£12.50

22 0514680 1

AF744385

EGE LIBRARY

*System Identification*

And when all is said, he that publishes a book runs a very great hazard, since nothing can be more impossible than to compose one that may secure the approbation of every reader.

Miguel de Cervantes Saavedra
*Don Quixote*, Vol. 2, Ch. 3.

Dr Healey

# System Identification

## Parameter and State Estimation


**Pieter Eykhoff**

*University of Technology*
*Eindhoven, the Netherlands*



*A Wiley–Interscience Publication*

JOHN WILEY & SONS

London · New York · Sydney · Toronto

Copyright © 1974 John Wiley & Sons Ltd.
All Rights Reserved. No part of this publication may be reproduced, stored in a retrieval system. or transmitted, in any form or by any means, electronic, mechanical photocopying, recording or otherwise, without the prior written permission of the Copyright owner.

Library of Congress Catalog Card No. 73-2781

ISBN 0 471 24980 7

UNIV. COLL. OF
LIBRARY
S. WALES & MON.
CARDIFF

Printed in Great Britain by J. W. Arrowsmith Ltd., Bristol

**Dedication:**

In memory of
**Professor ir. Léon H. M. Huydts (1890–1974)**
who served the University of Technology, Delft, the Netherlands, from 1914–1960;

an original thinker
an inspiring teacher
a devoted friend

# Preface

In recent years many aspects of system parameter and system state estimation have been discussed in many papers, at a number of conferences and in several university courses. The interest in this subject apparently has widely differing roots:

the desire of practising engineers in process industries to obtain better knowledge of specific plants in order to obtain improved control and thus lower operating cost or increased efficiency.

the study of high-performance aerodynamic and space vehicles and not incidentally the study of more down to earth systems such as railway carriages and hydrofoils.

the study of the human being in tracking action situations and in other types of control or learning processes.

the investigation of biological systems, e.g. neuromuscular systems such as the eye pupil control system, extremity (arm, leg) control, heart rate control system, etc.

This interest in estimation theory has not arisen solely from the need for increased system performance (or increased knowledge of system dynamics) but it is also a reflection of the dramatically changed possibilities of applying estimation theory brought about by the advent of modern computer hardware and software. The need for such computing facilities will become apparent when we come to discuss specific estimation algorithms.

More or less independently of the interest in estimation theory from the above sources under the general heading 'engineering sciences', there has also been considerable interest and activity from econometricians and statisticians who have been working on dynamic models (econometric models) of either some portion or the whole of social activity. This latter work has lead to occasional cross fertilization with the engineering sciences and is now culminating in the development of a general mathematical theory of systems applicable to systems of whatever type, be they social or physical. The net result of this development is a large number of publications, either accentuating a particular type of approach or describing a certain case study.

In this book we have tried to cast the estimation problem in a coherent form from a rather broad frame of reference. It is written with regard for the many different types of scientists and engineers concerned with dynamic systems—be they production processes, aircraft, biological objects, economic and other types of models.

Due to the nature of the subject, this book has to deal with statistics, system dynamics, stochastic signals, analogue, digital and hybrid simulation, and optimization. The limited treatment of some subjects (brought about by the constraint of a finite number of pages of text) may be supplemented by reference to the extensive bibliography of pertinent literature.

A few remarks on the organization of the book are in order here. When contemplating the theme, one may come up with a variety of legitimate distinctions to be used for the scheme of organization, e.g.

type of application;

statistical basis: least-squares, Markov, maximum likelihood, Bayes' estimates;

linearity and non-linearity of the model with respect to parameters (note that this is not the same as linearity with respect to dynamic behaviour);

signal representation: sampled and non-sampled;

implementation: analogue, digital, hybrid;

type of approach: open-loop and closed-loop estimation methods.

In order to provide the reader with convenient means of reference and to accentuate the coherence of the field the following approach has been chosen:

| | Chapters |
|---|---|
| The chapters **Introduction** and **Basic Approaches** serve as an explanation of the most important approaches toward the solution of estimation problems. | 1, 2 |
| The part **Prolegomena** summarizes the knowledge that is needed; the reader will find some condensation of knowledge | 3 |
| about signals, dynamic systems (processes) and topics from | 4 |
| probability theory, statistics and optimization theory. | 5 |
| The part **Estimation of Parameters** is organized along the following lines: | |

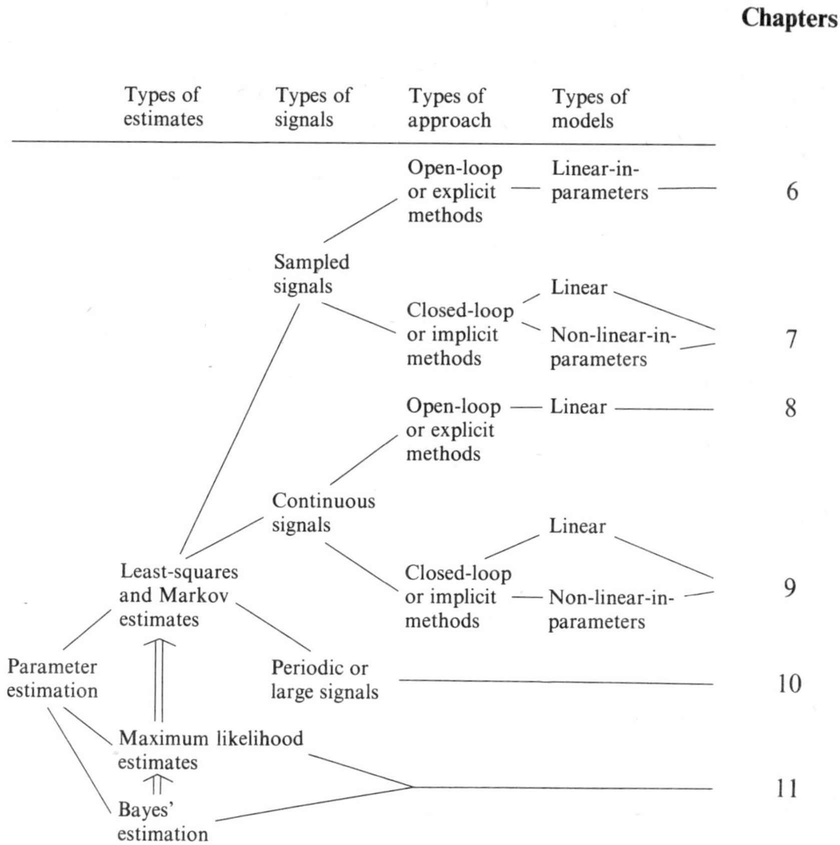


This way of presentation leads from the simple to the more complicated, from the particular to the more general case. In spite of the reversed trend in many research papers we feel that, for the first approach, the engineering student and research worker is more open to inductive reasoning, even where he recognizes the great beauty and merits of the deductive approach. Consequently in the organization of the material the pedagogical aspects have been weighed quite heavily. It is hoped and expected that the reader will profit from this emphasis by an enhanced insight into the problems under discussion and a clear perspective of possible applications.

By Chapter 5 is introduced already the relation between the different types of estimates.

Spectrum analysis techniques are referred to in those chapters where transformation to the frequency domain is practical.

**Chapters**

The estimation assignment is more complex, if not only parameters, but also (state) time functions in the system under study have to be determined. The description of system behaviour in terms of 'state variables' is reviewed in Chapter 4.

The part **Simultaneous Estimation of Parameters and States** reviews first the state-estimation problem. Here the deductive approach is chosen as it is expected that the reader has been exposed to this problem before or that, from the parallelism with the parameter-estimation problem, he can readily find his way. 12

The next chapter is devoted to the combined parameter and state-estimation techniques. 13

The last part of the book, **Applications,** refers to specific estimation cases that have been discussed in the literature, e.g. in control, in automatic industrial measurements, in biological systems and in learning or pattern recognition problems. 14

The book is written with potential users in mind, e.g. a graduate engineer or research worker, a postgraduate student or a college lecturer. With respect to the background needed for reading this text, the following prerequisites will be sufficient:

graduate level mathematics (advanced calculus, differential equations, complex-variable theory, some acquaintance with linear algebra and matrices); cf. Appendix C;

some basic knowledge of feedback control theory;

some knowledge of probability theory, elementary statistics and stochastic signal theory; the essential background is summarized in Appendix B;

some knowledge of digital- and analogue-computer techniques.

For easy reference and definition of notation, part of the necessary background knowledge is summarized in appendices. A list of symbols used in this text is given in Appendix A.

When teaching a course it is highly desirable to have the students reinforce their reading comprehension by working simple assignments. As in this field only examples of utmost simplicity can be calculated by hand one has to resort to the use of computers. Consequently those assignments will be very much dependent on the equipment and subroutines available. Therefore no attempt has been made to include problems;—anybody teaching such a course will not find it difficult to formulate such problems and 'hints'.

In organizing the manuscript it was a difficult decision whether or not to include digital computer programmes of estimation procedures. The choice of language (Fortan, Algol, PL1, etc.) and 'dialect', the availability of particular subroutines, the type of peripheral equipment and storage capacity, the speed of development of new estimation procedures, the size of the book—all these aspects had to be weighed in this decision. As a result no such programmes have been included; instead the potential user is strongly recommended to contact those authors who have reported on computer programmes that may be useful for a particular type of application in mind. Many references have been given.

Much information on identification and estimation can be found in the preprints/proceedings of specialized symposia:

*1st IFAC Symp.—Identification in Automatic Control Systems, Prague*, (1967); preprints (2 volumes) published by Academia, Prague.
*2nd IFAC Symp.—Identification and Process Parameter Estimation, Prague*, (1970); preprints (2 volumes) published by Academia, Prague.
*3rd IFAC Symp.—Identification and System Parameter Estimation, The Hague/Delft*, (1973); proceedings (2 volumes) published by North-Holland Publishing Comp., Amsterdam.

Many references are made also to IFAC congresses:
*1st IFAC Congr., Moscow*, (1960); proceedings titled 'Automatic and remote control' (4 volumes), published by Butterworths, London, 1961.
*2nd IFAC Congr., Basle*, (1963); proceedings titled 'Automatic and remote control' (2 volumes), published by Butterworths, London, and Oldenbourg, Munich, 1964.
*3rd IFAC Congr., London*, (1966); proceedings titled 'Automatic and remote control III' (4 volumes), published by the Inst. Mech. Eng., London.
*4th IFAC Congr., Warsaw*, (1969); preprints (50 volumes) published by Naczelna Organizacja Technicza, Warszawa, Poland.
*5th IFAC Congr., Paris*, (1972); preprints published by Instrument Society of America, Pittsburgh, Penna., U.S.A.

Besides the periodical literature valuable sources of information are also the preprints of the *Joint Automatic Control Conferences* (*JACC*), sponsored by the American Automatic Control Council, and many other conferences.

# Acknowledgement

As it is the case in almost all parts of human endeavour so also has the development of the field of identification and system parameter estimation been carried on by engineers and scientists all over the world.

It can be regarded as a duty to express the appreciation for so much relevant, interesting and outstanding work to which ample reference is made in this book. Yet I regret that, for reasons of publishing space and time, only part of that literature could be mentioned. The report literature, especially, could be referred to only incidentally.

It is more than a duty—it is a pleasure indeed—to express my thanks for the numerous professional contacts with colleagues in many countries. Mentioning names is risky as again completeness is virtually impossible. The friendships developed from these professional contacts I consider as belonging to the very valuable experiences in life.

The project of writing a manuscript can hardly be done without stimuli from the environment. For this I owe much thanks to colleagues, coworkers and students.

I am grateful for many opportunities; the time spent at the EE Department of the University of California, Berkeley (under the Visiting Research Scientists Programme of the U.S. National Academy of Sciences) and the EE Department of the University of Waterloo, Ontario, Canada, as well as the numerous visits to other institutes and industries.

Apart from the contributions by members of the Group 'Measurement and Control', EE Department, University of Technology, Eindhoven, the Netherlands that have been acknowledged by notes or references, I found great help in the friendly and cooperative atmosphere in this our group. For the typing, retyping and revisions I am much obliged to Mrs. M. J. C. P. M. Kanters-van Schaik, Mrs. J. E. M. van Bergen-Lucassen, Mrs. W. L. Mutsers-Akroyd and Miss E. Sanders.

Last, not least, the moral support and the acceptance of divided attention by my family is recognized with warm appreciation.

PIETER EYKHOFF

# Contents

PROLEGOMENA

ESTIMATION OF PARAMETERS

**13. Parameter and State Estimation**



APPLICATIONS

**14. Survey of Applications from Different Fields**



**Appendixes**

*CHAPTER 1*

# *Some Concepts; Model Building, Parameter and State Estimation*

## 1.1 Models

At the heart of the natural sciences and engineering there are the general notions of **observation** and **measurement.** Based on observation, the scientist builds a physical insight into the problem being studied and from that insight he formulates a theory by trial and error. This theory is a proposed concept of that aspect of nature which he is studying. Guided by this concept he designs new experiments. The observation of the results of these experiments either confirms the theory, dictates a change of the theory or rejects it completely. Although a concept may be beautiful and appealing to the mind of the scientist, the factual results nevertheless are dominant in this interaction of theory and experiment. From this point of view it is legitimate to say that in the fields of natural science and engineering, experiments and observations (measurements) are most fundamental.

Of almost equal importance is the idea of **model building** (Rosenblueth and Wiener, 1945). It can in fact hardly be separated from the observations and experiments mentioned before. The formulation of a theory (as a proposed concept of the aspect of nature being studied) may rightly be called 'model building'; the theory stands as a verbal or mathematical 'model' of reality. For our purposes, we will define a **model as a representation of the essential aspects of an existing system (or a system to be constructed) which presents knowledge of that system in a usable form.**

In this definition, **'system'** may be interpreted in a rather general way, e.g. as 'a collection of objects arranged in an ordered form, which is in some sense, purpose or goal directed'. Everything not belonging to the 'system' is part of the 'environment'. Depending on the borderline drawn between the system and its environment, one may characterize the system by input/output relations. What constitutes a 'system' depends on the point of view of the observer. The 'system' may be, for example, an amplifier consisting of electronic components, **or** a control loop including perhaps that amplifier as one of its many parts, **or** a chemical processing unit having many such loops, **or** a plant consisting of a number of units **or** a number of plants operating together as a system in the environment of a global economy.

The scope of the definition of **model** may be clarified by some remarks.

The model does not need to be a description of the actual mechanism of the system. It may simulate or 'mimic' the system behaviour. One may construct controlled arm prosthesis without understanding how the human being is able to utilize his limbs. It is the skill of the model maker to mask unexplained influences by more or less alchemistic properties such as 'crackability', 'spinning factor', etc.

Models may be **conceptual, physical** or **mathematical** (also called in that order: phenomenological, empirical and analytical) depending on: (1) the judgment of what is the essential aspect for the particular purpose, (2) the model building techniques that can be used and (3) the amount and quality of knowledge available. A clear and well-known example of the development of a model can be found in celestial mechanics. The Ptolemaic model of cycloidal motions stood as quite an adequate conceptual model for the observations of planetary motions. The Copernican model, which explained the observations as a result of both the motions of the earth (observer) and of the observed planet, can stand as a physical model. Kepler's laws provide a mathematical model that offers the possibility of numerical predictions. From his own experience the reader will be able to give examples of other models in the different areas cited.

Another distinction between models may be in terms of 'hardware' (instruments, apparatus) and of 'software' (mathematical expressions, computer programmes).

The object for which a model has to be made may be either a planned or an existing system. For engineers the importance of a model in the design stage of a process needs no further emphasis. For research scientists the model may provide a guide to further experiments.

The knowledge should be presented in a **usable** form. This is an essential aspect since the model must provide the background for further decisions. If the model is too complex its usefulness becomes questionable. The relative simplicity is a dominant feature of model construction. A model is a representation of reality with a reduced complexity. In many cases, the model complexity has to have a definite relation to the complexity of reality in order to be useful (e.g. for living systems).

Models can be found in many fields of human observation. These fields include physics, biology, astronomy, engineering, economics, sociology and psychology. In this text the attention will be focused mainly on the engineering fields, although the estimation techniques presented may be and are being used in the other fields mentioned as well. In engineering, one may distinguish among the following purposes of, or uses for models:

**Research.** One wants to give an interpretation of knowledge or measurements; the immediate use does not need to be clear beforehand but the gain in comprehension and consequently a concise formulation is of paramount importance (diagnostic situation). The model has to provide clues for further investigation.

**Design.** The knowledge of the components or subsystems to be used has to be expressed in a model, compatible with the design criteria (stability, error criterion, economic yield, intrinsic safety, etc.). In this context, the adaptation of an existing system due to higher demands or because a more economical performance is required has to be considered as part of a design too. Design essentially is a feedback action with the designer in the loop.

**Control.** The control actions on a system depend on the knowledge about that system. In this case one can distinguish:

(*a*) normal operating conditions, e.g. feedback or feedforward control, types of static optimization, dynamic optimization, adaptive control, batch control. Apparently there is a close relation between the usability of a 'difficult' dynamic system and the control means available.

(*b*) emergency situation, such as partial breakdown, when the control action depends on information about the type and degree of failure.

(*c*) start-up and shut-down situation, where certain steps of a programmed schedule may depend on the values of the system variables or parameters.

For the (engineering) scientist there is a real challenge in the attempt to analyse the common feature behind the immense number of applications of (automatic) control systems.

In the first place there is the wide spectrum of man-made control systems in engineering, ranging from the simplest forms of on–off control to complicated computer control in process industries. In addition to these, there are also the control mechanisms in society (e.g. the enforcement of the law) and the abundant number of control loops in living 'systems'. For such a wide variety of 'systems' only a rather loose notion can be expected to indicate the common factor behind the applications of control. Such a notion may be designated as **'uncertainty'.**

Since the celebrated application of Watt's regulator it has become apparent in innumerable cases that the introduction of some feedback mechanism can be used as an effective expedient for combating uncertainties. These uncertainties may be the result of unpredictable influences (disturbances) of the **environment** on a system, or they may originate **inside** the system under consideration (e.g. wear, aging, catalyst poisoning, etc.). Simple feedback can diminish the influence of uncertainty, or at least shift the adverse effect of uncertainty from an important system quantity to an unimportant quantity. Of course, this is no panacea and there are limitations to the application of this remedy. Such limitations are met in those cases where very large parameter variations occur; such a situation may lead to the use of **adaptive** control principles. Another limitation may be found in the requirements for optimizing some economic criterion, leading to **optimal** or **self-optimizing** systems.

In designing ordinary control the required knowledge of the behaviour of the process is limited to (possible) rather inaccurate data needed for stability considerations and to the data needed for evaluation of possible improvement by the application of feedback, i.e. the nature of loads and disturbances, constraints on the control variables, etc. Throughout the history of control theory it has been known that the knowledge of a system and its environment, required to design a system fully, is seldom available *a priori.* Even if the equations governing a system are known in principle, it often happens that knowledge of particular parameters is missing; it is not uncommon that the models which are available are much too complex. Such situations naturally occur in many other fields. There are, however, two facts which are always present in identification problems occurring in automatic control:

it is often possible to perform experiments on the system in order to obtain the knowledge which is lacking.

the purpose of the identification is to design a control strategy.

One of the factors which undoubtedly contributed very much to the great success of frequency response techniques in 'classical' control theory was the fact that the design methods were accompanied by a very powerful technique for systems identification, i.e. frequency analysis. This technique made it possible to determine the transfer functions accurately, and this is precisely what is needed to apply the synthesis methods based on logarithmic diagrams. The models used in 'modern' control theory are with a few exceptions **parametric** models in terms of state equations. The desire to determine such models from experimental data has naturally renewed the interests of control engineers in parameter estimation and related techniques.

For systems with large parameter variations (leading to adaptive control) and systems with strict 'economic' criteria (leading to optimal or self-optimizing control) there often is a definite need for more accurate and timely information on the systems, both with respect to parameters and state variables. Consequently, the necessity for process identification, modelling and parameter estimation is apparent. One could approach such a process as being a celebrated black box, i.e. a (sub)system with one or more 'inputs' and 'outputs' but of unknown structure. In most engineering situations, however, the black box approach is not a very realistic one. The experimenter in many cases, has derived some *a priori* knowledge from physical insight into the process under consideration. This may give information on the structure of a conceptual model for that process, and probably even an approximate knowledge of the values of the coefficients (parameters) in that model. Consequently, the box is more or less 'grey' or translucent. Even the fact that we know we want to consider a distillation column and not a lunar vehicle implies *a priori* knowledge (Pavlik).

Some types of problem characterization are provided by the answers to the following questions:

Does one want a model for static behaviour (e.g. for static process optimization) or for dynamic behaviour (e.g. for dynamic process optimization or for aircraft dynamics control) of a complete non-linear or a linearized model? The answer to this question may imply clues for the desired accuracy, the sophistication of the model, the mathematical or statistical approach, etc.

Should the model be assembled off-line, with paper and pencil from basic laws and some isolated experiments, or can we work on-line and are we allowed to experiment with the existing process?

What budget considerations have to be taken into account?

From this point on the number of questions increase rapidly and become more and more awkward. We shall mention just a few:

How is the quality of the model to be judged?

How is all the relevant knowledge to be incorporated?

What is the optimum strategy to obtain the missing knowledge?

How are non-linearities to be handled?

How can a complex system be approximated by a simple model?

The answer to these questions are strongly object oriented. Some general considerations are given in Chapter 4.

## 1.2 Model Building

In many cases model building starts with the application of basic physical laws (Newton's laws; Maxwell's laws; Kirchhoff's laws; mass balance; energy/heat balance; impulse balance; entropy balance) to the process being studied, e.g. a mechanical, an electrical or a thermodynamic process. From these laws, a number of relations between the variables at hand follow, e.g. ordinary or partial differential equations/difference equations. These are relations between 'forces' (intensive quantities) and 'currents' (extensive quantities), e.g.,

| | Forces | Currents |
|---|---|---|
| Function of time only | Temperature difference<br>Concentration difference<br>Electric potential difference | Heat flow<br>Mass flow<br>Electric current flow |
| Function of time and place | Temperature gradient<br>Concentration gradient<br>Electric potential gradient | (Vector) heat flow density<br>(Vector) mass flow density<br>(Vector) electric current flow density |

If all external and internal conditions of our system are known quantitatively and if our physical knowledge about the system is complete then, at least in principle, the numerical value of all coefficients (parameters) in those relations can be determined. This situation may occur for example in connexion with space-vehicle dynamics. Such a situation is, however, quite rare in down-to-earth (engineering) applications. Our *a priori* knowledge is restricted because of uncertainty as regards the environment, and because of an incomplete knowledge of the physical aspects inside the process.

Even if our knowledge of the physical effects in each 'elementary cell' of the process were perfect, then the devil of dimensionality would probably still force us to use a simplified description of the process as a whole. This may imply that the parameters and state of our model are less amenable to *a priori* calculation. For that reason one has to rely on (process) identification and estimation. Consequently, model building consists of the following steps:

selection of a model structure based on physical knowledge;

fitting of parameters to available data (estimation);

verification and testing of the model (diagnostic check);

application of the model to its given purpose.

In view of the wide field of disciplines and the great variety of applications no general rules can be given for these steps. The model structure is selected on the basis of initial (*a priori*) knowledge about the system and its purpose. Actually finding an appropriate model may be a very difficult problem even in a narrow field of application.

The main emphasis of this text is on the estimation aspect. The use of the word **'estimation'** is due to the fact that in almost all realistic situations the observations made on the system under study are contaminated with random influences (disturbances, errors). Consequently one has to call upon **statistical** methods to obtain a 'best' result from our measurements by 'filtering out' the influence of the disturbances.

The verification and the testing of the model is strongly related to the estimation; as a follow-up of the estimating procedure one should investigate to what extent the model really 'explains' the behaviour of our system. Conceptually this can be vizualized by driving the system and the model by the same 'input' signal and by studying the character and magnitude of the difference of both 'output' signals (errors or residuals).

The application of the model to its given purpose again is strongly object oriented.

For some examples of this procedure in process industry the reader is referred to Eykhoff, Van der Grinten, Kwakernaak and Veltman (1966). Examples in other fields can be found in the literature mentioned in Chapter 14. For another concise illustration of the (iterative) sequence of model-structure selection, fitting, diagnostic check and application (e.g. forecasting) cf. Box and Jenkins (1969; 1970).

## 1.3 Structure, Parameters and States

In system analysis a central problem is the determination of the output signal based on the input time history and the properties of the system. The problem discussed in this book is sometimes denoted by the **'inverse problem'** of system analysis; given an input and output time history determine the (differential) equations that describe the system behaviour.

**Identification** is defined by Zadeh (1962) as **the determination on the basis of input and output, of a system within a specified class of systems, to which the system under test is equivalent.** Using Zadeh's formulation it is necessary to specify a class of systems, $\mathscr{S} = \{S\}$, a class of input signals, $\mathscr{U}$, and the meaning of 'equivalent'. In the following we will call 'the system under test' simply the **process** and the elements of $\mathscr{S}$ will be called **models.** Equivalence is often defined in terms of a criterion, an error or a loss function, which is a functional of the process output $y$ and the model output $y_M$, i.e.

$$E = E(y, y_M)$$

Two models $M_1$ and $M_2$ are then said to be **equivalent** if the value of the loss function is the same for both models, i.e.

$$E(y, y_{M_1}) = E(y, y_{M_2})$$

There is a large freedom in the problem formulation; this freedom is reflected in literature on identification problems. The selection of the class of models, $\mathscr{S}$, the class of input signals, $\mathscr{U}$, and the criterion is influenced largely by the *a priori* knowledge of the process as well as by the purpose of the identification. It has to be recognized that a model represents three types of knowledge, viz.:

structure, expressed in terms of mathematical identities, block diagrams, flow diagrams or graphs, connexion matrices;

parameter values, i.e. quantities that show no dependence on the input or independent variables;

values of the dependent variables (state) at a certain instant or as a function of time.

The importance of the structure cannot be over-emphasized. Its choice is guided by the type of application of our model; the choice may be decisive for the ultimate success or failure of the estimation scheme.

Often identification has the connotation of starting from scratch without any prior knowledge about the process. As mentioned before, in the majority of engineering situations and in a number of cases in biology as well, this is not a realistic assumption; from the structure of the 'process' and at least a partial understanding of its operation, a certain amount of *a priori* knowledge will be available, e.g. the model structure. In such cases the knowledge to be derived is reduced to numerical values of a number of parameters (coefficients of differential equations governing the dynamics of the process, coefficients of a linear or non-linear 'model' of the process, etc.) and/or numerical values of the state variables. Consequently the identification problem is reduced to that of (process) parameter and/or state estimation.

**Parameter estimation** is defined as **the experimental determination of values of parameters that govern the dynamic and/or non-linear behaviour, assuming that the structure of the process model is known.** It should be pointed out that the distinction between knowledge of structure and parameter is not as straightforward as it may appear at first sight. The change from a parameter value $\beta \neq 0$ to $\beta = 0$ may represent a simplification in the structure as that 'branch' of the model may be deleted. On the other hand it is clear that in this way no 'additions' to the model can be made; the potential capabilities of the model are limited by the choice of the model. Little work on systematic improvement of models by adding new structural elements seems to have been done. At this state of the art, the human intelligence and intuition appear to be still the limiting factor.

In some cases, one is interested in a more detailed knowledge, e.g. a continuous information about the state of the process; this leads to **state estimation.** Roughly speaking the 'state' of a process is a variable (e.g. a vector) which, together with the subsequent input to the process, completely determines the subsequent behaviour. Consequently, if the process is left alone, i.e. if there is no input signal, then knowledge about the state at a certain time (together with knowledge about process structure and parameters) is sufficient to predict its future behaviour. As an example one may consider the initial conditions of a differential equation as a state vector; cf. Chapter 4. The assignment of determining the best estimate of the state can occur in modern control applications as, for example, optimal control. If there is in addition an uncertainty about a number of parameters then we have the combined problem of parameter and state estimation. From this description it is clear that in the majority of cases, state estimation is required to be done under normal operating conditions. For parameter estimation the determination can be based either on (active) experiments or (passive) observation of the process under study.

Figure 1.1* gives an illustration of the relation between '*a priori*' (structural) knowledge and '*a posteriori*' (measurement) knowledge in the model-building

* Based on discussions with Mr. A. U. Khan.

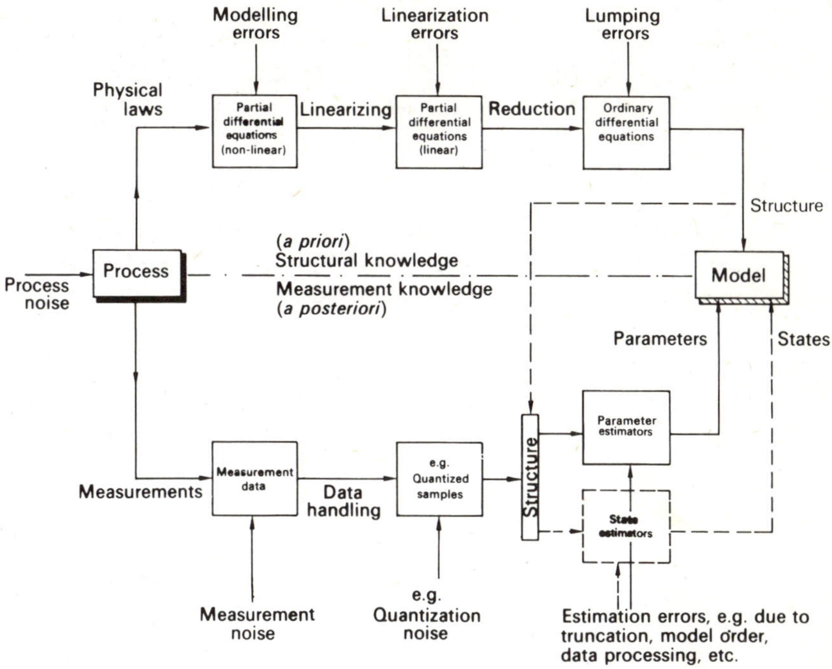


Figure 1.1

stage. The upper sequence indicates the 'model building' for a particular type of process, starting from physical laws, by applying linearization and reduction to ordinary differential equations. These resulting equations provide a model structure. In each step there are errors introduced. The lower sequence indicates the 'estimation procedure', starting from measurements, applying data handling and estimation algorithms. Here again different types of errors have to be recognized.

Note that the (parameter and/or state) estimation presupposes a model structure. Due to the unlimited variety of processes that might be considered and the wide variation of goals requiring particular types of models, Figure 1.1 can be considered only as an illustration; a complete discussion of the model building is outside the scope of this text. The emphasis will be on a coherent discussion of estimation techniques.

## 1.4 Representation of the Problem

Some simple representations of the class of problems at hand are given in Figure 1.2.

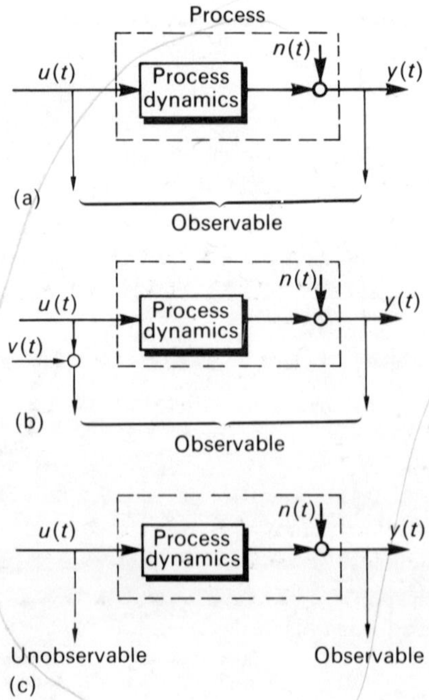


Figure 1.2

The quantities $u(t)$ and $y(t)$ are the input and the output of the process respectively. The differences between Figures 1.2a, b and c lie in the observability of the input signal, viz.: observable, observable but contaminated with noise, unobservable; cf. Kerr (1965). In all cases the output $y$ is contaminated with noise. From the observable signals and possible *a priori* knowledge about probability density functions the information on parameters and/or state has to be derived. In feedback and in multiple-input/multiple-output systems, combinations of these models may appear. The consequences of these different situations will be studied in Chapter 11.

The term 'process' can be used in the broadest sense of the word and may refer to an engineering system (e.g. aircraft control system, chemical plant, ionospheric transmission paths or other transmission systems), to a biological system (e.g. human tracking action, medical diagnosis), and perhaps also to an economic and sociological system.

Of course, the representation can be complicated by several factors. First of all there may be multiple-input/multiple-output situation where each input may affect a number of outputs. In this case the input and output may be represented by vector quantities $\mathbf{u}(t)$ and $\mathbf{y}(t)$ respectively, where each component of the vector is a time function corresponding to an input or output signal. Another type of complication is found in that situation where the noise is not of the additive type but where it is multiplicative (e.g. a changing gain or attenuation that is not a parameter of primary interest).

In Section 1.1 we emphasized the importance of models. The estimation problem can **conceptually** be stated in the following way. The process and a model are subjected to the same input signal. The process output, contaminated with noise, and the model output are compared. Then it is necessary to find an adjustment of the model that is optimal in some predefined sense.

The parameter values are not directly observable. Consequently the optimum has to be defined using a criterion with respect to the **output signal(s),** or a criterion related to the **expected error** of the estimated parameter value; cf. Section 2.3. The output signal is related to the parameter values by a functional relationship. This causes the complications of the process-parameter estimation and, generally speaking, makes the introduction of computer elements necessary. It was mentioned before that the criterion is often a minimization of a scalar loss function. In such a case it is natural to ask several questions:

Is the minimum achieved?

Is there a unique solution?

Is the uniqueness of the solution influenced by the choice of input signals?

If the solution is not unique, what is the character of the models which give the same value of the loss function and how should $\mathscr{S}$ be restricted in order to ensure uniqueness?

Answers to some of these problems have been given for a simple class of linear systems arising in biomedical applications by Bellman and Åström (1969). The loss function is chosen *ad hoc* when the identification problem is formulated as an optimization problem and it is a consequence of other assumptions when the problem is formulated as an estimation problem.

Often the criterion is expressed as a functional of an error, e.g.

$$E(y, y_M) = \int_0^T e^2(t)\, dt$$

where $y$ is the process output, $y_M$ the model output and $e$ the error; $y$, $y_M$ and $e$ are considered as functions defined on $(0, T)$. Notice that the criterion can be interpreted as a least-squares criterion for the error $e$. The case

$$e = y - y_M = y - M(u)$$

where $M(u)$ denotes the output of the model if its input is $u$, is referred to as the **output error.** It is the natural definition when the only disturbances are white noise errors in the measurement of the output (Figure 1.3a). The case

$$e = u - u_M = u - M^{-1}(y)$$

where $u_M$ denotes the input of the model which produces the output $y$, is called the **input error** (Figure 1.3b). The notation $M^{-1}$ implies the assumption

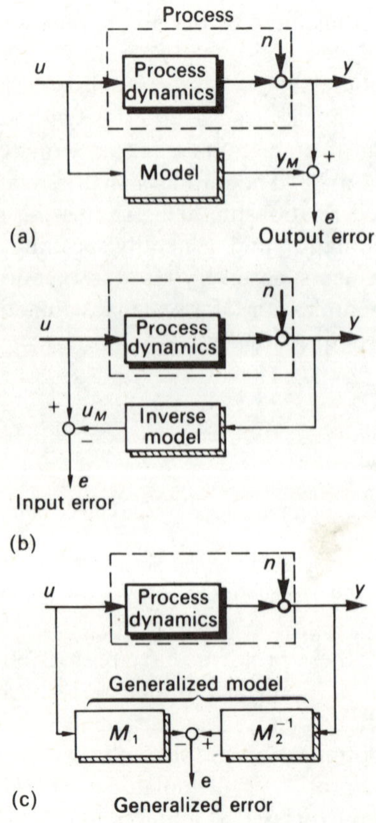


Figure 1.3

that the model is **invertible,** roughly speaking that it is always possible to find a unique input which produces a given output.*

From the point of view of estimation theory, the criterion with the error defined as the input error would be the natural criterion if the disturbances are white noise entering at the system input.

* Rigorous definitions of the concept of invertibility are discussed by Brockett and Mesarovic (1965), Silverman (1969), Sain and Massey (1969).

In a more general case the error can be defined as

$$e = M_2^{-1}(y) - M_1(u)$$

where $M_2$ represents an invertible model (Figure 1.3c). This type of model and error are referred to as **generalized model** and **generalized error** (Eykhoff, 1963). A special case of the generalized error is the 'equation error' introduced by Potts, Ornstein and Clymer (1961).

Another type of identification problem is obtained by imbedding the problem in a probabilistic framework. If $\mathscr{S}$ is defined as a parametric class, $\mathscr{S} = \{S_\beta\}$, where $\beta$ is a parameter, the identification problem then reduces to a parameter estimation problem. Such a formulation makes it possible to exploit the tools of estimation and decision theory. In particular it is possible to use special estimation methods, e.g. the maximum likelihood method or Bayes' method. It is possible to assign accuracies to the parameter estimates and to test various hypotheses.

However, in many probabilistic situations it turns out that the estimation problem can be reduced to an optimization problem. In such a case, the loss function is, however, given by the probabilistic assumptions. Conversely, to a given loss function it is often possible to find a probabilistic interpretation.

The estimation problem is closely related to the statistical theory of optimal systems (Pugachev, 1963). The optimum filtering problem (Figure 1.4a) belongs to the same category as does the prediction problem (Figures 1.4b and c) and combinations of these problems. It is also related to adaptation and learning problems (Tsypkin, 1966; 1967; 1968; 1970). The problem becomes still more complicated if the 'process' being studied is part of a closed loop. Fel'dbaum (1960, part 1) considers the situation where the control signal is used for a twofold purpose:

to **investigate** the characteristics of the process or the ways of controlling it.

to implement the regulation, to **direct** the process to the required state.

For the combination of these investigating and direction actions he coins the word **dual control.** He points out that these two functions are sometimes separated in frequency range (frequency-sharing operation) or in time intervals (time-sharing operation). This separation, however, is not necessary and a general theory on these dual-control systems is given by Fel'dbaum (1960, 1961; 1965). The essential problem is indicated by: 'In dual-control systems, there is a conflict between the two sides of the controlling process, the investigational and the directional. An efficient control can only be effected by a well-timed action on the object. A delayed action weakens the control process. But the control can only be effective when the properties of the object are sufficiently well known; one needs, however, more time to

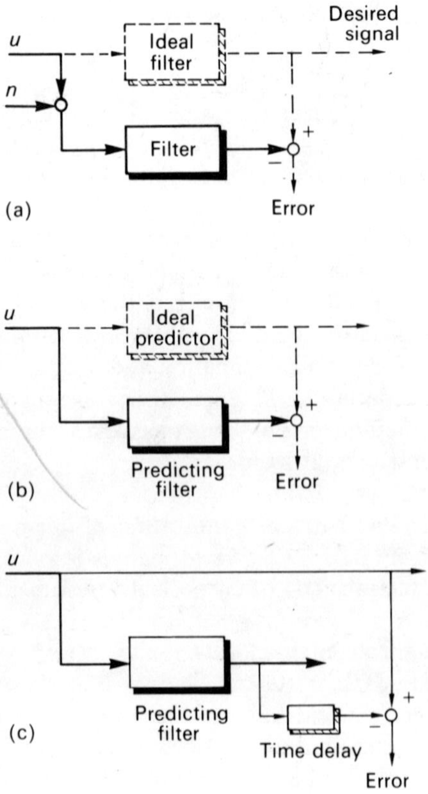


Figure 1.4

become familiar with them. A too 'hasty' controlling member will carry out the operational movement without making proper use of the results of trial investigations performed on the object. A too 'cautious' system will bide its time unnecessarily long and process the received information without directing the object to its required state at the right time. In each case, the control process may not prove the best one and may not even prove to be up to the mark. Our problem is to find out, one way or another, which combination of these two sides of the regulation would prove to be most suitable. The operations must be so selected as to maximize a criterion of the control's quality.

'Two factors should be taken into account by the controlling member which decides the specific amount of regulating movement at any given moment of time:

(*a*) The loss occurring in the value of the quality criterion due to the fact that the outcome of the operation at a given moment and at subsequent moments of time, will cause a deviation of the object either from the required state or from the best attainable one. The average value of this loss shall be called the **action risk.**

(*b*) The loss occurring in the value of the quality criterion due to the magnitude of the controlling action has not proved the best to obtain information on the characteristics of the object; in view of this, the subsequent actions will not be the best possible ones either. The average value of this loss shall be called the **investigation risk.**'

Another point of general interest is the relation between process parameters and signal parameters. Many types of deterministic signals can be generated, at least conceptually, as an impulse response or step response of a linear or non-linear filter. Many types of stochastic signals can be obtained by the filtering of white noise. This implies that in such cases signal characteristics can be expressed in filter (process) parameters and vice versa. Consequently there is a close tie between process parameter and signal parameter estimation.

## 1.5 Normal Operating Conditions

In many cases, the performance of an **experiment** implies that a number of important environmental conditions are preset or under control. This means that the experiment can be repeated. In other cases, e.g. celestial mechanics (before Sputnik), one had to be satisfied by making passive **observations** only.

There are cases in which the **duration** of the experiment is not an essential issue. In terms of accuracy, the amount of information that can be derived from the experiment is not restricted by time or by the number of trial runs. Restrictions may be induced by economic considerations or by other practical limitations. Of course, there also do exist fundamental limitations such as the uncertainty principle of Heisenberg, which states for example, that it is impossible to specify or determine simultaneously **both** the position and velocity of a particle as accurately as may be wished.

The situations where process parameter estimation is applicable are often unlike the experimental situation indicated before. The observation time is restricted, either through economic necessity or for fundamental engineering reasons. Included in the latter are process characteristics that change in time and yet have to be determined. Then, parameter estimation has to be performed under normal operating conditions. In some cases this has to be achieved through the use of (random) signals that may be present already; in other cases test signals of a special class with a sufficient level may be introduced.

It may be noted that in the previous lines the practical situation is clearly over-simplified. Many actual experiments do have aspects of the experimental as well as the normal operating conditions. A whole range of situations can be imagined. At one end of this range are the astronomical observations of celestial mechanics; this process is clearly under normal operating conditions. An example at the other end of the scale is the determination of transfer characteristics of a network in the laboratory using sinewave generators, etc.

Most cases, e.g. industrial-process measurements, will be in between these extremes because only part of the environment will be under control. It will often be desirable to make some kind of experiment, observe the process while using perturbations as input signals and observe the corresponding changes in process variables. In practice there are, however, often severe limitations on the experiments that can be performed. In order to get realistic models it is often necessary to carry out the experiments **during normal operation.** This means that if the system is perturbed, the perturbations must be small so that the production is hardly disturbed. It might be necessary to have several regulators in operation during the experiment in order to keep the process fluctuations within acceptable limits. This may have an important influence on the estimation results.

When carrying out identification experiments of this type there are many questions which arise naturally:

How should the experiment be planned? Should a sequential design be used, i.e. plan an experiment using the available *a priori* information, perform that experiment, plan a new experiment based on the results obtained, etc. When should the experimentation stop?

What kind of analysis should be applied to the results of the experiment in order to arrive at control strategies with desired properties? What confidence can be given to the results?

What type of perturbation signal should be used to obtain as good results as possible within the limits given by the experimental conditions?

If a digital computer is used, what is a suitable choice of the sampling interval?

In spite of the large amount of work that has been carried out in the area of system identification, we have at present practically no general answers to the problems raised above. In practice, most of these general problems are therefore answered in an *ad hoc* manner, leaving the analysis to more specific problems. It is worthwhile to recognize the fact, that the final purpose of identification is often the design of a control system, because this simple observation may resolve many of the ambiguities of an identification problem. A typical example is the discussion of whether the accuracy of an identification should be judged on the basis of deviations in the model **parameters** or in the **time response.** If the ultimate purpose is to design control systems, then it seems logical that the accuracy of an identification should be judged on the basis of the **performance** of the control system design from the results of the identification.

In any case, it is of vital importance that

the input signal should excite all modes of the process, i.e. the process has to be **controllable**, cf. Chapter 4, and the signals have to be of a class called **persistently exciting**, cf. Section 11.4;

the output signal contain sufficient information on 'what is going on in the process', Chapter 4: **observability** and **identifiability.**

Generally speaking, the time interval needed for the estimation has to be as short as possible in view of the need for actual information on the (slowly changing) parameters. On the other hand, the minimal time interval is bounded by the statistical effects (inherent variance of the measurements, influence of additive noise). In a sense, the time interval needed for the estimation can be related to the instrumentation costs. This can be explained by consideration of Figure 1.5.

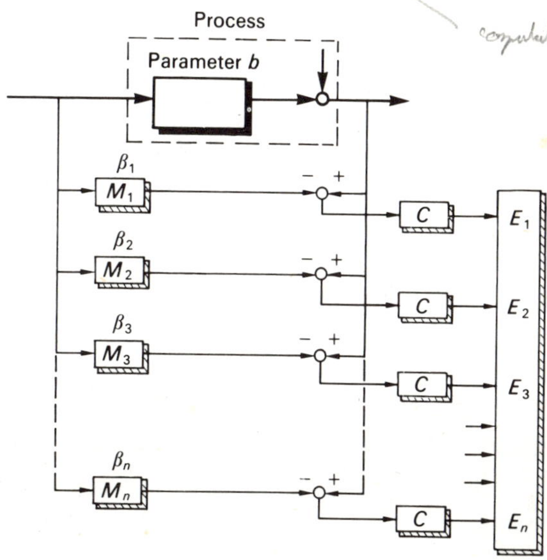


Figure 1.5

Assume that only one process parameter $b$ has to be determined. One could then use $n$ models, each with the same structure and a different parameter value $\beta_i$ ($i = 1, \ldots, n$). The output of the process is compared with that of each of the models. The difference or error is weighted by the boxes marked $C$ according to some criterion, e.g. by squaring and integrating over a time interval $T$. Now the value of $b$ is known after one such 'measuring time interval' $T$, for we know that $b \approx \beta_i$ if $E_i < E_j$ for each $j \neq i$.

Under stationary conditions, the same knowledge can be obtained using **one** model and setting in the $j$th measuring time interval $\beta = \beta_j$. Consequently in this case the desired information is obtained only after an interval of length $nT$; the instrumentation costs, however, are much less than in the previous case. For this reason one has to be careful in comparing different parameter-estimating schemes. For more parameters, say $m$, the total number of models or the total number of measurement intervals becomes extremely large, viz. $n^m$.

Although it was mentioned before that there exists a continuous range of cases, the situation indicated by 'under normal operating conditions' differs from the measurements made under test conditions. This difference consists of:

the information stored in the process (history);

the time requirements with respect to measurements and evaluation;

the choice of test signals to be used.

These differences necessitate a detailed study of the efficiency of the information-processing scheme that one wants to use for the estimation under consideration. The results of such an analysis have to be given in terms of:

the speed of convergence towards the asymptotic value;

the bias of the asymptotic value compared to the true parameter value;

the variance of the resulting parameter information as a function of time;

the influence of additive noise and of non-linearities in the process; etc.

These aspects have to be compared to the speed with which process parameters may change.

Let us consider briefly the complications mentioned above. The **information stored in the process** depends on the history of the input(s). For the elementary case of a linear time-invariant process, this effect can be seen from the convolution form of the input–output relation

$$y(t) = \int_{-\infty}^{t} u(\theta)h(t - \theta)\, d\theta$$

where $u$ is the process input, $y$ is the undisturbed process output and $h(t)$ is the process impulse response. Starting from a given time $t = T$ we wish to observe the input and output in order to obtain information on $h(t)$. The output $y(t)$ for $t > T$, is determined also by $u(t)$ for $t < T$. This contribution to the output can be written as:

$$y_1(t) = \int_{-\infty}^{T} u(\theta)h(t - \theta)\, d\theta$$

It acts as a disturbance to the output, since it is not related to the part of the input signal $u(t)$ for $t > T$ which we are observing. This difficulty can also be described by using state–space techniques, cf. Chapter 4. Roughly speaking, a state space of dimension $n$ can be assigned to each process described by a differential equation of order $n$. The coordinates of this space represent $n$ independent variables, e.g. the values of the derivatives of order 0 to $n - 1$. At each time $t$ the 'state' of the system is given by a point in the state space called the state vector $\mathbf{x}(t)$. This point moves under the influence of the equilibrium-restoring forces (dissipation of energy) and the forcing functions (input signals). At time $T$ the state $\mathbf{x}(T)$ contains all relevant information of the past history of the system. Under force-free conditions (no input signals) the state vector of a stable system will in time tend to an (some) equilibrium state position $\mathbf{x}_0$. This component of motion of the state vector, the free response, is added to that caused by the action of the input signals. Thus the free response acts as a disturbance signal if the input and output signals are observed say on a time interval starting at $t = T$. An extended form of the estimation problem is then: estimate the unknown parameters **and** the unknown state at $t = T$.

The **time requirements** for the measurement and evaluation are clearly dependent upon the fact that the parameter estimation has to use **signal characteristics** which cannot be obtained until after that signal has been observed over some interval of time. In general, there will be a close relationship between the length of the interval and the accuracy of the information obtained. **The choice of test signals** is also pertinent to the previous remark. In the experimental situation in the laboratory, large or periodic test signals can often be used. Under normal operating conditions, this is sometimes the case as well. If test signals are not allowed, then we have to make the best possible use of the available disturbances. If only a small signal may be added to the input then we have to consider the signal to noise ratio, the best choice of the predetermined characteristics of the test signal, the influence of the type of signal on the information processing operation, etc. If the process parameters are changing, then there is a fundamental uncertainty in the estimation results.

## 1.6 Some Fields of Application

The following diagram indicates the theories that are relevant for parameter and state estimation. It also indicates some fields of application.

**Theories**

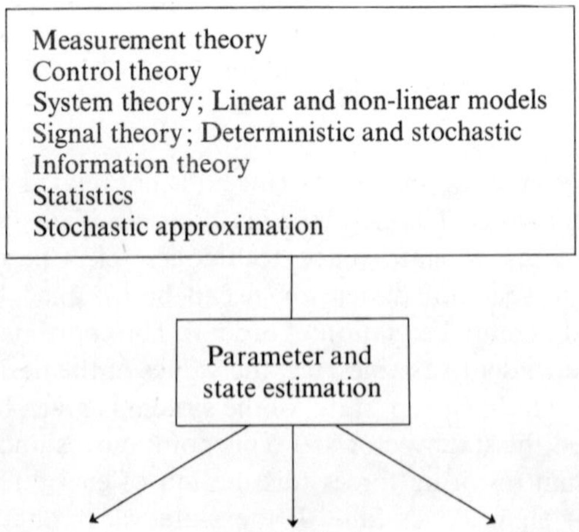


| Ways of application | Fields of application | Related problems |
|---|---|---|
| 'Diagnostic' applications | Communication engineering | Signal detection |
| Control applications | Power engineering | Optimal filtering |
| Automatic (industrial) measurements | Mechanical engineering | Prediction |
| Automatic (industrial) decisions | Aeronautical engineering | Learning |
| Automatic (industrial) adjustments | Chemical engineering | |
| Pattern recognition | Physics | |
| | Geology | |
| | Economics | |
| | Biology/medicine | |

Through the stochastic-signal theory, stochastic approximation, statistics and information theory a considerable use is made of probabilistic notions and results. The basic aspects needed, and assumed well-known, are summarized in Appendix B. The theoretical aspects of signal and model description are discussed in Chapter 3, 4 and 5.

The list of applications suggests that the potential uses of parameter estimation extend beyond those in control engineering. In each application, the goal of the estimation procedure has to be clear beforehand. This goal determines the accuracy needed and the effort permitted.

### *1.6.1 Estimates for 'Diagnostic' Applications*

In this category belong the measurements of transfer characteristics of a process, e.g. the measurements of a Bode plot, the determination of the impulse response, the determination of poles and zeros of a transfer function, the estimation of non-linear effects. The celebrated Fourier filtering method, i.e. the instrumentation of the equations:

$$\text{Re}\,[H(j\omega)] = \lim_{k\to\infty} \frac{1}{k}\int_0^{kT} y(t)u \sin \omega t \, dt$$

$$\text{Im}\,[H(j\omega)] = \lim_{k\to\infty} \frac{1}{k}\int_0^{kT} y(t)u \cos \omega t \, dt$$

where

$u(t) = u \sin \omega t$ is the process input signal

$y(t)$ is the process output signal

belongs to this class. Apart from applications in electrical and electro-mechanical systems one also finds applications in:

chemical engineering, e.g. the determinations of catalyst aging or poisoning; the determination of rate coefficients in chemical reactions;

industrial processes, e.g. the determination of heat-transfer coefficients;

physical measurements, e.g. the measurement of the electrical impedance of a mercury electrode (i.e. an electrode that consists of a mercury drop that grows at the end of a tube. After this reaches a certain volume then it falls and a new drop starts growing, etc. With such an arrangement there is always a 'clean' electrode surface available);

mechanical engineering, e.g. the measurement of the transfer characteristics of railroad wagons, textile machines;

nuclear reactors, e.g. the determination of reactivity coefficients;

aircraft and spacecraft engineering, e.g. the determination of the airplane dynamics during flight;

biology and medicine, e.g. the determination of the human characteristics when performing a 'tracking action'; a simple arrangement is shown in Figure 1.6, where an optical signal is given and where the assignment is to track the spot on the cathode ray tube as accurately as possible. Studies can be made of changes in the characteristics,

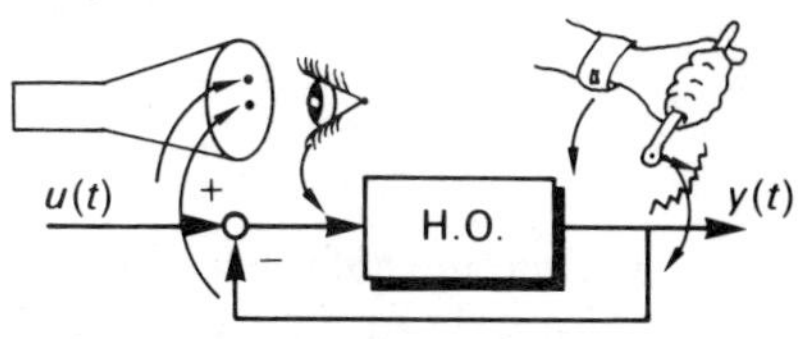


Figure 1.6

e.g. due to 'learning', fatigue, environment and drugs. Other examples from a large variety of possibilities: the pupil reaction on incident light; the determination of the parameters for a blood-circulation model;

economy.

In the following categories the accent lies on the type of application for which the estimate is to be used.

### *1.6.2 Estimates for Control Applications*

If the purpose of the identification is to design a control system, the character of the problem might vary widely depending on the nature of the control problem. A few examples:

Design a stable regulator.

Design a control programme for optimal transition from one state to another.

Design a regulator which minimizes the variations in process variables due to disturbances.

In the first case it might be sufficient to have a fairly crude model of the system dynamics. The second control problem might require a fairly accurate model of the system dynamics. In the third problem it is also necessary to have a model of the environment of the system.

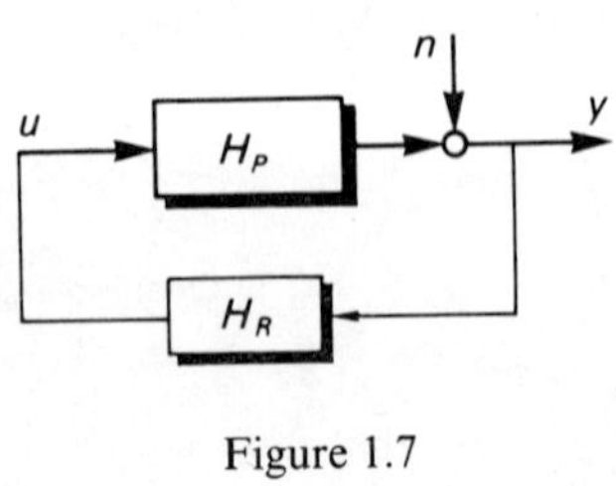


Figure 1.7

The danger of identifying systems under closed-loop control also deserves to be emphasized. Consider the classical example of Figure 1.7. An attempt to identify $H_P$ from measurements of $u$ and $y$ will give

$$\hat{H}_P = \frac{1}{H_R}$$

i.e. the inverse of the transfer functions of the feedback. In industrial applications the feedback can enter in very subtle ways, e.g. through the action of an operator who makes occasional adjustments. Fisher (1965) has shown the interesting result that the process may be identified if the feedback is made non-linear.

*Optimal Control Systems.* Here the optimality is based mostly on an economic criterion. In many cases the quality of the optimal control appears to be strongly dependent on the information that is available concerning the dynamical characteristics of the process that has to be controlled. If these characteristics change in time we have to determine them continuously. For optimal control the state of the process is also of great importance. This leads to a combination of parameter and state estimation. It is known (cf. Aoki, 1967), that only for a limited class of systems (viz. a linear process and using a quadratic cost function) the optimal control scheme can be divided into an optimal state estimator followed by an optimal controller; cf. Chapter 14. For other types of systems, this division into two separate problems leads to an approximation of the optimal control. Yet it may be an important aid to make the control problem tractable. As examples we can use the approach of and the landing of an airplane, the control of batch processes, the control of continuous industrial processes, where the optimal setpoints of the controllers are determined by use of a (partial) model of the process. Interesting examples are described in the case studies (Peschon and others, 1968):

on optimum control of electric-power systems;

of Kalman filtering in the C-5 aircraft navigation system;

in the paper-making industry;

on the optimal control of a flexible launch vehicle.

For other examples the reader is referred to Chapter 14.

*Self-optimizing and Self-adapting Systems.* For this class, the same holds as for the previous category. Now we want to adjust the parameters of the control system, based on the knowledge of the parameters of the process, in such a way that the transfer characteristics are optimal. An illuminating example of this is the airplane which, as a result of the difference in air density close to the earth surface and at 30,000 ft altitude, shows a great difference in dynamic behaviour.

Another illustration of adaptive control by process-parameter estimation is found in driving an automobile. Due to the weather conditions, there is a chance that the road is slippery. This may be detected by the introduction of small perturbations with the steering wheel. Evaluation of the results of these perturbations gives the information that is used in controlling the speed.

### *1.6.3 Estimates for Automatic (Industrial) Measurements*

On account of the economic aspects in this class, the speed of measurement is of great importance. As an example of a new approach to an old problem the automatic measurements of impedance can be used (cf. Figure 1.8). The impedance can be described by the differential equation

$$C\dot{u} + \frac{1}{R}u - i = 0$$

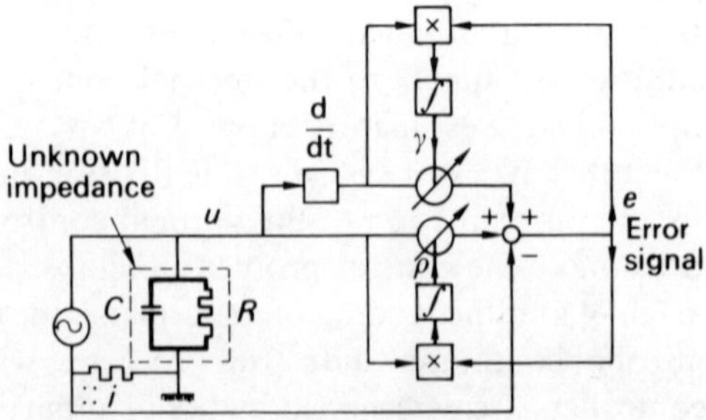


Figure 1.8

The model satisfies the equation

$$\gamma\dot{u} + \rho u - i = e$$

Due to the instrumentation shown in the diagram the model will be adjusted in such a way that the error signal approaches zero, that is to say its mean-squared value will be minimized. This principle can be extended in several directions, among which the determination of more complicated impedances or nonlinear devices, e.g. quartz crystals, electromechanical transducers.

### *1.6.4 Estimates for Automatic (Industrial) Decisions*

In these applications the accent also lies on the speed of information processing, particularly if this speed is so great that the human being is not able to do the job. As an example one may consider the decision of whether or not to switch off a generator in an electrical power station if there appears to be a short circuit in the distribution system. Here the criterion is whether or not the generator will fall out of step. This will be governed by the state variables of the generator and the power-distribution system.

### *1.6.5 Estimates for Automatic (Industrial) Adjustments*

An automatic adjustment can be considered as a series of repeated decisions on the continuation and the direction of an adjustment. The adjustment of band-pass filters during the manufacturing of radio and television sets can be interpreted as the adjustment of a model (the band-pass filter to be adjusted) according to a standard or reference circuit.

Another application is the compensation of a communication channel with regard to the intersymbol interference. These properties change if the cable configuration changes by interconnexions. Using a filter that consists of a delay line and potentiometers, one can obtain an automatic compensation (adaptive filter) that adjusts itself automatically with respect to the minimal interference.

### *1.6.6 Pattern Recognition*

This can be considered as the determination of a number of 'parameters' of the object under study together with the decision, based on these estimations, as to which class this object belongs. The most crucial question with respect to pattern recognition is: what parameters are the most important? As far as applications are concerned, one can think of character recognition, but also of signal classification (cardiographs etc.) and, for example, the classifications of data obtained by oil exploration.

An interesting list of possible applications of parameter estimation given by Clymer (1958) includes the following topics:

A. Aircraft and Automotive Industries.

determination of aerodynamic coefficients from
(*a*) flight tests involving complicated manoeuvres
(*b*) dynamic tests in wind tunnels with elastic mountings
(*c*) rocket sled tests;

determination of the flutter speed of an airplane or missile surface from subcritical tests of
(*a*) airplanes in flight
(*b*) wind-tunnel flutter models
(*c*) rocket sled models;

by determining the damping of the flutter mode as a function of frequency;

determination of the flexibility characteristics of any structure;

determination of servo internal characteristics and output impedance;

determination of the characteristics of the interior ballistics of guns, rocket capsules and the like;

determination of the dynamic characteristics of tires;

determination of the non-linear characteristics of airplane control sticks, rudder pedals and steering-wheel linkages;

determination of the dynamic characteristics of landing gears and automotive suspension systems;

determination of the dynamic characteristics of turbojet, turboprop, reciprocating, free-piston, and all other engines, compressors, etc.

determination of the parameters governing transient air flow through a porous material such as a fabric;

determination of the course and speed of a target as an aid to tracking of fire control;

determination of the characteristics of a 'satellite' and its upper-atmosphere environment, knowing the history of its motion (i.e. its orbit and perturbations);

design of aircraft landing-control systems, mid-air refuelling systems and other special-purpose computers using prediction.

## B. Process Industries (chemicals, fuels, materials, etc.).

determination of the chemical kinetic and other parameters which determine the reaction rates in batch or continuous processes;

determination of: the half-lives of the constituents, or the composition of a radio-chemical mixture;

determination of the dynamic rheological characteristics of materials undergoing deformation or flow;

determination of the lumped parameters in sets of ordinary differential questions used to approximate distributed-parameter systems such as heat exchangers;

determination of the characteristics of combustion processes;

determination of neutron process parameters in nuclear reactors;

determination of the dynamic parameters of instruments, tranducers and process-control system components.

## C. Miscellaneous.

determination of dynamic parameters in biological population systems;

determination of the characteristics of systems for which forecasts are to be made, such as dynamic economic systems (sales of product, growth of new industry, etc.);

determination of the dynamic characteristics of a human operator on various devices when he is a part of a loop;

determination of the dynamic characteristics of the regulatory system in human or animal physiology.

Summarizing, one may view the parameter and state estimation as an extension of the classical measurement theory to more complicated situations and tasks, made possible by the advent of modern computing equipment.

## 1.7 A Wider Perspective

In a number of fields of physical and engineering science, theorems have been derived with respect to the maximum result obtainable. Some examples:

thermodynamics: the maximal efficiency of energy conversion in a given situation is determined by the Carnot cycle;

physics: the uncertainty relations of Heisenberg indicate the limit of accuracy that can be obtained by simultaneous measurement of certain quantities;

communication theory: the Shannon relations indicate the (maximal) channel capacity.

In the fields of parameter and state estimation there is a need too for such limiting theorems. It would be most useful if it were possible to indicate the maximum amount of 'information' about a parameter or a state vector that could be derived in clearly specified situations. In that way different estimation techniques or schemes applied to a given situation, could be compared just as communication techniques can be evaluated by communication (information) theory. In fact, the comparison of parameter and state estimation with the communication problem suggests itself quite strongly. The values to be estimated are the messages sent by an information source; the transmitter, channel, receiver represent the information processing scheme. The (channel) noise stands for the disturbances in process and measurement devices. Appealing as this idea might be, a well-developed theory along these lines seems not yet to be available, in spite of some problem statements made (e.g. Zaborszky, 1966) and much material being available (Lindley, 1956; Kullback, 1959; Rényi, 1966; Barnes, 1968; Weidemann, 1969; Weidemann and Stear, 1970; Rajbman, 1970). Essential difficulties due to non-linearities are met if one tries to describe the development of crucial probability functions with respect to estimation time.

The best limiting theorem available at present is the Cramér–Rao inequality, cf. Chapter 11, which provides knowledge of the accuracy that can be obtained under certain prespecified conditions. It does not give, however, the possibility of taking into account the available *a priori* knowledge on the parameters before the estimation measurements and data processing.

## 1.8 Conclusion

It has been indicated that the field of identification/system-parameter estimation is rather complex. This complexity is due to the variety of applications, goals, conditions, etc. In practical situations, however, one may be guided by some basic requirements (Herles, 1970):

exact specification of the final aim to be attained by means of identification/parameter estimation;

selection of the 'important' properties;

analysis of the achievable precision.

These requirements and the present state of development of the field make that as yet each application exhibits elements of both 'art' and 'science'.

### References

Aoki, M. (1967). *Optimization of Stochastic Systems*, Academic Press, New York, 354 pp.

Barnes, J. (1968). 'Information theoretic aspects of feedback control systems', *Automatica*, **4,** 165–185.

Bellman, R. and K. J. Åström (1969). 'On structural identifiability', *Math. Biosci.*, **1,** 329–339.

Box, G. E. P. and G. M. Jenkins (1969). 'Discrete models for forecasting and control', in *Encyclopaedia of Linguistics, Information and Control* (Ed. A. R. Meetham), Pergamon Press, Oxford, pp. 162–168.

Box, G. E. P. and G. M. Jenkins (1970). *Time Series Analysis, Forecasting, and Control*, Holden Day, San Francisco.

Brockett, R. W. and M. D. Mesarovic (1965). 'The reproducibility of multivariable control systems', *JMAA*, **11,** 548–563.

Clymer, A. B. (1958). 'System synthesis by dynamic implicit computation', *Paper: Meeting Midwestern and Central Simulation Councils*, St. Louis (Mo.).

Eykhoff, P. (1963). 'Some fundamental aspects of process-parameter estimation', *IEEE Trans. Autom. Control*, **AC-8**, 347–357.

Eykhoff, P., P. M. E. M. Van der Grinten, H. Kwakernaak and B. P. Th. Veltman (1966). 'Systems modelling and identification', *3rd IFAC Congr., London*, survey paper. 83 references.

Fel'dbaum, A. A. (1960, 1961). 'Dual control theory', *Automn. Remote Control*, **21,** 874–880, 1033–1039; **22,** 1–12, 109–121.

Fel'dbaum, A. A. (1965). *Optimal Control Systems*, Academic Press, New York, 452 pp.

Fisher, E. E. (1965). 'Identification of linear systems', *Joint Autom. Control Conf.*, 473–475.

Herles, V. (1970). 'Limits of precision in identification procedures', *2nd IFAC Symp. Identification and Process Parameter Estimation, Prague*, paper 9.8.

Kerr, R. B. (1965). 'Statistical, structural, and regression models in linear process identification', *Joint Autom. Control Conf.*, 458–466.

Kullback, S. (1959). *Information Theory and Statistics*, Wiley, New York, 395 pp.

Lindley, D. V. (1956). 'On a measure of the information provided by an experiment', *Ann. Math. Stat.*, **27,** 986–1005.

Peschon, J. and others (1968). 'Case studies in system control', *Joint Autom. Control Conf., University of Michigan;* published by IEEE, New York.

Potts, T. F., G. N. Ornstein and A. B. Clymer (1961). 'The automatic determination of human and other system parameters', *Proc. West Joint Computer Conf., Los Angeles*, **19,** 645–660.

Pugachev, V. S. (1963). 'Statistical methods in automatic control', *Proc. 2nd IFAC Congr., Basle*, survey paper.

Rajbman, N. S. (1970). cf. references Chapter 2.

Rényi, A. (1966). 'On the amount of missing information and the Neyman–Pearson Lemma', in *Research Papers in Statistics* (Ed. F. N. David), Festschrift for J. Neyman, Wiley, London.

Rosenblueth, A. and N. Wiener (1945). 'The role of models in science', *Philosophy of Science*, **12,** 316–321.

Sain, M. K. and J. L. Massey (1969). 'Invertibility of linear, time-invariant dynamical systems', *IEEE Trans. Autom. Control*, **AC-14,** 141–149.

Silverman, L. M. (1969). 'Inversion of multivariable linear systems', *IEEE Trans. Autom. Control*, **AC-14**, 270–276.

Tsypkin, Ya. Z. (1966). 'Adaptation, learning and selflearning in control systems', *3rd IFAC Congr., London;* survey paper, also: 'Adaptation, training and self-organization in automatic systems', *Automn. Remote Control*, **27**, 16–51. 160 references.

Tsypkin, Ya. Z. (1967). 'Optimization, adaptation, and learning in automatic systems', *Computer and Information Sciences*, **2,** 15–32.

Tsypkin, Ya. Z. (1968). 'Self-learning—What is it?', *IEEE Trans. Autom. Control*, **AC-13,** 608–612.

Tsypkin, Ya. Z. (1970). 'Generalized algorithms of learning,' *Automn. Remote Control*, **31,** 97–104.

Weidemann, H. L. (1969). 'Entropy analysis of feedback control systems', in *Advances in Control Systems*, Vol. 7 (Ed. C. T. Leondes), Academic Press, New York, pp. 225–255.

Weidemann, H. L. and E. B. Stear (1970). 'Entropy analysis of estimating systems', *IEEE Trans. Information Theory*, **IT-16,** 264–270.

Zaborszky, J. (1966). 'An information theory viewpoint for the general identification problem (correspondence)', *IEEE Trans. Autom. Control*, **AC-11,** 130–131.

Zadeh, L. A. (1962). 'From circuit theory to system theory', *Proc. IRE*, **50,** 856–865.

*CHAPTER 2*

# *Basic Approaches to the Problem; Statistical and Engineering*

This chapter gives a **survey** of the parameter-estimation problem. The purpose is mainly to indicate the **relations** and **distinctions** between the different estimation techniques. This chapter should provide the reader with a clear picture of the organization of this text. There is nothing in this chapter that will not be discussed more fully and adequately in the chapters to follow. Consequently, the reader should not be alarmed by notions and discussions that as yet may be unfamiliar to him. On the other hand he is advised to return to this chapter once in a while to develop his overall picture of this field.

Some general aspects of the estimation problem, treated in Section 2.1, include process input signals, the *a priori* knowledge about the process and the choice of estimation schemes.

Attention is drawn to the existence of two classes of implementation in Section 2.2.

In Section 2.3, the statistical considerations that result in estimates providing explicit answers (expected value, variance or confidence interval) for the parameters to be determined are outlined.

In Section 2.4, attention is devoted to model-adjustment approaches. These imply the need of deriving partial derivatives of the error criterion with respect to the parameters to be adjusted. Some ways of obtaining these derivatives are given.

A particularly interesting situation arises if the error is a linear function or functional of the parameters to be adjusted. This case leads to the generalized model approach.

## 2.1 General Aspects of the Problem

### *2.1.1 Process Input Signals*

The estimating schemes are partly determined by the type of signals that are available. The signals may be either 'continuous in time' (non-sampled) or sampled. With the proper technical means the sampled and continuous forms can be converted into each other, provided that the sampling frequency

$f_s$ satisfies the Whittaker–Shannon relation: $f_s > 2f_h$ where $f_h$ is the highest frequency occurring in the band-limited continuous signal. Under that condition, the amount of information available does not decrease through the sampling process. Ways of describing signals and signal characteristics are discussed in **Chapter 3.**

In cases where the addition of a test signal is permissible, the amount of freedom in choosing such a signal is an important aspect. In those circumstances, one has to consider: the band width of the signal, the maximum permitted energy and amplitude, and the properties with respect to the generation and the information processing of the test signal (cf. **Chapter 10**).

### *2.1.2* A Priori *Knowledge About the Process*

In Chapter 1, the importance of initial knowledge of the process is indicated. That knowledge is represented in the structure and possible parameter values of the physical or mathematical (*a priori*) 'model' of the process. As regards implementation, the availability of such knowledge may amount to a difference in complexity and investment of an order of magnitude. The same is often true for the distinction between **non-linear** and **linear** aspects of the process. Some knowledge on the **speed of parameter variation** is a basic necessity as this may determine the feasibility of any type of implementation for the problem.

Although the 'black box approach' is attractive from the theoretical point of view it often leads to solutions which are highly wasteful due to neglecting some of the knowledge which is available. In many practical situations there may also be more measurable process quantities than merely one output. If no use is made of these signals it inevitably means a loss of available information (unless the dynamic relation between those signals and the output is known deterministically, e.g. a first-order low-pass filter with a time constant of $\tau$ seconds, and if no noise is added). The use of such additional 'outputs' has to be constantly kept in mind. Based on the knowledge of the process and the purpose of the estimation, it is mostly clear what must be demanded of a model. The parameters of the model must preferably be such that their values are a direct measure of the parameters of interest. Now, there are many different model representations available; such representations are discussed in **Chapter 4.** The choice of the type of representation is strongly case determined.

### *2.1.3 Estimation Scheme*

For each particular type of description the requirements posed on the implementation of the estimation problem will be to find the characteristics of the process:

accurately, not affected by additive noise and possible unwanted non-linearities in the system;

rapidly, yet under the condition that the system has to be always stable;

economically, using an instrumentation that is feasible and acceptable as regards costs.

The estimation scheme should preferably satisfy the following requirements:

it should be mathematically tractable;

it should be easily implemented;

it should be generally applicable;

it should yield an 'optimum' estimate;

it should yield an acceptable speed of convergence.

From the above list, which of course could be expanded, it will be obvious that the choice of an estimation scheme is governed by many considerations. Some of the general statistical and convergency notions are discussed in **Chapter 5.** For the reader who likes to review the notions from probability theory a survey is given in **Appendix B.**

## 2.2 Classes of Implementation

A distinction can be made with respect to the type of implementation used for solving the parameter-estimation problem. In principle one may distinguish two different classes (Eykhoff, 1963):

Implementation of mathematical relations that result in numerical quantities (e.g. coefficients of the differential equations; characteristic frequencies and damping ratios of poles and zeros; a number of points of the impulse response). This implementation will be denoted as: **'using explicit mathematical relations'.**

Implementation in the form of a (physical or mathematical) model of which the parameters are controlled in such a way, that the model characteristics approach the characteristics of the system under study in some predefined sense. This will be called: **'using a model-adjustment technique'.**

This difference is illustrated by the block diagrams of Figures 2.1a and b; one realization of these two classes is shown in Figures 2.1c and d.

In analytical terms the difference amounts to the following. Suppose the process output $y(t)$ is given by:

$$y(t) = f(u, \mathbf{b}, n) \tag{2.1a}$$

and the output $y_M$ of an (assumed) model by:

$$y_M(t) = f(u, \boldsymbol{\beta}, 0) \tag{2.1b}$$

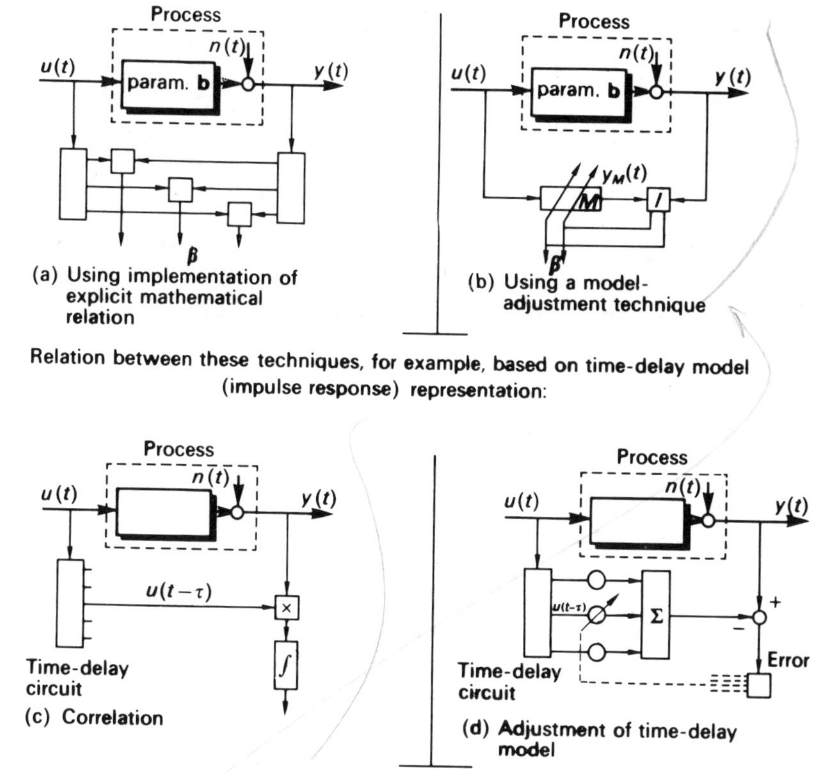


Figure 2.1

where

| | |
|---|---|
| **b** | is the $m+1$-dimensional vector of process parameters, i.e. a shorthand notation for the parameters $b_0, \ldots, b_m$; |
| $u = u(t)$ | is the process and model input; |
| $n = n(t)$ | is the noise (disturbance); |
| $\boldsymbol{\beta}$ | is the $m+1$-dimensional vector of parameter estimates; |
| $f$ | is a functional relationship between input and output quantities. |

This choice implies that we assume that the process and the model can be described by the same 'structure' and that the model does not suffer from an (appreciable) amount of noise.

As an example, let us assume that the correspondence between process and model is measured according to an error criterion:

$$E\{y, y_M; \boldsymbol{\beta}\} = \int_{t-T}^{t} \{y(t) - y_M(t; \boldsymbol{\beta})\}^2 \, dt. \tag{2.2}$$

Now one can follow two strategies:

I. Put

$$\frac{\partial E}{\partial \beta_i} \equiv 0 \qquad \text{for } i = 0, \ldots, m.$$

This is a necessary condition for obtaining the minimal error. These $m + 1$ equations with the $m + 1$ unknown estimates $(\beta_0, \ldots, \beta_m)$ can be solved for the $\beta$'s thus providing the **explicit mathematical relations.** (The additional requirement that $\partial^2 E/\partial \beta_i^2 > 0$ is tacitly assumed to be fulfilled.)

II. Put

$$\frac{\partial E}{\partial \beta_i} \to 0 \qquad \text{for } i = 0, \ldots, m.$$

A convergence towards zero can be obtained if $\partial E/\partial \beta_i$ can be determined by a suitable instrumentation and if these derivatives are used for the **adjustment of a physical model.**

Instead of these names chosen for the two classes, the following distinctions could also be used:

| **Class I** | **Class II** |
| --- | --- |
| using explicit mathematical relations | using model-adjustment techniques |
| or | or |
| explicit methods | implicit methods |
| or | or |
| open-loop methods | closed-loop methods |
| or | or |
| direct methods | iterative methods |
| or | |
| accumulative or 'one-shot' procedures | |
| where the estimate: | where the estimate: |
| is available after a **finite** number of elementary operations | is available after (in principle) an **infinite** number of elementary operations |
| requires considerable memory | requires less memory |
| is not available in an approximate form as an intermediate result | is available in an approximate form as an intermediate result |
| is open loop with respect to the estimate | is closed loop with respect to the estimate |
| | is found by a self-correcting procedure |

The reasoning behind these aspects will be clarified in

| | |
|---|---|
| **Chapter 6** for sampled signals | **Chapter 7** for sampled signals |
| **Chapter 8** for non-sampled signals | **Chapter 9** for non-sampled signals |

These differences are of great importance in engineering sense.

The number of operations and the memory required are relevant with respect to the computational means chosen for the implementation, viz. digital, analogue or hybrid. The availability of intermediate results may be essential for a particular type of application, e.g. in those cases where new measurements become available, and thus have to be used for on-line updating of the estimates (tracking); particular for control purposes such intermediate results are important. Consequently the distinction between **off-line** and **on-line** methods is relevant. The model offers the possibility of using *a priori* knowledge about the process (structure, initial estimates of parameters). The accuracy of the solution depends more strongly on the quality of some of the implementation components for class I than for class II, due to the properties of the (self-correcting) closed loop that can be recognized in class II.

The diagrams in Figure 2.1 are rather general ones; there is not only much freedom as regards the description of the signals, cf. Chapter 3, but they apply also to a wide variety of process descriptions (models), e.g. impulse-response model, transfer-function model, orthogonal series- expansion model, state-equation model. There is a large body of literature on these representations of processes, cf. Chapter 4.

## 2.3 Classes of Statistical Approaches

Among the problems studied in statistics are:

the condensation of information, given by a large set of data, to a few characteristic numbers (**data reduction**);

the sampling from a (conceptually) large population and the drawing of inferences about the properties of this large population from the properties of the sample taken. The goals of this procedure may be the following: (1) to test a statistical assumption, e.g. that a property of the population has a probability function of a certain class (**hypothesis testing**); (2) to find numerical values for the estimate and the variance of an unknown parameter in the distribution function of the population (**population parameters**);

the analysis of the relationship between two properties, e.g. to find numerical values for the estimate and its variance for coefficients in an assumed relationship between quantities $u$ and $y$ (**dependence parameters**).

The process-parameter estimation problem may be approached as a study of population parameters or of dependence parameters. The latter point of view is obvious if $u$ and $y$ represent the data available from the process input and output signals. The interpretation of a population-parameter study is arrived at by using the method of maximum likelihood estimates.

The results desired from the estimation procedure may be of a different kind, dictated by the type of application at hand and the available *a priori* knowledge. Consider a process represented in a state description form:*

$$\left.\begin{aligned}\dot{\mathbf{x}} &= \mathbf{f}(\mathbf{x}, u, \mathbf{b})\\ y &= g(\mathbf{x}, u, \mathbf{b}, n)\end{aligned}\right\} \tag{2.3a}$$

with

$$\begin{aligned} u = u(t) &= \text{input signal}\\ y = y(t) &= \text{output signal}\\ n = n(t) &= \text{noise}\\ \mathbf{x} = \mathbf{x}(t) &= \text{process state vector}\\ \mathbf{b} &= \text{process parameter vector}\end{aligned}$$

$$\mathbf{x}(t) = \begin{bmatrix} x_1(t)\\ \vdots \\ x_n(t)\end{bmatrix} \qquad \mathbf{b} = \begin{bmatrix} b_0\\ \vdots \\ b_m\end{bmatrix}$$

and a model description of the same form:

$$\left.\begin{aligned}\dot{\boldsymbol{\xi}} &= \mathbf{f}(\boldsymbol{\xi}, u, \boldsymbol{\beta})\\ y_M &= g(\boldsymbol{\xi}, u, \boldsymbol{\beta}, 0)\end{aligned}\right\} \tag{2.3b}$$

with

$$\begin{aligned}\boldsymbol{\xi} = \boldsymbol{\xi}(t) &= \text{model state vector}\\ \boldsymbol{\beta} &= \text{model parameter vector}\end{aligned}$$

A simplified linear case is sketched in Figure 2.2. Now the estimation problem may be formulated as a minimization of:

| | Parameter | State |
|---|---|---|
| the function(al) of a **measurable** quantity, viz. the difference of process and model output; | $E\{y(t;\mathbf{b}) - y_M(t;\boldsymbol{\beta})\}$ | $E\{y(\mathbf{x}) - y_M(\boldsymbol{\xi})\}$ |
| the expectation $\mathscr{E}[\ ]$† of the function(al) of an **unmeasurable** quantity, viz. the process parameters and/or state. | $\mathscr{E}[E\{\mathbf{b} - \boldsymbol{\beta}\}]$ | $\mathscr{E}[E\{\mathbf{x} - \boldsymbol{\xi}\}]$ |

* The reader unfamiliar with the state space description may turn to *Chapter 4* for a survey.
† cf. Appendix B.

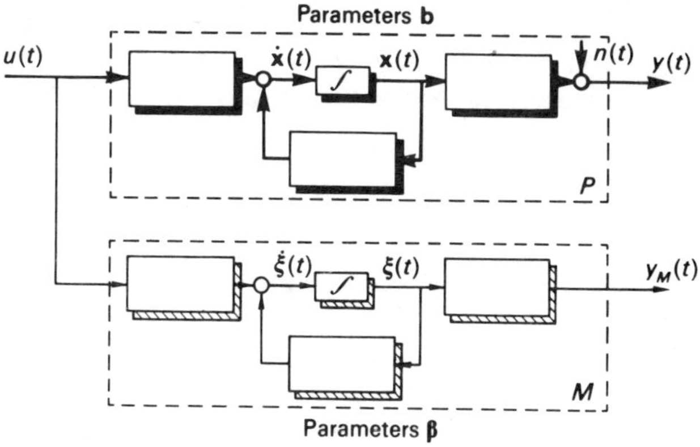


Figure 2.2

Some function(al)* of $e = y - y_M$, the 'error' between the output of the process and a model, can be minimized as $e(t)$ can be made measurable by proper implementation or by programming the model on a computer.

Depending on the model and on the input signal, this error provides some measure of the correspondence of the parameter and/or state vectors of process and model. In some cases, this correspondence of input–output relations is more important than parameter correspondence, particularly if the model is simpler, e.g. of a lower order, than the process. If we focus our attention on **b** and/or **x** directly then we notice that **b** and **x** are inaccessible for direct measurement. Consequently, one can minimize only the expectation of $E\{\mathbf{b} - \boldsymbol{\beta}\}$ or $E\{\mathbf{x} - \boldsymbol{\xi}\}$. This, however, is possible only if some *a priori* knowledge is available with respect to probability densities.

As in many other situations, we are tempted to look for the 'optimal' or 'best' estimation procedure. The definition of 'best', however, depends on the type of application, on the requirements as regards the properties of the estimate, on the *a priori* knowledge about the process and the additive noise. In the statistical literature, a number of different estimation procedures have been developed. These methods differ predominantly in the criteria used for defining optimality and in the use of available *a priori* knowledge. It is unfortunate that the choice between these criteria has aspects that are more or less subjective and that the mathematical approach is strongly dependent on these criteria. In **Chapter 5,** these estimation procedures and

*The distinction between function and functional can be illustrated by the simple example:

function $y(t) = a_1x(t) + a_2x^2(t) + a_3x^3(t)$, the value at $t$ is dependent only on $x(t)$;

functional $y(t) = \int_0^t x(t - \theta)h(\theta)\,d\theta$ the value at $t$ is dependent on $x$ over the interval $(0 - t)$.

their relations are discussed. Here a brief preview is in order. **Again it should be stressed that it suffices that from a first reading of this chapter the reader retains merely a general idea about the methods and the expressions that come into the picture.**

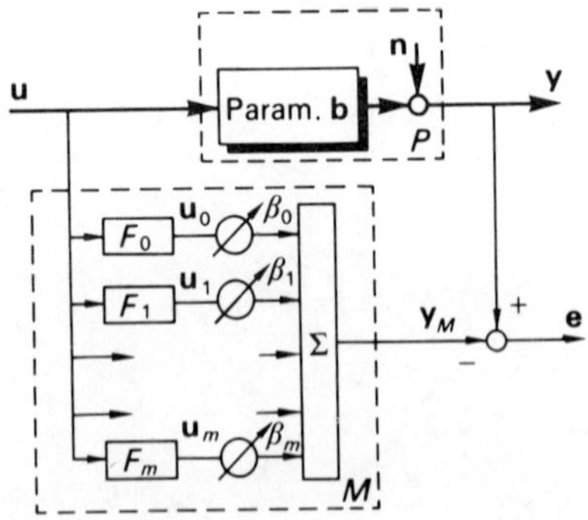


Figure 2.3

A simple approach can be represented by a block diagram according to Figure 2.3. The case of **sampled** time functions is considered. The quantities $\mathbf{u}$, $\mathbf{u}_i$, $\mathbf{y}$, $\mathbf{n}$ and $\mathbf{y}_M$ represent vectors consisting of samples, e.g.

$$\mathbf{y}' = [y(T_s), y(2T_s), \ldots, y(kT_s)]^* \tag{2.4}$$

The quantities $\mathbf{b}$ and $\boldsymbol{\beta}$ represent vectors consisting of the process parameters and model parameters respectively:

$$\mathbf{b}' = [b_0, b_1, \ldots, b_m]$$

$$\boldsymbol{\beta}' = [\beta_0, \beta_1, \ldots, \beta_m]$$

Consequently, one finds:

$$\mathbf{y} = \mathbf{y}(\mathbf{u}, \mathbf{b}, \mathbf{n}) \tag{2.5a}$$

$$\mathbf{y}_M = \mathbf{U}\boldsymbol{\beta} \tag{2.5b}$$

with

$$\mathbf{U} = [\mathbf{u}_0, \mathbf{u}_1, \ldots, \mathbf{u}_m] = \begin{bmatrix} u_0(1) & \ldots & u_m(1) \\ \vdots & & \vdots \\ u_0(k) & \ldots & u_m(k) \end{bmatrix} \tag{2.6}$$

* $\mathbf{y}'$ denotes the transpose of the (column) vector $\mathbf{y}$. In the remaining part of the text the sample interval $T_s$ will be constant and will be deleted.

In the matrix $\mathbf{U}$ each column $\mathbf{u}_i$ consists of sample values of the outputs of the blocks $F_i$. These blocks are chosen on the basis of the process representation needed, cf. **Chapter 4.**

Now we can define the problem as finding the estimate $\boldsymbol{\beta}$ of the process parameter vector $\mathbf{b}$, based on observations of $\mathbf{y}$. Based on the available *a priori* knowledge one can make a distinction between different kinds of estimates:

### 2.3.1 *Least-squares Estimation*

This does not require any additional knowledge. The error criterion is defined as:

$$E = \mathbf{e}'\mathbf{I}\mathbf{e} = \sum_{j=1}^{k} e^2(j) \tag{2.7}$$

with

$$\mathbf{e} = \mathbf{y} - \mathbf{y}_M \tag{2.8}$$

$\mathbf{I}$ is the identity matrix. From the minimization of the error criterion the so called 'normal equation' for the estimator* is found to be:

$$\hat{\boldsymbol{\beta}} = [\mathbf{U}'\mathbf{U}]^{-1}\mathbf{U}'\mathbf{y} \qquad \text{(least-squares estimator)} \tag{2.9}$$

where again $\mathbf{U}'$ is the transpose of $\mathbf{U}$.

### 2.3.2 *Markov or Generalized Least-squares Estimation*

Now it is assumed that we have some knowledge about the additive noise $\mathbf{n}$, viz. its covariance matrix:

$$\mathbf{N} = \mathscr{E}[\mathbf{n}\mathbf{n}'] = \begin{bmatrix} \mathscr{E}[n(1)n(1)] & \cdots & \mathscr{E}[n(1)n(k)] \\ \vdots & & \vdots \\ \mathscr{E}[n(k)n(1) & \cdots & \mathscr{E}[n(k)n(k)] \end{bmatrix} \tag{2.10}$$

It can be proved that if $\mathbf{N} \neq \mathbf{I}$ (identity matrix) a better linear estimator can be found by minimizing

$$E = \mathbf{e}'\mathbf{N}^{-1}\mathbf{e} \tag{2.11}$$

resulting in

$$\hat{\boldsymbol{\beta}} = [\mathbf{U}'\mathbf{N}^{-1}\mathbf{U}]^{-1}\mathbf{U}'\mathbf{N}^{-1}\mathbf{y} \qquad \text{(Markov or generalized least-squares estimator)} \tag{2.12}$$

* As in statistical literature we will denote by **estimator** a rule of estimation, by **estimate** the value to which it gives rise in a particular case.

### 2.3.3 *Maximum Likelihood Estimation*

It has already been mentioned that the estimation problem can also be approached as being a population-parameter study. Now the attention is focused on the output vector **y**.

*A priori* (before the measurement) we know that the samples of **y** are random variables, that have the joint probability function

$$p\{y(1), \ldots, y(k); \mathbf{b}\} \tag{2.13}$$

cf. Appendix B; **y** depends on **b**.

*A posteriori* (after the measurement) we **know** the observed sample values of these random variables, say $y(1) = c_1 \ldots, y(k) = c_k$. From this we want to determine an estimate $\boldsymbol{\beta}$. The functional relationship between $\mathbf{c}' = \{c_1, \ldots, c_k\}$ and $\boldsymbol{\beta}$ is the same as given in (2.13). In order to accentuate the dependence on the observed sample values we write this dependence as

$$L\{c_1, \ldots, c_k; \boldsymbol{\beta}\} \tag{2.14}$$

This is called the **likelihood function.** As the sample values are given by the measurements this $L$ is a function of $\boldsymbol{\beta}$ only. For the sake of convenience, $\ln L$ is usually considered; as the logarithmic function is monotonic, the maximum of $L$ and the maximum of $\ln L$ occur at the same value of $\boldsymbol{\beta}$. This value of $\boldsymbol{\beta}$ can be obtained by solving:

$$\nabla_{\boldsymbol{\beta}} \ln L\{\mathbf{c}; \boldsymbol{\beta}\}\bigg|_{\boldsymbol{\beta}=\hat{\boldsymbol{\beta}}} = \frac{\partial}{\partial \boldsymbol{\beta}} \ln L\{\mathbf{c}; \boldsymbol{\beta}\}\bigg|_{\boldsymbol{\beta}=\hat{\boldsymbol{\beta}}} = \mathbf{0} \tag{2.15}$$

Such a solution, $\hat{\boldsymbol{\beta}}$, is called a maximum likelihood estimate of **b**, if $\hat{\boldsymbol{\beta}}$ actually depends on the observed sample values $c_1, \ldots, c_k$. From this outline it follows that not only the covariance matrix of the noise has to be known but also the joint probability density function of its samples.

### 2.3.4 *Minimum Risk Estimation*

In this case, even more *a priori* information is needed. As outlined in Chapter 5, the probability density functions of **u**, **n** and **b** have to be given.

It is a matter of theoretical and engineering interest to consider how much available *a priori* knowledge will be used in practice. Generally speaking, the use of more *a priori* knowledge may lead to a better estimation scheme; the costs of implementation, however, may easily become prohibitive. Whatever estimation technique is being used, it has to be based on observations, which are random variables due to the additive noise. Consequently, any estimate, being a function of these observations (statistic), is a random variable too.

## 2.4 Model-adjustment Techniques

### *2.4.1 'Odd-function' and 'Even-function' Error*

The model-adjustment technique has been interpreted with Figure 2.1b as the task to adjust the model $M$ in such a way that the transfer characteristics of $M$ approach those of the process $P$ in some predefined sense. The crucial part of Figure 2.1b is the block denoted by $I$. This block processes the information that is available. The **error** may be chosen as $e(t) = y(t) - y_M(t)$ and the **error criterion** may be the minimization of any even-functional $E$ of this error, e.g.

$$E = \int_{t_1}^{t_2} e^2(t, b, \boldsymbol{\beta})\,dt \tag{2.16}$$

where again $b$ is the parameter of the process under consideration and $\beta$ is the corresponding model parameter. Such an even functional leads to the type of relation indicated in Figure 2.4 between $E$ and the value of the

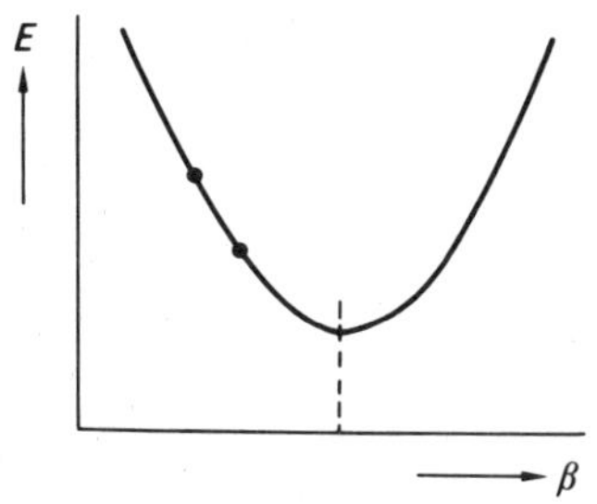


Figure 2.4

parameter that has to be adjusted. This brings us to the distinction between the so called **'odd-function'** and **'even-function'** control (Eykhoff, 1960). Let us look at the human eye, which shows both these different types of control (cf. Figure 2.5a).

The **iris** controls the illumination level of the retina in an ordinary control-system way; if the level is too high, then the error signal causes the iris to contract and vice versa. The **sign** of the error signal is a direct indication of the direction in which the iris has to change (cf. Figure 2.5b); the error signal goes to zero.

The **lens** controls the definition of the image on the retina. Now there is no error signal that changes sign when the lens accommodates from too far a distance to too close or vice versa. The function of definition versus amount of accommodation, and thus the error signal, has an extremum

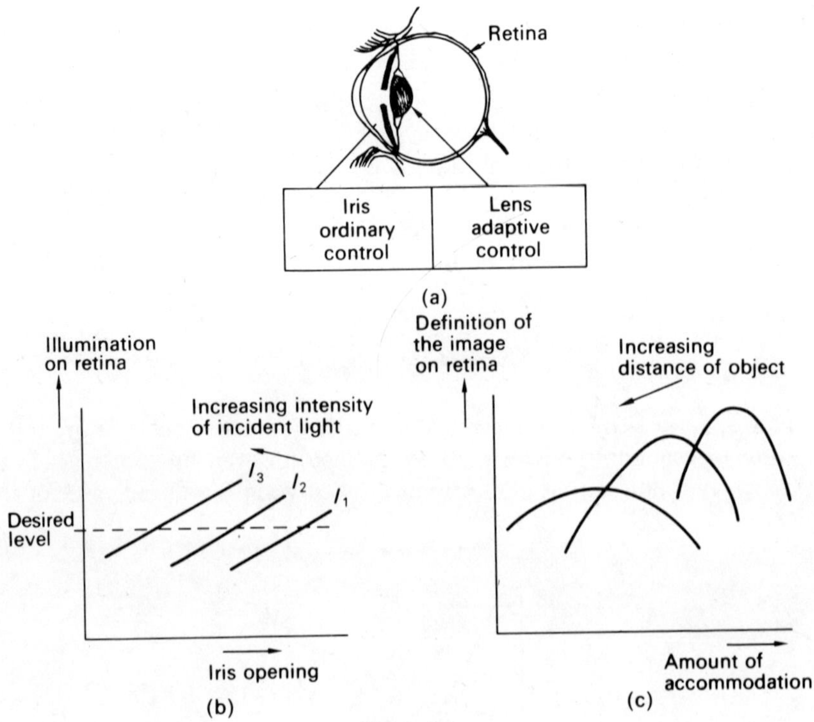


Figure 2.5

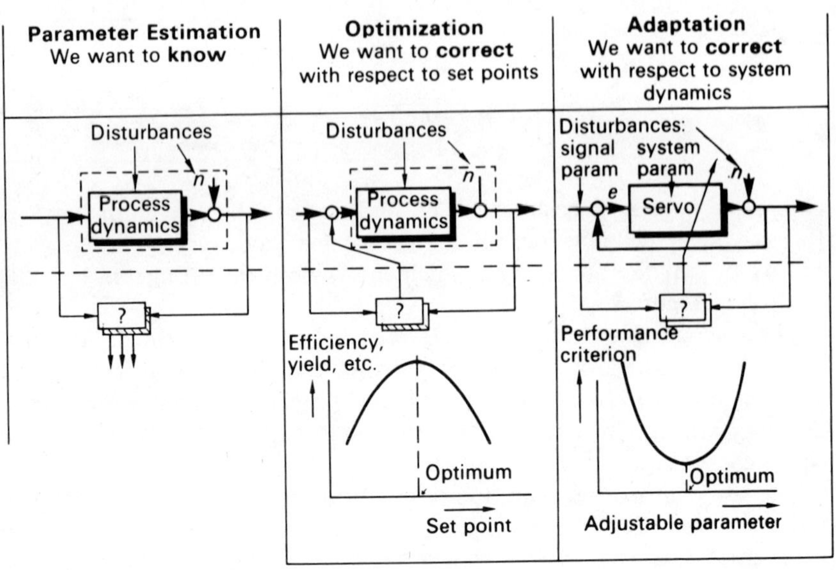


Figure 2.6

(cf. Figure 2.5c).* We face the same situation with the error defined for the model-adjustment problem. This situation seems to be characteristic of adaptive and optimizing feedback control systems as well. Figure 2.6 gives a comparison of the goals of parameter estimation, optimization and adaptation. Figure 2.7 shows the main difference between the ordinary control and this 'adaptive' type of control. The two types of feedback systems will be called with regard to the fundamental difference **'odd-function'** (error) systems and **'even-function'** (error) systems. From Figure 2.8 it is evident

| | I<br>**Ordinary feedback control** | II<br>**Adaptive and self optimizing feedback control** |
|---|---|---|
| General scheme | 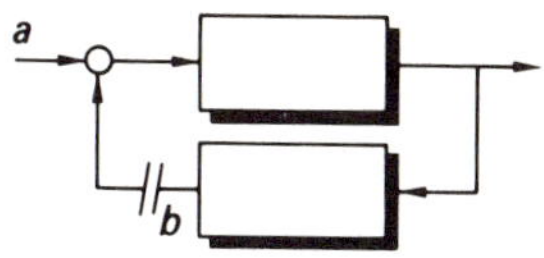 | Many different schemes |
| 'Error gain' function | 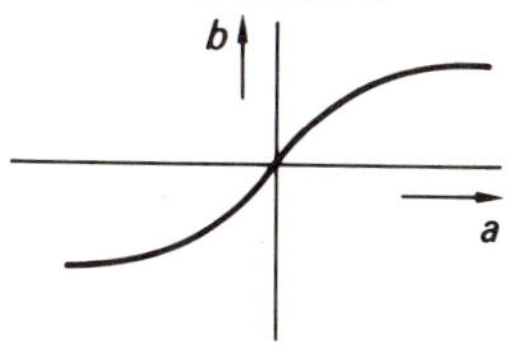<br>'Odd' function | Performance<br>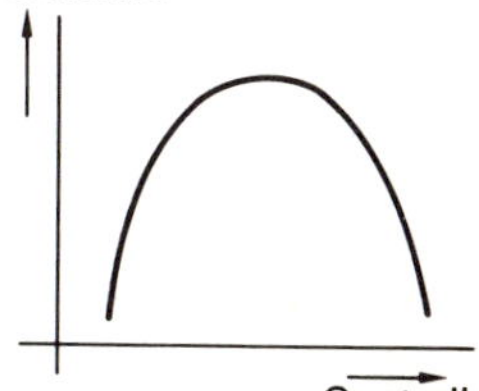<br>'Even' function |
| Principal property | 1st derivative does not change sign | 2nd derivative does not change sign |
| *A priori* knowledge | Nul exists;<br>Sign of 1st derivative known | Extreme exists;<br>Sign of 2nd derivative known |
| Sufficient information for feedback action | One observation | Minimum two observations (and data reduction) |

Figure 2.7

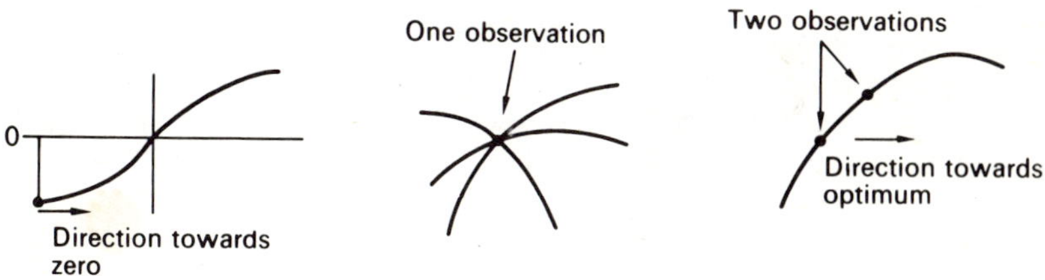


(a) (b) (c)

Figure 2.8

* In this situation a number of different clues are being used for the control of accommodation.

that when the sign of the second derivative is known (maximum or minimum of the function) for an even-function system at least **two** observations are required to locate the direction of the extremum as regards to these observations (cf. Figure 2.8c). In order to carry out control action we have to reduce the information of the two or more observations to one quantity. This data reduction changes the 'even-function' error into an 'odd-function' error. After this reduction has been applied, ordinary control principles based on this odd-function error can be used.

The direction towards the optimum adjustment is shown by the first derivative with respect to $\beta$. For **continuous (non-sampled) signals,** using expression (2.16) for the integral squared error one finds:

$$\nabla_{\boldsymbol{\beta}} E = \frac{\partial E}{\partial \boldsymbol{\beta}} = 2 \int_{t_1}^{t_2} e(t, b, \boldsymbol{\beta}) \frac{\partial e}{\partial \boldsymbol{\beta}} \, dt \tag{2.17}$$

Neither $\partial E/\partial \boldsymbol{\beta}$ nor $\partial e/\partial \boldsymbol{\beta} = -\partial y_M/\partial \boldsymbol{\beta}$ are measurable explicitly without additional instrumentation. Some techniques for obtaining the desired information are summarized in the following. If such partial derivatives can be determined then by 'hill climbing' techniques, cf. **Chapter 5,** on the function(al) $E$ one may find the 'best' value of $\boldsymbol{\beta}$.

For **sampled signals** one finds that, if the representation chosen is adequate for describing the process behaviour, (cf. Figure 2.3):

$$\begin{aligned} \mathbf{y} &= \mathbf{U}\mathbf{b} + \mathbf{n} \quad \text{Process} \\ \mathbf{y}_M &= \mathbf{U}\boldsymbol{\beta} \qquad\quad \text{Model} \end{aligned} \tag{2.18a}$$

and the error is found to be

$$\mathbf{e} = \mathbf{y} - \mathbf{y}_M = \mathbf{U}(\mathbf{b} - \boldsymbol{\beta}) + \mathbf{n} \tag{2.18b}$$

Consequently a quadratic form

$$E = \mathbf{e}'\mathbf{e} \tag{2.19}$$

will be minimized if $\mathbf{b} = \boldsymbol{\beta}$. Again $\nabla_{\boldsymbol{\beta}} E = \partial E/\partial \boldsymbol{\beta}$ has to be found. Assuming that $\partial E/\partial \boldsymbol{\beta}$ can be determined, then the task is to find an implementation that gives an automatic adjustment of the parameters $\boldsymbol{\beta}$. There are several possible ways of using this information for obtaining that parameter-adjustment scheme. The adjustment can be continuous or intermittent.

*Continuous Adjustment Scheme.* A favoured policy is obtained by choosing

$$\boldsymbol{\beta} = \Gamma \nabla_{\boldsymbol{\beta}} E \tag{2.20}$$

Such a choice leads to the gradient method. The coefficient $\Gamma$ represents a 'gain factor', governing the speed of the adjustment. We will find that this coefficient may be constant or time dependent; it may be a scalar $\Gamma$ viz. $\Gamma(t)$

or a matrix $\boldsymbol{\Gamma}$ viz. $\boldsymbol{\Gamma}(t)$. Strictly speaking, $\boldsymbol{\beta}$ has to be time invariant while the gradient is determined. This is not the case in continuous adjustment schemes. For that reason equation (2.20) offers an approximate description of the adjustment dynamics only if that adjustment is comparatively slow.

*Intermittent Adjustment Scheme.* The problem of the gradient determination just mentioned does not occur if one uses an intermittent scheme: measuring again, etc. This problem is studied in the theory of stochastic approximation (Tsypkin, 1966), which may lead to an adjustment algorithm as for example:

$$\boldsymbol{\beta}(i+1) = \boldsymbol{\beta}(i) - \Gamma(i)\nabla_{\boldsymbol{\beta}}E(i) \tag{2.21}$$

with

$\boldsymbol{\beta}(i)$ = the parameter vector of the model after the $i$th adjustment

$E(i)$ = the value of the error criterion after the $i$th sequence of error samples

$\Gamma(i)$ = a 'gain factor' governing the speed of convergence

A simple example may indicate again the connexion between model-adjustment techniques and the use of explicit mathematical relations. Assume that the adjustment criterion is:

$$\min \mathbf{e}'\mathbf{e}$$

After the $i$th model adjustment we know from equation (2.9) that the optimal parameter is given by:

$$\begin{aligned}\hat{\boldsymbol{\beta}}(i+1) &= [\mathbf{U}'\mathbf{U}]^{-1}\mathbf{U}'(\mathbf{y}_M + \mathbf{e}) \\ &= [\mathbf{U}'\mathbf{U}]^{-1}\mathbf{U}'\mathbf{U}\boldsymbol{\beta}(i) + [\mathbf{U}'\mathbf{U}]^{-1}\mathbf{U}'\mathbf{e} \\ &= \boldsymbol{\beta}(i) + [\mathbf{U}'\mathbf{U}]^{-1}\mathbf{U}'\mathbf{e}\end{aligned}$$

This is the best least-squares estimation, given the '*a priori*' model adjustment and the particular series of error samples. For consecutive model adjustments one finds; cf. equation (2.21):

$$\hat{\boldsymbol{\beta}}(i+1) = \hat{\boldsymbol{\beta}}(i) + [\mathbf{U}'\mathbf{U}]^{-1}\mathbf{U}'\mathbf{e}(i)$$

A measure for the 'uncertainty' of this estimate, cov $[\hat{\boldsymbol{\beta}}(i)]$, can also be determined.

In the same way, min $\mathbf{e}'\mathbf{N}^{-1}\mathbf{e}$ could be used as the model-adjustment criterion. Convergence schemes, that is the way in which the gradients are used by the implementation, are discussed in **Chapter 5.**

### 2.4.2 *Techniques for Obtaining the Partial Derivatives or Gradients*

In principle the following approaches are available:

(*a*) *The Use of Parameter Influence Coefficients or Parameter Sensitivity Functions.* In recent years this notion has gained much interest, cf. **Chapter 9.** A simple example following Meissinger (1960) may indicate the procedure. Let us assume the model to be of the first order having one unknown parameter $\beta$ and an output time function $y_M = w$:

$$\frac{dw}{dt} + \beta w = u \tag{2.22}$$

and the error to be $e = y - w$. Thus $\partial e/\partial\beta = -\partial w/\partial\beta$. As we are interested in the behaviour for $-\infty < t < \infty$ initial conditions do not concern us. Formally $w = w(t, \beta)$ and

$$w(t, \beta) = w(t, \beta_0) + \left.\frac{\partial w}{\partial \beta}\right|_{\beta_0} \Delta\beta + \frac{1}{2!}\left.\frac{\partial^2 w}{\partial \beta^2}\right|_{\beta_0} (\Delta\beta)^2 + \cdots \tag{2.23}$$

Only the first-order extrapolation will be used. The corresponding term can be obtained from the differential equation (2.22) by differentiation with respect to $\beta$:

$$\frac{\partial^2 w}{\partial\beta\partial t} + \beta\frac{\partial w}{\partial \beta} + w = 0 \tag{2.24}$$

Interchanging the order of differentiation and substituting $v = \partial w/\partial\beta$ one finds:

$$\frac{\partial v}{\partial t} + \beta v = -w \tag{2.25}$$

This is called the sensitivity equation with respect to $\beta$. For varying $\beta$ the quantity $v$ is an approximation to the gradient $\partial w/\partial\beta$ which is a function of time. Equations (2.22) and (2.25) can be instrumented according to Figure 2.9 where the time function $v$ is generated. Equations such as (2.17) and (2.20) or (2.21) can be instrumented to give a system which converges towards $b$ under a few restrictions. This is discussed in detail in **Chapter 9.**

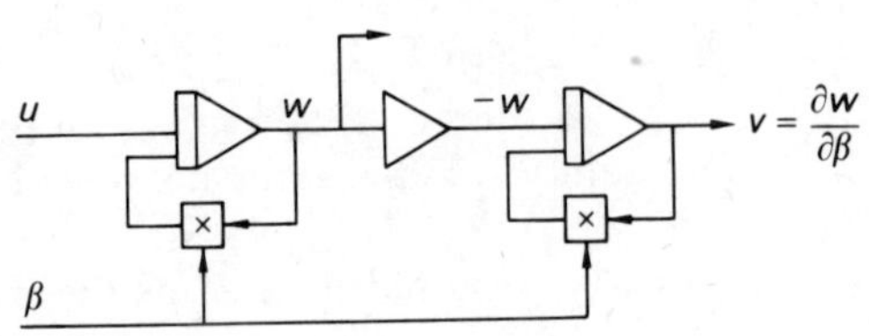


Figure 2.9

(*b*) *The Use of Two Models with Parameters* $\beta$ *and* $(\beta + \Delta\beta)$. The first derivative with respect to $\beta$ can be approximated by using two models with parameters $\beta$ and $(\beta + \Delta\beta)$ respectively. The difference of the outputs $w_1$ and $w_2$ of these models then does give a quantity $\Delta w/\Delta\beta = -\Delta e/\Delta\beta$ that can be used as an approximation of $\partial e/\partial\beta$ in an equation such as (2.17) and (2.20) or (2.21) (cf. Figure 2.10). This technique is discussed in **Chapter 9** as well.

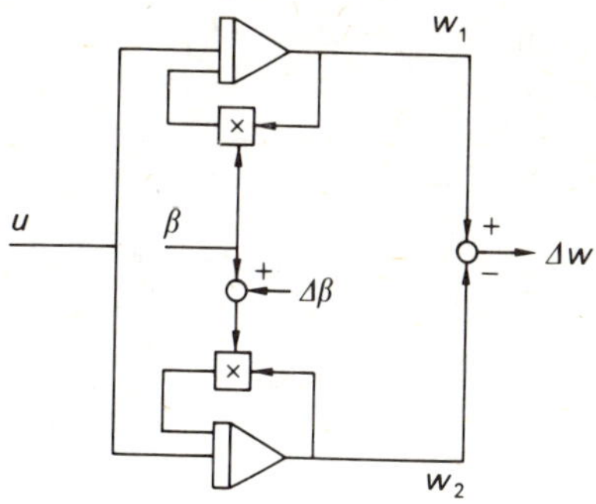


Figure 2.10

(*c*) *The Use of One Model with a Parameter Perturbation*. A third method is based on:

$$\frac{dE}{dt} = \frac{\partial E}{\partial\beta}\frac{d\beta}{dt} + \frac{\partial E}{\partial t} \tag{2.26}$$

From this expression it follows that if the model parameter is disturbed according to a known time function (test signal) and the resulting disturbance $dE/dt$ can be measured, then information on $\partial E/\partial\beta$ can be derived (cf. Figure 2.11). This technique is studied in **Chapter 9** too.

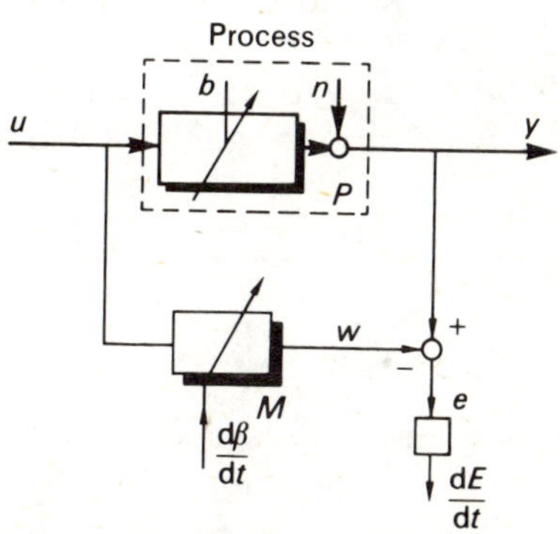


Figure 2.11

(*d*) *The Use of a Generalized Model.* It will be clear that the methods mentioned up to now for obtaining $\partial E/\partial \beta$, $\partial w/\partial \beta$ or an approximation of these quantities are putting quite heavy demands on the amount of implementation or computation. This is even more so if there are a number of parameters that have to be determined. Generally speaking, for each parameter to be estimated one needs another additional model or another model parameter perturbed by a suitably chosen test signal. For this reason it is worthwhile looking for an arrangement where the error is **linear-in-parameters** to be adjusted, for example:

$$e = \sum_{i=0}^{n} \alpha_i y_i + \sum_{j=0}^{m} \beta_j u_j \tag{2.27}$$

In that case the following holds:

$$\frac{\partial e}{\partial \alpha_i} = y_i \qquad \frac{\partial e}{\partial \beta_j} = u_j \tag{2.28}$$

i.e. the parameters sensitivity functions are already explicitly present in the model. Figure 2.12 gives an illustration of such a case. The input signal $u(t)$

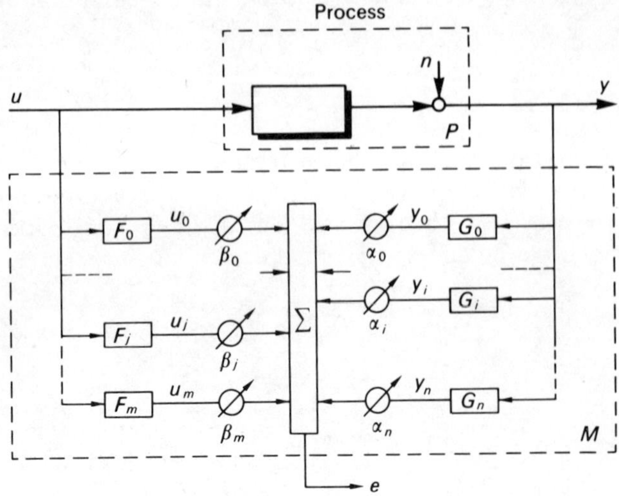


Figure 2.12

is stationary and in most cases considered to be stochastic. The blocks $F_0, \ldots, F_m$ and $G_0, \ldots, G_n$ are operators (e.g. linear transfer functions, non-linear operators). These blocks together with the 'potentiometers' $\alpha_0, \ldots, \alpha_n$ and $\beta_0, \ldots, \beta_m$ and the adder, form a **'generalized model'** of the

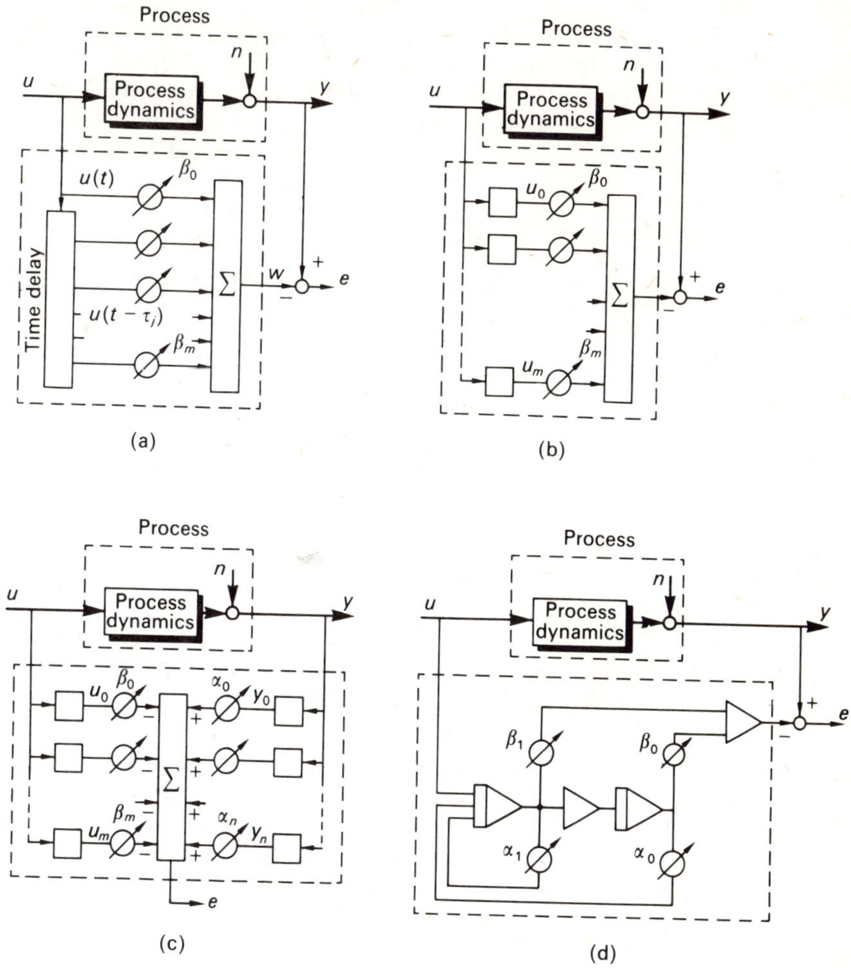


Figure 2.13

process. The 'potentiometers' can be adjusted to positive and negative coefficient values. The output $e(t)$ will be called a **'generalized error'.**

The reader will appreciate the convenience of a model that leads to an error which is linear in the parameters to be estimated. Figures 2.13a, b and c give examples of such situations:

(a) $$e(t) = y(t) - \sum \beta_j u(t - \tau_j)$$

(b) $$e(t) = y(t) - \sum \beta_j u_j(t)$$

(c) $$e(t) = \sum \alpha_i y_i(t) - \sum \beta_j u_j(t)$$

Note that (a) is a special case of (b) and (b) is a special case of (c). This does not necessarily imply that the input–output relationship of the model has to be linear. In connection with Figure 2.12, it has already been pointed out that the blocks $F_i$ and $G_j$ may be non-linear. Figure 2.13d gives an example where the error is non-linear in the parameters to be estimated:

(d) $$e(t) = y(t) - \mathcal{O}[u(t), \boldsymbol{\alpha}, \boldsymbol{\beta}]$$

where $\mathcal{O}[\ ]$ is the dynamic operator provided by the model and $\boldsymbol{\alpha}$ and $\boldsymbol{\beta}$ are the parameter vectors with components $\alpha_i$ and $\beta_j$.

## 2.5 Parameter and State Estimation

In quite a number of applications, the notion of state as mentioned in Chapter 1 plays an important part; this is particularly true for optimal control problems. As the state variables need not be measurable directly and as the quantities that are measurable will be subject to noise, the process state can be estimated only. This estimation problem is adequately solved for linear processes for which the dynamic properties are known completely (including the numerical values of the process parameters) (Kalman and Bucy, 1961). Many approximate schemes for non-linear processes have been proposed. These estimation techniques are summarized in **Chapter 12.**

The problem becomes more complicated if the parameter values of the process are not known beforehand but have to be estimated together with the state variables. This type of problem is discussed in **Chapter 13.**

**Chapter 14** is devoted to applications in different fields of engineering and other sciences.

## 2.6 Conclusion

Now the reader will be in a position to appreciate the organization of this book. He is strongly recommended to read again the preface, including the organization chart of the part on parameter estimation. In periodical literature, all kinds of estimation techniques are presented casewise. In studying the following parts of the book, the reader will notice many relations between the different estimation techniques. By having read this chapter the reader is well-stimulated to recognize such relations.

For survey papers in this field the reader may consult: Eykhoff (1964); Eykhoff, Van der Grinten, Kwakernaak and Veltman (1966); Cuenod and Sage (1967); Eykhoff (1967); Strobel (1967, 1968); Aleksandrovskii and Deich (1968); Strobel (1968); Young (1968); Bekey (1969); Balakrishnan and Peterka (1969); Åström and Eykhoff (1970).

Monographs of general interest are: Lee (1964); Davies (1970); Rajbman (1970); Sage and Melsa (1971); Richalet, Rault and Pouliquen (1971); Isermann (1971); Granpe (1972).

An extensive bibliography is: Nieman, Fisher and Seborg (1971).

## References

Aleksandrovskii, N. M. and A. M. Deich (1968). 'Determination of dynamic characteristics of nonlinear objects', (review), *Automn. Remote Control*, **29,** 142–160. 134 references.

Åström, K. J. and P. Eykhoff (1970). 'System identification—a survey', *2nd IFAC Symp. Identification and Process Parameter Estimation, Prague*, survey paper; also in *Automatica*, **7,** 123–162 (1971). 230 references.

Balakrishnan, A. V. and V. Peterka (1969). 'Identification in automatic control systems', *4th IFAC Congr., Warsaw*, survey paper 9; also in *Automatica*, **5,** 817–829 (1969). 125 references.

Bekey, G. A. (1969). 'Identification of dynamic systems by digital computers', *Joint Autom. Control Conf.* (Computer control workshop), 7–18.

Cuenod, M. and A. Sage (1967). 'Comparison of some methods used for process identification', *IFAC Symp. Identification in Autom. Control Systems, Prague*, survey paper; also in *Automatica*, **4,** 235–269 (1968). 79 references.

Davies, W. D. T. (1970). *System Identification for Self-Adaptive Control*, Wiley–Interscience, London, 380 pp.

Eykhoff, P. (1960). 'Adaptive and optimalizing control systems (correspondence)', *IRE Trans. Autom. Control*, **AC-5,** 148–151.

Eykhoff, P. (1963). cf. references Chapter 1.

Eykhoff, P. (1964). 'Process-parameter estimation', *Progress in Control Engineering*, Vol. 2 (Ed. R. H. Macmillan), Heywood, London.

Eykhoff, P., P. M. Van der Grinten, H. Kwakernaak and B. P. Veltman (1966). cf. references Chapter 1.

Eykhoff, P. (1967). 'Process parameter and state estimation', *IFAC Symp. Identification in Autom. Control Systems, Prague*, survey paper; also in *Automatica*, **4,** 205–233 (1968). 11 references.

Graupe, D. (1972). *Identification of Systems*, Van Nostrand, New York, 276 pp.

Isermann, R. (1971). *Experimental Analysis of Control System Dynamics* and *Theoretical Analysis* . . . (in German), Bibliographisches Inst., Mannheim, 276 and 122 pp.

Kalman, R. E. and R. S. Bucy (1961). cf. references Chapter 12.

Lee, R. C. K. (1964). *Optimal Estimation, Identification and Control*, M.I.T. Press, Cambridge Mass., 152 pp.

Meissinger, H. F. (1960). 'The use of parameter influence coefficients in computer analysis of dynamic systems', *Proc. West, Joint Computer Conf.*, San Francisco, 181–192.

Nieman, R. E., D. G. Fisher and D. E. Seborg (1971). 'A Review of process identification and parameter estimation techniques', *Int. J. Control*, **13,** 209–264.

Rajbman, N. S. (1970). *What is System Identification?* (in Russian), Nauka, Moscow.

Richalet, J., A. Rault and R. Pouliquen (1971). *Process Identification by the Model Method* (in French), Gordon and Breach, Paris, 361 pp.

Sage, A. P. and J. L. Melsa (1971). *System Identification*, Academic Press, New York, 221 pp.

Strobel, H. (1967, 1968). 'The approximation problem of the experimental system analysis' (in German), *Z. Messen. Steuern, Regeln.*, **10,** 460–464; **11,** 29–34; 73–77. 161 references.

Strobel, H. (1968). *System Analysis Using Deterministic Test Signals* (in German), V.E.B. Verlag Technik, Berlin, 219 pp.

Tsypkin, Ya. Z. (1966). cf. references Chapter 1.

Young, P. C. (1968). 'Process parameter estimation', *Control*, **12,** 931–936.

CHAPTER 3

# Signals; Deterministic and Stochastic

## 3.1 Classes of Signals

The basis of estimation and identification lies in the mathematical description of the relation between some (set of) 'input' and (set of) 'output' time functions of the process. In choosing the type of description which will be used for the signals, one has to weigh both aspects of signal representation:

the display of the information-bearing attributes of the **signals;**
the giving of means for studying the transmission properties of the **processes.**

Emphasis on either one aspect may lead to a different choice.

The signals occurring in estimation schemes may be distinguished as **'continuous'** (a better term would be non-sampled) and **sampled.** These refer to the representation of the information as a function of **time** (cf. Figure 3.1).

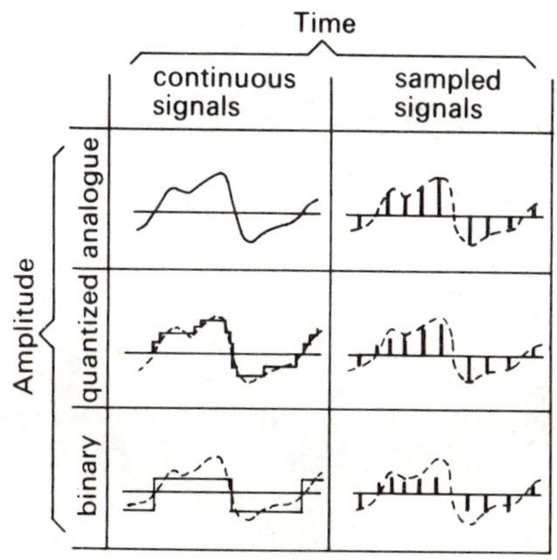


Figure 3.1

In numerous situations the sampling of signals may be useful and/or necessary. This is the case when the information processing is of an intermittent type (e.g. by digital computer) or when time sharing is used for an information-transportation channel. By sampling one generates from a time function $x(t)$ a sequence of sample values that can be represented as a vector:

$$\mathbf{x}' = [x(1), \ldots, x(k)]$$

The sampling operation is a linear one, for if the sampling operation is indicated by $\mathscr{S}[\ ]$ then

$$\mathscr{S}[\alpha x(t)] = \alpha \mathscr{S}[x(t)] \qquad \text{for each } \alpha$$

$$\mathscr{S}[\alpha x(t) + \beta y(t)] = \alpha \mathscr{S}[x(t)] + \beta \mathscr{S}[y(t)] \qquad \text{for each } \alpha \text{ and } \beta$$

provided that the sampling of both signals occurs simultaneously.

With respect to **amplitude,** a distinction can be made between **analogue, quantized** and **binary.** This is also illustrated by Figure 3.1. The binary signal is the limiting case of a quantized signal. The quantizing operation is non-linear; if the quantizing operator is indicated by $\mathscr{Q}[\ ]$ then the expressions

$$\mathscr{Q}[\alpha x(t) + \beta y(t)] \quad \text{and} \quad \alpha \mathscr{Q}[x(t)] + \beta \mathscr{Q}[y(t)]$$

for each $\alpha$ and $\beta$ are not necessarily equal. This non-linearity is responsible for many mathematical problems when dealing with quantized signals (cf. Section 8.2).

## 3.2 Orthogonal Functions

Functions of a rather general nature can be described by the development in a set of orthogonal functions $\{g_i(t)\}$. In this way, such an 'arbitrary' function can (approximately) be described by a **finite number of coefficients.** Another advantage of orthogonal expansions will appear later on in this book with respect to convergence properties of adjustment procedures.

A **discrete** set of orthogonal functions $g_1(t), g_2(t), \ldots,$ over the interval $(a, b)$ satisfies the conditions:

$$\int_a^b g_i(t) g_k(t)\,\mathrm{d}t = d_{ik} \qquad \begin{cases} = 0 & \text{if } i \neq k \\ \neq 0 & \text{if } i = k \end{cases} \tag{3.1}$$

If $d_{ik} = \delta_{ik}$ (Kronecker delta; $\delta_{ik} = 1$ for $i = k$) then the set is called **orthonormal.** Such a set is called **complete** if no non-zero function $y(t)$ can be found such that:

$$\int_a^b y(t) g_i(t)\,\mathrm{d}t = 0 \qquad \text{for all } i \tag{3.2}$$

If a set is complete then a time function $x(t)$ can be expressed in these orthonormal functions:

$$x(t) = \sum_{i=1}^{\infty} c_i g_i(t) \tag{3.3}$$

where

$$c_i = \int_a^b x(t) g_i(t)\, dt \tag{3.4}$$

The Fourier series is an example of the use of orthogonal functions.

A **continuous** set of orthonormal functions is given by:

$$\int_a^b g_\mu(t) g_\nu(t)\, dt = \delta(\mu - \nu) \qquad \begin{cases} \delta(\xi) = 0 & \text{for } \xi \neq 0 \\ \int_{-\infty}^{+\infty} \delta(\xi)\, d\xi = 1 \end{cases} \tag{3.5}$$

Dirac 'function'

A time function $x(t)$ can be expressed as follows:

$$x(t) = \int_{-\infty}^{+\infty} c_\nu g_\nu(t)\, d\nu \tag{3.6}$$

where

$$c_\nu = \int_a^b x(t) g_\nu(t)\, dt \tag{3.7}$$

One member of this family is the Fourier transform.

For discrete sets of orthonormal functions, the following properties are of interest:

The approximation of the signal by the sum of a finite number of orthonormal functions in the minimum mean-square error sense leads to the same coefficients as given by equation (3.4). Take a set of $m$ orthonormal functions:

$$E = \int_a^b \left\{ x(t) - \sum_{i=1}^{m} \gamma_i g_i(t) \right\}^2 dt \tag{3.8}$$

Minimization of $E$ with respect to $\gamma_k$ requires as a necessary condition:

$$0 = \left.\frac{dE}{d\gamma_k}\right|_{\gamma_k = \hat{\gamma}_k} = -2 \int_a^b \left\{ x(t) - \sum_{i=1}^{m} \gamma_i g_i(t) \right\} g_k(t)\, dt.$$

Consequently

$$\int_a^b x(t) g_k(t)\, dt = \sum_{i=1}^{m} \gamma_i \int_a^b g_i(t) g_k(t)\, dt = \hat{\gamma}_k \tag{3.9}$$

because of the orthonormality. A comparison of equation (3.9) with equation (3.4) shows that $\hat{\gamma}_i = c_i$. This implies that $\sum c_i g_i(t)$ gives a smaller mean-square error than any other linear combination of these orthonormal functions.

Note that the value of $\hat{\gamma}_i$ does not depend on $m$, the number of orthonormal functions in the set used. Consequently, an improvement by using $n$ ($n > m$) functions does not alter the coefficients $c_1, \ldots, c_m$. This is a desirable characteristic indeed.

By using $\hat{\gamma}_i = c_i$ in equation (3.8) it follows that:

$$\min E = \int_a^b x(t)^2 \, dt - \sum_{i=1}^{m} c_i^2 \geqslant 0$$

or

$$\sum_{i=1}^{m} c_i^2 \leqslant \int_a^b x(t)^2 \, dt \tag{3.10}$$

For $m \to \infty$ this is the **Bessel inequality**. Thus $c_i \to 0$ for $i \to \infty$ provided $\int_a^b x^2(t) \, dt < \infty$.

If the minimum error approaches zero for $m \to \infty$ then

$$\sum_{i=1}^{\infty} c_i^2 = \int_a^b x(t)^2 \, dt$$

which is called the **Parseval equality**.

Starting from a set of functions $\{f_i(t)\}$ defined over the interval $(a, b)$, one can derive a set of orthonormal functions $\{g_i(t)\}$ by applying the well-known Schmidt procedure. The first function is chosen as:

$$g_1(t) = \frac{f_1(t)}{\| f_1(t) \|} \tag{3.11}$$

with

$$\| f_1(t) \| \overset{\text{def}}{=} \sqrt{\int_a^b f_1(t)^2 \, dt} \tag{3.12}$$

As a result of this choice $\|g_1(t)\| = 1$, i.e. the function is 'normed'. If $f_2(t)$ is independent of $f_1(t)$, then one can choose a coefficient $\alpha$ in such a way that $f_2(t) - \alpha g_1(t)$ is orthogonal to $g_1(t)$:

$$\int_a^b \{f_2(t) - \alpha g_1(t)\} g_1(t) \, dt = 0$$

from which follows

$$\alpha = \int_a^b f_2(t) g_1(t) \, dt$$

and

$$g_2(t) = \frac{f_2(t) - \alpha g_1(t)}{\| f_2(t) - \alpha g_1(t) \|} \tag{3.13}$$

Similarly $g_i(t)$ can be generated so that this signal is orthogonal to $g_1(t) \ldots g_{i-1}$ and also $\|g_i(t)\| = 1$.

Table 3.1. Sets of Orthonormal Functions

| | Interval $(a, b)$ | Weighting function $w(t)$ | Functions |
|---|---|---|---|
| **Discrete sets** | | | |
| Fourier | $-\pi \leqslant t \leqslant \pi$ | 1 | $g_k(t) = \varepsilon^{jkt}$ |
| Legendre polynomials | $-1 \leqslant t \leqslant 1$ | 1 | $P_k(t) = \frac{\sqrt{2k+1}}{2^k\sqrt{2}k!}\frac{d^k}{dt^k}(t^2-1)^k$ |
| Tschebychef polynomials | $-1 \leqslant t \leqslant 1$ | $\frac{1}{\sqrt{1-t^2}}$ | $T_k(t) = \sqrt{\frac{2}{11}}\cos(k\cos^{-1} t)$ |
| Laguerre polynomials | $0 \leqslant t < \infty$ | $\varepsilon^{-t}$ | $L_k(t) = \frac{1}{k!}\varepsilon^{t}\frac{d^k}{dt^k}(t^k\varepsilon^{-t})$ |
| Laguerre functions | $0 \leqslant t < \infty$ | $\rightarrow$ | $l_k(t) = L_k(t)\,\varepsilon^{-t/2}$ |
| Hermite polynomials | $-\infty < t < \infty$ | $\varepsilon^{-t^2}$ | $H_k(t) = \frac{(-1)^k}{\sqrt{2^k k!\sqrt{\pi}}}\varepsilon^{t^2}\frac{d^k}{dt^k}(\varepsilon^{-t^2})$ |
| Hermite functions | $-\infty < t < \infty$ | $\rightarrow$ | $h_k(t) = H_k(t)\,\varepsilon^{-t^2/2}$ |
| **Continuous sets** | | | |
| Fourier | $-\infty < t < \infty$ | | $g_v(t) = \varepsilon^{jvt}$ |
| Dirac | $-\infty < t < \infty$ | | $g_v(t) = \delta(t - v)$ |

The notions of orthogonality and orthonormality can be extended by introducing a weighting function $w(t)$; this offers the possibility of emphasizing the contributions to the mean-square error in a predetermined way. The equivalent expressions for equations (3.1) and (3.8) are:

$$\int_a^b g_i(t)g_k(t)w(t)\,\mathrm{d}t = \delta_{ik} \tag{3.14}$$

$$E = \int_a^b \left\{x(t) - \sum_{i=1}^{m} c_i g_i(t)\right\}^2 w(t)\,\mathrm{d}t$$

Depending on the choice of the interval $(a, b)$ and the weighting function $w(t)$, one obtains the sets of orthogonal functions given in Table 3.1. For further aspects of orthogonal functions the reader is referred to Courant and Hilbert (1968); Sneddon (1961); Lee (1963). Examples of practical applications are found in the literature where, for example, covariance functions are identified as coefficients in orthogonal series expansions. E.g. Lampard (1955); Kitamori (1960); Barker and Hawley (1966); Roberts (1966) (1967); Schwarze (1969); Gorecki and Turowicz (1969); Donati and Milanese (1970).

## 3.3 Description of Deterministic Signals

The description of signals in the frequency domain by Fourier, Laplace or $z$ transform is assumed to be well known. The definitions given in Table 3.2* will be used. It is of importance to note that due to the Parseval theorem, the orthogonality in the time domain also implies orthogonality in the frequency domain.

One way of describing signals is by means of orthogonal components as indicated in the previous section. The coefficients of these components can be found, for example, from equation (3.4)

$$c_i = \int_a^b x(t)\,g_i(t)\,\mathrm{d}t \tag{3.15}$$

This can be instrumented in several ways:

Using a multiplication of the signal $x(t)$ with a member of the set of orthonormal functions and integrating the multiplier output over the interval $a \leqslant t \leqslant b$; Figure 3.2. The orthonormal function can be programmed in function generators or they can be obtained as the solutions of differential equations, programmed on a computer.

Using a set of stable filters. Figure 3.3 shows an example of filters that have orthogonal impulse responses, i.e. if the input of the filter is a $\delta(t)$ function, then the outputs are the respective time functions of an orthogonal set. However, as a filter leads to a convolution

* With kind permission this table has been reprinted from: *Hewlett–Packard Journal*, **19**, No. 2, Oct. 1967.

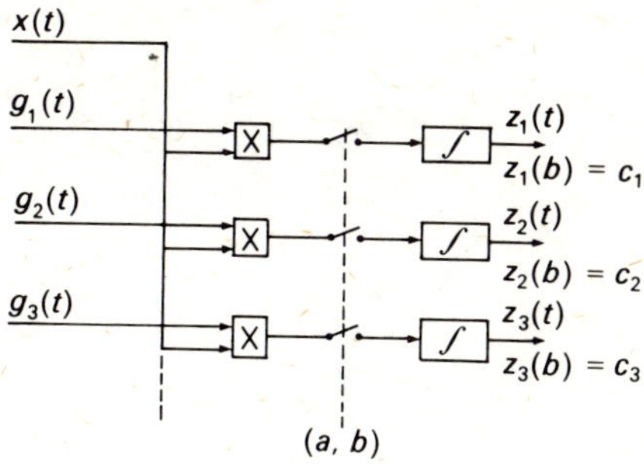


Figure 3.2

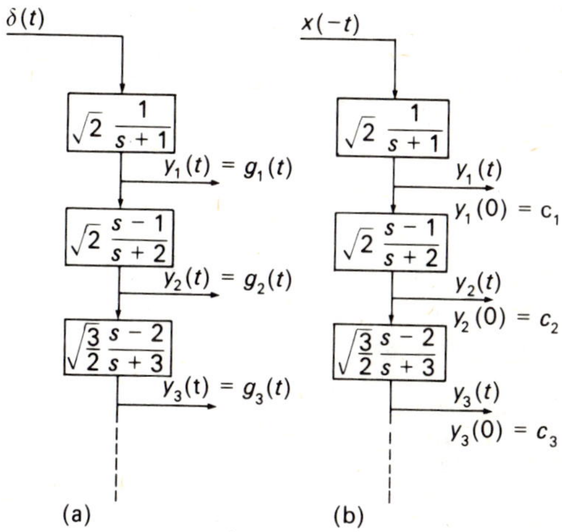


Figure 3.3

of input signal and the filter impulse response,

$$y_i(t) = \int_0^\infty x(t - \theta)g_i(\theta)\,d\theta \tag{3.16}$$

a straightforward instrumentation does not lead to the desired result. At $t = 0$ this equation reduces to

$$y_i(0) = \int_0^\infty x(-\theta)g_i(\theta)\,d\theta \tag{3.17}$$

Comparing this with equation (3.4) one observes that

$$y_i(0) = c_i \qquad \text{if } x(-\theta) = x(t)$$

Consequently the signal to be developed has to be recorded and played back in the reverse direction (Figure 3.3b). For parameter estimation this is, of course, an unwanted feature.

Table 3.2. Definition of Different Transforms with Their Inverses

| Fourier transform | Laplace transform (two-sided) | Laplace transform (one-sided) | $\mathcal{Z}$ Transform (two-sided) | $\mathcal{Z}$ Transform (one-sided) |
|---|---|---|---|---|
| $F(j\omega) = \int_{-\infty}^{\infty} f(t)\varepsilon^{-j\omega t}\,dt$ | $F(s) = \int_{-\infty}^{\infty} f(t)\varepsilon^{-st}\,dt$ | $F(s) = \int_{0}^{\infty} f(t)\varepsilon^{-st}\,dt$ | $f^{*}(t) = \sum_{n=-\infty}^{\infty} f(n)\delta(t - nT)$ | $f^{*}(t) = \sum_{n=0}^{\infty} f(n)\delta(t - nT)$ |
| $\omega$ real | $s = \sigma + j\omega$ | $s = \sigma + j\omega$ | $F(z) = \sum_{n=-\infty}^{\infty} f(n)z^{-n}$ | $F(z) = \sum_{n=0}^{\infty} f(n)z^{-n}$ |
| $F(j\omega)$ exists if | $F(s)$ exist if | $F(s)$ exists if | or | or |
| $\int_{-\infty}^{\infty} \lvert f(t)\rvert\,dt$ exists | $\int_{-\infty}^{\infty} \lvert f(t)\varepsilon^{-st}\rvert\,dt$ exists<br>This integral exists in some allowable region of convergence $\sigma_1 < \sigma < \sigma_2$ | $\int_{0}^{\infty} \lvert f(t)\varepsilon^{-st}\rvert\,dt$ exists<br>This integral exists in some allowable region of convergence $\sigma > \sigma_1$ | $F(z) = F^{*}(s)\rvert_{z=\varepsilon^{sT}}$<br>This exists in an allowable annulus of convergence $r_1 < \lvert z\rvert < r_2$ | $F(z) = F^{*}(s)\rvert_{z=\varepsilon^{sT}}$<br>This exists in an allowable region of convergence outside a circle centered at the origin. |
| $f(t) = \frac{1}{2\pi}\int_{-\infty}^{\infty} F(j\omega)\varepsilon^{j\omega t}\,d\omega$<br>which may be evaluated directly or from tables or | $f(t) = \frac{1}{2\pi j}\int_{\sigma_0 - j\infty}^{\sigma_0 + j\infty} F(s)\varepsilon^{st}\,ds$<br>$\sigma_1 < \sigma_0 < \sigma_2$ | $f(t) = \frac{1}{2\pi j}\int_{\sigma_0 - j\infty}^{\sigma_0 + j\infty} F(s)\varepsilon^{st}\,ds$<br>$\sigma_0 > \sigma_1$ | $f(n) = \frac{1}{2\pi j}\oint_{C} F(z)z^{n-1}\,dz$<br>where $C$ is a circle of radius $r_0$ such that $r_1 < r_0 < r_2$. | $f(n) = \frac{1}{2\pi j}\oint_{C} F(z)z^{n-1}\,dz$<br>where $C$ encloses all the poles of $F(z)z^{n-1}$ |
| $f(t) = \frac{1}{2\pi j}\int_{-j\infty}^{j\infty} F(s)\varepsilon^{st}\,ds\Big\rvert_{s=j\omega}$ | | | | |
| For $t < 0$ | For $t < 0$ | For $t < 0$ | For $n < 0$ | For $n < 0$ |
| $f(t) = \frac{1}{2\pi j}\oint_{C_1} F(s)\varepsilon^{st}\,ds$<br>$= -$(Sum of residues of poles in right half plane.) | $f(t) = \frac{1}{2\pi j}\oint_{C_1} F(s)\varepsilon^{st}\,ds$<br>$= -$(Sum of residues of poles with real parts to right of $\sigma_0$.) | $f(t) = 0$ | $f(n) =$ sum of residues of $F(z)z^{n-1}$ outside $C$. | $f(n) = 0$ |

Table 3.2. Definition of Different Transforms with Their Inverses *contd.*

| Fourier transform | Laplace transform (two-sided) | Laplace transform (one-sided) | $\mathscr{Z}$ Transform (two-sided) | $\mathscr{Z}$ Transform (one-sided) |
|---|---|---|---|---|
| For $t > 0$<br>$f(t) = \frac{1}{2\pi j}\oint_{C_2} F(s)\varepsilon^{st}\,ds$<br>= Sum of residues of poles in left hand plane. | For $t > 0$<br>$f(t) = \frac{1}{2\pi j}\oint_{C_2} F(s)\varepsilon^{st}\,ds$<br>= Sum of residues of poles with real parts to left of $\sigma_0$. | For $t > 0$<br>$f(t) = \frac{1}{2\pi j}\oint_{C} F(s)\varepsilon^{st}\,ds$<br>= Sum of residues of poles of $F(s)$. [$C$ encloses all poles of $F(s)$.] | For $n \geqq 0$<br>$f(n)$ = sum of residues of $F(z)z^{n-1}$ inside $C$. | For $n \geqq 0$<br>$f(n)$ = sum of the residues of $F(z)z^{n-1}$. |
| | Note: Residues are defined for counterclockwise contours. Contour $C_1$ is clockwise. The integral around $C_1$ is, therefore, **minus** the sum of the residues of the enclosed poles. | Note: As a consequence of the definition of the one-sided transform an acceptable $\sigma_0$ is automatically to the right of the poles of $F(s)$. | Note: the formula for $f(n)$ is the formula for the coefficients of the Laurent series for $F(z)$ (see footnotes). | Note: the formula for $f(n)$ is the formula for the coefficients of the Laurent series for $F(z)$ (see footnotes). |
| 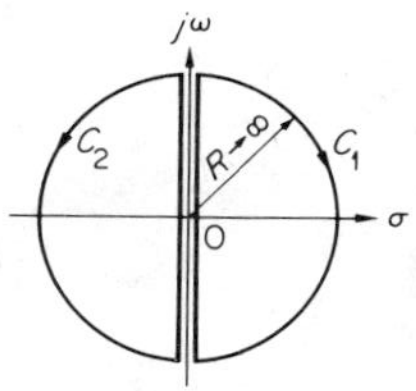 | 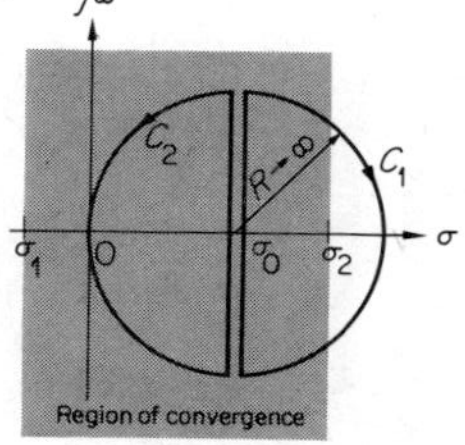 | 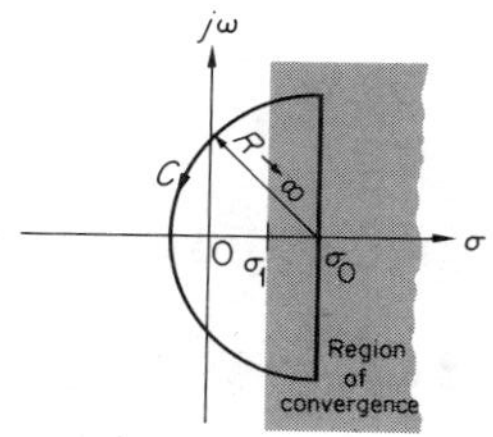 | 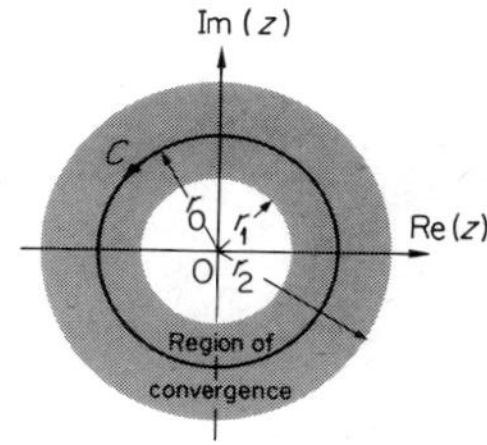 | 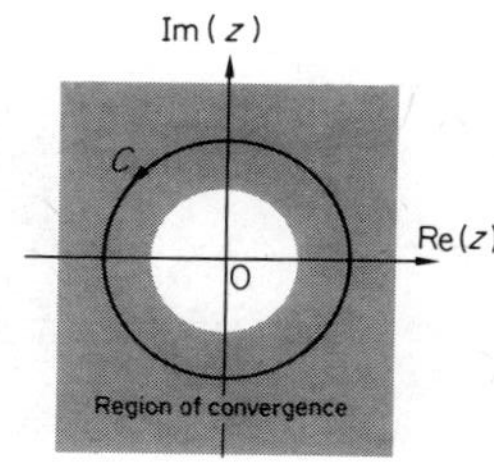 |











**Footnotes.** Complex-variable Formulas

a. $\oint_C f(z)\,dz = 0$ if $f(z)$ is analytic on and inside the closed contour $C$.

b. $\oint_C g(z)\,dz = 2\pi j \sum$ [residues of poles of $g(z)$ inside $C$].

If $g(z)$ has an $n$th-order pole at $z_0$ then residue of $z_0 = \frac{1}{(n-1)!}\left[\frac{d^{n-1}}{dz^{n-1}}\{(z-z_0)^n g(z)\}\right]_{z=z_0}$

c. If $f(z)$ is analytic on $C_1$ and $C_2$ and in the region between them, then $f(z)$ can be represented in that region by a Laurent Series.

$$f(z) = \sum_{n=-\infty}^{\infty} A_n(z-z_0)^n$$

where $A_n = \frac{1}{2\pi j}\oint_C \frac{f(z)}{(z-z_0)^{n+1}}\,dz.$

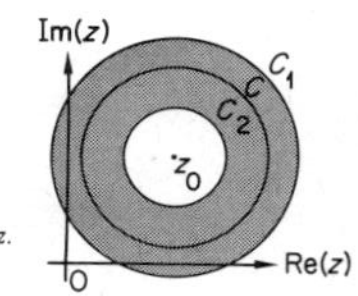



$C$ is any curve between $C_1$ and $C_2$ and enclosing $z_0$.

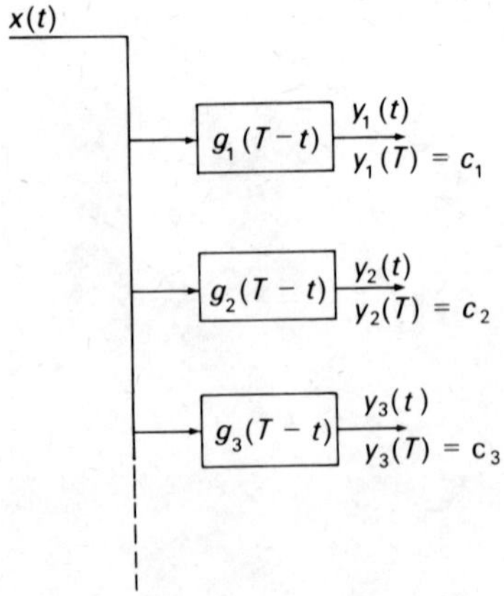


Figure 3.4

Using a set of unstable filters. Equation (3.15) can also be written as:

$$y_i(T) = \int_0^T x(\theta)g_i(T - \theta)\,d\theta$$

if $x(t) = 0$ for $t < 0$ and if one chooses for convolution the time interval $(0, T)$ (Figure 3.4). Comparing this again with equation (3.4) shows that

$$y_i(T) = c_i; \qquad \text{if } g_i(T - \theta) = g_i(t)$$

The filters to be used are unstable ones. This can be realized by analogue computer programming, using suitable clamping techniques.

Another way of describing signals in the time domain is by decomposing them in terms of the form $a \exp(\alpha t)$ and $b \exp(\beta t) \sin(\omega t + \varphi)$, etc. These derive their importance from the fact that such terms occur in the solution of linear homogeneous differential equations with constant coefficients.

## 3.4 Description of Stochastic Signals

For stochastic time functions, Figure 3.5 shows a few relations together with the symbols used for the quantities under consideration. These notions are summarized in Appendix B.

The most important ones are:

the **correlation function,** general definition:

$$\psi_{xy}(t_1, t_2) = \int_{-\infty}^{+\infty}\int_{-\infty}^{+\infty} x(t_1)y(t_2)p(x, t_1; y, t_2)\,dx\,dy \tag{3.18}$$

for stationary signals:

$$\psi_{xy}(\tau) = \int_{-\infty}^{+\infty}\int_{-\infty}^{+\infty} x(t)y(t - \tau)p(x, y; \tau)\,dx\,dy \tag{3.19}$$

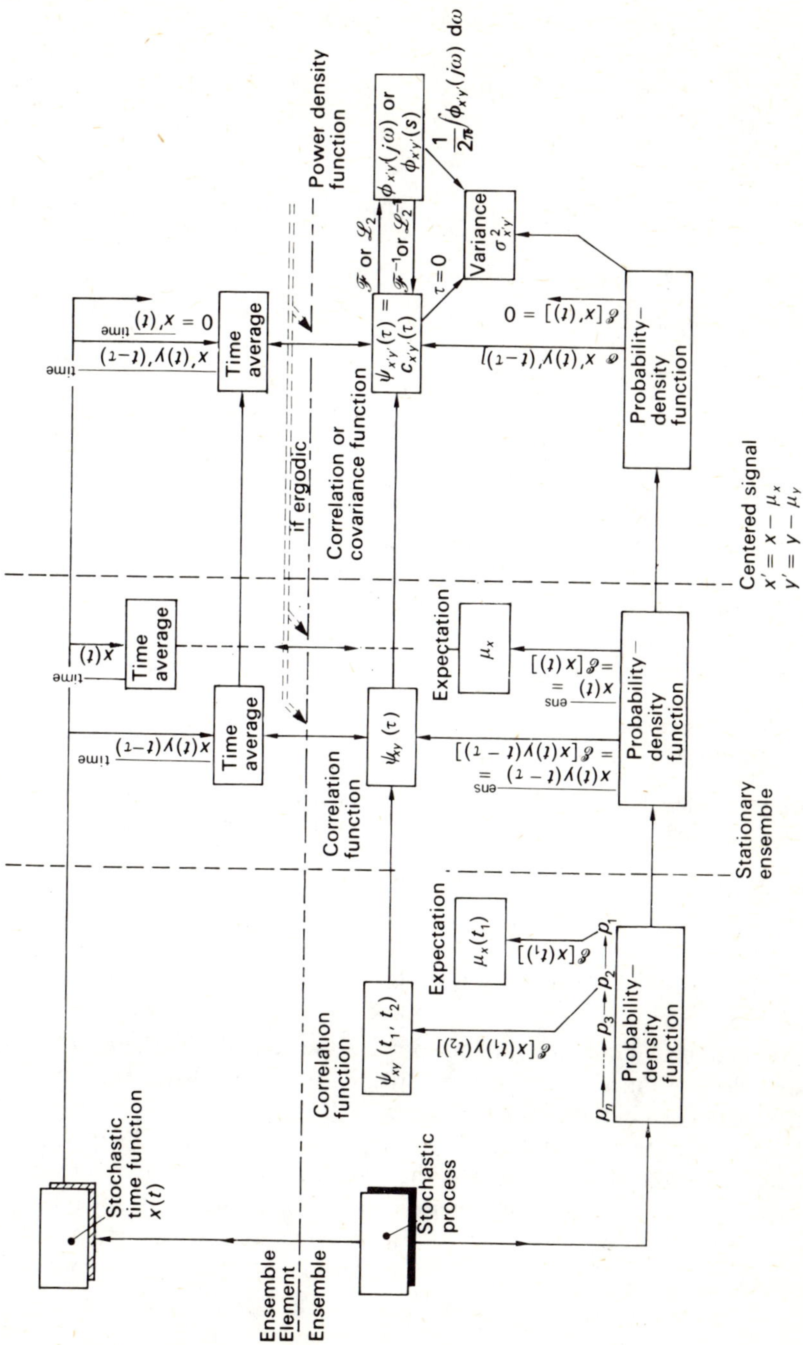
Ensemble
Element
Ensemble
Stochastic time function x(t)
Stochastic process
Probability-density function
Correlation function
Expectation
Stationary ensemble
Time average
Correlation or covariance function
if ergodic
Centered signal
Variance
Power density function


Figure 3.5

for ergodic signals this correlation function can also be found as a time average:

$$\psi_{xy}(\tau) = \lim_{T\to\infty} \frac{1}{T}\int_0^T x(t)y(t-\tau)\,dt \tag{3.20}$$

for finite-interval observation:

$$\tilde{\psi}_{xy}(\tau; T) = \frac{1}{T}\int_0^T x(t)y(t-\tau)\,dt \tag{3.21}$$

which is, of course, a random variable.

the **power density functions**

for stationary signals:

$$\Phi_{xy}(s) = \mathscr{L}_2[\psi_{xy}(\tau)] \tag{3.22}$$

or

$$\Phi_{xy}(j\omega) = \mathscr{F}[\psi_{xy}(\tau)] \quad * \tag{3.23}$$

For studying random signals in systems the assumption of a Gaussian (normal) distribution of the random processes is used in the majority of cases because:

the origin of the random signal, via the central-limit theorem, warrants this assumption;

passing a Gaussian random signal through a linear element results in a random signal with a Gaussian distribution;

other probability-density distributions can be thought of as being derived from a Gaussian random signal by passing it through a non-linear element;

a one-dimensional Gaussian distribution is completely determined if only the statistical mean $\mu$ and the variance $\sigma^2$ are given:

$$p(x) = \frac{1}{\sqrt{2\pi}\sigma} \exp\left[-\tfrac{1}{2}\left(\frac{x-\mu}{\sigma}\right)^2\right] \tag{3.24}$$

For a stationary Gaussian random process the second-order (joint) probability density function for $x(t)$ and $y(t) = x(t-\tau)$ is also Gaussian and can be expressed in terms of $\psi_{xx}(0)$ and $\psi_{xx}(\tau)$.

Similar statements hold for higher-order probability-density functions.

For stationary random signals, the effect upon the signal means and variances because of dynamic operations (transfer functions), adders and multipliers are given in Table 3.3. All these relations are assumed to be well-known, except perhaps that of the variance of the multiplier output; this relationship is derived in Appendix D.

* Note the use of the $j$ operator in the argument of the $\mathscr{F}$ transformed variable.

Table 3.3.

| | Time function | $\mathscr{F}$-Transform | $\mathscr{L}$-Transform | Correlation function | Power spectral density function | Probability-density function | Characteristic function | General moments | Central moments |
|---|---|---|---|---|---|---|---|---|---|
| $x$<br>$y$<br>$G$ | $x(t)$<br>$y(t)$<br>$g(t)$ | $X(j\omega)$<br>$Y(j\omega)$<br>$G(j\omega)$ | $X(s)$<br>$Y(s)$<br>$G(s)$ | $\psi_{xx}(\tau)\quad \psi_{xy}(\tau)$<br>$\psi_{yy}(\tau)\quad \psi_{yx}(\tau)$<br>$g(\tau)$ | $\Phi_{xx}(s)\quad \Phi_{xy}(s)$<br>$\Phi_{yy}(s)\quad \Phi_{yx}(s)$<br>$G(s)$ | $p_X(x)$<br>$p_Y(y)\quad p_{XY}(x, y)$ | $\varphi_X(u)$<br>$\varphi_Y(v)$ | $\alpha_{10}\quad \alpha_{20}\quad \alpha_{30}$<br>$\alpha_{11}\quad \alpha_{21}$<br>$\alpha_{12}$<br>$\alpha_{01}\quad \alpha_{02}\quad \alpha_{03}$ | $\beta_{10}\quad \beta_{20}\quad \beta_{30}$<br>$\beta_{21}$<br>$\beta_{11}\quad \beta_{12}$<br>$\beta_{01}\quad \beta_{02}\quad \beta_{03}$ |
| Dynamic operator 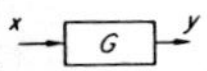 | $y(t) = g(t) * x(t)$<br>$= \int g(t-\tau)x(\tau)\,d\tau$<br>$= \int g(\tau)x(t-\tau)\,d\tau$ | $Y(j\omega) = G(j\omega)X(j\omega)$ | $Y(s) = G(s)X(s)$ | $\psi_{xy}(\tau) = g(\tau) * \psi_{xx}(\tau)$<br>$\psi_{yx}(\tau) = g(-\tau) * \psi_{xx}(\tau)$ | $\Phi_{xy}(s) = G(s)\Phi_{xx}(s)$<br>$\Phi_{yx}(s) = G(-s)\Phi_{xx}(s)$<br>$\Phi_{yy}(s) = G(-s)G(s)\Phi_{xx}(s)$ | | | $\alpha_{10} = \mu_x$<br>$\alpha_{01} = \mu_y$<br>$\mu_y = G(0)\mu_x$ | $\beta_{10} = \beta_{01} = 0$<br>$\beta_{20} = \sigma_x^2\quad \beta_{11} = \sigma_{xy}^2$<br>$\beta_{02} = \sigma_y^2$<br>$\sigma_y^2 = \psi_{yy}(0)$ |
| Non-linear operator 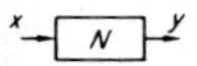 | $y(t) = f[x(t)]$ | | | Gaussian:<br>$\psi_{xy}(\tau) = \dfrac{\psi_{xx}(\tau)}{\psi_{xx}(0)}\psi_{xy}(0)$<br>$\psi_{xy}(0) = f'\psi_{xx}(0) + \dfrac{f'''}{3!}\psi_{xx^3}(0) + \cdots$ | | | | | |
| Adder 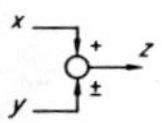 | $z(t) = x(t) \pm y(t)$ | $Z(j\omega) = X(j\omega) \pm Y(j\omega)$ | $Z(s) = X(s) \pm Y(s)$ | $\psi_{zz}(\tau) = \psi_{xx}(\tau) + \psi_{yy}(\tau) \pm \psi_{xy}(\tau) \pm \psi_{yx}(\tau)$ | $\Phi_{zz}(s) = \Phi_{xx}(s) + \Phi_{yy}(s) \pm \Phi_{xy}(s) \pm \Phi_{yx}(s)$ | $x, y$ independent<br>$p_Z(z) = p_X(x) * p_Y(y)$ | $x, y$ independent<br>$\varphi_Z(w) = \varphi_X(u)\varphi_Y(v)$ | $\mu_z = \mu_x \pm \mu_y$ | $\sigma_z^2 = \sigma_x^2 + \sigma_y^2 \pm 2\sigma_{xy}^2$ |
| Multiplier 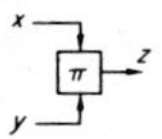 | $z(t) = x(t)y(t)$ | $Z(j\omega) = X(j\omega) * Y(j\omega)$ | $Z(s) = X(s) * Y(s)$ | Gaussian:<br>$\psi_{zz}(\tau) = \psi_{xy}^2(0) + \psi_{xx}(\tau)\psi_{yy}(\tau) + \psi_{xy}(\tau)\psi_{yx}(\tau)$ | Gaussian; $\psi_{xy}(0) = 0$:<br>$\Phi'_{zz}(s) = \Phi_{xx}(s) * \Phi_{yy}(s) + \Phi_{xy}(s) * \Phi_{yx}(s)$ | | | $\mu_z = \psi_{xy}(0)$ | Gaussian:<br>$\sigma_z^2 = \sigma_x^2\sigma_y^2 + \sigma_{xy}^4$ |

$$X(j\omega) = \int_{-\infty}^{+\infty} \varepsilon^{-j\omega t}x(t)\,dt \quad \xrightarrow{\quad} \quad \mathscr{F}\text{-transform}$$

$$x(t) = \frac{1}{2\pi}\int_{-\infty}^{+\infty} \varepsilon^{j\omega t}X(j\omega)\,d\omega \quad \xleftarrow{\quad} \quad \mathscr{F}^{-1}\text{-transform}$$

$$X(s) = \int_{0}^{\infty} \varepsilon^{-st}x(t)\,dt \quad \mathscr{L}\text{-transform} \quad \xrightarrow{\quad}$$

$$x(t) = \frac{1}{2\pi j}\int_{c-j\infty}^{c+j\infty} \varepsilon^{st}X(s)\,ds \quad \mathscr{L}^{-1}\text{-transform} \quad \xleftarrow{\quad}$$

$$\psi(\tau) = \lim_{T\to\infty}\frac{1}{T}\int_{0}^{+T} x(t)y(t-\tau)\,dt \quad \text{correlation} \quad \xrightarrow{\quad}$$

(time average; if ergodic)

$$\Phi_{xy}(s) = \int_{-\infty}^{+\infty} \varepsilon^{-s\tau}\psi_{xy}(\tau)\,d\tau \quad \xrightarrow{\quad} \quad \mathscr{L}_2\text{-transform}$$

$$\psi_{xy}(\tau) = \frac{1}{2\pi j}\int_{-\infty}^{+\infty} \varepsilon^{s\tau}\Phi_{xy}(s)\,ds \quad \xleftarrow{\quad} \quad \mathscr{L}_2^{-1}\text{-transform}$$

$$\psi_{xy} = \frac{\partial^2}{\partial u\,\partial v}\varphi(u, v)\Big|_{u=v=0} \quad \xleftarrow{\quad}$$

$$\psi_{xy}(\tau) = \int_{-\infty}^{+\infty}\int_{-\infty}^{+\infty} x(t)y(t-\tau)p(x, y)\,dx\,dy = \mathscr{E}[x(t)y(t-\tau)] \quad \xleftarrow{\quad}$$

(ensemble average)

$$\varphi(u, v) = \mathscr{E}[\varepsilon^{j(ux+vy)}] \quad \xrightarrow{\quad} \quad \mathscr{F}^{-1}\text{-transform}$$

$$p(x, y) = \frac{1}{(2\pi)^2}\int_{-\infty}^{+\infty}\int_{-\infty}^{+\infty} \varepsilon^{-j(ux+vy)}\varphi(u, v)\,du\,dv \quad \xleftarrow{\quad}$$

$$\alpha_n = j^{-n}\varphi^{(n)}(u)\big|_{u=0} \quad \xrightarrow{\quad}$$

$$\xleftarrow{\quad} \quad \beta_{11} = \alpha_{11} - \alpha_{10}\alpha_{01}$$
$$\beta_{20} = \alpha_{20} - \alpha_{10}^2 \text{ etc.}$$

$$\alpha_{i,k} = \mathscr{E}[x^i(t)y^k(t)] \quad \xrightarrow{\quad}$$

$$\beta_{n0} = \sum_{i=0}^{n}(-1)^r\binom{n}{r}\alpha_{10}^r\alpha_{(n-r)0} \quad \xrightarrow{\quad}$$

$$\beta_{i,k} = \mathscr{E}[\{x(t) - \bar{x}\}^i\{y(t) - \bar{y}\}^k] \quad \xrightarrow{\quad}$$

## 3.5 Signal Processing; FFT

The utility of the different types of signal descriptions is highly dependent on the means and procedures available to implement the necessary signal processing. The description in the frequency domain by means of the Fourier transform, as defined in Table 3.2, if of great importance in spite of its drawbacks:

the infinite (time) interval over which the operation extends;

the integral type of operation.

It is obvious that in practical situations the first aspect is a real problem as an infinite measurement interval is highly impractical. What happens if the measurement interval is limited is illustrated by Figure 3.6. Here the

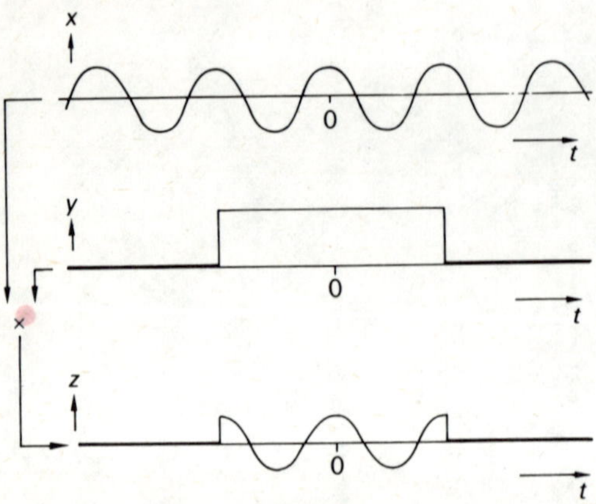


Figure 3.6

observed (truncated) part of a cosine wave $z(t)$ is considered as the product of cosine function $x(t)$ and a block function $y(t)$:

$$z(t) = x(t) \times y(t) \tag{3.25}$$

As a multiplication in the time domain corresponds with a convolution in the frequency domain the resulting $\mathscr{F}$-transform will be:

$$Z(j\omega) = X(j\omega) * Y(j\omega) \tag{3.26}$$

This is illustrated by Figure 3.7. As the width of the sinc function $Y(j\omega)$ is inversely proportional to the length $T$ of the block function the limitations with respect to resolution and the 'deformation' of the frequency spectrum are apparent.

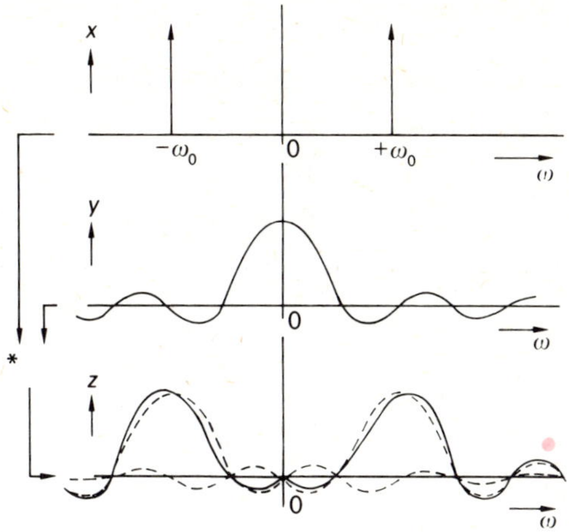


Figure 3.7

The sinc function can be circumvented by choosing some smoother 'window function' for $y(t)$. For these and other aspects of $\mathscr{F}$-transform see e.g., Blackman and Tukey (1959); Richards (1967); Bendat and Piersol (1971).

The second drawback mentioned of the Fourier transform was the integral type of operation. For approximation of this transform using a digital computer the procedure has to be written in a discrete form.

### 3.5.1 *Discrete Fourier Transform (DFT)*

This DFT for $N$ sample points and its inverse can be written as

$$X(k) = \sum_{n=0}^{N-1} x(n) \exp\left(-j\frac{2\pi kn}{N}\right) \qquad k = 0, \ldots, N-1 \tag{3.27}$$

and

$$x(n) = \frac{1}{N} \sum_{n=0}^{N-1} X(k) \exp\left(j\frac{2\pi nk}{N}\right) \qquad n = 0, \ldots, N-1 \tag{3.28}$$

Note that both expressions have a strong resemblance, which implies that by one procedure both transforms can be realized. Note, however, also that the execution of such a transform requires $N^2$ multiplications. For a (time) series with a large number of samples this implies quite a burden with respect to computer time.

### 3.5.2 *Fast Fourier Transform (FFT)*

With the development of the FFT the practical use of frequency domain representations has been greatly enhanced. FFT provides an appreciable reduction of computation time and diminishes the round-off errors. Also it can be organized in such a way that it requires comparatively little memory space.

FFT is based on the recognition of the fact that the number of multiplications with the terms

$$\exp\left(-j\frac{2\pi kn}{N}\right) = W^{kn}$$

with

$$W = \exp\left(-j\frac{2\pi}{N}\right)$$

and

$$N = 2^l \qquad l \text{ integer}$$

can be made far less than $N^2$ due to:

the occurrence of only $N$ different values of $W^{kn}$; this is illustrated for $N = 8; k = 0, \ldots, 7$; in Figure 3.8;

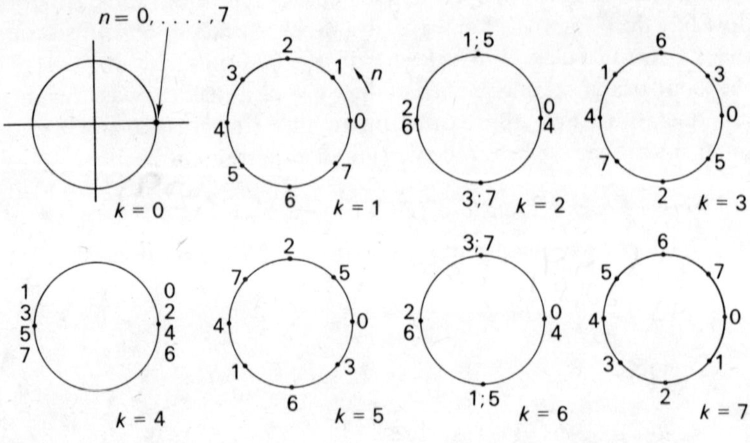


Figure 3.8

the symmetry in those values as indicated in Figure 3.9;

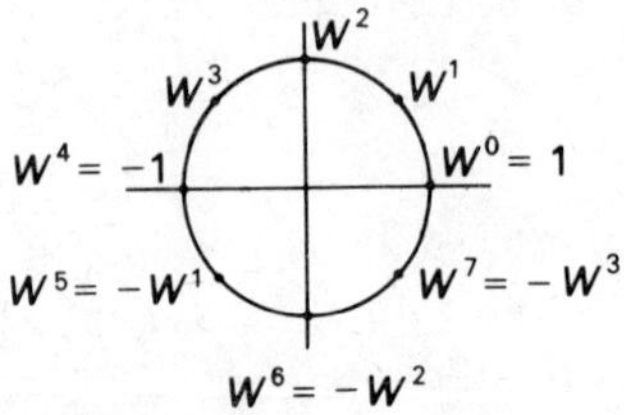


Figure 3.9

the observation that the same product may occur several times in the transformation scheme, and consequently, that terms may be combined in an efficient way.

It can be shown that the number of multiplications can be reduced to

$$N \log_2 N \qquad \text{if } N = 2^l, \quad \text{i.e.}$$

a power of 2. Then for $N = 1024 = 2^{10}$ the computer time may be reduced to the order of

$$\frac{N \log_2 N}{N^2} = \frac{10}{1024} \approx 1\%$$

For introductions to FFT the reader is referred to Cochran et al. (1967) and to Brigham and Morrow (1967). Other interesting publications on these techniques are: Cooley and Tukey (1965); Gentleman and Sande (1966); Fast Fourier Transform (1967) (1969).

*Example.* The use of FFT on a small digital computer is illustrated by Figures 3.10 and 3.11.* In Figure 3.10 in each of the photographs the upper

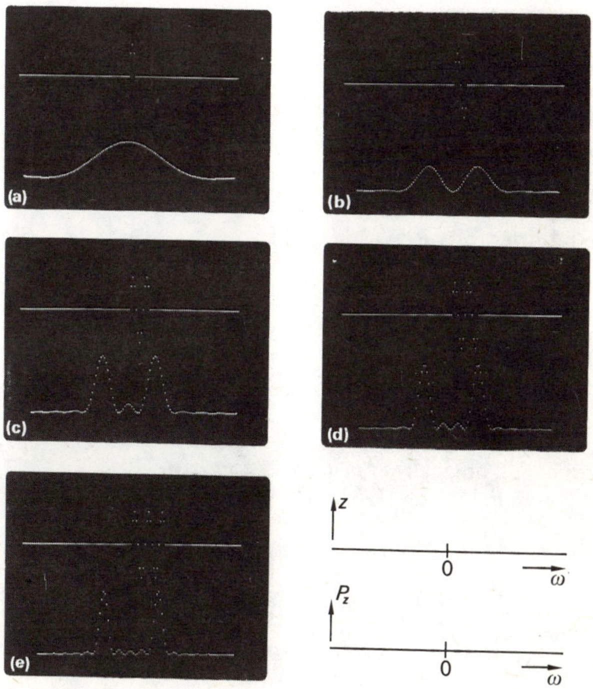


Figure 3.10

* The calculations are performed on a PDP-8 computer; the author is indebted to Mr. N. G. M. Kouwenberg, University of Technology, Eindhoven, Netherlands for these illustrations.

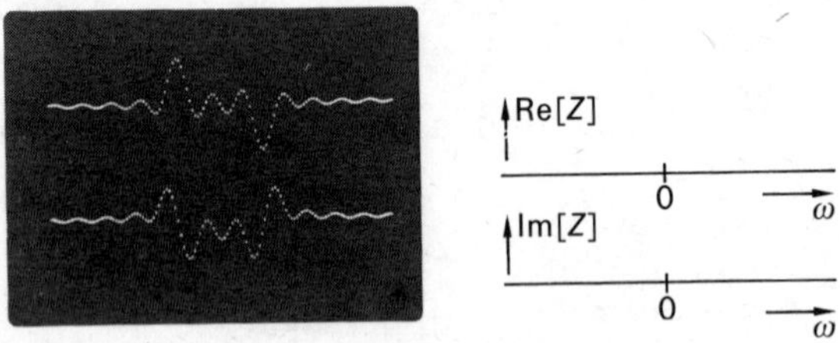


Figure 3.11

trace indicates the time samples $z(k)$ that have been processed, the lower trace shows the resulting power spectrum $P_z = Z^*(j\omega)Z(j\omega)$. The figures (a) . . . (e) show the effect of truncation as discussed earlier in this section. As an extreme case of truncation, parts of a sine wave have been chosen, viz. half a period to two and a half periods. From these pictures the 'development' of the spectrum and the relationship between 'spectral resolution' and 'measurement interval' is clear; cf. Figures 3.6 and 3.7 where $Z$ was given instead of $P_z$. In Figure 3.11 the real and imaginary parts of $Z$ are given for the case corresponding to Figure 3.10c.

## 3.6 Conclusion

The main function of this chapter is to summarize some important aspects of signals and to define the notation used in this text. In the References and Additional Literature a selection of relevant publications are mentioned. An extensive bibliography is Wold (1965).

## References

Barker, H. A. and D. A. Hawley (1966). 'Orthogonalising and optimization', *3rd IFAC Congr., London*, paper 36A.

Bendat, J. S. and A. G. Piersol (1971). *Random Data Analysis and Measurement Procedures*, Wiley–Interscience, New York, 407 pp.

Blackman, R. B. and J. W. Tukey (1959). *The Measurement of Power Spectra*, Dover, New York, 190 pp.; originally in *Bell Syst. Tech. J.*, **37** (1958).

Brigham, E. O. and R. E. Morrow (1967). 'The fast Fourier transform', *IEEE Spectrum*, **4,** No. 12, 63–70.

Cochran, W. T. et al. (1967). 'What is Fast Fourier Transform?', *Proc. IEEE*, **55,** 1664–1674.

Cooley, J. W. and J. W. Tukey (1965). 'An algorithm for the machine calculation of complex Fourier series', *Math. of Comput.*, **19,** 297–301.

Courant, R. and D. Hilbert (1968). *Methods of Mathematical Physics*, Part I (in German), Springer, Berlin, 3rd ed., 496 pp.

Donati, F. and M. Milanese (1970). 'System identification with approximated models', *2nd IFAC Symp. Identification and Process Parameter Estimation, Prague*, paper 1.9.

Fast Fourier Transform (1967). Special issue *IEEE Trans. on Audio and Electro-acoustics*, **AU-15,** No. 2, June 1967.

Fast Fourier Transform (1969). Special issue *IEEE Trans. on Audio and Electroacoustics*, **AU-17,** No. 2, June 1969.

Gentleman, W. M. and G. Sande (1966). 'Fast Fourier transforms—for fun and profit', *Proc. Fall Joint Computer Conf.*, 563–578.

Gorecki, H. and A. Turowicz (1969). 'The approximation method of identification', *4th IFAC Congr., Warsaw*, paper 5.5.

Kitamori, T. (1960). 'Application of orthogonal functions to the determination of process dynamic characteristics and to the construction of self-optimizing control systems', *1st IFAC Congr., Moscow*, 613–618.

Lampard, D. G. (1955). 'A new method of determining correlation functions of stationary time series', *Proc. IEE*, **102C,** 35–41.

Lee, Y. W. (1963). *Statistical Theory of Communication*, Wiley, New York, 3rd print, 509 pp.

Richards, P. I. (1967). 'Computing reliable power spectra', *IEEE Spectrum*, **4,** No. 1, 83–90.

Roberts, P. D. (1966). 'The application of orthogonal functions to self-optimizing control systems containing a digital compensator', *3rd IFAC Congr., London*, paper 36C.

Roberts, P. D. (1967). 'Orthogonal transformations applied to control system identification and optimization', *IFAC Symp. Identification in Autom. Control Systems, Prague*, paper 5.12.

Schwarze, G. (1969). 'Determination of models for identification with quality characteristics in the time domain', *4th IFAC Congr., Warsaw*, paper 19.3.

Sneddon, I. N. (1961). *Special Functions of Mathematical Physics and Chemistry*, Oliver and Boyd, Edinburgh, 184 pp.

Wold, H. O. A. (1965). *Bibliography on Time Series and Stochastic Processes*, Oliver and Boyd, Edinburgh, 516 pp.

## Additional Literature

Anderson, W. W. (1965). 'Optimum estimation of the mean of a Gaussian random process' (correspondence). *Proc. IEEE*, **53,** 1640–1641.

Arato, M. (1962). 'Estimation of the parameters of a stationary Gaussian Markov process' (in Russian), *Dokl. Akad. Nauk. SSSR*, **145,** 13.

Balch, H. T. and others (1966). 'Estimation of the mean of a stationary random process by periodic sampling', *Bell. Syst. Tech. J.*, **45,** 733–741.

Bendat, J. S. and A. G. Piersol (1966). *Measurement and Analysis of Random Data*, Wiley, New York, 390 pp.

Berger, E. R. (1964). 'The increase of information during observation of the relative frequency distribution of an ergodic process' (in German), *Nachrtech. Z.*, **17,** 590–594.

Berndt, H. and G. R. Cooper (1965). 'An optimum observation time for estimates of time-varying correlation functions' (correspondence), *IEEE Trans. Information Theory*, **IT-11,** 307–310.

Berndt, H. and G. R. Cooper (1966). 'Estimates of correlation functions of nonstationary random processes', *IEEE Trans. Information Theory*, **IT-12,** 70–72.

Bingulac, S. (1965). 'The orthonormalisation of functions' (correspondence), *IEEE Trans. Autom. Control*, **AC-10,** 374–375.

Bocharov, I. N. and R. I. Stakhovskii (1962). 'Random process probability distribution density analyser', *Automn. Remote Control*, **23,** 154–160.

Boulton, P. I. and R. J. Kavanagh (1963). 'A method of producing multiple non-correlated random signals from a single Gaussian noise source', *IEEE Trans. Applic. and Ind.*, **82,** No. 65, 46–52.

Brown Jr., J. L. (1964). 'Generating uncorrelated random outputs by nonlinear processing of a single noise source', *IEEE Trans. Applic. and Ind.*, **83,** No. 75, 408–410.

Brubaker, T. A. and G. A. Korn (1961). 'Accurate amplitude distribution analyzer combining analog and digital logic', *Rev. Scient. Instrum.*, **32,** 317.

Bruni, C. (1964). 'Analysis of approximation of linear and time-invariant systems pulse response by means of Laguerre finite term expansion' (correspondence), *IEEE Trans. Autom. Control*, **AC-9,** 580–581.

Davenport Jr., W. B. and W. L. Root (1958). *An Introduction to the Theory of Random Signals and Noise*, McGraw-Hill, New York, 393 pp.

Eier, R. (1966). 'Signal analysis using Laguerre polynomials' (in German), *Arch. Elekt. Ubertr.*, **20,** 185–194.

Fine, T. and N. Johnson (1965). 'On the estimation of the mean of a random process' (correspondence), *Proc. IEEE*, **53,** 187–188.

Gorskii, A. A. (1965). 'Formulas for computing the mean and the variance of estimates of the spectral density of a stationary random process', *Automn. Remote Control*, **26,** 1650–1655.

Hänel, H. (1962). 'Possibilities of producing very low frequencies', *Z. Messen, Steuern, Regeln*, **5,** 7–13.

Havel, J. (1960). 'An electronic generator of random sequences', *Trans. 2nd Prague Conf.—Information Theory, Statistical Decision Functions, Random Processes*, 219–229.

Herschel, R. (1961). 'Determination of the coefficients of expansions into orthogonal functions by analog computers' (in German), *Elektron. Rechenanl.*, **3,** 212–217.

Hofstetter, E. M. (1964). 'Construction of time-limited functions with specified autocorrelation functions', *IEEE Trans. Information Theory*, **IT-10**, 119–126.

Ivanenko, V. I. and O. A. Hohel (1967). 'Stabilization problems of parameters of artificially generated random processes', *IFAC Symp. Identification in Autom. Control Systems, Prague*, paper 5.13.

Janáč, K. (1962). 'Direct determination of correlation function for generators of random processes', *Proc. 3rd Internat. Conf. on Analog Computers, Opatija.*

Kliger, I. (1964). 'On the determination of Laguerre's spectrum from the Laplace transform of a given function' (correspondence), *IEEE Trans. autom. Control*, **AC-9,** 192–193.

Korn, G. A. (1966). *Random-Process Simulation and Measurements*, McGraw-Hill, New York, 234 pp.

Lee, Y. W. (1950). 'Application of statistical methods to communication problems', *Tech. Report 181, Res. Lab. of Electron., M.I.T.*, Cambridge, Mass.

März, K. (1965). 'The mean error of short-interval correlation functions' (in German), *Nachrtech. Z.*, **18,** 485–487.

Pugachev, V. S. (1965). cf. references Chapter 8.

Reissmann, H. C. (1961). 'Generation of a noise spectrum for lowest frequencies by using radio-active decay processes' (in German), *Wiss. Z. Hochsch. Elektrotech. Ilmenau*, D.D.R., **7,** 185–191.

Rice, S. O. (1944, 1945). 'Mathematical analysis of random noise', part 1–4, *Bell Syst. Tech. J.*, **23,** 282–332; **24,** 46–156.

Rowe, I. H. (1965). 'A method for generating a large number of uncorrelated random signals from a single noise source' (correspondence), *IEEE Trans. Autom. Control*, **AC-10,** 478–479.

Thelliez, S. F. (1967). 'Note on the experimental design of a random symbol generator', *IFAC Symp. Identification in Autom. Control Systems, Prague*, paper 6.2.

Wax, N. (1954). *Selected Papers on Noise and Stochastic Processes*, Dover, New York, 337 pp.

Werner, G. W. and H. C. Reissmann (1963). 'The generation of very low frequency noise spectra and methods for their measurement in control engineering' (in German), *Z. Messen, Steuern, Regeln*, **6,** 328–334.

Yakovlev, V. P. (1965). 'Obtaining a set of non-correlated random processes via non-linear transformations of a single random signal', *Automn. Remote Control*, **26,** 1086–1092.

Young, T. Y. and W. H. Huggins (1962). 'Discrete orthonormal exponentials', *Proc. nat. Electron. Conf.*, **18,** 10–18.

Young, T. Y. and W. H. Huggins (1962). 'Complementary signals and orthogonalized exponentials', *IRE Trans. Circuit Theory*, **CT-9,** 362–370.

CHAPTER 4

# Process Models; Linear, Varylinear, Non-linear

In Chapter 1, the importance of 'model building' was stressed. In this chapter the field of possible model choices is scanned.

First the classes of process models are distinguished according to the well-known aspects linear/varylinear/non-linear and non-parametric/parametric. These types of description are summarized in Sections 4.2, 4.3 and 4.4. For identification techniques the concepts of controllability, observability and identifiability, as summarized in Section 4.5, are relevant. In Section 4.6 attention is focussed on the concept 'linearity-in-the-parameters' which concept is of importance in estimation procedures, and on canonical forms having the least number of parameters.

## 4.1 Classes of Process Models

Stimulated by the theory of classical mechanics, an important method of interpreting the behaviour of processes has been by means of differential equations, expressed in terms of input and output quantities $u = u(t)$ and $y = y(t)$, respectively:

$$a_0 \frac{d^n y}{dt^n} + \cdots + a_{n-1} \frac{dy}{dt} + a_n y = b_0 \frac{d^n u}{dt^n} + \cdots + b_{n-1} \frac{du}{dt} + b_n u + c \qquad (4.1)$$

together with the initial conditions $d^i y/dt^i$ at $t = 0$ for $i = 1, \ldots, n$. For processes that include a pure time delay $u = u(t - \tau)$. One may arbitrarily choose one of the coefficients, e.g. $a_0 = 1$.

In a **linear** case the coefficients $a_i$, $b_j$ do not depend on $u$ and $y$ and their derivatives. If, in addition, they do not depend on time either, we have a case of **constant coefficients.** This is the most tractable case. These coefficients may, however, depend on time; in that case the equation is called **time varying** or **varylinear.** If any $a_i$ or $b_j$ does depend on $u$, $y$, or their derivatives, the process is **non-linear.**

The dominating distinction between linear and non-linear processes is the principle of superposition which holds only for the former. This principle implies that if

$y_1$ is the output due to input $u_1$

and

$$y_2 \text{ is the output due to input } u_2$$

then

$$\alpha y_1 + \beta y_2 \text{ is the output due to input } \alpha u_1 + \beta u_2$$

A simple example may serve as an illustration:

**Linear differential equation**

$$\dot{y} + ay = u$$

consequently

$$\begin{array}{cccccc} \dot{y}_1 & + & ay_1 & \equiv & u_1 \\ \dot{y}_2 & + & ay_2 & \equiv & u_2 \\ \hline \end{array}$$

$$(\dot{y}_1 + \dot{y}_2) + a(y_1 + y_2) \equiv u_1 + u_2$$

if

$$u_1 + u_2 = u_3$$

$$y_1 + y_2 = y_3$$

and from this

$$\dot{y}_1 + \dot{y}_2 = \dot{y}_3$$

then

$$\dot{y}_3 + ay_3 \equiv u_3$$

**Non-linear differential equation**

$$\dot{y} + ay^3 = u$$

consequently

$$\begin{array}{cccccc} \dot{y}_1 & + & ay_1^3 & \equiv & u_1 \\ \dot{y}_2 & + & ay_2^3 & \equiv & u_2 \\ \hline \end{array}$$

$$(\dot{y}_1 + \dot{y}_2) + a(y_1^3 + y_2^3) \equiv u_1 + u_2$$

if

$$u_1 + u_2 = u_3$$

$$y_1 + y_2 = y_3$$

and from this

$$\dot{y}_1 + \dot{y}_2 = \dot{y}_3$$

then

$$(y_1^3 + y_2^3) \not\equiv (y_1 + y_2)^3 = y_3^3$$

and

$$\dot{y}_3 + ay_3^3 \not\equiv u_3$$

In the linear case, the signal sums obey the same differential equation as the original signals; in the non-linear case this does not hold. The principle of superposition holds as well for the output signal due to initial conditions (general solution) as for the output signal due to an input (particular solution). Note that in the linear case the superposition (linearity) is preserved even if the coefficient $a$ of the differential equation is a function of time. In the engineering sense, however, such time-varying coefficients (parameters) make an important difference. For that reason this case will be discussed separately under the heading **varylinear** processes (Section 4.3).

The difficulties of solving **non-linear** differential equations are well known. So is the incompleteness of the collection of methods for solving them. The 'classical' approaches for studying such systems, viz. by phase plane analysis and by describing function approximation, are of little use for our problem of process representation. As one of the possible approaches for a class of non-linear systems, the Volterra series is discussed in Section 4.4.

For processes, important types of relations are: (1) those between input and output signals and (2) those expressed in state variables. Let us consider the first category through the expression:

$$F\{u, \dot{u}, \ldots, \overset{(m)}{u}; y, \dot{y}, \ldots, \overset{(n)}{y}\} = 0 \tag{4.2}$$

where $F$ is an analytic function in all variables $u, \dot{u}, \ldots; y, \dot{y}, \ldots$. As both the input and the output signals are found only implicitly in the differential equation the analysis will in general be devoted to obtaining an **explicit** relation between $u(t)$ and $y(t)$. Such a relation will be denoted by

$$y(t) = \mathcal{O}[u(\tau)]^{t}_{-\infty} \tag{4.3}$$

In mathematical terms the operator $\mathcal{O}$ can be a functional, i.e. a rule which makes a value correspond to each function over a given interval. As indicated by (4.3), the interval of interest for $u(\tau)$ may be $-\infty < \tau \leqslant t$. In the following we will consider the operator $\mathcal{O}$ for a number of simple cases (Van Harten, 1966).

(1) $$F = y - bu = 0$$

In this trivial case there is a linear relation between $u(t)$ and $y(t)$; the process does not include dynamical elements that accumulate energy (**linear-without-memory**). In this case the functional $\mathcal{O}[u(\tau)]$ reduces to a linear function of $u$:

$$y(t) = \mathcal{O}[u(\tau)]^{t}_{-\infty} = bu(t)$$

(2) $$F = \ddot{y} + a_1\dot{y} + a_2 y - \dot{u} - b_2 u = 0 \qquad a_1, a_2, b_2 \text{ constant}$$

From the theory of linear differential equations it follows that $y(t)$ can be written explicitly by the functional

$$y(t) = \mathcal{O}[u(\tau)]^{t}_{-\infty} = \int_{-\infty}^{t} h(t - \tau)u(\tau)\,d\tau$$

The process includes dynamical elements that provide some kind of memory; consequently, there is a dependence on previous values of $u(t)$ (**linear-with-memory**). One of the ways of determining the weighting function or impulse response $h(t)$ from the differential equation is by use of the Laplace transform theory;

$$h(t) = \mathcal{L}^{-1}\left[\frac{s + b_2}{s^2 + a_1 s + a_2}\right]$$

Another approach for obtaining $h(t)$ is indicated in Section 4.2.

(3) $$F = y + ay^3 - u = 0 \qquad a \text{ is constant}$$

This represents a non-linear relation between process input and output (e.g. a saturating amplifier; we restrict the discussion to a monotonic function). There are no dynamical elements that provide a memory (**non-linear-without-memory**). Consequently the functional again reduces to a function:

$$y(t) = \mathcal{O}[u(\tau)]^{t}_{-\infty} = g\{u(t)\}$$

This function can be obtained by substitution or iteration procedures or by a simple Taylor development. For this example it follows that

$$y = u - au^3 + 3a^2u^5 - 12a^3u^7 + \cdots$$

This equation holds only for $u$ inside the region of convergence. Although the implicit relation is very simple the explicit one is rather awkward. In some cases a graphical determination has advantages (McFee, 1961).

(4) $$F = \ddot{y} + a_1\dot{y} + a_2 y + f(y) - u = 0 \qquad a_1, a_2 \text{ constant}$$

This equation describes the behaviour of the process indicated in Figure 4.1 which is non-linear and has dynamic elements (**non-linear-with-memory**).

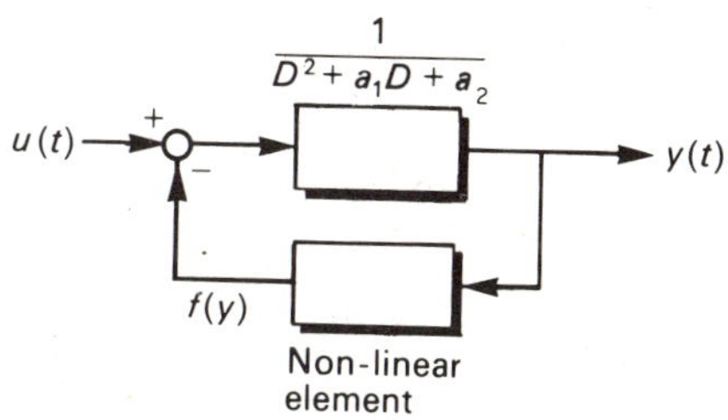


Figure 4.1

The non-linearity itself is assumed to have no memory; this restriction excludes hysteresis type of non-linearities. In Section 4.4 it will be shown that for this case an explicit relation can be written of the following type (Volterra series)

$$\begin{aligned} y(t) &= \mathcal{O}[u(\tau)]^{t}_{-\infty} \\ &= \int_{-\infty}^{t} h_1(t-\tau)u(\tau)\,\mathrm{d}\tau + \int_{-\infty}^{t}\int_{-\infty}^{t} h_2(t-\tau_1, t-\tau_2)u(\tau_1)u(\tau_2)\,\mathrm{d}\tau_1\,\mathrm{d}\tau_2 \\ &\quad + \int_{-\infty}^{t}\int_{-\infty}^{t}\int_{-\infty}^{t} h_3(t-\tau_1, t-\tau_2, t-\tau_3)u(\tau_1)u(\tau_2)u(\tau_3)\,\mathrm{d}\tau_1\,\mathrm{d}\tau_2\,\mathrm{d}\tau_3 \\ &\quad + \cdots \end{aligned}$$

These examples may also serve as illustrations of a discussion pertinent to the use of models as given by McGhee (1963). He coins the terms 'function space description' and 'parameter space description'.

The first uses the idea of transformation defined over a **function space.** This function space provides a representation of the process input signal. The representation may be such that the signal over a fixed time interval is given by **one point** in space, e.g. a Hilbert space. A representation can also be found for the signal using an interval of time that 'slides' along the $t$ axis; in that case one finds a curve in the function space. Examples of such representations can be found in e.g. the Fourier series expansion, the signal space expansion (using time samples), the Laguerre function expansion. The transformation that is defined over this space follows from the dynamics of the process. The process output signal may be represented on a similar space or on the one-dimensional amplitude space. In these terms the estimation problem is to find what transformation from input function space to output function space characterizes the process. As no information about the physical structure of the process or its assumed mathematical equations is used, this approach is of the black box identification type.

The **parameter space** description starts from an assumed mathematical description of the process dynamics. This description is a parametric model of finite dimensionality. The coordinates of the parameter space are the numerical values of the quantities that determine the 'output' of the model. If, for example, the assumed description is an ordinary differential equation, then the coordinates may be the value of the coefficients and the values of the initial conditions. If there is no forcing function, then from this one point in the parameter space one can predict the process output. If there is a forcing function then the unknown parameters of that signal increase the dimensionality of the parameter space. The dimensionality remains finite, while in principle an infinite number of parameters had to be determined in the function space description.

Consequently, one may distinguish:

**non-parametric models,** for example:

impulse responses;

covariance functions;

spectral densities;

Volterra series;

**parametric models,** for example:

differential equations of predetermined form and order;

state models.

The parametric models can give results with large errors if the order of the model does not agree with the order of the process. The non-parametric representations have the advantage that it is not necessary to specify the order of the process explicitly. These representations are, however, intrinsically infinite dimensional which means that it is frequently possible to obtain a model such that its output agrees exactly with the process output.

Interesting aspects of parametric versus non-parametric models are found in the literature on time series analysis. See for example Mann and Wald (1943); Whittle (1963); Grenander and Rosenblatt (1957); Jenkins and Watts (1968). Needless to say, the model must of course finally be judged with respect to the ultimate aim and the use that is going to be made of it. A very comprehensive discussion of dynamical models can be found in MacFarlane (1970).

## 4.2 Linear Models

For systems operating on continuous-time functions, Figure 4.2 gives a number of relations between different types of descriptions. It also shows several testing procedures and the type of information that is obtained from these tests. In the following sections some of these relations are reviewed and expanded.

### *4.2.1 Deterministic Signals; Continuous (Non-sampled)*

*Input–Output Description.* As a result of the superposition principle, the dynamic behaviour of a linear process can be described by means of characteristic time functions: **impulse-response function** and/or **step-response function.** These are the solutions of the differential equation (4.1) when all initial conditions are zero and when $u$ is defined respectively as:

$$\left.\begin{aligned} u(t) &= \delta(t) \\ \delta(t) &= 0 \qquad \text{for } t \neq 0 \\ \int_{-\infty}^{+\infty} \delta(t)\,\mathrm{d}t &= 1 \end{aligned}\right\} \text{(unit) impulse 'function'} \tag{4.4}$$

and

$$\left.\begin{aligned} u(t) &= 0 \qquad \text{for } t < 0 \\ u(t) &= 1 \qquad \text{for } t \geqslant 0 \end{aligned}\right\} \text{(unit) step function} \tag{4.5}$$

Physically each time function $u$ can formally be considered to consist of such impulse or step functions. This leads to the **convolution integral** which, for constant coefficients and zero initial conditions, can be written as:

$$y(t) = \int_0^{\infty} h(\tau)u(t-\tau)\,\mathrm{d}\tau = \int_{-\infty}^{t} h(t-\tau)u(\tau)\,\mathrm{d}\tau \tag{4.6}$$

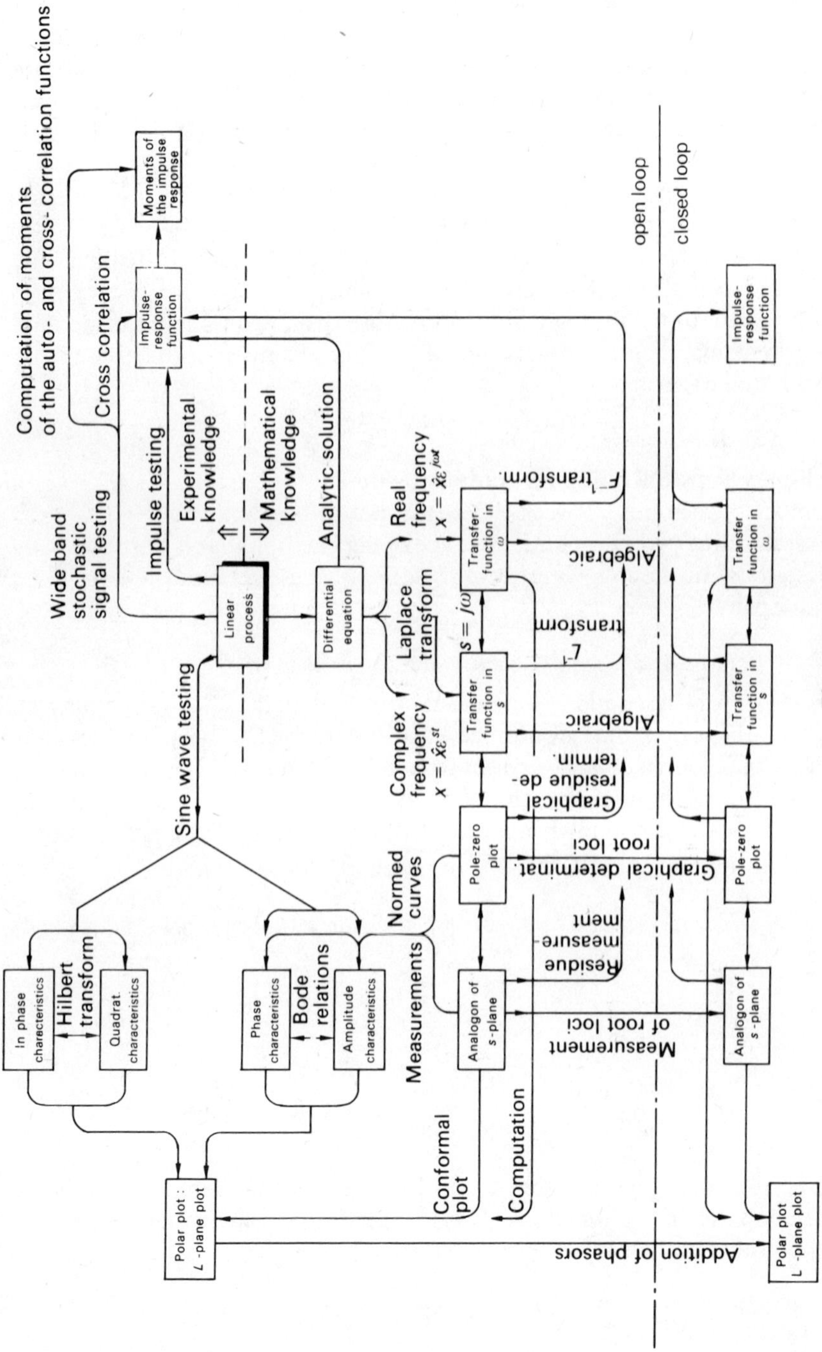

Computation of moments of the auto- and cross-correlation functions
Moments of the impulse response
Cross correlation
Impulse-response function
Wide band stochastic signal testing
Impulse testing
Experimental knowledge
Mathematical knowledge
Analytic solution
Linear process
Differential equation
Sine wave testing
Laplace transform
Real frequency
Complex frequency
$s = j\omega$
Transfer function in $\omega$
Transfer function in $s$
$F^{-1}$ transform.
$L^{-1}$ transform
Algebraic
Graphical residue determin.
Graphical determinat. root loci
Residue measurement
Measurement of root loci
Normed curves
Measurements
Pole-zero plot
Analogon of $s$-plane
In phase characteristics
Hilbert transform
Quadrat. characteristics
Phase characteristics
Bode relations
Amplitude characteristics
Conformal plot
Computation
Polar plot: $L$-plane plot
Addition of phasors
open loop
closed loop


Figure 4.2

If $u(t) = \delta(t)$ then

$$y(t) = \int_0^\infty h(\tau)\delta(t-\tau)\,d\tau = h(t) \tag{4.6a}$$

which explains the name impulse response (weighting function) for $h(t)$. For physical realizability $h(t) \equiv 0$ for $t < 0$; no response before excitation. This has already been used in the integration limits of (4.6). Instead of solving the differential equation for $y$ if $u = \delta(t)$ in order to obtain $h(t)$ one can also use the following procedure. Start with equation (4.6) as an assumption. Then differentiate this equation to form all derivatives that are needed in equation (4.1). Substitution of these expressions in (4.1) and the equating of the coefficients gives an expression for $h(t)$.

*Example.* Consider the simple first-order differential equation:

$$\dot{y} + ay = u \qquad y(0) = 0 \qquad u(t) = 0 \qquad \text{for } t < 0$$

with

$$y(t) = \int_0^t h(t-\tau)u(\tau)\,d\tau$$

Consequently, using Leibniz's rule*:

$$\dot{y}(t) = h(0)u(t) + \int_0^t \frac{dh(t-\tau)}{dt}u(\tau)\,d\tau$$

These expressions for $y$ and $\dot{y}$ are substituted in the differential equation. As the resulting equation has to hold for all $t$ and as this must be the case for arbitrary $u(t)$ it follows that the corresponding parts of the left- and right-hand sides have to be equated:

$$h(0)u(t) = u(t) \quad \Rightarrow h(0) = 1 \tag{4.7}$$

$$\int_0^t \left\{\frac{dh(t-\tau)}{dt} + ah(t-\tau)\right\}u(\tau)\,d\tau = 0$$

Again this equation has to be true for all time functions $u(t)$. It follows that:

$$\frac{dh(t)}{dt} + ah(t) = 0 \quad \Rightarrow h(t) = c\varepsilon^{-at} \tag{4.8}$$

Combining (4.7) and (4.8) leads us to

$$h(t) = \varepsilon^{-at}$$

* If

$$y(t) = \int_{a(t)}^{b(t)} f\{\tau, t\}\,d\tau$$

where $a$ and $b$ are differentiable functions of $t$ and where $f\{\tau, t\}$ and $\partial f\{\tau, t\}/\partial t$ are continuous in both $t$ and $\tau$, then is

$$\frac{dy}{dt} = f\{b(t), t\}\frac{db(t)}{dt} - f\{a(t), t\}\frac{da(t)}{dt} + \int_{a(t)}^{b(t)} \frac{\partial f\{\tau, t\}}{\partial t}\,d\tau$$

In this way we have derived the well-known result without using the properties of the symbolic impulse 'function' $\delta(t)$ as the independent variable. The same type of reasoning holds for $y(0) = b \neq 0$.

*State–Space Description.* In modern system and control theory the concept of state plays an important role (Zadeh and Desoer, 1963; Tou, 1964; Elgerd, 1967). In Table 4.1 some of the well-known relationships between time and frequency domain state-description are summarized and compared to the input–output relationships. Disturbances (noise signals) are not included in these descriptions.

The state and the output equation are schematically represented in Figure 4.3, where possible disturbances are represented by $\mathbf{v}(t)$ and $\mathbf{n}(t)$. The matrices are named:

$\mathbf{A}$ = system matrix
$\mathbf{B}$ = distribution matrix
$\mathbf{C}$ = output or measurement matrix
$\mathbf{D}$ = input–output matrix

The disturbances are named:

$\mathbf{v(t)}$ = process noise
$\mathbf{n(t)}$ = observation noise

The direct solution of the state equation in the time domain, neglecting $\mathbf{v}(t)$ and $\mathbf{n}(t)$, goes along the following lines:

$$\dot{\mathbf{x}} - \mathbf{Ax} = \mathbf{Bu}$$

$$\frac{\mathrm{d}}{\mathrm{d}t}[\varepsilon^{-\mathbf{A}t}\mathbf{x}] = \varepsilon^{-\mathbf{A}t}[\dot{\mathbf{x}} - \mathbf{Ax}] = \varepsilon^{-\mathbf{A}t}\,\mathbf{Bu}$$

For the definition of the matrix exponential cf. Appendix C. Consequently, integration over $(0, t)$ gives:

$$\varepsilon^{-\mathbf{A}t}\mathbf{x}(t) - \mathbf{x}(0) = \int_0^t \varepsilon^{-\mathbf{A}\tau}\mathbf{Bu}(\tau)\,\mathrm{d}\tau$$

or

$$\mathbf{x}(t) = \varepsilon^{\mathbf{A}t}\mathbf{x}(0) + \int_0^t \varepsilon^{\mathbf{A}(t-\tau)}\mathbf{Bu}(\tau)\,\mathrm{d}\tau \tag{4.9}$$

The state concept is closely related to the **mode** concept. For an illustration of this point consider only the set of homogeneous equations

$$\dot{\mathbf{x}} = \mathbf{Ax} \tag{4.10}$$

where $\mathbf{A}$ is an $(n \times n)$ matrix. Equation (4.10) has the solution

$$\mathbf{x}(t) = \varepsilon^{\mathbf{A}t}\mathbf{x}(0)$$

Table 4.1

| Time domain | $\xrightarrow{\mathscr{L}[\ ]}$ $\mathscr{L}^{-1}[\ ]\leftarrow$ | Frequency domain |
|---|---|---|
| **Input–output relationships** (2nd-order example) | | |
| $\ddot{y} + a_1\dot{y} + a_2 y = u$<br>$\dot{y}(0) \quad y(0)$<br>$\Downarrow$ Direct solution<br>$y(t) = \underbrace{h_0(t)y(0) + h_1(t)\dot{y}(0)}_{\text{General solution}} + \underbrace{\int_0^t h_1(t-\tau)u(\tau)\,d\tau}_{\text{Particular solution}}$ | $\rightarrow$ / $\leftarrow$ | $Y(s) = \frac{s + a_1}{s^2 + a_1 s + a_2}y(0) + \frac{1}{s^2 + a_1 s + a_2}\dot{y}(0) + \underbrace{\frac{1}{s^2 + a_1 s + a_2}}_{\text{Transfer function}}U(s)$ |
| **State relationships** (general example) | | |
| $\dot{\mathbf{x}} = \mathbf{Ax} + \mathbf{Bu}$ State equation<br>$\mathbf{y} = \mathbf{Cx} + \mathbf{Du}$ Output equation<br>$\mathbf{x}(0)$ Initial condition<br>$\Downarrow$ Direct solution<br>$\mathbf{x}(t) = \underbrace{\boldsymbol{\phi}(t)\mathbf{x}(0)}_{\text{General solution}} + \underbrace{\int_0^t \boldsymbol{\phi}(t-\tau)\mathbf{Bu}(\tau)\,d\tau}_{\text{Particular solution}}$ | $\rightarrow$ / $\leftarrow$ | $\mathbf{X}(s) = [s\mathbf{I} - \mathbf{A}]^{-1}\mathbf{x}(0) + [\mathbf{sI} - \mathbf{A}]^{-1}\mathbf{BU}(s)$<br>$\downarrow$ |
| $\boldsymbol{\phi}(t)$ = State transition matrix or principal matrix | $\leftarrow$ | $[s\mathbf{I} - \mathbf{A}]^{-1}$ |
| $\boldsymbol{\phi}(t) = \varepsilon^{At}$ | | |

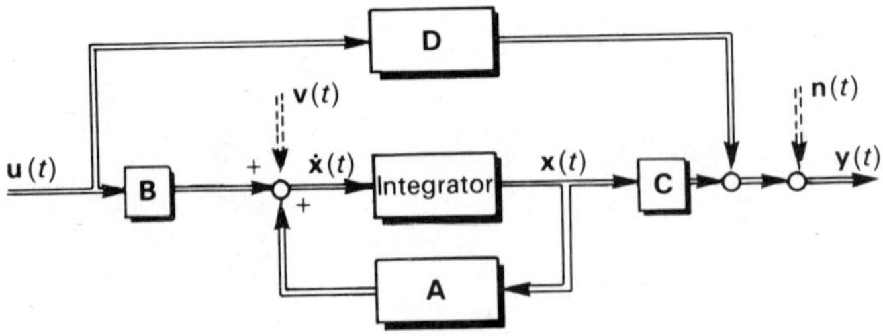


Figure 4.3

Instead of this formal solution we can look for special initial conditions under which one and only one mode is excited, i.e. where

$$\mathbf{x}(t) = \mathbf{r}\varepsilon^{\lambda t}$$

This substituted in equation (4.10) gives:

$$\lambda \mathbf{r}\varepsilon^{\lambda t} = \mathbf{A}\mathbf{r}\varepsilon^{\lambda t}$$

or

$$[\mathbf{A} - \lambda \mathbf{I}]\mathbf{r} = \mathbf{0} \tag{4.11}$$

A non-zero solution to this set of algebraic equations will exist if the determinant equals zero, that is if

$$|\mathbf{A} - \lambda \mathbf{I}| = 0 \tag{4.12}$$

The roots of this algebraic equation are called the **characteristic or eigenvalues** $\lambda_1, \ldots, \lambda_n$. Let us assume that all $\lambda$'s are distinct, then to each eigenvalue $\lambda_i$ there corresponds a **characteristic or eigenvector** $\mathbf{r}_i$ which is found by solving equation (4.11). A simple example is given in Table 4.2.

If the eigenvalues are distinct, then it is possible to choose a set of state–space coordinates such that the differential equations in these new coordinates become independent. For the theory on multiple eigenvalues and the generalized eigenvectors associated with these eigenvalues, cf. Friedman (1956).

We still have to show the equivalence of the state-transition matrix $\boldsymbol{\phi}(t)$, as given in Table 4.1, and the formal solution $\varepsilon^{\mathbf{A}t}$. This is done for the example in Table 4.2. Making use of the Cayley–Hamilton theorem (Friedman, 1956; p. 121) one can write for this second-order process, provided the eigenvalues are distinct:

$$\varepsilon^{\mathbf{A}t} = \alpha \mathbf{I} + \beta \mathbf{A}$$

where $\alpha$ and $\beta$ have to be determined. As a result of that theorem, this can be done using the eigenvalues instead of $\mathbf{A}$.

$$\varepsilon^{-t} = \alpha - \beta$$

$$\varepsilon^{-3t} = \alpha - 3\beta$$

Table 4.2

| Time domain | $\rightarrow\mathscr{L}[\ ]$ / $\mathscr{L}^{-1}[\ ]\leftarrow$ | Frequency domain |
|---|---|---|
| $\boldsymbol{\phi}(t) = \begin{cases} \varepsilon^{\mathbf{A}t} \\ \begin{bmatrix} h_0(t) & h_1(t) \\ h_0^0(t) & h_1^0(t) \end{bmatrix} \end{cases}$ (second-order case) | | |
| Example | | |
| $\mathbf{A} = \begin{bmatrix} 0 & 1 \\ -3 & -4 \end{bmatrix}$ | | $s\mathbf{I} - \mathbf{A} = \begin{bmatrix} s & -1 \\ 3 & s+4 \end{bmatrix}$ |
| $\boldsymbol{\phi}(t) = \frac{1}{2}\begin{bmatrix} 3\varepsilon^{-t} - \varepsilon^{-3t} & \varepsilon^{-t} - \varepsilon^{-3t} \\ -3\varepsilon^{-t} + 3\varepsilon^{-3t} & -\varepsilon^{-t} + 3\varepsilon^{-3t} \end{bmatrix}$ | $\leftarrow$ | $[s\mathbf{I} - \mathbf{A}]^{-1} = \begin{bmatrix} \frac{s+4}{s^2+4s+3} & \frac{1}{s^2+4s+3} \\ \frac{-3}{s^2+4s+3} & \frac{s}{s^2+4s+3} \end{bmatrix}$ |
| Eigenvalues and eigenvectors: | | |
| $\lvert\mathbf{A} - \lambda\mathbf{I}\rvert = \lambda^2 + 4\lambda + 3 = 0$ | | |
| $\lambda_1 = -1 \Rightarrow \begin{bmatrix} 1 & 1 \\ -3 & -3 \end{bmatrix}\mathbf{r}_1 = 0 \Rightarrow \mathbf{r}_1 = \alpha_1\begin{bmatrix} 1 \\ -1 \end{bmatrix}$ | | $\mathbf{x}(0) = \begin{cases} \mathbf{r}_1 \Rightarrow \mathbf{X}(s) = \alpha_1 \frac{1}{s+1}\mathbf{r}_1 \\ \mathbf{r}_2 \Rightarrow \mathbf{X}(s) = \alpha_2 \frac{1}{s+3}\mathbf{r}_2 \end{cases}$ |
| $\lambda_2 = -3 \Rightarrow \begin{bmatrix} 3 & 1 \\ -3 & -1 \end{bmatrix}\mathbf{r}_2 = 0 \Rightarrow \mathbf{r}_2 = \alpha_2\begin{bmatrix} 1 \\ -3 \end{bmatrix}$ | | |
| No input, arbitrary $\mathbf{x}(0)$: | | |
| $\mathbf{x}(t) = \gamma_1\mathbf{r}_1\varepsilon^{-t} + \gamma_2\mathbf{r}_2\varepsilon^{-3t}$ | | |
| if | | |
| $\mathbf{x}(0) = \gamma_1\mathbf{r}_1 + \gamma_2\mathbf{r}_2$ | | |

and it follows:

$$\alpha = \tfrac{1}{2}\{3\varepsilon^{-t} - \varepsilon^{-3t}\}$$

$$\beta = \tfrac{1}{2}\{\varepsilon^{-t} - \varepsilon^{-3t}\}$$

Consequently:

$$\varepsilon^{\left[\begin{smallmatrix} 0 & 1 \\ -3 & -4 \end{smallmatrix}\right]t} = \tfrac{1}{2}\{3\varepsilon^{-t} - \varepsilon^{-3t}\}\begin{bmatrix} 1 & 0 \\ 0 & 1 \end{bmatrix} + \tfrac{1}{2}\{\varepsilon^{-t} - \varepsilon^{-3t}\}\begin{bmatrix} 0 & 1 \\ -3 & -4 \end{bmatrix}$$

$$= \tfrac{1}{2}\begin{bmatrix} 3\varepsilon^{-t} - \varepsilon^{-3t} & \varepsilon^{-t} + 3\varepsilon^{-3t} \\ -3\varepsilon^{-t} + 3\varepsilon^{-3t} & -\varepsilon^{-t} + 3\varepsilon^{-3t} \end{bmatrix}$$

which is the same as $\phi(t)$ given in Table 4.2.

### 4.2.2 *Deterministic Signals; Sampled*

For the description of the dynamic behaviour of processes in which there are only sample values available of input and output signals one can use difference equations instead of differential equations. Such situations arise naturally when:

certain types of measuring instruments are used (e.g. those performing a relatively slow chemical analysis);

using measuring instruments and/or communication channels in a time-sharing fashion;

using digital data reduction or communication equipment;

the data are of a discrete nature (as e.g. in economics).

Another application may be found in continuous processes and signals, where a sampled data approach is used for mathematical simplicity. For sampled signals a relation scheme analogously to Figure 4.2 can be given. In this case difference equations, unit-sample response functions and $\mathscr{Z}$ transforms are used. Note that we restrict ourselves to that situation where both process input and output are given in sampled form (cf. Elgerd, 1967, p. 393).

*Input–Output Description.* Let the sample values of input and output signals be denoted by $u(k) = u(kT_s)$ and $y(k) = y(kT_s)$ respectively. Then the difference equation may be of the form:

$$a_0 y(k) + a_1 y(k+1) + \cdots + a_n y(k+n) = b_0 u(k) + \cdots + b_m u(k+m) \quad (4.13)$$

plus initial conditions, where again $a_0 = 1$ may be chosen arbitrarily. (Note that the difference equation also can be expressed in terms of $a_i y(k-i)$, etc.)

Again the linear, constant-coefficient case will be reviewed. Superposition holds because of the linearity, and the dynamic behaviour of the process can be described by a sampled time function $h(k)$, the **unit-sample-response-function.** The convolution integral goes over into a summation:

$$y(k) = \sum_{i=-\infty}^{k} h(k-i)u(i) \tag{4.14}$$

Again $h(k)$ can be derived by a method analogous to that presented for the continuous case.*

*Example.* The following simple difference equation is taken:

$$y(k+1) + ay(k) = u(k) \qquad y(0) = 0$$

For equation (4.14):

$$\begin{aligned} y(k+1) &= h(0)u(k+1) + h(1)u(k) + h(2)u(k-1) + \cdots \\ ay(k) &= \qquad\qquad\qquad\quad ah(0)u(k) + ah(1)u(k-1) + \cdots \end{aligned}$$

---

$$\begin{aligned} u(k) = y(k+1) + ay(k) &= h(0)u(k+1) + \{h(1) + ah(0)\}u(k) \\ &\quad + \{h(2) + ah(1)\}u(k-1) + \cdots \end{aligned}$$

This equation has to hold for all sample sequences $u(i)$. Consequently

$$\begin{aligned} &h(0) = 0 \\ &h(1) = 1 \\ &\cdots\cdots\cdots\cdots \\ &h(i+1) + ah(i) = 0 \qquad \text{for } i = 1, 2, 3, \ldots \end{aligned}$$

The solution to this set of equations is:

$$\begin{aligned} h(1) &= && 1 \\ h(2) &= -ah(1) = && -a \\ h(3) &= -ah(2) = && a^2 \\ &\cdots\cdots\cdots\cdots \\ h(i) &= && (-a)^{i-1} \qquad \text{for } i = 1, 2, 3, \ldots \end{aligned}$$

For stable processes, only a limited number of points of the unit-sample-response-function have to be given if $h(i) \approx 0$ for $i > L$. Consequently

$$y(k) \approx \sum_{i=0}^{L} h(i)u(k-i) \qquad k \geqslant L \tag{4.14a}$$

* Please note that this $h(k)$ is not necessarily identical with $h(t)$ at $t = kT_s$ for the continuous case, if this one is being approximated by a sampled version. The same symbol is used to accentuate the **similarities** between both cases.

The system behaviour is now characterized by the set of values $\{h(0), \ldots, h(L)\}$. Introducing the symbol $z$ as a shift operator, i.e.

$$zy(k) = y(k+1)$$

and

$$Y(z) = \sum_{k=0}^{\infty} y(k)z^{-k}$$

to the difference equation (4.13) leads to:

$$\begin{aligned} &\{a_0 z^k + a_1 z^{k+1} + \cdots + a_n z^{k+n}\} Y(z) \\ &\quad = \{b_0 z^k + b_1 z^{k+1} + \cdots + b_m z^{k+m}\} U(z) \end{aligned} \tag{4.15}$$

where $z^k$ can be divided out. The first $n$ values of $y$ are assumed to be zero. From equation (4.15) follows:

$$H(z) = \frac{Y(z)}{U(z)} = \frac{b_0 + b_1 z + \cdots + b_m z^m}{a_0 + a_1 z + \cdots + a_n z^n} \tag{4.16}$$

which is called the **sample transfer function.** By long division one obtains:

$$H(z) = h_0 + h_1 z^{-1} + h_2 z^{-2} + \cdots \tag{4.17}$$

where the coefficient $h_i = h(i)$ as in equation (4.14).

*State–Space Description.* Instead of the difference equation (4.13) one may use again a vector difference equation. In Table 4.3, some relationships between the time and $z$-domain description are summarized and compared to the input–output relationship. The direct solution of the state equation in the time domain is straight forward:

$$\mathbf{x}(k+1) = \mathbf{A}\mathbf{x}(k) + \mathbf{B}\mathbf{u}(k)$$

leads to:

$$\begin{aligned} &\mathbf{x}(1) = \mathbf{A}\mathbf{x}(0) + \mathbf{B}\mathbf{u}(0) \\ &\mathbf{x}(2) = \mathbf{A}\mathbf{x}(1) + \mathbf{B}\mathbf{u}(1) = \mathbf{A}^2\mathbf{x}(0) + \mathbf{A}\mathbf{B}\mathbf{u}(0) + \mathbf{B}\mathbf{u}(1) \\ &\cdots\cdots\cdots\cdots\cdots\cdots\cdots\cdots\cdots\cdots \\ &\mathbf{x}(k) = \mathbf{A}^k\mathbf{x}(0) + \sum_{i=0}^{k-1} \mathbf{A}^{k-1-i}\mathbf{B}\mathbf{u}(i) \end{aligned}$$

For a simple example cf. Table 4.4.

### 4.2.3 *Stochastic Signals*

In principle, the description given in the previous sections also holds for stochastic signals. With such signals there is little sense in asking for the

Table 4.3

| Time domain | $\xrightarrow{} \mathscr{Z}[\ ]$ / $\mathscr{Z}^{-1}[\ ] \leftarrow$ | z-Domain |
|---|---|---|
| **Input–output relationships** (2nd order example) | | |
| $y(k+2) + a_1 y(k+1) + a_0 y(k) = u(k)$; $y(1)$, $y(0)$ | → | $Y(z) = \frac{z^2 + a_1 z}{z^2 + a_1 z + a_0} y(0) + \frac{z}{z^2 + a_1 z + a_0} y(1) + \underbrace{\frac{1}{z^2 + a_1 z + a_0}}_{\text{Sample transfer function}} U(z)$ |
| ⇓ Direct solution | | |
| $y(k) = \underbrace{h_0(k) y(0) + h_1(k) y(1)}_{\text{General solution}} + \underbrace{\sum_{i=0}^{k} h_u(k-i) u(i)}_{\text{Particular solution}}$ | ← | |
| **State relationships** (general example) | | |
| $\mathbf{x}(k+1) = \mathbf{Ax}(k) + \mathbf{Bu}(k)$ State equation; $\mathbf{y}(k) = \mathbf{Cx}(k) + \mathbf{Du}(k)$ Output equation; $\mathbf{x}(0)$ Initial condition | → | $\mathbf{X}(z) = z[z\mathbf{I} - \mathbf{A}]^{-1}\mathbf{x}(0) + [z\mathbf{I} - \mathbf{A}]^{-1}\mathbf{BU}(z)$ |
| ⇓ Direct solution | | ↓ |
| $\mathbf{x}(k) = \underbrace{\boldsymbol{\phi}(k)\mathbf{x}(0)}_{\text{General solution}} + \underbrace{\sum_{i=0}^{k-1} \boldsymbol{\phi}(k-1-i)\mathbf{Bu}(i)}_{\text{Particular solution}}$ | ← | |
| $\boldsymbol{\phi}(k)$ Transition or principal matrix | ← | $z[z\mathbf{I} - \mathbf{A}]^{-1}$ |
| $\boldsymbol{\phi}(k) = \mathbf{A}^k$ | | |

Table 4.4

| Time domain | $\xrightarrow{\mathscr{Z}[\ ]}$ $\mathscr{Z}^{-1}[\ ]\leftarrow$ | z-Domain |
|---|---|---|
| $\phi(k) = \begin{cases} \mathbf{A}^k \\ \begin{bmatrix} h_0(k) & h_1(k) \\ h_0(k+1) & h_1(k+1) \end{bmatrix} \end{cases}$ (second-order case) | | |
| Example | | $z\mathbf{I} - \mathbf{A} = \begin{bmatrix} z & -1 \\ 3 & z+4 \end{bmatrix}$ |
| $\mathbf{A} = \begin{bmatrix} 0 & 1 \\ -3 & -4 \end{bmatrix}$ | | |
| $\phi(k) = \frac{1}{2}\begin{bmatrix} 3(-1)^k - (-3)^k & (-1)^k - (-3)^k \\ -3(-1)^k + 3(-3)^k & -(-1)^k + 3(-3)^k \end{bmatrix}$ | $\longleftarrow$ | $z[z\mathbf{I} - \mathbf{A}]^{-1} = \begin{bmatrix} \dfrac{z^2+4z}{z^2+4z+3} & \dfrac{z}{z^2+4z+3} \\ \dfrac{-3z}{z^2+4z+3} & \dfrac{z^2}{z^2+4z+3} \end{bmatrix}$ |
| $= \begin{bmatrix} \{1, 0, -3, \ldots\} & \{0, 1, -4, \ldots\} \\ \{0, -3, 12, \ldots\} & \{1, -4, 13, \ldots\} \end{bmatrix}$ | | |
| Unstable! | | |

Just as in the continuous case a development using eigenvalues and eigenvectors can be given, together with the corresponding modes.

instantaneous value of the amplitude. Consequently, the most useful questions are those with respect to the statistical properties and parameters, e.g. spectral density, expected value, variance, correlation function, etc. (cf. Appendix B and Section 3.4).

The input signal of a linear system is $u(t)$, its output signal $y(t)$. These signals are related through the convolution integral:

$$y(t) = \int_{-\infty}^{+\infty} h(\theta)u(t-\theta)\,d\theta \tag{4.18}$$

with

$$h(t) = 0 \qquad \text{for } t < 0$$

We find the **first** moment by averaging equation (4.18) over the ensemble (taking the expectation):

$$\mathscr{E}[y(t)] = \int_{-\infty}^{+\infty} h(\theta)\mathscr{E}[u(t-\theta)]\,d\theta$$

i.e.

$$\int_{-\infty}^{+\infty}\int_{-\infty}^{+\infty} y(t)p(u, y)\,du\,dy = \int_{-\infty}^{+\infty} h(\theta)\left\{\int_{-\infty}^{+\infty}\int_{-\infty}^{+\infty} u(t-\theta)p(u, y)\,du\,dy\right\} d\theta$$

or

$$\mu_y(t) = \int_{-\infty}^{+\infty} h(\theta)\mu_u(t-\theta)\,d\theta$$

for a general (non-stationary) stochastic process;

$$\mu_y = \mu_u \int_{-\infty}^{+\infty} h(\theta)\,d\theta$$

if $u(t)$ is a stationary process;

$$\mu_y = 0$$

if $u(t)$ is a centered process ($\mu_u = 0$).

For the **second** moment we can multiply equation (4.18) with a suitably chosen time function and take the expectation:

$$\mathscr{E}[u(t-\tau)y(t)] = \int_{-\infty}^{+\infty} h(\theta)\mathscr{E}[u(t-\tau)u(t-\theta)]\,d\theta \tag{4.19}$$

or

$$\psi_{uy}(t-\tau, t) = \int_{-\infty}^{+\infty} h(\theta)\psi_{uu}(t-\tau, t-\theta)\,d\theta$$

for a general (non-stationary) stochastic process;

$$\psi_{uy}(\tau) = \int_{-\infty}^{+\infty} h(\theta)\psi_{uu}(\tau - \theta)\,\mathrm{d}\theta \overset{\text{def}}{=} h(\tau) * \psi_{uu}(\tau)$$

for a stationary process. Along these lines one can find the following relations:

$$\begin{aligned} \psi_{uy}(\tau) &= h(\tau) * \psi_{uu}(\tau) \\ \psi_{yu}(\tau) &= h(-\tau) * \psi_{uu}(\tau) \\ \psi_{yy}(\tau) &= h(\tau) * \psi_{yu}(\tau) \\ \psi_{yy}(\tau) &= h(-\tau) * \psi_{uy}(\tau) \end{aligned} \tag{4.20}$$

which can be represented by the following diagram.

$$\begin{array}{ccccc} & & \psi_{uy}(\tau) & & \\ & {\scriptstyle * h(\tau)} \nearrow & & \searrow {\scriptstyle * h(-\tau)} & \\ \psi_{uu}(\tau) & & & & \psi_{yy}(\tau) \\ & {\scriptstyle * h(-\tau)} \searrow & & \nearrow {\scriptstyle * h(\tau)} & \\ & & \psi_{yu}(\tau) & & \end{array}$$

Note that for $\mu_u = 0$:

$$\psi_{uu}(0) = \sigma_u^2 \quad \text{and} \quad \psi_{yy}(0) = \sigma_y^2$$

By $\mathscr{L}_2$ transformation, one obtains the corresponding relations in the frequency domain:

$$\begin{aligned} \Phi_{uy}(s) &= H(s)\Phi_{uu}(s) \\ \Phi_{yu}(s) &= H(-s)\Phi_{uu}(s) \\ \Phi_{yy}(s) &= H(s)\Phi_{yu}(s) \\ \Phi_{yy}(s) &= H(-s)\Phi_{uy}(s) \end{aligned} \tag{4.21}$$

which is represented in the following diagram.

$$\begin{array}{ccccc} & & \Phi_{uy}(s) & & \\ & {\scriptstyle H(s)} \nearrow & & \searrow {\scriptstyle H(-s)} & \\ \Phi_{uu}(s) & & & & \Phi_{yy}(s) \\ & {\scriptstyle H(-s)} \searrow & & \nearrow {\scriptstyle H(s)} & \\ & & \Phi_{yu}(s) & & \end{array}$$

From the auto- and cross-correlation function the process impulse response can be determined by 'deconvolution'. Generally speaking it is not a trivial problem to determine $h(\tau)$ from equation (4.20) if $\psi_{uy}(\tau)$ and $\psi_{uu}(\tau)$ are given by correlation measurements (cf. Section 8.1).

Instead of using the impulse response one may also characterize a process by the time moments of that impulse response, defined as

$$m_i = \int_0^{\infty} t^i h(t)\,\mathrm{d}t$$

(cf. Goodman and Hillsley (1958)).

From the power spectral density function the transfer function follows as:

$$H(j\omega) = \frac{\Phi_{uy}(j\omega)}{\Phi_{uu}(j\omega)} \tag{4.22}$$

## 4.3 Varylinear Models

As mentioned in Section 4.1, the time dependence of coefficients of a (linear) differential or difference equation does not destroy the linearity. Consequently, the types of description given in the previous section remain valid. The simple Laplace transform, however, does not offer much help. The useful relations can be summarized as follows:

### *4.3.1 Input–Output Description*

$$y(t) = \int_{-\infty}^{t} h(t, \tau)u(\tau)\,\mathrm{d}\tau \tag{4.23}$$

where $h(t, \tau)$ can be found through substitution of $y$, $\dot{y}$, etc. in the differential equation and equating coefficients. For a survey of techniques used in describing time-varying networks cf. Zadeh (1961), where 138 references to the literature are given.

### *4.3.2 State–Space Description*

$$\begin{aligned} \dot{\mathbf{x}}(t) &= \mathbf{A}(t)\mathbf{x}(t) + \mathbf{B}(t)\mathbf{u}(t) \\ \mathbf{y}(t) &= \mathbf{C}(t)\mathbf{x}(t) + \mathbf{D}(t)\mathbf{u}(t) \end{aligned} \tag{4.24}$$

from which:

$$\mathbf{x}(t) = \boldsymbol{\phi}(t, t_0)\mathbf{x}(t_0) + \int_{t_0}^{t} \boldsymbol{\phi}(t, \tau)\mathbf{B}(\tau)\mathbf{u}(\tau)\,\mathrm{d}\tau \tag{4.25}$$

with

$$\boldsymbol{\phi}(t_0, t_0) = \mathbf{I} \qquad [\boldsymbol{\phi}(t, \tau)]^{-1} = \boldsymbol{\phi}(\tau, t)$$

$$\frac{\mathrm{d}}{\mathrm{d}t}\boldsymbol{\phi}(t, \tau) = \mathbf{A}(t)\boldsymbol{\phi}(t, \tau) \tag{4.26}$$

$$\frac{\mathrm{d}}{\mathrm{d}\tau}\boldsymbol{\phi}(t, \tau) = -\mathbf{A}'(\tau)\boldsymbol{\phi}'(t, \tau) \quad \text{adjoint equation} \tag{4.26a}$$

$$\boldsymbol{\phi}(t, \tau)\boldsymbol{\phi}(\tau, \theta) = \boldsymbol{\phi}(t, \theta) \tag{4.27}$$

For sample signals the state–space representation may be given as:

$$\left.\begin{aligned} \mathbf{x}(k+1) &= \mathbf{A}(k)\mathbf{x}(k) + \mathbf{B}(k)\mathbf{u}(k) \\ \mathbf{y}(k) &= \mathbf{C}(k)\mathbf{x}(k) + \mathbf{D}(k)\mathbf{u}(k) \end{aligned}\right\} \tag{4.28}$$

and corresponding expressions for the transition matrix $\boldsymbol{\phi}(k, i)$.

## 4.4 Non-linear Models

In Section 4.1, the importance of linearity and the associated principle of superposition has been stressed. In such a case the knowledge of one characteristic process time function (e.g. impulse response) is sufficient to determine the process output for arbitrary input signals. Such a procedure has advantages which we would also like to have for the description of non-linear processes (George, 1959) viz:

it gives an explicit input–output relationship;

it facilitates the discussion of combinations of systems;

it allows the consideration of random inputs.

For some classes of non-linear processes these requirements are fulfilled by the Volterra series.

### 4.4.1 *Volterra Representation** (Volterra, 1959)

Through the use of a Volterra series, whose kernels in addition to the ordinary linear impulse response are higher-order impulse responses, it is possible to characterize a non-linear process in a manner which yields a good deal of physical insight into the problem. Moreover, the method of solution has the great conceptual advantage of presenting a non-linear system as a rather straightforward generalization of the linear case, although the actual mechanics may be much more involved. Said in a converse manner, the Volterra series method treats the linear case as a subclass of the non-linear case, a very desirable property indeed (Zadeh, 1953). This advantage can hardly be overstressed because of the great deal of effort and time engineers have spent on solving linear systems in both the time and frequency domains; the Volterra series makes it possible to utilize these concepts and way of thinking for non-linear processes.

Let us consider a non-linear process with memory. We will assume that the input $u(t) = 0$ for $t < 0$. Moreover we will assume that the memory is **finite,** that is to say, it is granted that the value $u(t_0 - \tau)$ with $\tau$ sufficiently large no longer adds to the response $y(t_0)$. These assumptions follow naturally as a result of stability as well as physical realizability of the process. (An

* Much of the material in this section is taken from Van Harten (1966).

ideal integrator, however, has an infinite memory.) The input quantity can be approximated through rectangular pulses with finite dimensions. The samples from $u(t)$ for $\tau \leqslant t_0$, taken periodically with time intervals $\Delta t$, will be called $u_1, u_2, \ldots, u_N$ respectively (cf. Figure 4.4). $N$ will be taken so large

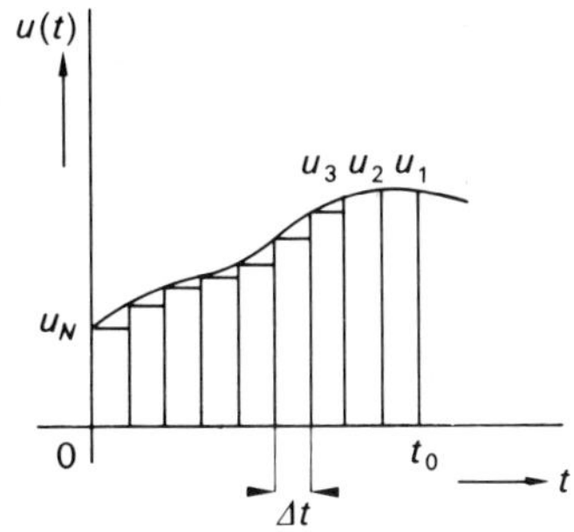


Figure 4.4

that $u_n$ for $n > N$ will no longer add to the value of $y(t)$. The response $\tilde{y}(t_0) = f(u_1, u_2, \ldots, u_N)$ is thus approximated as function of $N$ variables. A multi-dimensional Taylor development gives us (cf. Gabor et al., 1961):

$$\begin{aligned}\tilde{y}(t) &= (a_1u_1 + a_2u_2 + \cdots + a_Nu_N)\\ &\quad +(a_{11}u_1^2 + a_{12}u_1u_2 + \cdots + a_{NN}u_N^2)\\ &\quad +(a_{111}u_1^3 + a_{123}u_1u_2u_3 + \cdots + a_{NNN}u_N^3) + \cdots\\ &= \sum_{i=1}^{N} a_iu_i + \sum_{i=1}^{N}\sum_{j=1}^{N} a_{ij}u_iu_j + \sum_{i=1}^{N}\sum_{j=1}^{N}\sum_{k=1}^{N} a_{ijk}u_iu_ju_k + \cdots\\ &= \tilde{y}_{\text{lin}} + \tilde{y}_{\text{quadr}} + \tilde{y}_{\text{cub}} + \cdots \end{aligned} \tag{4.29}$$

(Kolmogorov–Gabor polynomial). Here it is assumed that the initial conditions are zero. Under the previously named conditions and if $\Delta t$ is small in comparison to the 'time constant' of the system $\tilde{y}(t)$ will be a close approximation of $y(t)$.

Should all $a_{ij}$ and $a_{ijk}$ be nought then only $y_{\text{lin}} = \sum_{i=1}^{N} a_iu_i$ will remain and this is an approximation of the convolution integral for linear processes. The latter is shown as follows:

We introduce $a_i = h_i\,\Delta t$ where $h_i$ is the height and $\Delta t$ is the breadth of the pulse. Therefore:

$$\tilde{y}_{\text{lin}}(t_0) = h_1u_1\,\Delta t + h_2u_2\,\Delta t + \cdots = \sum_{i=1}^{N} h_iu_i\,\Delta t \tag{4.30}$$

Taking into account the direction of the numbering of the $u_i$, cf. Figure 4.4, this sum for $\Delta t \to 0$ and $N \to \infty$ and $N\,\Delta t = t_0$ is converted into:

$$y_{\text{lin}}(t_0) = \int_0^{t_0} h_1(\tau)u(t_0 - \tau)\,d\tau \tag{4.31}$$

Should two pulses be applied to the system then the contribution to the linear sum follows the superposition principle: $h_i u_i \Delta t + h_j u_j \Delta t$. However, in sums that are quadratic, cubic, etc., the interaction between the pulses is also evident. Suppose $a_{ij} = h_{ij}\,\Delta t\,\Delta t$, then there is a contribution to $\tilde{y}_{\text{quadr}}$

$$h_{ii}u_i^2\,\Delta t\,\Delta t + 2h_{ij}u_i u_j\,\Delta t\,\Delta t + h_{jj}u_j^2\,\Delta t\,\Delta t \tag{4.32}$$

In the limit $\tilde{y}_{\text{quadr}}$ becomes:

$$y_{\text{quadr}}(t_0) = \int_0^{t_0}\int_0^{t_0} h_2(\tau_1, \tau_2)u(t_0 - \tau_1)u(t_0 - \tau_2)\,d\tau_1\,d\tau_2 \tag{4.33}$$

In this way

$$\tilde{y}(t) = \tilde{y}_{\text{lin}}(t) + \tilde{y}_{\text{quadr}}(t) + \tilde{y}_{\text{cub}}(t)$$

is converted into the Volterra series

$$\begin{aligned} y(t) = \int_0^t h_1(\tau)u(t-\tau)\,d\tau + \int_0^t\int_0^t h_2(\tau_1, \tau_2)u(t-\tau_1)u(t-\tau_2)\,d\tau_1\,d\tau_2 \\ + \int_0^t\int_0^t\int_0^t h_3(\ldots) \end{aligned} \tag{4.34}$$

The description using the Volterra series is therefore a direct generalization of the description of the linear processes using a convolution integral. The impulse response $h(\tau)$ of the linear system gives place to the impulse responses (kernels)

$$h_1(\tau), \qquad h_2(\tau_1, \tau_2), \qquad h_3(\tau_1, \tau_2, \tau_3), \quad \text{etc.}$$

The term $h_i$ is called an impulse response of the $i$th degree. The integral

$$\int_0^t\int_0^t \cdots \int_0^t h_i(\tau_1, \tau_2, \ldots, \tau_i)u(t-\tau_1)\ldots u(t-\tau_i)\,d\tau_1\ldots d\tau_i$$

is called the functional of the $i$th degree.

The Volterra series (4.34) is known as a functional power series or a functional development. The correspondence with the power series is apparent. If a process has no dynamic elements or if, in the case of an harmonic input signal, the frequency approaches zero then the series (4.34) gives place to a power series in $u$. Let us suppose for example that there are no dynamics so that

$$h_1(\tau) = c_1\delta(\tau)$$

$$h_2(\tau_1, \tau_2) = c_2\delta(\tau_1)\delta(\tau_2)$$

then the functional series takes the following form:

$$y(t) = c_1u(t) + c_2u(t)u(t) + \cdots$$

We shall see later that the behaviour of the power series gives us an insight into the kind of non-linearity. In certain cases it is possible to investigate the convergence of the Volterra series using a power series. If the Volterra series continues infinitely then according to Fréchet's theorem the functional power series can represent the input–output relation of all continuous non-linear processes. The functional series is unique and converges for $|u(t)| < r$; $r$ being the radius of convergence. As it seldom happens that the series can be successfully brought into a closed form, in most cases only a series can be used which is truncated after several terms, or which gives a satisfactory estimate after a few terms.

The kernels of the Volterra series for physical realizable processes have the following properties

(a) $h_n(\tau_1, \tau_2, \ldots, \tau_n) = 0$ for $\tau_i < 0 \qquad i = 1, 2, 3, \ldots, n;$

(b) $\lim\limits_{\tau_i \to \infty} h_n(\tau_1, \ldots, \tau_n) = 0 \qquad i = 1, 2, 3, \ldots, n;$

(c) $h_n(\tau_1, \ldots, \tau_n)$ are or can be made symmetrical.

The first two properties are trivial; the output of a physical realizable system depends merely on the past of the input. Every stable process has a finite memory so that the impulse responses become zero for large values of $\tau_i$. Note that an integrator is an example of a process that is on the border of stability and has an infinite memory.

The following applies to the third property: The kernels are symmetrical or can be made symmetrical through the cyclic changing of the arguments of the asymmetrical kernel. We shall illustrate this in the two-dimensional case, using

$$h_{2_{\text{sym}}}(\tau_1, \tau_2) \quad \text{and} \quad \tfrac{1}{2}\{h_{2_{\text{asym}}}(\tau_1, \tau_2) + h_{2_{\text{asym}}}(\tau_2, \tau_1)\}$$

The left-hand term and the right-hand term are not equal but since integration takes place with equal integration boundaries in the first quadrant of the $(\tau_1, \tau_2)$ plane the integral

$$\int_0^t \int_0^t h_{2_{\text{sym}}}(\tau_1, \tau_2)u(t - \tau_1)u(t - \tau_2)\,\mathrm{d}\tau_1\,\mathrm{d}\tau_2 \tag{4.35}$$

has the same value as the integral

$$\frac{1}{2}\int_0^t\int_0^t \{h_{2_{\text{asym}}}(\tau_1,\tau_2) + h_{2_{\text{asym}}}(\tau_2,\tau_1)\}u(t-\tau_1)u(t-\tau_2)\,\mathrm{d}\tau_1\,\mathrm{d}\tau_2 \tag{4.36}$$

The general symmetry properties thus become

$$h_{n_{\text{sym}}}(\tau_1,\tau_2,\ldots,\tau_n) \Leftrightarrow \frac{1}{n!}\{h_{n_{\text{asym}}}(\tau_1,\ldots,\tau_n) + \cdots + h_{n_{\text{asym}}}(\tau_n,\ldots,\tau_1)\}$$

*The Determination of the Kernels from the Differential Equation.* In principle, this can be done by substituting the Volterra series for $y(t)$ and its derivatives into the differential equation, just as was done for the linear case in Section 4.2. This procedure has been indicated by Flake (1963) for differential equations of the form:

$$\mathcal{O}[y(t)] + \mathcal{K}[y(t), \dot{y}(t), \ldots, \overset{(n-1)}{y}(t)] = u(t) \tag{4.37}$$

where $\mathcal{O}[\ ]$ is a linear time-independent dynamic operator, e.g. $\mathcal{O}[y(t)] = \overset{(n)}{y} + \cdots + a_1\dot{y} + a_0 y$ and $\mathcal{K}[\ ]$ is a multinomial in $y$ and its derivatives. The highest derivative appears in $\mathcal{O}$, the order of $\mathcal{K}$ is two or higher. The method consists of determining $\dot{y}, \ddot{y}, \ldots, \overset{(n)}{y}$ by differentiation of the Volterra series (4.34) and by substituting these expressions in (4.37). Recognition that the resulting expression has to hold for arbitrary forcing functions, leads to equating to zero the coefficient of each functional of $u(t)$. This results in a set of partial differential equations that may be solved successively, yielding Volterra kernels. Flake illustrates this substitution procedure for a differential equation

$$\dot{y} + y + ky^2 = u(t)$$

This procedure is quite cumbersome, particularly for more complicated differential equations. For that reason there is an advantage in introducing the multidimensional Laplace transform method. Therefore we will not repeat this exercise.

*The Use of Laplace Transform Techniques.* The multidimensional Laplace transform is defined as:

$$H_n(s_1,\ldots,s_n) = \underbrace{\int_{-\infty}^{+\infty}\cdots\int_{-\infty}^{+\infty}}_{n} h_n(\tau_1,\ldots,\tau_n)\exp\left[-\sum_{i=1}^{n} s_i\tau_i\right]\mathrm{d}\tau_1\ldots\mathrm{d}\tau_n \tag{4.38}$$

This is called a Laplace transform of the $n$th degree. To transform the functional

$$y_n(t) = \int_0^t\cdots\int_0^t h_n(\tau_1,\ldots,\tau_n)u(t-\tau_1)\ldots u(t-\tau_n)\,\mathrm{d}\tau_1\ldots\mathrm{d}\tau_n$$

we introduce new variables $t_1, t_2, \ldots, t_n$ instead of the single variable $t$. If

$$h_n(\tau_1, \ldots, \tau_n) \stackrel{\cdots\cdots}{\underset{\cdots\cdots}{=\!=}} H_n(s_1, \ldots, s_n) \quad * \tag{4.39}$$

and

$$u(t) \div U(s) \tag{4.40}$$

then

$$Y_n(s_1, \ldots, s_n) = H_n(s_1, \ldots, s_n) U(s_1) \ldots U(s_n) \tag{4.41}$$

Now the time function $y_n(t)$ can be obtained by applying the inverse transform to $Y_n(s_1, s_2, \ldots, s_n)$ and then substituting $t_1 = t_2 = \cdots = t_n = t$. This is quite laborious; a better way is by 'association of variables', introduced by George (1959), which provides a relation between $Y_n(s_1, s_2, \ldots, s_n)$ and $Y_n(s)$.

As an example take

$$y_2(t_1, t_2) = \frac{1}{(2\pi j)^2} \int_{c_1 - j\infty}^{c_1 + j\infty} \int_{c_2 - j\infty}^{c_2 + j\infty} Y_2(s_1, s_2) \varepsilon^{s_1 t_1 + s_2 t_2} \, ds_1 \, ds_2 \tag{4.42}$$

Put $t_1 = t_2 = t$ and $s_1 + s_2 = s$, then:

$$y_2(t, t) = \frac{1}{(2\pi j)^2} \int_{c_1 - j\infty}^{c_1 + j\infty} \int_{c_2 - j\infty}^{c_2 + j\infty} Y_2(s - s_2, s_2) \, ds_2 \varepsilon^{st} \, ds \tag{4.43}$$

Consequently

$$y_2(t) \div Y_2(s) \qquad \text{i.e. } y_2(t) = \frac{1}{2\pi j} \int_{c_2 - j\infty}^{c_2 + j\infty} Y_2(s) \varepsilon^{st} \, ds$$

$$Y_2(s) = \frac{1}{2\pi j} \int_{c_1 - j\infty}^{c_1 + j\infty} Y_2(s - s_2, s_2) \, ds_2 \tag{4.44}$$

$Y_n(s_1, s_2, \ldots, s_n)$ can be reduced to $Y_n(s)$ by applying the relation (4.44) successively to $s_1, s_2; s_2, s_3$; etc. Association of variables can also be done by inspection.

In the $s$-domain the symmetry properties of the kernels are given by:

$$H_{n_{\text{sym}}}(s_1, s_2, \ldots, s_n) \Leftrightarrow \frac{1}{n!} \{ H_{n_{\text{asym}}}(s_1, s_2, s_3, \ldots, s_n) + \cdots + H_{n_{\text{asym}}}(s_n, s_{n-1}, \ldots, s_1) \} \tag{4.45}$$

*Processes with Linear Elements and Multipliers.* These can very well be described using the Volterra series. The transforms of the kernels can be derived directly from the structure of such a non-linear process. These considerations will be helpful to our goal, i.e. determination of the kernels

* $\stackrel{\cdot\cdot}{=}$ indicates a Laplace transform relation of the second degree, $\stackrel{\cdots}{\equiv}$ of the third degree, etc.

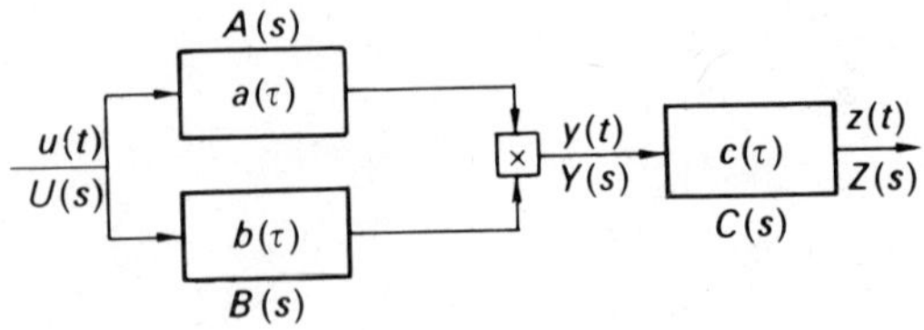


Figure 4.5

that pertain to a differential equation. They will also be useful in instrumenting these kernels in a model of such a non-linear process.

Let us consider the transform properties of the diagram in Figure 4.5.

$$y(t) = \int_0^t a(\tau)u(t-\tau)\,\mathrm{d}\tau \int_0^t b(\tau)u(t-\tau)\,\mathrm{d}\tau$$

$$= \int_0^t \int_0^t a(\tau_1)b(\tau_2)u(t-\tau_1)u(t-\tau_2)\,\mathrm{d}\tau_1\,\mathrm{d}\tau_2$$

$$z(t) = \int_0^t c(\tau_3)y(t-\tau_3)\,\mathrm{d}\tau_3$$

$$= \int_0^t c(\tau_3) \int_0^{t-\tau_3} \int_0^{t-\tau_3} a(\tau_1)b(\tau_2)u(t-\tau_3-\tau_1)u(t-\tau_3-\tau_2) \times \mathrm{d}\tau_1\,\mathrm{d}\tau_2\,\mathrm{d}\tau_3$$

This is a functional of the second degree in $u(t)$. The representation in the $s$-domain is found by:

$$a(\tau_1)b(\tau_2) \doteqdot A(s_1)B(s_2) \qquad a(\tau_1) \div A(s_1) \quad b(\tau_2) \div B(s_2)$$

$$y(t_1, t_2) \doteqdot Y(s_1, s_2) = A(s_1)B(s_2)U(s_1)U(s_2)$$

Using the equation (4.44), the association of variables leads us to:

$$Y(s) = \frac{1}{2\pi j} \int_{c-j\infty}^{c+j\infty} A(s-s_2)B(s_2)U(s-s_2)U(s_2)\,\mathrm{d}s_2$$

Consequently

$$Z(s) = C(s)Y(s)$$

$$= C(s)\frac{1}{2\pi j} \int_{c-j\infty}^{c+j\infty} A(s-s_2)B(s_2)U(s-s_2)U(s_2)\,\mathrm{d}s_2$$

$$= \frac{1}{2\pi j} \int_{c-j\infty}^{c+j\infty} A(s-s_2)B(s_2)C(s-s_2+s_2)U(s-s_2)U(s_2)\,\mathrm{d}s_2$$

This leads to

$$Z(s_1, s_2) = A(s_1)B(s_2)C(s_1+s_2)U(s_1)U(s_2)$$

and

$$z(t) = \int_0^t \int_0^t h(\tau_1, \tau_2)u(t - \tau_1)u(t - \tau_2)\,d\tau_1\,d\tau_2$$

with

$$h(\tau_1, \tau_2) \risingdotseq A(s_1)B(s_2)C(s_1 + s_2)$$

This can be generalized to the following statements:

if we consider linear networks and if we multiply the outputs of these networks then the transformation of the kernel of this process is given by

$$A_1(s_1)A_2(s_2)\cdots A_n(s_n) \tag{4.46}$$

if a linear network with transfer function $B(s)$ is put in series with the system just mentioned then the transformation of the kernel of the total process is:

$$A_1(s_1)A_2(s_2)\cdots A_n(s_n)B(s_1 + s_2 + \cdots s_n) \tag{4.47}$$

This is a function of $n$ variables which corresponds to a time-domain representation of a functional of the $n$th degree.

The time-domain representation leads to complicated integrals. The use of Laplace transformation offers advantages as the transforms of the kernels have a simple structure and can be 'translated' into a network. Summarizing the rules, cf. Figure 4.6:

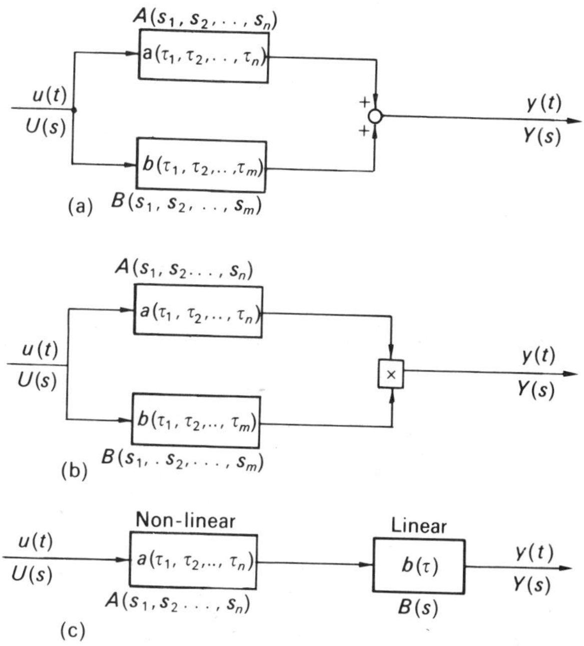


Figure 4.6

(*a*) If $m \neq n$ then the system can be described by:

$$y(t) = \underbrace{\int_0^t \cdots \int_0^t}_{n} a(\tau_1, \ldots, \tau_n) u(t - \tau_1) \ldots u(t - \tau_n)\, d\tau_1 \ldots d\tau_n$$
$$+ \underbrace{\int_0^t \cdots \int_0^t}_{m} b(\tau_1, \ldots, \tau_m) u(t - \tau_1) \ldots u(t - \tau_m)\, d\tau_1 \ldots d\tau_m \tag{4.48}$$

If $m = n$:

$$a(\tau_1, \tau_2, \ldots, \tau_n) + b(\tau_1, \tau_2, \ldots, \tau_n) \;\vdots\!\!=\!\!\vdots\; A(s_1, s_2, \ldots, s_n) + B(s_1, s_2, \ldots, s_n) \tag{4.49}$$

(*b*) $$y(t) = \int_0^t \cdots \int_0^t a(\tau_1, \tau_2, \ldots, \tau_n) b(\tau_{n+1}, \tau_{n+2}, \ldots, \tau_{m+n}) u(t - \tau_1) \ldots$$
$$\ldots u(t - \tau_{m+n})\, d\tau_1 \ldots d\tau_{m+n} \tag{4.50}$$

with

$$a(\tau_1, \ldots, \tau_n) b(\tau_{n+1}, \ldots, \tau_{m+n}) \;\vdots\!\!=\!\!\vdots\; A(s_1, \ldots, s_n) B(s_{n+1}, \ldots, s_{m+n})$$

(*c*) $$y(t) = \int_0^t \cdots \int_0^t h(\tau_1, \tau_2, \ldots, \tau_n) u(t - \tau_1) u(t - \tau_2) \ldots u(t - \tau_n)$$
$$\times\, d\tau_1 \ldots d\tau_n \tag{4.51}$$

with

$$h(\tau_1, \tau_2, \ldots, \tau_n) \;\vdots\!\!=\!\!\vdots\; A(s_1, s_2, \ldots, s_n) B(s_1 + s_2 + \cdots + s_n)$$

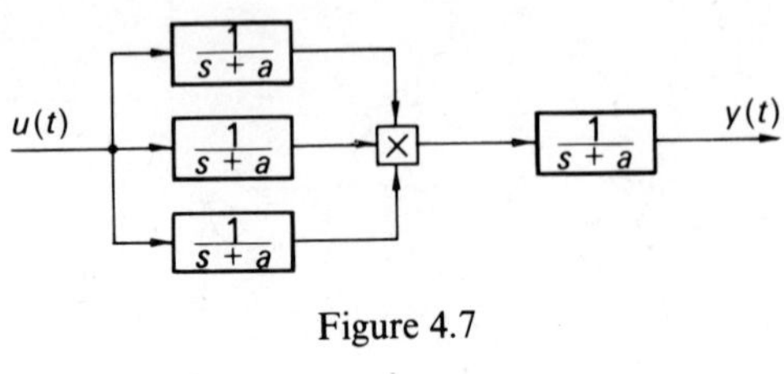


Figure 4.7

*Example 1.* According to Figure 4.7:

$$y(t) = \int_0^t \int_0^t \int_0^t h_3(\tau_1, \tau_2, \tau_3) u(t - \tau_1) u(t - \tau_2) u(t - \tau_3)\, d\tau_1\, d\tau_2\, d\tau_3$$

with

$$H_3(s_1, s_2, s_3) = \frac{1}{s_1 + a} \times \frac{1}{s_2 + a} \times \frac{1}{s_3 + a} \times \frac{1}{s_1 + s_2 + s_3 + a} \doteqdot h_3(\tau_1, \tau_2, \tau_3)$$

Use has been made of rules (*b*) and (*c*).

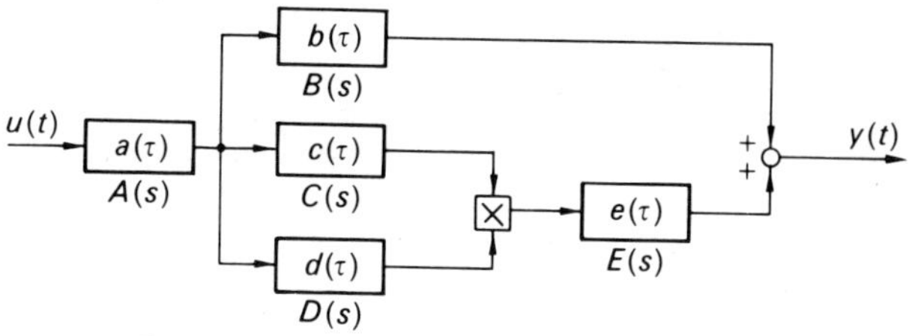


Figure 4.8

*Example 2*. In Figure 4.8 is represented a process consisting of a linear part, and a non-linear part. The dynamics can be described by:

$$y(t) = \int_0^t h_1(\tau)u(t - \tau)\,d\tau + \int_0^t \int_0^t h_2(\tau_1, \tau_2)u(t - \tau_1)u(t - \tau_2)d\tau_1\,d\tau_2$$

with

$$h_1(\tau) \doteqdot A(s)B(s)$$

$$h_2(\tau_1, \tau_2) \doteqdot A(s_1)C(s_1)A(s_2)D(s_2)E(s_1 + s_2)$$

This example illustrates the elegance of the multidimensional Laplace transform; the expression for $h(\tau_1, \tau_2)$ is quite complicated, the transformation of $h(\tau_1, \tau_2)$, however, is of a simple form and can be interpreted easily. The simplicity of the expressions in the frequency domain can be exploited when determining the Volterra kernels from a differential equation.

*Example 3*. We shall choose the equation:

$$\dot{y} + y + ay^2 = u(t)$$

The differentiation of the Volterra series is a time-consuming activity. Therefore, it is advisable to circumvent differentiation by applying the corresponding operation to the transformed expressions. For the one-dimensional Laplace transform this reduces to multiplication by $s$, for the $n$-dimensional transform it implies multiplication by $(s_1 + s_2 + \cdots + s_n)$. In our example the second-order kernel $H_2(s_1, s_2) = \mathscr{L}[h_2(\tau_1, \tau_2)]$ changes through the effect of differentiation to $(s_1 + s_2)H_2(s_1, s_2)$. Substitution of this result in the equation yields:

$$\underbrace{(s_1 + s_2)H_2(s_1, s_2)}_{\dot{y}} + \underbrace{H_2(s_1, s_2)}_{y} + \underbrace{a\left(\frac{1}{s_1 + 1}\right)\left(\frac{1}{s_2 + 1}\right)}_{\text{Quadratic term}} = 0$$

contribution from: $\dot{y}$ $y$ Quadratic term

or

$$H_2(s_1, s_2) = -a\frac{1}{s_1 + s_2 + 1} \times \frac{1}{s_1 + 1} \times \frac{1}{s_2 + 1}$$

The inverse transformation yields:

$$h_2(\tau_1, \tau_2) = a\{\varepsilon^{-(\tau_1+\tau_2)} - \varepsilon^{-\frac{1}{2}\{(\tau_1+\tau_2)-|\tau_1-\tau_2|\}}\}$$

For $\tau_1 = \tau_2 = t$

$$h_2(t) = a\{\varepsilon^{-2t} - \varepsilon^{-t}\}$$

The higher-order terms can be calculated along the same lines. Without some *a priori* knowledge of the process it is impossible to indicate how many terms will be needed for an adequate description. This example illustrates the fact that for some types of differential equations the Volterra series can be determined in a rather simple way. The kernels $h_1(\tau)$, $h_2(\tau_1, \tau_2)$, ... become available through their transforms. The Volterra series can be instrumented using linear elements and multipliers. On the other hand, the transfer characteristics can be determined for non-linear processes consisting of linear elements and multipliers.

From the examples given, it will be clear that models based on the Volterra representation can easily be programmed using analogue computer elements. Figure 4.9 shows two photographs of a kernel of the second

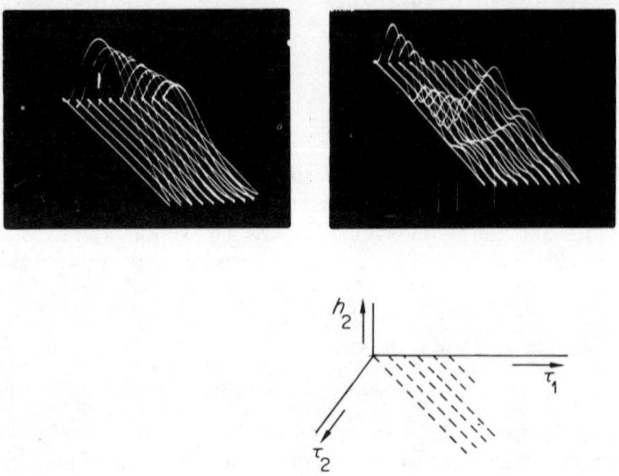


Figure 4.9

degree for a simple non-linear process. This 'impulse response' was determined experimentally (Van Harten, 1966). Further notes on theory and application of Volterra series can be found in references M.I.T. (1963–1965), Roy and Sherman (1967). Flake (1963) also discusses the case of non-zero initial conditions. For non-linear sampled-data processes one can use a discrete form of the Volterra series. By multidimensional $\mathscr{Z}$-transform one obtains a scheme analogous to the one given before for non-sampled processes (cf. Alper, 1965).

### 4.4.2 *Other Representations*

Besides the Volterra series representation there also exist several other representations which will be considered here only very briefly. These methods of description originated more or less as a result of the research carried out into purposeful **identification** methods for non-linear processes. Lee and Schetzen (1965) constructed a functional development with orthogonal properties. The input signal to be applied is noise with a flat power spectrum ('white' noise). The development takes the following form:

$$y(t) = \sum_{1}^{n} G_i[g_i, u(t)] \tag{4.52}$$

$G_i[g_i, u(t)]$ is a functional in $u(t)$, $g_i$ is a function of $n$ variables, which can be compared to the kernels of the Volterra series but are not equal to those kernels. If the input signal $u(t)$ is white then $G_j$ is a functional of the $j$th degree which is orthogonal to all functionals $G_i$ with $i < j$:

$$\overline{G_i[g_i, u(t)] \times G_j[g_j, u(t)]} = 0 \qquad \text{for } i < j \tag{4.53}$$

The line indicates the time average.

The first few terms of the functional development are the following:

$$G_1[g_1, u(t)] = \int_{-\infty}^{+\infty} g_1(\tau_1)u(t-\tau_1)\,d\tau_1$$

$$G_2[g_2, u(t)] = \int_{-\infty}^{+\infty}\int_{-\infty}^{+\infty} g_2(\tau_1, \tau_2)u(t-\tau_1)u(t-\tau_2)\,d\tau_1\,d\tau_2$$

$$- k\int_{-\infty}^{+\infty} g_2(\tau_1, \tau_1)\,d\tau_1 \tag{4.54}$$

$$G_3[g_3, u(t)] = \int_{-\infty}^{+\infty}\int_{-\infty}^{+\infty}\int_{-\infty}^{+\infty} g_3(\tau_1, \tau_2, \tau_3)u(t-\tau_1)u(t-\tau_2)u(t-\tau_3)$$

$$\times\, d\tau_1\,d\tau_2\,d\tau_3 - 3k\int_{-\infty}^{+\infty}\int_{-\infty}^{+\infty} g_3(\tau_1, \tau_2, \tau_2)u(t-\tau_1)\,d\tau_1\,d\tau_2$$

$k$ is the average power of the white noise. The orthogonal properties of the series lead to 'simple' techniques for the identification of completely unknown non-linear processes. The kernels $g_j(\tau_1, \tau_2, \ldots, \tau_j)$ can be successfully determined using correlation techniques.

Wiener (1958) developed another representation of non-linear processes. Here too a method of description developed out of the need to identify an unknown process. Wiener made use of orthogonal Laguerre functions and Hermite polynomials. The orthogonality properties of these polynomials lead to a simple general model. He develops $u(t)$ in Laguerre functions. This development takes the form

$$u(-t) = \sum_{i=1}^{\infty} a_i l_i(t) \qquad t \geqslant 0 \tag{4.55}$$

where $l_i(t)$ are Laguerre functions. The past of $u(t)$, therefore is known if the coefficients $a_i$ are given (Figure 4.10). If $u(t)$ is developed into orthogonal

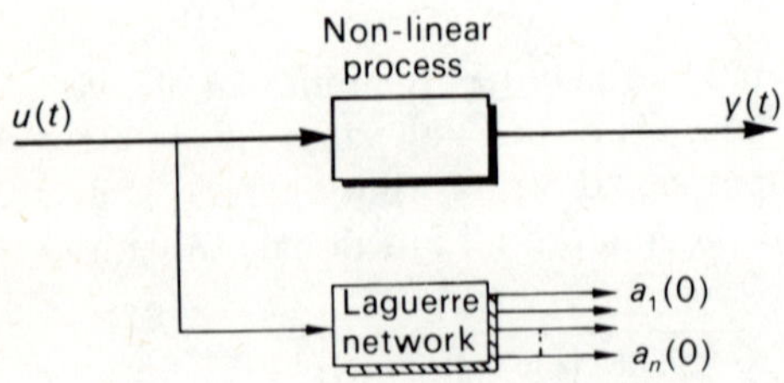


Figure 4.10

polynomials, the response $y(t)$ takes the following form

$$y(t) = \lim_{n \to \infty} F(a_1, a_2, \ldots, a_n) \tag{4.56}$$

in which $a_i$ are the coefficients of the development of $u(t)$ at time $t$. The next step is to develop $F$ into the Hermite polynomials $H_k(x)$ so that

$$\begin{aligned} y(t) = \lim_{n \to \infty} \sum_{i=1} \sum_{j=1} \cdots \sum_{k=1} b_{ij,\ldots,k} H_i(a_1) H_j(a_2) \cdots \\ \cdots H_k(a_n) \exp\left[-\frac{a_1^2 + a_2^2 + \cdots + a_n^2}{2}\right] \end{aligned} \tag{4.57}$$

As this development represents the input–output relation of a general non-linear process we can identify an unknown process by using a model that has been built according to this development. If the input quantity is white noise the orthogonal properties will give a model that (in principle) is easily adjusted.

A model based on this description can be outlined as follows. With a slight change with respect to Table 3.1 the Laguerre functions can be defined as:

$$l_k(t) = \frac{1}{k!} \frac{d^k}{dt^k}(t^k \varepsilon^{-t}) \tag{4.58}$$

which has the Laplace transform

$$L_k(s) = \frac{s^k}{(s+1)^{k+1}} \tag{4.59}$$

The determination of the Laguerre coefficients can take place with the use of a fairly simple linear ladder network, or by analogue computer elements.

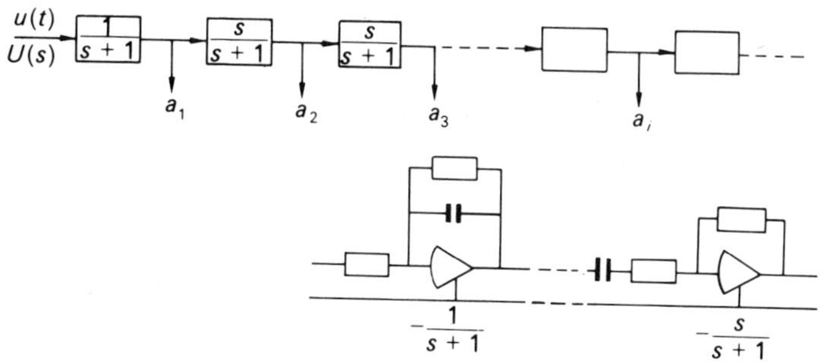


Figure 4.11

In Figure 4.11, such a filter and a realization of its sections is outlined. The Laguerre coefficients are found to be

$$a_i(t) = \int_0^\infty u(t-\tau)l_i(\tau)\,d\tau$$

or at $t = 0$

$$a_i(0) = \int_0^\infty u(-\tau)l_i(\tau)\,d\tau$$

Consequently, the coefficients $a_i(0)$ determine the past of $u(t)$.

As the last example of representations we will give the development using Tchebycheff polynomials (Smets, 1960). This development can be used only if the input quantity is bounded. For a non-linear process without memory

$$y(t) = f[u(t)] \tag{4.60}$$

applies and the development

$$y(t) = \sum_{i=0}^{\infty} a_i T_i\left(\frac{u(t)}{u_0}\right) \tag{4.61}$$

in which $u_0$ is the maximum value of the input signal. This development is well adapted to the case of sinusoidal input signals. If $u = u_0 \cos \omega t$ then the following applies for a system without a memory:

$$y(t) = \sum_{i=1}^{n} a_i T_i(\cos \omega t) = \sum_{i=1}^{n} a_i \cos i\omega t \tag{4.62}$$

The coefficients of the polynomial development are similar to the amplitudes of the higher harmonics of the input signal. The above-mentioned relation comes into being as a result of the recurrent relation of Tchebycheff polynomials.

$$T_k(u) = 2uT_{k-1}(u) - T_{k-2}(u) \tag{4.63}$$

Suppose that $T_{k-1}$ and $T_{k-2}$ are of the form $\cos(k-1)\omega t$ and $\cos(k-2)\omega t$. Then

$$T_k(u) = 2\cos\omega t\cos(k-1)\omega t - \cos\{(k-1)\omega t - \omega t\}$$

$$T_k(u) = \cos\omega t\cos(k-1)\omega t - \sin(k-1)\omega t\sin\omega t = \cos k\omega t$$

The substitution of $u = \cos\omega t$ in $T_0(u)$ and $T_1(u)$ shows that the relation for $k = 0$ and $k = 1$ is correct, and therefore for all $k$.

For the purpose of analysis the Volterra representation is most satisfactory. The orthogonal developments have advantages in the identification of unknown systems. They give fairly simple models with which a system can be simulated. A disadvantage of the development, however, is that the characteristic quantities are difficult to define in terms of the quantities of the Volterra series.

*State–Space Description.* Along the same lines as in the previous sections, the state–space representation for nonlinear processes can be chosen as:

$$\begin{aligned}\dot{\mathbf{x}}(t) &= \mathbf{f}[\mathbf{x}(t), \mathbf{u}(t), \mathbf{a}(t), \mathbf{v}(t), t]\\ \mathbf{y}(t) &= \mathbf{g}[\mathbf{x}(t), \mathbf{u}(t), \mathbf{n}(t), t]\end{aligned} \tag{4.64}$$

with

$\mathbf{x}(t)$ = process state vector
$\mathbf{u}(t)$ = measurable process input vector
$\mathbf{a}(t)$ = process parameter vector
$\mathbf{v}(t)$ = process noise vector (unmeasurable input)
$\mathbf{n}(t)$ = observation (or measurement) noise vector
$\mathbf{y}(t)$ = observable process output vector

## 4.5 Controllability, Observability, Identifiability

Since the introduction of these concepts by Kalman (1960) they have been discussed by many authors, e.g. Elgerd (1967); Lee (1964). The reader is referred to the publications cited at the end of this chapter for a thorough discussion of this subject. Here we only summarize the results for processes described by (stationary) linear difference equations. For time-varying linear processes, cf. Kreindler and Sarachik (1964). The process under discussion is described by:

$$\left.\begin{aligned} \mathbf{x}(k+1) &= \mathbf{A}\mathbf{x}(k) + \mathbf{B}\mathbf{u}(k) \\ \mathbf{y}(k) &= \mathbf{C}\mathbf{x}(k) \end{aligned}\right\} \tag{4.65}$$

where $\mathbf{x}$ is an $n$-dimensional vector.

A process is called **controllable** if it is possible to find an (unconstrained) control vector which brings the system from any initial state to any specified final state in finite time. Consequently, we ask for the condition under which we can determine the control necessary to drive the system from $\mathbf{x}(0)$ to a preassigned state $\mathbf{x}(n)$.

$$\mathbf{x}(1) = \mathbf{A}\mathbf{x}(0) + \mathbf{B}\mathbf{u}(0)$$

$$\mathbf{x}(2) = \mathbf{A}\mathbf{x}(1) + \mathbf{B}\mathbf{u}(1) = \mathbf{A}^2\mathbf{x}(0) + \mathbf{A}\mathbf{B}\mathbf{u}(0) + \mathbf{B}\mathbf{u}(1)$$

$$\cdots\cdots\cdots\cdots\cdots\cdots\cdots\cdots\cdots\cdots\cdots\cdots$$

$$\mathbf{x}(n) = \mathbf{A}^n\mathbf{x}(0) + \mathbf{A}^{n-1}\mathbf{B}\mathbf{u}(0) + \cdots + \mathbf{B}\mathbf{u}(n-1)$$

or

$$\mathbf{x}(n) - \mathbf{A}^n\mathbf{x}(0) = [\mathbf{A}^{n-1}\mathbf{B} \,\vdots\, \mathbf{A}^{n-2}\mathbf{B} \,\vdots\, \cdots \,\vdots\, \mathbf{B}] \begin{bmatrix} \mathbf{u}(0) \\ \vdots \\ \mathbf{u}(n-1) \end{bmatrix} \tag{4.66}$$

As $\mathbf{A}$, $\mathbf{x}(0)$ and $\mathbf{x}(n)$ are known the left-hand side of equation (4.66) is determined. A unique solution of the $\mathbf{u}$ vector can be found if the matrix

$$[\mathbf{A}^{n-1}\mathbf{B} \,\vdots\, \mathbf{A}^{n-2}\mathbf{B} \,\vdots\, \cdots \,\vdots\, \mathbf{B}]$$

has rank $n$. In that case $\mathbf{A}$, $\mathbf{B}$ is called a controllable pair.

A process is called **observable** if from the measurements of the output it is possible to determine the state of that process. Consequently, we ask for the condition under which we can determine $\mathbf{x}(0)$ by measuring $\mathbf{y}$

$$\mathbf{y}(0) = \mathbf{C}\mathbf{x}(0)$$

$$\mathbf{y}(1) = \mathbf{C}\mathbf{x}(1) = \mathbf{C}\mathbf{A}\mathbf{x}(0)$$

$$\cdots\cdots\cdots\cdots\cdots\cdots\cdots\cdots$$

$$\mathbf{y}(n-1) = \mathbf{C}\mathbf{A}^{n-1}\mathbf{x}(0)$$

or taking the transpose

$$[\mathbf{y}'(0) \,\vdots\, \mathbf{y}'(1) \,\vdots\, \cdots \,\vdots\, \mathbf{y}'(n-1)] = \mathbf{x}'(0)[\mathbf{C}' \,\vdots\, \mathbf{A}'\mathbf{C}' \,\vdots\, \cdots \,\vdots\, \mathbf{A}'^{n-1}\mathbf{C}'] \tag{4.67}$$

As the $\mathbf{y}$ vectors are known, a unique solution for $\mathbf{x}(0)$ can only be found if the matrix

$$[\mathbf{C}' \,\vdots\, \mathbf{A}'\mathbf{C}' \,\vdots\, \cdots \,\vdots\, \mathbf{A}'^{n-1}\mathbf{C}']$$

has rank $n$. In that case **A**, **C** is called an observable pair. It has been shown by Brockett (1965) that the introduction of state variable feedback to a process may change the observability of that process.

A process is called **identifiable** if from the measurements of the state variables, it is possible to determine the system matrix **A**.

$$\mathbf{x}(1) = \mathbf{A}\mathbf{x}(0)$$
$$\mathbf{x}(2) = \mathbf{A}^2\mathbf{x}(0)$$
$$\cdots\cdots\cdots\cdots\cdots$$
$$\mathbf{x}(n) = \mathbf{A}^n\mathbf{x}(0)$$

or

$$[\mathbf{x}(1) \,\vdots\, \mathbf{x}(2) \,\vdots\, \cdots \,\vdots\, \mathbf{x}(n)] = \mathbf{A}[\mathbf{x}(0) \,\vdots\, \mathbf{A}\mathbf{x}(0) \,\vdots\, \cdots \,\vdots\, \mathbf{A}^{n-1}\mathbf{x}(0)] \tag{4.68}$$

As the **x** vectors are known a unique solution for **A** can be found only if the following matrix has rank $n$:

$$[\mathbf{x}(0) \,\vdots\, \mathbf{A}\mathbf{x}(0) \,\vdots\, \cdots \,\vdots\, \mathbf{A}^{n-1}\mathbf{x}(0)]$$

The interested reader will find in the literature more elaborate studies of these concepts and finer distinctions, e.g. for controllability, which may be either: state or output, completely or totally, strongly or weakly. The observability may be either completely or totally (Kreindler and Sarachik, 1964).

Fischer and Vámos (1967) have extended the controllability concept to what they call $\delta$-controllability, in which case an initial state $\mathbf{x}(0)$ can be driven to the $\delta$-environment of a desired given state. Aoki (1967) has extended the definitions of controllability, observability and identifiability to stochastic situations. The question of identifiability has also been discussed by Balakrishnan (1969) and by Staley and Yue (1970). Closely related to identifiability is the concept of determinable classes (Root, 1965; Prosser and Root, 1966; Root, 1970).

*Examples.* For illustrating these concepts a simple second-order system will be chosen:

$$\mathbf{x}(k+1) = \mathbf{A}\mathbf{x}(k) + \mathbf{B}\mathbf{u}(k)$$
$$\mathbf{y}(k) = \mathbf{C}\mathbf{x}(k)$$

cf. Figure 4.12 with

$$\mathbf{A} = \begin{bmatrix} a_{11} & a_{12} \\ a_{21} & a_{22} \end{bmatrix} \qquad \mathbf{B} = \begin{bmatrix} 1 & 0 \\ 0 & 0 \end{bmatrix} \qquad \mathbf{C} = \begin{bmatrix} 1 & 0 \\ 0 & 0 \end{bmatrix}$$

Controllability:

$$[\mathbf{AB} \,\vdots\, \mathbf{B}] = \begin{bmatrix} a_{11} & 0 & 1 & 0 \\ a_{21} & 0 & 0 & 0 \end{bmatrix}$$

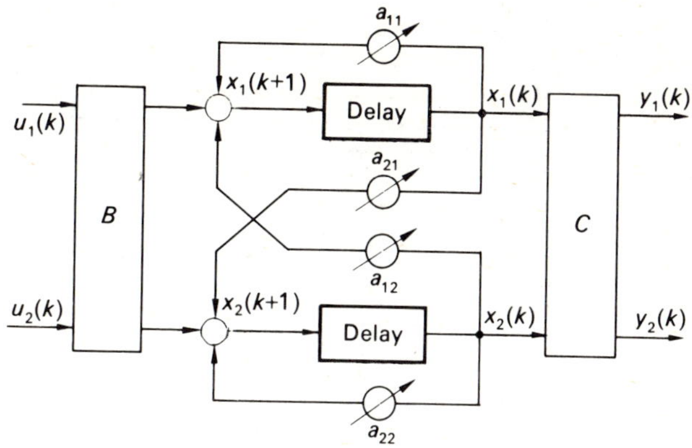


Figure 4.12

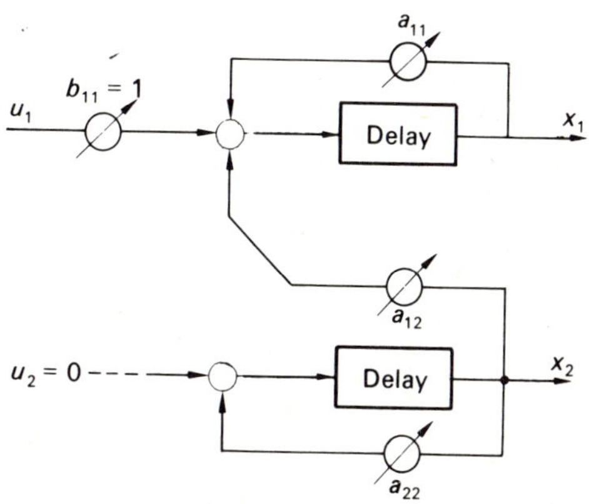


Figure 4.13

Controllable if rank = 2; uncontrollable if $a_{21} = 0$. From Figure 4.13, it is clear that if $a_{21} = 0$, we do not have control over the state $x_2$. Note that all other parameters $a_{ij}$ may be zero for the system to remain controllable.
Observability:

$$[\mathbf{C}' \vdots \mathbf{A}'\mathbf{C}'] = \begin{bmatrix} 1 & 0 & a_{11} & 0 \\ 0 & 0 & a_{12} & 0 \end{bmatrix}$$

Observable if rank = 2; unobservable if $a_{12} = 0$. From Figure 4.14, it is clear that if $a_{12} = 0$, the output variable **y** does not provide any knowledge about $x_2$. Note again that all other parameters $a_{ij}$ may be zero for the system to remain observable.
Identifiability:

$$[\mathbf{x}(0) \vdots \mathbf{A}\mathbf{x}(0)] = \begin{bmatrix} x_1(0) & a_{11}x_1(0) + a_{12}x_2(0) \\ x_2(0) & a_{21}x_1(0) + a_{22}x_2(0) \end{bmatrix}$$

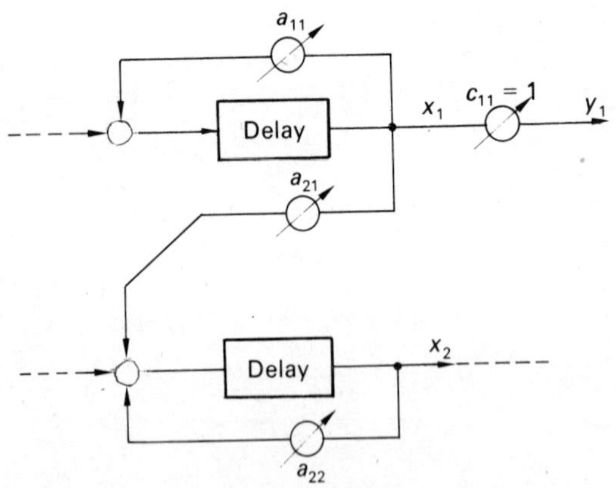


Figure 4.14

Identifiable if rank = 2; unidentifiable, if the determinant of this matrix equals zero. This is the case if both columns of the matrix are linearly dependent. We can distinguish:

a trivial case: $x_1(0) = x_2(0) = 0$, that is, the process at rest cannot be identified;

a non-trivial case:

$$\lambda \begin{bmatrix} x_1(0) \\ x_2(0) \end{bmatrix} = \begin{bmatrix} a_{11} & a_{12} \\ a_{21} & a_{22} \end{bmatrix} \begin{bmatrix} x_1(0) \\ x_2(0) \end{bmatrix} \quad \text{or} \quad [\mathbf{A} - \lambda \mathbf{I}] \begin{bmatrix} x_1(0) \\ x_2(0) \end{bmatrix} = 0$$

This is the approach for finding the eigenvalues $\lambda_1$ and $\lambda_2$ and the corresponding eigenvectors $\mathbf{r}_1$ and $\mathbf{r}_2$. If $\mathbf{x}(0) = \alpha \mathbf{r}_1$ then only the mode $\exp[\lambda_1 t]$ of the process is excited by $\mathbf{x}(0)$ and the mode $\exp[\lambda_2 t]$ is not identifiable. If $\mathbf{x}(0) = \beta \mathbf{r}_2$ then only the mode $\exp[\lambda_2 t]$ can be identified. Consequently, the process is identifiable only if all modes of the process are excited by $\mathbf{x}(0)$.

In Chapter 11 other notions of identifiability are given.

## 4.6 Models Based on These Descriptions

In the previous sections, several types of description of process characteristics have been mentioned. Now the models or simulations based on these descriptions will be considered briefly. Only general ideas are presented, as the choice of a model is strongly object and goal oriented.

### *4.6.1 Linearity-of-the-dynamics versus Linearity-in-the-parameters*

In Section 4.1, a distinction between linear and non-linear has been made on the criterion of dynamic behaviour, i.e. the relation between the **dependent**

**and the independent time-variables.** For parameter estimation another distinction between linearity and non-linearity is of as much importance, viz. with respect to the relation between the **dependent variables and the parameters**. A model is said to be linear-in-the-parameters if the (generalized) error is linear-in-the-parameters.

Apparently, these two notions of linearity have no immediate relation as can be seen from the following examples. We assume a process with input signal $u$ and output signal $y$. Then the 'model' may be chosen to form an 'error' $e$ between process and model output in the following way:

| | | Model; in-the-parameters: | | Process |
|---|---|---|---|---|
| | | Linear | Non-linear | |
| Dynamic behaviour | Linear | $e = \dot{y} + \alpha y - u$ | $e = y - w = y - \dfrac{u}{D + \alpha}$<br>$\dot{w} + \alpha w = u$ | $\dot{y} + ay = u$ |
| | Non-linear | $e = \dot{y} + \alpha y^3 - u$ | $e = y - w = y - \mathcal{O}[u, \alpha]$<br>$\dot{w} + \alpha w^3 = u$ | $\dot{y} + ay^3 = u$ |

The two different uses of the terms linear and non-linear may cause some confusion. This is due to the mixing of concepts from the fields of system theory and regression analysis*. Henceforth we will use the term **'linear'** for the dynamic behaviour and use **'linear-in-the-parameters'** for the other type.

In connexion with estimation schemes the great importance of linearity-in-the-parameters will become clear. A typical example is the identification of a sampled linear system when the output is disturbed with white measurement noise. The representation of the system by the coefficients of the pulse transfer function leads to a non-linear regression problem while the representation of the model by coefficients of a generalized model or by the ordinates of the weighting function leads to an estimation problem which is 'linear-in-the-parameters'. It also pays to try to find transformations of the variables to obtain such a linearity if possible. Some simple examples may

* Note that the term 'order' may cause confusion also. In regression analysis this term refers to the highest degree of the independent variable:

$$\left.\begin{aligned} y &= \beta_0 + \beta_1 u_1 + n \\ y &= \beta_0 + \beta_1 u_1 + \beta_2 u_2 + \cdots + \beta_m u_m + n \end{aligned}\right\} \text{Model of the first order}$$

$$y = \beta_0 + \beta_1 u_1 + \beta_2 u_1^2 \qquad \text{Model of the second order}$$

illustrate this.

$$z = \frac{\alpha_2 x_2 + x_1}{\alpha_1 x_1 x_2} \Rightarrow \left\{ \begin{matrix} z = y & \\ \dfrac{1}{x_1} = u_1 & \dfrac{\alpha_2}{\alpha_1} = \beta_2 \\ \dfrac{1}{x_2} = u_2 & \dfrac{1}{\alpha_1} = \beta_1 \end{matrix} \right\} \Rightarrow y = \beta_1 u_1 + \beta_2 u_2$$

Reciprocal transformation

$$z = c x_1^{\alpha_1} x_2^{\alpha_2} \Rightarrow \left\{ \begin{matrix} \log z = y & \log c = \beta_0 \\ \log x_1 = u_1 & \alpha_1 = \beta_1 \\ \log x_2 = u_2 & \alpha_2 = \beta_2 \end{matrix} \right\} \Rightarrow y = \beta_0 + \beta_1 u_1 + \beta_2 u_2$$

Logarithmic transformation

Such non-linear expressions that can be made linear-in-the-parameters through transformation are called **intrinsically linear.** If such a linearization is not possible then the term **intrinsically non-linear** is used. It may pay to use transformations even if the system is intrinsically non-linear (see e.g. Diskind, 1969).

### 4.6.2 *Linear Input–Output Models*

Linear models naturally represent the most extensively developed area in the field of identification. Now we will consider linear processes as well as 'linear environments', that is environments that can be characterized by linear stochastic models. In most control problems the properties of the environment will be just as important as the system dynamics, because e.g. it is the presence of disturbances that creates a control problem. If classical design techniques are to be used, the model can be characterized by a transfer function or by an impulse response. Many recently developed design methods will however require a state model, i.e. a parametric model.

Several problems arise naturally:

Suppose the impulse response is desired. Should this be identified directly or is it 'better' to identify a parametric model and then compute the impulse response?

Assume that a parametric model is desired. Should this be fitted directly or is it 'better' to first determine the impulse response and then fit a parametric model to that?

Since a parametric model contains the order of the system explicitly, what happens if the wrong order is assumed in the problem formulation?

There are not yet any general answers to these problems. Special cases have been investigated by Gustavsson (1969) in connexion with identification of nuclear reactor and distillation tower dynamics as well as on simulated data.

For linear processes the well known analogue computer techniques provide a straightforward way of modelling ordinary differential equations such as

$$\frac{d^n y}{dt^n} + \cdots + a_{n-1}\frac{dy}{dt} + a_n y = b_0\frac{d^n u}{dt^n} + \cdots + b_{n-1}\frac{du}{dt} + b_n u$$

This leads to a transfer function

$$H(s) = \frac{Y(s)}{U(s)} = \frac{b_0 s^n + b_1 s^{n-1} + \cdots + b_{n-1}s + b_n}{s^n + a_1 s^{n-1} + \cdots + a_{n-1}s + a_n} \tag{4.69}$$

and a computer diagram as given in Figure 4.15. If the state variables are chosen as $x_{n-1}, \ldots, x_0$ then those quantities are directly measurable in this

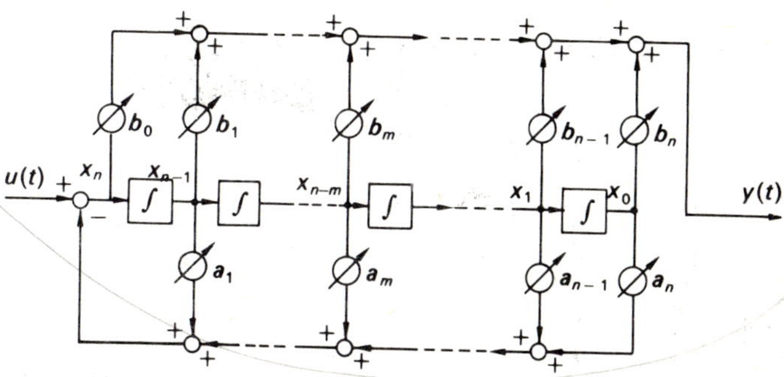


Figure 4.15

model. If $y$ and its derivatives are chosen as state variables then they can be formed as linear combinations of these $x$ variables. Note that $H(s)$ is non-linear-in-the-parameters $a_i$. The transfer function of a linear(ized) process, e.g.

$$H(s) = \frac{(s + d_1)(s + d_2)}{(s + e_1)(s + e_2)(s + e_3)} \tag{4.70}$$

can be decomposed in several ways, cf. Figure 4.16:

as a series combination of simple transfer functions having only one or two poles and perhaps a zero of the overall transfer function

$$H(s) = \frac{s + d_1}{s + e_1} \cdot \frac{s + d_2}{s + e_2} \cdot \frac{1}{s + e_3} \tag{4.71}$$

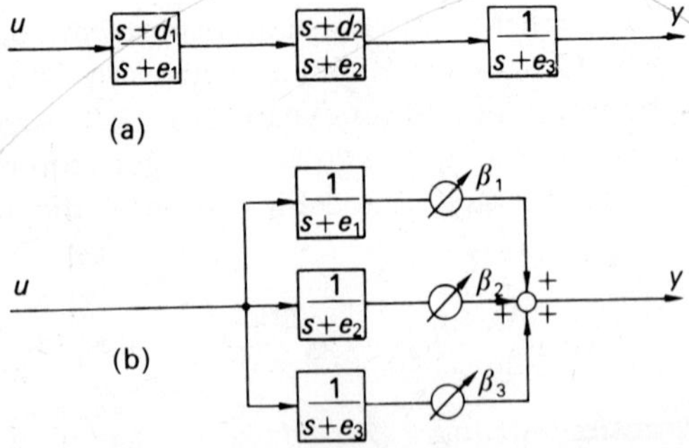


Figure 4.16

as a parallel combination of simple transfer functions, each weighted by a particular coefficient $\beta_i$

$$H(s) = \frac{\beta_1}{s + e_1} + \frac{\beta_2}{s + e_2} + \frac{\beta_3}{s + e_3} \tag{4.72}$$

as a (parallel) combination of orthogonal transfer functions

$$H(s) = \sum_{i=1}^{n} \beta_i F_i(s) \tag{4.73}$$

Example: the Laguerre representation

$$F_i(s) = \frac{1}{s + \alpha}\left(\frac{s - \alpha}{s + \alpha}\right)^{i-1}$$

which leads to the block diagram of Figure 4.17.

The modelling technique mentioned before can also be applied to systems in which information of the signals is available only as samples at discrete

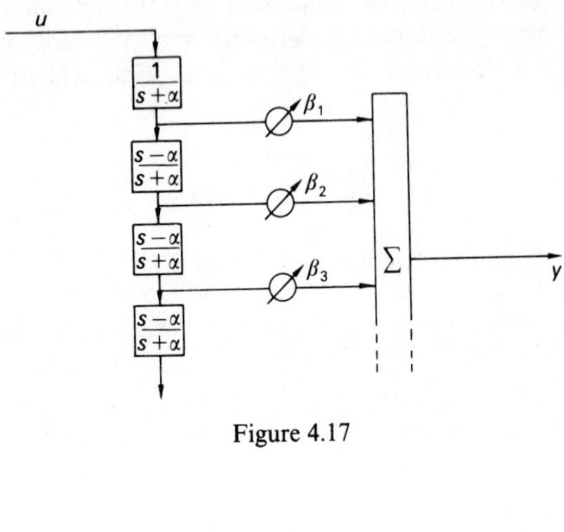


Figure 4.17

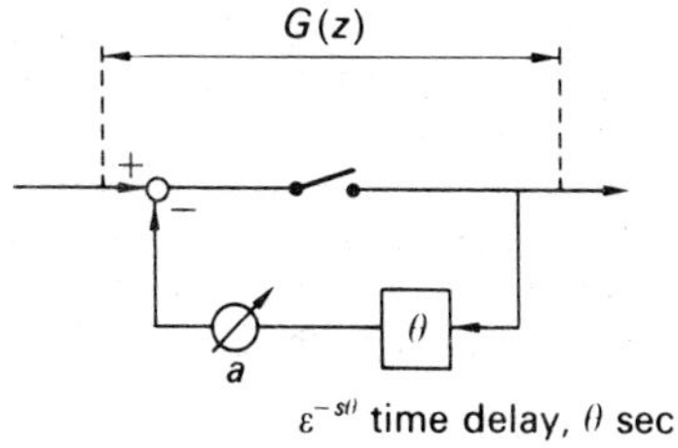


Figure 4.18

intervals of time. Figure 4.18 gives one example of a simple first-order system with the transfer function:

$$H(z) = \frac{z}{z + a} \qquad \text{with } z = \varepsilon^{s\theta}$$

A higher-order transfer function

$$H(z) = \frac{b_0 + b_1 z + \cdots + b_n z^n}{1 + a_1 z + \cdots + a_n z^n} \tag{4.74}$$

can be instrumented using delay units analogously to Figure 4.15, and in addition it can be decomposed analogously to Figure 4.16. A pure time delay can easily be incorporated:

$$G(z) = z^{-k} H(z)$$

This type of process description can well be used for control purposes. If the process is contaminated with non-white additive noise, then the noise properties can also be described (and estimated) in terms of such a transfer function with a white-noise input.

The impulse response $h(t)$ provides a general, non-parametric representation for a linear process; it requires no more *a priori* information than that the process is linear.

A pure time delay can easily be accommodated. It can be instrumented in several ways. A condition, however, is that the model enables the convolution of $h(t)$ with $u(t)$. This leads to the following possibilities:

$h(t)$ given as a photographic mask and $u(t)$ as a film track;
$h(t)$ given as a series of coefficients of orthonormal terms; cf. Chapter 3:

$$h(t) = \sum \beta_i g_i(t) \tag{4.75}$$

with

$$\int_a^b g_i(t) g_j(t)\, dt = \delta_{ij}$$

where $\delta_{ij} = 1$ if $i = j$ and $\delta_{ij} = 0$ if $i \neq j$. The usefulness of the representation as given by equation (4.75) depends to a large extent on the 'match' between the impulse response and the type of (orthonormal) functions chosen. In a favourable case, only a few parameters $\beta_i$ unequal to zero may give a satisfactory approximation; otherwise it may take a rather large number. It has been proposed (Wolff and Dietz, 1962; Izawa and Furuta, 1967) that the functions $g_i(t)$ be made dependent on other parameters, e.g. $|g_i(t;\alpha)$. In that case, one needs to determine the best values of the $\beta$ and $\alpha$ parameters, complicating the estimation procedure. Attempts have been made to extend the orthogonal models to multivariable processes (Barker, 1965).

$h(t)$ given as a number of points $\beta_0 = h(0)$, $\beta_1 = h(\theta)$, $\beta_2 = h(2\theta), \ldots$. This leads to the delay-line synthesizer (Goodman and Reswick, 1956).

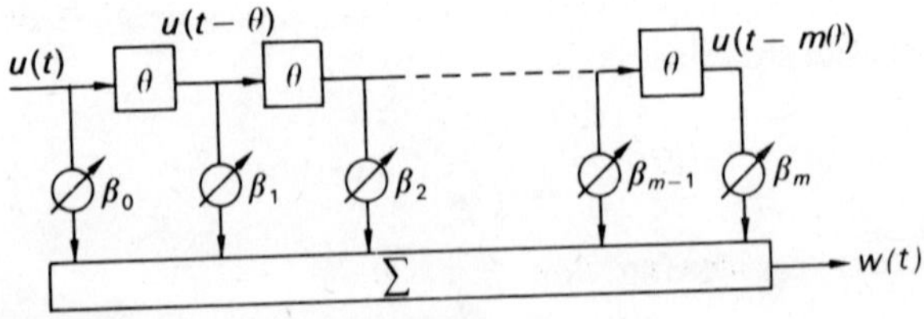


Figure 4.19

Instrumented according to Figure 4.19, the convolution is approximated by

$$w(t) = \sum_{i=0}^{n} \beta_i u(t - i\theta) \tag{4.76}$$

A disadvantage is the rather large number of points (parameters) needed. The approximation of the impulse response may take several forms, cf. Figure 4.20:

$h(t)$ as a (continuous) function

$$h_1(t) = \sum h(t)\delta(t - i\theta)$$

$$h_2(t) = h(i\theta) \qquad \text{for } 0 \leqslant t - i\theta < \theta$$

$$h_3(t) = h(i\theta) + \frac{t - i\theta}{\theta}[h\{(i+1)\theta\} - h\{i\theta\}] \qquad \text{for } 0 \leqslant t - i\theta < \theta$$

Linearity-in-the-parameters enables us to represent the sampled-data case in the form

$$\mathbf{y} = \mathbf{U}\mathbf{b} \tag{4.77}$$

with

$$\mathbf{y} = \begin{bmatrix} y(1) \\ \vdots \\ y(k) \end{bmatrix} \qquad \mathbf{U} = \begin{bmatrix} u_0(1) & \cdots & u_m(1) \\ \vdots & & \vdots \\ u_0(k) & \cdots & u_m(k) \end{bmatrix} \qquad \mathbf{b} = \begin{bmatrix} b_0 \\ \vdots \\ b_m \end{bmatrix}$$

Note that

$$\mathbf{U} = [\mathbf{u}_0, \ldots, \mathbf{u}_m]$$

and that no assumption has yet been made with respect to the relations between $u_i$ and $u_j$. They may represent the same sample values shifted in

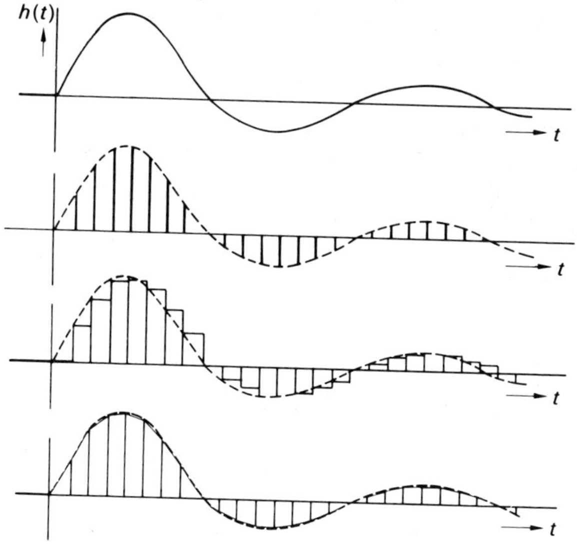


Figure 4.20

time (delay-line outputs), but they may as well stand for output samples of non-linear filters. If the 'delay-line' representation is chosen then

$$\mathbf{U} = \begin{bmatrix} u(1) & 0 & 0 & \cdots & 0 \\ u(2) & u(1) & 0 & & 0 \\ u(3) & u(2) & u(1) & & \vdots \\ \vdots & & & & u(1) \\ & & & & \vdots \\ u(k) & u(k-1) & & \cdots & u(k-m) \end{bmatrix} \tag{4.78}$$

where $u(i) = 0$ for $i \leqslant 0$ is assumed. Also the following type of notation can be used:

$$\mathbf{y} = \mathbf{B}\mathbf{u} \tag{4.79}$$

with

$$\mathbf{y} = \begin{bmatrix} y(1) \\ \vdots \\ y(k) \end{bmatrix} \qquad \mathbf{B} = \begin{bmatrix} b_0 & 0 & 0 & \cdots & 0 \\ b_1 & b_0 & 0 & & \\ b_2 & b_1 & b_0 & & \\ \vdots & & & \ddots & \vdots \\ b_m & & & & \\ 0 & b_m & & & \\ 0 & 0 & b_m & \cdots & b_0 \end{bmatrix} \qquad \mathbf{u} = \begin{bmatrix} u(1) \\ \vdots \\ u(k) \end{bmatrix}$$

This permits a direct notation of the covariance of **y**, viz.

$$\operatorname{cov}[\mathbf{y}] = \mathbf{B}\operatorname{cov}[\mathbf{u}]\mathbf{B}' \tag{4.80}$$

### 4.6.3 *Linear State Models; Canonical Forms**

Finally, our attention will be focussed on the important type of state model $S(\mathbf{A}, \mathbf{B}, \mathbf{C}, \mathbf{D})$ defined by:

$$\begin{aligned} \dot{\mathbf{x}} &= \mathbf{A}\mathbf{x} + \mathbf{B}\mathbf{u} \\ \mathbf{y} &= \mathbf{C}\mathbf{x} + \mathbf{D}\mathbf{u} \end{aligned} \tag{4.81}$$

where **x** is an $n$-vector, the input **u** is a $p$-vector and the output **y** is an $r$-vector. It is well known that the systems $A(\mathbf{A}, \mathbf{B}, \mathbf{C}, \mathbf{D})$ and $S(\mathbf{TAT}^{-1}, \mathbf{TB}, \mathbf{CT}^{-1}, \mathbf{D})$ where **T** is a non-singular matrix are equivalent in the sense that they have the same input–output relation. It is also easy to verify that the systems $S(\mathbf{A}, \mathbf{B}, \mathbf{C}, \mathbf{D})$ and $S(\tilde{\mathbf{A}}, \tilde{\mathbf{B}}, \tilde{\mathbf{C}}, \tilde{\mathbf{D}})$ are equivalent in the sense that they have the same input–output relation if

$$\begin{aligned} \mathbf{D} &= \tilde{\mathbf{D}} \\ \mathbf{C}\mathbf{A}^k\mathbf{B} &= \tilde{\mathbf{C}}\tilde{\mathbf{A}}^k\tilde{\mathbf{B}} \qquad k = 0, 1, \ldots, n \end{aligned} \tag{4.82}$$

The relations between the different representations were clarified by Kalman's work (Kalman, 1963). The impulse response and the transfer function represent only the part of the system $S$ which is completely controllable. It is thus clear that only the completely controllable and completely observable part of a state model $S(\mathbf{A}, \mathbf{B}, \mathbf{C}, \mathbf{D})$ can be determined from input–output measurements. The impulse response and the transfer function are easily obtained from the state description. The problem of determining a state model from the impulse response is more subtle, even if we disregard the fact that only the controllable and observable subsystem can be determined from the impulse response. The problem of assigning a state model of the lowest possible order which has a given impulse response has been solved by Ho and Kalman (1966). See also Kalman, Falb and Arbib (1969) and Bruni, Isidori and Ruberti (1969). Again the solution is not unique. The model $S(\mathbf{A}, \mathbf{B}, \mathbf{C}, \mathbf{D})$ contains

$$N_1 = n^2 + np + nr + pr \tag{4.83}$$

parameters. The fact that the input–output relation is invariant under a linear transformation of the state variables implies that all $N_1$ parameters

* The following part is taken from Åström and Eykhoff (1970).

cannot be determined from input–output measurements. To obtain unique solutions as well as to be able to construct efficient algorithms, it is therefore of great interest to find representations of the system which contain the smallest number of parameter, i.e. **canonical representations.**

*Canonical Forms for Linear Deterministic Models.* Canonical forms for linear systems are discussed, for example, by Kalman (1963). If the matrix **A** has distinct eigenvalues, canonical forms can be obtained as follows. By a suitable choice of coordinates the matrix **A** can be brought into diagonal form.

$$\dot{\mathbf{x}} = \begin{bmatrix} \lambda_1 & 0 & \cdots & 0 \\ 0 & \lambda_2 & \cdots & 0 \\ \vdots & \vdots & & \vdots \\ 0 & 0 & \cdots & \lambda_n \end{bmatrix} \mathbf{x} + \begin{bmatrix} b_{11} & b_{12} & \cdots & b_{1p} \\ b_{21} & b_{22} & \cdots & b_{2p} \\ \vdots & \vdots & & \vdots \\ b_{n1} & b_{n2} & \cdots & b_{np} \end{bmatrix} \mathbf{u}$$

$$\mathbf{y} = \begin{bmatrix} c_{11} & c_{12} & \cdots & c_{1n} \\ c_{21} & c_{22} & \cdots & c_{2n} \\ \vdots & \vdots & & \vdots \\ c_{r1} & c_{r2} & \cdots & c_{rn} \end{bmatrix} \mathbf{x} + \begin{bmatrix} d_{11} & d_{12} & \cdots & d_{1p} \\ d_{21} & d_{22} & \cdots & d_{2p} \\ \vdots & \vdots & & \vdots \\ d_{r1} & d_{r2} & \cdots & d_{rp} \end{bmatrix} \mathbf{u} \tag{4.84}$$

This representation contains $n + np + nr + pr$ parameters of which $n$ are redundant since all state variables can be scaled without affecting the input–output relations. The input–output relation can thus be characterized by

$$N_2 = n(p + r) + pr \tag{4.85}$$

parameters. Since the system is completely controllable and observable there is at least one non-zero element in each row of the **B** matrix and of each column of the **C** matrix. The redundancy in (4.84) can thus be reduced by imposing conditions such as

$$\max_j b_{ij} = 1 \qquad i = 1, 2, \ldots, n \tag{4.86}$$

$$\sum_j |b_{ij}| = 1 \qquad i = 1, 2, \ldots, n \tag{4.87}$$

or similar conditions on the **C**-matrix. When the matrix **A** has multiple eigenvalues the problem of finding a minimal parameter representation is much more complex. If **A** not only has multiple eigenvalues but is cyclic as well (i.e. there exists a vector **x** such that the vectors $\mathbf{x}, \mathbf{A}\mathbf{x}, \mathbf{A}^2\mathbf{x}, \ldots, \mathbf{A}^{n-1}\mathbf{x}$ span the $n$-dimensional space), the matrix can be transformed to companion form and a minimal parameter representation is then given by

$$\dot{\mathbf{x}} = \begin{bmatrix} -a_1 & 1 & 0 & \cdots & 0 \\ -a_2 & 0 & 1 & \cdots & 0 \\ \vdots & \vdots & \vdots & & \vdots \\ -a_{n-1} & 0 & 0 & \cdots & 1 \\ -a_n & 0 & 0 & \cdots & 0 \end{bmatrix} \mathbf{x} + \begin{bmatrix} b_{11} & \cdots\cdots\cdots & b_{1p} \\ \vdots & & \vdots \\ \vdots & & \vdots \\ \vdots & & \vdots \\ b_{n1} & \cdots\cdots\cdots & b_{np} \end{bmatrix} \mathbf{u} \tag{4.88}$$

$$\mathbf{y} = \begin{bmatrix} c_{11} & \cdots\cdots\cdots & c_{1n} \\ \vdots & & \vdots \\ \vdots & & \vdots \\ \vdots & & \vdots \\ c_{r1} & \cdots\cdots\cdots & c_{rn} \end{bmatrix} \mathbf{x} + \begin{bmatrix} d_{11} & \cdots\cdots\cdots & d_{1p} \\ \vdots & & \vdots \\ \vdots & & \vdots \\ \vdots & & \vdots \\ d_{r1} & \cdots\cdots\cdots & d_{rp} \end{bmatrix} \mathbf{u}$$

where $n$ additional conditions, e.g. of the form (4.86) or (4.87) are imposed on the elements of the matrices **B** and **C**.

In the case of processes with one output, the additional conditions are conveniently introduced by specifying all elements of the vector **C**, e.g. $\mathbf{C}' = [1 \quad 0 \quad \cdots \quad 0]$. The canonical form then becomes

$$\begin{aligned} Y(s) = & \left[ d_{11} + \frac{b_{11}s^{n-1} + b_{21}s^{n-2} + \cdots + b_{n1}}{s^n + a_1 s^{n-1} + \cdots + a_n} \right] U_1(s) \\ & + \cdots + \\ & + \left[ d_{1p} + \frac{b_{1p}s^{n-1} + b_{2p}s^{n-2} + \cdots + b_{np}}{s^n + a_1 s^{n-1} + \cdots + a_n} \right] U_p(s) \end{aligned} \tag{4.89}$$

where $Y$ and $U_i$ denote the Laplace transforms of $y$ and $u_i$. A canonical representation of a process of the $n$th order with $p$ inputs and one output can thus be written as

$$\begin{aligned} \frac{d^n y}{dt^n} + a_1 \frac{d^{n-1} y}{dt^{n-1}} + \cdots + a_n y = & \left[ b'_{01} \frac{d^n u_1}{dt^n} + \cdots + b'_{n1} u_1 \right] \\ & + \cdots + \\ & + \left[ b'_{0p} \frac{d^n u_p}{dt^n} + \cdots + b'_{np} u_p \right] \end{aligned} \tag{4.90}$$

An analogous form for systems with several outputs is

$$\frac{d^n\mathbf{y}}{dt^n} + \mathbf{A}_1\frac{d^{n-1}\mathbf{y}}{dt^{n-1}} + \cdots + \mathbf{A}_n\mathbf{y} = \left[\mathbf{B}_{01}\frac{d^n u_1}{dt^n} + \cdots + \mathbf{B}_{n1}u_1\right] + \cdots + \left[\mathbf{B}_{0p}\frac{d^n u_p}{dt^n} + \cdots + \mathbf{B}_{np}u_p\right] \tag{4.91}$$

This form has been used among others by Wong et al. (1968) and Rowe (1968). The determination of the order of the process (4.91), which in general is different from $n$, as well as the reduction of (4.91) for state form has been done by Tuel (1966). Canonical forms for linear multivariable systems have also been studied by Luenberger (1967) and Bucy (1968). Analogous results hold for discrete time systems.

When the matrix $\mathbf{A}$ has multiple eigenvalues and is not cyclic it is not clear what a 'minimal parameter representation' means. The matrix $\mathbf{A}$ can of course always be transformed to Jordan canonical form (Friedman, 1956). Since the eigenvalues of $\mathbf{A}$ are not distinct the matrix $\mathbf{A}$ can, strictly speaking, be characterized by fewer than $n$ parameters. The ones in the superdiagonal of the Jordan form can, however, be arranged in many different ways depending on the internal couplings which leads to many different structures.

*Canonical Forms for Linear Stochastic Models.* We will now discuss canonical forms for stochastic systems. To avoid the technical difficulties associated with continuous-time white noise we will present the results for discrete-time systems. The analogous results are, however, true also for continuous-time systems. Consider the system

$$\begin{aligned} \mathbf{x}(k+1) &= \mathbf{A}\mathbf{x}(k) + \mathbf{B}\mathbf{u}(k) + \mathbf{v}(k) \\ \mathbf{y}(k) &= \mathbf{C}\mathbf{x}(k) + \mathbf{D}\mathbf{u}(k) + \mathbf{n}(k) \end{aligned} \tag{4.92}$$

where $k$ takes integer values. The state vector $\mathbf{x}$, the input $\mathbf{u}$ and the output $\mathbf{y}$ have dimensions $n$, $p$ and $r$; $\{v(k)\}$ and $\{n(k)\}$ are sequences of independent random vectors with zero mean values and covariance $\mathbf{V}$ and $\mathbf{N}$. Since the covariance matrices are symmetric the model (4.92) contains

$$\begin{aligned} N_3 &= n^2 + np + nr + pr + \frac{1}{2}n(n+1) + \frac{1}{2}r(r+1) \\ &= n\left(\frac{3}{2}n + \frac{1}{2} + p + r\right) + r\left(p + \frac{r}{2} + \frac{1}{2}\right) \end{aligned} \tag{4.93}$$

parameters. Two models of the type (4.92) are said to be equivalent if: (1) their input–output relations are the same when $n = 0$ and $v = 0$ and (2) the stochastic properties of the outputs are the same when $u = 0$. The parameters of $\mathbf{A}$, $\mathbf{B}$ and $\mathbf{C}$ can be reduced by the techniques applied previously. It still remains to reduce the parameters representing the disturbances. This is accomplished, e.g. by the Kalman filtering theorem. It follows from this that the output process can be represented as

$$\begin{aligned} \hat{\mathbf{x}}(k+1) &= \mathbf{A}\hat{\mathbf{x}}(k) + \mathbf{B}\mathbf{u}(k) + \mathbf{K}\mathbf{r}(k) \\ \mathbf{y}(k) &= \mathbf{C}\hat{\mathbf{x}}(k) + \mathbf{D}\mathbf{u}(k) + \mathbf{r}(k) \end{aligned} \tag{4.94}$$

where $\hat{\mathbf{x}}(k)$ denotes the conditional mean of $\mathbf{x}(k)$ given $\mathbf{y}(k-1), \mathbf{y}(k-2), \ldots$, and $\{\mathbf{r}(k)\}$ is a sequence of independent equally distributed random variables with zero mean values and covariance $\mathbf{R}$.

The single output version of the model (4.94) was used in Åström (1964). Kailath and Frost (1968) call (4.94) an **innovations representation** of the process. A detailed discussion is given in Åström (1970). The model (4.94) is also used by Mehra (1969).

Notice that the state of the model (4.94) has a physical interpretation as the conditional mean of the state of (4.92). If $\mathbf{A}$ is chosen to be in diagonal form and if conditions such as (4.86) are introduced on $\mathbf{B}$ and $\mathbf{C}$ the model (4.94) is a canonical representation in which the number of parameters is:

$$N_4 = n(p + 2r) + r\left(p + \frac{r}{2} + \frac{1}{2}\right) \tag{4.95}$$

For systems with one output, where the additional conditions are $\mathbf{C}' = [1 \quad 0 \quad \cdots \quad 0]$, the equation (4.94) then reduces to

$$\begin{aligned} y(k) &+ a_1 y(k-1) + \cdots + a_n y(k-n) \\ &= [b'_{01}u_1(k) + \cdots + b'_{n1}u_1(k-n)] \\ &+ \cdots + \\ &+ [b'_{0p}u_p(k) + \cdots + b'_{np}u_p(k-n)] \\ &+ r(k) + c_1 r(k-1) + \cdots + c_n r(k-n) \end{aligned} \tag{4.96}$$

By introducing the shift operator $z$ defined by

$$zy(k) = y(k+1) \tag{4.97}$$

the polynomials

$$\begin{aligned} A(z) &= \quad z^n + a_1 z^{n-1} + \cdots + a_n \\ B_i(z) &= b'_{0i}z^n + b'_{1i}z^{n-1} + \cdots + b'_{ni} \qquad i = 1, 2, \ldots, p \\ C(z) &= \quad z^n + c_1 z^{n-1} + \cdots + c_n \end{aligned} \tag{4.98}$$

and the corresponding reciprocal polynomials

$$\begin{aligned} A^*(z^{-1}) &= 1 \quad + a_1 z^{-1} + \cdots + a_n z^{-n} = z^{-n} A(z) \\ B_i^*(z^{-1}) &= b'_{0i} + b'_{1i} z^{-1} + \cdots + b'_{ni} z^{-n} = z^{-n} B_i(z) \\ C^*(z^{-1}) &= 1 \quad + c_1 z^{-1} + \cdots + c_n z^{-n} = z^{-n} C(z) \end{aligned} \tag{4.99}$$

the equation (4.96) can be written as

$$A^*(z^{-1}) y(k) = \sum_{i=1}^{p} B_i^*(z^{-1}) u_i(k) + C^*(z^{-1}) r(k) \tag{4.96a}$$

or

$$A(z) y(k) = \sum_{i=1}^{p} B_i(z) u_i(k) + C(z) r(k) \tag{4.96b}$$

This canonical form of an $n$th order system was introduced in Åström, Bohlin and Wensmark (1965) and has since then been used extensively. The corresponding form for multivariable systems is obtained by interpreting $y$ and $u_i$ as vectors and $A$, $B_i$ and $C$ as polynomials whose coefficients are matrices. Such models have been discussed by Eaton (1967), Kashyap (1970), Rowe (1968), Vališ (1970).

The following canonical form

$$\begin{aligned} y(k) = {} & \frac{B_1(z)}{A_1(z)} u_1(k) + \frac{B_2(z)}{A_2(z)} u_2(k) + \cdots + \\ & + \frac{B_p(z)}{A_p(z)} u_p(k) + \frac{C(z)}{A(z)} r(k) \end{aligned} \tag{4.100}$$

has been used by Bohlin (1968) and Steiglitz and McBride (1965) as an alternative to (4.96).

The choice of model structure can greatly influence the amount of work required to solve a particular problem. We illustrate this by a filtering example.

*Example.* Assume that the final goal of the identification is to design a predictor using Kalman filtering. If the process is modelled by

$$\begin{aligned} \mathbf{x}(k+1) &= \mathbf{A}\mathbf{x}(k) + \mathbf{v}(k) \\ \mathbf{y}(k) &= \mathbf{C}\mathbf{x}(k) + \mathbf{n}(k) \end{aligned} \tag{4.101}$$

where $\{n(k)\}$ and $\{v(k)\}$ are discrete-time white noise with covariances $\mathbf{N}$ and $\mathbf{V}$, the likelihood function for the estimation problem can be written as

$$\begin{aligned} -\log L = {} & \tfrac{1}{2} \sum_{k=1}^{n} [\mathbf{v}'(k)\mathbf{V}^{-1}\mathbf{v}(k) + \mathbf{n}'(k)\mathbf{N}^{-1}\mathbf{n}(k)] \\ & + \frac{n}{2} \log(\det \mathbf{V})(\det \mathbf{N}) + \text{const.} \end{aligned} \tag{4.102}$$

where the system equations are considered as constraints. The evaluation of gradients of the loss function leads to two point boundary-value problems. In addition, when the identification is done, the solution of the Kalman filtering problem requires the solution of a Riccati equation. Assume instead that the process is identified using the structure

$$\begin{aligned} \mathbf{w}(k+1) &= \mathbf{A}\mathbf{w}(k) + \mathbf{K}\mathbf{r}(k) \\ \mathbf{y}(k) &= \mathbf{C}\mathbf{w}(k) + \mathbf{r}(k) \end{aligned} \tag{4.103}$$

the likelihood function then becomes

$$-\log L = \tfrac{1}{2} \sum_{k=1}^{n} \mathbf{r}'(k)\mathbf{R}^{-1}\mathbf{r}(k) + \frac{n}{2} \log \det \mathbf{R} \tag{4.104}$$

The evaluation of gradients of the loss function in this case is done simply as an initial value problem. When the identification is done the steady-state Kalman filter is simply given by

$$\hat{\mathbf{x}}(k+1) = \mathbf{A}\hat{\mathbf{x}}(k) + \mathbf{K}[\mathbf{y}(k) - \mathbf{A}\hat{\mathbf{x}}(k)] \tag{4.105}$$

Hence if the model with the structure (4.103) is known there is no need to solve a Riccati equation in order to obtain the steady-state Kalman filter.

The choice of the particular type of (canonical) form to be used is dictated by a number of considerations, e.g. the interest in the eigenvalues, the coefficients of difference equation, certain critical parameters, the minimum number of parameters, etc. Some examples of models for industrial processes are given by Van der Grinten in the IFAC survey paper (Eykhoff et al., 1966), cf. the reference at the end of Chapter 1. From this reference is the following quotation from Kwakernaak:

'To show that the notion of state can be advantageously employed in models of varying complexity we quote some examples which are picked from different areas of application in a rather random fashion.

Models of electrical networks have as their state variables capacitance voltages and inductance currents.

Mechanical systems can be characterized by models having positions and velocities as their state variables.

Thermal systems with material flow have enthalpies as their state variables.

A simple model for the design of a longitudinal guidance system for aircraft landing employs as state variables pitch rate, altitude and rate of ascent of the aircraft.

The model of a multimachine power system has as its state variables magnetic fluxes, rotor angles and machine speeds.

In process engineering also examples are known in which the concept of state is applied. A lumped-parameter model of a superheater includes as its state variables the output steam enthalpy and the tube wall temperature.

A model of a distillation column has as its state variables the liquid composition and the saturated liquid at each tray. Numerical methods can be employed to obtain verification.

Also in other fields the notion of state finds application in dynamic model building:

Simulation of traffic on a "microscopic" scale can be done with a model in which the positions and speeds of individual automobiles are recorded.

Dynamic models in industrial engineering also employ the notion of state. A model of a production–distribution system developed by Forrester has as its state variables (denoted as "levels") "inventories, number of employees, unfilled orders, stocks of capital equipment, bank balances, orders in transit in communication channels, goods in shipment and unfilled orders for new employees".

It appears that the notion of state can be advantageously employed for the description of systems of a widely varying nature. It is of great help in understanding the operation of the system and in conceiving the model. Digital computers are very suitable to carry out the numerical manipulations necessary for simulation'.

### *4.6.4 Non-linear Models*

For a wide class of non-linear processes a rather general, non-parametric representation is given by the Volterra series, using impulse responses of increasing degree.

Approximation of these functions by a finite number of points leads to a model that is linear-in-the-parameters (Eykhoff, 1963; Alper, 1965; Roy and Sherman, 1967). For most practical cases, the number of parameters needed for this description is too large. Assume that for a process the impulse response of the first degree, $h_1(\tau)$, can be approximated by say 10 sample values (parameters). Then taking into account the symmetry $h_2(\tau_1, \tau_2)$ would need $\frac{1}{2} \times 10 \times (10 + 1) = 55$ parameters, $h_3(\tau_1, \tau_2, \tau_3)$ would need 220 parameters, etc., if these impulse responses of higher degree have to have the same 'resolution'. Apparently the dimensionality of the parameter-estimation problem becomes excessive. Moreover the physical interpretation of these parameters probably will hardly be possible. Consequently, more than for linear processes, for which non-parametric models are feasible, one needs to choose a parametric model when studying non-linear processes. The use of *a priori* knowledge due to physical insight into the process is of high importance.

Yet a scheme for deriving a parametric model without *a priori* knowledge from observations has been proposed by Ivakhnenko (1968; 1970), Ivakhnenko and Koppa (1970), under the name 'Heuristic Self-Organization' or 'Group Method of Data Handling' (GMDH).

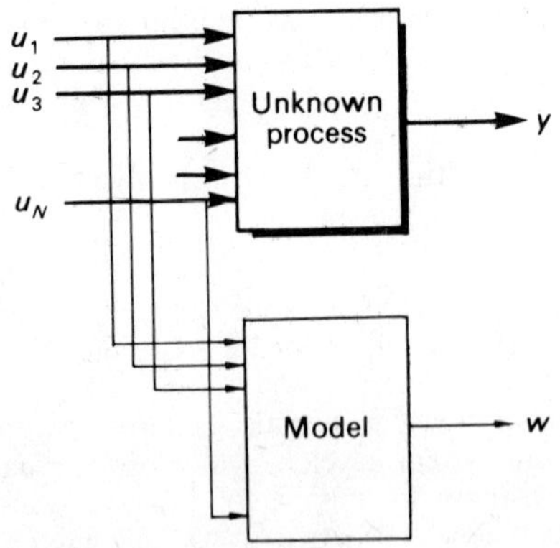


Figure 4.21

In Figure 4.21 the unknown process is characterized by some functional relationship between the output variable $y$ and the set of input variables $\{u_1, \ldots, u_N\}$, i.e.:

$$y = \mathcal{O}[u_1, \ldots, u_N]$$

A model has to be constructed without *a priori* knowledge about $\mathcal{O}[\ ]$ in such a way that $w$ is a good approximation of $y$. This could be done by using the Kolmogorov–Gabor polynomial (equation 4.29); its problems are:

the dimensionality problem caused by the large number of coefficients to be estimated;

the large number of data points of the input and output variables needed for the estimation.

Ivakhnenko's method consists of choosing a hierarchy of 'partial' models instead of one 'complete' model. There are several 'columns' of partial models as indicated in Figure 4.22. In the first column each partial model has as inputs a few of the process variables $u_i$ and only one output variable

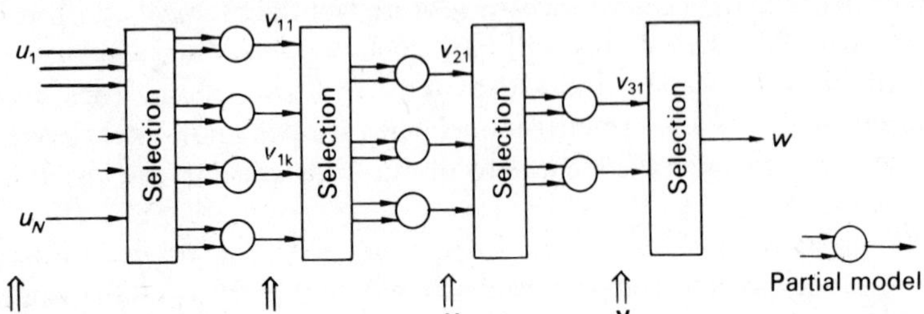


Figure 4.22

$v_{1k}$. The parameters of the partial models are chosen in such a way that each $v_{1k}$ approaches $y$ as well as possible. In the second column each partial model has as inputs a few of the variables $v_{1i}$ and only one output variable $v_{2k}$. The parameters of the partial models in the second column are chosen in such a way that each $v_{2k}$ approaches $y$ as well as possible. This structure can be repeated a number of times. Consequently the GMDH method is an algorithm according to which a model is built from identical partial models; these are arranged in an hierarchial structure. By repeatedly generating new variables one attempts to obtain finally a variable $w$ that is a good approximation of $y$. The new variables are formed by a combination algorithm from the variables of the previous column ('level'). At each level only the best variables are selected for further manipulation.

Two aspects remain to be discussed:

the choice of the type of partial models;

the way of selecting variables for further manipulation.

The partial model can be of a simple kind: e.g. with two inputs, viz. $u_1$ and $u_2$:

(1) $v = \beta_0 + \beta_1 u_1 + \beta_2 u_2$
a linear combination with 3 parameters

(2) $v = \beta_0 + \beta_1 u_1 + \beta_2 u_2 + \beta_{11} u_1^2 + \beta_{22} u_2^2 + \beta_{12} u_1 u_2$
a quadratic polynomial with 6 parameters

(3) $v = \beta_0 + \beta_1 u + \beta_2 u_2 + \beta_{12} u_1 u_2$
4 parameters

(4) $v = \beta_1 u_1 u_2 + \beta_2 \bar{u}_1 u_2 + \beta_3 u_1 \bar{u}_2 + \beta_4 \bar{u}_1 \bar{u}_2$
for binary variables with 4 parameters etc.

Partial models of type (1) do not increase the degree of the model; in such a case the higher powers and cross terms of the process input variables have to be generated as components of the model input vector. The selection of variables at each level is necessary due to the many combinations of $(u_i, u_j)$ out of the set $(u_1, \ldots, u_N)$, viz. $N!$ For further manipulation the 'best' variables are selected on the basis of: (1) the closeness of that variable to $y$ in a least-squares sense and (2) the diversity of the group of constituting variables. The first criterion is straightforward and goal-directed. The second criterion results from the possibility that variables which need not be promising on the basis of a squared error compared to $y$ still may contain essential information. This may become apparent in combination with other variables, at a 'higher' level. There are no general rules yet that give a more precise recipe on the selection and on the percentage of intermediate variables that has to be chosen for further processing at the next level.

As each of the partial models has a small number of parameters only a comparatively small number of data points is needed for estimating their values. For non-linear processes that have memory the model input variables $u_i$ include also delayed process input variables.

In his publications Ivakhnenko gives several examples of applications with very good correspondence between process and model.

## 4.7 Conclusion

In spite of the length of presentation the reader has to be aware of the fact that only a number of aspects of process models could be summarized and discussed. Obvious omissions are the discussion of:

time-lag processes (cf. Choksy (1962));
distributed parameter processes (cf. Phillipson (1971)).

Particularly for non-linear processes it is highly desirable to base the model building on the *a priori* knowledge available. This implies a wide variety of models directed to particular processes and specific goals. Examples of model building for specific applications can be found in the literature given in Chapter 14.

## References

Alper, P. (1965). 'A consideration of the discrete Volterra series', *IEEE Trans. Autom. Control*, **AC-10,** 322–327.

Aoki, M. (1967). *Optimization of Stochastic Systems; Topics in Discrete-Time Systems*, Academic Press, New York, 354 pp.

Åström, K. J. (1964). 'Control problems in papermaking', *Proc. IBM Scientific Computing Symp. on Control Theory and Applic.*, Yorktown Heights, New York, 135.

Åström, K. J., T. Bohlin and S. Wensmark (1965). 'Automatic construction of linear stochastic dynamic models for industrial processes with random disturbances using operation records', *Report IBM Nordic Lab., Stockholm.*

Åström, K. J. (1970). *Introduction to Stochastic Control Theory*, Academic Press, New York, 299 pp.

Åström, K. J. and P. Eykhoff (1970). cf. references Chapter 2.

Balakrishnan, A. V. (1969). *Techniques of System Identification*, Lecture notes, Institute Electrotecnio, University of Rome.

Barker, H. A. (1965). 'Multivariable models for control systems', *IFAC Symp.—Theory of Self-adaptive Control Systems*, Teddington, 112–117.

Bohlin, T. (1968). 'Real-time estimation of time-variable processes characteristics', *Report, IBM Nordic Lab., Stockholm.*

Brockett, R. W. (1965). 'Poles, zeros and feedback: state space interpretation', *IEEE Trans. Autom. Control*, **AC-10,** 129–135.

Bruni, C., A. Isidori and R. Ruberti (1969). 'Order and factorization of the impulse response matrix', *4th IFAC Congr., Warsaw*, paper 19.4.

Bucy, R. S. (1968). 'Canonical forms for multivariable systems' (short paper), *IEEE Trans. Autom. Control*, **AC-13,** 567–569.

Choksy, N. H. (1962). 'Time lag systems', in *Progress in Control Engineering*, Vol. 1 (Ed. R. H. Macmillan), Heywood, London, pp. 19–38.

Diskind, T. (1969). 'Parameter evaluation by complex domain analysis', *Joint Autom. Control Conf.*, 770–781.

Eaton, J. (1967). 'Identification for control purposes', paper, *IEEE Winter meeting*, New York.

Elgerd, O. I. (1967). *Control Systems Theory*, McGraw-Hill, New York, 562 pp.

Eykhoff, P. (1963). 'Some fundamental aspects of process-parameter estimation', *IEEE Trans. Autom. Control*, **AC-8,** 347–357.

Fischer, P. and T. Vámos (1967). 'A generalization of the controllability and observability concept from the point of view of systems identification', *IFAC Symp.—Identification in Autom. Control Systems, Prague*, paper 1.10.

Flake, R. H. (1963). 'Volterra series representation of nonlinear systems', *AIEE Trans.*, **81** part II (Applications and Industry, 64), 330–335.

Friedman, B. (1956). *Principles and Techniques of Applied Mathematics*, Wiley, New York, 315 pp.

Gabor, D., W. P. L. Wilby and R. Woodcock (1961). 'A universal nonlinear filter, predictor and simulator which optimizes itself by a learning process', *Proc. IEE*, **108B,** 422–438.

George, D. A. (1959). 'Continuous nonlinear systems', *Tech. Rep. 355. Res. Lab. of Electron. M.I.T.*, Cambridge, Mass.

Goodman, T. P. and R. H. Hillsley (1958). 'Continuous measurement of the characteristics of systems with random inputs: a step towards self optimizing control', *Trans. ASME*, **80,** 1839–1848.

Goodman, T. P. and J. B. Reswick (1956). cf. references Chapter 8.

Grenander, U. and M. Rosenblatt (1957). *Statistical Analysis of Stationary Time Series*, Wiley, New York, 300 pp.

Gustavsson, I. (1969). 'Maximum likelihood identification of the Ågesta reactor and comparison with results of spectral analysis', Report 6903, *Div. of Aut. Control Lund Inst. of Technol., Lund, Sweden.*

Ho, B. L. and R. E. Kalman (1966). 'Effective construction of linear state-variable models from input/output functions', *Regelungstechnik*, **14,** 545–548.

Ivakhnenko, A. G. (1968). 'The group method of data handling—a rival of the method of stochastic approximation', *Soviet Autom. Control*, **13,** 43–55; English translation of the Ukrainian journal *Avtomatika*.

Ivakhnenko, A. G. and Ju. V. Koppa (1970). 'The group method of data handling for the solution of the various interpolation problems of cybernetics', *2nd IFAC Symp.—Identification and Process Parameter Estimation, Prague*, paper 2.1.

Ivakhnenko, A. G. (1970). 'Heuristic self-organization in problems of engineering cybernetics', *Automatica*, **6,** 207–219.

Izawa, K. and K. Furuta (1967). 'Process identification using correlation coefficient. *IFAC Symp.—Identification in Autom. Control Systems, Prague*, paper 3.9.

Jenkins, G. and D. Watts (1968). *Spectral Analysis and its Applications*, Holden Day, San Francisco, 525 pp.

Kailath, T. and P. Frost (1968). 'An innovations approach to least-squares estimation', part I, *IEEE Trans. Autom. Control*, **AC-13,** 646–660.

Kalman, R. E. (1960). 'On the general theory of control systems', *1st IFAC congress, Moscow*, 481–492.

Kalman, R. E. (1963). 'Mathematical description of linear dynamical systems', *SIAM J. Control*, **1,** 152–192.

Kalman, R. E., P. L. Falb and M. A. Arbib (1969). *Topics in Mathematical System Theory*, McGraw-Hill, New York.

Kashyap, R. L. (1970). 'Maximum likelihood identification of stochastic linear systems', *IEEE Trans. Autom. Control*, **AC-15,** 25–34.

Kreindler, E. and P. E. Sarachik (1964). 'On the concepts of controllability and observability of linear systems', *IEEE Trans. Autom. Control*, **AC-9,** 129–136.
Lee, R. C. K. (1964). *Optimal Estimation, Identification, and Control*, M.I.T. Press, Cambridge, Mass., 152 pp. (Research Monograph 28).
Lee, Y. W. and M. Schetzen (1965). 'Measurement of the Wiener kernels of a nonlinear system by cross correlation', *Int. J. Control*, **2,** 237–254.
Luenberger, D. G. (1967). 'Canonical forms for linear multivariable systems' (short paper), *IEEE Trans. Autom. Control*, **AC-12,** 290–293.
MacFarlane, A. G. J. (1970). *Dynamical System Models*, Harrap, London, 503 pp.
Mann, H. B. and A. Wald (1943). 'On the statistical treatment of linear stochastic difference equations', *Econometrica*, **11,** Nos. 3 and 4.
McFee, R. (1961). 'Determining the response of nonlinear systems to arbitrary inputs', *AIEE Trans.*, **80,** part II (Applications and Industry, 56), 189–193.
McGhee, R. B. (1963). 'Identification of nonlinear-dynamic systems by regression analysis methods', *Ph.D. Dissert., E.E. Dept., University of Southern California*, Los Angeles, Cal., 217 pp.
Mehra, R. K. (1969). 'Identification of stochastic linear dynamic systems', *Proc. 1969 IEEE Symp. on Adaptive Processes*, Pennsylvania State University.
M.I.T. (1963–'65). Several references and notes. *Q. Prog. Report. Res. Lab. of Electron. M.I.T.*, Cambridge, Mass., e.g. Nos. 71, 75, 76, 77.
Phillipson, G. A. (1971). *Identification of Distributed Systems*, Elsevier, Amsterdam, 154 pp.
Prosser, R. T. and W. L. Root (1966). 'Determinable classes of channels', *J. Math. Mech.*, **16,** 365–398.
Root, W. L. (1965). 'On system measurement and identification', *Symp. on System Theory*, Polytechnic Institute of Brooklyn.
Root, W. L. (1970). 'On the structure of a class of system identification problems', *2nd IFAC Symp.—Identification and Process Parameter Estimation, Prague*, paper 8.1.
Rowe, I. H. (1968). 'A statistical model for the identification of multivariable stochastic systems', *Proc. IFAC Symp. on Multivariable Systems, Düsseldorf.*
Roy, R. and J. Sherman (1967). 'System identification and pattern recognition', *IFAC Symp.—Identification in Autom. Control Systems, Prague*, paper 1.6.
Smets, H. B. (1960). 'Analysis and synthesis of nonlinear systems', *IRE Trans. Circuit Theory*, **CT-7,** 459–469.
Staley, R. M. and P. C. Yue (1970). 'On system parameter identifiability', *Information Sci.*, **2,** 127–138.
Steiglitz, K. and L. E. McBride (1965). 'A technique for the identification of linear systems' (short paper), *IEEE Trans. Autom. Control*, **AC-10,** 461–464.
Tou, J. T. (1964). *Modern Control Theory*, McGraw-Hill, New York, 427 pp.
Tuel Jr., W. G. (1966). 'Canonical forms for linear systems I and II', *IBM Res. Report* RJ 375 and 390.
Vališ, J. (1970). 'On-line identification of multivariable-linear systems of unknown structure from input output data', *2nd IFAC Symp.—Identification and Process Parameter Estimation, Prague*, paper 1.5.
Van Harten, J. (1966). 'The description of non-linear systems' (in Dutch), *Thesis, E.E. Dept., University of Technology, Eindhoven, Netherlands.*
Volterra, V. (1959). *Theory of Functionals and of Integral and Integro-Differential Equations*, Dover, New York, 226 pp.
Whittle, P. (1963). *Prediction and Regulation*, Van Nostrand, Princeton, N.J.
Wiener, N. (1958, 1962). *Nonlinear Problems in Random Theory*, M.I.T. Press, Cambridge, Mass., 140 pp.

Wolff, A. A. and J. H. Dietz (1962). 'An adaptive correlator for underwater measurements', *Marine Sciences Instrumentation*, **1,** 347–354.
Wong, K. Y., K. M. Wiig and E. J. Allbritton (1968). 'Computer control of the Clarksville cement plant by state space design method', *IEEE Cement Ind. Conf., St. Louis, Mo.*
Zadeh, L. A. (1953). 'A contribution to the theory of nonlinear systems', *J. Franklin Inst.*, **255,** 387–408.
Zadeh, L. A. (1961). 'Time-varying networks', *Proc. IRE*, **49,** 1488–1503.
Zadeh, L. A. and C. A. Desoer (1963). *Linear System Theory; The State Space Approach*, McGraw-Hill, New York, 628 pp.

## Additional Literature

### *4.2 Linear Models*

Anderson, B. D. O., and others (1966). 'Equivalence of linear time-invariant dynamical systems', *J. Franklin Inst.*, **281,** 371–378.
Balakrishnan, A. V. (1967). 'Foundations of state-space theory of continuous systems', *J. Computer and System Sciences*, **1,** 91–116.
Barker, H. A. and D. A. Hawley (1966). cf. references Chapter 3.
Bolte, W. (1965). 'An approximation procedure for determining the transient response from the frequency response' (in German), *Z. Messen, Steuern, Regeln*, **13,** 248–249.
Brown, B. M. (1961). *The Mathematical Theory of Linear Systems*, Chapman and Hall, London, 282 pp.
Budin, M. A. (1971). 'Minimal realization of discrete linear systems from input–output observations', *IEEE Trans. Autom. Control*, **AC-16,** 395–401.
Chen, C. F. and B. L. Philip (1963). 'Graphical determination of transfer function coefficients of a system from its frequency response', *IEEE Trans. Applic. and Ind.*, **82,** No. 65, 42–45.
Davies, W. D. T. and W. Fishwick (1964). 'The synthesis of dynamic models of plants and processes', *Proc. Instn. Mech. Engrs.*, **179,** pt 3H (Advances in Automatic Control), 29–36.
DeRusso, P. M., R. J. Roy and C. M. Close (1965). *State Variables for Engineers*, Wiley, New York, 608 pp.
Douce, J. L. and P. D. Roberts (1964). cf. references Chapter 6.
Kitamori, T. (1960). cf. references Chapter 3.
Koenig, J. F. (1964). 'A method for obtaining the time response of any linear system' (short paper), *IEEE Trans. Autom. Control*, **AC-9,** 556–565.
Krishnamurthy, E. V. (1964). 'Solution of linear systems as eigen value problems' (correspondence), *Proc. IEEE*, **52,** 1065–1066.
Kuh, E. S. and R. A. Rohrer (1965). 'The state-variable approach to network analysis', *Proc. IEEE*, **53,** 672–686.
Lovering, W. F. (1965). 'Analog computer simulation of transfer functions', *Proc. IEEE*, **53,** 306–307.
Lubbock, J. K. and H. A. Barker (1964). 'A solution of the identification problem', *IEEE Trans. Applic. and Ind.*, **83,** No. 72, 166–173.
Milsum, J. H. (1960). 'Determining transfer functions with analog computers', *ISA Journal*, **7,** 54–57.
Nelepin, R. A. (1965). 'Investigation of a class of control in parameter space systems', *Engng. Cybernetics*, **3,** 123–132.
Ogata, K. (1967). *State Space Analysis of Control Systems*, Prentice–Hall, Englewood Cliffs, N.J., 596 pp.

Peterka, V. and P. Vidinčev (1967). 'Rational-fraction approximation of transfer functions', *IFAC Symp.—Identification in Autom. Control Systems, Prague*, paper 2.1.

Polonnikov, D. E. (1965). 'Method for determining system phase-frequency characteristics from the logarithmic amplitude-frequency characteristic with arbitrary-slopes of its segments', *Automn. Remote Control*, **26,** 716–719.

Radkte, M. (1966). 'On the approximation of linear aperiodic transient responses' (in German), *Z. Messen, Steuren, Regeln*, **9,** 192–196.

Roberts, P. D. (1966). cf. references Chapters 3 and 6.

Roberts, P. D. (1967). cf. references Chapter 3.

Sauer, W. (1966). 'A survey of system theoretical parameters' (in German), *Nachr. technik*, **16,** 194–197.

Seide, W. (1963). 'Determination of characteristic values from the weight function' (in German), *Z. Messen, Steuern, Regeln*, **6,** 323–325.

Storer, J. E. (1957). *Passive Network Synthesis*, McGraw-Hill, New York, 329 pp.

Thal-Larsen, H. (1955). 'Frequency response from experimental nonoscillatory transient response data', *Trans. AIEE*, **74,** part II (Applicat. and Industry), 109–114.

Timothy, L. K. and B. E. Bona (1968). *State Space Analysis; an Introduction*, McGraw-Hill, New York, 406 pp.

Wang, P. K. C. (1961). 'Mathematical models for time-domain design of electrohydraulic servomechanisms', *Trans. AIEE*, **80,** part II (Applicat. and Industry, No. 57), 252–260.

Woodside, C. M. (1967). 'The interpretation of time domain models in the frequency domain' (correspondence), *IEEE Trans. Autom. Control*, **AC-12,** 110–111.

Zakian, V. (1965). 'Relation between system parameters and response duration', *Electron. Letters*, **1,** 293–294.

Zorn, J. (1963). 'Methods of evaluating Fourier transforms with application to control engineering', *Doct. Dissert., University of Technology, Delft*, Netherlands.

*4.3 Varylinear Models*

Bishop, K. A., C. M. Sliepcevich and T. H. Puckett (1962). 'Techniques for identification of linear and linear time varying processes', *Joint Autom. Control Conf.*

Calahan, D. A. (1964). 'Equivalence of time-varying systems' (correspondence), *Proc. IEEE*, **52,** 1372.

Litovchenko, Ts. G. (1961). 'Analytical solutions of linear equations describing one class of dynamical systems with variable parameters', *Automn. Remote Control*, **22,** 390–408.

Meerov, M. V. (1965). 'Methods of synthesis of structures equivalent to self-adjusting systems in plants with variable parameters', *IFAC Symp.—Theory of Self-adaptive Control Systems, Teddington*, 186–194.

Peschel, M. (1963). 'On the transfer characteristics of time varying linear blocks' (in German), *Z. Messen, Steuern, Regeln*, **6,** 310–315.

Schweizer, G. (1965). 'A general method for studying systems with periodically changing parameters' (in German), *Arch. Elekt. Ubertr.*, **19,** 469–482.

Solodov, A. V. (1958). 'Conversion of output initial conditions in a linear system with variable parameters into an equivalent input signal', *Automn. Remote Control*, **19,** 645–651.

Solodov, A. V. (1966). *Linear Automatic Control Systems with Varying Parameters*, Blackie, London, 288 pp.

Tucker, D. G. (1964). *Circuits with Periodically-Varying Parameters*, MacDonald, London, 175 pp.

Wunsch, G. (1962). 'The mathematical treatment of linear systems with variable parameters', *Nachrtechnik*, **11**, 221–222.
Wunsch, G. (1966). 'Analysis of networks with changing parameters by means of Z transformation' (in German), *Arch. Elekt. Ubertr.*, **20**, 289–293.
Zadeh, L. A. (1950). 'Correlation functions and power spectra in variable networks', *Proc. IRE*, **38**, 1342–1345.
Zadeh, L. A. (1950). 'The determination of the impulsive response of variable networks', *J. Appl. Phys.*, **21**, 642–645.

### *4.4 Non-linear Models*

Alper, P. (1964). 'Higher-dimensional Z-transforms and non-linear discrete systems', *Revue A*, **6**, 199–212.
Balakrishnan, A. V. (1966). 'State-space theory of nonlinear systems', in *Functional Analysis and Optimization* (Ed. E. Cavaniello), Academic Press, New York, 225 pp.
Barrett, J. F. (1957). 'The use of functionals in the analysis of non-linear physical systems', *Report Statistical Advisory Unit*. No. 1/57. Ministry of Supply, England; also in *J. Electron. Control*, **15**, 567–617 (1963).
Bose, A. G. (1956). 'A theory of nonlinear systems', *Tech. Report 309*, *Res. Lab. of Electron. M.I.T.*, Cambridge, Mass.
Brilliant, M. B. (1958). 'Theory of the analysis of nonlinear systems', *Tech. Report 345*, *Res. Lab. of Electron. M.I.T.*, Cambridge, Mass.
Bruni, C., G. Di Pillo and G. Koch (1971). 'On the mathematical models of bilinear systems', *Ricerche di Automatica*, **2**, 11–26.
Cameron, R. H. and W. T. Martin (1947). 'The orthogonal development of nonlinear functionals in series of Fourier–Hermite functionals', *Ann. Math. Stat.*, **48**, 385–389.
Flake, R. H. (1963). 'Volterra series representation of time varying nonlinear systems', *2nd IFAC Congr., Basle*, 91–99.
Frank, P. and R. McFee (1963). 'Determining input–output relationships of nonlinear systems by inversions', *IEEE Trans. Circuit Theory*, **CT-10**, 168–180.
Hemani, H. and R. L. Cosgriff (1966), 'Graph theory in the analysis of non-linear systems', *IEEE Trans. Autom. Control*, **AC-11**, 550–552.
Kohr, R. H. (1963). cf. references Chapter 9.
Ku, Y. H. and A. A. Wolf (1966). 'Volterra–Wiener functionals for the analysis of nonlinear systems', *J. Franklin Inst.*, **281**, 9.
Lubbock, J. K. (1960). 'Optimization of a class of nonlinear systems', *1st IFAC Congr., Moscow*, 761–769.
Oppenheim, A. V. (1965). 'Superposition in a class of nonlinear systems', *Tech. Report 432*, *Res. Lab. of Electron, M.I.T.*, Cambridge, Mass.
Pugachev, V. S. (1960). 'Statistical theory of systems reducible to linear', *IRE Trans. Circuit Theory*, **CT-7**, 506–513.
Radtke, M. (1966). 'On the determination of the dynamic characteristic values of non-linear transmission elements from graphically represented transition functions', *3rd IFAC Congr., London*, paper 12E.
Roy, R. J. and P. M. DeRusso (1962). 'A digital orthogonal model for non-linear processes with two-level inputs', *IRE Trans. Autom. Control*, **AC-7**, 93–101.
Simpson, H. R. (1965). 'A sampled data non-linear filter', *Proc. IEE*, **112**, 1187–1196.
Singleton, H. B. (1950). 'The theory of nonlinear transducers', *Tech. Report 160*, *Res. Lab. of Electron. M.I.T.*, Cambridge, Mass.,
Van Trees, H. L. (1962). *Synthesis of Optimum Non-linear Control Systems*, M.I.T. Press, Cambridge, Mass.

Voelker, D. and G. Doetsch (1950). *The Two-dimensional Laplace Transform* (in German). Birkhäuser Publ. Co. Zürich.
Wolf, A. A. (1961). 'Some recent advances in the analysis and synthesis of non-linear systems', *ATEE Trans.*, **80,** part II (Applications and Industry), 289–300.
Zadeh, L. A. (1953). 'Nonlinear multipoles', *Proc. Nat. Acad. Sci.*, Wash., D.C., **39,** 274–280.
Zadeh, L. A. (1953). 'Optimum nonlinear filters', *J. Appl. Phys.*, **24,** 396–404.
Zadeh, L. A. (1956). 'On the identification problem', *IRE Trans. Circuit Theory*, **CT-3,** 277–281.
Zadeh, L. A. (1957). 'On the representation of nonlinear operators', *IRE Conv. Rec.*, part II, 105–113.

### *4.5 Controllability Observability, Identifiability*

Chen, C. T. and others (1966). 'Simplified conditions for controllability and observability of linear time-invariant systems', (correspondence), *IEEE Trans. Autom. Control*, **AC-11,** 613–614.
Chen, C. T. and C. A. Desoer (1967). 'Controllability and observability of composite systems', *IEEE Trans. Autom. Control*, **AC-12,** 402–409.
Gilbert, E. G. (1963). 'Controllability and observability in multivariable control systems', *SIAM J. Control*, **1,** 128–151.
Sekiguchi, T. (1969). 'Observability of linear dynamic measuring systems and some applications', *4th IFAC Congr., Warsaw*, paper 12.5.
Tse, E. and J. J. Anton (1972). 'On the identifiability of parameters', *IEEE Trans. Autom. Control*, **AC-17**, 637–646.
Wang, P. K. C. (1964). 'Control of distributed parameter systems', in *Advances in Control Systems*, Vol. 1 (Ed. C. T. Leondes), Academic Press, New York, 75–172.
Weiss, L. and R. E. Kalman (1965). 'Contributions to linear system theory', *Internat. J. Engng. Sci.*, **3,** 141–171.

### *4.6 Models Based on These Descriptions*

Davison, E. J. (1968). 'A new method for simplifying large linear dynamic systems' (correspondence), *IEEE Trans. Autom. Control*, **AC-13,** 214–215.
Kashyap, R. L. (1970). 'A new method of recursive estimation in discrete linear systems', *IEEE Trans. Autom. Control*, **AC-15,** 34–43.
Koepcke, R. W. (1963). 'Solution to sampled, minimum-time problem', *Joint Autom. Control Conf.*, 94–100; also in *Trans. ASME* (*J. Basic Engng*), **86D,** 145–150 (1964).

CHAPTER 5

# *Some Probabilistic Notions; Estimation Theory, Convergence Schemes, Stochastic Approximation*

In this chapter, some ideas are discussed that are of great importance for parameter estimation. First the estimators are considered from the statistical point of view. Evidently there is a close relation between the amount of knowledge of the process that is initially available and the type of 'optimum' procedure that can be used to estimate the parameters. The characteristics of different kinds of estimators are considered:

| | |
|---|---|
| the least-squares | estimator |
| the Markov | estimator |
| the maximum likelihood | estimator |
| the Bayes' | estimator |

The amount of assumed initial knowledge available on the process increases in this order. For the least-squares estimator, the only assumption is that the dynamics of the process can be approximated sufficiently by the 'model' chosen. The Markov estimator requires as well the knowledge of the co-variance matrix of the noise.

For the maximum likelihood estimator, one needs to know the probability density function of the stochastic process from which the sample values are taken.

The Bayes', minimum cost or minimum risk procedure requires a knowledge of the *a priori* probability density functions of the unknown parameters, and of the costs of making errors. Properties such as efficiency, asymptotic efficiency, consistency, bias, etc., are indicated.

Starting from the Bayes' estimator one can derive the other estimators as special cases if there is less *a priori* knowledge available. In this way, relations between the different estimators are derived in this chapter. In many cases, for example, in the implicit or model-adjustment methods, the estimation techniques really are optimization methods. Such an optimization may be the search for a minimum of a cost function, a maximum probability, a minimum squared error, etc. For that reason optimum seeking or convergence schemes are discussed in this chapter as well. Depending on the absence or presence of stochastic disturbances these schemes can be categorized as deterministic, Section 5.2, or stochastic, Section 5.3.

## 5.1 Estimation Theory

The cases to be considered can be illustrated by Figure 5.1. In this section only the central ideas are summarized; a more detailed discussion is given

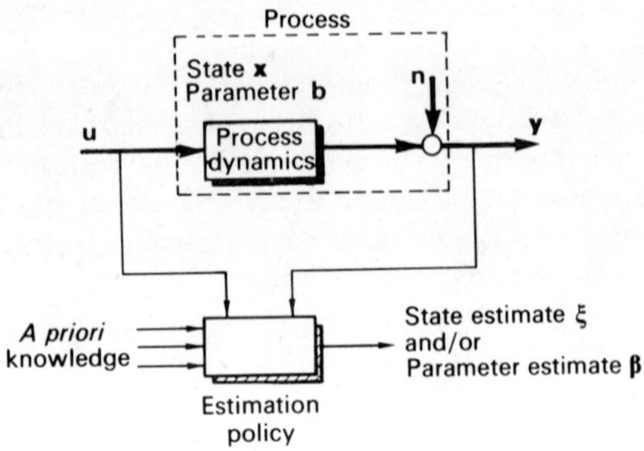


Figure 5.1

in Chapter 11. Here we restrict the attention to parameter estimation. We want to derive a rule (an **estimator**), i.e. a relationship

$$\beta = \beta\{u(1), \ldots, u(k), y(1), \ldots, y(k)\} = \beta(\mathbf{u}', \mathbf{y}') \tag{5.1}$$

such that a numerical value (**estimate**) $\beta$ can be assigned for the process parameter $b$ under consideration. This estimate depends on the sequences of observations. When a number of estimates $[\beta_0, \ldots, \beta_m] = \boldsymbol{\beta}'$ have to be found for the process parameters $\mathbf{b}$, then the relationship is indicated by the vector expression:

$$\boldsymbol{\beta} = \boldsymbol{\beta}(\mathbf{u}', \mathbf{y}')$$

### 5.1.1 *Characteristics of Estimators*

While estimating we are interested in the development of our knowledge about the parameters of interest. That knowledge can be expressed in terms of the probability density function $p(\boldsymbol{\beta}; T)$ or $p(\boldsymbol{\beta}; k)$. This function depends on the length of the observation interval $T$ or, equivalently, on the number $k$ of signal samples processed. This is indicated in Figure 5.2 for a one-dimensional case; $T_1 < T_2 < T_3$. This probability-density function is the

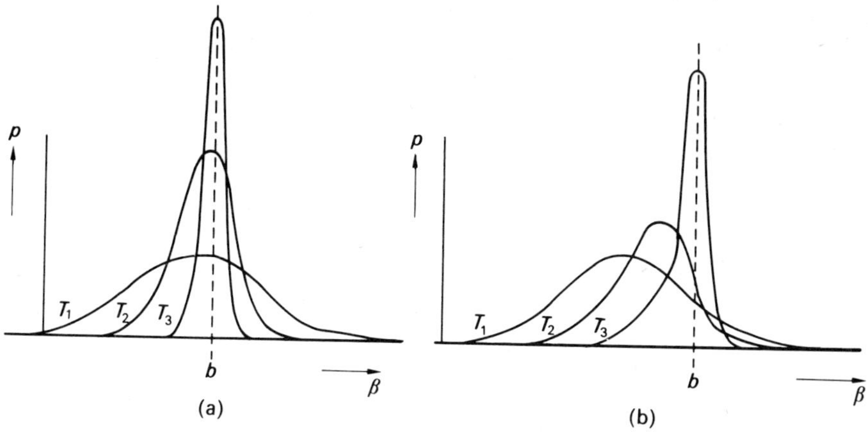


Figure 5.2

most complete type of knowledge that can be derived through application of statistical techniques. Working with this particular type of knowledge, however, is quite complicated and impractical, particularly if the dimensionality of $\boldsymbol{\beta}$ is higher than two; a multidimensional presentation would be needed to give the results to the experimenter. For that reason, in most cases the interest is reduced from the complete probability-density function to its most significant characteristics:*

| | |
|---|---|
| **expected value** | $\mathscr{E}[\boldsymbol{\beta}]$ |
| **bias** | $\mathscr{E}[\boldsymbol{\beta}] - \mathbf{b}$ |
| **covariance** | $\operatorname{cov}[\boldsymbol{\beta}] = \mathscr{E}[\{\boldsymbol{\beta} - \mathscr{E}[\boldsymbol{\beta}]\}\{\boldsymbol{\beta} - \mathscr{E}[\boldsymbol{\beta}]\}']$ |

The reader will appreciate this simplification from functions to numerical constants by application of this reduction to the examples of Figures 5.2a and b. In this way he will also get an idea about the 'information' that is preserved by this simplification. Note that if the probability density function were (multivariate) Gaussian then no 'information' is lost by restricting to $\mathscr{E}[\boldsymbol{\beta}]$ and $\operatorname{cov}[\boldsymbol{\beta}]$, as Gaussian functions are completely characterized by their first and second moments.

* If necessary the reader will find an interpretation of probabilistic notions in Appendix B.

In statistical literature, e.g. Deutsch (1965), Nahi (1969), some desirable properties are defined for the estimators $\boldsymbol{\beta}$ of the parameters $\mathbf{b}$:

**unbiased estimator** if for each $k$:

$$\mathscr{E}[\boldsymbol{\beta}] = \mathbf{b}$$

**consistent estimator** if for increasing $k$:

$$\lim_{k\to\infty} P[|\boldsymbol{\beta} - \mathbf{b}| > \epsilon] = 0$$

with $\epsilon$ arbitrary small; i.e. $\boldsymbol{\beta}$ converges in probability (or stochastically) to the true value $\mathbf{b}$.

**efficient estimator** if for all unbiased estimators $\boldsymbol{\gamma}$

$$\operatorname{cov}[\boldsymbol{\beta}] = \mathscr{E}[\{\boldsymbol{\beta} - \mathbf{b}\}\{\boldsymbol{\beta} - \mathbf{b}\}'] \leqslant \mathscr{E}[\{\boldsymbol{\gamma} - \mathbf{b}\}\{\boldsymbol{\gamma} - \mathbf{b}\}'] = \operatorname{cov}[\boldsymbol{\gamma}]$$

or

$$\det[\operatorname{cov}[\boldsymbol{\gamma}] - \operatorname{cov}[\boldsymbol{\beta}]] \geqslant 0$$

**sufficient estimator** if for all other estimators $\boldsymbol{\gamma}$ it is true that $p(\boldsymbol{\gamma}|\boldsymbol{\beta})$ is not dependent on $\mathbf{b}$.

If the first and third property hold only for $k \to \infty$ then they are called asymptotic unbiasedness and efficiency. Referring again to Figure 5.2 we note that (a) suggests an unbiased case and (b) a biased, but possibly asymptotically unbiased, case.

Note that another important characteristic of these ways of description is, whether or not they provide us with a possibility of introducing *a priori* knowledge including our 'confidence' in this knowledge, which might be available from physical insight or from prior measurements.

*Some Types of Estimators.* As a starting point for deriving estimators, we will choose a situation where much *a priori* knowledge is available (Maslov, 1963), viz.:

(*a*) The probability density function of the noise $\mathbf{n}$. From this function the probability density of the measurements $\mathbf{y}$ follows; this function is dependent on the process parameters $\mathbf{b}$, and is denoted as $p(\mathbf{y}|\mathbf{b})$. It is understood, of course, that $p(\mathbf{y})$ also depends on $\mathbf{u}$.

(*b*) The probability-density functions of the parameter values $\mathbf{b}$. This function is written as $q(\mathbf{b})$.

(*c*) The cost of choosing the value $\boldsymbol{\beta}$ for the estimate if the true value of the process parameters is $\mathbf{b}$. This **cost** or **loss function** $C(\boldsymbol{\beta}, \mathbf{b})$ has a minimum for $\boldsymbol{\beta} = \mathbf{b}$.

After a discussion of this most elaborate case we will consider successively the effect of dropping the assumptions (*c*), (*b*) and (*a*).

### 5.1.2 *Bayes' Estimators*

| *a priori* knowledge: |
|---|
| $p(\mathbf{y}\|\mathbf{b})$ |
| $q(\mathbf{b})$ |
| $C(\boldsymbol{\beta}, \mathbf{b})$ |

The essence of the method is centered around Bayes' rule:

$$p(\mathbf{y}|\mathbf{b})q(\mathbf{b}) = p(\mathbf{y}, \mathbf{b}) = p(\mathbf{b}|\mathbf{y})p(\mathbf{y})$$

For our purpose we consider

$$p(\mathbf{b}|\mathbf{y}) = \frac{p(\mathbf{y}, \mathbf{b})}{p(\mathbf{y})} = \frac{p(\mathbf{y}|\mathbf{b})q(\mathbf{b})}{p(\mathbf{y})} \tag{5.2}$$

where

$$p(\mathbf{y}) = \int_{m+1} p(\mathbf{y}, \mathbf{b})\, d^{m+1}\mathbf{b}$$

$\int_{m+1}$ indicates an $(m + 1)$fold integral

$d^{m+1}\mathbf{b}$ indicates $db_0, db_1 \ldots db_m$

Consequently, all terms in the right-hand part follow from our *a priori* knowledge.

This conditional probability function $p(\mathbf{b}|\mathbf{y})$ can be interpreted as: **the** (*a posteriori*) **probability-density function of the parameters b, given the results of the measurements on y,** say $\mathbf{y} = \mathbf{c}$.

*Example.* It may be helpful to illustrate these ideas with an example of utmost simplicity. Consider the situation of Figure 5.3 where an estimate $\beta$ has to be found for parameter $b$.

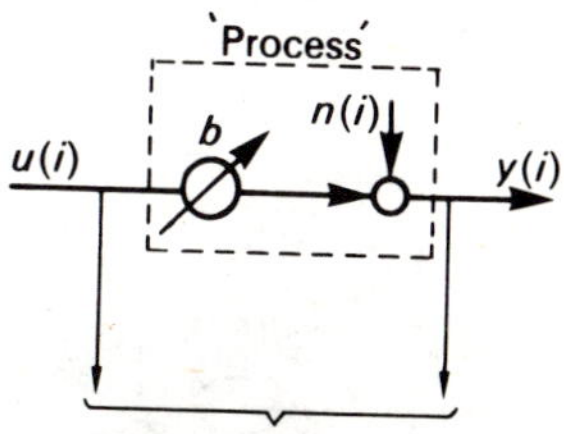


Figure 5.3

This estimate has to be derived from a number of signal samples $u(i)$ and $y(i)$ where

$$y(i) = bu(i) + n(i) \qquad i = 1, \ldots, k$$

and where the average (expected) value of $n(i)$ is zero. Let the probability-density functions of $n$ and $b$ be given by $p_n$ and $q$. If $n$ and $b$ are statistically independent then

$$p(y, b) = p(y - ub, b)\underbrace{\left|\frac{\partial(y - ub)}{\partial y}\right|}_{1} = p_n(y - ub)q(b)$$

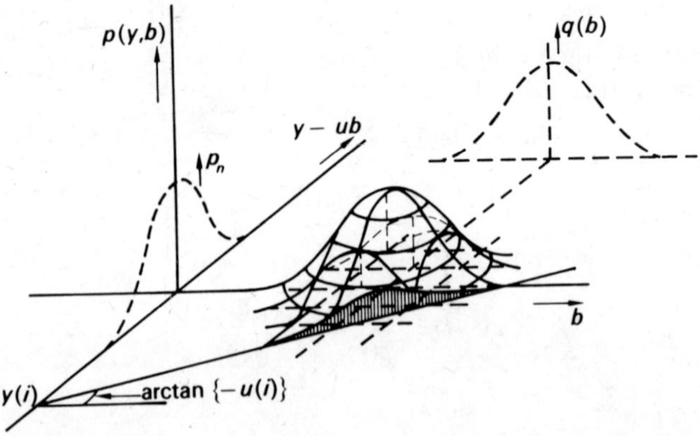


Figure 5.4

This is indicated in Figure 5.4. Note that this is the *a priori* knowledge, available before the measurements of $u(i)$ and $y(i)$ are made. After these measurements are made they provide us with a 'cut' through the probability-density function, from which the *a posteriori* probability function for $b$ follows. This new probability function now may be used as the *a priori* knowledge for the following measurement. In this way the development of $q(b)$ with increasing number of observations can be followed (cf. Figure 5.5). Note that $p_n$ does not change as the additive noise is assumed stationary.

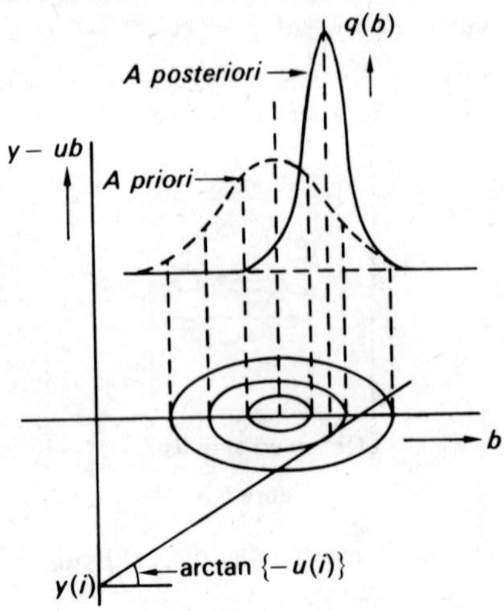


Figure 5.5

The reader is invited to consider special cases such as $u(k) = 0$ and $u(k) = \rightarrow \infty$. Again the method can be generalized to more parameters; its visualization has severe limitations, however. Note that the knowledge on $b$ is given in terms of a **function.**

Based on the *a posteriori* probability-density function $p(\mathbf{b}|\mathbf{y})$ one has to decide on the question what value will be assigned as the estimate. Application of a cost function $C(\beta, b)$ is illustrated in Figure 5.6 for a one-dimensional

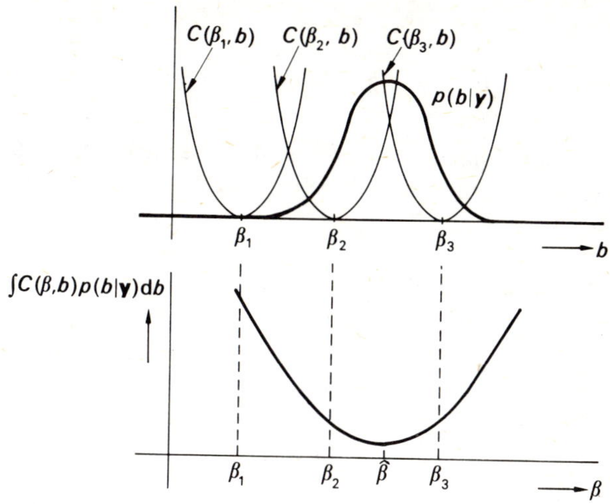


Figure 5.6

case and a quadratic function $C(\beta, b) = (\beta - b)^2$. If $\beta_1$ should be chosen as the estimate, then the expected cost or loss will be

$$\int_{-\infty}^{+\infty} (\beta_1 - b)^2 p(b|\mathbf{y})\, \mathrm{d}b \qquad \text{with } \mathbf{y} = \mathbf{c} \tag{5.3}$$

This is a scalar, and can be plotted as a function of $\beta$. This is shown in Figure 5.6. Consequently, then the best estimate is that value $\hat{\beta}$ for which this function has a minimum value.

A similar line of thought proceeds in the following manner. The **conditional risk** of choosing $\boldsymbol{\beta}(\mathbf{y})$ if the true process parameter value is $\mathbf{b}$ can be written as the expectation* of the cost function with respect to the observations $\mathbf{y}$:

$$\mathscr{E}_{\mathbf{y}|\mathbf{b}}[C(\boldsymbol{\beta}, \mathbf{b})] = \int_k C(\boldsymbol{\beta}, \mathbf{b}) p(\mathbf{y}|\mathbf{b})\, \mathrm{d}^k\mathbf{y} \tag{5.4}$$

* Note that:

$$\mathscr{E}_{\mathbf{y}}[\ldots] = \int_k \ldots p(\mathbf{y})\, d^k\mathbf{y}$$

$$\mathscr{E}_{\mathbf{y}|\mathbf{b}}[\cdots] = \int_k \ldots p(\mathbf{y}|\mathbf{b})\, d^k\mathbf{y}$$

and

$$\int_k p(\mathbf{y}|\mathbf{b})\, d^k\mathbf{y} = 1$$

The **average risk** for this estimating situation is the expectation with respect to the probability of the value of the process parameter **b**

$$R(\boldsymbol{\beta}) = \mathscr{E}_{\mathbf{b}}[\mathscr{E}_{\mathbf{y}|\mathbf{b}}[C(\boldsymbol{\beta}, \mathbf{b})]] \\ = \int_{m+1} \int_{k} C(\boldsymbol{\beta}, \mathbf{b}) p(\mathbf{y}|\mathbf{b}) q(\mathbf{b}) \, \mathrm{d}^k\mathbf{y} \, \mathrm{d}^{m+1}\mathbf{b} \tag{5.5}$$

The estimate that minimizes this expression is called the **minimum risk** or **minimum cost** estimate. Because of the Bayes' relationships, this average risk can be written as:

$$R(\boldsymbol{\beta}) = \int_k \mathrm{d}^k\mathbf{y} p(\mathbf{y}) \int_{m+1} C(\boldsymbol{\beta}, \mathbf{b}) p(\mathbf{b}|\mathbf{y}) \, \mathrm{d}^{m+1}\mathbf{b} \tag{5.6}$$

As $p(\mathbf{y}) \geqslant 0$ the average risk $R(\boldsymbol{\beta})$ can be minimized by making the second integral as small as possible for the observed $\mathbf{y} = \mathbf{c}$:

$$\min_{\boldsymbol{\beta}} \int_{m+1} C(\boldsymbol{\beta}, \mathbf{b}) p(\mathbf{b}|\mathbf{c}) \, \mathrm{d}^{m+1}\mathbf{b} \tag{5.7}$$

A necessary condition for such a minimum is simply:

$$\frac{\partial}{\partial \boldsymbol{\beta}} \int_{m+1} C(\boldsymbol{\beta}, \mathbf{b}) p(\mathbf{b}|\mathbf{c}) \, \mathrm{d}^{m+1}\mathbf{b} \bigg|_{\boldsymbol{\beta} = \hat{\boldsymbol{\beta}}} = \mathbf{0}$$

Due to the use made of Bayes' theorem, the resulting estimate is also called **Bayes' estimate.**

*Example.* As a simple illustration consider again the case where we want to estimate one parameter $\beta$ using a quadratic cost function $C(\beta, b) = (\beta - b)^2$. Then equation (5.7) results in

$$\hat{\boldsymbol{\beta}} = \frac{\int bp(b|\mathbf{c}) \, db}{\int p(b|\mathbf{c}) \, db} = \int bp(b|\mathbf{c}) \, db = \mathscr{E}_{b|\mathbf{c}}[b] \tag{5.8}$$

i.e. the optimum estimate is the (conditional) expected value of $b$.

As a suitable cost function $C(\boldsymbol{\beta}, \mathbf{b})$ has a minimum for $\boldsymbol{\beta} = \mathbf{b}$, as it presumably has small values for values of $\boldsymbol{\beta}$ in the vicinity of $\mathbf{b}$ and as

$$\int_{m+1} p(\mathbf{b}|\mathbf{y}) \, \mathrm{d}^{m+1}\mathbf{b} = 1$$

it will be clear that equation (5.7) is satisfied if $\boldsymbol{\beta}$ is chosen in the neighbourhood of that $\mathbf{b}$ where the conditional probability $p(\mathbf{b}|\mathbf{y})$ is maximum (cf. Figure 5.6). Now we will drop the assumption (c), i.e. the knowledge about an adequate cost or loss function $C(\boldsymbol{\beta}, \mathbf{b})$. In that case it is reasonable to chose $\boldsymbol{\beta}$ at that value of $\mathbf{b}$ for which $p(\mathbf{b}|\mathbf{y})$ is a maximum. As according to

| *a priori* knowledge |
|---|
| $p(\mathbf{y}|\mathbf{b})$ |
| $q(\mathbf{b})$ |

Bayes' rule:

$$p(\mathbf{b}|\mathbf{y}) = \frac{p(\mathbf{y}|\mathbf{b})q(\mathbf{b})}{p(\mathbf{y})}$$

with

$$p(\mathbf{y}) = \int_{m+1} p(\mathbf{y}|\mathbf{b})q(\mathbf{b})\, \mathrm{d}^{m+1}\mathbf{b}$$

both assumptions ($a$) and ($b$) are still being used.

### 5.1.3 *Maximum Likelihood Estimators*

Next we will consider the consequences of dropping both assumptions ($b$) and ($c$); the *a priori* probability $q(\mathbf{b})$ of the process parameter $\mathbf{b}$ is also unknown. This ignorance can be expressed by assuming a uniform distribution $q(\mathbf{b}) = \text{const.}$ over the interval of consideration. In that case for any $\mathbf{y}$:

$$\max_{\mathbf{b}} p(\mathbf{b}|\mathbf{y}) = \frac{\text{const.}}{p(\mathbf{y})} \max_{\mathbf{b}} p(\mathbf{y}\,;\mathbf{b}) \tag{5.9}$$

Now $\mathbf{b}$ is no longer a random variable but an unknown constant parameter; this is expressed by a semicolon instead of a bar in the argument of $p$. The probability-density function (*a priori*) is given by

$$p\{y(1), \ldots, y(k); \mathbf{b}\}$$

*A posteriori*, after the measurement with the observed sample values $y(1) = c_1, \ldots, y(k) = c_k$ or $\mathbf{y} = \mathbf{c}$ this function is noted as

$$L\{c_1, \ldots, c_k; \boldsymbol{\beta}\}$$

the **likelihood function;** cf. Section 2.3.

*Example.* Again the simple example of Figure 5.3 can be used as an illustration. We have to know $p_n$, the probability-density function of $n(i)$. In that case the measurement of $u(i)$ provides us with the knowledge sketched in Figure 5.7, i.e. the *a priori* probability density of $y(i)$ with $b$ as parameter. Now the measurement $y(i) = c_i$ brings us to the situation indicated in Figure 5.8, the *a posteriori* knowledge. From the likelihood function we have to assign an estimate $\hat{\beta}$. A reasonable and popular choice is to take that value $\hat{\beta}$, for which $L$ has its maximal value.

*a priori* knowledge
$p(\mathbf{y}\,;\mathbf{b})$

For the higher-dimensional case the following necessary condition for finding the maximum can be given:

$$\left.\frac{\partial}{\partial \boldsymbol{\beta}} L(\mathbf{c}\,;\boldsymbol{\beta})\right|_{\boldsymbol{\beta}=\hat{\boldsymbol{\beta}}} = \mathbf{0} \tag{5.10a}$$

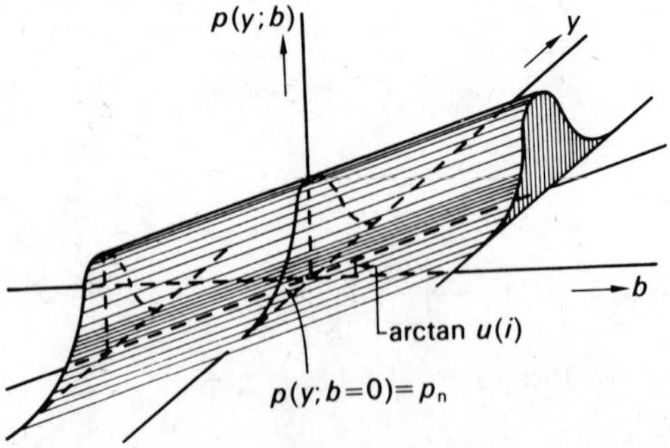


Figure 5.7

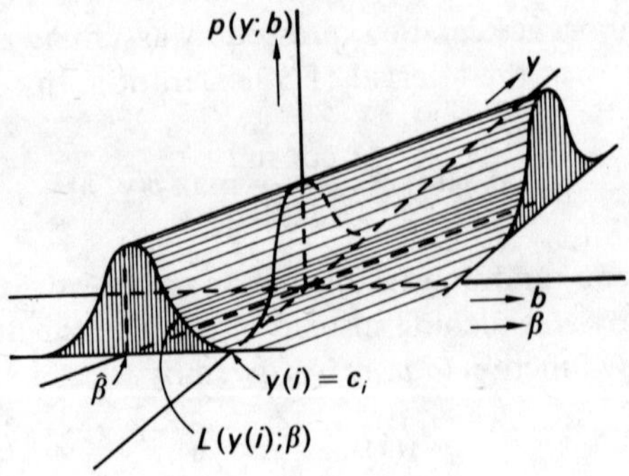


Figure 5.8

or as the logarithmic function is monotonic

$$\left.\frac{\partial}{\partial \boldsymbol{\beta}} \ln L(\mathbf{c}; \boldsymbol{\beta})\right|_{\boldsymbol{\beta}=\hat{\boldsymbol{\beta}}} = \mathbf{0} \tag{5.10b}$$

This expression is called the likelihood equation. Picking the root of this set of equations which yields the largest value for $L(\mathbf{c}; \boldsymbol{\beta})$ or for $\ln L(\mathbf{c}; \boldsymbol{\beta})$ we are given the celebrated **maximum likelihood estimate** (M.L.E.). The M.L.E. has been extensively discussed in literature. Some of its interesting properties are the following:

**asymptotic normality**: i.e. $p(\hat{\boldsymbol{\beta}}; \mathbf{b})$ approaches a normal distribution for $k \to \infty$; for the meaning of $k$ cf. equation (5.1);

**asymptotic unbiasedness**: i.e. $\mathscr{E}[\hat{\boldsymbol{\beta}}] = \mathbf{b}$ for $k \to \infty$;

**asymptotic efficiency**: i.e. approaching the best accuracy or minimum variance as given by the Cramér–Rao (in)equality, cf. Chapter 11;

**consistency:** as defined before;

**invariance:** i.e. if $\hat{\boldsymbol{\beta}}$ is a M.L.E. of $\mathbf{b}$ then $g(\hat{\boldsymbol{\beta}})$ is a M.L.E. of $g(\mathbf{b})$.

### *5.1.4 Markov and Least-squares Estimators*

It remains to be shown how the Markov and least-squares estimators, mentioned in Section 2.2 and discussed in Chapter 6, are related to those

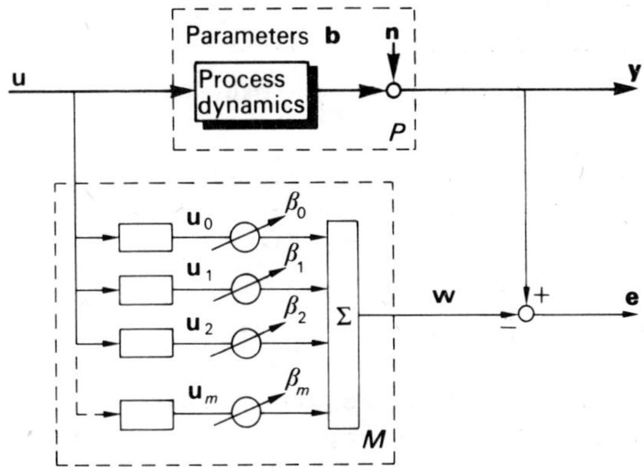


Figure 5.9

derived in this section. Let us refer to Figure 5.9 and assume that the components of $\mathbf{n}$ have a $k$-variate **Gaussian (normal) distribution,** i.e.

$$p(\mathbf{n}) = \frac{1}{(2\pi)^{k/2}|\mathbf{N}|^{\frac{1}{2}}} \exp\left[-\tfrac{1}{2}(\mathbf{n}'\mathbf{N}^{-1}\mathbf{n})\right] \tag{5.11}$$

with

$$\mathscr{E}[\mathbf{n}] = \mathbf{0} \quad \text{and} \quad \mathscr{E}[\mathbf{n}\mathbf{n}'] = \mathbf{N}$$

Note that $\mathbf{n}$ is the only stochastic part of $\mathbf{e} = \mathbf{y} - \mathbf{w} = \mathbf{y} - \mathbf{U}\boldsymbol{\beta}$. Consequently we can write for the logarithm of the probability-density function of $\mathbf{e}$:

$$\ln p(\mathbf{y} - \mathbf{U}\boldsymbol{\beta}) = -\tfrac{1}{2}\ln\{(2\pi)^k|\mathbf{N}|\} - \tfrac{1}{2}\{(\mathbf{y} - \mathbf{U}\boldsymbol{\beta})'\mathbf{N}^{-1}(\mathbf{y} - \mathbf{U}\boldsymbol{\beta})\}$$

Maximizing this functions leads to:

$$\frac{\partial}{\partial\boldsymbol{\beta}}(\mathbf{y} - \mathbf{U}\boldsymbol{\beta})'\mathbf{N}^{-1}(\mathbf{y} - \mathbf{U}\boldsymbol{\beta})\bigg|_{\boldsymbol{\beta}=\hat{\boldsymbol{\beta}}} = \mathbf{0} \tag{5.12}$$

or

$$\mathbf{U}'\mathbf{N}^{-1}\mathbf{U}\hat{\boldsymbol{\beta}} - \mathbf{U}'\mathbf{N}^{-1}\mathbf{y} = \mathbf{0} \tag{5.13}$$

| a priori knowledge |
| --- |
| $\mathbf{N} = \mathscr{E}[\mathbf{nn}']$ |

or if $\mathbf{U}'\mathbf{N}^{-1}\mathbf{U}$ has an inverse then equation (5.13) can be written

$$\hat{\boldsymbol{\beta}} = [\mathbf{U}'\mathbf{N}^{-1}\mathbf{U}]^{-1}\mathbf{U}'\mathbf{N}^{-1}\mathbf{y} \tag{5.14}$$

This is the expression for the **Markov estimate.** It has the following properties:

**linearity:** i.e. $\hat{\boldsymbol{\beta}} = \mathbf{Qy}$

**unbiasedness:** i.e. $\mathscr{E}[\hat{\boldsymbol{\beta}}] = \mathbf{b}$;

**minimum variance:** of all linear unbiased estimates. This variance follows from:

$$\text{cov}\,[\hat{\boldsymbol{\beta}}] = \mathscr{E}[(\hat{\boldsymbol{\beta}} - \mathbf{b})(\hat{\boldsymbol{\beta}} - \mathbf{b})']$$

| a priori knowledge |
| --- |
| nil |

If knowledge of the covariance matrix of the noise is also lacking it is appropriate to choose $\mathbf{N}^{-1} = \sigma^{-2}\mathbf{I}$, where $\sigma^2$ is the variance of the noise. This implies the assumption that the noise is white. Consequently:

$$\hat{\boldsymbol{\beta}} = [\mathbf{U}'\mathbf{U}]^{-1}\mathbf{U}'\mathbf{y} \tag{5.15}$$

This is the expression for the **least-squares estimator.**

The Markov and least-squares estimators have been 'derived' from the maximum likelihood estimate under the assumption of Gaussian noise. (This assumption of a Gaussian distribution is not necessary. In Chapter 6 this estimator is derived irrespective of the Gaussian assumption.)

*Example.* Such a derivation can be illustrated by using again the simple example of Figure 5.3. Employing the least-squares method, the estimate is chosen in such a way that the loss function, defined as

$$E(\beta) = \sum_{i=1}^{k} \{y(i) - u(i)\beta\}^2 = [\mathbf{y} - \mathbf{u}\beta]'[\mathbf{y} - \mathbf{u}\beta]$$

is a minimum.

In Figure 5.10 the differences between the observations $y$ and the 'model outputs' $u\beta$ are indicated. The minimization can be pursued analytically; a necessary condition

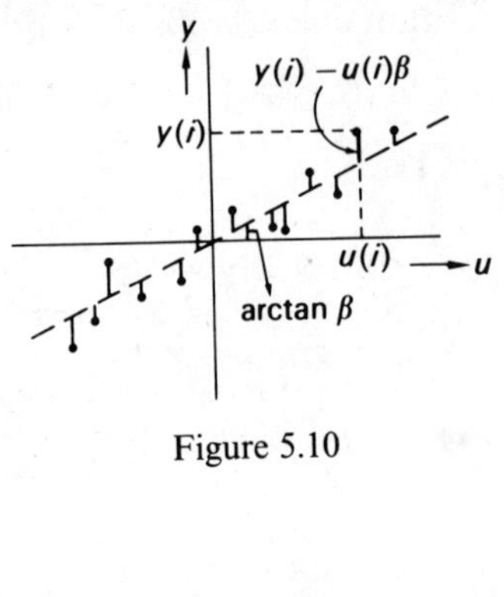


Figure 5.10

for the minimum is

$$\frac{d}{d\beta}\sum_{i=1}^{k}\{y(i)-u(i)\beta\}^2\Big|_{\beta=\hat\beta}=0$$

or

$$\sum_{i=1}^{k}u(i)\{y(i)-u(i)\hat\beta\}=0 \tag{5.16}$$

or

$$\hat\beta=\frac{\sum_{i=1}^{k}u(i)y(i)}{\sum_{i=1}^{k}u(i)u(i)} \tag{5.17}$$

$\hat\beta$ is the optimal estimate under the conditions given. Note from (5.16) that the terms $y(i)-u(i)\beta$ are weighted with respect to $u(i)$; quite naturally the larger the input signal, the more importance is assigned to the deviation between observation $y(i)$ and 'prediction' $u(i)\beta$. Equation (5.17) refers to the correlation methods.

### 5.1.5 *Some Other Aspects*

The simple example of Figure 5.3 also serves very well to illustrate the so-called 'problem of input noise'. Consider the system illustrated by the block diagram in Figure 5.11 where neither the input nor the output can be observed exactly.

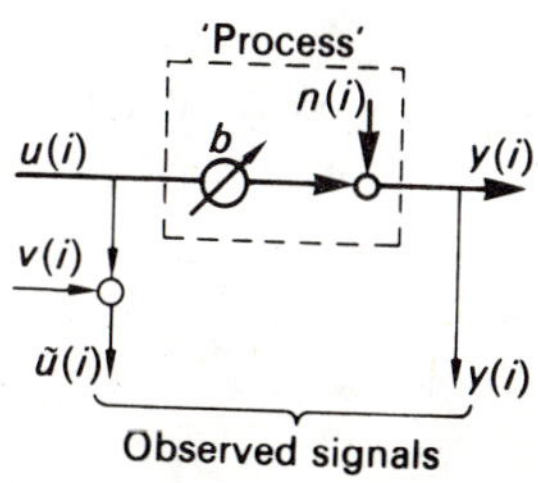


Figure 5.11

It is well known in statistics (see e.g. Lindley, 1965) that, unless specific assumptions are made concerning the variations in $u$, $n$ and $v$ it is not possible to estimate the dynamics of the process. Consider, for example,

$$y(i)=bu(i)+n(i)$$
$$\tilde u(i)=u(i)+v(i)$$

Assume that $v(i)$ and $n(i)$ are independent stochastic variables with zero mean values. If $v(i)=0$ and $n(k)\not\equiv 0$ we find that the estimate of $b$ as given by (5.17) is

$$\hat{\beta} = \frac{\sum_{i=1}^{k} \tilde{u}(i)y(i)}{\sum_{i=1}^{k} \tilde{u}(i)\tilde{u}(i)}$$

However, if $n(k) = 0$ and $v(k) \not\equiv 0$ we find by the same argument that the estimate of $b$ is

$$\hat{\beta} = \frac{\sum_{i=1}^{k} y(i)y(i)}{\sum_{i=1}^{k} y(i)\tilde{u}(i)}$$

This corresponds to choosing $\beta$ such that the difference between the observations $\tilde{u}(i)$ and the predictions $y(i)/\beta$ are as small as possible in the least-squares sense (cf. Figure 5.12). Without additional information it is of course impossible to tell which estimate to choose.

From the expression (5.17) for the least-squares estimator, which provides an explicit mathematical or **accumulative** solution, it is quite simple to derive a model-adjustment or **iterative** solution. We write the expressions for $k + 1$ and for $k$ (pairs of) observation samples:

$$\left\{\sum_{i=1}^{k+1} u^2(i)\right\} \beta(k + 1) = \sum_{i=1}^{k+1} u(i)y(i)$$

$$\left\{\sum_{i=1}^{k} u^2(i)\right\} \beta(k) = \sum_{i=1}^{k} u(i)y(i)$$

These expression for the estimates $\hat{\beta}(k)$ and $\hat{\beta}(k + 1)$ can be rewritten as:

$$\left\{\sum_{i=1}^{k+1} u^2(i)\right\} \beta(k + 1) = \sum_{i=1}^{k} u(i)y(i) + u(k + 1)y(k + 1)$$

$$\left\{\sum_{i=1}^{k+1} u^2(i)\right\} \beta(k) = \sum_{i=1}^{k} u(i)y(i) + u^2(k + 1)\hat{\beta}(k)$$

$$\left\{\sum_{i=1}^{k+1} u^2(i)\right\} \{\beta(k + 1) - \beta(k)\} = u(k + 1)\{y(k + 1) - u(k + 1)\hat{\beta}(k)\}$$

or

$$\hat{\beta}(k + 1) = \hat{\beta}(k) + \left\{\sum_{i=1}^{k+1} u^2(i)\right\}^{-1} u(k + 1)\underbrace{\{y(k + 1) - u(k + 1)\hat{\beta}(k)\}}_{e(k+1)}$$

The last term, $e(k + 1)$, is the error between the observation $y(k + 1)$ and the model output $u(k + 1)\hat{\beta}(k)$. The new estimate consists of the old estimate,

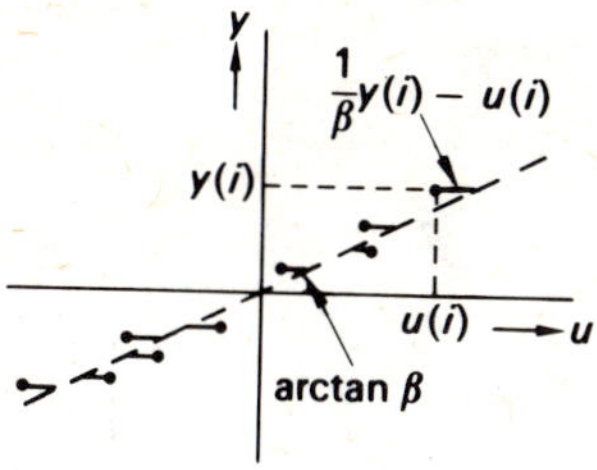


Figure 5.12

corrected with a proportion of the error $e(k + 1)$. What 'weight' is given to the error depends on the magnitude of $u(k + 1)$ and on the sum of squares of all previous input samples ('energy' measure of the input signal).

Some of the relations between the different types of estimates as derived in this section are summarized in Table 5.1. This discussion has been restricted to sampled signals. By increasing the number of samples and decreasing the sampling interval the corresponding expressions can be derived for continuous signals.

## 5.2 Deterministic Convergence Schemes*

As pointed out in Chapter 2, the estimation problem can be formulated as a minimization of an error defined according to some criterion. This error criterion necessarily has to be such that both positive and negative errors contribute in a positive sense to the criterion value. Loosely speaking this was called an 'even-function error criterion'. For example:
Let the criterion be given as a minimization of the loss function

$$E\{\boldsymbol{\beta}\} = \tfrac{1}{2}\int_0^T (\mathbf{y} - \mathbf{w})'\mathbf{G}(\mathbf{y} - \mathbf{w})\,\mathrm{d}t \tag{5.18}$$

where

$\mathbf{y} = \mathbf{y}(\mathbf{b}, t)$ is the process output;

$\mathbf{w} = \mathbf{w}(\boldsymbol{\beta}, t)$ is the corresponding model output;

$\mathbf{G}$ is a positive definite matrix, weighing the relative reliabilities of $y_i$'s

In model-adjustment approaches for parameter estimation, just as in cases of optimal and self-optimizing control, the extremum of a function or functional has to be found. Assume the error criterion to be a minimization of:

$$E = E(\beta_0, \beta_1, \ldots, \beta_m) = E(\boldsymbol{\beta}) \tag{5.19}$$

* This section has been reworked and expanded by Mr. A. A. H. Damen, University of Technology, Eindhoven, Netherlands.

TABLE 5.1

| | Bayes' estimation | Unconditional maximum likelihood estimation | Conditional maximum likelihood estimation | Markov estimation | Least-squares estimation |
|---|---|---|---|---|---|
| *A priori* knowledge | $p(\mathbf{y}\|\mathbf{b})$<br>$q(\mathbf{b})$<br>$C(\boldsymbol{\beta}, \mathbf{b})$ | $p(\mathbf{y}\|\mathbf{b})$<br>$q(\mathbf{b})$ | $p(\mathbf{y}; \mathbf{b})$ | $\mathbf{N} = \mathscr{E}[\mathbf{nn}']$ | Nil |
| Conditions | $\min_{\boldsymbol{\beta}} \int_{m+1} C(\boldsymbol{\beta}, \mathbf{b}) p(\mathbf{b}\|\mathbf{y})\, \mathrm{d}^{m+1}\mathbf{b}$ | $\max p(\mathbf{b}\|\mathbf{y}) = \max \frac{p(\mathbf{y}\|\mathbf{b})q(\mathbf{b})}{p(\mathbf{y})}$ | $\max_{\mathbf{b}} p(\mathbf{y}; \mathbf{b})$ or $\max_{\mathbf{b}} \ln p(\mathbf{y}; \mathbf{b})$ | $\hat{\boldsymbol{\beta}} = [\mathbf{U}'\mathbf{N}^{-1}\mathbf{U}]^{-1}\mathbf{U}'\mathbf{N}^{-1}\mathbf{y}$ | $\hat{\boldsymbol{\beta}} = [\mathbf{U}'\mathbf{U}]^{-1}\mathbf{U}'\mathbf{y}$ |
| | Observations $\mathbf{y} = \mathbf{c}$ | ⇔Same | ⇔Same | ⇔Same | ⇔Observations $\mathbf{y} = \mathbf{c}$ |
| Properties | minimum risk or cost | | Asympt. normality<br>Asympt. unbiasedness<br>Asympt. efficiency | Linearity<br>Unbiasedness<br>(minimum variance) | Linearity<br>Unbiasedness |
| Relations | cf. Text (Bayes' ↔ Unconditional) | cf. Text (Unconditional ↔ Conditional) | If gaussian noise (Conditional ↔ Markov) | If white noise (Markov ↔ Least-squares) | |

where $\beta_j$ are the model parameters to be adjusted. In Chapter 2 some methods have been indicated to determine the partial derivatives with respect to these parameters. These derivatives can be used in a variety of ways. A rather general representation follows from the following reasoning. The error function given by equation (5.19) can be considered as a hypersurface defined by the expression:

$$E(\boldsymbol{\beta}) - E_0 \tag{5.20}$$

where $E_0$ is the minimum of the error function, that we want to find: $E_0 = E|_{\boldsymbol{\beta}=\mathbf{b}}$. The error function can be expressed as a Taylor expansion around the minimum:

$$E(\boldsymbol{\beta}) = E_0 + \left.\frac{\partial E}{\partial \boldsymbol{\beta}'}\right|_{\boldsymbol{\beta}=\mathbf{b}} (\boldsymbol{\beta} - \mathbf{b}) + \frac{1}{2}(\boldsymbol{\beta} - \mathbf{b})' \left.\frac{\partial^2 E}{\partial \boldsymbol{\beta}\, \partial \boldsymbol{\beta}'}\right|_{\boldsymbol{\beta}=\mathbf{b}} (\boldsymbol{\beta} - \mathbf{b}) + \text{higher-order terms} \tag{5.21a}$$

Because $E$ has a minimum in $\boldsymbol{\beta} = \mathbf{b}$, the following holds:

$$\left.\frac{\partial E}{\partial \boldsymbol{\beta}}\right|_{\boldsymbol{\beta}=\mathbf{b}} = \mathbf{0} \quad \text{and} \quad \left.\frac{\partial^2 E}{\partial \boldsymbol{\beta}\, \partial \boldsymbol{\beta}'}\right|_{\boldsymbol{\beta}=\mathbf{b}} \quad \text{is positive definite.}$$

Moreover in most cases the higher-order terms can be neglected (if not, cf. method (*h*) mentioned below). Consequently,

$$E(\boldsymbol{\beta}) \approx E_0 + \frac{1}{2}(\boldsymbol{\beta} - \mathbf{b})' \left.\frac{\partial^2 E}{\partial \boldsymbol{\beta}\, \partial \boldsymbol{\beta}'}\right|_{\boldsymbol{\beta}=\mathbf{b}} (\boldsymbol{\beta} - \mathbf{b}) \tag{5.21b}$$

After substituting (5.21b), equation (5.20) now represents a hyperparaboloid (cf. Figure 5.13).

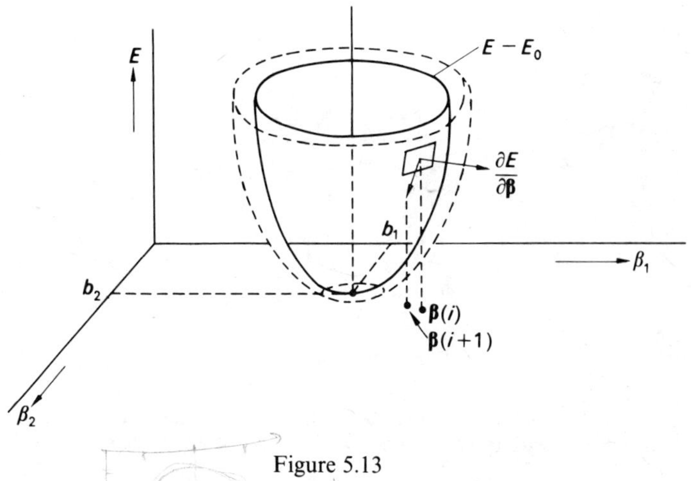


Figure 5.13

For the convergence study of the adjustment schemes that we will discuss, some knowledge about the iso-criterion curves $E =$ constant is important. In the above quadratic case these are all hyperellipsoids. Before discussing details of the adjustment schemes, an example of a quadratic error function is given.

*Example.* For a simple case a rough approximation leads to the following description. Assume a process describable by the differential equation

$$(a_nD^n + \cdots + a_1D + 1)y(t) = (b_mD^m + \cdots + b_1D + b_0)u(t) \tag{5.22}$$

with $D \overset{\text{def}}{=} d/dt$. A model is given by the equation:

$$(\alpha_nD^n + \cdots + \alpha_1D + 1)w(t) = (\beta_mD^m + \cdots + \beta_1D + \beta_0)u(t) \tag{5.23}$$

These equations can be written explicitly as $y(t) = \ldots$ and $w(t) = \ldots$ In the vicinity of the optimum $e(t) = y(t) - w(t)$ is small, if no noise is present. If $e(t)$ is sufficiently small then $w(t)$ can be approximated by:

$$\tilde{w}(t) = -(\alpha_nD^n + \cdots + \alpha_1D)y(t) + (\beta_mD^m + \cdots + \beta_1D + \beta_0)u(t) \tag{5.24}$$

Consequently,

$$\begin{aligned} e(t) \approx y(t) - \tilde{w}(t) = {} & \{(\alpha_n - a_n)D^n + \cdots + (\alpha_1 - a_1)D\}y(t) \\ & -\{(\beta_m - b_m)D^m + \cdots + (\beta_1 - b_1)D + (\beta_0 - b_0)\}u(t) \end{aligned} \tag{5.25}$$

In the neighbourhood of the optimum $e(t)$ is approximately a linear function of $\alpha$ and $\beta$. By renaming*

$$\left.\begin{aligned} \alpha_1 - a_1 &= \gamma_1 \\ &\vdots \\ \alpha_n - a_n &= \gamma_n \\ \beta_0 - b_0 &= \gamma_{n+1} \\ &\vdots \\ \beta_m - b_m &= \gamma_{n+m+1} \end{aligned}\right\} \text{ and } \left\{\begin{aligned} Dy(t) &= v_1(t) \\ &\vdots \\ D^ny(t) &= v_n(t) \\ u(t) &= v_{n+1}(t) \\ Du(t) &= v_{n+2}(t) \\ &\vdots \\ D^mu(t) &= v_{n+m+1}(t) \end{aligned}\right.$$

we find the simple expressions:

$$e(t) \approx \sum_{i=1}^{N} \gamma_i v_i \qquad N = n + m + 1$$

$$e^2(t) \approx \sum_i^N \sum_j^N \gamma_i \gamma_j v_i v_j$$

$$E = \int_{t-T}^{t} e^2(t)\,dt = \sum_i^N \sum_j^N \gamma_i \gamma_j r_{ij} \tag{5.26}$$

with

$$r_{ij} = \int_{t-T}^{t} v_i(t)v_j(t)\,dt = r_{ji}$$

* In practice differentials are difficult to obtain. However, an analogous scheme can be derived for integrals or semi-integrals (using $1/(s + \sigma)$ instead of $1/s$, where $s$ is Laplace operator); cf. Chapter 9.

This also can be written as:

$$E = \boldsymbol{\gamma}'\mathbf{R}\boldsymbol{\gamma} \tag{5.27}$$

with

$$\boldsymbol{\gamma} = \begin{pmatrix} \gamma_1 \\ \vdots \\ \gamma_N \end{pmatrix} \qquad \mathbf{R} = \begin{pmatrix} r_{11} & \dots & r_{1N} \\ \vdots & & \\ r_{N1} & \dots & r_{NN} \end{pmatrix}$$

Comparison of equation (5.27) with (5.21b) shows, that they are equivalent for $E_0 = 0$ and

$$\mathbf{R} = \frac{1}{2} \frac{\partial^2 E}{\partial \boldsymbol{\beta}\, \partial \boldsymbol{\beta}'}\bigg|_{\boldsymbol{\beta}=\mathbf{b}}$$

The adjustment schemes can be distinguished as **direct-search methods** and **gradient methods.** The first class does not require an explicit evaluation of the gradient of the error or loss function, whereas the second class is based on the assumption that the gradient can be determined. For surveys of optimization techniques cf. Box, Davies and Swann (1969), Wilde (1964), Bekey and McGee (1964), Wilde and Beightler (1967), Kowalik and Osborne (1968), Powell (1970), Dixon (1972).

The **direct-search methods** can be divided into two subclasses:

α. **Tabulating methods.** Evaluating the error function in several points spread out in the parameter space, e.g. at the junctions of a regular grid or randomly distributed.

β. **Exploratory methods.** Those algorithms follow implicitly the gradient path by stepwise looking for a descending direction. An example is the simplex method starting with $n$ points at equal distances in the $n$-dimensional parameter space. Each iteration is made by 'mirroring' the point with the largest function value onto the new point, which has also the same distances to the other ones. After some iterations the distances have to be diminished in order to prevent finding the same points again. Another approach is the creeping random-search method, where, starting from the last iteration point, a random direction is examined for smaller function values. In these methods there is a great variety of combinations between earlier iterations in the parameters. A survey can be found in White (1970). The creeping random-search methods are especially important when the number of parameters is greater than about twenty.

The **gradient methods** can be either

continuous-adjustment schemes or

intermittent-adjustment schemes

where intermittent refers to the alternation of the measurement and the model-adjustment periods.

As a **continuous**-adjustment technique the method of steepest descent is quite popular. In this case the model will be taken along a trajectory which, for a certain speed of adjustment, gives the most rapid reduction of the error. It turns out that this trajectory is at any point normal to the curves for constant $E$. For example, in the two-dimensional case $E$ = constant implies:

$$\mathrm{d}E = \frac{\partial E}{\partial \beta_1}\mathrm{d}\beta_1 + \frac{\partial E}{\partial \beta_2}\mathrm{d}\beta_2 = 0 \tag{5.28}$$

The equation for the tangent to the curve at the point $P(\beta_{1p}, \beta_{2p})$ is given by:

$$\left(\frac{\partial E}{\partial \beta_1}\right)_P(\beta_1 - \beta_{1p}) + \left(\frac{\partial E}{\partial \beta_2}\right)_P(\beta_2 - \beta_{2p}) = 0 \tag{5.29}$$

The vector

$$\left(\frac{\partial E}{\partial \beta_1}, \frac{\partial E}{\partial \beta_2}\right)'$$

is perpendicular to the tangent and consequently also to the curve. For the steepest-descent method the parameters are adjusted according to the equation

$$\begin{pmatrix}\dot{\beta}_1\\ \dot{\beta}_2\end{pmatrix} = -\Gamma\begin{pmatrix}\dfrac{\partial E}{\partial \beta_1}\\[2mm] \dfrac{\partial E}{\partial \beta_2}\end{pmatrix} = -\Gamma\frac{\partial E}{\partial \boldsymbol{\beta}} \tag{5.30}$$

Thus, $P$ will follow a trajectory which is perpendicular everywhere to the curves $E$ = constant (cf. Figure 5.14). Here $\Gamma$ is a constant which, together

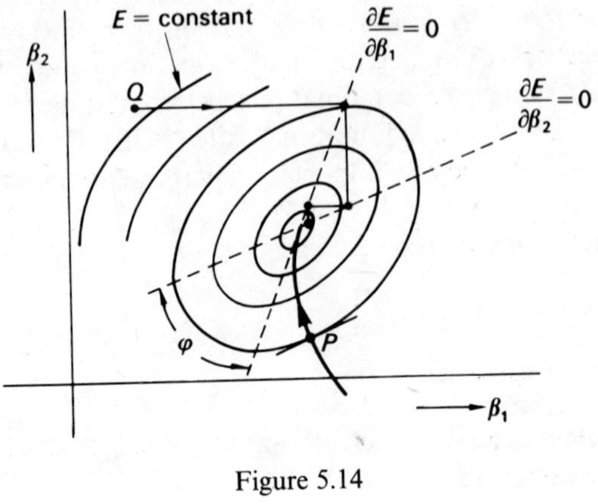


Figure 5.14

with the partial derivatives, determines the rate of variation of the parameters. The 'measurement' of the partial derivatives is not instantaneous and, therefore, a limitation on the rate of variation must be maintained. If too high a gain $\Gamma$ is used, the measurement of the derivatives will degenerate and consequently, a steepest-descent trajectory will not be followed. For a still higher gain, the system even may turn unstable. For arbitrary systems there is no generally applicable technique yet for determining stability, because our system is both time varying and non-linear. For simple systems, however, the technique developed by Liapunov can be applied successfully. For a low gain, the approach to the optimum will be slow.

The **intermittent** adjustment schemes can be such that:

the parameters are adjusted one by one

all parameters are adjusted simultaneously.

(*a*) **Cyclic adjustments of the parameters one by one.** Say that parameter $\beta_1$ is first adjusted. One possibility is then to let the adjustment system give $\beta_1$ such a value that $\partial E/\partial \beta_1$ is equal to zero. The same procedure will then be repeated for all parameters of the model. If necessary more cycles have to be made to obtain a satisfactory model match. In the two-dimensional case with parameters $\beta_1$ and $\beta_2$ the $(\beta_1, \beta_2)$-plane can be mapped by curves for constant $E$. As pointed out, near the optimum these will in general be concentric ellipses, the principal axes of which may be of arbitrary orientation (cf. Figure 5.14). Therefore, we must expect interaction between the model parameters and generally more cycles are needed to bring the model to the neighbourhood of the optimum. If the system has been made orthogonal through a transformation then one adjustment for each parameter may be sufficient. The problem, however, is that we do not know exactly which linear combination to use for orthogonalization. A feature of the ellipses is the angle between the lines, which represents the geometrical position of all points satisfying either

$$\frac{\partial E}{\partial \beta_1} = 0 \quad \text{or} \quad \frac{\partial E}{\partial \beta_2} = 0 \tag{5.31}$$

The smaller this angle the more cycles are necessary to obtain a given accuracy. In Figure 5.14 a possible adjustment path is sketched ($Q$). If the iso-criterion curves are not such simple ellipses but a 'ridge', then this method has the disadvantage that the adjustment may stop at a value $E > E_0$.

All other adjustment schemes considered will be with simultaneous parameter adjustments. For a more elaborate discussion of some of these schemes cf. Rosenbrock and Storey (1966).

(*b*) **Steepest descent.** The adjustments are made according to the algorithm

$$\boldsymbol{\beta}(i+1) = \boldsymbol{\beta}(i) - \Gamma \left.\frac{\partial E}{\partial \boldsymbol{\beta}}\right|_{\boldsymbol{\beta}=\boldsymbol{\beta}(i)} \qquad \Gamma > 0$$

where $\Gamma$ is a constant. Depending on the value chosen for $\Gamma$ we obtain a slow and smooth approach to the optimum ($\Gamma$ small, cf. path $P_1$ in Figure 5.15) or an oscillatory one (path $P_2$). If the gradient cannot be calculated

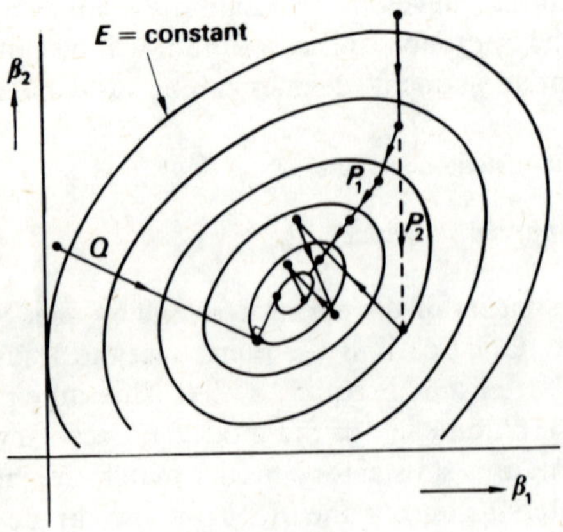


Figure 5.15

then it has to be determined by 'experiments', i.e. by (small) excursions around $\boldsymbol{\beta}(i)$.

(*c*) **Steepest descent with minimization along a line.** This is governed by the algorithm

$$\boldsymbol{\beta}(i+1) = \boldsymbol{\beta}(i) - \Gamma(i) \left.\frac{\partial E}{\partial \boldsymbol{\beta}}\right|_{\boldsymbol{\beta}=\boldsymbol{\beta}(i)} \qquad \Gamma(i) > 0$$

Now $\Gamma(i)$ is chosen so that the system moves along the gradient direction until in that direction the minimum is reached. After that a new gradient determination is performed, etc. Path $Q$ in Figure 5.15 shows an example of such an adjustment. Note that necessarily the successive adjustment paths are orthogonal. Near the minimum the convergence can be slow. It makes sense to ask under what condition one would obtain a one-step procedure to the optimum, assuming elliptic curves. For that case, it follows from equation (5.26) in the two-dimensional case

$$E = r_{11}\gamma_1^2 + 2r_{12}\gamma_1\gamma_2 + r_{22}\gamma_2^2 \tag{5.32}$$

$$\frac{\partial E}{\partial \beta_i} = \frac{\partial E}{\partial \gamma_i} \quad \text{as} \quad \gamma_i = \beta_i - b_i$$

that

$$\begin{cases} \dfrac{\partial E}{\partial \gamma_1} = 2r_{11}\gamma_1 + 2r_{12}\gamma_2 \approx 2r_{11}\gamma_1 & \text{if } r_{12} \ll r_{11} \\ \dfrac{\partial E}{\partial \gamma_2} = 2r_{12}\gamma_1 + 2r_{22}\gamma_2 \approx 2r_{22}\gamma_2 & \text{if } r_{12} \ll r_{22} \end{cases}$$

For a one-step approach to hold it is necessary that

$$\left.\begin{aligned} -\gamma_1 = \Delta\gamma_1 = -\Gamma_1 \frac{\partial E}{\partial \gamma_1} \approx -2\Gamma_1 r_{11}\gamma_1 \quad \text{or} \quad \Gamma_1 = \frac{1}{2r_{11}} \\ -\gamma_2 = \Delta\gamma_2 = -\Gamma_2 \frac{\partial E}{\partial \gamma_2} \approx -2\Gamma_2 r_{22}\gamma_2 \quad \text{or} \quad \Gamma_2 = \frac{1}{2r_{22}} \end{aligned}\right\} \tag{5.33}$$

$\Gamma_1 = \Gamma_2 = \Gamma$ is acceptable only if $r_{11} \approx r_{22}$, otherwise the 'gain coefficients' have to be adapted to the $r$ elements.

Returning to the hypersurface interpretation it is clear, that the assumptions $r_{12} = r_{21} = 0$ and $r_{11} = r_{22}$ involve, that the iso-criterion lines are circles. In general the steepest-descent method will be optimal, when the iso-criterion surfaces are hyperspheres. According to formula (5.21b) the iso-criterion surfaces near the minimum are approximately hyperellipsoids, so it is obvious to try to transform these ellipsoids into spheres in another space and to use the steepest-descent method in that other space. Because

$$\left.\frac{\partial^2 E}{\partial \boldsymbol{\beta}\, \partial \boldsymbol{\beta}'}\right|_{\boldsymbol{\beta}=\mathbf{b}}$$

is symmetric and positive definite, the following holds:

$$\left.\frac{\partial^2 E}{\partial \boldsymbol{\beta}\, \partial \boldsymbol{\beta}'}\right|_{\boldsymbol{\beta}=\mathbf{b}} = \mathbf{S}\boldsymbol{\Lambda}\mathbf{S}' \tag{5.34}$$

where $\boldsymbol{\Lambda}$ is a diagonal matrix containing the eigenvalues, and the columns of matrix $\mathbf{S}$ are the orthonormal eigenvectors. Substitution of (5.34) in (5.21b) leads to

$$E(\mathbf{z}^*) = E_0 + \tfrac{1}{2}\mathbf{z}^{*\prime}\boldsymbol{\Lambda}\mathbf{z}^* \tag{5.35}$$

where

$$\mathbf{z}^* = \mathbf{S}'(\boldsymbol{\beta} - \mathbf{b})$$

Define

$$\mathbf{z} = \boldsymbol{\Lambda}^{\frac{1}{2}}\mathbf{z}^* \quad \text{and} \quad \mathbf{z}^* = \boldsymbol{\Lambda}^{-\frac{1}{2}}\mathbf{z}$$

which results in

$$E(\mathbf{z}) = E_0 + \tfrac{1}{2}\mathbf{z}'\mathbf{z} \tag{5.36}$$

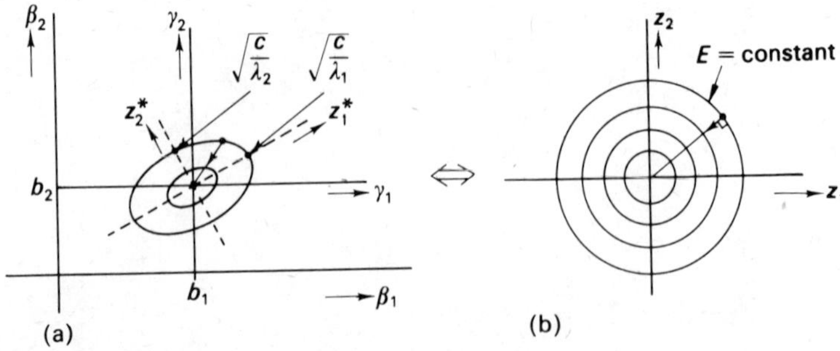


Figure 5.16

In the **z**-space the iso-criterion surfaces are spheres (cf. Figure 5.16). Performance of a steepest-descent iteration in the **z**-space with $\Gamma = 1$:

$$\mathbf{z}(i+1) = \mathbf{z}(i) - \left.\frac{\partial E}{\partial \mathbf{z}}\right|_{\mathbf{z}=\mathbf{z}(i)} = \mathbf{z}(i) - \mathbf{z}(i) = \mathbf{0}$$

and because $\boldsymbol{\beta} - \mathbf{b} = \mathbf{S}\boldsymbol{\Lambda}^{-\frac{1}{2}}\mathbf{z}$ with **S** and $\boldsymbol{\Lambda}$ positive definite, $\boldsymbol{\beta} - \mathbf{b} = \mathbf{0}$ also holds.

Transformation back to the $\boldsymbol{\beta}$-space yields:

$$\boldsymbol{\beta}(i+1) = \boldsymbol{\beta}(i) - \left[\frac{\partial^2 E}{\partial \boldsymbol{\beta}\, \partial \boldsymbol{\beta}'}\right]^{-1}_{\boldsymbol{\beta}=\mathbf{b}} \left.\frac{\partial E}{\partial \boldsymbol{\beta}}\right|_{\boldsymbol{\beta}=\boldsymbol{\beta}(i)} \tag{5.37}$$

Only in purely quadratic error functions the second derivative will be independent of $\boldsymbol{\beta}$; in many other cases, however, it can be assumed to be almost independent of $\boldsymbol{\beta}$, which results in:

(*d*) **Newton–Raphson method.** The algorithm is given by:

$$\boldsymbol{\beta}(i+1) = \boldsymbol{\beta}(i) - \left[\frac{\partial^2 E}{\partial \boldsymbol{\beta}\, \partial \boldsymbol{\beta}'}\right]^{-1}_{\boldsymbol{\beta}=\boldsymbol{\beta}(i)} \left.\frac{\partial E}{\partial \boldsymbol{\beta}}\right|_{\boldsymbol{\beta}=\boldsymbol{\beta}(i)} \tag{5.38}$$

This means that a Newton–Raphson iteration is directed to the center of a hyperellipsoid, as it uses the second derivative at $\boldsymbol{\beta}(i)$. This can also be interpreted as going to the minimum of a hyperparaboloid, which is the second-order approximation of the hypersurface given by (5.20) (cf. Figure 5.17). This can be shown by means of a Taylor expansion around $\boldsymbol{\beta}(i)$:

$$E(\boldsymbol{\beta}) = E(\boldsymbol{\beta}(i)) + \left.\frac{\partial E}{\partial \boldsymbol{\beta}'}\right|_{\boldsymbol{\beta}=\boldsymbol{\beta}(i)} (\boldsymbol{\beta} - \boldsymbol{\beta}(i)) + \frac{1}{2}(\boldsymbol{\beta} - \boldsymbol{\beta}(i))' \left.\frac{\partial^2 E}{\partial \boldsymbol{\beta}\, \partial \boldsymbol{\beta}'}\right|_{\boldsymbol{\beta}=\boldsymbol{\beta}(i)} (\boldsymbol{\beta} - \boldsymbol{\beta}(i)) + \cdots \tag{5.39}$$

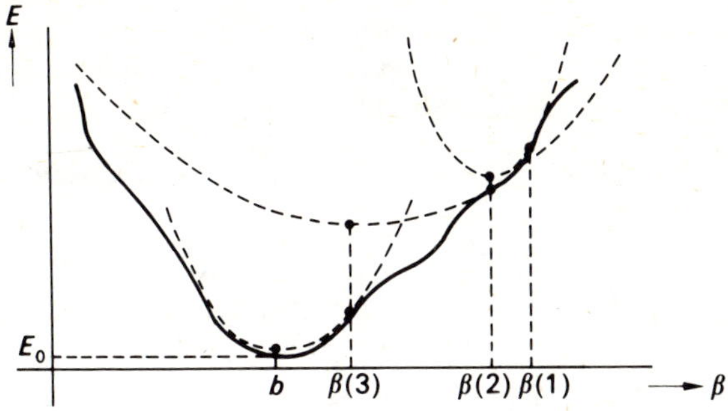


Figure 5.17

The minimum $\boldsymbol{\beta} = \boldsymbol{\beta}(i+1)$ is found by setting $\partial E/\partial\boldsymbol{\beta}|_{\boldsymbol{\beta}=\boldsymbol{\beta}(i+1)} = \mathbf{0}$. Consequently

$$\left.\frac{\partial E}{\partial \boldsymbol{\beta}}\right|_{\boldsymbol{\beta}=\boldsymbol{\beta}(i)} + \left.\frac{\partial^2 E}{\partial \boldsymbol{\beta}\,\partial \boldsymbol{\beta}'}\right|_{\boldsymbol{\beta}=\boldsymbol{\beta}(i)} (\boldsymbol{\beta}(i+1) - \boldsymbol{\beta}(i)) = \mathbf{0}$$

which results in equation (5.38).

Evaluation of the second derivative is quite a troublesome job, but some simplification is possible, if $E$ is a least-squares criterion like (5.18). In that case

$$\frac{\partial E}{\partial \boldsymbol{\beta}} = -\int_0^T \frac{\partial \mathbf{w}'}{\partial \boldsymbol{\beta}} \mathbf{G}(\mathbf{y} - \mathbf{w})\,dt$$

$$\frac{\partial^2 E}{\partial \boldsymbol{\beta}\,\partial \boldsymbol{\beta}'} = \int_0^T \frac{\partial \mathbf{w}'}{\partial \boldsymbol{\beta}} \mathbf{G} \frac{\partial \mathbf{w}}{\partial \boldsymbol{\beta}'}\,dt - \int_0^T \frac{\partial^2 \mathbf{w}}{\partial \boldsymbol{\beta}\,\partial \boldsymbol{\beta}'} \mathbf{G}(\mathbf{y} - \mathbf{w})\,dt^*$$

Close to the minimum $\mathbf{y} \approx \mathbf{w}$, so that the second term can be neglected. Substitution in (5.38) gives us the

(*e*) **Gauss–Newton method.** The algorithm is:

$$\boldsymbol{\beta}(i+1) = \boldsymbol{\beta}(i) + \left[\int_0^T \frac{\partial \mathbf{w}'}{\partial \boldsymbol{\beta}} \mathbf{G} \frac{\partial \mathbf{w}}{\partial \boldsymbol{\beta}'}\,dt\right]^{-1}_{\boldsymbol{\beta}=\boldsymbol{\beta}(i)} \left[\int_0^T \frac{\partial \mathbf{w}'}{\partial \boldsymbol{\beta}} \mathbf{G}(\mathbf{y} - \mathbf{w})\,dt\right]_{\boldsymbol{\beta}=\boldsymbol{\beta}(i)} \tag{5.40}$$

It is easy to verify, that this is a simple least-squares estimate, if $\mathbf{w}$ is linear in $\boldsymbol{\beta}$. The Gauss–Newton method is preferable to the steepest descent on account of the quadratic convergence, although, contrary to the steepest-descent method, the convergence is not guaranteed. So a combination of the two methods incorporating both properties would be advisable. The Marquardt method satisfies this condition.

* The notation is not correct, as $\partial^2\mathbf{w}/\partial\boldsymbol{\beta}\,\partial\boldsymbol{\beta}'$ is a tensor.

(*f*) **Marquardt method.** Let the center of a hypersphere in the parameter space be given by $\boldsymbol{\beta}(i)$. We will search for the minimum on that hypersphere according to Lagrange minimizing $E(\boldsymbol{\beta})$ under the restriction:

$$\Delta\boldsymbol{\beta} = \boldsymbol{\beta}(i+1) - \boldsymbol{\beta}(i) \qquad \Delta\boldsymbol{\beta}'\Delta\boldsymbol{\beta} = r^2 = \text{constant}$$

This results in:

$$\frac{\partial E}{\partial \Delta\boldsymbol{\beta}} + \mu\Delta\boldsymbol{\beta} = \mathbf{0} \tag{5.41}$$

Handling now a Taylor expansion of $\mathbf{w}$ implies that the assumed hypersphere need not be a pure hypersphere:

$$\mathbf{w}(i+1) = \mathbf{w}(i) + \left.\frac{\partial \mathbf{w}}{\partial \boldsymbol{\beta}'}\right|_{\boldsymbol{\beta}=\boldsymbol{\beta}(i)} \Delta\boldsymbol{\beta} \quad \text{or} \quad \frac{\partial \mathbf{w}(i+1)}{\partial \Delta\boldsymbol{\beta}'} \approx \left.\frac{\partial \mathbf{w}}{\partial \boldsymbol{\beta}'}\right|_{\boldsymbol{\beta}=\boldsymbol{\beta}(i)}$$

Consequently, equation (5.41) will be:

$$-\int_0^T \left.\frac{\partial \mathbf{w}'}{\partial \boldsymbol{\beta}}\right|_{\boldsymbol{\beta}=\boldsymbol{\beta}(i)} \mathbf{G}\left(\mathbf{y} - \mathbf{w}(i) - \left.\frac{\partial \mathbf{w}}{\partial \boldsymbol{\beta}'}\right|_{\boldsymbol{\beta}=\boldsymbol{\beta}(i)} \Delta\boldsymbol{\beta}\right) dt + \mu\Delta\boldsymbol{\beta} = \mathbf{0}$$

which implies:

$$\boldsymbol{\beta}(i+1) = \boldsymbol{\beta}(i) + \left[\int_0^T \frac{\partial \mathbf{w}'}{\partial \boldsymbol{\beta}}\mathbf{G}\frac{\partial \mathbf{w}}{\partial \boldsymbol{\beta}'} + \mu\mathbf{I}\right]^{-1} \int_0^T \left.\frac{\partial \mathbf{w}'}{\partial \boldsymbol{\beta}}\right|_{\boldsymbol{\beta}=\boldsymbol{\beta}(i)} \mathbf{G}(\mathbf{y} - \mathbf{w}(i))\, dt \tag{5.42}$$

Further consideration of this formula shows:

$\mu = 0 \Rightarrow$ pure Gauss–Newton

$\mu = \infty \Rightarrow$ pure steepest descent, step size zero, cf. Figure 5.18.

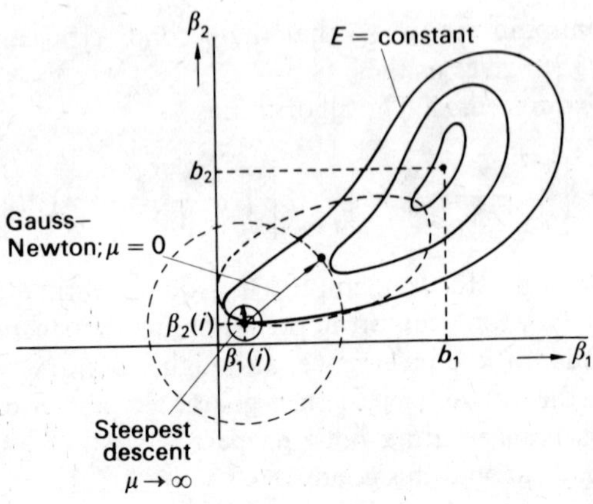


Figure 5.18

By changing $\mu$ for each iteration one can control the convergence properties. A disadvantage is still the evaluation of the (approximated) second derivative of $E$ and the inversion of this matrix, especially, when the number of parameters is large. It would be advisable to compose this inverted matrix from information obtained during iterations of simpler methods. This leads us to:

(g) **Conjugate direction methods.** For the sake of convenience we first observe, that (5.21b) can be written as:

$$E = E^* + \mathbf{a}'\boldsymbol{\beta} + \tfrac{1}{2}\boldsymbol{\beta}'\mathbf{A}\boldsymbol{\beta} \tag{5.43}$$

where

$$\mathbf{A} = \left.\frac{\partial^2 E}{\partial\boldsymbol{\beta}\,\partial\boldsymbol{\beta}'}\right|_{\boldsymbol{\beta}=\mathbf{b}} \qquad \mathbf{a} = -\mathbf{A}\mathbf{b} \qquad E^* = E_0 + \tfrac{1}{2}\mathbf{b}'\mathbf{A}\mathbf{b}$$

so

$$\left.\frac{\partial E}{\partial\boldsymbol{\beta}}\right|_{\boldsymbol{\beta}=\boldsymbol{\beta}(i)} = \mathbf{g}(i) = \mathbf{a} + \mathbf{A}\boldsymbol{\beta}(i)$$

Two directions **u** and **v** are defined to be conjugate with respect to the positive definite matrix **A** if

$$\mathbf{u}'\mathbf{A}\mathbf{v} = 0$$

(note that orthogonal vectors are conjugate with respect to the identity matrix).

An $n \times n$ matrix **A** (positive definite) has at least one set of $n$ independent vectors that are mutually conjugate, as the eigenvectors of **A** form such a set. So in the parameter space the vector $\gamma$ from the initial point to the minimum can be built up by $n$ independent conjugate directions $\mathbf{d}_i$:

$$\gamma = \sum_{i=1}^{n} \alpha_i \mathbf{d}_i$$

Modification of the steepest-descent method by following a conjugate direction instead of the gradient will, as we shall see, guarantee us to reach the minimum in $n$ iterations for an $n$-dimensional quadratic error function. The iteration formula becomes:

$$\boldsymbol{\beta}(i+1) = \boldsymbol{\beta}(i) + \lambda_i \mathbf{d}(i) \tag{5.44}$$

where $\lambda_i$ is found by line minimization along $\mathbf{d}(i)$, which implies:

$$\mathbf{g}'(i+1)\mathbf{d}(i) = 0 \tag{5.45a}$$

As

$$\boldsymbol{\beta}(i+1) = \boldsymbol{\beta}(k+1) + \sum_{j=k+1}^{i} \lambda_j \mathbf{d}(j)$$

expression

$$\mathbf{g}'(i+1)\mathbf{d}(k) = 0 \qquad k = 1, 2, \ldots, i \tag{5.45b}$$

holds too, because:

$$\mathbf{g}(i+1) = \mathbf{A}\boldsymbol{\beta}(i+1) + \mathbf{a} = \mathbf{A}\left\{\boldsymbol{\beta}(k+1) + \sum_{j=k+1}^{i} \lambda_j \mathbf{d}(j)\right\} + \mathbf{a}$$

$$= \mathbf{g}(k+1) + \sum_{j=k+1}^{i} \lambda_j \mathbf{A}\mathbf{d}(j)$$

and

$$\mathbf{d}'(k)\mathbf{g}(i+1) = \mathbf{d}'(k)\mathbf{g}(k+1) + \sum_{j=k+1}^{i} \lambda_j \mathbf{d}'(k)\mathbf{A}\mathbf{d}(j) = 0$$

Consequently the gradient in the last iteration is perpendicular to all already used conjugate directions. When the next conjugate direction $\mathbf{d}(i+1)$ incorporates gradient $\mathbf{g}(i+1)$, we are sure to explore in a new dimension, which explains the hypothesis:

$$\mathbf{d}(i+1) = -\mathbf{g}(i+1) + \gamma_{i+1}\mathbf{d}(i) \tag{5.46}$$

By substitution this hypothesis can easily be verified, which results in:

$$\gamma_{i+1} = \frac{\mathbf{g}'(i+1)\mathbf{g}(i+1)}{\mathbf{g}'(i)\mathbf{g}(i)} \tag{5.47}$$

Starting with the gradient as the first direction after $n$ iterations one possesses $n$ independent conjugate directions. In this last iteration point the gradient has to be perpendicular to all these independent directions according to (5.45b), which is impossible in the $n$-dimensional space. Consequently, the gradient is zero and the minimum is reached. The above algorithm has been developed by Fletcher and Reeves (1964).

Another approach, more frequently applied, is the algorithm of Davidon, (Fletcher and Powell, 1963). The concept of this method is an updating of the inverse second derivative $\mathbf{A}$ in each iteration step. An explicit form for the inverse matrix can be given in terms of the conjugate directions:

$$\mathbf{H} = \sum_{i=1}^{n} \frac{\mathbf{d}(i)\mathbf{d}'(i)}{\mathbf{d}'(i)\mathbf{A}\mathbf{d}(i)} \Rightarrow \mathbf{H}\mathbf{A}\mathbf{d}(k) = \mathbf{d}(k) \qquad \text{for } k = 1, 2, \ldots, n$$

or

$$\mathbf{H}\mathbf{A} = \mathbf{I} \Rightarrow \mathbf{H} = \mathbf{A}^{-1}$$

It can be proved then, that if

$$\mathbf{d}(i) = -\mathbf{H}(i-1)\mathbf{g}(i) \tag{5.48}$$

and

$$\mathbf{H}(i) = \mathbf{H}(i-1) + \lambda_i \frac{\mathbf{d}(i)\mathbf{d}'(i)}{\mathbf{g}'(i)\mathbf{H}(i-1)\mathbf{g}(i)} - \frac{\mathbf{H}(i-1)\mathbf{y}(i)\mathbf{y}'(i)\mathbf{H}(i-1)}{\mathbf{y}'(i)\mathbf{H}(i-1)\mathbf{y}(i)}$$

with

$$\mathbf{y}(i) = \mathbf{g}(i+1) - \mathbf{g}(i)$$

in $n$ iterations $\mathbf{H} = \mathbf{A}^{-1}$, which means that the minimum is reached for a quadratic criterion. A disadvantage of this method is the line minimization that we need to perform in each iteration. Davidon (1968), however, introduced a similar method without line minimization. This algorithm normally reaches the minimum in the same number of iterations, but uses only one function and one gradient per iteration.

All methods mentioned above behave theoretically very well on almost quadratic surface, but there are many other 'smooth' surfaces, where we also can expect a successful application of gradient methods. Such a category is the set of homogeneous functions:

$$E(\boldsymbol{\beta}) = \frac{1}{\gamma}(\boldsymbol{\beta} - b)'\mathbf{g}(\boldsymbol{\beta}) + E_0^* \tag{5.49}$$

where

$$\mathbf{g}(\boldsymbol{\beta}) = \frac{\partial E}{\partial \boldsymbol{\beta}}$$

A promising algorithm for minimizing these homogeneous functions is given by the following method.

(*h*) **Jacobson and Oksman's method (1970).** The idea itself is astonishingly simple: Suppose we know the function and gradient values in $n + 2$ points $\boldsymbol{\beta}(i)$, where (5.49) can be written as:

$$\mathbf{g}'(\boldsymbol{\beta}(i))\mathbf{b} + E(\boldsymbol{\beta}(i))\gamma - \gamma E_0^* = \mathbf{g}'(\boldsymbol{\beta}(i))\boldsymbol{\beta}(i) \qquad i = 1, 2, \ldots, n+2$$

which are $n + 2$ (or more) linear equations in $n + 2$ parameters $\mathbf{b}$, $\gamma$ and $\gamma E_0^*$. When the coefficient matrix can be made regular by well-chosen $\boldsymbol{\beta}(i)$'s, then the solution is obtained by matrix inversion (or pseudo-inversion). Jacobson and Oksman transformed this idea into an updating scheme. For many test functions they claim better results than the Powell, Fletcher, Davidon method.

Such a more rigorous test for convergence algorithms is Rosenbrock's valley, given by the expression

$$E = 100(\beta_1^2 - \beta_2)^2 + (1 - \beta_1)^2$$

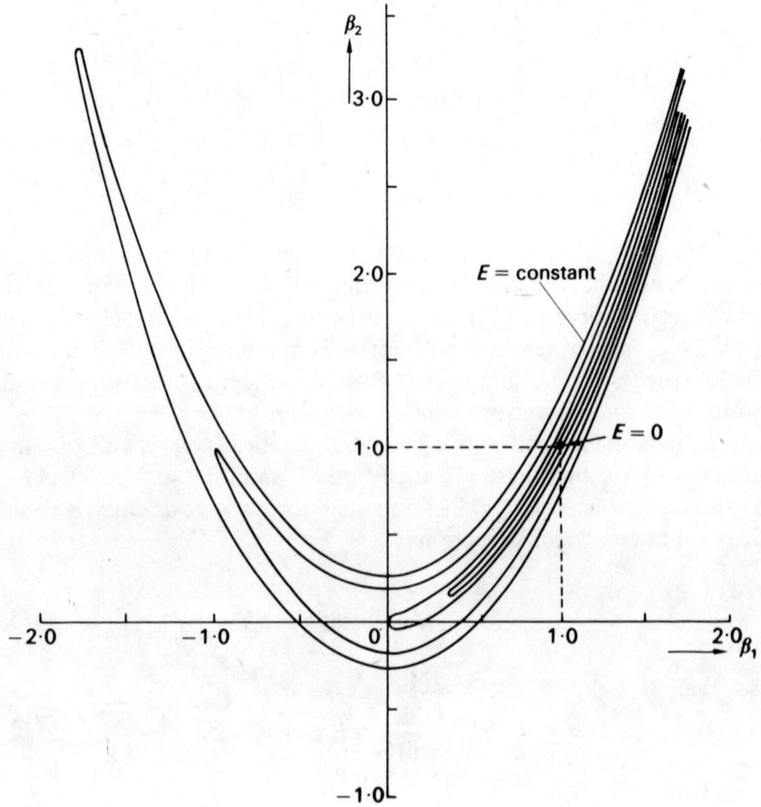


Figure 5.19

and sketched in Figure 5.19 (cf. Rosenbrock and Storey, 1966). In this reference a comparison can be found of several algorithms applied to a number of different 'valleys'.

*Examples.* In order to give the reader a visual impression of the different methods a simulation has been made of a number of these schemes on a small digital computer.* Both a purely quadratic function

$$E = \tfrac{1}{2}\{5\beta_1^2 + 6\beta_1\beta_2 + 5\beta_2^2\}$$

and Rosenbrock's valley have been tested. Every algorithm was allowed to execute 20 iterations. In the Figures 5.20–5.29 the (a) part represents the quadratic function, the (b) part Rosenbrock's valley.

* The author is indebted to Mrs. J. Kregting-Jansen and Messrs. A. A. H. Damen, P. Damman and C. de Groot for this simulation.

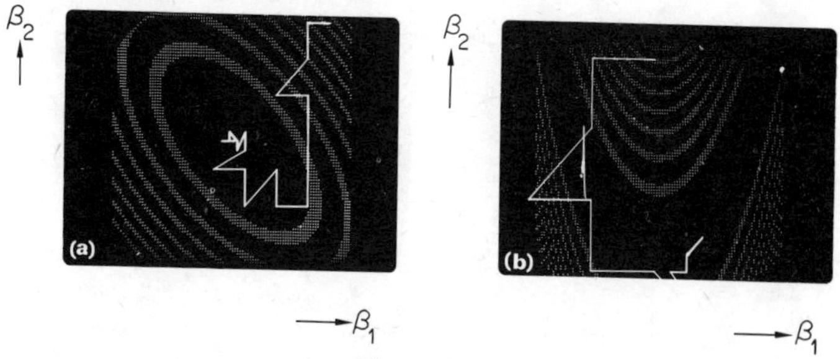


Figure 5.20

*Hooke and Jeeves* (for explanation see Kowalik and Osborne, 1968) initial values:

Figure 5.20a $\left\{\begin{array}{l}\beta_1 = 10\\ \beta_2 = 10\end{array}\right.$ Figure 5.20b $\left\{\begin{array}{ll}\beta_1 = -1{\cdot}2 & 0{\cdot}\\ \beta_2 = 1{\cdot} & 3{\cdot}\end{array}\right.$

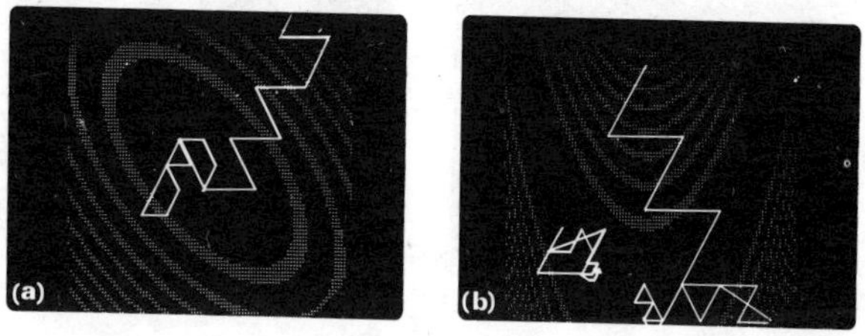


Figure 5.21

*Simplex Method* (for explanation see Box, Davies and Swann, 1969) initial values:

Figure 5.21a $\left\{\begin{array}{l}\beta_1 = 8{\cdot}\\ \beta_2 = 12{\cdot}\end{array}\right.$ Figure 5.21b $\left\{\begin{array}{ll}\beta_1 = -1{\cdot}2 & 0{\cdot}\\ \beta_2 = 1{\cdot} & 3{\cdot}\end{array}\right.$

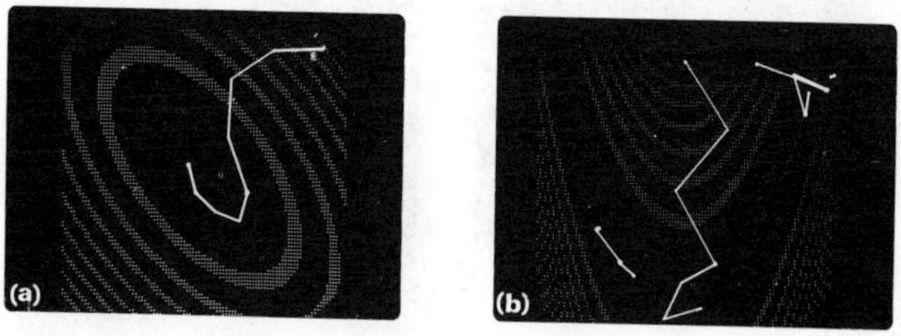


Figure 5.22

*Creeping Random Search* (for explanation see White, 1970) initial values:

Figure 5.22a $\left\{\begin{array}{l}\beta_1 = 10{\cdot}\\ \beta_2 = 8{\cdot}\end{array}\right.$ Figure 5.22b $\left\{\begin{array}{lll}\beta_1 = -1{\cdot}2 & 0{\cdot} & 1{\cdot}\\ \beta_2 = 1{\cdot} & 3{\cdot} & 3{\cdot}\end{array}\right.$

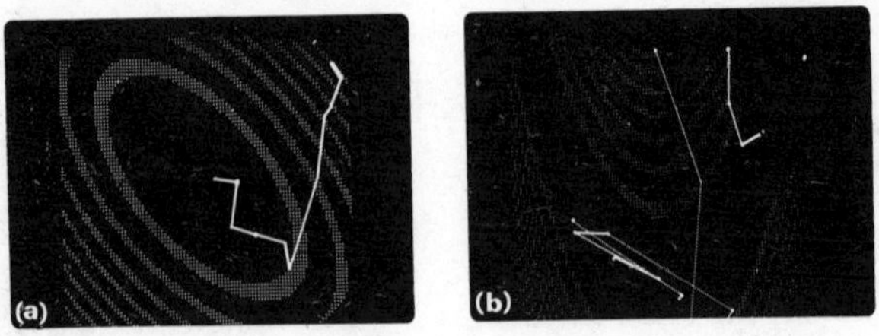


Figure 5.23

*Adaptive Random Search* (for explanation see White, 1970)
initial values:

$$\text{Figure 5.23a}\begin{cases}\beta_1 = 10\cdot \\ \beta_2 = 8\cdot\end{cases} \qquad \text{Figure 5.23b}\begin{cases}\beta_1 = -1{\cdot}2 \quad 0\cdot \quad 1\cdot \\ \beta_2 = 1\cdot \qquad 3\cdot \quad 3\cdot\end{cases}$$

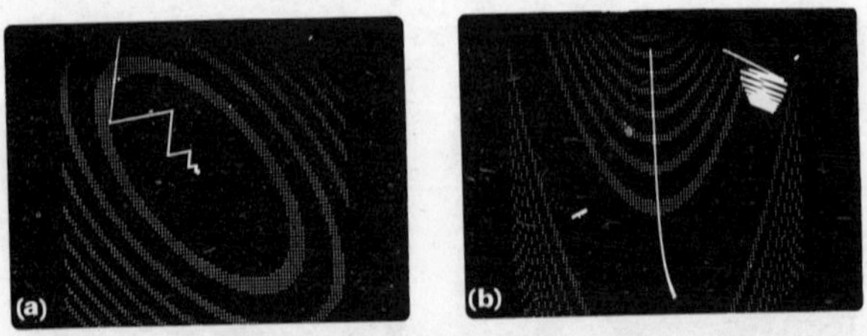


Figure 5.24

*Gradient Methods, Intermittent-adjustment Schemes*
*Steepest Descent*
initial values:

$$\text{Figure 5.24a}\begin{cases}\beta_1 = -6\cdot \\ \beta_2 = 12\cdot\end{cases} \qquad \text{Figure 5.24b}\begin{cases}\beta_1 = -1{\cdot}2 \quad 0\cdot \quad 1\cdot \\ \beta_2 = 1\cdot \qquad 3\cdot \quad 3\cdot\end{cases}$$

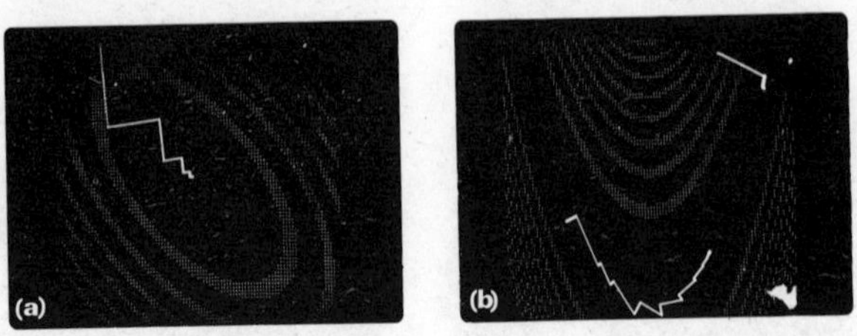


Figure 5.25

*Steepest Descent with Line Minimization*
initial values:

$$\text{Figure 5.25a}\begin{cases}\beta_1 = -8\cdot \\ \beta_2 = 12\cdot\end{cases} \qquad \text{Figure 5.25b}\begin{cases}\beta_1 = -1{\cdot}2 \quad 1\cdot^{*} \\ \beta_2 = 1\cdot \qquad 3\cdot\end{cases}$$

* The successive steps are not perpendicular, because the line minimization was done by third-order interpolation and, in this case, consequently not exact.

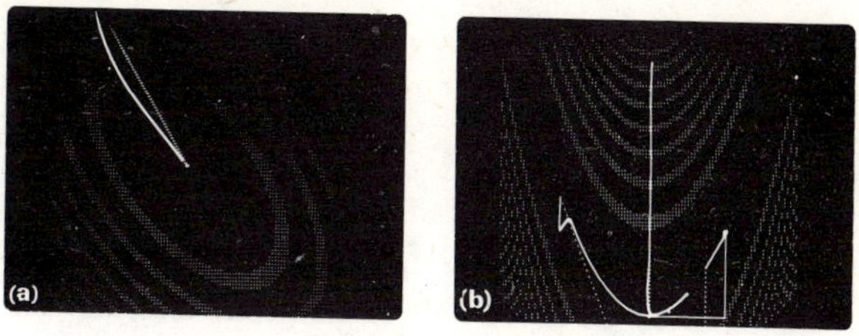


Figure 5.26

*Marquardt Method*
initial values:

$$\text{Figure 5.26a}\begin{cases}\beta_1 = -8\cdot & -8\cdot \\ \beta_2 = 12\cdot & 12\cdot \\ \mu = 100\cdot & 0\cdot^{*}\end{cases} \qquad \text{Figure 5.26b}\begin{cases}\beta_1 = -1\cdot 2 & 0\cdot & -1\cdot 2 & 0\cdot \\ \beta_2 = 1\cdot & 3\cdot & 1\cdot & 3\cdot \\ \mu = 100\cdot & 1000\cdot & 0\cdot\dagger & 0\cdot\dagger\end{cases}$$

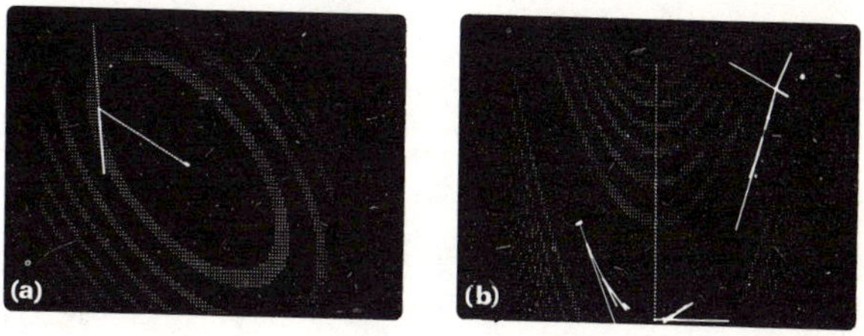


Figure 5.27

*Fletcher Powell method* (for explanation see Fletcher and Powell, 1963)
initial values:

$$\text{Figure 5.27a}\begin{cases}\beta_1 = -8\cdot \\ \beta_2 = 12\cdot\end{cases} \qquad \text{Figure 5.27b}\begin{cases}\beta_1 = -1\cdot 2 & 0\cdot & 1\cdot \\ \beta_2 = 1\cdot & 3\cdot & 3\cdot\end{cases}$$

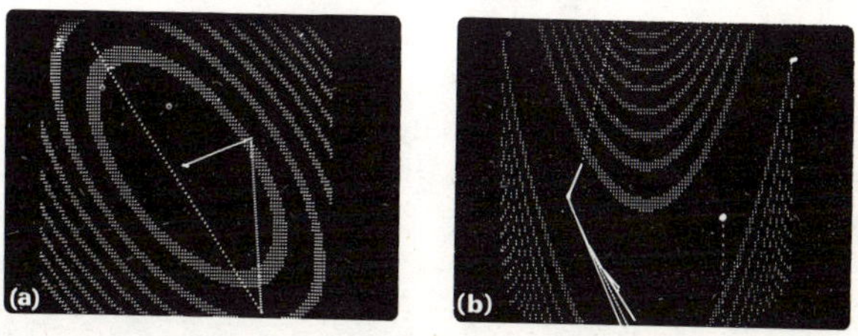


Figure 5.28

*Davidon method* (for explanation see Davidon, 1968)
initial values:

$$\text{Figure 5.28a}\begin{cases}\beta_1 = -8\cdot \\ \beta_2 = 8\cdot 5\end{cases} \qquad \text{Figure 5.28b}\begin{cases}\beta_1 = -1\cdot 2 & 1\cdot \\ \beta_2 = 1. & 3\cdot\end{cases}$$

* The successive steps are not perpendicular, because the line minimization was done by third-order interpolation and, in this case, consequently not exact.

† $\mu = 0$ implies Newton iteration!

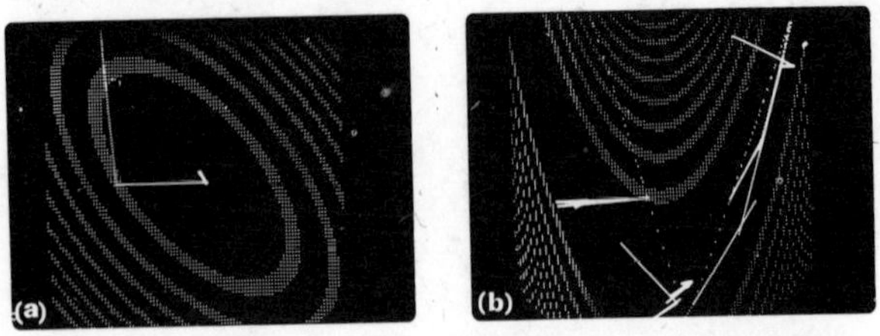


Figure 5.29

*Method of Jacobson and Oksman* (1970)
initial values:

Figure 5.29a $\begin{cases}\beta_1 = -8\cdot \\ \beta_2 = 12\cdot\end{cases}$ Figure 5.29b $\begin{cases}\beta_1 = -1\cdot 2 & 1\cdot \\ \beta_2 = 1\cdot & 3\cdot\end{cases}$

The reader is advised to go over the discussion of adjustment schemes again, using the illustrations provided by the previous figures.

## 5.3 Stochastic Convergence Schemes; Stochastic Approximation

The discussion of convergence schemes in Section 5.2 did not take into account the disturbances introduced by additive noise in the determination of the gradient. Stochastic approximation, as discussed in this section, can be used for several types of problems connected with optimization, adaptation, model adjustment and learning under the influence of noise (Tsypkin, 1966).

Robbins and Monro (1951) introduced an approach to the problem of finding a point of a regression function. They start with the conditional probability function $p(y|x)$ or distribution function $F(y|x)$ and assume that:

the regression function $f(x)$ exists;

the equation $f(x) = \lambda$ has a unique solution $x = \theta$.

The latter condition implies that $f(x)$ is a monotonic function.

Now the following iterative algorithm is proposed for determining the root $x = \theta$ of the equation $f(x) = \lambda$:

$$x(i+1) = x(i) - \Gamma(i)\{f(x(i)) - \lambda\} \tag{5.50}$$

where $x(i)$ is the $i$th estimate and the observation $f(x(i))$ are contaminated with an **additive** random disturbance; cf. Figure 5.30; $\Gamma(i)$ with $i = 1, 2, \ldots,$

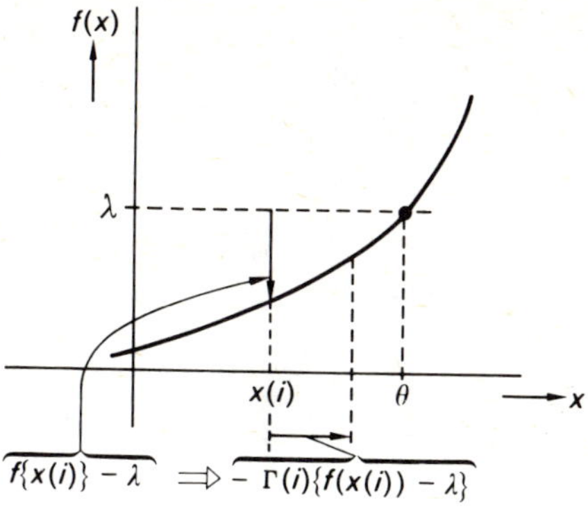


Figure 5.30

is a sequence of numbers with properties that still have to be determined. This sequence has to be chosen in such a way that $x(i)$ converges in probability to $\theta$, i.e. for all $\epsilon > 0$:

$$\lim_{i \to \infty} P[|x(i) - \theta| \geqslant \epsilon] = 0 \tag{5.51}$$

The following reasoning indicates the restrictions that result from this condition. Consider $\{x(i+1) - \theta\}^2$ together with equation (5.50) and take the expectation:

$$\begin{aligned} \mathscr{E}[\{x(i+1) - \theta\}^2] = \mathscr{E}[\{x(i) - \theta\}^2] + \Gamma^2(i)\mathscr{E}[\{f(x(i)) - \lambda\}^2] \\ - 2\Gamma(i)\mathscr{E}[\{x(i) - \theta\}\{f(x(i)) - \lambda\}] \end{aligned} \tag{5.52}$$

or

$$b(i+1) = b(i) + \Gamma^2(i)e(i) - 2\Gamma(i)d(i) \tag{5.52a}$$

with

$$\begin{aligned} b(i) &= \mathscr{E}[\{x(i) - \theta\}^2] \geqslant 0 \\ e(i) &= \mathscr{E}[\{f(x(i)) - \lambda\}^2] \geqslant 0 \\ d(i) &= \mathscr{E}[\{x(i) - \theta\}\{f(x(i)) - \lambda\}] \\ \Gamma(i) &> 0 \end{aligned}$$

For the existence of $e(i)$ it is required that $f(x(i))$ is finite except possibly at a finite number of points. We require $b(i+1) < b(i)$. So $d(i) > 0$. This implies that $f(x)$ has to be monotonic at least in the region of $\theta$. Writing the expressions for $b(2), b(3), \ldots, b(i+1)$ yields:

$$b(i+1) = b(1) + \sum_{j=1}^{i} \Gamma^2(j)e(j) - 2\sum_{j=1}^{i} \Gamma(j)\,d(j) \tag{5.53}$$

or

$$\begin{aligned}\sum_{j=1}^{i} \Gamma(j)d(j) &= \frac{1}{2}\left[b(1) + \sum_{j=1}^{i} \Gamma^2(j)e(j) - b(i+1)\right] \\ &\leqslant \frac{1}{2}\left[b(1) + \sum_{j=1}^{i} \Gamma^2(j)e(j)\right]\end{aligned} \tag{5.54}$$

Under the condition that $e(j)$ is finite and

$$\sum_{j=1}^{\infty} \Gamma^2(j) < \infty \quad \text{one finds} \quad \sum_{j=1}^{\infty} \Gamma(j)d(j) < \infty \tag{5.55}$$

If $\Gamma(j)$ is chosen such that

$$\sum_{j=1}^{\infty} \Gamma(j) = \infty \quad \text{then} \quad \lim_{j\to\infty} d(j) = 0 \tag{5.56}$$

Consequently

$$\lim_{i\to\infty} \mathscr{E}[\{x(i) - \theta\}\{f(x(i)) - \lambda\}] = 0 \tag{5.57}$$

Development of $f(x(i))$ in the vicinity of $x = \theta$ yields:

$$f(x(i)) = f(\theta) + \frac{f'(\theta)}{1!}\{x(i) - \theta\} + 0\{x(i) - \theta\}^2 \tag{5.58}$$

or

$$f(x(i)) - \lambda = \frac{f'(\theta)}{1!}\{x(i) - \theta\} + 0\{x(i) - \theta\}^2 \tag{5.59}$$

By substitution in equation (5.57)

$$\lim_{i\to\infty} \mathscr{E}[f'(\theta)\{x(i) - \theta\}^2] = 0 \tag{5.60}$$

or

$$\lim_{i\to\infty} P[|x(i) - \theta| \geqslant \epsilon] = 0 \tag{5.61}$$

Sequences that obey the conditions (5.55) and (5.56) are $\Gamma(j) = c/j^{\alpha}$ with $0{\cdot}5 < \alpha \leqslant 1$.

As mentioned this method was developed originally for determining the root of an equation, consisting of a monotonic function. The method has been applied by Kiefer and Wolfowitz (1952) to determine the extremum of a regression function. As the regression function itself it unknown the derivative is estimated by

$$\nabla_x f(x(i)) \approx \frac{f(x(i) + a(i)) - f(x(i) - a(i))}{2a(i)} \tag{5.62}$$

By this operation the 'even' function is replaced by an 'odd' function, cf. Chapter 2. The algorithm for the determination of a minimum is:

$$x(i + 1) = x(i) - \Gamma(i)\nabla_x f(x(i)) \tag{5.63}$$

This algorithm converges if all conditions with respect fo $f(x)$ and $\Gamma(i)$ are satisfied. Kiefer and Wolfowitz also have extended this method to more-dimensional and continuous cases.

The multidimensional case can be formulated in the following way. The function

$$F(\mathbf{x}) = F(x_1, x_2, \ldots, x_N) \tag{5.64}$$

has to be maximized (or minimized) by a suitable choice of $\mathbf{x}$. Assume the maximum occurs for $\mathbf{x} = \hat{\mathbf{x}}$. If $F(\mathbf{x})$ is known and if there are no extremal boundary values, saddle points etc., then $\hat{\mathbf{x}}$ follows from

$$\nabla_{\mathbf{x}} F(\mathbf{x}) = \left\{ \frac{\partial F(\mathbf{x})}{\partial x_1}, \ldots, \frac{\partial F(\mathbf{x})}{\partial x_N} \right\}' = \mathbf{0} \qquad \text{for } \mathbf{x} = \hat{\mathbf{x}}$$

For determining $\hat{\mathbf{x}}$ the following algorithm is suitable:

$$\mathbf{x}(i + 1) = \mathbf{x}(i) - \Gamma(i)\nabla_{\mathbf{x}} F(\mathbf{x}(i)) \tag{5.65}$$

The choice of $\Gamma(i)$ determines the iteration method. If the convergence conditions are satisfied then

$$\mathbf{x}(i) \to \hat{\mathbf{x}} \qquad \text{for } i \to \infty$$

In cases that $F(\mathbf{x})$ is not known the gradient can be estimated by:

$$\begin{aligned} \nabla_{\mathbf{x}} F(\mathbf{x}) &\approx \Delta F'(\mathbf{x}, \pm a) \\ &= \left\{ \frac{F(\mathbf{x} + a\mathbf{e}_1) - F(\mathbf{x} - a\mathbf{e}_1)}{2a}, \ldots, \frac{F(\mathbf{x} + a\mathbf{e}_N) - F(\mathbf{x} - a\mathbf{e}_N)}{2a} \right\}' \end{aligned} \tag{5.66}$$

where $\mathbf{e}_1, \ldots, \mathbf{e}_N$ are the basis vectors and $a$ is a small scalar value; $a$ can be made dependent on the iteration number and on the vector component $x_j$. If in addition $F(\mathbf{x})$ is of a stochastic nature then this stochastic aspect can be expressed by the function $F(\mathbf{n}|\mathbf{x})$ where $\mathbf{n}$ is a stochastic vector. The best thing to do in such cases, is to maximize (or minimize) the expected value

$$\int_{\mathbf{n}} F(\mathbf{n}|\mathbf{x})p(\mathbf{n})\,d\mathbf{n} = \mathscr{E}_{\mathbf{n}}[F(\mathbf{n}|\mathbf{x})] \tag{5.67}$$

Then $\mathbf{x} = \hat{\mathbf{x}}$ follows from

$$\nabla_{\mathbf{x}}\mathscr{E}_{\mathbf{n}}[F(\mathbf{n}|\mathbf{x})]|_{\mathbf{x}=\hat{\mathbf{x}}} = \mathbf{0}$$

and the algorithm is

$$\mathbf{x}(i+1) = \mathbf{x}(i) - \Gamma(i)\nabla_{\mathbf{x}}\mathscr{E}_{\mathbf{n}}[F(\mathbf{n}|\mathbf{x}(i))] \tag{5.68}$$

If $p(\mathbf{n})$ is not given beforehand and if $F(\mathbf{x})$ is unknown then we have to apply an estimate of the gradient as used in the next expression:

$$\mathbf{x}(i+1) = \mathbf{x}(i) - \Gamma(i)\,\frac{F(\mathbf{n}|\mathbf{x}(i) + \mathbf{a}(i)) - F(\mathbf{n}|\mathbf{x}(i) - \mathbf{a}(i))}{2\mathbf{a}(i)} \tag{5.69}$$

Now $\mathbf{x}(i) \to \hat{\mathbf{x}}$, i.e.

$$\lim_{i\to\infty} \mathscr{E}[|\mathbf{x}(i) - \hat{\mathbf{x}}|^2] = 0 \tag{5.70}$$

if

$$\sum_{i=1}^{\infty} \Gamma(i) = \infty \qquad \sum_{i=1}^{\infty} \frac{\Gamma^2(i)}{a^2(i)} < \infty \qquad \sum_{i=1}^{\infty} \Gamma(i)a(i) < \infty$$

These conditions imply that $\Gamma(i)$ has to decrease fast enough, in order that the variance disappears. On the other hand $\Gamma(i)$ may not decrease too fast, because otherwise not a sufficient amount of data is being processed. If the process is deterministic, then one can drop the requirements with respect to $\Gamma(i)$ and $\Gamma$ can be chosen to be constant.

*Example.* As an illustration of the capabilities of the stochastic approximation approach, the results of a simple experiment* are shown in the Figures 5.31 and 5.32. The function $F$ has been chosen to be one dimensional;

$$F(n(i)|x(i)) = 1 + x^2(i) + \lambda n(i)$$

$$p(n) = 1 \qquad \text{for } 0 \leqslant n \leqslant 1$$

$$a(i) = a = 0{\cdot}5$$

For a low noise level, $\lambda = 0{\cdot}1$, the convergence to the right value, $x = 0$, is shown in Figure 5.31. $\Gamma = 0{\cdot}5$ results in a one-step approach; $\Gamma = 0{\cdot}9$ leads to an oscillatory behaviour; $\Gamma = 1/i$ shows a convergence in two steps. In case of a high noise level, $\lambda = 10$, two approaches are given in Figures 5.32a and b for $\Gamma = 0{\cdot}5$ viz. $\Gamma = 1/i$. Each figure shows the behaviour for two different noise sequences $n(i)$. In case of the constant $\Gamma$ there is no convergence to $x = 0$. The stochastic approximation algorithm with $\Gamma = 1/i$ converges satisfactorily in spite of the same noise.

* By Mr. J. P. C. M. de Weerd, University of Technology, Eindhoven, Netherlands.

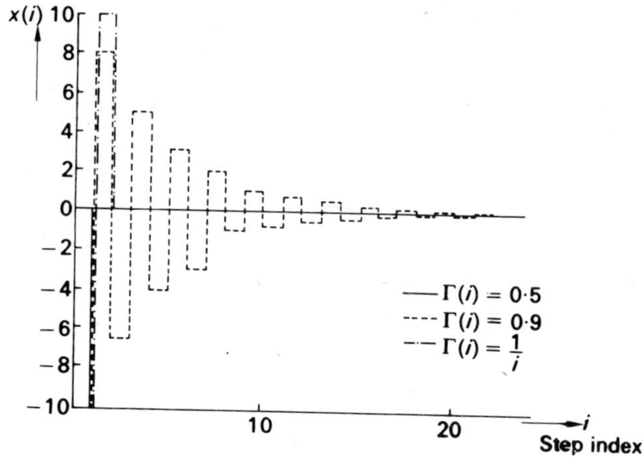


Figure 5.31

Some of these algorithms can be made continuous in time:

$$\frac{\mathrm{d}\mathbf{x}(t)}{\mathrm{d}t} = -\Gamma(t)\nabla_{\mathbf{x}}F\{\mathbf{x}(t)\} \tag{5.71}$$

for the non-stochastic case and

$$\frac{\mathrm{d}\mathbf{x}(t)}{\mathrm{d}t} = -\Gamma(t)\nabla_{\mathbf{x}}F\{\mathbf{n}(t)|\mathbf{x}(t)\} \tag{5.72}$$

for the stochastic case.

For more information on stochastic approximation and on different types of application the reader is referred to the excellent surveys available. See e.g. Albert and Gardner (1967) and Tsypkin (1966). Some applications are given by Sakrison (1967), Saridis and Stein (1968a, 1968b), Holmes (1968), Elliott and Sworder (1969), Neal and Bekey (1969).

## 5.4 Conclusion

In this chapter some essential notions have been summarized. In each of the sections the emphasis has been laid on presenting a coherent picture and not on an elaborate exposition including mathematical intricacies. With the references and additional literature given the reader may dig deeper if he so desires.

This chapter, together with the previous ones on Signals and Process Models constitute the *Prolegomena*, the knowledge that is desirable as a prerequisite for discussing estimation procedures. At this point we are in a fair position to start that discussion.

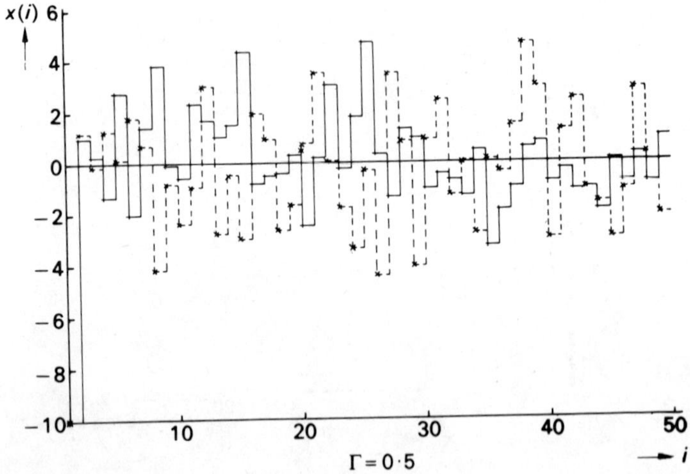


Γ = 0·5

(a)

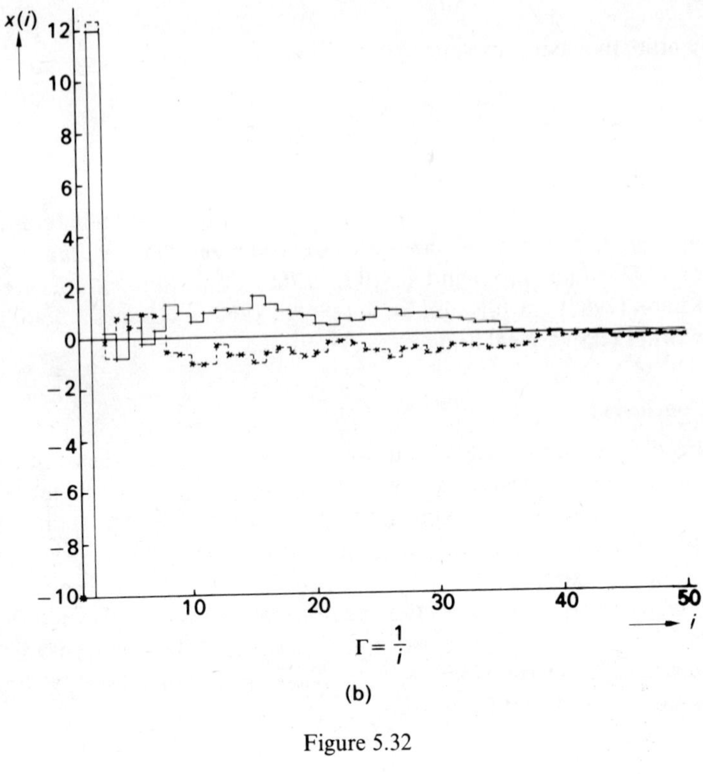


$\Gamma = \frac{1}{i}$

(b)

Figure 5.32

## References

Albert, A. E. and L. A. Gardner, Jr. (1967). *Stochastic Approximation and Nonlinear Regression*, M.I.T. Press, Cambridge, Mass., 204 pp.

Bekey, G. A. and R. B. McGee (1964). 'Gradient methods for the optimization of dynamic system parameters by hybrid computation', in: *Computing methods in Optimization Problems* (Eds. A. V. Balakrishnan and L. W. Neustadt), Academic Press, New York, pp. 305–327.

Box, M. J., Davies and W. H. Swann (1969). *Non-linear Optimization Techniques*, Oliver and Boyd, Edinburgh, 60 pp.

Davidon, W. C. (1968). 'Variance algorithm for minimization', *Computer J.*, **10**, 406–410.

Deutsch, R. (1965). *Estimation Theory*, Prentice-Hall, Englewood Cliffs, N.J., 269 pp.

Dixon, L. C. W. (1972). *Nonlinear Optimization*, English Universities Press, London, 224 pp.

Elliott, D. F. and D. D. Sworder (1969). 'Applications of a simplified multidimensional stochastic approximation algorithm', *Joint Autom. Control Conf.*, 148–154.

Fletcher, R. and M. J. D. Powell (1963). 'A rapidly convergent descent method for minimization', *Computer J.*, **6,** 163–168.

Fletcher, R. and C. M. Reeves (1964). 'Function minimization by conjugate gradients', *Computer J.*, **7,** 149–154.

Holmes, J. K. (1968). 'System identification from noise corrupted measurements', *J. Optimizat. Theory and Applic.*, **2,** 103–116.

Jacobson, D. H. and W. Oksman (1970). 'New algorithms for function minimization', *IEEE Symp. on Adaptive Processes, Decision and Control*, Univ. of Texas, Austin, December 7–9, page XXI, 1.1–4.

Kiefer, J. and J. Wolfowitz (1952). 'Stochastic estimation of the maximum of a regression function', *Ann. Math. Stat.*, **23,** 462–466.

Kowalik, J. and M. R. Osborne (1968). *Methods for Unconstrained Optimization Problems*, American Elsevier Publ., New York, 206 pp.

Lindley, D. V. (1965). *Introduction to Probability and Statistics, from a Bayesian Viewpoint*, Cambridge Univ. Press, London.

Maslov, E. P. (1963). 'Application of the theory of statistical decisions to the estimation of object parameters', *Automn. Remote Control*, **24,** 1214–1226.

Nahi, N. E. (1969). *Estimation Theory and Applications*, Wiley, New York, 280 pp.

Neal, C. B. and G. A. Bekey (1969). cf. references Chapter 7.

Powell, M. J. D. (1970). 'A survey of numerical methods for unconstrained optimization', *SIAM Rev.*, **11,** 79–97.

Robbins, H. and S. Monro (1951). 'A stochastic approximation method', *Ann. Math. Stat.*, **22,** 400–407.

Rosenbrock, H. H. and C. Storey (1966). *Computational Techniques for Chemical Engineers*, Pergamon, Oxford, 328 pp.

Sakrison, D. J. (1967). 'The use of stochastic approximation to solve the system identification problem', *IEEE Trans. Autom. Control*, **AC-12,** 563–567.

Saridis, G. N. and G. Stein (1968a). 'Stochastic approximation algorithms for linear discrete-time system identification', *IEEE Trans. Autom. Control*, **AC-13,** 515–523.

Saridis, G. N. and G. Stein (1968b). 'A new algorithm for linear system identification (correspondence)', *IEEE Trans. Autom. Control*, **AC-13,** 592–594.

Tsypkin, Y. Z. (1966). cf. references Chapter 1.

White, R. (1970). 'A survey of random methods for parameter optimization', *Report, E.E. Dept., University of Technology, Eindhoven*, Netherlands.

Wilde, D. J. (1964). *Optimum Seeking Methods*, Prentice-Hall, Englewood Cliffs, N.J., 202 pp.

Wilde, D. J. and C. S. Beightler (1967). *Foundations of Optimization*, Prentice-Hall, Englewood Cliffs, N.J., 480 pp.

## Additional Literature

### *Deterministic Convergence Schemes*

Bryson, A. E. and W. Denham (1962). 'A steepest ascent method for solving optimum programming problems', *J. Appl. Mech.*, **29**, 247–257.

Goldstein, A. A. (1965). 'On steepest descent', *SIAM J. Control*, **3**, 147–151.

Goldstein, A. A. (1966). 'Minimizing functionals on normed linear spaces', *SIAM J. Control*, **4**, 81–89.

Goodwin, G. C. (1969). 'Application of curvature methods to parameter and state estimation', *Proc. IEE*, **116**, 1107–1110.

Hestenes, M. R. and E. Stiefel (1952). 'Methods of conjugate gradients for solving linear systems', *J. Res. N.B.S.*, **49**, 409.

Kelley, H. J. (1962). 'Methods of gradients', in *Optimization Techniques* (Ed. G. Leitmann), Academic Press, New York.

Kopp, R. E. and R. McGill (1964). 'Several trajectory optimization techniques', in *Computation Methods in Optimization Problems* (Eds. A. V. Balakrishnan and L. W. Neustadt), Academic Press, New York, pp. 65–89.

Lasdon, L. S., S. K. Mitter and A. D. Warren (1967). 'Conjugate gradient method for optimal control problems', *IEEE Trans. Autom. Control*, **AC-12**, 132–138.

Leitmann, G. (Ed.) (1962). *Optimization Techniques*, Academic Press, New York.

Ortega, J. M. and W. C. Rheinholdt (1970). *Iterative Solution of Nonlinear Equations in Several Variables*, Academic Press, New York, 572 pp.

Rosenbrock, H. H. (1960). 'An automatic method for finding the greatest or least value of a function', *Computer J.*, **3**, 175–184.

Saaty, T. L. and J. Bram (1964). *Nonlinear Mathematics*, McGraw-Hill, New York, 381 pp.

Sorenson, H. W. (1969). 'Comparison of some conjugate direction procedures for function minimization', *J. Franklin Inst.*, **288**, 421–441.

Voronov, A. A. and M. B. Ignatjev (1963). 'On the searching of extrema of functions in automatic control systems', *2nd IFAC Congr., Basle*, 459–464.

### *Stochastic Approximation*

Blum, J. A. (1954). 'Multidimensional stochastic approximation methods', *Ann. Math. Stat.*, **25**, 737–744.

Blum, J. R. (1954). 'Approximation methods which converge with probability one', *Ann. Math. Stat.*, **25**, 382–386.

Bojanov, E. S. (1967). 'Application of stochastic approximation method for object characteristic regeneration', *Automn. Remote Control*, **28**, 95–103.

Chung, K. L. (1954). 'On a stochastic approximation method', *Ann. Math. Stat.*, **25**, 463–483.

Derman, C. (1956). 'An application of Chun's lemma to the Kiefer–Wolfowitz stochastic approximation procedure', *Ann. Math. Stat.*, **27**, 532–536.

Derman, C. and J. Sacks (1959). 'On Dvoretsky's stochastic approximation theorem', *Ann. Math. Stat.*, **30**, 601–606.

Driml, M. and T. Nedoma (1959). 'Stochastic approximations for continuous random processes', *Trans. of the Second Prague Conference on Information Theory, Statistical Decision Functions, Random Processes, Prague.*

Dvoretsky, A. (1956). 'On stochastic approximation', *Proc. Third Berkeley Symp. Math. Statist. Probab.* (Ed. J. Neyman), Univ. of California Press, Berkeley, Cal., pp. 39–55.

Gladysev, E. G. (1965). 'On the stochastic approximation', *The Theory of Probability and its Applications*, Vol. X, p. 2.

Holmes, J. K. (1967). 'System identification with noisy measurements', *IFAC Symp.—Identification in Autom. Control Systems, Prague*, paper 3.3.

Holmes, J. K. (1969). 'Two stochastic approximation procedures for identifying linear systems' (short paper), *IEEE Trans. Autom. Control*, **AC-14,** 292–295.

Imedadze, V. V. and Sh. G. Lelashvili (1967). 'Investigation of some algorithms of stochastic approximation in identification of multidimensional plants', *IFAC Symp.—Identification in Autom. Control Systems, Prague*, paper 3.1.

Jury, E. I. and K. G. Oza (1968). 'System identification and the principle of random contraction mapping', *SIAM J. Control*, **6,** 244–257.

Kesten, H. (1958), 'Accelerated stochastic approximation', *Ann. Math. Stat.*, **29,** 41–59.

Panuska, V. (1968). 'A stochastic approximation method for identification of linear systems using adaptive filtering', *Joint Autom. Control Conf.*, 1014–1021.

Popkov, Yu. S. (1967). 'Restoration of dynamic characteristics of nonlinear plants in the presence of noise', *IFAC Symp.—Identification in Autom. Control Systems, Prague*, paper 3.15.

Roberts, J. D. (1965). 'Extremum or hill climbing regulation: a statistical theory involving lags, disturbances and noise', *Proc. IEE*, **112,** 137–150, 1646–1650.

Sacks, J. (1958). 'Asymptotic distribution of stochastic approximation procedures', *Ann. Math. Stat.*, **29,** 373–405.

Sebestyen, G. S. (1962). *Decision-making Processes in Pattern Recognition*, Macmillan, New York, 162 pp.

Wolfowitz, J. (1952). 'On the stochastic approximation method of Robbins and Monroe', *Ann. Math. Stat.*, **23,** 457–461.

Wolfowitz, J. (1956). 'On stochastic approximation methods', *Ann. Math. Stat.*, **27,** 1151–1155.

Zhivoglyadov, V. P. and V. Kh. Kaipov (1966). 'On application of stochastic approximations method in the identification problem', *Automn. Remote Control*, **27,** 1702–1706.

*CHAPTER 6*

# *Sampled Signals; Explicit Methods*

In Chapter 2 the least-squares and the generalized least-squares estimators were introduced. In Chapter 5 they were 'derived' from the maximum likelihood approach under special conditions. This was done in order to show the relationship between the various estimation procedures.

In this chapter, these estimators are derived starting from the concepts of regression analysis. Properties such as linearity, unbiasedness and minimum variance are discussed (Section 6.1).

Next, the implementation of such estimators is considered, (Section 6.2), as well as the types of errors which occur in practical applications: noise error, truncation error, state-correspondence error, errors due to the implementation, sampling error (Section 6.3). In the estimation procedure the residuals can be used to derive more information, for example, about the noise variance and the number of process parameters; this procedure is treated in Section 6.4.

An attempt to use these explicit methods on the class of generalized models leads to estimates which generally are biased. An asymptotic expression is derived for this bias. A few methods for circumventing this problem are outlined in Section 6.5.

## 6.1 Regression Analysis

### *6.1.1 Regression Curves and Planes*

The elementary notions of regression analysis (Cramér, 1946; Aitken, 1952) can be given as follows using a two-dimensional case.

Let $x$ and $y$ be random variables with a joint probability-density function $p(x, y)$. If this function is continuous in $y$ then the conditional probability-density function of $y$ with respect to $x$ can be written as:

$$p(y|x) = \frac{p(x, y)}{\int_{-\infty}^{+\infty} p(x, y)\,dy} = \frac{p(x, y)}{p(x)} \tag{6.1}$$

From this function we want to determine some functional relation, for example, the expectation of $y$ for any value of $x$:

$$y_x = \mathscr{E}[y|x] = \frac{\int_{-\infty}^{+\infty} yp(x, y)\,dy}{p(x)} = f(x) \tag{6.2}$$

That function is called a **regression curve** of $y$ on $x$. It can be shown (Aitken, 1952, p. 96) that this regression curve gives the best possible estimation of $y$ in the mean-squared error sense.

In this way, one determines the functional relationship

$$y_x = f(x)$$

without restrictions on $f$. We may, however, want this function to belong to a certain class, for example, the class of all linear functions or the class of all functions of a given polynomial form. Choosing such a class and minimizing the expectation in the mean-squares sense leads to **least-squares regression curves.**

*Example.* Let us assume that we want to find the best linear function

$$w = \alpha + \beta x \tag{6.3}$$

This is called a **linear regression model;** $\alpha$ and $\beta$ are called the regression coefficients. Then we have to minimize

$$\mathscr{E}[(y - w)^2] = \mathscr{E}[(y - \alpha - \beta x)^2] \tag{6.4}$$

The following notation will be useful, cf. Appendix B:

$$\mu_x = \mathscr{E}[x]$$

$$\mu_y = \mathscr{E}[y]$$

$$\sigma_x^2 = \mathscr{E}[x^2] - \mathscr{E}^2[x]$$

$$\sigma_{xy}^2 = \mathscr{E}[xy] - \mathscr{E}[x]\mathscr{E}[y]$$

$$\rho = \frac{\sigma_{xy}^2}{\sigma_x \sigma_y} \quad \text{correlation coefficient}$$

Differentiation of equation (6.4) with respect to $\alpha$ and with respect to $\beta$ and equating to zero leads to

$$\left.\begin{aligned} -2\mathscr{E}[y - \alpha - \beta x] &= -2\mathscr{E}[y] + 2\alpha + 2\beta\mathscr{E}[x] = 0 \\ -2\mathscr{E}[x(y - \alpha - \beta x)] &= -2\mathscr{E}[xy] + 2\alpha\mathscr{E}[x] + 2\beta\mathscr{E}[x^2] = 0 \end{aligned}\right\} \tag{6.5}$$

From this it follows that:

$$\alpha = \mu_y - \mu_x \frac{\sigma_{xy}^2}{\sigma_x^2}$$

$$\beta = \frac{\sigma_{xy}^2}{\sigma_x^2}$$

Substitution of the expressions for $\alpha$ and $\beta$ in equation (6.3) yields the equation of the regression line:

$$w = \mu_y + \frac{\sigma_{xy}^2}{\sigma_x^2}(x - \mu_x) = \mu_y + \rho\frac{\sigma_y}{\sigma_x}(x - \mu_x) \tag{6.6}$$

The reader may wish to apply these ideas to a bivariate normal probability-density function $p(x, y)$.

These ideas can be generalized so as to find the best non-linear regression model, for example, the polynomial functions $w = \alpha + \beta x + \gamma x^2 + \cdots$. It may be useful to choose a set of orthonormal polynomials, cf. Aitken (1952), p. 115. These polynomials $\pi_i$ are of the degree $i$ in $x$ $(i = 0, 1, \ldots, m)$ and fulfil the conditions of orthonormality:

$$\int_{-\infty}^{+\infty} \pi_i(x)\pi_j(x)p(x)\,dx = \begin{cases} 1 & \text{for } i = j \\ 0 & \text{for } i \neq j \end{cases} \tag{6.7}$$

where

$$p(x) = \int_{-\infty}^{+\infty} p(x, y)\,dy$$

is the marginal probability density function of $x$. If now

$$w = \sum_{i=0}^{m} \beta_i \pi_i(x) \tag{6.8}$$

then we have to minimize

$$\mathscr{E}\left[\left\{y - \sum_{i=0}^{m} \beta_i \pi_i(x)\right\}^2\right] \tag{6.9}$$

By differentiation with respect to $\beta_j$ one obtains:

$$\begin{aligned}\mathscr{E}\left[\pi_j(x)\left\{y - \sum_{i=0}^{m} \beta_i \pi_i(x)\right\}\right] &= \int_{-\infty}^{+\infty}\int_{-\infty}^{+\infty} y\pi_j(x)p(x, y)\,dx\,dy \\ &\quad - \sum_{i=0}^{m} \beta_i \int_{-\infty}^{+\infty} \pi_j(x)\pi_i(x)p(x)\,dx\end{aligned} \tag{6.10}$$

This has to be equated to zero for $\beta_j = \hat{\beta}_j$. As a result of the orthonormality, this reduces to:

$$\hat{\beta}_j = \int_{-\infty}^{+\infty}\int_{-\infty}^{+\infty} y\pi_j(x)p(x, y)\,dx\,dy \tag{6.11}$$

Consequently, there is no need to solve a set of linear equations. This is particularly advantageous since the linear equations often become nearly singular as the order of the model increases (Hamming).

These ideas can be extended to the more-dimensional case where a conditional probability-density function of the form

$$p(y|x_1, x_2, \ldots, x_m) = p(y|\mathbf{x}) \tag{6.12}$$

is considered. Again the class of functions may be restricted, for example, to the linear ones. Then the choice of

$$w = \alpha + \beta_1 x_1 + \cdots + \beta_m x_m \tag{6.13}$$

leads to a **mean-square regression plane** for $y$ with respect to $x_1, \ldots, x_m$ if the following expectation is minimized:

$$\mathscr{E}[(y - w)^2] = \mathscr{E}[(y - \alpha - \beta_1 x_1 - \cdots - \beta_m x_m)^2] \tag{6.14}$$

If all random variables are assumed to be centred, i.e. with zero expectation, then the coefficient $\alpha$ can be dropped. Differentiating with respect to $\beta_1, \ldots, \beta_m$ respectively, and equating to zero for $\beta_1 = \hat{\beta}_1$, etc. leads to:

$$\begin{matrix} \hat{\beta}_1 \mathscr{E}[x_1^2] + \hat{\beta}_2 \mathscr{E}[x_1 x_2] + \cdots + \hat{\beta}_m \mathscr{E}[x_1 x_m] = \mathscr{E}[x_1 y] \\ \vdots \\ \hat{\beta}_1 \mathscr{E}[x_m x_1] + \hat{\beta}_2 \mathscr{E}[x_m x_2] + \cdots + \hat{\beta}_m \mathscr{E}[x_m^2] = \mathscr{E}[x_m y] \end{matrix} \tag{6.15}$$

If the matrix of expectations is not singular this set of linear equations can be solved for the regression coefficients $\hat{\beta}_1, \ldots, \hat{\beta}_m$. When $w$ is chosen as a more complicated function than polynomials (e.g. a set of exponentials), then the resulting expressions for the coefficients are often quite difficult and tedious to obtain.

### 6.1.2 *Estimation from a Finite Number of Observations*

So far we have assumed that we can determine all $\mathscr{E}[\ ]$ operations, i.e. that the joint probability-density function $p(x_1, \ldots, x_m, y)$ is known. This will seldom be the case. Then we will have to estimate the parameters using a **finite** number of observations, viz. the sample values. Consequently, the estimate has to be some function of these sample values, which are actually observed values of a random variable. This implies that the estimate is also a random variable and can be characterized by a probability-density function. The quality of the estimate depends on this function, for example, its expectation and its variance.

The methods discussed have a long history. K. Gauss already used them in 1795 for the study of planetary motion. In our time they are used for example, for determining parameters of satellite orbits. In passing, it has to be noted that apart from the ordinary type of regression model

$$y_i = \beta_1 x_{1i} + \cdots + \beta_m x_{mi} + n_i$$

with $n_i$ being a random variable, in statistical literature the **autoregression model**

$$y_i + \alpha_1 y_{i-1} + \cdots + \alpha_k y_{i-k} = n_i$$

and the **generalized regression model**

$$y_i + \alpha_1 y_{i-1} + \cdots + \alpha_k y_{i-k} = \beta_1 x_{1i} + \cdots + \beta_m x_{mi} + n_i$$

are studied as well (Durbin, 1960; Jenkins and Watts, 1968). General references are: Deutsch (1965), Draper and Smith (1966), Fraser (1958), Graybill (1961), Grenander and Rosenblatt (1957).

*Notation.* Now we will see how estimates can be derived. The following notation will be used. Observations are made on the process output signal $y$. As in Chapter 2, this $y$ is composed of: the response to the process input $\mathbf{u}$, and the noise $\mathbf{n}$ added in the process and in the measuring instrument. The output at the $j$th sampling instant is

$$y(j) = y(\mathbf{u}, \mathbf{b}, \mathbf{n}, j) \tag{6.16}$$

with $j = 1, \ldots, k$. The argument $\mathbf{b}$ of $y$ indicates that these samples depend on the parameters $b_0, b_1, \ldots, b_m$ which are the components of the process parameter vector $\mathbf{b}$. We define

$$\mathbf{y}' = [y(1), y(2), \ldots, y(k)] \tag{6.17}$$

The noise may be given by its mean and its covariance matrix

$$\mathscr{E}[\mathbf{n}] = \mathbf{0} \tag{6.18}$$

$$\mathscr{E}[\mathbf{nn}'] = \begin{bmatrix} \mathscr{E}[n(1)n(1)] & \cdots & \mathscr{E}[n(k)n(1)] \\ \vdots & & \vdots \\ \mathscr{E}[n(1)n(k)] & \cdots & \mathscr{E}[n(k)n(k)] \end{bmatrix} = \mathbf{N} \tag{6.19}$$

The objective is to determine a coefficient vector $\boldsymbol{\beta}$ which estimates $\mathbf{b}$. This is done by using the theoretical response vector $\mathbf{w}$, the model output, which is dependent on $\boldsymbol{\beta}$, the coefficient vector, which has the components $\beta_0, \beta_1, \ldots, \beta_m$. This functional relationship can be chosen in many different ways.

The simplest case is that in which there is a **linear** relation between $\mathbf{w}$ and $\boldsymbol{\beta}$, a model which is linear-in-the-parameters (cf. Section 4.6), i.e.

$$w(j) = \sum_{i=0}^{m} \beta_i u_i(j) \qquad j = 1, \ldots, k$$

where $u_i(j)$ are linearly independent known functions. The $\mathbf{w}$ can be written as:

$$\mathbf{w} = \mathbf{U}\boldsymbol{\beta} \tag{6.20}$$

with

$$\mathbf{U} = \begin{bmatrix} u_0(1) & \cdots & u_m(1) \\ \vdots & & \vdots \\ u_0(k) & \cdots & u_m(k) \end{bmatrix} \tag{6.21}$$

Note again, that this choice of a linear relationship between $\mathbf{w}$ and $\boldsymbol{\beta}$ does **not** imply that the input–output relation of the model has to be linear. The matrix $\mathbf{U}$ is assumed to be known perfectly, that is, it can be measured without errors. Moreover, it is assumed that there are more observations ($k$) than there are unknown parameters ($m + 1$).

### *6.1.3 Linear Unbiased Estimators*

The class of linear unbiased estimators is defined by:

$$\boldsymbol{\beta} = \mathbf{Qy} \qquad \textbf{linearity} \tag{6.22}$$

where $\mathbf{Q}$ is a matrix having $m + 1$ rows and $k$ columns, and

$$\mathscr{E}[\boldsymbol{\beta}] = \mathbf{b} \qquad \textbf{unbiasedness} \tag{6.23}$$

It is assumed that (6.20) may give a complete representation model of the process, that is:

$$\mathbf{y} = \mathbf{Ub} + \mathbf{n} \tag{6.24}$$

For the moment we will assume that $\mathbf{U}$ and $\mathbf{n}$ are statistically independent. Now an error vector $\mathbf{e}$ may be defined as

$$\mathbf{e} = \mathbf{y} - \mathbf{w} = \mathbf{y} - \mathbf{U}\boldsymbol{\beta} \tag{6.25}$$

An error or loss function can be taken as a positive definite form:

$$E = \mathbf{e}'\mathbf{R}\mathbf{e} \overset{\text{def}}{=} \|\mathbf{e}\|_{\mathbf{R}}^2 \tag{6.26}$$

where $\mathbf{R}$ is a matrix of weighting coefficients $r_{ij}$. Without loss of generality this matrix will be assumed symmetric. This error function can be written as:

$$\begin{aligned} E &= \|\mathbf{y} - \mathbf{U}\boldsymbol{\beta}\|_{\mathbf{R}}^2 = [\mathbf{y} - \mathbf{U}\boldsymbol{\beta}]'\mathbf{R}[\mathbf{y} - \mathbf{U}\boldsymbol{\beta}] \\ &= \mathbf{y}'\mathbf{R}\mathbf{y} + [\mathbf{U}\boldsymbol{\beta}]'\mathbf{R}[\mathbf{U}\boldsymbol{\beta}] - \mathbf{y}'\mathbf{R}[\mathbf{U}\boldsymbol{\beta}] - [\mathbf{U}\boldsymbol{\beta}]'\mathbf{R}\mathbf{y} \end{aligned} \tag{6.27}$$

Since $[\mathbf{U}\boldsymbol{\beta}]' = \boldsymbol{\beta}'\mathbf{U}'$ and since $\mathbf{R}$ is a symmetric matrix:

$$E = \mathbf{y}'\mathbf{R}\mathbf{y} + \boldsymbol{\beta}'\mathbf{U}'\mathbf{R}\mathbf{U}\boldsymbol{\beta} - 2\boldsymbol{\beta}'\mathbf{U}'\mathbf{R}y \tag{6.28}$$

Differentiation of equation (6.28) with respect to $\boldsymbol{\beta}$ gives (cf. Appendix $C$):

$$\frac{\partial E}{\partial \boldsymbol{\beta}} = \nabla_{\boldsymbol{\beta}} E = 2\mathbf{U}'\mathbf{R}\mathbf{U}\boldsymbol{\beta} - 2\mathbf{U}'\mathbf{R}\mathbf{y} \tag{6.29}$$

This can also be written as:

$$-2\mathbf{U}'\mathbf{R}[\mathbf{y} - \mathbf{U}\boldsymbol{\beta}] = -2\mathbf{U}'\mathbf{R}\mathbf{e}$$

For some vector $\hat{\boldsymbol{\beta}}$ equation (6.29) can be equated to zero, leading to an expression for $\hat{\boldsymbol{\beta}}$ at which an extremum of $E$ occurs:

$$\mathbf{U}'\mathbf{R}\mathbf{U}\hat{\boldsymbol{\beta}} = \mathbf{U}'\mathbf{R}\mathbf{y} \tag{6.30}$$

This is called the **'normal equation'.** If $\mathbf{U}'\mathbf{R}\mathbf{U}$ is non-singular then

$$\hat{\boldsymbol{\beta}} = [\mathbf{U}'\mathbf{R}\mathbf{U}]^{-1}\mathbf{U}'\mathbf{R}y \tag{6.31}$$

It is not difficult to show that for $\hat{\boldsymbol{\beta}}$ the error function $E$ actually has a minimum value. This value $E(\hat{\boldsymbol{\beta}})$ is called the **residual error** (based on $k$ observations).

Some remarks are in order here:

Of course equation (6.31) can also be derived through calculus of variations:

$$\left.\frac{d}{d\alpha}E(\hat{\boldsymbol{\beta}} + \alpha\Delta\boldsymbol{\beta})\right|_{\alpha=0} = 0$$

or

$$\Delta\boldsymbol{\beta}'\mathbf{U}'\mathbf{R}[\mathbf{y} - \mathbf{U}\hat{\boldsymbol{\beta}}] = 0$$

for arbitrary $\Delta\boldsymbol{\beta}$ (orthogonality principle).

A direct proof of the minimization of $E$ can be found by the well-known method of completing the square term in $\boldsymbol{\beta}$. From equation (6.27):

$$\begin{aligned}E &= \mathbf{y}'\mathbf{R}\mathbf{y} + \boldsymbol{\beta}'\mathbf{U}'\mathbf{R}\mathbf{U}\boldsymbol{\beta} - \mathbf{y}'\mathbf{R}\mathbf{U}\boldsymbol{\beta} - \boldsymbol{\beta}'\mathbf{U}'\mathbf{R}\mathbf{y}\\ &= [\boldsymbol{\beta} - [\mathbf{U}'\mathbf{R}\mathbf{U}]^{-1}\mathbf{U}'\mathbf{R}\mathbf{y}]'\mathbf{U}'\mathbf{R}\mathbf{U}[\boldsymbol{\beta} - [\mathbf{U}'\mathbf{R}\mathbf{U}]^{-1}\mathbf{U}'\mathbf{R}\mathbf{y}]\\ &\quad -\mathbf{y}'\mathbf{R}\mathbf{U}[\mathbf{U}'\mathbf{R}\mathbf{U}]^{-1}\mathbf{U}'\mathbf{R}\mathbf{U}[\mathbf{U}'\mathbf{R}\mathbf{U}]^{-1}\mathbf{U}'\mathbf{R}\mathbf{y} + \mathbf{y}'\mathbf{R}\mathbf{y}.\end{aligned}$$

Evidently this is a minimum if $\boldsymbol{\beta}$ satisfies equation (6.31)

As a mnemonic technique the reader may find it useful to remember

$$\mathbf{y} = \mathbf{U}\mathbf{b} + \mathbf{n}$$

multiply by $\mathbf{U}'\mathbf{R}$

$$\mathbf{U}'\mathbf{R}\mathbf{y} = \mathbf{U}'\mathbf{R}\mathbf{U}\mathbf{b} + \mathbf{U}'\mathbf{R}\mathbf{n}$$

As the last term is unknown, unmeasurable and as it is assumed that $\mathbf{U}$ and $\mathbf{n}$ are statistically independent, that term is discarded, leaving an estimate $\boldsymbol{\beta}$ instead of the true parameter $\mathbf{b}$ as shown in equation (6.30). Naturally this way of deriving equation (6.31) does not indicate in what sense the estimator is optimal.

This estimate fulfils the condition of equation (6.22) for **linearity** because

$$\hat{\boldsymbol{\beta}} = \mathbf{Q}\mathbf{y} \quad \text{with} \quad \mathbf{Q} = [\mathbf{U}'\mathbf{R}\mathbf{U}]^{-1}\mathbf{U}'\mathbf{R} \tag{6.32}$$

From equations (6.21) and (6.24) it follows that

$$\hat{\boldsymbol{\beta}} = [\mathbf{U}'\mathbf{R}\mathbf{U}]^{-1}\mathbf{U}'\mathbf{R}[\mathbf{U}\mathbf{b} + \mathbf{n}] = \mathbf{b} + [\mathbf{U}'\mathbf{R}\mathbf{U}]^{-1}\mathbf{U}'\mathbf{R}\mathbf{n}$$

As the input and noise signal are statistically independent:

$$\mathscr{E}[\hat{\boldsymbol{\beta}}] = \mathbf{b} + \mathscr{E}[[\mathbf{U}'\mathbf{R}\mathbf{U}]^{-1}\mathbf{U}'\mathbf{R}]\mathscr{E}[\mathbf{n}] \tag{6.33}$$

Furthermore, it was already assumed that $\mathscr{E}[\mathbf{n}] = \mathbf{0}$, consequently, the estimate is **unbiased**:

$$\mathscr{E}[\hat{\boldsymbol{\beta}}] = \mathbf{b}$$

Another characteristic of the estimate, $\hat{\boldsymbol{\beta}}$ of equation (6.31), which we want to determine is the **variance.** One may also be interested in the correlation between the elements of the vector $\hat{\boldsymbol{\beta}}$. This can be found by means of the covariance matrix:

$$\operatorname{cov}[\hat{\boldsymbol{\beta}}] = \mathscr{E}[[\hat{\boldsymbol{\beta}} - \mathbf{b}][\hat{\boldsymbol{\beta}} - \mathbf{b}]'] = \mathscr{E}[[\mathbf{Q}\mathbf{y} - \mathbf{b}][\mathbf{Q}\mathbf{y} - \mathbf{b}]'] \tag{6.34}$$

It it is again assumed that equation (6.24) holds and that $\mathbf{U}$ and $\mathbf{n}$ are statistically independent then one finds, using (6.32):

$$\begin{aligned}\operatorname{cov}[\hat{\boldsymbol{\beta}}] &= \mathscr{E}[[\mathbf{QUb} + \mathbf{Qn} - \mathbf{b}][\mathbf{QUb} + \mathbf{Qn} - \mathbf{b}]'] \\ &= \mathscr{E}[[\mathbf{Qn}][\mathbf{Qn}]'] = \mathbf{Q}\mathscr{E}[\mathbf{nn}']\mathbf{Q}' \\ &= \mathbf{QNQ}'\end{aligned} \tag{6.35}$$

Consequently

$$\operatorname{cov}[\hat{\boldsymbol{\beta}}] = [\mathbf{U}'\mathbf{R}\mathbf{U}]^{-1}\mathbf{U}'\mathbf{RNRU}[\mathbf{U}'\mathbf{R}\mathbf{U}]^{-1}$$

We will find that in several cases of practical interest this expression is simplified considerably. Estimates of the variances of the parameter estimates are obtained from the main diagonal of this matrix.

### 6.1.4 *Least-squares Estimators* (Levin, 1959; 1960)

The least-squares estimators imply a minimization of

$$E = \mathbf{e}'\mathbf{e} = \|\mathbf{e}\|^2 \tag{6.36}$$

Consequently in (6.26) and following equations:

$$\mathbf{R} = \mathbf{I}$$

and from equations (6.30), (6.31) and (6.35)

$$\mathbf{U}'\mathbf{U}\hat{\boldsymbol{\beta}}_S = \mathbf{U}'\mathbf{y} \tag{6.37}$$

or

$$\hat{\boldsymbol{\beta}}_S = [\mathbf{U}'\mathbf{U}]^{-1}\mathbf{U}'\mathbf{y} \tag{6.38}$$

and

$$\operatorname{cov}[\hat{\boldsymbol{\beta}}_S] = [\mathbf{U}'\mathbf{U}]^{-1}\mathbf{U}'\mathbf{N}\mathbf{U}[\mathbf{U}'\mathbf{U}]^{-1} \tag{6.39}$$

If **U** is a square matrix, i.e. if the number of samples is equal to the number of parameters to be determined, and if **U** has an inverse, then:

$$[\mathbf{U}'\mathbf{U}]^{-1}\mathbf{U}' = \mathbf{U}^{-1} \tag{6.40}$$

and

$$\hat{\boldsymbol{\beta}}_S = \mathbf{U}^{-1}\mathbf{y} \tag{6.41}$$

i.e. a solution of a set of linear equations. From the engineering point of view this case is of little interest due to the disturbance by the noise. For reduction of the influence of the noise, the number of samples chosen has to be much larger than the number of parameters.

If equation (6.37) is written in terms of $u$'s then:

$$\begin{bmatrix} \sum_{j=1}^{k} u_0^2(j) & \cdots & \sum_{j=1}^{k} u_0(j)u_m(j) \\ \vdots & & \vdots \\ \sum_{j=1}^{k} u_m(j)u_0(j) & \cdots & \sum_{j=1}^{k} u_m^2(j) \end{bmatrix} \begin{bmatrix} \hat{\beta}_0 \\ \vdots \\ \hat{\beta}_m \end{bmatrix} = \begin{bmatrix} \sum_{j=1}^{k} u_0(j)y(j) \\ \vdots \\ \sum_{j=1}^{k} u_m(j)y(j) \end{bmatrix} \tag{6.42}$$

Choice of test signals $u_i(j)$ which are orthogonal or orthonormal may introduce desirable simplification; cf. Chapter 4. In the case of orthonormality then:

$$\mathbf{U}'\mathbf{U} = \mathbf{I}, \text{ the identity matrix}$$

and

$$\hat{\boldsymbol{\beta}} = \mathbf{U}'\mathbf{y}$$

or

$$\hat{\beta}_i = \sum_{j=1}^{k} u_i(j)y(j) \qquad i = 0, \ldots, m$$

A simple geometrical interpretation of the least-squares estimate can be given as follows; cf. Figure 6.1 for the case where $\boldsymbol{\beta}$ is two dimensional. The length of

$$\mathbf{e} = \mathbf{y} - \mathbf{w}$$

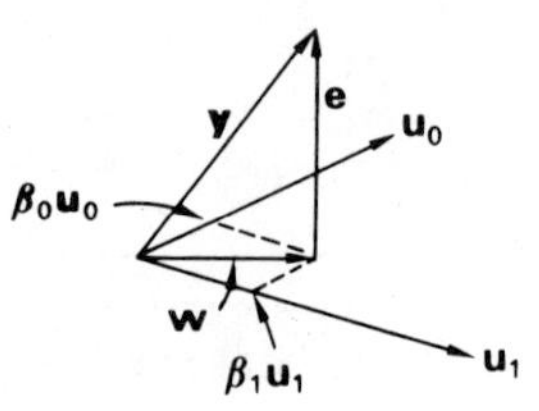


Figure 6.1

has to be minimized. This is true if $\mathbf{e}$ is orthogonal to $\mathbf{u}_1$ and $\mathbf{u}_2$;

$$\left.\begin{aligned}\mathbf{u}_1'\mathbf{e} = 0\\ \mathbf{u}_2'\mathbf{e} = 0\end{aligned}\right\} \quad \text{or} \quad \mathbf{U}'\mathbf{e} = \mathbf{0}$$

Consequently

$$\mathbf{U}'(\mathbf{y} - \mathbf{w}) = \mathbf{U}'(\mathbf{y} - \mathbf{U}\hat{\boldsymbol{\beta}}_S) = \mathbf{0}$$

or

$$\mathbf{U}'\mathbf{U}\hat{\boldsymbol{\beta}}_S = \mathbf{U}'\mathbf{y}$$

which is equation (6.37).

### *6.1.5 Markov Estimators*

The Markov estimators imply a minimization of

$$E = \mathbf{e}'\mathbf{N}^{-1}\mathbf{e} = \|\mathbf{e}\|^2_{\mathbf{N}^{-1}} \tag{6.43}$$

Consequently

$$\mathbf{R} = \mathbf{N}^{-1}$$

and from equations (6.30), (6.31) and (6.35):

$$\mathbf{U}'\mathbf{N}^{-1}\mathbf{U}\hat{\boldsymbol{\beta}}_M = \mathbf{U}'\mathbf{N}^{-1}\mathbf{y} \tag{6.44}$$

or

$$\hat{\boldsymbol{\beta}}_M = [\mathbf{U}'\mathbf{N}^{-1}\mathbf{U}]^{-1}\mathbf{U}'\mathbf{N}^{-1}\mathbf{y} \tag{6.45}$$

and

$$\begin{aligned}\operatorname{cov}[\hat{\boldsymbol{\beta}}_M] &= [\mathbf{U}'\mathbf{N}^{-1}\mathbf{U}]^{-1}\mathbf{U}'\mathbf{N}^{-1}\mathbf{N}\mathbf{N}^{-1}\mathbf{U}[\mathbf{U}'\mathbf{N}^{-1}\mathbf{U}]^{-1}\\ &= [\mathbf{U}'\mathbf{N}^{-1}\mathbf{U}]^{-1}[\mathbf{U}'\mathbf{N}^{-1}\mathbf{U}][\mathbf{U}'\mathbf{N}^{-1}\mathbf{U}]^{-1}\\ &= [\mathbf{U}'\mathbf{N}^{-1}\mathbf{U}]^{-1}\end{aligned} \tag{6.46}$$

As the covariance matrix (6.46) is part of the estimator (6.45), one obtains both estimate and covariance at the same time. The diagonal elements of (6.46) represent the variances of the parameter estimates, the off-diagonal elements provide the correlation between the errors of the different parameters. If $k \to \infty$ and $\sum_{j=1}^{k} u_i^2(j) \to \infty$ the variances approach zero and consequently the estimates are **consistent,** cf. Section 5.1.

Using Schwarz's inequality for matrices (cf. AppendixC), it can quite easily be proved that this linear unbiased estimator has the property of minimum variance. Schwarz's inequality:

$$[\mathbf{B}'\mathbf{B}] \geqslant [\mathbf{B}'\mathbf{A}][\mathbf{A}'\mathbf{A}]^{-1}[\mathbf{A}'\mathbf{B}]$$

Choose: $\mathbf{A} = \mathbf{N}^{-\frac{1}{2}}\mathbf{U}$ and $\mathbf{B}' = \mathbf{Q}\mathbf{N}^{\frac{1}{2}}$, then

$$\begin{aligned}\mathbf{Q}\mathbf{N}^{\frac{1}{2}}\mathbf{N}^{\frac{1}{2}'}\mathbf{Q}' &= \mathbf{Q}\mathbf{N}\mathbf{Q}' \\ &\geqslant [\mathbf{Q}\mathbf{N}^{\frac{1}{2}}\mathbf{N}^{-\frac{1}{2}}\mathbf{U}][\mathbf{U}'\mathbf{N}^{-\frac{1}{2}'}\mathbf{N}^{-\frac{1}{2}}\mathbf{U}]^{-1}[\mathbf{U}'\mathbf{N}^{-\frac{1}{2}'}\mathbf{N}^{\frac{1}{2}'}\mathbf{Q}']\end{aligned}$$

Because $\mathbf{QU} = \mathbf{I}$ then

$$\mathbf{QNQ}' \geqslant [\mathbf{U}'\mathbf{N}^{-1}\mathbf{U}]^{-1}$$

or

$$\text{cov}\,[\boldsymbol{\beta}] \geqslant \text{cov}\,[\hat{\boldsymbol{\beta}}_M] \tag{6.47}$$

i.e.

$$\text{cov}\,[\boldsymbol{\beta}] - \text{cov}\,[\hat{\boldsymbol{\beta}}_M]$$

is non-negative definite. Here $\boldsymbol{\beta}$ is any other linear estimate. This implies that the estimator is **efficient**; cf. Section 5.1.

From the previous sections, it is clear that if the additive noise is 'white' that is

$$\mathbf{N} = \mathscr{E}[\mathbf{nn}'] = \sigma_n^2\mathbf{I} \tag{6.48}$$

then

$$\boldsymbol{\beta}_S = \hat{\boldsymbol{\beta}}_M \tag{6.49}$$

and

$$\text{cov}\,[\hat{\boldsymbol{\beta}}_M] = \sigma_n^2[\mathbf{U}'\mathbf{U}]^{-1} \tag{6.50}$$

As was indicated in Chapter 5, for Gaussian noise the least-squares and Markov estimates coincide with the maximum likelihood ones, each having the minimum variance of all unbiased estimates.

The class of linear unbiased estimates was derived by minimizing the positive definite form

$$E = \|\mathbf{e}\|_{\mathbf{R}}^2$$

Subsequently, the choices $\mathbf{R} = \mathbf{I}$ and $\mathbf{R} = \mathbf{N}^{-1}$ led to the least-squares and the Markov estimates. Other choices would be possible as well, e.g. a weighting matrix taking into account a possible non-stationary additive noise. In that case one has to minimize the error function

$$E = \sum_{j=1}^{k} \frac{1}{\sigma_n^2(j)}\{y(j) - \mathbf{u}'(j)\boldsymbol{\beta}\}^2 \tag{6.51}$$

where $\sigma_n^2(j)$ represents the noise variance at time $j$.

## 6.2 Implementation of 'Open-loop' Estimation Schemes

In Section 6.1 we found the following explicit expressions:

**least-squares estimate**

$$\hat{\boldsymbol{\beta}}_S = [\mathbf{U}'\mathbf{U}]^{-1}\mathbf{U}'\mathbf{y} \tag{6.52}$$

$$\operatorname{cov}[\hat{\boldsymbol{\beta}}_S] = [\mathbf{U}'\mathbf{U}]^{-1}\mathbf{U}'\mathbf{N}\mathbf{U}[\mathbf{U}'\mathbf{U}]^{-1} \tag{6.53}$$

**Markov estimate**

$$\hat{\boldsymbol{\beta}}_M = [\mathbf{U}'\mathbf{N}^{-1}\mathbf{U}]^{-1}\mathbf{U}'\mathbf{N}^{-1}\mathbf{y} \tag{6.54}$$

$$\operatorname{cov}[\hat{\boldsymbol{\beta}}_M] = [\mathbf{U}'\mathbf{N}^{-1}\mathbf{U}]^{-1} \tag{6.55}$$

In this section we will discuss some aspects of the implementation of these expressions.

### 6.2.1 *Least-squares Estimation*

We will start with the simplest case, the least-squares estimation, following from

$$\min_{\beta} E \quad \text{with} \quad E = \mathbf{e}'\mathbf{e} = \sum_{j=1}^{k} e^2(j)$$

resulting in equation (6.52). The implementation of this equation leads to the celebrated correlation technique.

*Example.* Let us assume that only parameter $\beta_i$ has to be determined. Equation (6.52) then reduces to:

$$\beta_i = (\mathbf{u}_i'\mathbf{u}_i)^{-1}\mathbf{u}_i'\mathbf{y} = \frac{\sum_{j=1}^{k} u_i(j)y(j)}{\sum_{j=1}^{k} u_i^2(j)} \tag{6.56}$$

The numerator and the denominator can be recognized as the time-averaged products that approximate the respective cross and autocorrelations for $k \to \infty$ (cf. Figure 6.2). The errors that may occur will be discussed later in this section.

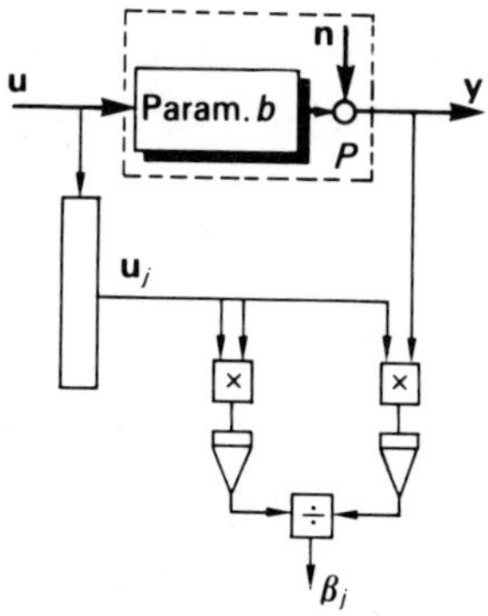


Figure 6.2

For multiple-parameter case equation (6.52) can be written as:

$$\hat{\boldsymbol{\beta}} = \begin{bmatrix} \hat{\beta}_0 \\ \vdots \\ \hat{\beta}_m \end{bmatrix} = \begin{bmatrix} \mathbf{u}_0'\mathbf{u}_0 & \cdots & \mathbf{u}_0'\mathbf{u}_m \\ \vdots & & \vdots \\ \mathbf{u}_m'\mathbf{u}_0 & \cdots & \mathbf{u}_m'\mathbf{u}_m \end{bmatrix}^{-1} \begin{bmatrix} \mathbf{u}_0'\mathbf{y} \\ \vdots \\ \mathbf{u}_m'\mathbf{y} \end{bmatrix} \tag{6.57}$$

Some remarks are in order here:

The matrix $\mathbf{U}'\mathbf{U}$ is symmetric about its main diagonal because $\mathbf{u}_i'\mathbf{u}_j = \mathbf{u}_j'\mathbf{u}_i$. It also has real elements, consequently it is self-adjoint and can be brought in diagonal form by a similarity transformation.

Among the practical problems there is the computation of the inverse matrix. (Westlake, 1968; Peterka and Šmuk, 1969). Using available matrix-inversion subroutines, it is a simple matter to write a straightforward estimation procedure for the parameters. If desired, the residuals and the value of the error function can also be determined. The covariance matrix can be computed if the noise covariance matrix $\mathbf{N}$ is known.

If $\mathbf{u}_i'\mathbf{u}_j = \delta_{ij}$, the Kronecker delta, then $\mathbf{U}'\mathbf{U} = \mathbf{I}$, the identity matrix. Such conditions of orthonormality can be approximated by a suitable choice of the input signal $\mathbf{u}$ and appropriate transfer functions $G_i$ (cf. Figure 6.3). The combination $\mathbf{u}$ = white noise,

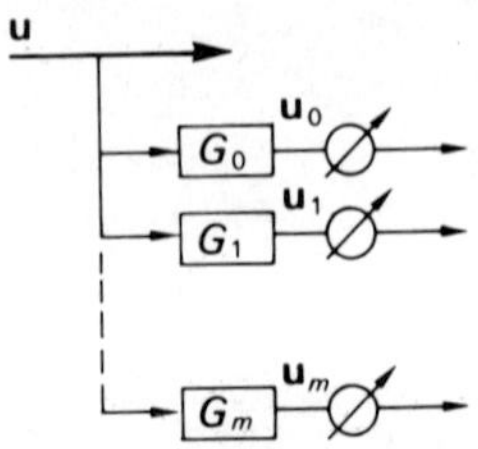


Figure 6.3

$G_i$ = time-delay elements is most frequently used. The use of orthonormal filters for $G_i$ (e.g. Laguerre filters, cf. Chapter 4) is quite often proposed. The calculation of filters that provide orthogonal components $\mathbf{u}_j$ for a particular input signal $\mathbf{u}$ has been studied by Douce and Roberts (1964) and by Roberts (1967).

Instead of the operation on sampled signals, one may use the corresponding operations on continuous (non-sampled) signals (cf. Chapter 8).

Instead of operating on signals with a continuous amplitude scale one may work with quantized signals which can have only a limited number of predetermined amplitude values. The limiting case of quantization gives a binary signal (cf. Chapter 8).

If one may use a test signal, then $\mathbf{u}$ or $u(t)$ can be chosen with full emphasis on its properties. Well-known choices are test signals composed of a number of sine waves and pseudo-random binary sequences. Both types of signals can be given interesting properties with respect to orthogonality and generation (cf. Chapter 10).

In the derivation of the estimate no assumption was made that the process $P$ has to be linear. The operations $G_i$ in Figure 6.3 may be non-linear. The only requirements imposed by the condition $\mathbf{w} = \mathbf{U}\boldsymbol{\beta}$ as a model for $\mathbf{y} = \mathbf{U}\mathbf{b}$ is the linearity of $\mathbf{w}$ with respect to the parameters to be determined. Consequently classes of non-linear systems may be amenable to correlation techniques: cf. Wiener's characterization of non-linear systems, Volterra expansions, Pugashev's systems reducible to linear (cf. Chapter 4).

If

$$\mathbf{u}_i' = [u_i(1), u_i(2), \ldots, u_i(k)] \qquad \text{for } i = 0, 1, \ldots, m$$

then

$$\mathbf{u}_i'\mathbf{u}_j = \sum_{l=1}^{k} u_i(l)u_j(l) = k\tilde{\psi}_{u_iu_j}(0, k) \tag{6.58}$$

where $\tilde{\psi}(0, k)$ is an approximation of one point of a correlation function, based on $k$ observations. If $\mathbf{u}_i$ and $\mathbf{u}_j$ are delayed versions of a stationary $\mathbf{u}$ then:

$$\mathbf{u}_i'\mathbf{u}_j = \sum_{l=1}^{k} u(l-i)u(l-j) = k\tilde{\psi}_{uu}(j-i, k)$$

For ergodic (stationary) time series

$$\tilde{\psi}_{uu}(j, k) \to \psi_{uu}(j) \qquad \text{for } k \to \infty$$

### 6.2.2 *Markov Estimation*

The next degree of complication is found in the instrumentation of the **Markov estimate,** following from

$$\min_{\beta} E \quad \text{with} \quad E = \mathbf{e}'\mathbf{N}^{-1}\mathbf{e} \tag{6.59}$$

resulting in equation (6.54). The matrix $\mathbf{N}^{-1}$ can be separated into a lower triangular matrix $\mathbf{D}$ and its transpose:

$$\mathbf{N}^{-1} = \mathbf{D}'\mathbf{D}$$

Using this notation, equations (6.59) and (6.54) can be written as:

$$E = [\mathbf{D}\mathbf{e}]'\mathbf{D}\mathbf{e}$$

$$\hat{\boldsymbol{\beta}} = [[\mathbf{D}\mathbf{U}]'\mathbf{D}\mathbf{U}]^{-1}[\mathbf{D}\mathbf{U}]'\mathbf{D}\mathbf{y}$$

The matrix $\mathbf{D}$ represents a 'noise-whitening' filter; given $\mathbf{n}$ as the input sequence, the output of that filter is white noise. Figure 6.4 indicates how these filters may be introduced into the instrumentation of the correlation technique. As was mentioned before, this provides the minimum variance estimate (of all linear unbiased estimates that can be found). In spite of this interesting property and the relative simplicity of the instrumentation it seems that it has found little application for lack of *a priori* knowledge. The same remarks that have been made regarding $[\mathbf{U}'\mathbf{U}]^{-1}$ in the least-squares case, may be made as regards the matrix $[[\mathbf{D}\mathbf{U}]'\mathbf{D}\mathbf{U}]^{-1}$.

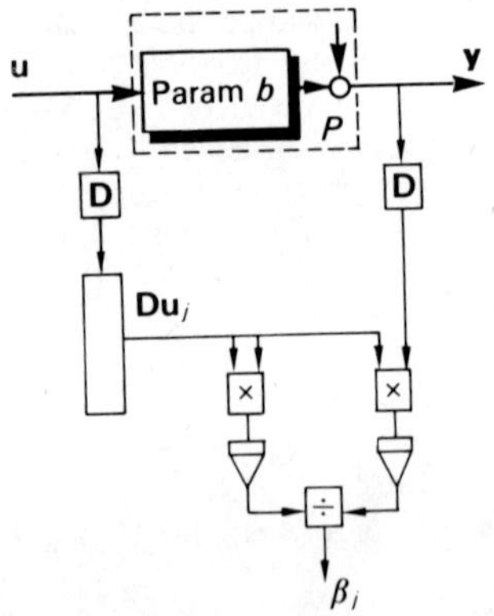


Figure 6.4

In practice, the available *a priori* knowledge of the disturbances is seldom sufficient for using the Markov estimating scheme. Consequently, one may want to estimate the statistics of the disturbance in terms of autoregression of the residual. From that knowledge one may derive the filters **D** as shown in Figure 6.4.

One such technique can be indicated as follows. Determine

$$\hat{\boldsymbol{\beta}} = [\mathbf{U}'\mathbf{U}]^{-1}\mathbf{U}'\,\mathbf{y}$$

$$\mathbf{e} = \mathbf{y} - \mathbf{U}\hat{\boldsymbol{\beta}}$$

Now the residuals **e** are considered as an approximation of the disturbances **n**. Assume that these disturbances are generated from 'white' noise through an $m$th-order filter. From this interpretation follows that one wants to minimize the square of:

$$\mathbf{e} - \mathbf{E}\boldsymbol{\delta}$$

or

$$\begin{bmatrix} e(m+1) \\ \vdots \\ e(k) \end{bmatrix} - \begin{bmatrix} e(m) & \cdots & e(1) \\ \vdots & & \vdots \\ e(k-1) & \cdots & e(k-m) \end{bmatrix} \begin{bmatrix} \delta_1 \\ \vdots \\ \delta_m \end{bmatrix}$$

From this follows, via the normal equations,

$$\boldsymbol{\delta} = [\mathbf{E}'\mathbf{E}]^{-1}\mathbf{E}'\mathbf{e}$$

With these estimates one may adjust (approximately) the model $\mathscr{D}$ of filter **D**. Then a new estimate for $\hat{\boldsymbol{\beta}}$ is determined, after that a new estimate of $\boldsymbol{\delta}$ is made, etc. (cf. Section 7.2).

A brute force method of obtaining observations without correlation when the noise is non-white, is by waiting long enough between successive samples. In this way, however, the observation time may become excessively long. For other approaches to this problem the reader is referred to Durbin (1960), Mayne (1967) and Chapter 7. Åström's work, cf. Chapter 11, is directed to the estimation of both process parameters and noise statistics.

### 6.2.3 General Considerations

Up to this point, we did not specify the model except by the condition that its output has to be linearly related to the parameters to be determined. In the majority of cases $\mathbf{u}_i$ is chosen as a delayed version of the input signal $\mathbf{u}$. When using a digital computer, such a choice is of course very convenient. In that case, the parameter $\beta_i$ is an estimate of one point $h(i)$ of the process impulse response. Figure 6.5 indicates this type of model, using a 'delay line'.

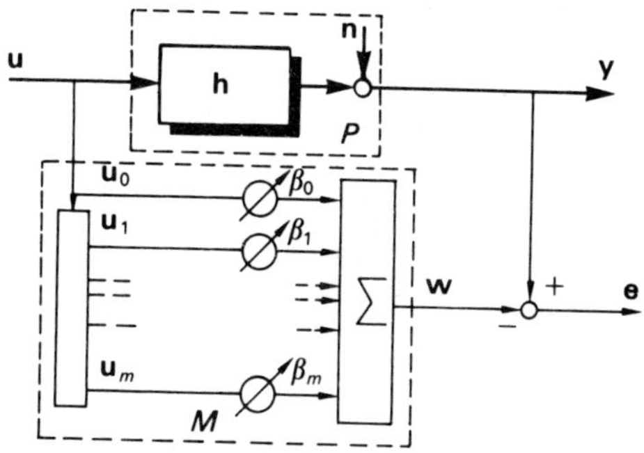


Figure 6.5

In that case, we assume that the process output $y(j)$ can be found from a convolution sum:

$$y(j) = \sum_i u(i)h(j - i) + n(j) \qquad j = 1, \ldots, k$$

or

$$\mathbf{y} = \mathbf{Uh} + \mathbf{n} = [\mathbf{u}_0, \mathbf{u}_1, \ldots, \mathbf{u}_m]\mathbf{h} + \mathbf{n}$$

In that case equation (6.57) can be written as

$$\hat{\boldsymbol{\beta}} = \begin{bmatrix} \hat{\beta}_0 \\ \vdots \\ \hat{\beta}_m \end{bmatrix} = \begin{bmatrix} \tilde{\psi}_{uu}(0, k) & \cdots & \tilde{\psi}_{uu}(m, k) \\ \vdots & & \vdots \\ \tilde{\psi}_{uu}(m, k) & \cdots & \tilde{\psi}_{uu}(0, k) \end{bmatrix}^{-1} \begin{bmatrix} \tilde{\psi}_{uy}(0, k) \\ \vdots \\ \tilde{\psi}_{uy}(m, k) \end{bmatrix} \tag{6.60}$$

where each $\tilde{\psi}$ is a summation over $k$ products. If $k \to \infty$ and the signal $u$ is ergodic $\tilde{\psi}_{uu}(j, k)$ approaches $\psi_{uu}(j)$, the autocorrelation function. For an input time series with the character of 'white' noise

$$\psi_{uu}(j) = \begin{cases} \psi_{uu}(0) = \sigma_u^2 & \text{for } j = 0 \\ 0 & \text{for } j \neq 0 \end{cases}$$

the parameter estimates follow from

$$\hat{\beta}_j = \frac{\psi_{uy}(j)}{\psi_{uu}(0)}$$

*Example.* At this point the reader may wish to stop for a moment and to try out on one of the most simple examples the ideas developed so far. The process is given by:

$$y(k) = a + bu(k) + n(k) \quad \text{or} \quad \mathbf{y} = a\mathbf{1} + b\mathbf{u} + \mathbf{n} \qquad *$$

The model is selected to be:

$$w(k) = \alpha + \beta u(k) \quad \text{or} \quad \mathbf{w} = \alpha\mathbf{1} + \beta\mathbf{u}$$

The error is:

$$e(k) = y(k) - w(k) \quad \text{or} \quad \mathbf{e} = \mathbf{y} - \mathbf{w}$$

The error criterion to be minimized is chosen to be:

$$\sum_{j=1}^{k} e^2(j) \quad \text{or} \quad \mathbf{e}'\mathbf{e}$$

The parameters to be estimated are $[a; b] = [4; 2]$; the estimates are indicated by $[\alpha; \beta]$. With the proper interpretation one may use directly equation (6.52). For reviewing purposes, however, let us go over the minimization procedure again:

$$\frac{\partial}{\partial \alpha}(\mathbf{e}'\mathbf{e}) = -\mathbf{1}'[\mathbf{y} - \hat{\alpha}\mathbf{1} - \hat{\beta}\mathbf{u}] = 0$$

$$\frac{\partial}{\partial \beta}(\mathbf{e}'\mathbf{e}) = -\mathbf{u}'[\mathbf{y} - \hat{\alpha}\mathbf{1} - \hat{\beta}\mathbf{u}] = 0$$

or

$$\begin{bmatrix} \mathbf{1}'\mathbf{y} \\ \mathbf{u}'\mathbf{y} \end{bmatrix} = \begin{bmatrix} \mathbf{1}'\mathbf{1} & \mathbf{1}'\mathbf{u} \\ \mathbf{u}'\mathbf{1} & \mathbf{u}'\mathbf{u} \end{bmatrix} \begin{bmatrix} \hat{\alpha} \\ \hat{\beta} \end{bmatrix}$$

or

$$\begin{bmatrix} \sum_{j=1}^{k} y(j) \\ \sum_{j=1}^{k} u(j)y(j) \end{bmatrix} = \begin{bmatrix} k & \sum_{j=1}^{k} u(j) \\ \sum_{j=1}^{k} u(j) & \sum_{j=1}^{k} u^2(j) \end{bmatrix} \begin{bmatrix} \hat{\alpha} \\ \hat{\beta} \end{bmatrix}$$

This corresponds with the expressions found before. The reader is invited either to write a simple computer programme that generates $u(k)$ and $n(k)$, e.g. as random numbers, that simulates the process and estimates the parameters (The estimates might be plotted, e.g. for $k = 3; 10; 30; 100$);

* The symbol **1** is used to indicate a vector having all elements equal to 1.

or to perform this experiment by hand in using, e.g. a die or a table of random numbers for generating $u(k)$ and $n(k)$, making the estimation calculations by hand.

When performing a computer simulation it pays to do the same experiment a number of times, chosing different starting values of the random-number generator in each experiment. By plotting the results of those experiments one may obtain a visual impression of the rate of convergence.

One might also plot the observations $\{u(k), y(k)\}$ in the $u, y$ plane, draw the 'process' representation without noise: $y = 4 + 2u$ and the regression line found from the estimates $\hat{\alpha}$ and $\hat{\beta}$.

## 6.3 Accuracy; Some Error Causes

For practical use it is of prime importance to recognize the possible error causes. We will distinguish:

6.3.1 noise error;

6.3.2 truncation error;

6.3.3 state-correspondence error;

6.3.4 errors due to simplification of the implementation;

6.3.5 sampling error.

Some general references are: Levin (1960), Kwakernaak (1963), Melis (1967), Van den Boom and Melis (1969).

### *6.3.1 Noise Error*

This error is due to the noise $\mathbf{n}$, added to the process output. We assume that the model may represent the process completely (cf. Figure 6.5):

$$\mathbf{y} = \mathbf{U}\mathbf{b} + \mathbf{n}$$

$$\mathbf{w} = \mathbf{U}\boldsymbol{\beta}$$

with $\mathbf{b}$ and $\boldsymbol{\beta}$ of the same dimension. For that situation, we have already derived the least-squares equation (6.53):

$$\begin{aligned} \operatorname{cov}[\hat{\boldsymbol{\beta}}] &= \mathscr{E}[[\hat{\boldsymbol{\beta}} - \mathbf{b}][\hat{\boldsymbol{\beta}} - \mathbf{b}]'] \\ &= [\mathbf{U}'\mathbf{U}]^{-1}\mathbf{U}'\mathbf{N}\mathbf{U}[\mathbf{U}'\mathbf{U}]^{-1} \end{aligned} \tag{6.61}$$

with

$$\mathbf{N} = \mathscr{E}[\mathbf{n}\mathbf{n}']$$

Note that equation (6.61) tells the complete story with respect to the noise error, based on *a priori* knowledge about the noise and *a posteriori* knowledge (measurements) of the process input signal $\mathbf{u}$. The experimenter can obtain a clear picture of the accuracy of his estimates and the correlation between the errors in the components of $\hat{\boldsymbol{\beta}}$. If necessary, an implementation of equation (6.61) on the digital computer together with the estimator is possible.* For the sake of conceptual clearness we will consider some cases using a few simplifying assumptions.

*Example.* If the noise is white and $\mathscr{E}[\mathbf{n}] = \mathbf{0}$ then

$$\mathbf{N} = \psi_{nn}(0)\mathbf{I} = \sigma_n^2\mathbf{I}$$

and

$$\operatorname{cov}[\hat{\boldsymbol{\beta}}] = \sigma_n^2[\mathbf{U}'\mathbf{U}]^{-1} \tag{6.62}$$

If in addition the noise is Gaussian, then this least-squares estimation is identical with the maximum likelihood estimation (cf. Chapter 5). Consequently the properties of the M.L.E. as summarized in that chapter hold; for example, $\hat{\boldsymbol{\beta}}$ is asymptotically normal with mean $\mathbf{b}$ and a covariance given by (6.62).

*Example.* Now we assume further that the input $\mathbf{u}$ of the process has the character of stationary 'white' noise with the *a priori* knowledge:

$$\mathscr{E}[\mathbf{u}] = \mathbf{0}$$

$$\mathscr{E}[\mathbf{U}'\mathbf{U}] = k\psi_{uu}(0)\mathbf{I} = k\sigma_u^2\mathbf{I}$$

In this situation, the expectation operator refers to the stochastic variable $\mathbf{u}$. Again $k$ samples are considered. This leads to

$$\operatorname{cov}[\hat{\boldsymbol{\beta}}] \approx \frac{\psi_{nn}(0)}{k\psi_{uu}(0)}\mathbf{I} = \frac{\sigma_n^2}{k\sigma_u^2}\mathbf{I} \tag{6.63}$$

For each parameter $\beta_j$, one finds a standard deviation (spread):

$$\sigma_{\beta_j} \approx \frac{\sigma_n}{\sqrt{k}\sigma_u}$$

For $k \to \infty$ the spread of the probability-density curve goes to zero. It must be noticed, however, that due to the square root, the convergence is rather slow. $\sigma_u^2$ and $\sigma_n^2$ can be considered as a measure of the respective 'powers'. This expression for $\sigma_\beta$ is independent of the number of parameters to be estimated.

Another approach to finding the error due to additive noise is the following. As can be seen from equation (6.38) and noting that $\mathbf{y} = \mathbf{U}\mathbf{b} + \mathbf{n}$:

$$\hat{\boldsymbol{\beta}} = [\mathbf{U}'\mathbf{U}]^{-1}\mathbf{U}'\mathbf{y} = \mathbf{b} + [\mathbf{U}'\mathbf{U}]^{-1}\mathbf{U}'\mathbf{n}$$

or

$$\Delta\boldsymbol{\beta} = \hat{\boldsymbol{\beta}} - \mathbf{b} = [\mathbf{U}'\mathbf{U}]^{-1}\mathbf{U}'\mathbf{n} = \mathbf{P}\mathbf{U}'\mathbf{n} \quad \text{with} \quad \mathbf{P} = k[p_{ij}]$$

Consequently

$$\Delta\beta_j = p_{0j}\tilde{\psi}_{un}(0, k) + \cdots + p_{jj}\tilde{\psi}_{un}(j, k) + \cdots + p_{mj}\tilde{\psi}_{un}(m, k)$$

* Of course if $\mathbf{N}$ is available as *a priori* knowledge then a Markov estimator could be implemented.

If $u$ has a white-noise character then $p_{jj} = \tilde{\psi}_{uu}^{-1}(0, k) \approx \sigma_u^{-2}$ and $p_{ij} \approx 0$ for $i \neq j$. Therefore the dominant term will be that with $p_{jj}$ and

$$\Delta\beta_j \approx \frac{\tilde{\psi}_{un}(j, k)}{\tilde{\psi}_{uu}(0, k)} \qquad (k \text{ large}) \tag{6.64}$$

Of course an extra noise error manifests itself in a situation where the additive noise is not white if the ordinary least-squares estimation is used.

### 6.3.2 *Truncation Error*

If only part of the process parameters can be taken care of by the (conceptual) model, then for a good grasp of the problem it is advisable to consider the process to consist of two parts (cf. Figure 6.6).

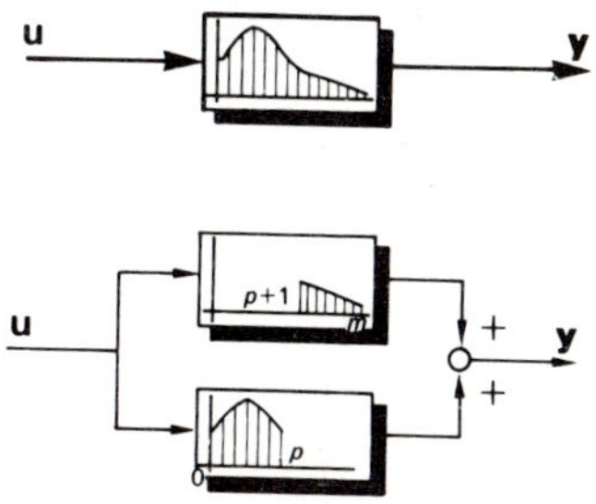


Figure 6.6

The parameters (points of the impulse response) $b_0, \ldots, b_p$ can be estimated, the parameters $b_{p+1}, \ldots, b_m$ are not considered. This situation may occur:

if the initial guess of the length of the impulse response is too restricted;

if we are interested only in the initial (or any other restricted) part of the response;

if the impulse response extends to infinity, i.e. if the process contains a pure integrating action.

In Figure 6.7 the model parameters $\boldsymbol{\beta}' = [\beta_0, \ldots, \beta_p]$ correspond to $\mathbf{b}_1' = [b_0, \ldots, b_p]$. The process output $\mathbf{y}$ consists of two parts

$$\mathbf{y} = \mathbf{y}_1(\mathbf{b}_1) + \mathbf{y}_2(\mathbf{b}_2)$$

As far as the estimating procedure is concerned $\mathbf{y}_2$, due to the 'truncated' part of the process response, acts as additive disturbances. Clearly $\mathbf{y}_1$ and $\mathbf{y}_2$ may be correlated as they both originate from $\mathbf{u}$.

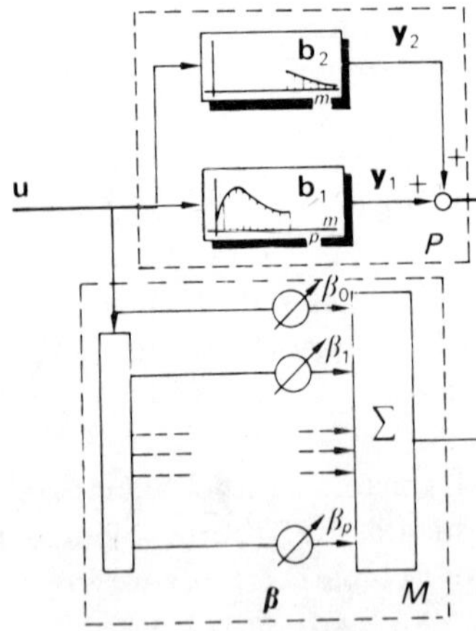


Figure 6.7

For studying the effect of the truncation we put:

$$\mathbf{y} = \mathbf{U}\mathbf{b} = [\mathbf{U}_1 \,\vdots\, \mathbf{U}_2]\begin{bmatrix}\mathbf{b}_1 \\ \hdashline \mathbf{b}_2\end{bmatrix} \quad \text{with} \quad \mathbf{b}_1 = \begin{bmatrix} b_0 \\ \vdots \\ b_p \end{bmatrix} \quad \text{and} \quad \mathbf{b}_2 = \begin{bmatrix} b_{p+1} \\ \vdots \\ b_m \end{bmatrix}$$

or

$$\begin{aligned} \mathbf{y} &= \mathbf{U}_1\mathbf{b}_1 + \mathbf{U}_2\mathbf{b}_2 \\ \mathbf{U}_1'\mathbf{y} &= \mathbf{U}_1'\mathbf{U}_1\mathbf{b}_1 + \mathbf{U}_1'\mathbf{U}_2\mathbf{b}_2 \end{aligned} \tag{6.65}$$

The estimator for the 'truncated' part of the process is

$$\hat{\boldsymbol{\beta}} = [\mathbf{U}_1'\mathbf{U}_1]^{-1}\mathbf{U}_1'\mathbf{y} \quad \text{or} \quad \mathbf{U}_1'\mathbf{y} = \mathbf{U}_1'\mathbf{U}_1\hat{\boldsymbol{\beta}}$$

Combining this expression with equation (6.65) leads to:

$$\Delta\boldsymbol{\beta} = \hat{\boldsymbol{\beta}} - \mathbf{b}_1 = [\mathbf{U}_1'\mathbf{U}_1]^{-1}\mathbf{U}_1'\mathbf{U}_2\mathbf{b}_2 \tag{6.66}$$

If the input **u** has the character of **white noise** then $\mathscr{E}[\mathbf{U}_1'\mathbf{U}_2] = \mathbf{0}$ and $\mathscr{E}[\Delta\boldsymbol{\beta}] = \mathbf{0}$. For other input signals this expectation may be unequal to zero and the estimates can be biased.

For the covariance of $\Delta\boldsymbol{\beta}$:

$$\operatorname{cov}[\Delta\boldsymbol{\beta}] = \mathscr{E}[[\mathbf{U}_1'\mathbf{U}_1]^{-1}\mathbf{U}_1'\mathbf{U}_2\mathbf{b}_2\mathbf{b}_2'\mathbf{U}_2'\mathbf{U}_1[\mathbf{U}_1'\mathbf{U}_1]^{-1}] \tag{6.67}$$

From these equations and from physical insight it is clear that $\mathscr{E}(\Delta\boldsymbol{\beta}]$ and cov $[\Delta\boldsymbol{\beta}]$ are independent of the level ('power') of the input signal.

A case of special interest is that in which the impulse response of the process has an asymptotic value $h(\infty) \neq 0$ (cf. Figure 6.8). The parameters

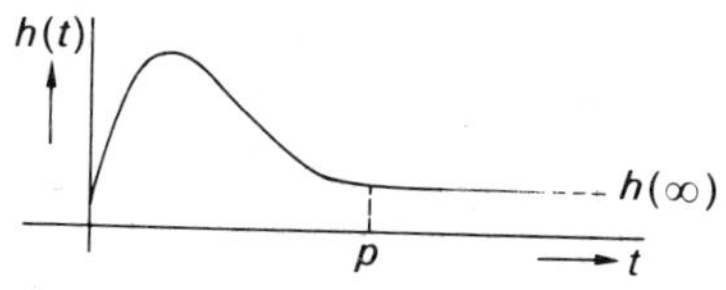


Figure 6.8

$\mathbf{b}_1$ ought to be determined from

$$\hat{\boldsymbol{\beta}} = [\mathbf{U}_1'\mathbf{U}_1]^{-1}\mathbf{U}_1'\mathbf{y}_1 = [\mathbf{U}_1'\mathbf{U}_1]^{-1}\mathbf{U}_1'(\mathbf{y} - \mathbf{y}_2) \tag{6.68}$$

However, $\mathbf{y}_2$ is not known beforehand and $\mathbf{y}$ has to be used. Consequently, errors will occur. No matter how many samples are being used in the estimation, the uncertainty with respect to these estimates never converges to zero. It turns out, that for each parameter $\beta_j$ with $j > p$ holds (Melis, 1967):

$$\sigma_{\beta_j} = \frac{h(\infty)}{\sqrt{2}} \tag{6.69}$$

If we are going to use $\mathbf{y}$ then the last one of the estimated parameters, $\beta_p$ in Figure 6.8, may provide us with some knowledge about $h(\infty)$, albeit with an error due to truncation. It was assumed that $\beta_j = \beta_p$ for $j \geqslant p$. Therefore, $\mathbf{y}_2$ is known with some uncertainty. This implies that we (partially) can compensate $\mathbf{y}$ by $\mathbf{y}_2$ and derive a better estimate for $\beta_j$, etc.

Another way to handle this situation is indicated in Figure 6.9, where an integrator (summer) is chosen to represent the 'tail' of the impulse response. In an analogous way one might take care of impulse-response components due to large time constants by including in the (conceptual) model a low-pass filter instead of the integrator. Note the difference between this scheme and that given by Levin (1959) who places an integrator before the delay line.

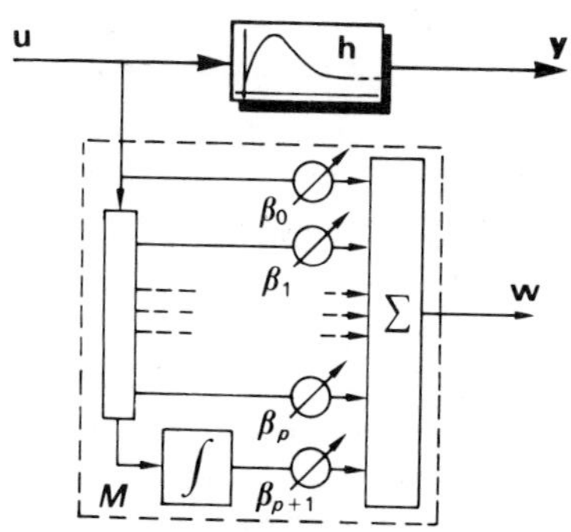


Figure 6.9

The truncation errors in the estimated parameters are highly correlated and almost equal in magnitude and sign. This holds especially for large $k$. From this it follows that for large $k$ the shape of the impulse response is measured adequately, but that it cannot be indicated how much each estimated point differs from the corresponding point of the true impulse response (Melis, 1967).

### 6.3.3 *State-correspondence Error*

When starting an observation on the input $\mathbf{u}$ and the output $\mathbf{y}$ of a process the state vector need not be zero. This results in contributions to the first (few) samples of $\mathbf{y}$ which are not the result of the input samples $\mathbf{u}$. Choosing again the delay-line representation, one finds

$$\text{process:} \quad \mathbf{y} = \mathbf{U}_P\mathbf{b} + \mathbf{n}$$

$$\text{model:} \quad \mathbf{w} = \mathbf{U}_M\boldsymbol{\beta}$$

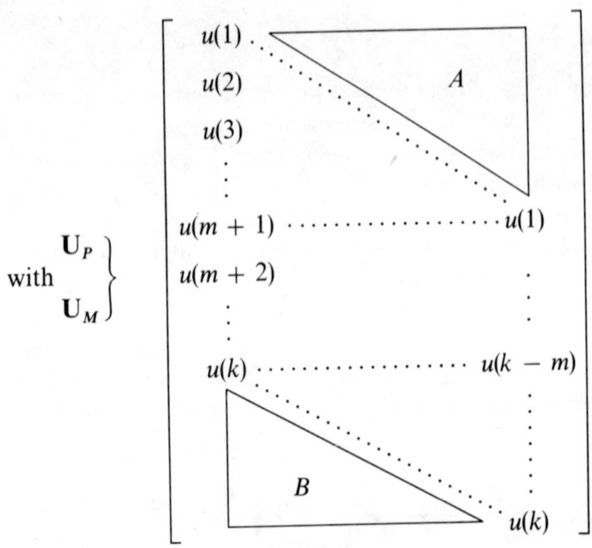


(6.70)

If at $t = 0$ the state of the process is zero, then all elements of $\mathbf{U}_P$ in the triangle $A$ are zero; if not then they contribute to $y(1), \ldots, y(m)$. The same holds for the end of the observation period; if we have control over the input signal $u$ and we would take $u(j) = 0$ for $j > k$, then the triangle $B$ will consist of zeros. If this control on the input signal is not available then the elements in $B$ add contributions to the last few samples of $y$, which are not relevant for our estimation. This results in an extra variance of the parameter estimates. Of course this error is only of some importance when handling **small** series of samples for estimating purposes.

This effect can be diminished in the following way. The least-square estimation is based on minimizing

$$E = \mathbf{e}'\mathbf{e} = \sum e^2(j)$$

When starting the observation we want to take the first few error samples 'less seriously' then the samples later on. Near the end of the observations we want to do the same with the last few error samples. This can be effected by minimizing (Draper and Smith, 1966):

$$E = \mathbf{e}'\mathbf{R}\mathbf{e} \quad \text{with} \quad \mathbf{R} = \begin{bmatrix} r_1 & & & & 0 \\ & r_2 & & & \\ & & r_3 & & \\ & & & \ddots & \\ 0 & & & & r_k \end{bmatrix}$$

with $r_1 < r_2 < r_3, \ldots, r_i = r_j, \ldots, r_{k-2} > r_{k-1} > r_k$. This results in an estimate given by

$$\hat{\boldsymbol{\beta}} = [\mathbf{U}'\mathbf{R}\mathbf{U}]^{-1}\mathbf{U}'\mathbf{R}\,\mathbf{y} \tag{6.71}$$

Another approach to this problem is taken by estimating the parameters and the state at $t = 0$ simultaneously.

### 6.3.4 *Errors Due to Simplification of the Implementation*

We have seen under what conditions the estimate

$$\hat{\boldsymbol{\beta}} = [\mathbf{U}'\mathbf{U}]^{-1}\mathbf{U}'\mathbf{y} \tag{6.72}$$

is unbiased. The matrix inversion and its multiplication with a matrix is unpleasant from the computational point of view. Consequently it is of interest to know whether it is permissible to simplify equation (6.72) to

$$\boldsymbol{\beta} = \frac{\mathbf{U}'\mathbf{y}}{k\tilde{\psi}_{uu}(0, k)} \quad \text{or} \quad \beta_j = \frac{\tilde{\psi}_{uy}(j, k)}{\tilde{\psi}_{uu}(0, k)} \tag{6.73}$$

particularly as we know from Section 6.2.3 that this expression is valid when $u$ has the character of white noise. From equation (6.72)

$$\begin{bmatrix} \tilde{\psi}_{uu}(0, k) & \cdots\cdots\cdots\cdots & & & \tilde{\psi}_{uu}(m, k) \\ \vdots & & & & \vdots \\ \cdots\cdots & \tilde{\psi}_{uu}(1, k) & \tilde{\psi}_{uu}(0, k) & \tilde{\psi}_{uu}(1, k) & \cdots\cdots \\ \vdots & & & & \vdots \\ \tilde{\psi}_{uu}(m, k) & \cdots\cdots\cdots\cdots & & & \tilde{\psi}_{uu}(0, k) \end{bmatrix} \begin{bmatrix} \hat{\beta}_0 \\ \vdots \\ \hat{\beta}_j \\ \vdots \\ \hat{\beta}_m \end{bmatrix} = \begin{bmatrix} \tilde{\psi}_{uy}(0, k) \\ \vdots \\ \tilde{\psi}_{uy}(j, k) \\ \vdots \\ \tilde{\psi}_{uy}(m, k) \end{bmatrix}$$

it follows that

$$\hat{\beta}_j = \frac{\tilde{\psi}_{uy}(j, k)}{\tilde{\psi}_{uu}(0, k)} - \frac{1}{\tilde{\psi}_{uu}(0, k)}\{(\hat{\beta}_{j+1} + \hat{\beta}_{j-1})\tilde{\psi}_{uu}(1, k) + \cdots\} \tag{6.74}$$

As $\beta_j$ determined without using matrix inversion gives only the first term, it follows that the second term indicates the bias due to this simplified instrumentation (Van den Boom and Melis, 1969).

Another point of interest is to compare the variance of the estimates determined with and without matrix inversion. With matrix inversion and 'white' additive noise $\mathbf{n}$ we have already found that:

$$\text{cov}\,[\hat{\boldsymbol{\beta}}] = [\mathbf{U'U}]^{-1}\mathbf{U'NU}[\mathbf{U'U}]^{-1} = \sigma_n^2[\mathbf{U'U}]^{-1} \tag{6.75}$$

Without matrix inversion the estimator is:

$$\boldsymbol{\beta} = \frac{\mathbf{U'y}}{k\tilde{\psi}_{uu}(0, k)} = \frac{\mathbf{U'Ub} + \mathbf{U'n}}{k\tilde{\psi}_{uu}(0, k)}$$

In general this will be biased, as in most cases:

$$\mathscr{E}[\boldsymbol{\beta}] = \frac{\mathbf{U'U}}{k\tilde{\psi}_{uu}(0, k)}\mathbf{b} \neq \mathbf{Ib} \tag{6.76}$$

Consequently the covariance is:

$$\text{cov}\,[\boldsymbol{\beta}] = \mathscr{E}\left[\left\|\boldsymbol{\beta} - \frac{\mathbf{U'Ub}}{k\tilde{\psi}_{uu}(0, k)}\right\|^2\right] = \frac{\mathbf{U'NU}}{k^2\tilde{\psi}^2_{uu}(0, k)} \tag{6.77}$$

and for white noise:

$$\text{cov}\,[\boldsymbol{\beta}] = \frac{\sigma_n^2\mathbf{U'U}}{k^2\tilde{\psi}^2_{uu}(0, k)} \tag{6.78}$$

If the input $\mathbf{u}$ has the character of white noise then for large $k$ the covariance has the same expression as in the case with matrix inversion; cf. equation (6.63). Consequently, a comparison of the explicit method with and without matrix inversion results in the following list of properties:

| With matrix inversion | Without matrix inversion |
|---|---|
| (*a*) The estimate is linear. $\hat{\boldsymbol{\beta}} = \mathbf{Qy}$ | Same. |
| (*b*) The estimate is unbiased. $\mathscr{E}[\hat{\boldsymbol{\beta}}] = \mathbf{b}$ <br> (*c*) $\lim_{k\to\infty} \text{cov}\,[\hat{\boldsymbol{\beta}}] = 0\mathbf{I}$ <br> (*d*) Suited for all types of input signals, provided all modes are excited. | The estimate is biased, unless the input signal is white and $k$ large enough; equation (6.76). |

| | |
|---|---|
| (*e*) Suited for all types of additive noise uncorrelated with $u$. | Same |
| (*f*) Requires matrix inversion, matrix–matrix multiplication and two matrix–vector multiplications. | Requires only a matrix–vector multiplication if $u$ is white and $\tilde{\psi}_{uu}(0, k)$ is approximately known beforehand. |
| (g) With $u$ white, $n$ white then $$\operatorname{cov}[\hat{\boldsymbol{\beta}}] \approx \frac{\sigma_n^2}{k\sigma_u^2}$$ | Same. |

### 6.3.5 *Sampling Error*

If the process under consideration is continuous (non-sampled), then one has to be aware of the influence of the choice of the sampling-interval length (cf. Chapter 7).

## 6.4 Residuals, Noise Properties and Model Order

In the previous sections we derived estimators for the parameters and for their covariance matrices. In practice, however, there are often more aspects of interest that may not be known beforehand, e.g. the properties of the noise (disturbances) and the number of the model parameters needed. For such type of knowledge we may study the residuals:

$$\mathbf{e} = \mathbf{y} - \mathbf{U}\hat{\boldsymbol{\beta}} = \mathbf{U}(\mathbf{b} - \hat{\boldsymbol{\beta}}) + \mathbf{n} \tag{6.79}$$

The residuals carry information on both the parameter–vector correspondence and the additive noise.

If it is not known beforehand how many parameters are significant, then it is of interest to have available an estimation procedure that is **recursive in the number of parameters.** This means that, starting with an estimate of the parameter vector $\mathbf{b}'_m = [b_0, \ldots, b_{m-1}]$, it is not necessary to do all the calculations again for estimating the parameter vector $\mathbf{b}'_p = [b_0, \ldots, b_{p-1}]$ with $p > m$.

Using the notation $\mathbf{b}'_p = [\mathbf{b}'_m \;\vdots\; \mathbf{b}'_{p-m}]$ we find

$$\mathbf{y} = [\mathbf{U}_m \;\vdots\; \mathbf{U}_{p-m}]\left[\begin{array}{c} \mathbf{b}'_m \\ \hline \mathbf{b}'_{p-m} \end{array}\right] + \mathbf{n} \tag{6.80}$$

which leads to the following least-squares equations:

$$\left[\begin{array}{c|c} \mathbf{U}'_m\mathbf{U}_m & - \\ \hline - & - \end{array}\right]\left[\begin{array}{c} \hat{\boldsymbol{\beta}}_{m;m} \\ \hline 0 \end{array}\right] = \left[\begin{array}{c} \mathbf{U}'_m\mathbf{y} \\ \hline - \end{array}\right]$$

and

$$\left[\begin{array}{c|c} \mathbf{U}'_m\mathbf{U}_m & \mathbf{U}'_m\mathbf{U}_{m-p} \\ \hline \mathbf{U}'_{m-p}\mathbf{U}_m & \mathbf{U}'_{m-p}\mathbf{U}_{m-p} \end{array}\right]\left[\begin{array}{c} \hat{\boldsymbol{\beta}}_{m;p} \\ \hline \hat{\boldsymbol{\beta}}_{p-m;p} \end{array}\right] = \left[\begin{array}{c} \mathbf{U}'_m\mathbf{y} \\ \hline \mathbf{U}'_{m-p}\mathbf{y} \end{array}\right] = \left[\begin{array}{c} \mathbf{U}'_m\mathbf{U}_m\hat{\boldsymbol{\beta}}_{m;m} \\ \hline \mathbf{U}'_{p-m}\mathbf{y} \end{array}\right]$$

$\hat{\boldsymbol{\beta}}_{m;m}$ and $\hat{\boldsymbol{\beta}}_{m;p}$ are the old and the new estimates of $b_0, \ldots, b_{m-1}$, respectively, that is, before and after the increase of the number of parameters from $m$

to $p$, using the same set of observations. These equations can be rewritten as:

$$\mathbf{U}'_m\mathbf{U}_m[\hat{\boldsymbol{\beta}}_{m;p} - \hat{\boldsymbol{\beta}}_{m;m}] + \mathbf{U}'_m\mathbf{U}_{m-p}\hat{\boldsymbol{\beta}}_{m-p;p} = \mathbf{0}$$

$$\mathbf{U}'_{m-p}\mathbf{U}_m[\hat{\boldsymbol{\beta}}_{m;p} - \hat{\boldsymbol{\beta}}_{m;m}] + \mathbf{U}'_{m-p}\mathbf{U}'_{m-p}\hat{\boldsymbol{\beta}}_{m-p;p} = \mathbf{U}'_{m-p}[\mathbf{y} - \mathbf{U}_m\hat{\boldsymbol{\beta}}_{m;m}]$$

Solving for $\hat{\boldsymbol{\beta}}_{m;p}$ and $\hat{\boldsymbol{\beta}}_{m-p,p}$ leads to:

$$\left.\begin{aligned}\hat{\boldsymbol{\beta}}_{m;p} &= \hat{\boldsymbol{\beta}}_{m;m} - \mathbf{BCU}'_{m-p}[\mathbf{y} - \mathbf{U}_m\hat{\boldsymbol{\beta}}_{m;m}]\\ \hat{\boldsymbol{\beta}}_{m-p;p} &= \mathbf{CU}'_{m-p}[\mathbf{y} - \mathbf{U}_m\hat{\boldsymbol{\beta}}_{m;m}]\end{aligned}\right\} \qquad (6.81)$$

with

$$\mathbf{A} = [\mathbf{U}'_m\mathbf{U}_m]^{-1}$$

$$\mathbf{B} = \mathbf{AU}'_m\mathbf{U}_{m-p}$$

$$\mathbf{C} = [\mathbf{U}'_{m-p}\mathbf{U}_{m-p} - \mathbf{U}'_{m-p}\mathbf{U}_m\mathbf{AU}'_m\mathbf{U}_{m-p}]^{-1}$$

The term $[\mathbf{y} - \mathbf{U}_m\hat{\boldsymbol{\beta}}_{m;m}]$ represents the **residuals** of the initial model with parameters $\beta_0, \ldots, \beta_{m-1}$. Note that $\mathbf{A} = [\mathbf{U}'_m\mathbf{U}_m]^{-1}$ is known from the previous estimation and that the matrix to be inverted is only of the order $(m - p)$. Consequently, $\hat{\boldsymbol{\beta}}_{m;p}$ and $\hat{\boldsymbol{\beta}}_{m-p;p}$ can be computed quite easily. For the next estimation one also needs the inverse of the augmented $\mathbf{U}'\mathbf{U}$ matrix. This can be shown to satisfy the relationship.

$$\begin{bmatrix}\mathbf{U}'_m\mathbf{U}_m & \mathbf{U}'_m\mathbf{U}_{m-p}\\ \mathbf{U}'_{m-p}\mathbf{U}_m & \mathbf{U}'_{m-p}\mathbf{U}_{m-p}\end{bmatrix} = \begin{bmatrix}\mathbf{A} - \mathbf{BCB}' & -\mathbf{BC}\\ -\mathbf{C}'\mathbf{B}' & \mathbf{C}\end{bmatrix} \qquad (6.82)$$

Consequently, it consists of the matrices that have been determined already.

In this way, it is possible to calculate the estimates for models with different numbers of parameters. We have, however, no answer yet to the question what number of parameters (what model order) actually is needed. Generally speaking, an increase of the number of model parameters will result in a decrease of the error or loss function $E = \mathbf{e}'\mathbf{e}$. Actually $E$ may become zero if the number of parameters equals $k$, the number of samples from input and output signals. In that case, the noise contribution is incorporated completely in the model; this, of course, is undesirable. Therefore, we have to use a statistical test for determining whether a decrease of $E$, when increasing the number of parameters, is significant.

The idea to view the test of order as a decision problem has been discussed by Anderson (1962). It is also a standard tool in regression analysis. The following derivation of such a test is taken from Åström (1968). Assume the process to be

$$\mathbf{y} = \mathbf{U}_m\mathbf{b}_m + \mathbf{n} \qquad (6.83)$$

where $\mathbf{y}$ consists of $k$ sample values, $\mathbf{b}_m$ has $m$ parameters (e.g. $b_0, \ldots, b_{m-1}$) and where $\mathbf{n}$ is independent normal $(0, \sigma)$. Based on a model of order $p$ the estimate $\hat{\boldsymbol{\beta}}_p$ is:

$$\hat{\boldsymbol{\beta}}_p = [\mathbf{U}_p'\mathbf{U}_p]^{-1}\mathbf{U}_p'\mathbf{y} = \left[\frac{\mathbf{b}_m}{\mathbf{0}_{p-m}}\right] + [\mathbf{U}_p'\mathbf{U}_p]^{-1}\mathbf{U}_p'\mathbf{n} \tag{6.84}$$

and the residuals:

$$\begin{aligned}\mathbf{e}_p &= \mathbf{y} - \mathbf{U}_p\hat{\boldsymbol{\beta}}_p \\ &= \mathbf{U}_m\mathbf{b}_m + \mathbf{n} - [\mathbf{U}_m \;\vdots\; \mathbf{U}_{p-m}]\left[\frac{\mathbf{b}_m}{\mathbf{0}_{p-m}}\right] - \mathbf{U}_p[\mathbf{U}_p'\mathbf{U}_p]^{-1}\mathbf{U}_p'\mathbf{n} \qquad (6.85) \\ &= \mathbf{n} - \mathbf{U}_p[\mathbf{U}_p'\mathbf{U}_p]^{-1}\mathbf{U}_p'\mathbf{n} = [\mathbf{I} - \mathbf{A}]\mathbf{n}\end{aligned}$$

$\mathbf{A}$ is a symmetric idempotent matrix, i.e.

$$\mathbf{A} = \mathbf{A}^2 = \mathbf{A}^3 = \cdots$$

the same holds for the matrix $[\mathbf{I} - \mathbf{A}]$; cf. Appendix C. Consequently:

$$E_p = \mathbf{e}_p'\mathbf{e}_p = \mathbf{n}'[\mathbf{I}_k - \mathbf{U}_p[\mathbf{U}_p'\mathbf{U}_p]^{-1}\mathbf{U}_p']\mathbf{n} \tag{6.86}$$

Now if we consider two models with $p$ and $q$ ($p < q$) as the numbers of parameters, then the following trivial identity can be considered:

$$E_q + (E_p - E_q) + (\mathbf{n}'\mathbf{I}_k\mathbf{n} - E_p) = \mathbf{n}'\mathbf{I}_k\mathbf{n}$$

The quantities at the left-hand side of the equation are non-negative definite quadratic forms in $\mathbf{e}_q$, $\mathbf{e}_p$ and $\mathbf{n}$. Using thc knowledge about the corresponding matrices:

symmetric, non-negative definite;

which, by an orthogonal transformation can be brought to diagonal form having only 0's and 1's on the diagonal:

for which the rank = trace (invariant):

cf. Appendix C, one notices:

$$\begin{aligned}&\text{rank}\,[\mathbf{I}_k - \mathbf{U}_q[\mathbf{U}_q'\mathbf{U}_q]^{-1}\mathbf{U}_q'] = k - q \\ &\text{rank}\,[\mathbf{U}_q[\mathbf{U}_q'\mathbf{U}_q]^{-1}\mathbf{U}_q' - \mathbf{U}_p[\mathbf{U}_p'\mathbf{U}_p]^{-1}\mathbf{U}_p'] = q - p \\ &\text{rank}\,[\mathbf{U}_p[\mathbf{U}_p'\mathbf{U}_p]^{-1}\mathbf{U}_p'] = p\end{aligned}$$

By Cochran's theorem (cf. Kendall and Stuart, 1961, Vol. 2, p. 360 ff), each of $E_q$, $E_p - E_q$ and $\mathbf{n}'\mathbf{I}_k\mathbf{n} - E_p$ is distributed as $\chi^2$ and they are mutually independent. Consequently, if $q > p \geqslant m$ then $E_q$ and $E_p - E_q$ are independent random variables having $\chi^2$ distributions with $k - q$ and $q - p$ degrees of freedom, respectively.

Now a test of whether the error or loss function is significantly reduced by increasing the number of parameters from $p$ to $q$ can be made by

$$t = \frac{(E_p - E_q)/(q - p)}{E_q/(k - q)} \tag{6.87}$$

This has an $F_{k-q}^{q-p}$ distribution, which is tabulated extensively; e.g. Fraser (1958, p. 390 ff). From such tables, at the 5% level:

$$F_{120}^{1} = 3{\cdot}92 \qquad F_{\infty}^{1} = 3{\cdot}84$$

$$F_{120}^{2} = 3{\cdot}07 \qquad F_{\infty}^{2} = 3{\cdot}00$$

$$F_{120}^{3} = 2{\cdot}68 \qquad F_{\infty}^{3} = 2{\cdot}60$$

i.e.

$$P(F_{120}^{1} < 3{\cdot}92) = 0{\cdot}95$$

Consequently, if $t$ is greater than the $F$ value for the case under consideration a higher-order model has to be used. In engineering situations, one is interested in cases where $k \gg q$ and thus, it is of interest to remember that for $k \to \infty$ $F_{k-q}^{q-p}$ is asymptotically $\chi^2$, i.e.

$$F_{k-q}^{q-p} \Rightarrow \frac{\chi_{q-p}^2}{q - p}$$

Consequently, tables of the $\chi^2$ distribution can be used as well.

A perfect matching of model and process would result in $\mathbf{e} = \mathbf{n}$. If one has *a priori* assumptions about the statistics of the noise then it is possible to test the residuals with respect to that assumption. If $\mathbf{n}$ is assumed to be 'white', i.e. the sample values are statistically independent, then that property can be tested on $\mathbf{e}$ by correlation techniques; if $\mathbf{n}$ is assumed to be gaussian, then a $\chi^2$ test can be used as verification.

For further discussions of the order-determination problem the reader is referred to: Woodside (1970); Van den Boom and Van den Enden (1973); Unbehauen and Göhring (1973).

If the variance of the noise $\sigma_n$ is not known beforehand, then this can be estimated by

$$s_n^2 = \frac{1}{k - m}\mathbf{e}'\mathbf{e} \tag{6.88}$$

where $k$ is the number of samples and $m$ is the number of parameters to be determined. This can be shown as follows (Rosenberg and Shen, 1963;

Åström 1968). From equation (6.86) it follows that:

$$\begin{aligned}\mathscr{E}[\mathbf{e}'_m\mathbf{e}_m] &= \mathscr{E}[\mathbf{n}'[\mathbf{I}_k - \mathbf{U}_m[\mathbf{U}'_m\mathbf{U}_m]^{-1}\mathbf{U}'_m]\mathbf{n}]\\ &= \mathscr{E}[\text{trace }\mathbf{n}'[\mathbf{I}_k - \mathbf{U}_m[\mathbf{U}'_m\mathbf{U}_m]^{-1}\mathbf{U}'_m]\mathbf{n}]\\ &= \mathscr{E}[\text{trace }[\mathbf{I}_k - \mathbf{U}_m[\mathbf{U}'_m\mathbf{U}_m]^{-1}\mathbf{U}'_m]\mathbf{n}\mathbf{n}']\\ &= \sigma_n^2\{\text{trace }\mathbf{I}_k - \text{trace }\mathbf{U}_m[\mathbf{U}'_m\mathbf{U}_m]^{-1}\mathbf{U}'_m\} = \sigma_n^2(k-m)\end{aligned}$$

Consequently, an unbiased estimate of $\sigma_n^2$ is given by equation (6.88), i.e. by squaring the residuals and dividing by the number of degrees of freedom.

Note that in statistical literature, expressions can also be found for confidence intervals of parameters $\beta_i$ and for tests of the hypothesis $\beta_i = 0$.

*Some Further Points of Interest*

Hoppe (1965) describes a method in which the point by point measurements (samples) are replaced by the results of integration over short periods of time. This should be advantageous in situations in which the process output is corrupted by additive noise, the frequency of which is high compared to the sampling frequency. If one has control over the input $\mathbf{u}$ then the interesting question arises how to choose $\mathbf{U}$ in order to make the expected errors as small as possible. This problem is briefly considered by Leonov (1967).

## 6.5 Extension to Generalized Models and Processes with Feedback

### *6.5.1 The Generalized Model*

In view of the interesting properties and the convenient instrumentation (linearity-in-the-parameters and a quadratic error or loss function) it is tempting to try to extend the least-squares or Markov estimation to the generalized model as introduced in Chapter 2 and shown again in Figure 6.10. In this case

$$e(k) = y(k) + \sum_{j=1}^{n} \alpha_j y_j(k) - \sum_{j=0}^{m} \beta_j u_j(k)$$

or

$$\mathbf{e} = \mathbf{y} + \mathbf{Y}\boldsymbol{\alpha} - \mathbf{U}\boldsymbol{\beta} = \mathbf{y} - [\mathbf{U} \mid \mathbf{Y}]\begin{bmatrix}\boldsymbol{\beta}\\ \hline -\boldsymbol{\alpha}\end{bmatrix} \tag{6.89}$$

$$= \mathbf{y} - \quad \boldsymbol{\Omega} \quad\; \boldsymbol{\theta}$$

In the sequel $m = n$ will be used as the number of parameters in $\boldsymbol{\alpha}$ and $\boldsymbol{\beta}$. Each expression can easily be adapted to the situation where $m \neq n$. For the sake of convenience, the term $\beta_0$ will be discarded as well. Along the same lines as before by minimizing $E = \mathbf{e}'\mathbf{e}$ the estimate is found to be:

$$\hat{\boldsymbol{\theta}} = [\boldsymbol{\Omega}'\boldsymbol{\Omega}]^{-1}\boldsymbol{\Omega}'\mathbf{y} \tag{6.90}$$

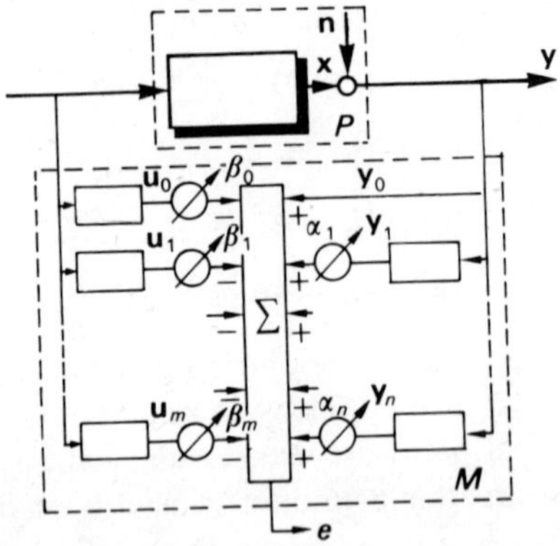


Figure 6.10

We will soon notice, that this description has a severe disadvantage due to a bias in the estimates caused by the additive noise. Yet it pays to study this case a bit further, as it covers one of the most important representations of a linear, time-invariant, discrete-time process with one input and one output. This is obtained by choosing $u_i(k) = u(k - i)$ and $y_j(k) = y(k - j)$:

process:

$$\begin{aligned} &x(k) + a_1 x(k-1) + \cdots + a_m x(k-m) = b_1 u(k-1) + \cdots + b_m u(k-m) \\ &y(k) = x(k) + n(k) \end{aligned} \tag{6.91}$$

model:

$$\begin{aligned} y(k) &+ \alpha_1 y(k-1) + \cdots + \alpha_m y(k-m) - \beta_1 u(k-1) - \cdots - \beta_m u(k-m) \\ &= e(k) \end{aligned} \tag{6.92}$$

By introduction of the shift operator

$$zy(k) = y(k+1)$$

as before, then the process can be described by:

$$A(z)x(k) = B(z)u(k) \quad \begin{cases} A(z) = z^m + a_1 z^{m-1} + \cdots + a_m \\ B(z) = \qquad\quad b_1 z^{m-1} + \cdots + b_m \end{cases}$$

or

$$A^*(z^{-1})x(k) = B^*(z^{-1})u(k) \quad \begin{cases} A^*(z^{-1}) = 1 + a_1 z^{-1} + \cdots + a_m z^{-m} \\ B^*(z^{-1}) = \qquad b_1 z^{-1} + \cdots + b_m z^{-m} \end{cases}$$

with

$$y(k) = x(k) + n(k)$$

For the model the corresponding expressions $\mathscr{A}$, $\mathscr{B}$, $\mathscr{A}^*$ and $\mathscr{B}^*$ with $\alpha$ and $\beta$ parameters will be used. Using the $\mathscr{A}^*$, $\mathscr{B}^*$ notation, the generalized error can be interpreted by different kinds of models:

(1) a 'generalized' model:

$$e(k) = \mathscr{A}^*(z^{-1})y(k) - \mathscr{B}^*(z^{-1})u(k) \tag{6.93}$$

which is represented schematically in Figure 6.11;

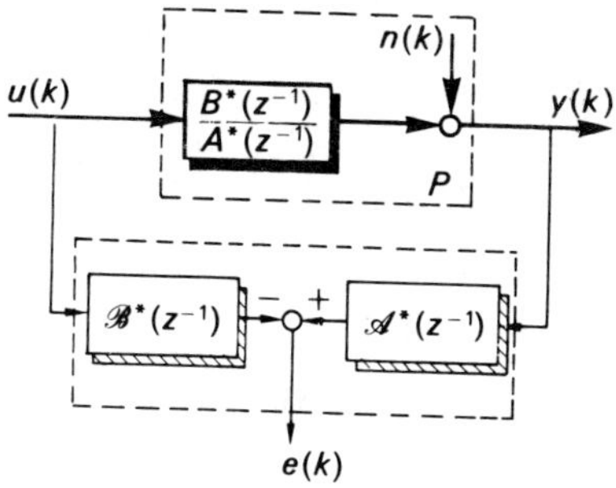


Figure 6.11

(2) a 'prediction' model

$$e(k) = y(k) - w(k) \tag{6.94}$$

where

$$w(k) = \{1 - \mathscr{A}^*(z^{-1})\}y(k) + \mathscr{B}^*(z^{-1})u(k) \tag{6.95}$$

This model is also linear-in-the-parameters. Its output can be interpreted as a prediction of $y(k)$ based on previous observations: $w(k) = \hat{y}(k|k-1)$ cf. Figure 6.12. Consequently, the generalized error can be considered as the

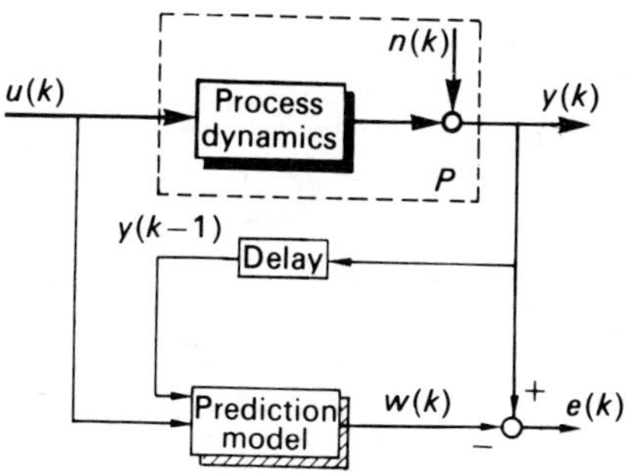


Figure 6.12

difference between the actual output and its prediction according to equation (6.95). Note that this type of description also holds for situations where there are no inputs, e.g. parametric time series analysis in the sense of fitting an autoregression. This has been discussed by Wold (1938) and Whittle (1963). Applications to EEG analysis have been given by Gersch (1970).

(3) an 'ordinary' model (and filtering of the error signal):

$$e(k) = \mathscr{A}^*(z^{-1})[y(k) - y_M(k)] \tag{6.96}$$

where

$$y_M(k) = \frac{\mathscr{B}^*(z^{-1})}{\mathscr{A}^*(z^{-1})}u(k) \tag{6.97}$$

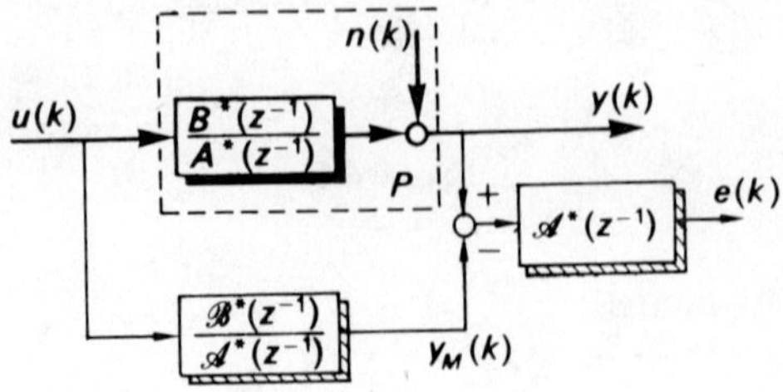


Figure 6.13

cf. Figure 6.13. Note that process and model can be represented as consisting of a forward part $B^*$, viz. $\mathscr{B}^*$, and a backward part $1 - A^*$, viz. $1 - \mathscr{A}^*$, as shown in Figure 6.14. Note also in this figure that the noise can be assumed

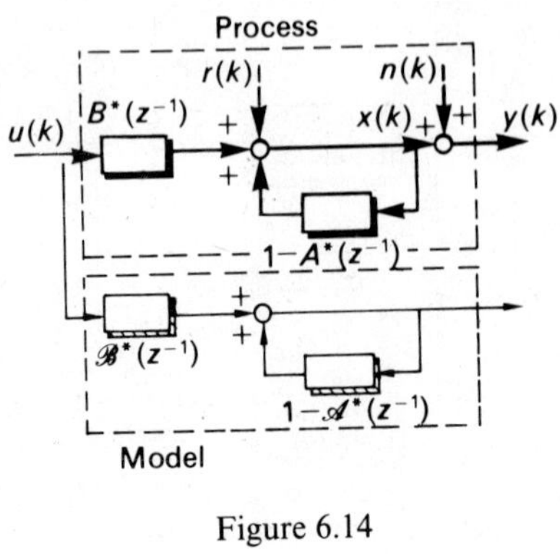


Figure 6.14

to affect the process in different ways, e.g. as $r(k)$ or as $n(k)$. This implies as process descriptions:

| | $r(k) \not\equiv 0 \quad n(k) \equiv 0$ | $r(k) \equiv 0 \quad n(k) \not\equiv 0$ |
|---|---|---|
| in $u$ and $x$ | $A^*x(k) = B^*u(k) + r(k)$ | $A^*x(k) = B^*u(k)$ |
| | $y(k) = x(k)$ | $y(k) = x(k) + n(k)$ |
| in $u$ and $y$ | $A^*y(k) = B^*u(k) + r(k)$ | $A^*y(k) = B^*u(k) + A^*n(k)$ |

Depending on the type of model and type of application, one may prefer one of these descriptions above the other.

For the interpretations just given, the estimator:

$$\hat{\boldsymbol{\theta}} = [\boldsymbol{\Omega}'\boldsymbol{\Omega}]^{-1}\boldsymbol{\Omega}'\mathbf{y} \tag{6.98}$$

consists of the following components:

$$\hat{\boldsymbol{\theta}}' = [\hat{\boldsymbol{\beta}}' \,\vdots\, -\hat{\boldsymbol{\alpha}}'] = [\hat{\beta}_1, \quad \ldots, \quad \hat{\beta}_m \,\vdots\, -\hat{\alpha}_1, \quad \ldots, \quad -\hat{\alpha}_m]$$

$$\mathbf{y} = \begin{bmatrix} y(1) \\ y(2) \\ \vdots \\ y(k) \end{bmatrix}$$

$$\boldsymbol{\Omega} = \left[\begin{array}{cccc|cccc} u(0) & u(-1) & \cdots & u(1-m) & y(0) & y(-1) & \cdots & y(1-m) \\ u(1) & u(0) & \cdots & u(2-m) & y(1) & y(0) & \cdots & y(2-m) \\ \vdots & & & \vdots & \vdots & & & \vdots \\ u(k-1) & \ldots & \ldots & u(k-m) & y(k-1) & y(k-2) & \ldots & y(k-m) \end{array}\right]$$

and the corresponding expressions for $\boldsymbol{\Omega}'\boldsymbol{\Omega}$ and $\boldsymbol{\Omega}'\mathbf{y}$. These expressions may be programmed in a straightforward way for a digital computer. Starting from the observations $\mathbf{u}$ and $\mathbf{y}$, such a programme results in numerical values for $\hat{\boldsymbol{\theta}}$ and, possibly, for the covariance matrix. As indicated before, this is an explicit or 'one-shot' method where all observations are processed in one sequence.

There is a close relation between this method and the well-known correlation technique for determining points of the impulse response. In that case the following (approximations of) correlation coefficients are calculated:

$$\left.\begin{aligned}
\tilde{\psi}_{uu}(j) &= \frac{1}{k-j}\sum_{i=1}^{k-j} u(i)u(i+j) \\
\tilde{\psi}_{yy}(j) &= \frac{1}{k-j}\sum_{i=1}^{k-j} y(i)y(i+j) \\
\tilde{\psi}_{yu}(j) &= \frac{1}{k-j}\sum_{i=1}^{k-j} y(i)u(i+j) \\
\tilde{\psi}_{uy}(j) &= \frac{1}{k-j}\sum_{i=1}^{k-j} u(i)y(i+j)
\end{aligned}\right\} \tag{6.99}$$

Comparing with the least-squares parameter estimation we find that the elements of the matrices $\mathbf{\Omega}'\mathbf{\Omega}$ and $\mathbf{\Omega}'\mathbf{y}$ consist essentially of auto and cross-correlations, if the beginning and the end of the observation series are neglected.

Notice that the technique of this section can immediately be applied to the identification of non-linear processes which are linear-in-the-parameters, e.g.

$$y(k) + ay(k-1) = b_1 u(k-1) + b_2 u^2(k-1)$$

### 6.5.2 *The Bias Problem*

It was mentioned previously that the least-squares estimation, using a generalized model, suffers from a bias in the estimates. This is due to the fact, that in the $\mathbf{\Omega}$ matrix of observations the $y$ elements are contaminated with noise:

$$\mathbf{\Omega} = [\mathbf{U} \,\vdots\, \mathbf{Y}] = [\mathbf{U} \,\vdots\, \mathbf{X}] + \left[\begin{array}{c|ccc} & n(0) & \cdots & n(1-m) \\ \mathbf{0} & \vdots & & \vdots \\ & n(k-1) & \cdots & n(k-m) \end{array}\right]$$

or in an obvious shorthand notation

$$\mathbf{\Omega} = \mathbf{\Omega}_{u,y} = \mathbf{\Omega}_{u,y} + \mathbf{\Omega}_{0,n} \tag{6.100}$$

This contamination is contrary to the assumptions under which the least-squares estimator was derived. The reader may convince himself that a bias is bound to occur, except in very special cases, by considering the following simple example of Figure 6.15.

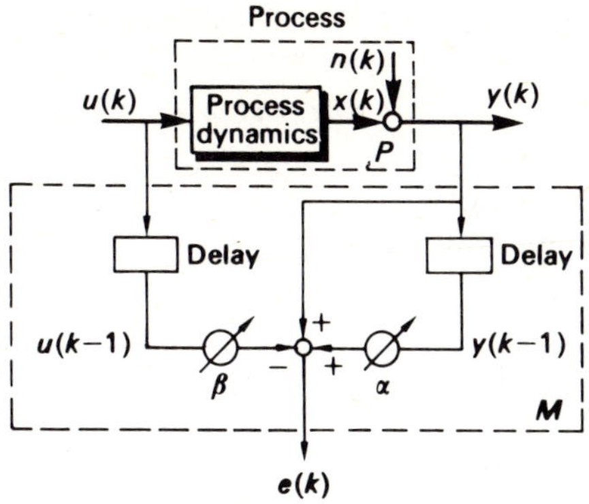


Figure 6.15

*Example*. This is described by:

$$\left.\begin{aligned} x(k+1) + ax(k) &= bu(k) \\ y(k) &= x(k) + n(k) \end{aligned}\right\} \text{process}$$

$$e(k) = y(k) + \alpha y(k-1) - \beta u(k-1) \quad \text{model}$$

The estimate follows from a (conceptual) model adjustment, giving a minimization of the loss function

$$E = \sum_{i=1}^{k} e^2(i)$$

with the necessary conditions

$$\frac{\partial E}{\partial \alpha} = \sum_{i=1}^{k} e(i)y(i-1) = 0,$$

$$\frac{\partial E}{\partial \beta} = \sum_{i=1}^{k} e(i)u(i-1) = 0$$

This can be written as

$$\left.\begin{aligned} \sum[\{x(i) + \alpha x(i-1) - \beta u(i-1) + n(i) + \alpha n(i-1)\}\{x(i-1) + n(i-1)\}] = 0 \\ \sum[\{x(i) + \alpha x(i-1) - \beta u(i-1) + n(i) + \alpha n(i-1)\}u(i-1)] = 0 \end{aligned}\right\} \quad (6.101)$$

These equations are fulfilled for $\alpha = a$ and $\beta = b$, i.e. **unbiased estimates**, if:

(*a*) $n(i) = 0 \qquad i = 1, \ldots, k$

i.e. there is no noise present.

(*b*) $\sum_i n(i)n(i-1) + a\sum_i n^2(i-1) = 0$

which is approximated by the respective points of the autocorrelation function:

$\psi_{nn}(1) + a\psi_{nn}(0) = 0$

i.e. there is noise which, by passing through the part of the generalized model, provides 'white' residuals.

In all other cases there will be a **bias.** To see how this works out consider an illustration: Assume that the process is actually described by the model

$$y(k) + ay(k-1) = bu(k-1) + n(k) + 0{\cdot}1\,n(k-1)$$

where $n(k)$ is a sequence of independent normal (0, 1) random variables, but that the system is identified using the least-squares method under the assumption that the residuals are uncorrelated. Below we give a typical result obtained from 500 pairs of inputs and outputs.

| Process parameters | Estimates |
|---|---|
| $a = -0{\cdot}5$ | $\hat{\alpha} = -0{\cdot}643 \pm 0{\cdot}029$ |
| $b = 1{\cdot}0$ | $\hat{\beta} = 1{\cdot}018 \pm 0{\cdot}062$ |

We are consequently in a very bad situation; not only is the estimate $\hat{\alpha}$ wrong but we have also a great deal of confidence in the wrong result. (The true value deviates from the estimate by about $5\sigma$.)

Now the properties of the estimate have to be studied in some detail.* Substitution of the relations

$$\mathbf{\Omega}_{u,x} = \mathbf{\Omega}_{u,y} - \mathbf{\Omega}_{0,n}$$

and

$$\mathbf{y} = \mathbf{\Omega}_{u,x}\mathbf{t} + \mathbf{n}$$

where $\mathbf{t}' = [\mathbf{b}' \,\vdots\, -\mathbf{a}']$ is the total process parameter vector, in the expression for the estimator (6.98) leads to:

$$\begin{aligned}\hat{\boldsymbol{\theta}} &= [\mathbf{\Omega}'_{u,y}\mathbf{\Omega}_{u,y}]^{-1}\mathbf{\Omega}'_{u,y}\mathbf{y} \\ &= [\mathbf{\Omega}'_{u,y}\mathbf{\Omega}_{u,y}]^{-1}\mathbf{\Omega}'_{u,y}[\mathbf{\Omega}_{u,y}\mathbf{t} - \mathbf{\Omega}_{0,n}\mathbf{t} + \mathbf{n}]\end{aligned} \tag{6.102}$$

The expectation is:

$$\mathscr{E}[\hat{\boldsymbol{\theta}}] = \mathbf{t} - \mathscr{E}[[\mathbf{\Omega}'_{u,y}\mathbf{\Omega}_{u,y}]^{-1}[\mathbf{\Omega}'_{u,y}\mathbf{\Omega}_{0,n}\mathbf{t} - \mathbf{\Omega}'_{u,y}\mathbf{n}]] \tag{6.103}$$

The second right-hand term indicates the bias of the estimate. Now we study the asymptotic properties of this estimator for an increasing number of observations; $k \to \infty$. Consider

$$\left[\frac{\mathbf{\Omega}'_{u,y}\mathbf{\Omega}_{u,y}}{k}\right]^{-1}\left[\frac{\mathbf{\Omega}'_{u,y}\mathbf{\Omega}_{0,n}}{k}\mathbf{t} - \frac{\mathbf{\Omega}'_{u,y}}{k}\mathbf{n}\right]$$

where the bias term is multiplied and divided by $k$. If $\mathbf{u}$ and $\mathbf{n}$ are well-behaved stationary random processes, then the sample averages will be consistent estimators of the auto- and cross-correlations. Using the notion

* Shaw, 1970, personal communication.

of 'limit-in-probability' (p lim)* and a theorem due to Slutsky† it follows from equation (6.102) that:

$$\underset{k\to\infty}{\text{p lim}}\,\hat{\boldsymbol{\theta}} = \mathbf{t} - \left[\underset{k\to\infty}{\text{p lim}}\,\frac{\boldsymbol{\Omega}'_{u,y}\boldsymbol{\Omega}_{u,y}}{k}\right]^{-1} \times \left[\underset{k\to\infty}{\text{p lim}}\,\frac{\boldsymbol{\Omega}'_{u,y}\boldsymbol{\Omega}_{0,n}}{k}\mathbf{t} - \underset{k\to\infty}{\text{p lim}}\,\frac{\boldsymbol{\Omega}'_{u,y}}{k}\mathbf{n}\right] \tag{6.104}$$

where:

$$\underset{k\to\infty}{\text{p lim}}\,\frac{\boldsymbol{\Omega}'_{u,y}\boldsymbol{\Omega}_{u,y}}{k} = \mathbf{Q}$$

$$\underset{k\to\infty}{\text{p lim}}\,\frac{\boldsymbol{\Omega}'_{u,y}\boldsymbol{\Omega}_{0,n}}{k} = \left[\begin{array}{c|ccc} \mathbf{0} & & \mathbf{0} & \\ \hline & \psi_{nn}(0) & \cdots & \psi_{nn}(1-m) \\ \mathbf{0} & \vdots & & \vdots \\ & \psi_{nn}(1-m) & \cdots & \psi_{nn}(0) \end{array}\right] = \left[\begin{array}{c|c} \mathbf{0} & \mathbf{0} \\ \hline \mathbf{0} & \mathbf{N} \end{array}\right]$$

$$\underset{k\to\infty}{\text{p lim}}\,\frac{\boldsymbol{\Omega}'_{u,y}\mathbf{n}}{k} = \left[\begin{array}{c} 0 \\ \vdots \\ 0 \\ \hline \psi_{nn}(1) \\ \vdots \\ \psi_{nn}(m) \end{array}\right]$$

*The non-random vector **c** is the p lim of $\mathbf{x}(k)$ for $k \to \infty$, i.e.

$$\text{p lim}\,\mathbf{x}(k) = \mathbf{c}$$

if

$$\lim_{k\to\infty} P(|\mathbf{x}(k) - \mathbf{c}| \geqslant \epsilon) = 0 \qquad \text{for each } \epsilon \geqslant \mathbf{0}$$

This implies

$$\lim_{k\to\infty} \mathscr{E}[\mathbf{x}(k)] = \mathbf{c}$$

†If the elements of $\mathbf{A}(k)$ and $\mathbf{B}(k)$ converge in probability then

$$\text{p lim}\,[\mathbf{A}^{-1}(k)\mathbf{B}(k)] = [\text{p lim}\,\mathbf{A}(k)]^{-1}[\text{p lim}\,\mathbf{B}(k)]$$

cf. Elgerd (1967, p. 118).

Assuming that the covariance matrix $\mathbf{Q}$ is positive definite:

$$\lim_{k\to\infty} \mathscr{E}[\hat{\boldsymbol{\theta}}] = \text{p}\lim_{k\to\infty} \hat{\boldsymbol{\theta}}$$

$$= \mathbf{t} - \mathbf{Q}^{-1}\left[\left[\begin{array}{c|c} \mathbf{0} & \mathbf{0} \\ \hline \mathbf{0} & \mathbf{N} \end{array}\right]\begin{bmatrix} b_1 \\ \vdots \\ b_m \\ \hline -a_1 \\ \vdots \\ -a_m \end{bmatrix} - \begin{bmatrix} 0 \\ \vdots \\ 0 \\ \hline \psi_{nn}(1) \\ \vdots \\ \psi_{nn}(m) \end{bmatrix}\right] \tag{6.105}$$

Consequently, **there is an asymptotic bias** unless the term in brackets is zero. This asymptotic bias does not exist if:

$$\mathbf{Na} = -\begin{bmatrix} \psi_{nn}(1) \\ \vdots \\ \psi_{nn}(m) \end{bmatrix} \tag{6.106}$$

i.e. if $n(k)$ is generated from white noise $w(k)$ by

$$n(k) + a_1 n(k-1) + \cdots + a_m n(k-m) = w(k)$$

This can be seen from multiplication by $n(k-i)$ and taking the expectation:

$$\mathscr{E}[n(k)n(k-1) + a_1 n^2(k-1) + \cdots + a_m n(k-m)n(k-1)]$$
$$= \mathscr{E}[w(k)n(k-1)]$$

$$\mathscr{E}[n(k)n(k-m) + a_1 n(k-1)n(k-m) + \cdots + a_m n^2(k-m)]$$
$$= \mathscr{E}[w(k)n(k-m)]$$

Here the right-hand side is zero, as $n(k-i)$ can depend only on previous samples of $w(k)$. Therefore, this expression is identical to equation (6.106). This condition for an unbiased estimate corresponds to the condition of 'white residuals', as in the notation of equation (6.93) for the right parameter values

$$\mathscr{A}^*(z^{-1}) = A^*(z^{-1}) \quad \text{and} \quad \mathscr{B}^*(z^{-1}) = B^*(z^{-1}):$$

$$\begin{aligned}
\text{model:} \quad & e(k) = A^*(z^{-1})x(k) + A^*(z^{-1})n(k) - B^*(z^{-1})u(k) \\
-\ \text{process:} \quad & \ \ \ \ 0 = A^*(z^{-1})x(k) \qquad\qquad\qquad\quad - B^*(z^{-1})u(k) \\
\hline
& e(k) = \qquad\qquad\qquad A^*(z^{-1})n(k) \\
& \quad = n(k) + a_1 n(k-1) + \cdots + a_m n(k-m) = w(k)
\end{aligned}$$

This phenomenon can be expressed by: the moving average

$$n(k) + a_1 n(k-1) + \cdots + a_m n(k-m)$$

whitens the noise generated by

$$n(k) + a_1 n(k-1) + \cdots + a_m n(k-m) = w(k)$$

This is illustrated in Figure 6.16.

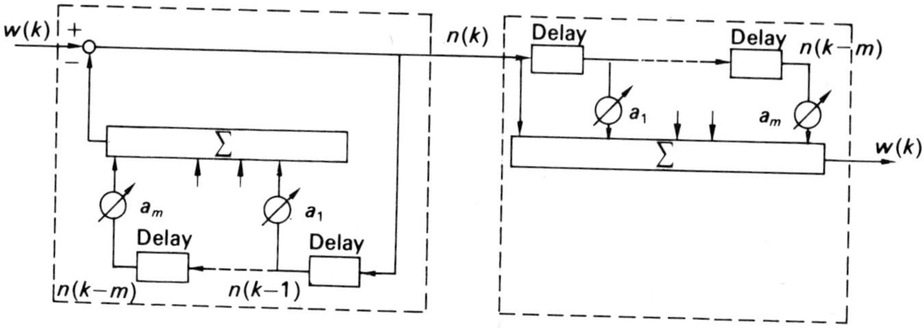


Figure 6.16

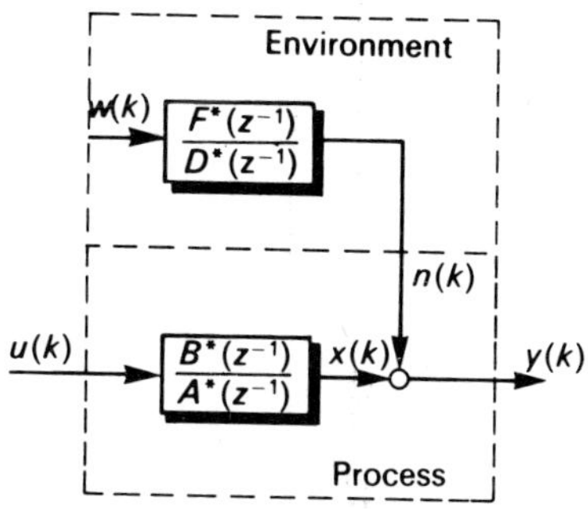


Figure 6.17

The condition of 'white residuals' can also be approached along the following lines. Consider the representation of process and environment given in Figure 6.17

$$A^*(z^{-1})x(k) = B^*(z^{-1})u(k)$$

$$y(k) = x(k) + n(k)$$

Consequently,

$$A^*(z^{-1})y(k) = B^*(z^{-1})u(k) + A^*(z^{-1})n(k) \tag{6.107}$$

If $n(k)$ is derived from a 'white' noise $w(k)$ through

$$D^*(z^{-1})n(k) = F^*(z^{-1})w(k) \tag{6.108}$$

then substitution of (6.108) in (6.107) with $C^*(z^{-1}) = A^*(z^{-1})F^*(z^{-1})$ results in:

$$A^*(z^{-1})y(k) = B^*(z^{-1})u(k) + \frac{C^*(z^{-1})}{D^*(z^{-1})}w(k) \tag{6.109}$$

By introducing

$$\tilde{y}(k) = \frac{D^*(z^{-1})}{C^*(z^{-1})}y(k)$$

$$\tilde{u}(k) = \frac{D^*(z^{-1}}{C^*(z^{-1})}u(k)$$

equation (6.107) can be written as:

$$A^*(z^{-1})\tilde{y}(k) = B^*(z^{-1})\tilde{u}(k) + w(k) \tag{6.110}$$

Thus, by using a generalized model operating on $\tilde{u}(k)$ and $\tilde{y}(k)$ the residuals are white and the estimates will be unbiased.

### 6.5.3 *Processes with Feedback*

Related problems occur when the process to be identified is part of a closed loop (cf. Figure 6.18). Note that the noise is inside the loop. The

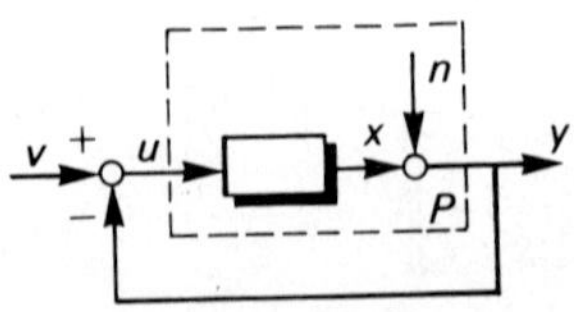


Figure 6.18

characteristics of $P$ are of interest. One has the freedom to consider either $v$ or $u$ as 'input signals'. In the first case, an estimation procedure based on $v$ and $y$, the overall characteristics of the closed-loop system are determined. The parameter determination for the process $P$ from these overall characteristics is not a trivial problem; e.g. the error propagation from the closed-loop description to the open loop one may lead to an accumulation of estimation errors.

In the second case, when the estimation procedure is based on $u$ and $y$, one has to face another problem. Due to the feedback, the 'input signal' $u = v - y$ is not statistically independent of the disturbance $n$, which was recognized to be a sufficient condition for obtaining unbiased estimates.

*Example* For the sampled case consider the determination of the sampled impulse response. The process is described by:

$$x(k) = b'_0 u(k) + b'_1 u(k-1) + \cdots + b'_m u(k-m) = \sum_{i=0}^{m} b'_i u(k-i) \tag{6.111}$$

or

$$x(k) = \sum_{i=0}^{m} b'_i v(k-i) - \sum_{i=0}^{m} b'_i y(k-i) \tag{6.112}$$

Because

$$y(k) = x(k) + n(k)$$

the output can be found explicitly as:

$$y(k) = \sum_{i=0}^{m} b_i v(k-i) - \sum_{i=1}^{m} b_i y(k-i) + \frac{1}{1+b'_0} n(k) \tag{6.113}$$

where $b_i = b'_i/(1 + b'_0)$. The condition of 'white residuals' can be studied as before.

Through the efforts to preserve the nice properties of the generalized model, that is the linearity-in-the-parameters, but to circumvent the detrimental effects of the additive noise, several techniques have been developed, viz. (cf. Åström and Eykhoff, 1970):

repeated least-squares (Section 6.5.4);

instrumental variables (Section 6.5.5);

tally principle (Section 6.5.6);

Levin's method (Section 6.5.7).

generalized least-squares and related schemes (cf. Chapter 7);

maximum likelihood (cf. Chapter 11).

### 6.5.4 *Repeated Least-Squares*

This method consists of increasing the order of the model

$$e(k) = \mathscr{A}^*(z^{-1})y(k) - \mathscr{B}^*(z^{-1})u(k) \tag{6.114}$$

in successive estimation runs. The correspondence of process and model is judged, based on the residual error. If an increase of model order does not have a significant improvement on the residual error then the estimation stops. It has been illustrated by Åström (1967) for a simple first-order system that a division of $\mathscr{A}^*(z^{-1})$ by $\mathscr{B}^*(z^{-1})$ results in a good model and that the rest of the division can be neglected. No formal proof of the validity of this method seems to be available yet.

### 6.5.5 *Instrumental Variables*

A straightforward solution to the problem of bias, due to observations which are contaminated with noise, can be indicated in a simple way. From the process equation

$$\mathbf{y} = \mathbf{\Omega b} + \mathbf{n} \tag{6.115}$$

follows through multiplication by $\mathbf{\Omega}'$

$$\mathbf{\Omega}'\mathbf{y} = \mathbf{\Omega}'\mathbf{\Omega b} + \mathbf{\Omega}'\mathbf{n} \tag{6.116}$$

If $\mathbf{\Omega}$ and $\mathbf{n}$ are statistically independent and if the inverse matrix exists, then the normal equation is

$$\hat{\boldsymbol{\beta}} = [\mathbf{\Omega}'\mathbf{\Omega}]^{-1}\mathbf{\Omega}'\mathbf{y} \tag{6.117}$$

Now assume that there is a matrix $\mathbf{V}$ available which is correlated with $\mathbf{u}$ but not with $\mathbf{n}$:

$$\text{p lim}\,[\mathbf{V}'\mathbf{n}] = \mathbf{0} \qquad \text{p lim}\,[\mathbf{V}'\mathbf{\Omega}] \quad \text{positive definite}$$

then

$$\mathbf{V}'\mathbf{y} = \mathbf{V}'\mathbf{\Omega b} + \mathbf{V}'\mathbf{n} = \mathbf{V}'\mathbf{\Omega}\hat{\boldsymbol{\beta}}$$

as $\mathbf{n}$ is unmeasurable, or:

$$\hat{\boldsymbol{\beta}} = [\mathbf{V}'\mathbf{\Omega}]^{-1}\mathbf{V}'\mathbf{y} \tag{6.118}$$

which guarantees an asymptotically unbiased estimate. The matrix $\mathbf{V}$ represents the so-called **'instrumental variables'** (cf. Figure 6.19). A particular

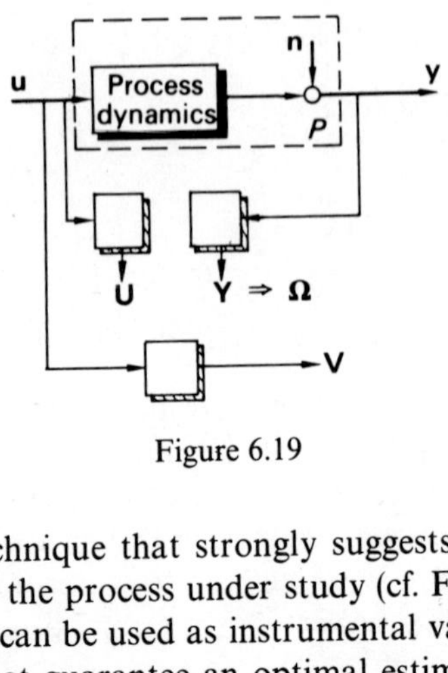


Figure 6.19

application of this technique that strongly suggests itself is the case of a feedback path around the process under study (cf. Figure 6.18). As $v$ is not correlated with $n$ this can be used as instrumental variable. Of course such an arbitrary $\mathbf{V}$ does not guarantee an optimal estimate. It is also possible to find instrumental variables such that the estimate has optimal properties. Moreover, this type of estimator can be formulated in a recursive version.

As references, cf. Reiersoll (1941), Kendall and Stuart (1961, Vol. 2, p. 397 ff), Joseph, Lewis and Tou (1961), Andeen and Shipley (1963), Wong and Polak (1967), Smets (1970).

The schemes proposed by Peterka and Šmuk (1969), Hsia and Landgrebe (1967) are closely related to the instrumental-variable technique. Vušković, Bingulac and Djorović (1970) indicate the use of pseudo-sensitivity functions as instrumental variables.

### 6.5.6 *Tally Principle*

This estimation procedure as described by Peterka and Hallousková (1970) starts from the canonical form

$$\begin{aligned} y(k) + a_1 y(k-1) + \cdots + a_m y(k-m) \\ = b_1 u(k-1) + \cdots + b_m u(k-m) + n(k) \end{aligned} \tag{6.119}$$

where

$$n(k) = v(k) + c_1 v(k-1) + \cdots + c_m v(k-m) \tag{6.120}$$

and $\{v(k)\}$ is a sequence of uncorrelated random variables. It is derived that:

$$\left.\begin{aligned} \mathscr{E}[n(i)u(i-j)] &= 0 \qquad \text{for } j > 0 \\ \mathscr{E}[n(i)y(i-j)] &= 0 \qquad \text{for } j > m \end{aligned}\right\} \tag{6.121}$$

For any estimate of the unknown parameters one finds a sequence $\{n^*(i)\}$. Now for a sequence of $k$ observations

$$\left.\begin{aligned} &\frac{1}{k}\sum_{i=1}^{k} n^*(i)u(i-j) \\ \text{and} \\ &\frac{1}{k}\sum_{i=1}^{k} n^*(i)y(i-j) \end{aligned}\right\} \tag{6.122}$$

have to 'tally' (duplicate) the corresponding expectations as well as possible in a least-square sense.

### 6.5.7 *Levin's Method*

A particular method for estimating the bias due to correlated residuals has been proposed by Levin (1959, 1960, 1964) for the case of a deterministic system with independent observation errors. Levin's results are based on a technique due to Koopmans and it gives the estimates in terms of an eigenvalue problem for the matrix $\mathbf{\Omega}'\mathbf{\Omega}$. A careful analysis of Levin's method which includes convergence proofs and error estimates has been done by Aoki and Yue (1970). The method has also been used by Smith (1968).

*Generalized Least-squares and Related Schemes*

These are discussed in Chapter 7.

*Maximum Likelihood*

This is discussed in Chapter 11.

**References**

Aitken, A. C. (1952). *Statistical Mathematics*, Oliver and Boyd, Edinburgh, 153 pp.

Andeen, R. E. and P. P. Shipley (1963). 'Digital adaptive flight control systems for aerospace vehicles', *AIAA J.*, **1,** 1105.

Anderson, T. W. (1962). 'The choice of the degree of a polynomial regression as a multiple decision problem', *Ann. Math. Stat.*, **33,** 255–265.

Aoki, M. and P. C. Yue (1970). 'On certain convergence questions in system identification', *SIAM J. Control*, **8,** 239–256.

Åström, K. J. (1967). Cf. references Chapter 11.

Åström, K. J. (1968). 'Lectures on the identification problem—the least squares method' report, *Lund Inst. Technol. Div. Autom. Control*, Sweden.

Åström, K. J. and P. Eykhoff (1970). Cf. references Chapter 2.

Cramér, H. (1946). *Mathematical Methods of Statistics*, Princeton University Press, Princeton, 575 pp.

Deutsch, R. (1965). Cf. references Chapter 5.

Douce, J. L. and P. D. Roberts (1964). 'Orthogonal functions', *Control*, **8,** 457–461; see also *Control*, **8,** 555.

Draper, N. R. and H. Smith (1966). *Applied Regression Analysis*, Wiley, New York, 407 pp.

Durbin, J. (1960). 'Estimation of parameters in time-series regression models', *J.R. Statist. Soc. Ser. B*, **22,** 139–153.

Elgerd, O. I. (1967). Cf. references Chapter 4.

Fraser, D. A. S. (1958). *Statistics, An Introduction*, Wiley, New York, 398 pp.

Gersch, W. (1970). Cf. references Section 14.7.

Graybill, F. A. (1961). *An Introduction to Linear Statistical Models*, McGraw-Hill, New York.

Grenander, U. and M. Rosenblatt (1957). Cf. references Chapter 4.

Hoppe, S. G. (1965). 'A least squares technique for the identification of a linear time-in-variant plant in a sampled data system' (correspondence), *IEEE Trans. Autom. Control*, **AC-10,** 490.

Hsia, T. C. and D. A. Landgrebe (1967). 'On a method for estimating power spectra', *IEEE Trans. Instrum. and Measurem.*, **IM-16,** 255–257.

Jenkins, G. and D. Watts (1968). Cf. references Chapter 4.

Joseph, P., J. Lewis and J. Tou (1961). 'Plant identification in the presence of disturbances and application to digital adaptive systems', *AIEE Trans.*, **80,** part II (Applications and Industry), 18–24.

Kendall, M. G. and A. Stuart (1961, 1962). *The Advanced Theory of Statistics*, Vol. 1 (1962) 2nd ed., Vol. 2 (1961), Griffin, London.

Kwakernaak, H. (1963). 'Some observation on the error in the estimation of impulse response' (correspondence), *IEEE Trans. Circuit Theory*, **CT-10,** 458–460.

Leonov, Yu, P. (1967). 'On estimate of parameters of linear dynamic systems', *IFAC Symp.—Identification in Autom. Control Systems, Prague*, paper 1.2.

Levin, M. J. (1959). 'Estimation of the characteristics of linear systems in the presence of noise', *Sc.D. Thesis, E.E. Dept. Columbia Univ.*, New York, N.Y., 170 pp.

Levin, M. J. (1960). 'Optimum estimation of impulse response in the presence of noise', *IRE Trans. Circuit Theory*, **CT-7,** 50–56.

Levin, M. J. (1964). 'Estimation of system pulse transfer function in the presence of noise', *IEEE Trans. Autom. Control*, **AC-9,** 229–235; cf. also same journal, **AC-9,** 214–215.

Mayne, D. Q. (1967). 'A method for estimating discrete time transfer functions', *IEE Convention—Advances in Computer Control*, April, 1967.

Melis, J. H. A. M. (1967). 'Explicit estimation of process parameters using a digital computer' (in Dutch), *M.Sc. Thesis, E.E. Dept., University of Technology, Eindhoven*, Netherlands.

Peterka, V. and K. Šmuk (1969). 'On-line estimation of dynamic model parameters from input–output data', *4th IFAC Congr., Warsaw*, paper 26.1.

Peterka, V. and A. Halousková (1970). 'Tally estimate of Åström model for stochastic systems', *2nd IFAC Symp.—Identification and Process Parameter Estimation*, Prague, paper 2.3.

Reiersoll, O. (1941). 'Confluence analysis by means of lag moments and other methods of confluence analysis', *Econometrica*, **9,** 1–23.

Roberts, P. D. (1967). 'Self-adjustable orthogonal digital filters for system identification and optimization', *Proc. IEE*, **114,** cf. also *Electron Letters*, **2,** 187–188 (1966).

Rosenberg, A. E. and D. W. C. Shen (1963). 'Regression analysis and its application to the system-identification problem', *Joint Autom. Control Conf.*, 446

Smets, A. J. (1970). 'The instrumental variable method and related identification schemes', *Report, E.E. Dept., University of Technology, Eindhoven, Netherlands.*

Smith, F. W. (1968). 'System Laplace-transform estimation from sampled data', *IEEE Trans. Autom. Control*, **AC-13,** 37–44.

Unbehauen, H. and B. Göring (1973). 'Application of different statistical tests for the determination of the most accurate order of the model in parameter estimation', *Proc. 3rd IFAC Symp.—Identification and System Parameter Estimation*, 917–928, North-Holland Publ. Co., Amsterdam.

Van den Boom, A. J. W. and J. H. A. M. Melis (1969). 'A comparison of some process parameter estimating schemes', *4th IFAC Congr., Warsaw*, paper 26.5.

Van den Boom, A. J. W. and A. W. M. van den Enden (1973). 'The determination of the order of process and noise dynamics', *Proc. 3rd IFAC Symp.—Identification and System Parameter Estimation*, 929–938, North Holland Publ. Co., Amsterdam.

Vuškovič, M. I., S. P. Bingulac and M. Djorovič (1970). 'Applications of the pseudo sensitivity functions in linear dynamic systems identification', *2nd IFAC Symp.—Identification and Process Parameter Estimation, Prague*, paper 2.5.

Westlake, J. R. (1968). *A Handbook of Numerical Matrix Inversion and Solution of Linear Equations*, Wiley, New York, 171 pp.

Whittle, P. (1963). Cf. references Chapter 4.

Wold, H. (1938). *A Study in the Analysis of Stationary Time Series*, Almqvist and Wiksell, Uppsala, 236 pp.

Wong, K. Y. and E. Polak (1967). 'Identification of linear discrete time systems using the instrumental variable method', *IEEE Trans. Autom. Control*, **AC-12,** 707–718.

Woodside, C. M. (1970). 'Estimation of the order of linear systems', *2nd IFAC Symp.—Identification and Process Parameter Estimation, Prague*, paper 1.7.

## Additional Literature


Balakrishnan, A. V. (1963). 'Determination of nonlinear system from input–output data', Paper: *Conference on Identification and Representation Problems*, Princeton Univ., Princeton, N.J.

Bigelow, S. C. and H. Ruge (1961). 'An adaptive system using periodic estimation of the pulse transfer function', *IRE Int. Conv. Rec.*, **9,** part 4, 25–38.

Diamessis, J. E. (1965). 'On the determination of the parameters of certain nonlinear systems' (correspondence), *Proc. IEEE*, **53,** 319–320.

Durbin, J. and G. S. Watson (1950). 'Testing for serial correlation in least squares regression', *Biometrika*, **37,** 409–428.

Efroymson, M. A. (1960). 'Multiple regression analysis', in: *Mathematical Methods for Digital Computers* (Eds. A. Ralston and H. Wilf), Wiley, New York, 191–203.

Fagin, S. L. (1964). 'Recursive linear regression theory, optimal filter theory and error analysis of optimal systems', *IEEE Con. Rec.*, **12,** part 1.

Fantanzzi, G. (1968). 'The relation between the sampling time and stochastic error for the impulsive response of linear time-independent systems' (short paper), *IEEE Trans. Autom. Control*, **AC-13,** 426–428.

Girshick, M. A. and L. J. Savage (1957). 'Bayes and minimax estimation for quadratic loss functions', *2nd Berkeley Symp. Math. Statist. Probab.* (Ed. J. Neyman), 53–73.

Graupe, D., B. H. Swanick and G. R. Cassir (1968). 'Reduction and identification of multivariable processes using regression analysis' (short paper), *IEEE Trans. Autom. Control*, **AC-13,** 564–567.

Graybill, F. A. (1969). 'On a method of correlation analysis for multivariate systems', *4th IFAC Congr., Warsaw*, paper 33.1.

Hanůs, B. and J. Vaníček (1967). 'Evaluation of transient characteristics from measured values', *IFAC Symp.—Identification in Autom. Control Systems, Prague*, paper 4.8.

Hsia, T. C. (1967). 'On power spectrum identification methods' (correspondence), *IEEE Trans. Atom. Control*, **AC-12,** 781–782.

Kerr, R. B. and W. H. Surber Jr. (1961). 'Precision of impulse response identification based on short, normal operating records', *IRE Trans. Autom. Control*, **AC-6,** 173–182.

Kichatov, Yu. F. (1965). 'Determination of the nonlinear characteristics of control plants under gaussian inputs', *Automn. Remote Control*, **26,** 465–470.

Koopmans, T. C. (1950). *Statistical Inference in Dynamic Economic Models*, Cowless Commission Monograph, 10, Wiley.

Koszelnik, M., J. Malkiewicz and St. Trybula (1969). Cf. references Section 14.4.

Kutikov, L. M. (1965). 'Inversion of correlation matrices and some problems in self-adjustment', *Engng. Cybernetics*, **3,** 35–39.

Lindenlaub, J. C. and G. R. Cooper (1963). 'Noise limitations of system identification techniques', *IEEE Trans. Autom. Control*, **AC-8,** 43–48.

Linnik, J. W. (1961). *The Least-Squares Method in a Modern Presentation* (in German), Deutscher Verlag der Wissenschaften, Berlin, 314 pp.

Loeb, J. and G. Cahen (1963). 'Derivation of the dynamic parameters from recorded measurements' (in French), *Automatisme*, **8,** 479–486.

Magnes, T. A. and J. B. McGuire (1962). 'Comparison of least square and minimum variance estimate of regression parameters', *Ann. Math. Stat.*, **33,** 462.

Malinvaud, E. (1966). *Statistical Methods of Econometrics, Studies in Mathematical and Managerial Economics*, Vol. 6, McNalley, Chicago.

Mann, H. B. and A. Wald (1943). Cf. references Chapter 4.

Parzen, E. (1961). 'Mathematical considerations in the estimation of spectra', *Technometrics*, **3,** 167–190.
Perel'man, I. I. (1967). 'Estimation accuracy of impulse response of a linear dynamic plant', *IFAC Symp.—Identification in Autom. Control Systems, Prague*, paper 1.9.
Placket, R. L. (1960). *Principles of Regression Analysis*, Oxford Univ. Press, New York.
Popov, E. P., G. M. Loskutov and R. M. Yusupov (1963). 'One self-adjusting control system without test disturbance signals', *2nd IFAC Congr., Basle*, 446.
Rubin, O. (1971). 'Process identification by generalized correlation' (short paper), *IEEE Trans. Autom. Control*, **AC-16,** 361–363.
Shaw, L. and G. Robinson (1967). 'Invariant estimation of stochastic system parameters', *IFAC Symp.—Identification in Autom. Control Systems, Prague*, paper 3.16.
Tretter, S. A. and K. Steiglitz (1967). Power spectrum identification in terms of rational models', *IEEE Trans. Autom. Control*, **AC-12,** 185–188.
Voronova, L. I. and Yu. V. Krementulo (1966). 'New method of complex dynamic system characteristics determination', in *Complex Control System*, Naukova Dumka, Kiev.
Voronova, L. I. (1967). 'Some methods of controlled plants dynamic characteristics determination and investigation of their accuracy', *IFAC Symp.—Identification in Autom. Control Systems, Prague*, paper 3.19.
Wegrzijn, S., G. Denis and J. Delisle (1969). 'The identification of processes by minimization of the distance between sets of signals', *4th IFAC Congr. Warsaw*, paper 15.1.
Williams, E. J. (1959). *Regression Analysis*, Wiley, New York, 179 pp.

*CHAPTER 7*

# *Sampled Signals; Implicit or Model-adjustment Techniques*

In the previous chapter we derived the 'normal equations' for the least-squares and the Markov estimators. These can readily be programmed for execution on a digital computer. Some disadvantages, however, are:

matrix inversion is a (computer-)time-consuming operation;

all sample values of input and output signals have to be stored in a fast memory in order to obtain acceptable computing time; this requires much memory space;

this method does not offer the possibilities of 'on-line' computation, i.e. taking into account new observations that may become available as time progresses and perhaps 'forgetting' old observations.

Based on this third consideration the **explicit** method, discussed in the previous chapter, can also be characterized as an accumulative, **'one-shot'** or **'one-step'** procedure.

In many applications it is highly desirable to obtain the result of the identification as the process develops. For example it might be of interest to proceed until a specified parameter accuracy is achieved. The problem is then as follows. Assume that an estimate $\hat{\boldsymbol{\beta}}(i)$ is obtained based on $i$ pairs of input–output samples. Is it necessary to repeat the whole identification procedure from the beginning, using the whole string of input–output data in order to obtain $\hat{\boldsymbol{\beta}}(i + 1)$ or is it possible to rearrange the computations using $\hat{\boldsymbol{\beta}}(i)$? An identification scheme which does not require that the whole string of input–output data is brought in at each step is called **recursive.** Such schemes may be considered for applications where one wants to process observations as soon as they become available; **on-line methods.** Apart from being of practical interest, this point of view on identification problems will also make it possible to establish connexions with other fields, e.g. non-linear filtering, stochastic approximation, learning and adaptation.

If the parameters of the process are truly time varying it is of course meaningless to do anything else but to track the parameters in real time. This is called **real-time identification.**

There are many different ways to obtain algorithms for implicit estimation techniques. Practically all methods, however, will yield algorithms with a strong similarity and be of the form

$$\hat{\boldsymbol{\beta}}(i+1) = \hat{\boldsymbol{\beta}}(i) + \boldsymbol{\Gamma}(i)\mathbf{e}(i)$$

where $\mathbf{e}(i)$ is the (generalized) error discussed earlier and $\boldsymbol{\Gamma}(i)$ is a gain factor which can be of varying complexity.

For models which are linear-in-the-parameters it is quite simple to determine the gradient and to base the adjustment scheme on those gradients. This is discussed in Section 7.1. When trying to use these methods for generalized models one meets with the problem of bias as discussed in the previous chapter. This implies the need for the extensive estimation schemes discussed in Section 7.2. Examples are given in Section 7.3. For models that are non-linear-in-the-parameters special gradient-determination schemes have to be devised or direct-search methods have to be used (cf. Section 7.4).

When studying this chapter, and especially the discussion of model adjustments, it should be kept in mind that the deterministic and the stochastic convergence schemes as discussed in Sections 5.2 and 5.3 have a direct applicability to this problem.

## 7.1 Models, Linear-in-the-parameters

The recursive or iterative solution to an estimation problem can be characterized by the task of updating an old estimate, say $\hat{\boldsymbol{\beta}}(i)$. The presence of this estimate implies the need for a 'model (e.g. programmed on a digital computer) as shown in Figure 7.1a. The input is simultaneously fed to the

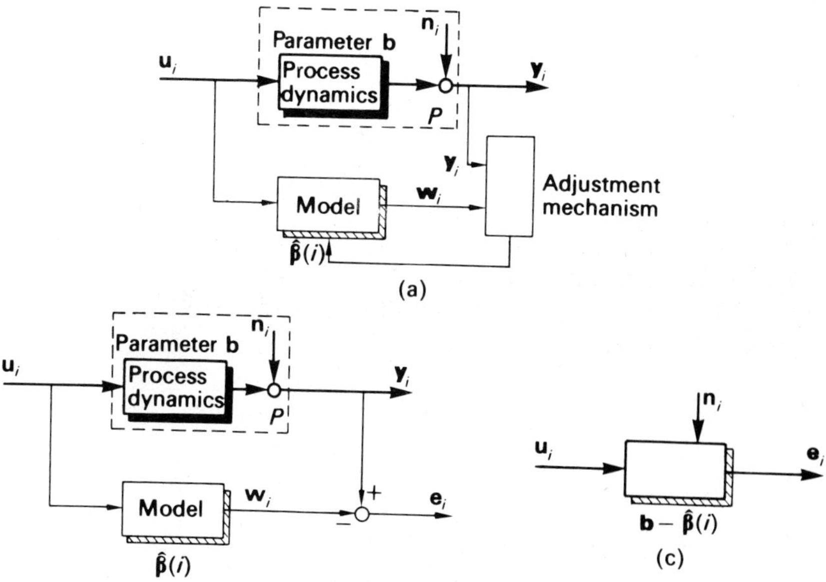


Figure 7.1

process and to a model with adjustable parameters. The adjustable parameters are changed by an adjustment mechanism which receives the process output $\mathbf{y}_i$ and the model output $\mathbf{w}_i$ as inputs. This formulation of the on-line identification problem was considered by Whitaker *et al* (1958). The essential problem is to determine the adjustment mechanism such that the model parameters in some sense will be close to the process parameters. Note again that in this approach the $\hat{\boldsymbol{\beta}}(i)$ is imbedded in a closed loop, which implies self-correcting properties for several types of errors (cf. Chapter 2).

As indicated in Chapter 4, the linearity-in-the-parameters leads to a simple case, even if the process has non-linear dynamics. This section will be devoted to that type of model structure:

$$\mathbf{w}(k) = \sum \beta_j \mathbf{u}_j(k) \quad \text{and} \quad \mathbf{w} = \mathbf{U}\boldsymbol{\beta}$$

or

$$\mathbf{e} = \mathbf{y} - \mathbf{U}\boldsymbol{\beta}$$

Iterative schemes may be derived at different levels of complexity; this will be discussed successively, viz:

corrections proportional to the gradient (Section 7.1.1);

least-squares approach to strings of observations (Section 7.1.2);

least-squares approach to single (pairs of) observations (Section 7.13);

stochastic approximation (Section 7.1.4);

contraction mapping (Section 7.1.5).

### 7.1.1 *Corrections Proportional to the Gradient*

Assume we are given the diagram of Figure 7.1b. The number of parameters to be estimated is again known beforehand. **U** and **n** are statistically independent. The vector $\hat{\boldsymbol{\beta}}(i)$ represents the constant model parameters during the $i$th measuring interval. The next parameter vector can be chosen as the old vector corrected with a quantity proportional to the gradient of the error function:

$$\hat{\boldsymbol{\beta}}(i+1) = \hat{\boldsymbol{\beta}}(i) - \frac{1}{2}\Gamma(i)\nabla_{\boldsymbol{\beta}} E_i \tag{7.1}$$

with

$$E_i = \mathbf{e}_i'\mathbf{R}\mathbf{e}_i$$

$$\mathbf{e}_i = \mathbf{y}_i - \mathbf{U}_i\hat{\boldsymbol{\beta}}(i)$$

$$\nabla_{\boldsymbol{\beta}} E_i = -2\mathbf{U}_i'\mathbf{R}\mathbf{e}_i \qquad \text{cf. Appendix C}$$

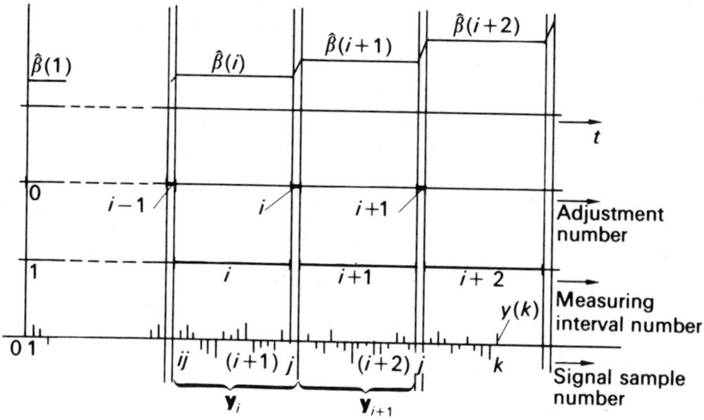


Figure 7.2

and where $\mathbf{y}_i$ and $\mathbf{U}_i$ have to be interpreted as the observations made during the $i$th measurement interval (cf. Figure 7.2). Consequently,

$$\hat{\boldsymbol{\beta}}(i + 1) = \hat{\boldsymbol{\beta}}(i) + \Gamma(i)\mathbf{U}_i'\mathbf{R}[\mathbf{y}_i - \mathbf{U}_i\hat{\boldsymbol{\beta}}(i)] \tag{7.2}$$

For the weighting matrix possible choices again include $\mathbf{R} = \mathbf{I}$ and $\mathbf{R} = \mathbf{N}^{-1}$. The 'gain factor' $\Gamma(i)$ can be chosen constant or as a function of $i$. The effects of this choice are explained in the following.

### 7.1.2 *Least-squares Approach to Strings of Observations*

In the situation given by Figure 7.1b one might estimate the difference between the process and the model-parameter vector:

$$\Delta\hat{\boldsymbol{\beta}}(i) = \mathbf{b} - \hat{\boldsymbol{\beta}}(i) = [\mathbf{U}_i'\mathbf{R}\mathbf{U}_i]^{-1}\mathbf{U}_i'\mathbf{R}\mathbf{e}_i$$

i.e. by use of equation (6.31) operating on $\mathbf{e}_i$ instead of $\mathbf{y}$ (Figure 7.1*c*). This implies that the process and the model partially compensate each other and that the parameters of the resulting dynamic system are estimated. Hence:

$$\hat{\boldsymbol{\beta}}(i + 1) = \hat{\boldsymbol{\beta}}(i) + \Gamma(i)\Delta\hat{\boldsymbol{\beta}}(i)$$

which yields:

$$\hat{\boldsymbol{\beta}}(i + 1) = \hat{\boldsymbol{\beta}}(i) + \Gamma(i)[\mathbf{U}_i'\mathbf{R}\mathbf{U}_i]^{-1}\mathbf{U}_i'\mathbf{R}[\mathbf{y}_i - \mathbf{U}_i\hat{\boldsymbol{\beta}}(i)] \tag{7.3}$$

Compare this result with equation (7.2). Again $\mathbf{R} = \mathbf{I}$ and $\mathbf{R} = \mathbf{N}^{-1}$ are among the possible choices. Unless otherwise stated $\mathbf{R} = \mathbf{I}$ will be used in the following part. The need of matrix inversion in each step of the iteration makes this algorithm still comparatively unpleasant for numerical solutions, although a much smaller number of samples is used per iteration than in the accumulative type of solution. The size of the $\mathbf{U}'\mathbf{U}$ matrix, however, is not reduced.

If there is no noise **n** present, then an adjustment scheme referred to as 'learning identification' with good convergence properties can be used. In that case equation (7.3) may be reduced to:

$$\hat{\boldsymbol{\beta}}(i+1) = \hat{\boldsymbol{\beta}}(i) + \Gamma \frac{u_i}{\|u_i\|^2}[\mathbf{y}_i - \mathbf{U}_i\hat{\boldsymbol{\beta}}(i)]$$

cf. Nagumo and Noda (1967), Bélanger (1968). The effect of working with a quantized signal sgn $u_i$ is considered by Crum and Wu (1968).

In real-life estimating situations the additive noise is the limiting factor. Several properties of the algorithm equation (7.3) have been derived by Van den Boom (1967); cf. also Van den Boom and Melis (1969).

The **development in time of the expectation** can be studied. Keeping in mind that $\mathbf{y}_i = \mathbf{U}_i\mathbf{b} + \mathbf{n}_i$, it follows from equation (7.3):

$$\mathscr{E}[\hat{\boldsymbol{\beta}}(i+1)] = \Gamma(i)\mathbf{b} + \{1 - \Gamma(i)\}\mathscr{E}[\hat{\boldsymbol{\beta}}(i)] \qquad i = 1, 2, \ldots \tag{7.4}$$

as **U** and **n** have been assumed to be statistically independent. Consequently

$$\mathscr{E}[\hat{\boldsymbol{\beta}}(2)] = \Gamma(1)\mathbf{b} + \{1 - \Gamma(1)\}\mathscr{E}[\hat{\boldsymbol{\beta}}(1)]$$
$$\mathscr{E}[\hat{\boldsymbol{\beta}}(3)] = \Gamma(2)\mathbf{b} + \{1 - \Gamma(2)\}\Gamma(1)\mathbf{b} + \{1 - \Gamma(2)\}\{1 - \Gamma(1)\}\mathscr{E}[\hat{\boldsymbol{\beta}}(1)]$$

As $\hat{\boldsymbol{\beta}}(1)$ is the (known) initial model setting we may put $\mathscr{E}[\hat{\boldsymbol{\beta}}(1)] = \hat{\boldsymbol{\beta}}(1)$. If no *a priori* knowledge about the process parameters is available a legitimate choice is $\hat{\boldsymbol{\beta}}(1) = \mathbf{0}$.

Also of interest is the **development in time of the covariance.** For the sake of simplicity we choose $\mathbf{R} = \mathbf{I}$.

$$\hat{\boldsymbol{\beta}}(i+1) = \Gamma(i)\mathbf{b} + \{1 - \Gamma(i)\}\hat{\boldsymbol{\beta}}(i) + \Gamma(i)[\mathbf{U}_i'\mathbf{U}_i]^{-1}\mathbf{U}_i'\mathbf{n}$$
$$\mathscr{E}[\hat{\boldsymbol{\beta}}(i+1)] = \Gamma(i)\mathbf{b} + \{1 - \Gamma(i)\}\mathscr{E}[\hat{\boldsymbol{\beta}}(i)] + \mathbf{0}$$
$$- \overline{\hat{\boldsymbol{\beta}}(i+1) - \mathscr{E}[\hat{\boldsymbol{\beta}}(i+1)] = \{1 - \Gamma(i)\}[\hat{\boldsymbol{\beta}}(i) - \mathscr{E}[\hat{\boldsymbol{\beta}}(i)]] + \Gamma(i)[\mathbf{U}_i'\mathbf{U}_i]^{-1}\mathbf{U}_i'\mathbf{n}}$$

If the noise **n** has a wide band character then $\hat{\boldsymbol{\beta}}(i)$ and $\mathbf{n}_i$ can also be considered independent in most practical cases. Consequently

$$\operatorname{cov}[\hat{\boldsymbol{\beta}}(i+1)] = \{1 - \Gamma(i)\}^2 \operatorname{cov}[\hat{\boldsymbol{\beta}}(i)] + \Gamma^2(i)[\mathbf{U}_i'\mathbf{U}_i]^{-1}\mathbf{U}_i'\mathscr{E}[\mathbf{n}_i\mathbf{n}_i']\mathbf{U}_i[\mathbf{U}_i'\mathbf{U}_i]^{-1} \tag{7.5}$$

i.e. a varylinear difference equation. The last term reduces to

$$\Gamma^2(i)\sigma_n^2[\mathbf{U}_i'\mathbf{U}_i]^{-1} \qquad \text{for 'white' additive noise } n \tag{7.6}$$

$$\approx \Gamma^2(i)\frac{\sigma_n^2}{j\sigma_u^2}\mathbf{I} \qquad \text{for 'white' additive noise } n \text{ and 'white' input signal } u \text{ of sufficient length } (j \text{ samples}) \tag{7.7}$$

Again $\hat{\boldsymbol{\beta}}(1)$ is the initial model setting and is consequently known. This implies $\text{cov}\,[\hat{\boldsymbol{\beta}}(1)] = 0\mathbf{I}$. From this initial condition the development of the covariance in time can be studied.

From equations (7.4) and (7.5) it is clear that the choice of $\Gamma(i)$ is of crucial importance for the covergence properties. This 'gain factor' may be chosen to be either a constant or a function of the adjustment number. In both cases a choice $\Gamma(i) = 1$ for any $i = l$ implies according to equation (7.4) that

$$\mathscr{E}[\hat{\boldsymbol{\beta}}(i + 1)] = \mathbf{b} \tag{7.8}$$

for all $i \geqslant l$. Consequently after that $l$th adjustment the estimate is unbiased irrespective of the initial model setting!

*Implementation of 'Closed-loop' Estimation Schemes.* Following Van den Boom and Melis (1969) we will consider the following cases:

(*a*) $\Gamma(i) = 1/i$

(*b*) $\Gamma(i) = c = \text{constant}$

(*a*) This case reminds us of the algorithms found for stochastic approximation (cf. Chapter 5). From equations (7.5) and (7.7) it follows that for 'white' noise $n$ and 'white' input signal $u$:

$$\text{cov}\,[\hat{\boldsymbol{\beta}}(i + 1)] = \frac{\sigma_n^2}{ij\sigma_u^2}\mathbf{I} \tag{7.9}$$

and

$$\lim_{i\to\infty} \text{cov}\,[\hat{\boldsymbol{\beta}}(i + 1)] = 0\mathbf{I} \tag{7.10}$$

This can easily be shown by recognizing the pattern in the successive covariance expressions for $\hat{\boldsymbol{\beta}}(1), \hat{\boldsymbol{\beta}}(2), \ldots$, etc. The term $ij$ in equation (7.9) implies the total number of samples that have been used in all $i$ measuring intervals together. A comparison of this expression with the one for the explicit method (equation 6.63) shows that the covariance for both methods is equal. For rather short lengths of samples ($j$ small) the relation

$$[\mathbf{U}_i'\mathbf{U}_i]^{-1} \approx \frac{1}{j\sigma_u^2}\mathbf{I}$$

does not necessarily hold. For such cases, experimental results show a shift towards greater variances (cf. Figure 7.3).

(*b*) The case $\Gamma(i) = c$ leads to

$$\mathscr{E}[\hat{\boldsymbol{\beta}}(i + 1)] = c\mathbf{b} + (1 - c)\mathscr{E}[\hat{\boldsymbol{\beta}}(i)] \tag{7.11}$$

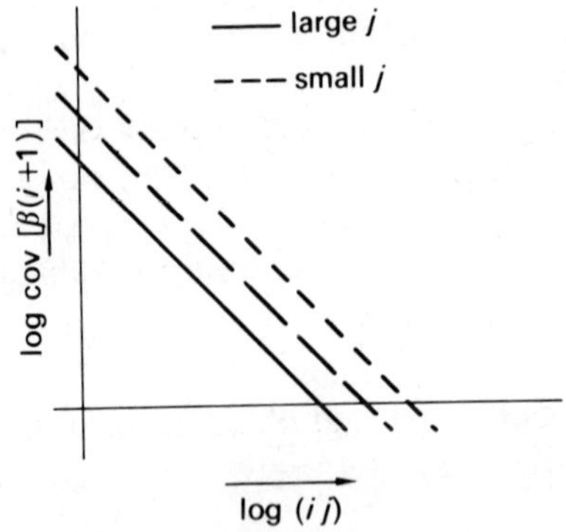


Figure 7.3

which can be written as:

$$\mathscr{E}[\hat{\boldsymbol{\beta}}(i+1)] = \{1-(1-c)^i\}\mathbf{b} + (1-c)^i\boldsymbol{\beta}(1) \tag{7.12}$$

Consequently,

$$\lim_{i\to\infty} \mathscr{E}[\hat{\boldsymbol{\beta}}(i+1)] = \mathbf{b} \qquad \text{for } 0 < c < 2 \tag{7.13}$$

i.e. the estimate is asymptotically unbiased (cf. Figure 7.4). The successive covariance expressions for $\hat{\boldsymbol{\beta}}(2)$, $\hat{\boldsymbol{\beta}}(3)$, . . . , etc. lead to

$$\lim_{i\to\infty} \operatorname{cov}[\hat{\boldsymbol{\beta}}(i+1)] = \frac{c}{2-c}\frac{\sigma_n^2}{j\sigma_u^2}\mathbf{I} \tag{7.14}$$

For this case of a constant 'gain factor' the covariance does **not** reduce to zero even for an infinite observation interval!

Apparently method (*a*) is usable if the parameters to be determined are constant. Such parameters can be found with any desired accuracy if the observation interval is of sufficient length. Note that because $\Gamma(i) = 1/i$ new

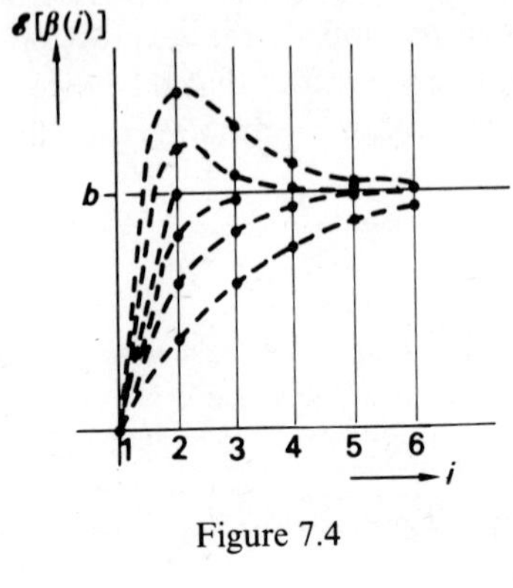


Figure 7.4

observations add less and less to the estimate as $i$ increases. The method (*b*) gives to new observations as much weight as to old observations. In cases where the parameters are slowly varying (parameter tracking) one would care for 'gradually-forgetting-the-past'. The implementation of that feature is discussed further on in this section.

Just as in Chapter 6 we may ask what will be the effect of simplifying the instrumentation and specially what will happen if the **matrix inversion is discarded.** It turns out that:

for $\Gamma(i) = 1/i$ the covariance is about the same as in the case when matrix inversion is used, with some deviation for small values of $ij$, the total number of samples (cf. Figure 7.5);

for $\Gamma(i) = c$ for large values of $ij$ the covariance has the same asymptotic value as with matrix inversion given by equation (7.14).

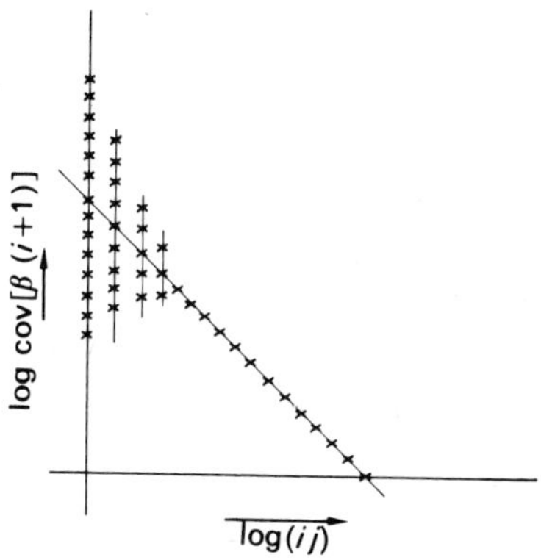


Figure 7.5

### *7.1.3 Least-squares Approach to Single (Pairs of) Observations*

For our next iterative scheme we change the notation of $\mathbf{U}$ to $\mathbf{U}_k$, indicating the last observation incorporated (Lee, 1964; Aoki, 1967).

$$\mathbf{U}_k = \begin{bmatrix} u_0(1) & \cdots & u_m(1) \\ \vdots & & \vdots \\ u_0(k) & \cdots & u_m(k) \end{bmatrix}$$

$$\mathbf{U}_{k+1} = \begin{bmatrix} u_0(1) & \cdots & u_m(1) \\ \vdots & & \vdots \\ u_0(k) & \cdots & u_m(k) \\ \hline u_0(k+1) & \cdots & u_m(k+1) \end{bmatrix} = \begin{bmatrix} \mathbf{U}_k \\ \hline \mathbf{u}'(k+1) \end{bmatrix} \tag{7.15}$$

where $\mathbf{u}'(k+1)$ represents the new observations. We define

$$\mathbf{P}^{-1}(k) = \mathbf{U}_k'\mathbf{U}_k \qquad \text{non-singular}$$
$$\mathbf{P}^{-1}(k+1) = \mathbf{U}_{k+1}'\mathbf{U}_{k+1} = [\mathbf{U}_k' \mid \mathbf{u}(k+1)]\left[\begin{array}{c}\mathbf{U}_k \\ \hline \mathbf{u}'(k+1)\end{array}\right] \tag{7.16}$$

or

$$\mathbf{P}^{-1}(k+1) = \mathbf{P}^{-1}(k) + \mathbf{u}(k+1)\mathbf{u}'(k+1) \tag{7.17}$$

Note that the dimensions of $\mathbf{P}$ do not change with increasing $k$. By using the matrix identities given in Appendix C one finds:

$$\begin{aligned}\mathbf{P}(k+1) &= \mathbf{P}(k) - \mathbf{P}(k)\mathbf{u}(k+1)[\mathbf{u}'(k+1)\mathbf{P}(k)\mathbf{u}(k+1) + 1]^{-1}\mathbf{u}'(k+1)\mathbf{P}(k) \\ &= \mathbf{P}(k) - \gamma(k)\mathbf{P}(k)\mathbf{u}(k+1)\mathbf{u}'(k+1)\mathbf{P}(k)\end{aligned} \tag{7.18}$$

where

$$\gamma(k) = [\mathbf{u}'(k+1)\mathbf{P}(k)\mathbf{u}(k+1) + 1]^{-1}$$

is a scalar. Using $\mathbf{y}_k' = [y(1), \ldots, y(k)]$ we find the estimator for $\mathbf{b}$ from equation (6.31):

$$\hat{\boldsymbol{\beta}}(k) = \mathbf{P}(k)\mathbf{U}_k'\mathbf{y}_k \tag{7.19}$$

$$\hat{\boldsymbol{\beta}}(k+1) = \mathbf{P}(k+1)\mathbf{U}_{k+1}'y_{k+1} = \mathbf{P}(k+1)[\mathbf{U}_k'\mathbf{y}_k + \mathbf{u}(k+1)y(k+1)] \tag{7.20}$$

Substituting equation (7.18) into (7.20) yields:

$$\begin{aligned}\hat{\boldsymbol{\beta}}(k+1) = {} & \mathbf{P}(k)\mathbf{U}_k'\mathbf{y}_k \\ & + \mathbf{P}(k)\mathbf{u}(k+1)\gamma(k)\gamma^{-1}(k)y(k+1) \\ & - \mathbf{P}(k)\mathbf{u}(k+1)\gamma(k)\mathbf{u}'(k+1)\mathbf{P}(k)[\mathbf{U}_k'\mathbf{y}_k + \mathbf{u}(k+1)y(k+1)]\end{aligned} \tag{7.21}$$

where the form $\gamma(k)\gamma^{-1}(k) = 1$ is introduced to facilitate further manipulation. Using (7.19) this can be reworked to the following expressions:

$$\hat{\boldsymbol{\beta}}(k+1) = \hat{\boldsymbol{\beta}}(k) + \gamma(k)\mathbf{P}(k)\mathbf{u}(k+1)\{y(k+1) - \mathbf{u}'(k+1)\hat{\boldsymbol{\beta}}(k)\} \tag{7.22}$$

or

$$\hat{\boldsymbol{\beta}}(k+1) = \hat{\boldsymbol{\beta}}(k) + \mathbf{P}(k+1)\mathbf{u}(k+1)\{y(k+1) - \mathbf{u}'(k+1)\hat{\boldsymbol{\beta}}(k)\} \tag{7.23}$$

The equivalence of the last two expressions can be shown by direct substitution of $\mathbf{P}(k+1)\mathbf{u}(k+1)$ or by use of the matrix identities given in Appendix C. The estimate after using the new observations $\mathbf{u}(k+1)$ and $y(k+1)$ equals the old estimate plus a correction term. Note that $\gamma$ acts as a normalizing factor. The term $\{y(k+1) - \mathbf{u}'(k+1)\hat{\boldsymbol{\beta}}(k)\}$ represents that part of the new output observation that is not accounted for by the input observations and the last parameter estimate. Note that there is no need for matrix inversion and that only 'old' knowledge and new observations are used for the new estimate. The pair $(\hat{\boldsymbol{\beta}}, \mathbf{P})$ thus represents the smallest number of variables, characterizing the input–output data, which are necessary to carry along in the computations. This is indicated in Figure 7.6. Each estimation iteration

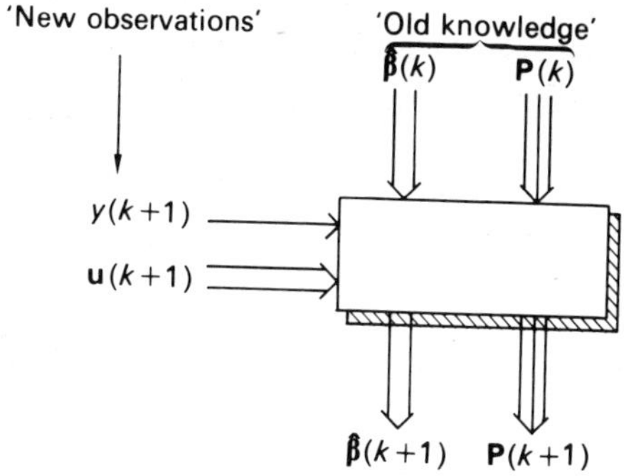


Figure 7.6

is given by the equations (7.18) and (7.22) or (7.23). Figure 7.7 shows schematically the operations that are needed for each such iteration. $P_k$ is a symmetrical matrix, hence only a triangular part needs to be updated. Notice that $\hat{\boldsymbol{\beta}}(k+1)$, as given by equations (7.22) or (7.23), is the least-square estimate;

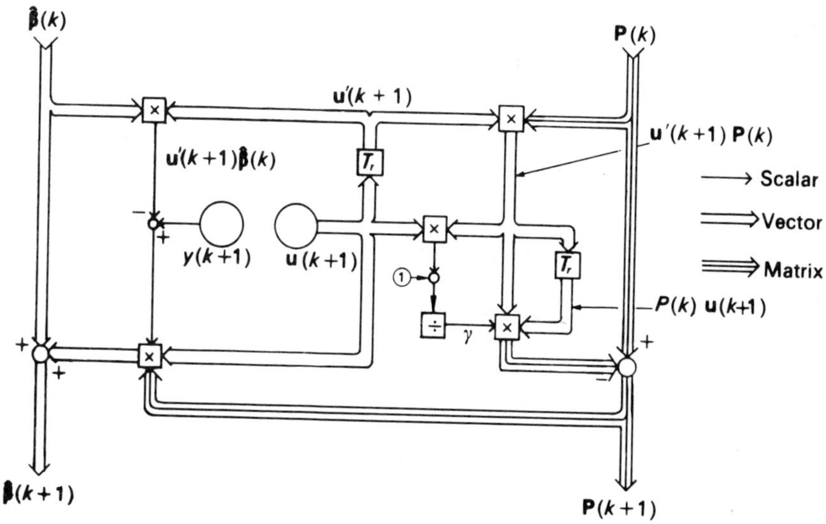


Figure 7.7

the convergence of the iterative scheme follows directly from the consistency proof of the least-squares estimate. Another way to write equation (7.22) is:

$$\hat{\boldsymbol{\beta}}(k+1) = \{1 - \gamma(k)\mathbf{P}(k)\mathbf{u}(k+1)\mathbf{u}'(k+1)\}\hat{\boldsymbol{\beta}}(k) + \gamma(k)\mathbf{P}(k)\mathbf{u}(k+1)y(k+1) \tag{7.24}$$

which indicates that the new estimate consists of a linear combination of the old estimate and an estimate following from the new observations only. The 'weighting factors' in this linear combination represent the 'credibility' of the respective parts.

As our estimator is just an iterative version of the accumulative type derived in Chapter 6 we know that

$$\mathscr{E}[\hat{\boldsymbol{\beta}}(k)] = \mathbf{b} \tag{7.25}$$

$$\operatorname{cov}[\hat{\boldsymbol{\beta}}(k)] = [\mathbf{U}_k'\mathbf{U}_k]^{-1}\mathbf{U}_k'\mathbf{N}\mathbf{U}_k[\mathbf{U}_k'\mathbf{U}_k]^{-1} = \mathbf{P}(k)\mathbf{U}_k'\mathbf{N}\mathbf{U}_k\mathbf{P}(k) \tag{7.26}$$

If $\mathbf{n}$ is white noise with $\mathbf{N} = \sigma_n^2\mathbf{I}$

$$\operatorname{cov}[\hat{\boldsymbol{\beta}}(k)] = \sigma_n^2[\mathbf{U}_k'\mathbf{U}_k]^{-1} = \sigma_n^2\mathbf{P}(k) \tag{7.27}$$

Consequently $\mathbf{P}(k)$ and $\mathbf{P}(k+1)$ are proportional to the covariance matrices of the estimates of $k$ and $k+1$ respectively. It can be shown for many types of input signals that $\mathbf{P}(k)$ is positive definite and that

$$\lim_{k\to\infty} \mathbf{P}(k) = 0\mathbf{I}$$

*Example.* In order to get a good and easy grasp of the significance of the different terms the reader is invited to work a simple example. Consider the simplification to a one-dimensional case:

$$\hat{\beta}(k+1) = \hat{\beta}(k) + P(k+1)u(k+1)\{y(k+1) - u(k+1)\hat{\beta}(k)\} \tag{7.28}$$

$$P^{-1}(k+1) = P^{-1}(k) + u^2(k+1) = P^{-1}(0) + \sum_{j=1}^{k+1} u^2(j) \tag{7.29}$$

If at the start little *a priori* knowledge is available on $\beta$, then a proper choice for $P^{-1}(0)$ is 0. If $\mathbf{u}$ is a white stationary signal with $\mathscr{E}[\mathbf{u}] = \mathbf{0}$, and $\operatorname{var}[\mathbf{u}] = \sigma_u^2\mathbf{I}$ then

$$\mathscr{E}[P(k+1)] = \frac{1}{(k+1)\sigma_u^2} \tag{7.30}$$

It is interesting to compare this weighting factor with the condition derived for the convergence in probability for stochastic approximation (cf. Section 5.4).

As an exercise consider the following case:

The process is given by

$$y(k) = bu(k) + n(k)$$

The model is selected using the estimated value found in the previous iteration:

$$w(k) = \beta(k-1)u(k)$$

The error is

$$e(k) = y(k) - w(k)$$

For reviewing purposes it pays to derive equations (7.28) and (7.29) in a direct way (cf. also Section 5.1).

The reader is invited:

either to write a simple computer programme that generates $u(k)$ and $n(k)$, e.g. as random numbers, that simulates the process and iteratively estimates the parameters;

or to perform this experiment by hand using, for example a die or a table of random numbers for generating $u(k)$ and $n(k)$, making the estimation calculations by hand.

Choose $\beta(0) = 0$ viz. $\beta(0) = b$. When performing a computer simulation it pays again to do the same experiment a number of times, choosing different starting values in each experiment. Plotting the results of those experiments provides a visual impression of the rate of convergence. This can be compared with the actual value of $P(k)$ and with $\mathscr{E}[P(k)]$.

By a straightforward manipulation along the lines given in this section the following relations can be derived as well:

$$\hat{\boldsymbol{\beta}}(k+1) = \hat{\boldsymbol{\beta}}(k) + \mathbf{P}(k+1)\mathbf{u}(k+1)r_{k+1}\{y(k+1) - \mathbf{u}'(k+1)\hat{\boldsymbol{\beta}}(k)\} \tag{7.31}$$

$$\mathbf{P}(k+1) = \mathbf{P}(k) - \mathbf{P}(k)\mathbf{u}(k+1)\left[\mathbf{u}'(k+1)\mathbf{P}(k)\mathbf{u}(k+1) + r_{k+1}\right]^{-1} \times \mathbf{u}'(k+1)\mathbf{P}(k) \tag{7.32}$$

where $r_{k+1}$ is the corresponding element of the diagonal matrix $\mathbf{R}$ in the error function $\mathbf{e}'\mathbf{Re}$.

For a more rigorous discussion of the iterative least-squares estimators the reader is referred to the paper by Albert and Sittler (1965).

*Starting the Estimation Procedure.* This estimation scheme, using an iteration after each new observation, can be started by several techniques (cf. Klinger, 1968).

(1) By using possibly available *a priori* knowledge about $\boldsymbol{\beta}$ and $\mathbf{P}$. If one knows that the input signal $\mathbf{u}$ has the character of stationary 'white' noise with variance $\sigma_u^2$ then this knowledge can be used in choosing:

$$\mathbf{P}(l) = \mathscr{E}[\mathbf{U}_l'\mathbf{U}_l]^{-1} = \frac{\mathbf{I}}{l\sigma_u^2}$$

Then the iterative procedure starts with iteration number $l + 1$.

(2) By determination from the explicit method:

$$\mathbf{P}(l) = [\mathbf{U}_l'\mathbf{U}_l]^{-1}$$

$$\boldsymbol{\beta}(l) = \mathbf{P}(l)\mathbf{U}_l'\mathbf{y}_l$$

using $l$ samples together for an initial estimate; after this calculation the iterative scheme can be used. This way of starting has the disadvantage that it needs a matrix inversion.

(3) By using $\mathbf{P}(0) = a\mathbf{I}$, where $a$ is a very large scalar and $\hat{\boldsymbol{\beta}}(0)$ is chosen arbitrarily. After one iteration:

$$\mathbf{P}^{-1}(1) = \mathbf{P}^{-1}(0) + \mathbf{u}(1)\mathbf{u}'(1)$$

$$\begin{aligned}\hat{\boldsymbol{\beta}}(1) &= \hat{\boldsymbol{\beta}}(0) + \mathbf{P}(1)\mathbf{u}(1)\{y(1) - \mathbf{u}'(1)\hat{\boldsymbol{\beta}}(0)\} \\ &= \mathbf{P}(1)\{\mathbf{u}(1)y(1) + \mathbf{P}^{-1}(0)\hat{\boldsymbol{\beta}}(0)\}\end{aligned}$$

cf. equations (7.17) and (7.23). In the same way follows:

$$\mathbf{P}^{-1}(l) = \mathbf{P}^{-1}(0) + \mathbf{U}_l'\mathbf{U}_l$$

$$\hat{\boldsymbol{\beta}}(l) = \mathbf{P}(l)\{\mathbf{U}_l\mathbf{y}_l + \mathbf{P}^{-1}(0)\hat{\boldsymbol{\beta}}(0)\}$$

If $\mathbf{P}^{-1}(0) = \dfrac{1}{a}\mathbf{I}$ and $a \to \infty$ then

$$\mathbf{P}^{-1}(l) \to \mathbf{U}_l'\mathbf{U}_l$$

$$\hat{\boldsymbol{\beta}}(l) \to \mathbf{P}(l)\mathbf{U}_l'\mathbf{y}_l$$

i.e. the expressions corresponding to the explicit estimation method. Now the iterative estimation procedure can start with the first observation samples.

*Exponential Weighting of Past Data.* As mentioned before the algorithms just given are iterative versions of the accumulative (one-shot) type of estimator. This implies that all $k$ samples processed for the estimate $\hat{\boldsymbol{\beta}}(k)$ get an equal weight, irrespective of the 'age' of those samples. For the estimation of (slowly) changing parameters this obviously is undesirable; one needs a method to 'forget' gradually the past. This can be done as follows:

$$\mathbf{U}_{k+1}'\mathbf{U}_{k+1} = [c\mathbf{U}_k]'[c\mathbf{U}_k] + \mathbf{u}(k+1)\mathbf{u}'(k+1)$$

or

$$c^{-2}\mathbf{P}^{-1}(k+1) = \mathbf{P}^{-1}(k) + \mathbf{u}(k+1)c^{-2}\mathbf{u}'(k+1) \tag{7.33}$$

with $0 < c \leqslant 1$. With $\rho = c^2$ this leads to the following equations (cf. Appendix C):

$$\hat{\boldsymbol{\beta}}(k+1) = \hat{\boldsymbol{\beta}}(k) + \gamma(k)\mathbf{P}(k)\mathbf{u}(k+1)\{\mathbf{y}(k+1) - \mathbf{u}'(k+1)\hat{\boldsymbol{\beta}}(k)\} \tag{7.34}$$

$$\mathbf{P}(k+1) = \rho^{-1}[\mathbf{P}(k) - \gamma(k)\mathbf{P}(k)\mathbf{u}(k+1)\mathbf{u}'(k+1)\mathbf{P}(k)] \tag{7.35}$$

$$\gamma(k) = \{\mathbf{u}'(k+1)\mathbf{P}(k)\mathbf{u}(k+1) + \rho\}^{-1} \tag{7.36}$$

Recognizing that the weighting factors on the previous samples change with the sample number $j$ as $c^{k-j}$, with $j \leqslant k$, it is quite simple to determine the number of samples that effectively contribute to the estimate for a given value of $c$. For noisy observations in general $c$ will have to be close to one, in order to obtain an appreciable effective observation length.

### 7.1.4 *Stochastic Approximation*

This method was discussed in Section 5.3. The algorithm that may be used is

$$\hat{\boldsymbol{\beta}}(k+1) = \hat{\boldsymbol{\beta}}(k) - \tfrac{1}{2}\Gamma(k)\nabla_{\boldsymbol{\beta}} E \tag{7.37}$$

where $\Gamma(k)$ has to fulfil the conditions

$$\Gamma(k) \geqslant 0 \qquad \sum_{k=0}^{\infty} \Gamma(k) = \infty \qquad \sum_{k=0}^{\infty} \Gamma^2(k) < \infty$$

For an iteration after each observation the following holds:

$$E = e^2(k+1)$$

$$e(k+1) = y(k+1) - \mathbf{u}'(k+1)\hat{\boldsymbol{\beta}}(k)$$

Consequently equation (7.37) can be rewritten as:

$$\hat{\boldsymbol{\beta}}(k+1) = \hat{\boldsymbol{\beta}}(k) + \Gamma(k)\mathbf{u}(k+1)\{y(k+1) - \mathbf{u}'(k+1)\hat{\boldsymbol{\beta}}(k)\} \tag{7.38}$$

This estimate converges in probability to the true parameter values (cf. Albert and Gardner, 1967; Holmes, 1968). The conditions of $\Gamma(k)$ can be fulfilled by the sequence:

$$\Gamma(k) = \frac{c}{k^{\alpha}} \qquad \text{with } c > 0 \qquad 0{\cdot}5 < \alpha \leqslant 1$$

Comparing equation (7.23) with equation (7.38), one notices that the only difference is that instead of $\mathbf{P}_{k+1}$ we have $\Gamma(k)$; that is the $\mathbf{P}$ matrix, that is closely related to the covariance matrix of the estimate, cf. equation (7.27), is replaced by a diagonal matrix $\Gamma(k)\mathbf{I}$. Consequently stochastic approximation provides a **simplified** algorithm that needs less computational effort because the $\mathbf{P}$ matrices need not to be calculated. On the other hand as the $\mathbf{P}$ matrix is not available, the stochastic approximation method does not provide a measure of the accuracy of the estimate. In the least-squares error sense it is suboptimal compared to the algorithms (7.23) and (7.18). Some references on applications of the stochastic approximation procedure are given on page 175; cf. also Neal and Bekey (1969).

### 7.1.5 *Contraction Mapping*

A technique of constructing recursive algorithms has been suggested by Oza and Jury (1968; 1969). We will explain the technique in connexion with the least-squares problem. Instead of solving the equation

$$\mathbf{U}_k'\mathbf{U}_k\hat{\boldsymbol{\beta}}(k) = \mathbf{U}_k'\mathbf{y}_k$$

for each $k$ and showing that $\hat{\boldsymbol{\beta}}(k)$ satisfies a recursive equation, Oza and Jury introduce the mapping

$$\mathscr{T}_k[\boldsymbol{\beta}] = \boldsymbol{\beta} - \gamma[\mathbf{U}_k'\mathbf{U}_k\boldsymbol{\beta} - \mathbf{U}_k'\mathbf{y}_k] \tag{7.39}$$

where $\gamma$ is a scalar. It is then shown that the sequence

$$\hat{\boldsymbol{\beta}}(k+1) = \mathscr{T}_k[\boldsymbol{\beta}(k)] \tag{7.40}$$

under suitable conditions converges to the true parameters as $k \to \infty$. When applied to the ordinary least-squares problems the algorithm (7.39) is not efficient in contrast with the recursive least-squares method. To obtain an efficient algorithm it is necessary to make $\gamma$ a matrix. With the choice

$$\boldsymbol{\gamma} = [\mathbf{U}_k'\mathbf{U}_k]^{-1}$$

the algorithm becomes equivalent to the least squares.

The method of Oza and Jury can be applied to more general cases than the least squares. It was actually proven for the case where there are errors in the measurements of both inputs and outputs, provided the covariance function of the measurement errors are known. The assumption of known covariances of the measurement errors severely limits the practical applicability of the method.

## 7.2 Extension to Generalized and State–Space Models

The iterative schemes given in the previous section can also be applied to the generalized model given by equation (6.83) and Figure 7.8:

$$\mathbf{e} = \mathbf{y} + \mathbf{Y}\boldsymbol{\alpha} - \mathbf{U}\boldsymbol{\beta} = \mathbf{y} - [\mathbf{U} \,\vdots\, \mathbf{Y}]\begin{bmatrix} \boldsymbol{\beta} \\ \hline -\boldsymbol{\alpha} \end{bmatrix} \stackrel{\text{def}}{=} \mathbf{y} - \boldsymbol{\Omega}\boldsymbol{\theta} \tag{7.41}$$

Note that this model is linear-in-the-parameters. Of course we will meet with the same problems as when using the explicit method for this case, i.e.

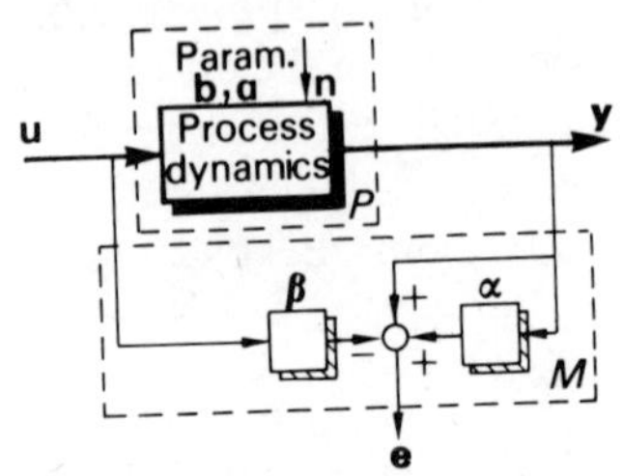


Figure 7.8

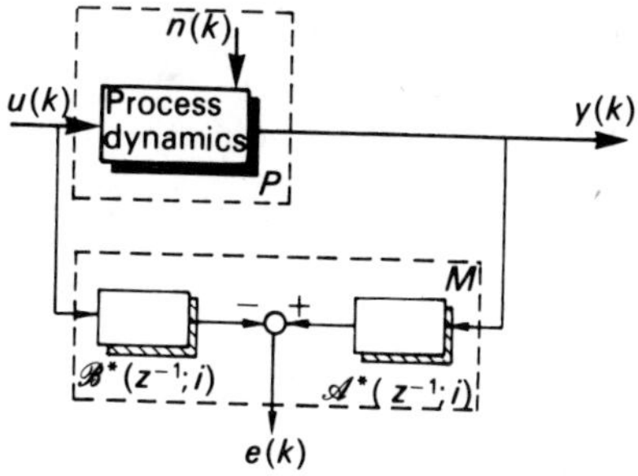


Figure 7.9

a bias of the parameters due to the noise components in **Y** (cf. Section 6.5). A specific generalized model can be given by Figure 7.9, a pulse-transfer-function representation, where:

$$\left.\begin{aligned} \mathscr{A}^*(z^{-1};i) &= 1 + \alpha_1(i)z^{-1} + \cdots + \alpha_m(i)z^{-m} \\ \mathscr{B}^*(z^{-1};i) &= \qquad \beta_1(i)z^{-1} + \cdots + \beta_m(i)z^{-m} \end{aligned}\right\} \tag{7.42}$$

Here the symbol $i$ refers to the parameter values at the $i$th iteration. For such a scheme Kalman (1958) proposed a simple estimation scheme leading to asymptotically biased estimates.

Techniques to circumvent the detrimental effects of additive noise have been indicated in Chapter 6. The discussion of one of those techniques, **generalized least-squares and related schemes,** was deferred to this chapter, as it implies model-adjustment procedures. In this category the following approaches will be discussed:

successive linear regression and filtering;

generalized least-squares (Markov) estimation;

extended matrix method.

### 7.2.1 *Successive Linear Regression and Filtering*

Instead of choosing the generalized error according to Figure 7.9. as:

$$e(k) = \mathscr{A}^*(z^{-1};i)y(k) - \mathscr{B}^*(z^{-1};i)u(k) \tag{7.43}$$

a proposal by Steiglitz and McBride (1965) implies the use of the error expression:

$$\begin{aligned} e'(k) &= \frac{e(k)}{\mathscr{A}^*(z^{-1};i-1)} = \frac{\mathscr{A}^*(z^{-1};i)}{\mathscr{A}^*(z^{-1};i-1)}y(k) - \frac{\mathscr{B}^*(z^{-1};i)}{\mathscr{A}^*(z^{-1};i-1)}u(k) \\ &= \left\{ y(k) - \frac{\mathscr{B}^*(z^{-1};i)}{\mathscr{A}^*(z^{-1};i)}u(k) \right\} \frac{\mathscr{A}^*(z^{-1};i)}{\mathscr{A}^*(z^{-1};i-1)} \end{aligned} \tag{7.44}$$

We still have a generalized model, linear with parameters $\alpha(i)$ and $\beta(i)$, but the term between brackets represents an ordinary model (Figure 7.10). For further improvement of the estimate it is also suggested that, after the

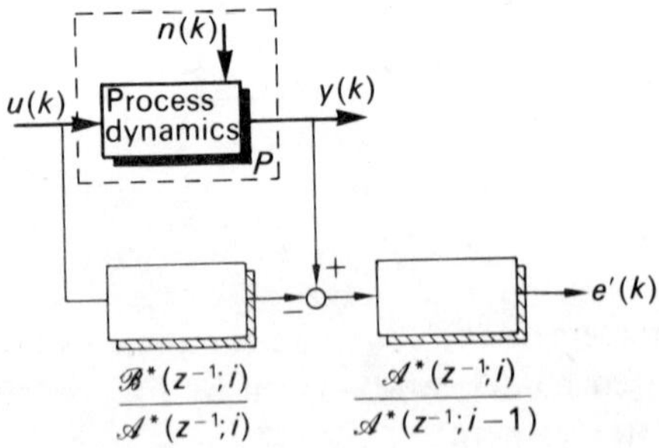


Figure 7.10

convergence has proceeded far enough, the scheme be switched from this generalized model to an ordinary model:

$$e(k) = y(k) - \frac{\mathscr{B}^*(z^{-1}; i)}{\mathscr{A}^*(z^{-1}; i)} u(k) \tag{7.45}$$

This has the advantage that the additive noise does not cause biased estimates. Near the optimum adjustment the non-linearity-in-the-parameters of $\mathscr{A}^*$ hardly has a detrimental effect on the convergence. The non-linearity-in-the-parameters can also be handled by means of quasilinearization (cf. Schulz, 1968).

The iterative procedure is as follows: As the first step, a linear regression gives

$$\hat{\boldsymbol{\theta}}(1) = \left[\begin{array}{c} \boldsymbol{\beta}(1) \\ \hline -\boldsymbol{\alpha}(1) \end{array}\right] = [\boldsymbol{\Omega}'\boldsymbol{\Omega}]^{-1}\boldsymbol{\Omega}'\mathbf{y} \tag{7.46}$$

where $\boldsymbol{\alpha}(1)$ and $\boldsymbol{\beta}(1)$ are the parameters of $\mathscr{A}^*(z^{-1}; i)$ and $\mathscr{B}^*(z^{-1}; i)$. The next step is to filter process input and output signals with the filter $\mathscr{A}^{*-1}(z^{-1}; i)$. Based on these filtered signals an estimate $\hat{\boldsymbol{\theta}}(2)$ is found, etc. No proof is known for the convergence of this scheme.

### 7.2.2 *Generalized Least-squares (Markov) Estimation Method*

The basic idea is as follows. Let the process be governed by

$$A^*(z^{-1})y(k) - B^*(z^{-1})u(k) + r(k) \tag{7.47}$$

where $A^*$ and $B^*$ are polynomials and $\{r(k)\}$ a sequence of correlated random variables. Suppose that the correlations of $r(k)$ are known. Say that they can be represented as

$$r(k) = G^*(z^{-1})w(k) \tag{7.48}$$

where $\{w(k)\}$ is a sequence of uncorrelated random variables and $G^*$ a pulse transfer function. The equation describing the process can be then written as

$$A^*(z^{-1})y(k) = B^*(z^{-1})u(k) + G^*(z^{-1})w(k) \tag{7.49}$$

where $A^*$ and $B^*$ characterize the process and $G^*$ characterizes the 'environment', or:

$$A^*(z^{-1})\tilde{y}(k) = B^*(z^{-1})\tilde{u}(k) + w(k) \tag{7.50}$$

where

$$\tilde{y}(k) = \frac{1}{G^*(z^{-1})}y(k) \quad \text{and} \quad \tilde{u}(k) = \frac{1}{G^*(z^{-1})}u(k)\dagger$$

Hence if the signals $\tilde{u}$ and $\tilde{y}$ are considered as the inputs and outputs we have an ordinary least-squares problem. Compare with (6.110). We thus find that the generalized least squares can be interpreted as a least-squares identification problem where the criterion is chosen as

$$\min \sum e^2(k)$$

with the generalized error defined as:

$$\begin{aligned} e(k) &= \frac{\mathscr{A}^*(z^{-1})}{G^*(z^{-1})}y(k) - \frac{\mathscr{B}^*(z^{-1})}{G^*(z^{-1})}u(k) \\ &= \mathscr{A}^*(z^{-1})\left[\frac{y(k)}{G^*(z^{-1})}\right] - \mathscr{B}^*(z^{-1})\left[\frac{u(k)}{G^*(z^{-1})}\right] \end{aligned} \tag{7.51}$$

Compare with the block diagram of Figure 7.11. This shows how the generalized error can be obtained from the process input and output and the model parameters $\alpha$ and $\beta$ in the generalized least-squares method.

The correlation of the residuals and the pulse transfer function $G^*$ are seldom known in practice. Clarke (1967) has proposed an iterative procedure to determine $G^*$ which has been tested on simulated data as well as on practical measurements (distillation-column identification). The procedure consists of the following steps. (In order to accentuate the link between pulse-transfer function representation and the matrix notation we will use both in parallel.)

† Note that even if the number of terms of $G^*(z^{-1})$ is quite limited, the number of significant terms of $\{G^*(z^{-1})\}^{-1}$ may be very high. This implies that many samples of $u$ are needed for $\tilde{u}$.

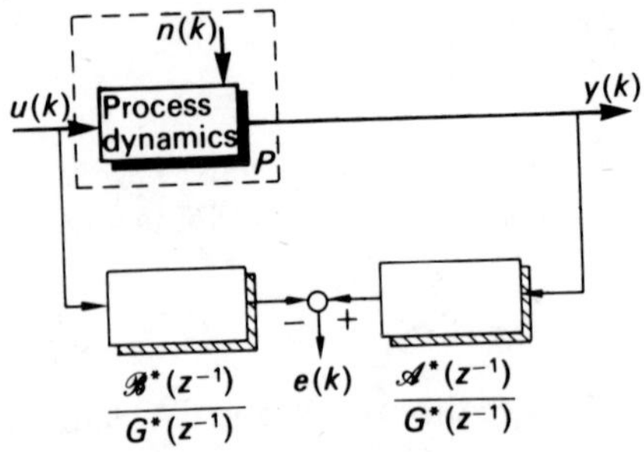


Figure 7.11

(1) Make an ordinary least-squares fit to the process input and output sequence

model

$$\mathscr{A}^*(z^{-1}; i)y(k) = \mathscr{B}^*(z^{-1}; i)u(k) + r(k)$$

model

$$\mathbf{e} = \mathbf{y} - \mathbf{\Omega\theta}(i)$$

estimate

$$\hat{\boldsymbol{\theta}}(i) = [\mathbf{\Omega}'\mathbf{\Omega}]^{-1}\mathbf{\Omega}'\mathbf{y}$$

(2) Now the error sequence is an approximation of the residuals. Analyze those residuals and fit an autoregression:

determine

$$\mathscr{D}^*(z^{-1}; i) = 1 + \delta_1(i)z^{-1} + \delta_2(i)z^{-2} + \cdots$$

in such a way that

$$\mathscr{D}^*(z^{-1}; i)\hat{r}(k) = \xi(k)$$

where $\xi(k)$ is discrete time white noise

residuals

$$\mathbf{e} = \hat{\mathbf{r}} = \mathbf{y} - \mathbf{\Omega}\hat{\boldsymbol{\theta}}(i)$$

model

$$\hat{\mathbf{r}} = -\hat{\mathbf{R}}\boldsymbol{\delta}(i) + \boldsymbol{\xi}$$

estimate

$$\hat{\boldsymbol{\delta}}(i) = -[\hat{\mathbf{R}}'\hat{\mathbf{R}}]^{-1}\hat{\mathbf{R}}'\hat{\mathbf{r}}$$

cf. Section 6.2

(3) Filter the process input and output sequence:

$$\tilde{u}(k) = \mathscr{D}^*(z^{-1}; i)u(k) \qquad \tilde{\mathbf{u}} = \mathbf{u} + \mathbf{U}\hat{\boldsymbol{\delta}}(i)$$

$$\tilde{y}(k) = \mathscr{D}^*(z^{-1}; i)y(k) \qquad \tilde{\mathbf{y}} = \mathbf{y} + \mathbf{Y}\hat{\boldsymbol{\delta}}(i)$$

(4) Make a new least-squares fit to the filtered process input and output sequence, i.e. repeat step 1.

(5) Repeat from step 2.

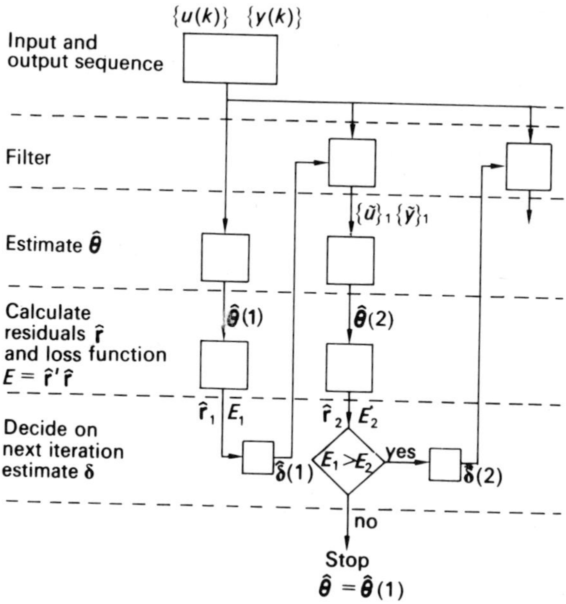


Figure 7.12

This recursive procedure is indicated in Figure 7.12. The stopping is governed by a comparison of the respective values of the loss function. If $\mathscr{A}^*(i) \to A^*$, $\mathscr{B}^*(i) \to B^*$, and $\mathscr{D}^*(i) \to \{G^*\}^{-1}$ then we will obtain

$$A^*(z^{-1})\{G^*(z^{-1})\}^{-1}y(k) = B^*(z^{-1})\{G^*(z^{-1})\}^{-1}u(k) + w(k) \quad (7.52)$$

i.e. a description with uncorrelated residuals. There do not seem to be available any convergence proofs as yet. It has, however, been shown to work very well in specific examples with reasonable *ad hoc* choices for the order of the process. The procedure has the disadvantage that there are no systematic rules (yet) for the choice of model order and for the order of the autoregression.

The procedure just described is not recursive in the process input–output data; a whole sequence of observations is handled in a one-shot manner.† A recursive version of this procedure has been described by Hasting-James and Sage (1969). Now the estimates of $\boldsymbol{\theta}$ and $\boldsymbol{\delta}$ are updated when new observations become available.

In the beginning, the estimates $\boldsymbol{\delta}$ will be poor and consequently the resulting samples $u(k)$ and $y(k)$ after filtering will not be optimal for estimating $\boldsymbol{\theta}$. Therefore, one has to introduce some kind of weighting in order to diminish

† Strictly speaking therefore this method could have been discussed in Chapter 6. It has found its place in this chapter for the sake of continuity in the discussion.

the influence of older input–output data on the estimation procedure. As the exponential weighting of past data has been discussed in the previous section, we only summarize the algorithms for this recursive estimation scheme:

$$\left.\begin{aligned} \boldsymbol{\Omega}_k &= \quad\quad = [\mathbf{U}_k \;\vdots\; \mathbf{Y}_k] \\ \boldsymbol{\Omega}_{k+1} &= \begin{bmatrix} \boldsymbol{\Omega}_k \\ \hdashline \boldsymbol{\omega}'_{k+1} \end{bmatrix} = \left[\begin{array}{c:c} \mathbf{U}_k & \mathbf{Y}_k \\ \hdashline \mathbf{u}'_{k+1} & \mathbf{y}'_{k+1} \end{array}\right] \end{aligned}\right\} \tag{7.53}$$

$$\left.\begin{aligned} \tilde{y}(k+1) &= y(k+1) + \mathbf{y}'(k)\hat{\boldsymbol{\delta}}(k) \\ \tilde{\omega}(k+1) &= \omega(k+1) + \boldsymbol{\omega}'(k)\hat{\boldsymbol{\delta}}(k) \\ \tilde{\boldsymbol{\omega}}(k+1) &= \boldsymbol{\omega}(k+1) + \boldsymbol{\Omega}_k\hat{\boldsymbol{\delta}}(k) \end{aligned}\right\} \tag{7.54}$$

$$\left.\begin{aligned} \hat{r}(k+1) &= y(k+1) - \boldsymbol{\omega}'(k+1)\hat{\boldsymbol{\theta}}(k) \\ \hat{\mathbf{r}}'(k+1) &= [\hat{r}(k), \hat{r}(k-1), \ldots, \hat{r}(k-p-1)] \end{aligned}\right\} \tag{7.55}$$

$$\left.\begin{aligned} \hat{\boldsymbol{\theta}}(k+1) &= \hat{\boldsymbol{\theta}}(k) + \gamma(k)\mathbf{P}(k)\tilde{\boldsymbol{\omega}}(k+1)\{\tilde{y}(k+1) - \tilde{\boldsymbol{\omega}}'(k+1)\hat{\boldsymbol{\theta}}(k)\} \\ \mathbf{P}(k+1) &= \rho^{-1}[\mathbf{P}(k) - \gamma(k)\mathbf{P}(k)\tilde{\boldsymbol{\omega}}(k+1)\tilde{\boldsymbol{\omega}}'(k+1)\mathbf{P}(k)] \\ \gamma(k) &= \{\tilde{\boldsymbol{\omega}}'(k+1)\mathbf{P}(k)\tilde{\boldsymbol{\omega}}(k+1) + \rho\}^{-1} \end{aligned}\right\} \tag{7.56}$$

$$\left.\begin{aligned} \hat{\boldsymbol{\delta}}(k+1) &= \hat{\boldsymbol{\delta}}(k) + \lambda(k)\mathbf{Q}(k)\hat{\mathbf{r}}(k+1)\{\hat{r}(k+1) - \mathbf{r}'(k+1)\hat{\boldsymbol{\delta}}(k)\} \\ \mathbf{Q}(k+1) &= \nu^{-1}[\mathbf{Q}(k) - \lambda(k)\mathbf{Q}(k)\hat{\mathbf{r}}(k+1)\hat{\mathbf{r}}'(k+1)\mathbf{Q}(k)] \\ \lambda(k) &= \{\hat{\mathbf{r}}'(k+1)\mathbf{Q}(k)\hat{\mathbf{r}}(k+1) + \nu\}^{-1} \end{aligned}\right\} \tag{7.57}$$

A choice has to be made for the weighting coefficients $\rho$ and $\nu$. When starting the estimation procedure the coefficients could be low, e.g. 0·975 in order to obtain a good convergence rate; in due course $\rho$ and $\nu$ could approach 1·0 for obtaining a good accuracy (small variance). No general theory is available yet that dictates the optimal choice of both $\rho$ and $\nu$. Some simulation results are given in Section 7.3.

### 7.2.3 *Extended Matrix Method*

The previous method was based on the assumption that the noise can be represented by

$$r(k) = \frac{1}{D^*(z^{-1})}w(k) \qquad D^*(z^{-1}) = 1 + d_1z^{-1} + d_2z^{-2} + \cdots \tag{7.58}$$

where $\{w(k)\}$ is a sequence of uncorrelated random variables (discrete 'white' noise). Depending on the situation at hand, the number of parameters to characterize the noise can be quite large. Consequently, a minimum number of parameters for such a characterization could be obtained by using:

$$r(k) = \frac{C^*(z^{-1})}{D^*(z^{-1})} w(k) \qquad \begin{aligned} C^*(z^{-1}) &= 1 + c_1 z^{-1} + c_2 z^{-2} + \cdots \\ D^*(z^{-1}) &= 1 + d_1 z^{-1} + d_2 z^{-2} + \cdots \end{aligned} \tag{7.59}$$

Consequently, it is of importance to develop an estimation scheme where the 'environment' of the process is characterized by estimates of the $c$ and $d$ parameters (cf. Figure 7.13).

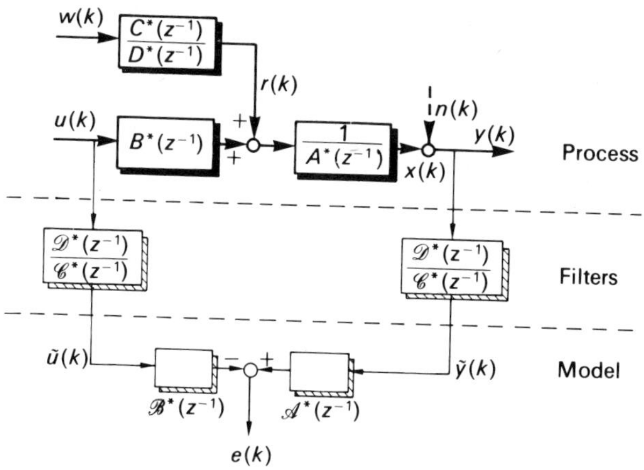


Figure 7.13

The process and the environment are described by

$$\left.\begin{aligned} \sum_{i=0}^{n} a_i x(k-i) &= \sum_{i=1}^{m} b_i u(k-i) && \text{Process dynamics} \\ y(k) &= x(k) + n(k) && \text{Process output} \end{aligned}\right\} \tag{7.60}$$

Consequently,

$$\sum_{i=0}^{n} a_i y(k-i) - \sum_{i=1}^{m} b_i u(k-i) = \sum_{i=0}^{n} a_i n(k-i) = r(k) \tag{7.61}$$

where the output noise is represented by $\{r(k)\}$ and $r(k)$ may be assumed to be derived from 'white'-noise samples $w(k)$ by:

$$r(k) + \sum_{i=1}^{p} d_i r(k-i) = w(k) + \sum_{i=1}^{q} c_i w(k-i) \tag{7.62}$$

Choosing again $a_0 = 1$ one finds:

$$\begin{aligned} y(k) = \sum_{i=1}^{m} b_i u(k-i) - \sum_{i=1}^{n} a_i y(k-i) \\ + \sum_{i=1}^{q} c_i w(k-i) - \sum_{i=1}^{p} d_i r(k-i) + w(k) \end{aligned} \tag{7.63}$$

and in the same way as before:

$$\mathbf{y} = [\mathbf{U} \mid \mathbf{Y} \mid \mathbf{W} \mid \mathbf{R}] \begin{bmatrix} \mathbf{b} \\ \hline -\mathbf{a} \\ \hline \mathbf{c} \\ \hline -\mathbf{d} \end{bmatrix} + \mathbf{w} \tag{7.64}$$

This, again, is a model that is linear in the process parameters and in the noise parameters, and is disturbed by white equation noise. As before, a consistent estimator is:

$$\begin{bmatrix} \boldsymbol{\beta} \\ \hline -\boldsymbol{\alpha} \\ \hline \boldsymbol{\gamma} \\ \hline -\boldsymbol{\delta} \end{bmatrix} = [[\mathbf{U} \mid \mathbf{Y} \mid \mathbf{W} \mid \mathbf{R}]'[\mathbf{U} \mid \mathbf{Y} \mid \mathbf{W} \mid \mathbf{R}]]^{-1}[\mathbf{U} \mid \mathbf{Y} \mid \mathbf{W} \mid \mathbf{R}]'\mathbf{y} \tag{7.65}$$

As we do not know, nor can measure, the elements of **R** and **W** these elements have to be replaced by their estimates $\hat{\mathbf{r}}$ and $\hat{\mathbf{w}}$:

$$\left.\begin{aligned} \hat{r}(k) &= y(k) + \sum_{i=1}^{n} \alpha_i y(k-i) - \sum_{i=1}^{m} \beta_i u(k-i) \\ \hat{w}(k) &= \hat{r}(k) + \sum_{i=1}^{p} \delta_i \hat{r}(k-i) - \sum_{i=1}^{q} \gamma_i \hat{w}(k-i) \end{aligned}\right\} \tag{7.66}$$

Consequently, after each estimation of the parameter vector, the elements of $\hat{\mathbf{R}}$ and $\hat{\mathbf{W}}$ are calculated. In the beginning of the estimation procedure, the accuracy will be poor, so that again a weighting factor has to be introduced in order to de-emphasize older information. This algorithm too can be written in recursive form along the same lines as before. With proper interpretation, e.g.

$$\left.\begin{aligned} \hat{\boldsymbol{\theta}}'(k+1) &= [\boldsymbol{\beta}'(k+1) \mid -\boldsymbol{\alpha}'(k+1) \mid \boldsymbol{\gamma}'(k+1) \mid -\boldsymbol{\delta}'(k+1)] \\ \boldsymbol{\omega}'(k+1) &= [\mathbf{u}'(k+1) \mid \mathbf{y}'(k+1) \mid \hat{\mathbf{w}}'(k+1) \mid \hat{\mathbf{r}}'(k+1)] \end{aligned}\right\} \tag{7.67}$$

etc. the algorithm given by equation (7.56) can be used. This method is discussed by Smets (1970) for $\boldsymbol{\gamma} = \mathbf{0}$, by Young (1968) for $\boldsymbol{\delta} = \mathbf{0}$ and by Talmon (1971), Talmon and Van den Boom (1973) for the general case. Some simulation results taken from the last reference, are given in Section 7.3.

Another approach to circumvent the bias problems of the generalized model is by means of the maximum likelihood method (Cf. Chapter 11).

The instrumental variable method (cf. Chapter 6) can also easily be arranged as an on-line iterative procedure (cf. Peterka and Šmuk, 1969; Young, 1969). An approximate on-line version of the maximum likelihood methods has been proposed by Panuska (1968). A discussion of on-line methods is also given in Leathrum (1969).

### 7.2.4 *State–Space Models*

The previous methods can also be applied for the estimation of the parameters, used in the state–space description of the process. A discussion of the estimation of **A** in

$$\left.\begin{aligned} \mathbf{x}(k+1) &= \mathbf{A}\mathbf{x}(k) \\ \mathbf{y}(k) &= \quad \mathbf{x}(k) + \mathbf{n}(k) \end{aligned}\right\} \tag{7.68}$$

is given in Ho and Whalen (1963). The situation where in addition, the input signal is measurable can be indicated as follows (cf. Figure 7.14):

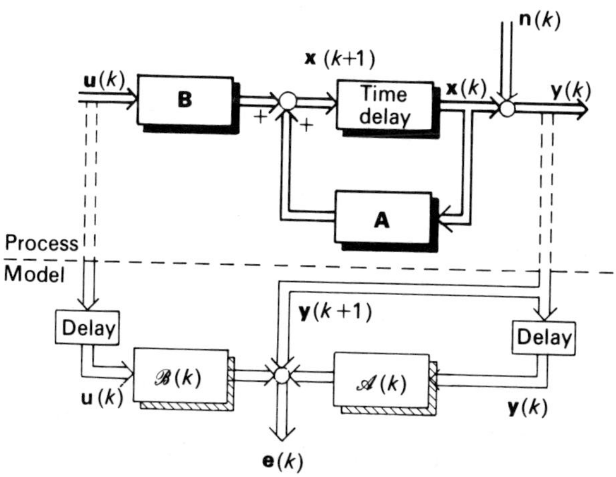


Figure 7.14

process:

$$\left.\begin{aligned} \mathbf{x}(k+1) &= \mathbf{A}\mathbf{x}(k) + \mathbf{B}\mathbf{u}(k) \\ \mathbf{y}(k) &= \quad \mathbf{x}(k) + \quad \mathbf{n}(k) \end{aligned}\right\} \tag{7.69}$$

model:

$$\mathbf{e}(k) = \mathbf{y}(k+1) - \mathscr{A}(k)\mathbf{y}(k) - \mathscr{B}(k)\mathbf{u}(k) \tag{7.70}$$

The criterion is chosen to minimize $E = \mathbf{e}'(k)\mathbf{e}(k)$ leading to

$$\left.\begin{aligned}\frac{\partial E}{\partial \mathscr{A}(k)} &= -2\mathbf{e}(k)\mathbf{y}'(k)\\ \frac{\partial E}{\partial \mathscr{B}(k)} &= -2\mathbf{e}(k)\mathbf{u}'(k)\end{aligned}\right\} \tag{7.71}$$

Consequently, the iterative estimation procedure can be chosen as:

$$\left.\begin{aligned}\mathscr{A}(k+1) &= \mathscr{A}(k) + \Gamma\mathbf{e}(k)\mathbf{y}'(k)\\ \mathscr{B}(k+1) &= \mathscr{B}(k) + \Gamma\mathbf{e}(k)\mathbf{u}'(k)\end{aligned}\right\} \tag{7.72}$$

The coefficient $\Gamma$ may again be chosen:

to be constant, if **A** and **B** are changing in time;

to be $1/k$ if **A** and **B** are constant; stochastic approximation.

Because of the additive noise the estimates will be biased.

Anderson et al. (1969) describe a procedure in order to obtain consistent estimates of the elements of the **A** matrix in

$$\left.\begin{aligned}\mathbf{x}(k+1) &= \mathbf{A}\mathbf{x}(k) + \mathbf{v}(k)\\ \mathbf{y}(k) &= \mathbf{x}(k) + \mathbf{n}(k)\end{aligned}\right\} \tag{7.73}$$

where $\mathbf{v}(k)$ and $\mathbf{n}(k)$ are white noises.

## 7.3 Some Computer Procedures and Results

As the methods discussed in Section 7.2 are promising with respect to practical application, the results of a number of experiments will be given to illustrate their properties. These illustrations are taken from Talmon (1971); cf. also: Talmon and Van den Boom (1973).

*Process*

In all cases given, the process has the pulse-transfer function.

$$H(z^{-1}) = \frac{B^*(z^{-1})}{A^*(z^{-1})} = \frac{z^{-1} + 0{\cdot}5z^{-2}}{1 - 1{\cdot}5z^{-1} + 0{\cdot}7z^{-2}} \tag{7.74}$$

and consequently:

$$\begin{matrix}[\mathbf{b}' \,\vdots\, -\mathbf{a}'] = [0, & 1, & 0{\cdot}5, & 1{\cdot}5, & -0{\cdot}7]\\ b_0 & b_1 & b_2 & -a_1 & -a_2\end{matrix}$$

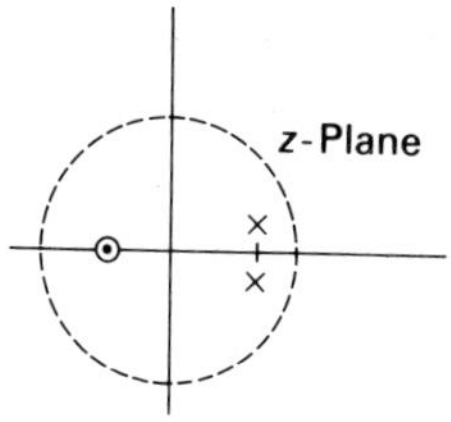


Figure 7.15

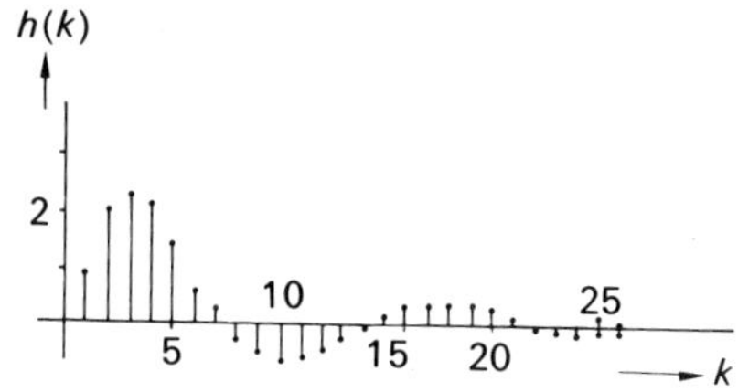


Figure 7.16

It has poles for $z = 0{\cdot}75 \pm 0{\cdot}36j$ and a zero at $z = -0{\cdot}5$. The pole-zero pattern is given in Figure 7.15, and the impulse response is shown in Figure 7.16. In all programmes, the input is chosen as a white-noise signal with a rectangular distribution between $-1$ and $+1$. If the white noise input signal has a power of $\sigma_u^2$ then the power of the output, $\sigma_x^2$, is given by:

$$\sigma_x^2 = \frac{\sigma_u^2}{2\pi j}\int_{|z|=1} H(z)H(z^{-1})\frac{dz}{z} \tag{7.75}$$

cf. Jury (1964). Åström, Jury and Agniel (1970) gave in their paper a fast method to calculate this integral. For the given process the ratio $\sigma_x^2/\sigma_u^2 = 18{\cdot}8$. The results given in the following sections are the averages of the estimates over ten runs of 1000 iterations. In addition, an average of the standard deviation, based on those ten runs, is given. Notes:

(1) In all tables, first the averages of the estimated values of the parameters over ten runs are given, unless otherwise indicated. Their standard deviations are given immediately below the averages.

(2) Consider a stochastic process $z(k) = \mu + x(k)$, in which $\mu$ is constant and $x(k)$ is a stochastic variable with $\mathscr{E}[x(k)] = 0$. The estimate of $\mu$ is given by $\bar{z} = 1/N \sum_{i=1}^{N} z(i)$. The confidence interval for $\mu$ is given by

$$\bar{z} - \frac{t_v(\tfrac{1}{2}\alpha)s}{\sqrt{N}} < \mu < \bar{z} + \frac{t_v(\tfrac{1}{2}\alpha)s}{\sqrt{N}} \tag{7.76}$$

where $s^2$ is the estimate of the variance of $z$ and where $v$ is the degree of freedom. In our case, $z(i)$ is the estimated value of a parameter in the $i$th run. We have 10 runs, so $\nu = 9$. When we want the confidence interval, with a chance of $\alpha = 5\%$ that $\mu$ lies outside this interval, then $t_9(\frac{1}{2}\alpha) = 2{\cdot}26$. Thus, we obtain:

$$\bar{z} - 0{\cdot}7s < \mu < \bar{z} + 0{\cdot}7s$$

(3) The programmes are written in ALGOL 60. The listings and the paper tapes of the programmes are available at the Group Measurement and Control, E.E. Dept., University of Technology, Eindhoven, the Netherlands.

*The Recursive Generalized Least-squares Estimation Method*

The algorithms used are those given by equations (7.56) and (7.57). The difference equation describing process and disturbances is:

$$y(k) = \sum_{i=0}^{2} b_i u(k-i) - \sum_{i=1}^{2} a_i y(k-i) - \sum_{i=1}^{2} d_i r(k-i) + \lambda w(k) \qquad (7.77)$$

$$b_0 = 0$$

$$b_1 = 1{\cdot}0 \qquad a_1 = -1{\cdot}5 \qquad d_1 = -1{\cdot}0$$

$$b_2 = 0{\cdot}5 \qquad a_2 = 0{\cdot}7 \qquad d_2 = 0{\cdot}2$$

cf. Figures 7.17a and b. $\{w(k)\}$ is a white-noise sequence with a rectangular amplitude distribution between $-1$ and $+1$. The transfer function that colours the white noise is:

$$\frac{1}{D^*(z^{-1})} = \frac{1}{1 - z^{-1} + 0{\cdot}2z^{-2}} \qquad (7.78)$$

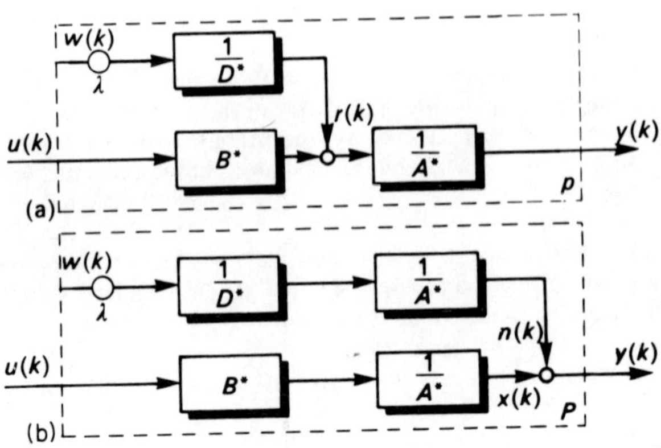


Figure 7.17

with poles at $z = 0{\cdot}725$ and $z = 0{\cdot}275$. For this case $\sigma_n^2/\sigma_w^2 = 98{\cdot}52$. Five process parameters and two noise parameters are estimated. After each estimation of $\mathbf{d}$ a new vector $\tilde{\boldsymbol{\omega}}(k+1)$ and a new value $\tilde{y}(k+1)$ are calculated. After each new estimation $\hat{\boldsymbol{\theta}}$ of the parameter vector a new vector $\hat{\mathbf{r}}(k+1)$ and a new value of $\hat{\mathbf{r}}(k+1)$ are calculated.

The following tables of experimental results are given:

| Table | $\lambda$ | Signal/noise ratio at process output $\frac{S}{N} = \frac{\sigma_x^2}{\sigma_n^2}$ | | Weighting: Coefficient $\rho = \nu =$ | Weighting: Number of samples in which the influence of model output errors is reduced to 1% |
|---|---|---|---|---|---|
| 7.1 | 0·25 | 3·2 | +5 dB | 0·9913 | 528 |
| 7.2 | 0·5 | 0·8 | −1 dB | Same | Same |
| 7.3 | 1 | 0·2 | −7 dB | Same | Same |
| 7.4 | 2 | 0·05 | −13 dB | Same | Same |
| 7.5 | 4 | 0·0125 | −19 dB | Same | Same |
| 7.6 | Same | Same | Same | 1·0 | ∞ |
| 7.7 | Same | Same | Same | 0·995 | 917 |
| 7.8 | Same | Same | Same | 0·990 | 458 |
| 7.9 | Same | Same | Same | 0·985 | 305 |
| 7.10 | Same | Same | Same | 0·980 | 228 |
| 7.11 | Same | Same | Same | 0·975 | 181 |

$\boxed{\rho = \nu = 0{\cdot}9913}$

Table 7.1. $S/N = 3{\cdot}2$ (+5 dB)

| Number of iterations | $\alpha_1$ | $\alpha_2$ | $\beta_0$ | $\beta_1$ | $\beta_2$ | $\delta_1$ | $\delta_2$ |
|---|---|---|---|---|---|---|---|
| 100 | −1·556 | ·752 | +·049 | 1·077 | ·521 | −0·995 | ·270 |
| | ·056 | ·082 | ·088 | ·098 | ·049 | ·162 | ·157 |
| 200 | −1·529 | ·730 | +·038 | 1·048 | ·514 | −1·060 | ·284 |
| | ·056 | ·079 | ·076 | ·082 | ·022 | ·113 | ·131 |
| 300 | −1·519 | ·720 | +·020 | 1·030 | ·505 | −1·068 | ·286 |
| | ·042 | ·061 | ·067 | ·069 | ·013 | ·092 | ·086 |
| 400 | −1·514 | ·715 | +·025 | 1·028 | ·504 | −1·003 | ·235 |
| | ·028 | ·038 | ·048 | ·046 | ·012 | ·066 | ·090 |
| 500 | −1·509 | ·708 | +·014 | 1·015 | ·501 | −0·971 | ·190 |
| | ·017 | ·024 | ·030 | ·034 | ·017 | ·101 | ·080 |
| 600 | −1·499 | ·700 | −·001 | 1·004 | ·504 | −0·985 | ·201 |
| | ·017 | ·020 | ·024 | ·027 | ·018 | ·076 | ·061 |
| 700 | −1·502 | ·703 | +·001 | 1·008 | ·501 | −0·950 | ·172 |
| | ·012 | ·017 | ·016 | ·016 | ·016 | ·077 | ·058 |
| 800 | −1·498 | ·702 | −·011 | 0·990 | ·489 | −0·960 | ·158 |
| | ·022 | ·022 | ·024 | ·031 | ·020 | ·057 | ·052 |
| 900 | −1·498 | ·708 | −·003 | 0·998 | ·493 | −0·977 | ·204 |
| | ·017 | ·019 | ·017 | ·018 | ·018 | ·066 | ·052 |
| 1000 | −1·502 | ·702 | +·001 | 1·007 | ·498 | −0·985 | ·196 |
| | ·016 | ·016 | ·016 | ·016 | ·016 | ·090 | ·087 |

$\rho = \nu = 0{\cdot}9913$

Table 7.2. $S/N = 0{\cdot}8$ ($-1$ dB)

| Number of iterations | $\alpha_1$ | $\alpha_2$ | $\beta_0$ | $\beta_1$ | $\beta_2$ | $\delta_1$ | $\delta_2$ |
|---|---|---|---|---|---|---|---|
| 100 | −1·565 | ·722 | +0·22 | 1·053 | ·510 | −0·925 | ·244 |
| | ·056 | ·056 | ·054 | ·085 | ·080 | ·132 | ·152 |
| ⋮ | ⋮ | ⋮ | ⋮ | ⋮ | ⋮ | ⋮ | ⋮ |
| 1000 | −1·504 | ·704 | +·011 | 1·014 | ·496 | −0·983 | ·196 |
| | ·028 | ·028 | ·031 | ·038 | ·035 | ·083 | ·080 |

Table 7.3. $S/N = 0{\cdot}2$ ($-7$ dB)

| Number of iterations | $\alpha_1$ | $\alpha_2$ | $\beta_0$ | $\beta_1$ | $\beta_2$ | $\delta_1$ | $\delta_2$ |
|---|---|---|---|---|---|---|---|
| 100 | −1·624 | ·763 | +·035 | 1·105 | ·507 | −0·849 | ·232 |
| | ·047 | ·053 | ·114 | ·190 | ·171 | ·102 | ·158 |
| ⋮ | ⋮ | ⋮ | ⋮ | ⋮ | ⋮ | ⋮ | ⋮ |
| 1000 | −1·517 | ·705 | +·007 | 1·038 | ·496 | −0·969 | ·208 |
| | ·039 | ·041 | ·068 | ·081 | ·079 | ·082 | ·082 |

Table 7.4. $S/N = 0{\cdot}05$ ($-13$ dB) (averaged over 9 runs)

| Number of iterations | $\alpha_1$ | $\alpha_2$ | $\beta_0$ | $\beta_1$ | $\beta_2$ | $\delta_1$ | $\delta_2$ |
|---|---|---|---|---|---|---|---|
| 100 | −1·672 | ·806 | +·043 | 1·106 | ·456 | −0·805 | ·245 |
| | ·049 | ·042 | ·225 | ·226 | ·258 | ·096 | ·167 |
| ⋮ | ⋮ | ⋮ | ⋮ | ⋮ | ⋮ | ⋮ | ⋮ |
| 1000 | −1·564 | ·722 | +·036 | 1·098 | ·490 | −0·926 | ·243 |
| | ·041 | ·042 | ·116 | ·140 | ·149 | ·069 | ·091 |

Table 7.5. $S/N = 0{\cdot}0125$ ($-19$ dB)

| Number of iterations | $\alpha_1$ | $\alpha$ | $\beta_0$ | $\beta_1$ | $\beta_2$ | $\delta_1$ | $\delta_2$ |
|---|---|---|---|---|---|---|---|
| 100 | −1·691 | ·816 | +·157 | 1·508 | ·636 | −0·776 | ·240 |
| | ·052 | ·047 | +·492 | 1·122 | ·769 | ·101 | ·155 |
| ⋮ | ⋮ | ⋮ | ⋮ | ⋮ | ⋮ | ⋮ | ⋮ |
| 1000 | −1·587 | ·731 | +·023 | 1·142 | ·470 | −0·905 | ·254 |
| | ·041 | ·043 | ·243 | ·260 | ·272 | ·087 | ·093 |

$S/N = 0{\cdot}0125\ (-19\ \text{dB})$

Table 7.6. $\rho = \nu = 1{\cdot}0$

| Number of iterations | $\alpha_1$ | $\alpha_2$ | $\beta_0$ | $\beta_1$ | $\beta_2$ | $\delta_1$ | $\delta_2$ |
|---|---|---|---|---|---|---|---|
| 100 | −1·698 | ·820 | +·142 | 1·103 | ·354 | −0·802 | ·301 |
| | ·094 | ·090 | ·395 | ·638 | ·581 | ·158 | ·094 |
| ⋮ | ⋮ | ⋮ | ⋮ | ⋮ | ⋮ | ⋮ | ⋮ |
| 1000 | −1·640 | ·764 | −·010 | 1·024 | ·458 | −0·858 | ·275 |
| | ·070 | ·053 | ·140 | ·168 | ·122 | ·067 | ·050 |

Table 7.7. $\rho = \nu = 0{\cdot}995$

| Number of iterations | $\alpha_1$ | $\alpha_2$ | $\beta_0$ | $\beta_1$ | $\beta_2$ | $\delta$ | $\delta_2$ |
|---|---|---|---|---|---|---|---|
| 100 | −1·676 | ·817 | +·138 | 1·102 | ·449 | −0·806 | ·300 |
| | ·092 | ·088 | ·367 | ·605 | ·538 | ·158 | ·097 |
| ⋮ | ⋮ | ⋮ | ⋮ | ⋮ | ⋮ | ⋮ | ⋮ |
| 1000 | −1·595 | ·731 | −·036 | 0·983 | ·490 | −0·886 | ·264 |
| | ·052 | ·036 | ·197 | ·237 | ·176 | ·065 | ·072 |

Table 7.8. $\rho = \nu = 0{\cdot}990$

| Number of iterations | $\alpha_1$ | $\alpha_2$ | $\beta_0$ | $\beta_1$ | $\beta_2$ | $\delta_1$ | $\delta_2$ |
|---|---|---|---|---|---|---|---|
| 100 | −1·667 | ·790 | +·285 | 1·335 | ·580 | −0·865 | ·334 |
| | ·103 | ·099 | ·293 | ·542 | ·483 | ·173 | ·071 |
| ⋮ | ⋮ | ⋮ | ⋮ | ⋮ | ⋮ | ⋮ | ⋮ |
| 1000 | −1·574 | ·731 | −·093 | 1·028 | ·551 | −0·926 | ·272 |
| | ·050 | ·040 | ·245 | ·270 | ·330 | ·071 | ·064 |

Table 7.9. $\rho = \nu = 0{\cdot}985$

| Number of iterations | $\alpha_1$ | $\alpha_2$ | $\beta_0$ | $\beta_1$ | $\beta_2$ | $\delta_1$ | $\delta_2$ |
|---|---|---|---|---|---|---|---|
| 100 | −1·662 | ·787 | +·265 | 1·327 | ·609 | −0·875 | ·337 |
| | ·102 | ·098 | ·313 | ·526 | ·455 | ·174 | ·077 |
| ⋮ | ⋮ | ⋮ | ⋮ | ⋮ | ⋮ | ⋮ | ⋮ |
| 1000 | −1·558 | ·722 | −·705 | 1·0681 | ·605 | −0·933 | ·266 |
| | ·058 | ·061 | ·329 | ·359 | ·409 | ·082 | ·080 |

S/N – 0·0125 (>19 dB)

Table 7.10. $\rho = \nu = 0{\cdot}980$

| Number of iterations | $\alpha_1$ | $\alpha_2$ | $\beta_0$ | $\beta_1$ | $\beta_2$ | $\delta_1$ | $\delta_2$ |
|---|---|---|---|---|---|---|---|
| 100 | −1·657<br>·101 | ·783<br>·098 | +·246<br>·344 | 1·318<br>·519 | ·632<br>·439 | −0·883<br>·177 | ·341<br>·086 |
| ⋮ | ⋮ | ⋮ | ⋮ | ⋮ | ⋮ | ⋮ | ⋮ |
| 1000 | −1·549<br>·073 | ·715<br>·081 | −·049<br>·413 | ·121<br>·448 | ·658<br>·469 | −0·935<br>·095 | ·264<br>·099 |

Table 7.11. $\rho = \nu = 0{\cdot}975$

| Number of iterations | $\alpha_1$ | $\alpha_2$ | $\beta_0$ | $\beta_1$ | $\beta_2$ | $\delta_1$ | $\delta_2$ |
|---|---|---|---|---|---|---|---|
| 100 | −1·652<br>·101 | ·779<br>·098 | +·230<br>·378 | 1·307<br>·521 | ·650<br>·435 | −0·891<br>·181 | ·345<br>·097 |
| ⋮ | ⋮ | ⋮ | ⋮ | ⋮ | ⋮ | ⋮ | ⋮ |
| 1000 | −1·543<br>·089 | ·711<br>·098 | −·020<br>·488 | 1·177<br>·526 | ·707<br>·517 | −0·936<br>·110 | ·264<br>·120 |

Consequently, from the experimental results one may judge:

(*a*) the influence of increasing noise level; tables 7.1–7.5;

(*b*) the effect of the weighting coefficients; tables 7.6–7.11.

(*a*) One notices that, with increasing noise power:

the standard deviations of $\beta_0, \beta_1$ and $\beta_2$ grow linearly with $\lambda$, i.e. proportionally with the square root of $S/N$;

the standard deviations of $\alpha_1$ and $\alpha_2$ go to a constant value;

the standard deviations of $\delta_1$ and $\delta_2$ remain nearly constant (cf. Figure 7.18).

With increasing noise power $b_i - \bar{\beta}_i(k)$ becomes larger for increasing $k$. In addition, it seems that with increasing noise power the speed of convergence decreases.

(*b*) One notices that, with decreasing weighting factors:

the means of the estimated value of the $\beta$'s vary more and more around the true values, whereas their standard deviations increase;

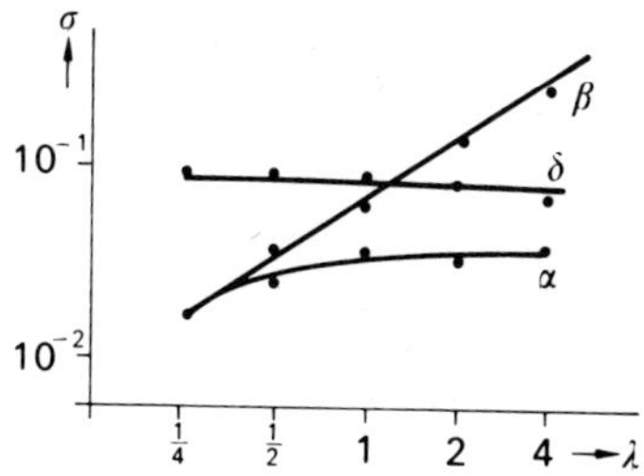


Figure 7.18

$a_i - \bar{\alpha}_i$ decrease constantly; the standard deviations diminish until $\rho = \nu = 0.990$ and then grow again;

the means of the $\delta$'s improve, while the standard deviations become a little bit larger.

These experimental results suggest that for low noise levels greater weighting factors can be chosen than for high noise levels. For such situations where the parameters remain constant it seems appropriate to increase the weighting factors gradually to one in course of the estimation procedure, in order to obtain both rapid convergence and small standard deviations.

*The Extended Matrix Method*

The process description and the estimation algorithms used are those indicated by equations (7.60) to (7.67). Using for the process description the same parameters as before one finds:

$$y(k) = \sum_{i=0}^{2} b_i u(k-i) - \sum_{i=1}^{2} a_i y(k-i) + r(k) \tag{7.79}$$

$$b_0 = 0$$

$$b_1 = 1 \qquad a_1 = -1.5$$

$$b_2 = 0.5 \qquad a_2 = 0.7$$

For the disturbances now a description with $c$ and $d$ parameters is chosen:

$$r(k) = \lambda \sum c_i w(k-i) - \sum d_i r(k-i) + \lambda w(k) \tag{7.80}$$

$$c_1 = 0.3 \qquad d_1 = -0.5$$

$$c_2 = 0.0 \qquad d_2 = 0.0$$

$$c_3 = 0.0 \qquad d_3 = 0.0$$

$$c_4 = 0.0$$

Again the factor $\lambda$ controls the noise level. The noise filter has a 'power gain' $\sigma_n^2/\sigma_w^2 = 43.62$.

Now one may study the effect on the process-parameter estimates when using the wrong number (and kind) of noise parameters. In the experimental results given, the weighting factor is increased to one in an exponential way:

$$\rho(k+1) = \rho(k) + \{1 - \rho(k)\}\Delta\rho = \{1 - \Delta\rho\}\rho(k) + \Delta\rho$$

$$\rho(0) = 0.9913 \tag{7.81}$$

The following tables and figures of experimental results are given:

| Table | $\lambda$ | Signal/noise ratio at process output | | $\Delta\rho$ | Estimated process parameters: $\alpha_1, \alpha_2, \beta_0, \beta_1, \beta_2$ Estimated noise parameters: | Figure |
|---|---|---|---|---|---|---|
| 7.12 | 1 | 0·432 | −3·7 dB | 0·001 | None | 7.19 |
| 7.13 | Same | Same | Same | Same | $\delta_1$ | 7.20 |
| 7.14 | Same | Same | Same | Same | $\delta_1\delta_2$ | 7.21 |
| 7.15 | Same | Same | Same | Same | $\delta_1\delta_2\delta_3$ | 7.22 |
| 7.16 | Same | Same | Same | Same | $\gamma_1$ | 7.23 |
| 7.17 | Same | Same | Same | Same | $\gamma_1\gamma_2$ | 7.24 |
| 7.18 | Same | Same | Same | Same | $\gamma_1\gamma_2\gamma_3$ | 7.25 |
| 7.19 | Same | Same | Same | Same | $\gamma_1\gamma_2\gamma_3\gamma_4$ | 7.26 |
| 7.20 | 0.25 | 6.92 | +8.3 dB | Same | $\delta_1\gamma_1$ | 7.27 |
| 7.21 | 1 | 0·432 | −3·7 dB | Same | Same | 7.28 |
| 7.22 | 4 | 0·0269 | −15·7 dB | Same | Same | 7.29 |
| 7.23 | 16 | 0·0017 | −27·7 dB | Same | Same | 7.30 |
| 7.24 | 1 | 0·432 | −3·7 dB | 0 | Same | |
| 7.25 | Same | Same | Same | 0·001 | Same | |
| 7.26 | Same | Same | Same | 0.005 | Same | |
| 7.27 | Same | Same | Same | 0·025 | Same | |

Table 7.12. Only Process Parameters Estimated

| Number of iterations | $\alpha_1$ | $\alpha_2$ | $\beta_0$ | $\beta_1$ | $\beta_2$ |
|---|---|---|---|---|---|
| 100 | −1·709<br>·027 | ·859<br>·044 | +·024<br>·103 | 1·042<br>·116 | ·386<br>·110 |
| 200 | −1·701<br>·030 | ·857<br>·026 | +·012<br>·065 | 1·029<br>·084 | ·386<br>·058 |
| 300 | −1·690<br>·031 | ·850<br>·027 | +·014<br>·072 | 1·026<br>·089 | ·394<br>·052 |
| 400 | −1·690<br>·029 | ·853<br>·021 | +·059<br>·055 | 1·057<br>·108 | ·398<br>·072 |
| 500 | −1·693<br>·015 | ·851<br>·018 | +·042<br>·106 | 1·035<br>·128 | ·392<br>·085 |
| 600 | −1·687<br>·011 | ·849<br>·015 | −·004<br>·077 | 0·068<br>·100 | ·357<br>·070 |
| 700 | −1·695<br>·011 | ·852<br>·019 | −0·11<br>·071 | 0·974<br>·094 | ·375<br>·050 |
| 800 | −1·689<br>·016 | ·853<br>·014 | −·029<br>·083 | 0·968<br>·090 | ·390<br>·044 |
| 900 | −1·688<br>·018 | ·849<br>·018 | −·011<br>·078 | 0·974<br>·089 | ·380<br>·063 |
| 1000 | −1·689<br>·013 | ·846<br>·014 | +·012<br>·040 | 0·998<br>·073 | ·410<br>·064 |

Table 7.13. Only One Backward Noise Parameter Estimated

| Number of iterations | $\alpha_1$ | $\alpha_2$ | $\beta_0$ | $\beta_1$ | $\beta_2$ | $\delta_1$ |
|---|---|---|---|---|---|---|
| 100 | −1·558<br>·080 | ·742<br>·073 | +·005<br>·064 | 1·038<br>·089 | ·482<br>·141 | −·625<br>·139 |
| ⋮ | ⋮ | ⋮ | ⋮ | ⋮ | ⋮ | ⋮ |
| 1000 | −1·481<br>·041 | ·718<br>·029 | −·011<br>·040 | 0·981<br>·051 | ·509<br>·065 | −·700<br>·051 |

Table 7.14. Two Backward Noise Parameters Estimated

| Number of iterations | $\alpha_1$ | $\alpha_2$ | $\beta_0$ | $\beta_1$ | $\beta_2$ | $\delta_1$ | $\delta_2$ |
|---|---|---|---|---|---|---|---|
| 100 | −1·623 | ·827 | −·019 | 0·983 | ·438 | −·661 | ·217 |
| | ·038 | ·044 | ·101 | ·118 | ·127 | ·166 | ·119 |
| ⋮ | ⋮ | ⋮ | ⋮ | ⋮ | ⋮ | ⋮ | ⋮ |
| 1000 | −1·512 | ·696 | +·005 | 0·998 | ·507 | −·770 | ·200 |
| | ·034 | ·033 | ·049 | ·055 | ·062 | ·068 | ·066 |

Table 7.15. Three Backward Noise Parameters Estimated

| Number of iterations | $\alpha_1$ | $\alpha_2$ | $\beta_0$ | $\beta_1$ | $\beta_2$ | $\delta_1$ | $\delta_2$ | $\delta_3$ |
|---|---|---|---|---|---|---|---|---|
| 100 | −1·583 | ·756 | +·025 | 1·006 | ·419 | −·720 | ·238 | −·104 |
| | ·045 | ·051 | ·135 | ·172 | ·173 | ·111 | ·186 | ·127 |
| ⋮ | ⋮ | ⋮ | ⋮ | ⋮ | ⋮ | ⋮ | ⋮ | ⋮ |
| 1000 | −1·511 | ·710 | +·010 | 1·008 | ·503 | −·783 | ·232 | −·069 |
| | ·037 | ·034 | ·052 | ·057 | ·060 | ·046 | ·067 | ·031 |

Table 7.16. One Forward Noise Parameter Estimated

| Number of iterations | $\alpha_1$ | $\alpha_2$ | $\beta_0$ | $\beta_1$ | $\beta_2$ | $\gamma_1$ |
|---|---|---|---|---|---|---|
| 100 | −1·643 | ·804 | −·009 | 1·026 | ·429 | ·642 |
| | ·034 | ·037 | ·060 | ·090 | ·129 | ·177 |
| ⋮ | ⋮ | ⋮ | ⋮ | ⋮ | ⋮ | ⋮ |
| 1000 | −1·602 | ·767 | −·014 | 0·995 | ·465 | ·677 |
| | ·020 | ·018 | ·033 | ·040 | ·053 | ·053 |

Table 7.17. Two Forward Noise Parameters estimated

| Number of Iterations | $\alpha_1$ | $\alpha_2$ | $\beta_0$ | $\beta_1$ | $\beta_2$ | $\gamma_1$ | $\gamma_2$ |
|---|---|---|---|---|---|---|---|
| 100 | −1·625 | ·794 | −·029 | 0·973 | ·435 | ·683 | ·237 |
| | ·042 | ·041 | ·096 | ·111 | ·126 | ·166 | ·173 |
| ⋮ | ⋮ | ⋮ | ⋮ | ⋮ | ⋮ | ⋮ | ⋮ |
| 1000 | −1·552 | ·726 | +·005 | 0·997 | ·487 | ·741 | ·297 |
| | ·034 | ·031 | ·048 | ·050 | ·058 | ·073 | ·056 |

Table 7.18. Three Forward Noise Parameters Estimated

| Number of iterations | $\alpha_1$ | $\alpha_2$ | $\beta_0$ | $\beta_1$ | $\beta_2$ | $\gamma_1$ | $\gamma_2$ | $\gamma_3$ |
|---|---|---|---|---|---|---|---|---|
| 100 | −1·597 | ·764 | +·039 | 1·020 | ·424 | ·760 | ·316 | ·124 |
| | ·046 | ·047 | ·141 | ·192 | ·161 | ·139 | ·215 | ·223 |
| ⋮ | ⋮ | ⋮ | ⋮ | ⋮ | ⋮ | ⋮ | ⋮ | ⋮ |
| 1000 | −1·524 | ·707 | +·007 | 1·009 | ·498 | ·779 | ·346 | ·151 |
| | ·030 | ·028 | ·052 | ·056 | ·059 | ·046 | ·065 | ·059 |

Table 7.19. Four Forward Noise Parameters Estimated

| Number of iterations | $\alpha_1$ | $\alpha_2$ | $\beta_0$ | $\beta_1$ | $\beta_2$ | $\gamma_1$ | $\gamma_2$ | $\gamma_3$ | $\gamma_4$ |
|---|---|---|---|---|---|---|---|---|---|
| 100 | −1·621 | ·788 | −·018 | 0·966 | ·411 | ·739 | ·241 | −·052 | +·012 |
| | ·046 | ·046 | ·124 | ·169 | ·103 | ·178 | ·183 | ·258 | ·166 |
| ⋮ | ⋮ | ⋮ | ⋮ | ⋮ | ⋮ | ⋮ | ⋮ | ⋮ | ⋮ |
| 1000 | −1·514 | ·706 | +·003 | 1·0006 | ·495 | ·790 | ·362 | +·175 | +·077 |
| | ·038 | ·033 | ·055 | ·069 | ·068 | ·047 | ·071 | ·096 | ·058 |

Table 7.20. $S/N = 6{\cdot}92$ (8·3 dB)

| Number of iterations | $\alpha_1$ | $\alpha_2$ | $\beta_0$ | $\beta_1$ | $\beta_2$ | $\delta_1$ | $\gamma_1$ |
|---|---|---|---|---|---|---|---|
| 100 | −1·519 | ·714 | −·009 | 0·989 | ·487 | −·452 | ·342 |
| | ·019 | ·025 | ·024 | ·028 | ·029 | ·172 | ·169 |
| ⋮ | ⋮ | ⋮ | ⋮ | ⋮ | ⋮ | ⋮ | ⋮ |
| 1000 | −1·500 | ·701 | +·000 | 0·999 | ·502 | −·502 | ·300 |
| | ·010 | ·011 | ·013 | ·014 | ·018 | ·060 | ·095 |

Table 7.21. $S/N = 0{\cdot}432$ (−3·7 dB)

| Number of iterations | $\alpha_1$ | $\alpha_2$ | $\beta_0$ | $\beta_1$ | $\beta_2$ | $\delta_1$ | $\gamma_1$ |
|---|---|---|---|---|---|---|---|
| 100 | −1·601 | ·777 | −·018 | 0·983 | ·433 | −·366 | ·332 |
| | ·039 | ·047 | ·101 | ·135 | ·139 | ·160 | ·170 |
| ⋮ | ⋮ | ⋮ | ⋮ | ⋮ | ⋮ | ⋮ | ⋮ |
| 1000 | −1·508 | ·703 | +·002 | 1·000 | ·504 | −·482 | ·308 |
| | ·042 | ·034 | ·046 | ·051 | ·059 | ·085 | ·092 |

Table 7.22. $S/N = 0{\cdot}0269$ ($-15{\cdot}7$ dB)

| Number of iterations | $\alpha_1$ | $\alpha_2$ | $\beta_0$ | $\beta_1$ | $\beta_2$ | $\delta_1$ | $\gamma_1$ |
|---|---|---|---|---|---|---|---|
| 100 | −1·627<br>·057 | ·793<br>·065 | −0·77<br>·405 | 0·915<br>·470 | ·415<br>·494 | −·327<br>·173 | ·349<br>·183 |
| ⋮ | ⋮ | ⋮ | ⋮ | ⋮ | ⋮ | ⋮ | ⋮ |
| 1000 | −1·515<br>·057 | ·706<br>·045 | +·005<br>·202 | 0·969<br>·218 | ·517<br>·220 | −·473<br>·104 | ·311<br>·094 |

Table 7.23. $S/N = 0{\cdot}0017$ ($-27{\cdot}7$ dB)

| Number of iterations | $\alpha_1$ | $\alpha_2$ | $\beta_0$ | $\beta_1$ | $\beta_2$ | $\delta_1$ | $\gamma_1$ |
|---|---|---|---|---|---|---|---|
| 100 | −1·623<br>·064 | ·788<br>·075 | −·329<br>1·621 | 0·603<br>1·956 | +·304<br>1·875 | −·342<br>·155 | ·330<br>·180 |
| ⋮ | ⋮ | ⋮ | ⋮ | ⋮ | ⋮ | ⋮ | ⋮ |
| 1000 | −1·516<br>·057 | ·707<br>·045 | −·013<br>·806 | 0·863<br>·870 | +·576<br>·883 | −·471<br>·106 | ·312<br>·097 |

Table 7.24. $\rho_0 = 0{\cdot}9913$, $\Delta\rho = 0{\cdot}0$

| Number of iterations | $\alpha_1$ | $\alpha_2$ | $\beta_0$ | $\beta_1$ | $\beta_2$ | $\delta_1$ | $\gamma_1$ |
|---|---|---|---|---|---|---|---|
| 100 | −1·605<br>·045 | ·778<br>·050 | −·021<br>·099 | 0·982<br>·118 | ·447<br>·127 | −·341<br>·193 | ·364<br>·169 |
| ⋮ | ⋮ | ⋮ | ⋮ | ⋮ | ⋮ | ⋮ | ⋮ |
| 1000 | −1·509<br>·051 | ·704<br>·040 | +·010<br>·081 | 1·010<br>·086 | ·513<br>·090 | −·474<br>·106 | ·333<br>·111 |

Table 7.25. $\rho_0 = 0{\cdot}9913$, $\Delta\rho = 0{\cdot}001$

| Number of iterations | $\alpha_1$ | $\alpha_2$ | $\beta_0$ | $\beta_1$ | $\beta_2$ | $\delta_1$ | $\gamma_1$ |
|---|---|---|---|---|---|---|---|
| 100 | −1·601<br>·039 | ·777<br>·047 | −·018<br>·101 | 0·983<br>·135 | ·433<br>·139 | −·366<br>·160 | ·332<br>·170 |
| ⋮ | ⋮ | ⋮ | ⋮ | ⋮ | ⋮ | ⋮ | ⋮ |
| 1000 | −1·508<br>·042 | ·703<br>·034 | +·002<br>·046 | 1·000<br>·051 | ·504<br>·059 | −·482<br>·085 | ·308<br>·092 |

Table 7.26. $\rho_0 = 0{\cdot}9913$, $\Delta\rho = 0{\cdot}005$

| Number of iterations | $\alpha_1$ | $\alpha_2$ | $\beta_0$ | $\beta_1$ | $\beta_2$ | $\delta_1$ | $\gamma_1$ |
|---|---|---|---|---|---|---|---|
| 100 | −1·603 | ·778 | −·018 | 0·984 | ·434 | −·366 | ·328 |
| | ·038 | ·047 | ·098 | ·133 | ·138 | ·160 | ·170 |
| ⋮ | ⋮ | ⋮ | ⋮ | ⋮ | ⋮ | ⋮ | ⋮ |
| 1000 | −1·519 | ·709 | −·005 | 0·985 | ·486 | −·482 | ·290 |
| | ·025 | ·026 | ·037 | ·055 | ·042 | ·064 | ·077 |

Table 7.27. $\rho_0 = 0{\cdot}9913$, $\Delta\rho = 0{\cdot}025$

| Number of iterations | $\alpha_1$ | $\alpha_2$ | $\beta_0$ | $\beta_1$ | $\beta_2$ | $\delta_1$ | $\gamma_1$ |
|---|---|---|---|---|---|---|---|
| 100 | −1·612 | ·785 | −·019 | 0·992 | ·447 | −·343 | ·348 |
| | ·043 | ·050 | ·091 | ·113 | ·124 | ·179 | ·148 |
| ⋮ | ⋮ | ⋮ | ⋮ | ⋮ | ⋮ | ⋮ | ⋮ |
| 1000 | −1·538 | ·721 | −·005 | 0·985 | ·475 | −·458 | ·294 |
| | ·022 | ·025 | ·038 | ·060 | ·045 | ·071 | ·073 |

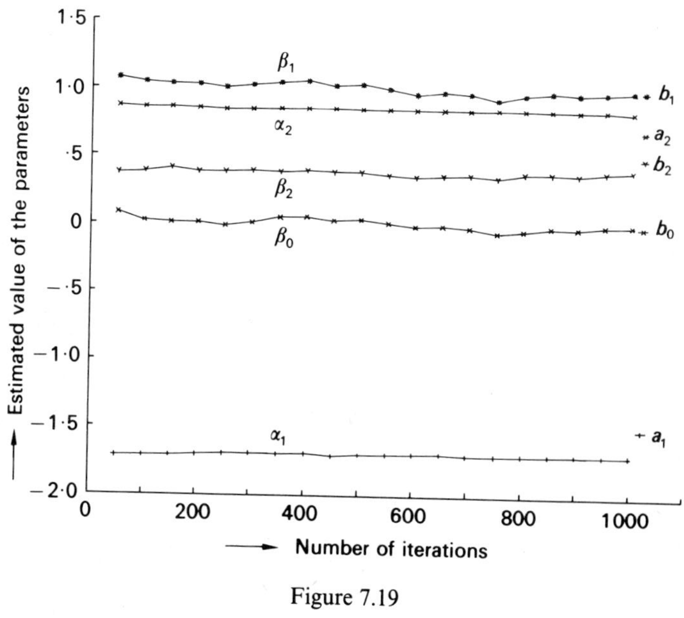


Figure 7.19

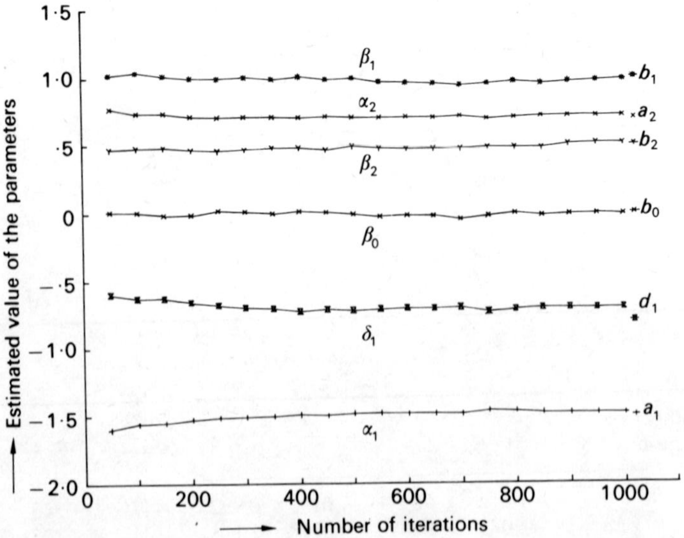


Figure 7.20

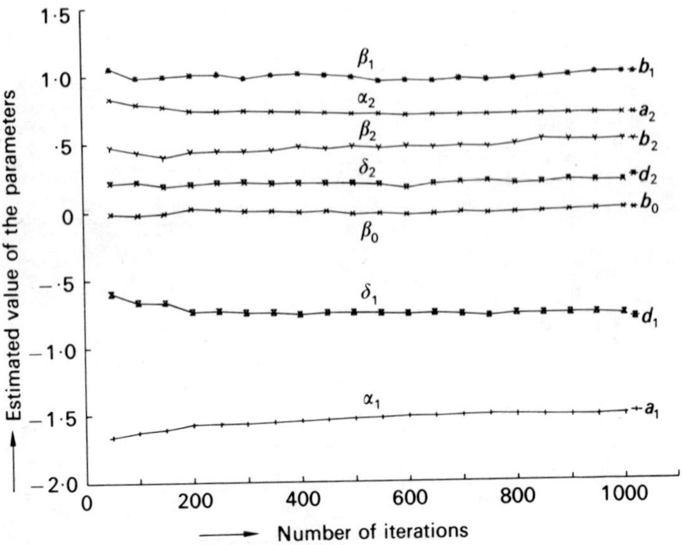


Figure 7.21

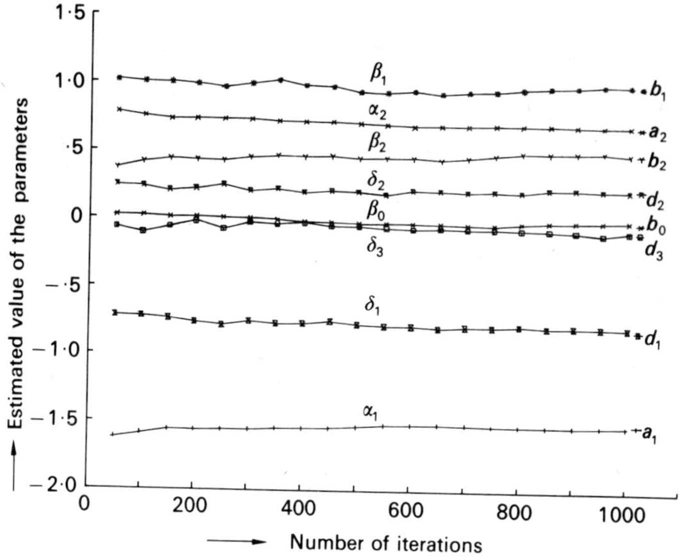


Figure 7.22

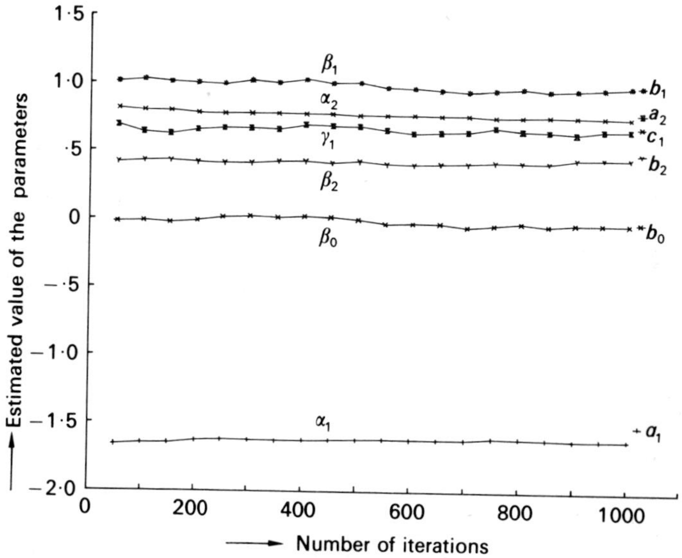


Figure 7.23

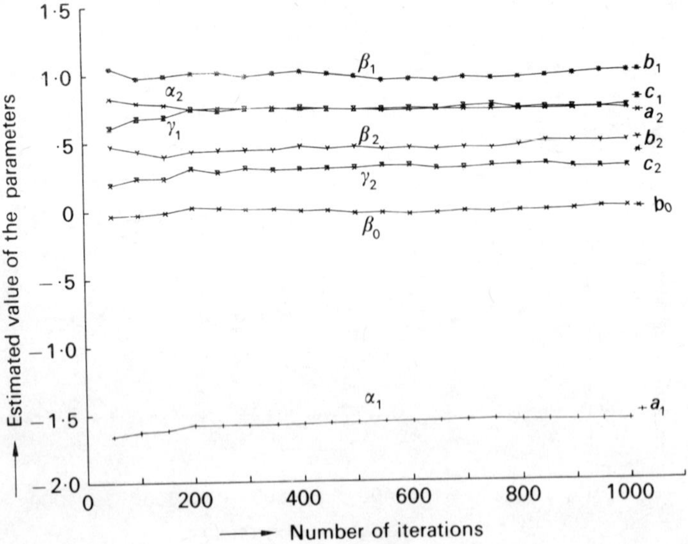


Figure 7.24

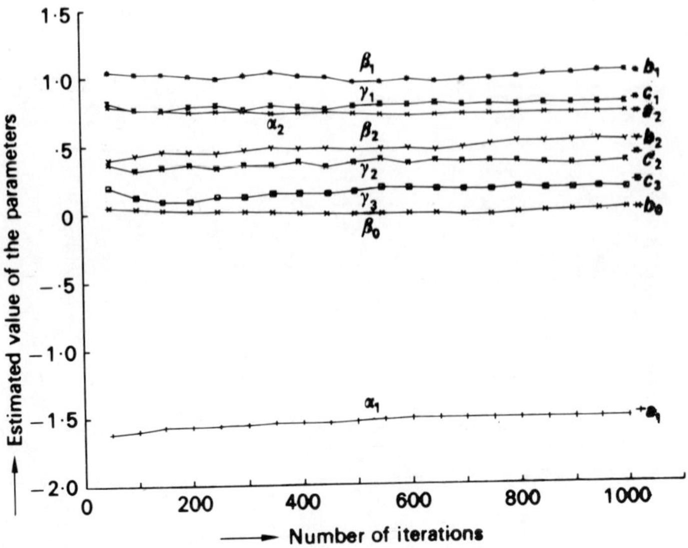


Figure 7.25

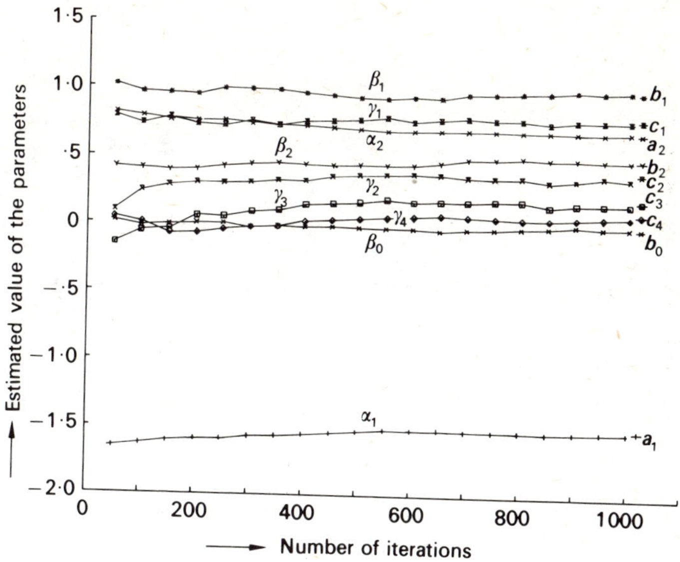


Figure 7.26

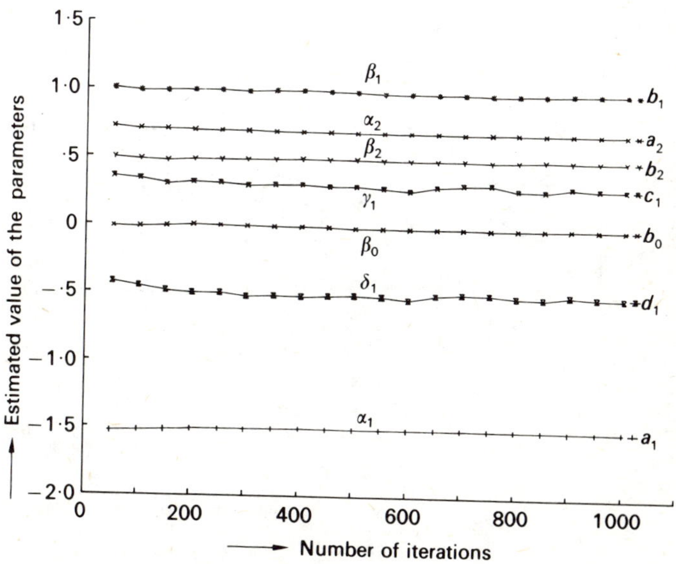


Figure 7.27

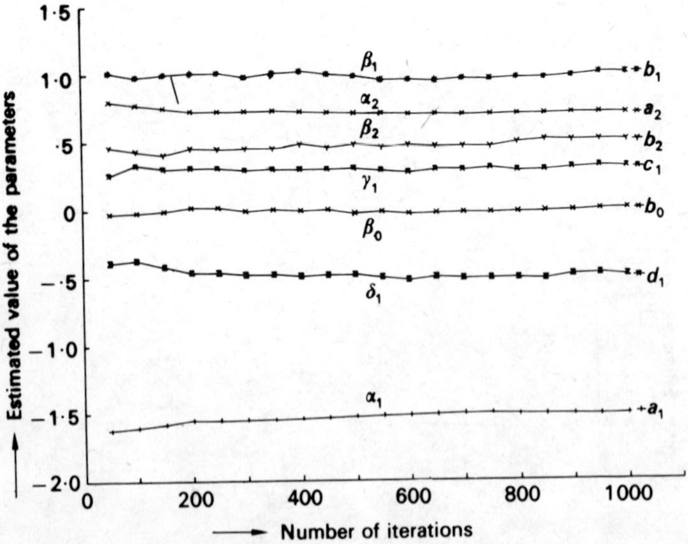


Figure 7.28

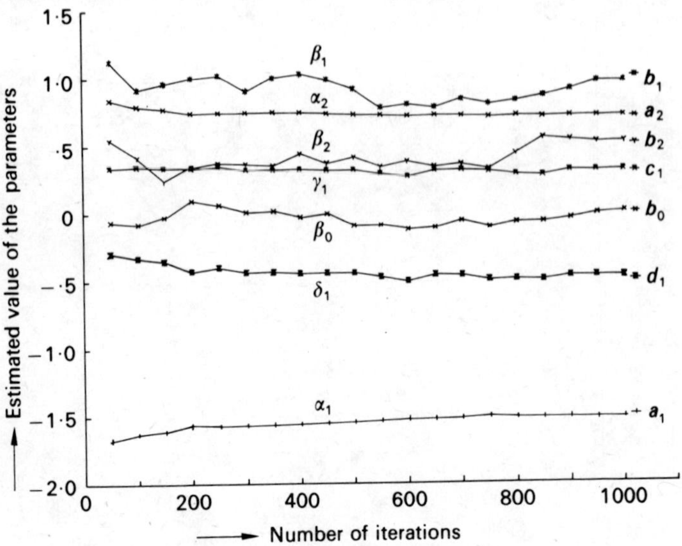


Figure 7.29

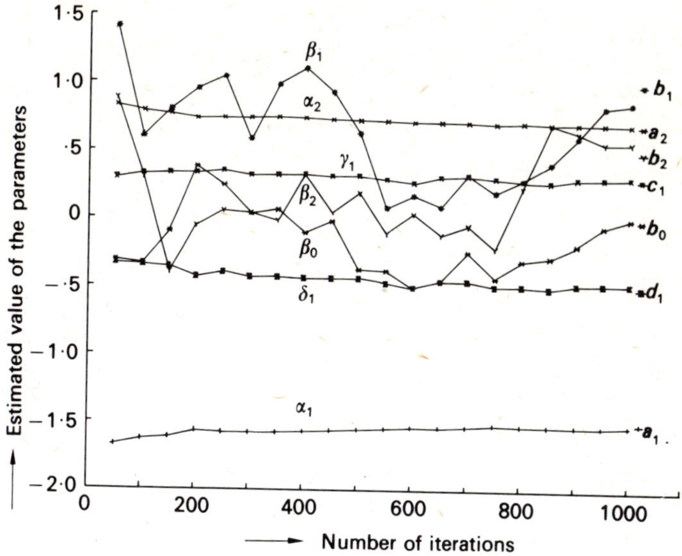


Figure 7.30

Table 7.12 and Figure 7.19 represent cases where only process parameters are estimated. As expected, there is an appreciable bias in some of the estimates such that the true value does not fall in the one standard-deviation interval.

Tables 7.13–7.15 and Figures 7.20–7.22 represent cases where the model interprets the noise by means of an autoregression expression ('backward' parameters), i.e. as if it were generated by:

$$r'(k) = -\sum d_i r'(k-i) + \lambda w(k)$$

or

$$r'(k) = \frac{\lambda w(k)}{D^*(z^{-1})} \tag{7.82}$$

The true noise-generation mechanism can be approximated in this way:

$$\begin{aligned} r(k) = \frac{C^*(z^{-1})}{D^*(z^{-1})}\lambda w(k) &= \frac{1 + 0{\cdot}3z^{-1}}{1 - 0{\cdot}5z^{-1}}\lambda w(k) \\ &= \frac{\lambda w(k)}{1 - 0{\cdot}8z^{-1} + 0{\cdot}24z^{-2} - 0{\cdot}072z^{-3} + \cdots} \end{aligned} \tag{7.83}$$

Increasing the number of parameters will result in a better representation of the noise and the process parameters; a comparison with Figure 7.19 indicates the improvement with respect to the bias. Note the true parameter values with respect to the estimated values $\pm$ the standard deviation.

Tables 7.16–7.19 and Figures 7.23–7.26 represent cases where the model interprets the noise by means of a moving-average model ('forward' parameters), i.e. as if it were generated by:

$$r''(k) = \lambda\{\sum c_i w(k-i) + w(k)\}$$

or

$$r''(k) = \lambda C^*(z^{-1})w(k) \tag{7.84}$$

The true noise-generation mechanism can be approximated in this way:

$$\begin{aligned} r(k) &= \frac{C^*(z^{-1})}{D^*(z^{-1})}\lambda w(k) = \frac{1+0{\cdot}3z^{-1}}{1-0{\cdot}5z^{-1}}\lambda w(k) \\ &= \{1 + 0{\cdot}8z^{-1} + 0{\cdot}4z^{-2} + 0{\cdot}2z^{-3} + \cdots\}\lambda w(k) \end{aligned} \tag{7.85}$$

Again increasing the number of parameters will result in a better representation of the noise and the process parameters; cf. with Figure 7.19. It turns out that one has to estimate more $c$ parameters than $d$ parameters in order to obtain results of the same quality. This is due to the fact that the values of $d_i$ decline more rapidly with increasing $i$ than the values of $c_i$.

Tables 7.20–7.23 and Figures 7.27–7.30 represent the case where the model interprets the noise by means of one $c$ and one $d$ parameter, i.e. a model that has the right topology and number of parameters. Table 7.21 and Figure 7.28 has the same $S/N$ ratio as the examples discussed before; the process and noise parameters are estimated very well.

These tables and figures give the results of experiments with decreasing signal to noise ratios at the process output; in each step the $S/N$ is deteriorated fourfold; in Figure 7.30 note that $S/N = 0{\cdot}0017$!

Study of these results indicates that:

the standard deviations of the $\beta$ parameters grow linearly with $\lambda$, i.e. proportionally with the square root of $S/N$;

the standard deviations of the $\alpha$ parameters tend to a constant value;

the standard deviations of the $\gamma$ and $\delta$ parameters remain constant.

Tables 7.24–7.27 represent cases where the speed of change with $k$ of the weighting factor $\rho$ changes. The cases chosen:

$$\Delta\rho = 0 \qquad 0{\cdot}001 \qquad 0{\cdot}005 \qquad 0{\cdot}025$$

correspond to a reduction of the difference between its value and 1 to half the initial value of that difference after the following number of iterations:

$$\infty \qquad 698 \qquad 128 \qquad 27$$

From the tables, one notices that with increasing $\rho$ the standard deviations tend to diminish. When increasing $\Delta\rho$ then in general the means of the estimates become worse, while the standard deviation tends to some lower bound.

For a complete insight into the properties of the different estimation schemes, more theoretical work and experiments have to be done. This holds particularly for the choice of the sequence of weighting factors. Of vital importance remains the problem of choosing or deriving the right number of $\alpha$, $\beta$, $\gamma$ and $\delta$ parameters.

### 7.4 Models, Non-linear-in-the-parameters

In Chapter 4, the importance of linearity-in-the-parameters has been stressed. The methods in the previous sections were devoted to such processes. For the simplicity of the estimating schemes this type of linearity proved to be of more importance than the linearity-of-the-dynamics. For special classes of non-linear systems, where the linear dynamics and the non-linearity-without-memory can be separated, the identification technique by means of an adjustable model is sometimes feasible (cf. Butler and Bohn, 1966; Narenda and Gallman, 1966).

Let us now consider the general case of non-linearity-in-the-parameters. The model output may be written as

$$\mathbf{w} = \mathbf{f}(\mathbf{U}, \boldsymbol{\beta}) \tag{7.86}$$

the error and error (loss) function may again be given as

$$E = \mathbf{e}'\mathbf{e} \qquad \text{with} \qquad \mathbf{e} = \mathbf{y} - \mathbf{w} = \mathbf{y} - \mathbf{f}(\mathbf{U}, \boldsymbol{\beta}) \tag{7.87}$$

From this point on the techniques mentioned in Section 5.2 for minimization of a function can be used. Note that linearity-in-the-parameters resulted in an error function which is quadratic-in-the-parameters. This neat property is lost in the cases that are considered now. Two approaches may serve as illustrations:

(*a*) *Gauss* or *Gauss–Newton Method*

This method is based on an expansion of $\mathbf{f}$ in a Taylor series:

$$\mathbf{w} = \mathbf{f}(\mathbf{U}, \boldsymbol{\beta} + \Delta\boldsymbol{\beta}) \approx \mathbf{f}(\mathbf{U}, \boldsymbol{\beta}) + \frac{\partial \mathbf{f}}{\partial \boldsymbol{\beta}'}\Delta\boldsymbol{\beta}$$

or

$$\mathbf{w} \approx \mathbf{f} + \mathbf{M}\Delta\boldsymbol{\beta} \tag{7.88}$$

Now $\mathbf{w}$ is again linear in $\Delta\boldsymbol{\beta}$; consequently, $\Delta\boldsymbol{\beta}$ can be found by the standard least-squares techniques discussed before. The model is adjusted according to $\Gamma\Delta\boldsymbol{\beta}$ and the same procedure is followed in the next iteration.

(*b*) *Gradient Method*

The gradient of the error function is determined:

$$\nabla_{\boldsymbol{\beta}} E = \left[\frac{\partial E}{\partial \beta_1}, \ldots, \frac{\partial E}{\partial \beta_m}\right]' \tag{7.89}$$

and the model adjustment is chosen as $\Gamma\nabla_{\boldsymbol{\beta}}E$.

In Section 2.4 the determination of partial derivatives or gradients has been discussed. It may be done by evaluating the effects of parameter perturbations of the 'model'. For such situations, a fast analogue computer model, combined with a digital machine, that is, a hybrid computer, may be a useful implementation.

Generally speaking, method (*a*) may suffer from divergence; method (*b*) may converge very slowly in the vicinity of the optimum. In Marquardt (1963) both methods are combined. For a further discussion of convergence schemes the reader is referred to Chapter 5.

The reason for using a model non-linear-in-the-parameters may be: the type of process representation needed for the application in mind; the properties of the estimates.

If our attention is focussed on the pulse-transfer function representation, then the generalized model is an adequate choice. It has the disadvantage, however, that the additive noise causes bias of some parameters. Therefore,

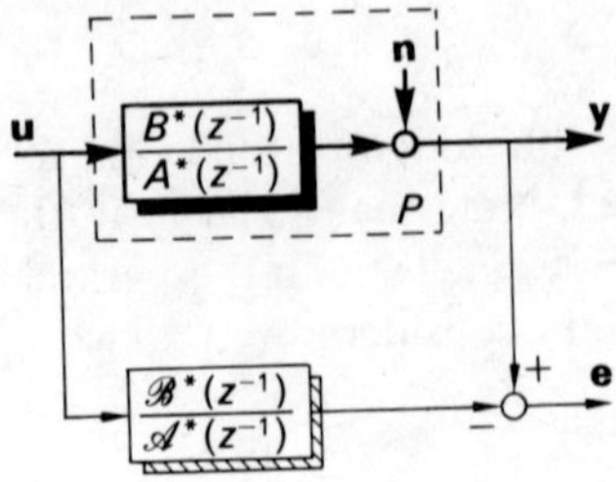


Figure 7.31

it may be necessary to consider the model according to Figure 7.31 and perform a minimization of

$$E = \sum e^2(k) \tag{7.90}$$

with

$$e(k) = y(k) - \frac{\mathscr{B}^*(z^{-1})}{\mathscr{A}^*(z^{-1})} u(k) \tag{7.91}$$

Using quasilinearization, Bellman and Kalaba (1965) have presented a solution for the non-linear optimization problem. Interesting applications of this technique are found in Buel, Kagiwada and Kalaba (1967), Schulz (1968), Buel and Kalaba (1969), Vaneček and Fessl (1970).

Another method for solving the non-linear optimization problem has been given by Taylor, Iliff and Powers (1969) in connexion with application to in-flight determination of stability derivatives. Again the error criterion can be given a probabilistic interpretation if it is assumed that the only disturbances are white-noise measurement errors. A technique which admits the measurement errors to be a stationary process with unknown rational spectral density has been proposed by Åström, Bohlin and Wensmark (1965). Due to specific assumptions that are made concerning the model structure, one might expect that serious mistakes can be made if these assumptions are not true. Results which prove or disprove this are not known.

This problem of adjustment of non-linear models is intimately related to the subjects discussed in Chapter 13, where more references may be found.

## References


Albert, A. and R. W. Sittler (1965). 'A method for computing least squares estimators that keep up with the data', *SIAM J. Control*, **3,** 384–417.

Albert, A. E. and L. A. Gardner Jr. (1967). Cf. references Chapter 5.

Anderson, W. M. et al. (1969). Cf. references Chapter 13.

Aoki, M. (1967). Cf. references Chapter 4.

Åström, K. J., T. Bohlin and S. Wensmark (1965). Cf. references Chapter 4.

Åstrom, K. J., E. I. Jury and R. G. Agniel (1970). 'A numerical method for the evaluation of complex integrals' (short paper), *IEEE Trans. Autom. Control*, **AC-15,** 468–471.

Bélanger, P. R. (1968). 'Comments on "A learning method for system identification' (correspondence), *IEEE Trans. Autom. Control*, **AC-13,** 207–208.

Bellman, R. E. and R. E. Kalaba (1965). Cf. references Chapter 13.

Buel, J. D., H. H. Kagiwada and R. E. Kalaba (1967). 'A proposed computational method for estimation of orbital elements drag coefficients and potential fields parameters from satellite measurements', *Ann. Géophys.* **23,** 35–39.

Buel, J. D. and R. E. Kalaba (1969). 'Quasilinearization and the fitting of nonlinear models of drug metabolism to experimental kinetic data', *Math. Biosci.*, **5,** 121–132.

Butler, R. E. and E. V. Bohn (1966). 'An automatic identification technique for a class of nonlinear systems' (short paper), *IEEE Trans. Autom. Control*, **AC-11,** 292–296.

Clarke, D. W. (1967). 'Generalized-least-squares estimation of the parameters of a dynamic model', *IFAC Symp.—Identification in Autom. Control Systems, Prague*, paper 3.17.

Crum, L. A. and S. H. Wu (1968). 'Convergence of the quantizing learning method for system identification' (correspondence), *IEEE Trans. Autom. Control*, **AC-13,** 297–298.

Hasting-James, R. and M. W. Sage (1969). 'Recursive generalised-least-squares procedure for on-line identification of process parameters', *Proc. IEE*, **116,** 2057–2062.

Ho, Y. C. and B. W. Whalen (1963). 'An approach to the identification and control of linear dynamic systems with unknown parameters' (correspondence), *IEEE Trans. Autom. Control*, **AC-8,** 255–256.

Holmes, J. K. (1968). Cf. references Chapter 5.

Jury, E. I. (1964). Cf. references Chapter 12.

Kalman, R. E. (1958). 'Design of a self-optimizing control system', *Trans. ASME*, **80,** 468–478.

Klinger, A. (1968). 'Prior information and bias in sequential estimation' (correspondence), *IEEE Trans. Autom. Control*, **AC-13,** 102–103.

Leathrum, J. F. (1969). 'On-line recursive estimation', *Joint Autom. Control Conf.*, 173–178.

Lee, R. C. K. (1964). Cf. references Chapter 4.

Marquardt, D. W. (1963). 'An algorithm for least-squares estimation of nonlinear parameters', *SIAM J.*, **11,** 431–441.

Nagumo, J. and A. Noda (1967). 'A learning method for system identification', *IEEE Trans. Autom. Control*, **AC-12,** 282–287. Cf. also same journal **AC-13,** 207–208.

Narenda, K. S. and P. G. Gallman (1966). 'An iterative method for the identification of nonlinear systems using a Hammerstein model' (short paper), *IEEE Trans. Autom. Control*, **AC-11,** 546–550.

Neal, C. B. and G. A. Bekey (1969). 'Estimation of the parameters of sampled data systems by means of stochastic approximation', *Joint Autom. Control Conf.*, 977–978.

Oza, K. G. and E. I. Jury (1968). 'System identification and the principle of contraction mapping', *SIAM J. Control*, **6,** 244–257.

Oza, K. G. and E. I. Jury (1969). 'Adaptive algorithms for identification problem', *4th IFAC Congr., Warsaw*, paper 26.4. Also: *Automatica*, **6,** 795–799 (1970).

Panuska, V. (1968). 'On the maximum likelihood estimation of rational pulse transfer function parameters' (correspondence), *IEEE Trans. Autom. Control*, **AC-13,** 304.

Peterka, V. and K. Šmuk (1969). Cf. references Chapter 6.

Schulz, E. R. (1968). 'Estimation of pulse transfer function parameters by quasi-linearization' (short paper), *IEEE Trans. Autom. Control*, **AC-13,** 424–426.

Smets, A. J. (1970). Cf. references Chapter 6.

Steiglitz, K. and L. E. McBride (1965). Cf. references Chapter 4.

Talmon, J. L. (1971). 'Approximated Gauss–Markov estimators and related schemes', *Report, Group Measurement and Control, E.E. Dept., University of Technology, Eindhoven*, Netherlands.

Talmon, J. L. and A. J. W. Van den Boom (1973). 'On the estimation of the transfer function parameters of process- and noise dynamics using a single-stage estimator', *Proc. 3rd IFAC Symp.—Identification and System Parameter Estimation*, 711–720, North-Holland Publ. Co., Amsterdam.

Taylor, L. W., K. W. Iliff and B. G. Powers (1969). 'A comparison of Newton-Raphson and other methods for determining stability derivatives from flight data'. *Paper, AIAA third Flight Test Simulation and Support Conf.*, Houston, Texas.

Van den Boom, A. J. W. (1967). 'Implicit estimation of process parameters using a digital computer' (in Dutch), *Thesis, E.E. Dept., University of Technology, Eindhoven*, Netherlands.

Van den Boom, A. J. W. and J. H. A. M. Melis (1969). Cf. references Chapter 6.

Vaneček, A. and J. Fessl (1970). 'A contribution to parameters and state estimation and synthesis via quasi-linearization', *2nd IFAC Symp.—Identification and Process Parameter Estimation, Prague*, paper 2.6.

Whitaker, H. P. et al. (1958). 'Design of a model reference adaptive control system for aircraft', *Report R-164, Instrumentation Lab., MIT*, Boston.

Young, P. C. (1968). 'The use of linear regression and related procedures for the identification of dynamic processes', *Proc. IEEE Symp. on Adapt. Processes*, Dec. 1968.

Young, P. C. (1969). 'An instrumental variable method for real-time identification of a noisy process', *4th IFAC Congr., Warsaw*, paper 26.6. Also: *Automatica*, **6,** 271–287 (1970).

## Additional Literature

Bellman, R. E., H. H. Kagiwada and R. E. Kalaba (1965). 'Identification of linear systems via numerical inversion of Laplace transforms' (correspondence), *IEEE Trans. Autom. Control*, **AC-10,** 111–112.

Budin, M. A. (1972) 'A new approach to system identification and state estimation', *IEEE Trans. Systems, Man. Cybernet.*, **SMC-2,** 396–402.

Durbin, J. (1960). 'Estimation of parameters in time-series regression models', *J.R. Statist. Soc., Ser. B*, **22B,** 139–153.

Farison, J. B., R. E. Graham and C. Shelton Rey (1967). 'Identification and control of linear discrete systems', *IEEE Trans. Autom. Control*, **AC-12,** 438–442.

Farison, J. B. (1967). 'Parameter identification for a class of linear discrete systems' (correspondence), *IEEE Trans. Autom. Control*, **AC-12,** 109.

Genin, Y. (1968). 'A note on linear minimum variance estimation problems' (correspondence), *IEEE Trans. Autom. Control*, **AC-13,** 103.

Holmes, J. K. (1967). 'System determination with noisy measurements', *IFAC Symp.—Identification in Autom. Control Systems, Prague*, paper 3.3.

Hsieh, H. C. (1964). 'An on-line identification scheme for multi-variable nonlinear systems', in: *Computing Methods in Optimization Problems* (Eds. A. V. Balakrishnan and L. W. Neustadt), Academic Press, New York, 193–210.

Hsieh, H. C. (1964). 'Least square estimation of linear and nonlinear weighting function matrices', *Information and Control*, **7,** 84.

Mintz, M. L. and J. S. Thorp (1966). 'A method for discrete system identification approximation', *J. Franklin Inst.*, **283,** 389–403.

Müller, J. A. (1968). 'Analysis of control loops by means of adaptive models' (in German), *Z. Messen. Steuern, Regeln*, **11,** 78–80, 146–152.

Nishimura, M., K. Fujii and Y. Suzuki (1969). 'On-line estimation of the process parameters and its applications to an adaptive control system', *4th IFAC Congr., Warsaw*, paper 26.3.

Ochem, W., A. Rault and H. Weingarten (1970). 'A new identification algorithm for linear discrete time systems', *2nd IFAC Symp.—Identification and Process Parameter Estimation, Prague*, paper 2.9.

Pearson, A. E. (1966). 'A regression analysis in function space for the identification of multivariable linear systems', *3rd IFAC Congr., London*, paper 46 B.

Pearson, J. D. (1967). 'Deterministic aspects of the identification of linear discrete dynamic systems' (short paper), *IEEE Trans. Autom. Control*, **AC-12,** 764–766.

Rault, A., R. Pouliquen and J. Richalet (1969). Cf. references Chapter 11.

Roy, R. and J. Sherman (1967). 'A learning technique for Volterra series representation' (short paper), *IEEE Trans. Autom. Control*, **AC-12,** 761–764.

Šmuk, K. (1970). 'Comparison of some algorithms for least squares updating', *2nd IFAC Symp.—Identification and Process Parameter Estimation, Prague*, paper 2.10.

Stankovič, S., D. Velasevič and M. Carapič (1970). 'An approach to on-line discrete-time system identification', *2nd IFAC Symp.—Identification and Process Parameter Estimation, Prague*, paper 4.2.

Wilkie, D. F. and W. R. Perkins (1969). 'Design of model following systems using the companion transformation', *4th IFAC Congr., Warsaw*, paper 19.2.

CHAPTER 8

# Continuous Signals; Explicit Methods

In this chapter correlation techniques are studied for signals which are continuous (non-sampled) time functions. In Chapter 2 it was stated that as regards to amplitude, one can distinguish between 'analogue', quantized and binary signals. For a quantized signal the amplitude is restricted to a limited number of prefixed levels; for a binary signal this number is two. The correlation techniques for 'analogue' signals are discussed in Section 8.1. Section 8.2 is devoted to the description of the (non-linear) quantizing operation. As a result of quantizing, the correlation multiplier can be of a simpler construction than the one used for analogue signals. This holds even more in the case of binary signals where a still simpler relay or logic circuit may be used for multiplication. This is discussed in Section 8.3. Figure 8.1 gives a schematic representation of these cases mentioned.

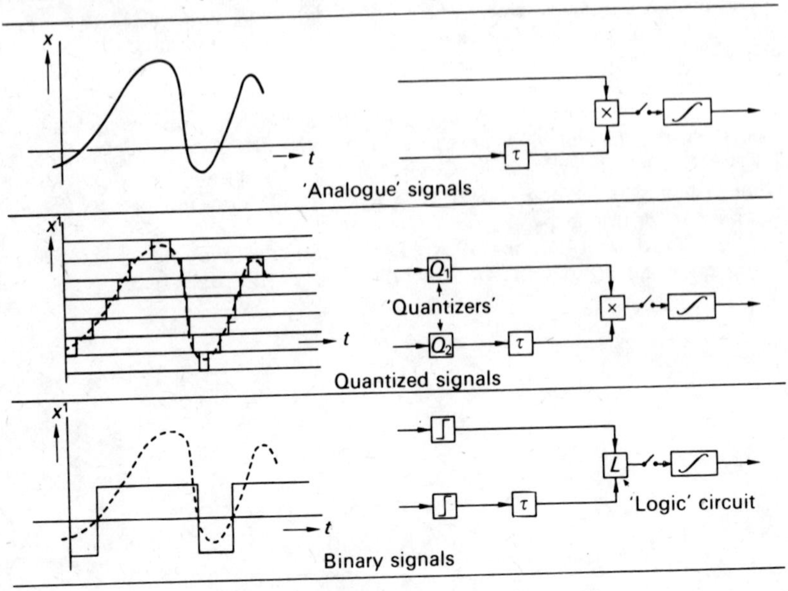


Figure 8.1

The essence of the explicit methods is the use of an operator (e.g. multiplication by a time function and integration of the product) to process input and output signals. Correlation techniques can be considered as a subclass of this 'equations-of-motion' method, just as it is the case with differential approximation. This is discussed in Section 8.4.

For the characterization of processes with non-linear dynamics higher-order correlation functions are just mentioned in Section 8.5. Another attempt to cope with non-linear processes is given in Section 8.6 by means of dispersion functions.

## 8.1 Operations on 'Analogue' Signals

### *8.1.1 General Properties of Correlation Measurements*

As a start it seems worthwhile to recall that correlation techniques have found application for a wide variety of purposes: e.g.

stochastic signal characterization;

system characterization;

signal detection;

and in a wide variety of fields: e.g.

communication

radioastronomy

radar and sonar receivers

engineering process studies

engineering measurement techniques

mechanical vibration studies

seismology

acoustics

brain research

Although the importance of correlation measurements was recognized for many years, it took a long time before commercial instruments became available. This may be attributed to the rather difficult and critical nature of some of the elements which constitute a correlator. Particularly this is true for the time-delay unit, the multipliers usable over an extensive range of frequencies and the integrators for long integration intervals. The advent of digital and hybrid techniques has alleviated many of these problems.

Correlation techniques may provide estimates of the unknown parameter in the sense of a minimum mean-squared error (cf. Chapter 2). This can easily be shown for continuous signals. We consider a linear time-invariant process $P$, stationary signals (ergodic random processes), and want to

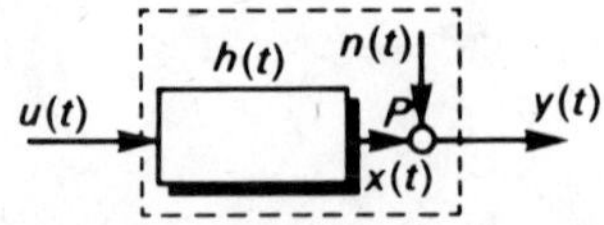


Figure 8.2

estimate the impulse response of $P$ (cf. Figure 8.2). Due to linearity, the **convolution integral** holds:

$$\begin{aligned} y(t) &= \int_0^\infty h(\theta)u(t-\theta)\,\mathrm{d}\theta + n(t) \\ &\stackrel{\text{def}}{=} h(t) * u(t) \quad + n(t) \end{aligned} \tag{8.1}$$

The signals $u(t)$ and $n(t)$ are stochastic; consequently, $y(t)$ is stochastic too. Stationarity of $u(t)$ implies that for processes with a 'memory' (impulse response $h(t)$) of finite duration $y(t)$ will be stationary also.

*Correlation Technique.* Multiplying both sides of equation (8.1) with $u(t-\tau)$ and taking the mathematical expectation we find:

$$\mathscr{E}[u(t-\tau)y(t)] = \int_0^\infty h(\theta)\mathscr{E}[u(t-\tau)u(t-\theta)]\,\mathrm{d}\theta + \mathscr{E}[u(t-\tau)n(t)] \tag{8.2}$$

or using the notation of Appendix B:

$$\begin{aligned} \psi_{uy}(\tau) &= \int_0^\infty h(\theta)\,\psi_{uu}(\tau-\theta)\,\mathrm{d}\theta + 0 \\ &= h(\tau) * \psi_{uu}(\tau) + 0 \end{aligned} \tag{8.3}$$

if $n(t)$ and $u(t)$ are independent. Equation (8.1) relates the **stochastic** signals $u(t)$ and $y(t)$; equation (8.3) is a relation between **deterministic** functions, describing important characteristics of those stochastic signals.

*Minimization of the Average-squared Error.* One wants to determine a function $h(t)$ as an estimate of the impulse response. Again the average-squared error is considered:

$$E = \frac{1}{T}\int_0^T e^2\,\mathrm{d}t \tag{8.4}$$

with

$$e(t) = y(t) - \int_0^\infty h(\theta)u(t - \theta)\,\mathrm{d}\theta \tag{8.5}$$

Assume that $\hat{h}(\theta)$ is the function that minimizes $E$. Then any $h(\theta)$ can be represented as

$$h(\theta) = \hat{h}(\theta) + \alpha h_a(\theta) \tag{8.6}$$

where $h_a$ is an arbitrary function satisfying $h_a(\theta) \equiv 0$ for $\theta < 0$. By definition $E$ will be larger for $\alpha \neq 0$ than for $\alpha = 0$. Consequently,

$$\left.\frac{\partial E}{\partial \alpha}\right|_{\alpha=0} = 0$$

or

$$\frac{2}{T}\int_0^T \left\{ y(t) - \int_0^\infty \hat{h}(\theta)u(t - \theta)\,\mathrm{d}\theta \right\}\left\{ \int_0^\infty h_a(\tau)u(t - \tau)\,\mathrm{d}\tau \right\}\mathrm{d}t = 0$$

This can be written as:

$$\int_0^\infty h_a(\tau)\left\{ \frac{1}{T}\int_0^T y(t)u(t - \tau)\,\mathrm{d}t - \int_0^\infty \hat{h}(\theta)\,\mathrm{d}\theta \frac{1}{T}\int_0^T u(t - \theta)u(t - \tau)\,\mathrm{d}t \right\}\mathrm{d}\tau = 0 \tag{8.7}$$

As $h_a(\tau)$ is an arbitrary function, we can infer that

$$\frac{1}{T}\int_0^T y(t)u(t - \tau)\,\mathrm{d}t = \int_0^\infty \hat{h}(\theta)\,\mathrm{d}\theta \frac{1}{T}\int_0^T u(t - \theta)u(t - \tau)\,\mathrm{d}t \tag{8.8}$$

or

$$\tilde{\psi}_{uy}(\tau, T) = \int_0^\infty \hat{h}(\theta)\tilde{\psi}_{uu}(\tau - \theta, T)\,\mathrm{d}\theta \tag{8.9}$$

In practical situations the correlation measurement has to be over a finite time interval, which results in an approximation (indicated by ˜) of the true correlation function. For $T \to \infty$ and using the assumption of ergodicity:

$$\psi_{uy}(\tau) = \int_0^\infty \hat{h}(\theta)\psi_{uu}(\tau - \theta)\,\mathrm{d}\theta \qquad \text{Wiener–Hopf equation} \tag{8.10}$$

Compare this result with equation (8.3)!

By studying $\partial^2 E/\partial\alpha^2$ one has to verify that $E$ really is minimized. Note that no restrictions have been introduced with respect to $\hat{h}(\theta)$ except that for physical realizability of the process

$$\hat{h}(\theta) \equiv 0 \qquad \text{for } \theta < 0$$

*Determination of* $h(\theta)$. Determination of the impulse response from the correlation functions $\psi_{uy}$ and $\psi_{uu}$ (**deconvolution**) can be done by means of analogue or digital techniques and along the lines of time-domain or frequency-domain operations. In the time domain the situation is quite simple if the bandwidth of $u(t)$ is large compared to the bandwidth of the process. For such practical cases of a 'white' input signal the following approximation can be used:

$$\psi_{uu}(\tau - \theta) \approx c\delta(\tau - \theta) \quad \text{with} \quad c = \int_{-\infty}^{+\infty} \psi_{uu}(\tau)\, \mathrm{d}\tau$$

and, using the properties of Dirac 'functions'

$$\psi_{uy}(\tau) \approx ch(\tau) \quad \text{or} \quad h(\tau) \approx \frac{1}{c}\psi_{uy}(\tau)$$

In such a case, the cross-correlation function is a good approximation of the process impulse response (cf. Figure 8.3).

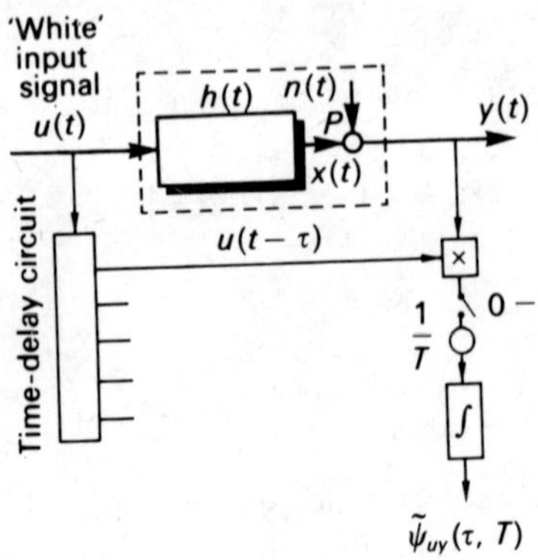


Figure 8.3

If the input signal cannot be considered to be 'white', then one has to face the problem of solving for $h(\theta)$ from equation (8.10). One type of solution is to adjust a model in such a way that $\psi_{uy}(t)$ is its output signal if $\psi_{uu}(t)$ is its input (cf. Wallman, 1950; Goodman and Reswick, 1956; Chang, Goodman and Reswick, 1956; Goodman, 1957).

Another approach is by approximating the convolution integral (8.10) as a convolution sum:

$$\psi_{uy}(\tau) = \int_0^{\infty} h(\theta)\psi_{uu}(\tau - \theta)\, d\theta$$

$$= \int_0^{\frac{1}{2}T_s} h(\theta)\psi_{uu}(\tau - \theta)\, d\theta + \sum_{i=1}^{\infty} \int_{(i-\frac{1}{2})T_s}^{(i+\frac{1}{2})T_s} h(\theta)\psi_{uu}(\tau - \theta)\, d\theta$$

or, as indicated in Figure 8.4:

$$\psi_{uy}(kT_s) \approx \frac{T_s}{2} h(0)\psi_{uu}(kT_s) + T_s \sum_{i=1}^{\infty} h(iT_s)\psi_{uu}((k - i)T_s) \qquad (8.11)$$

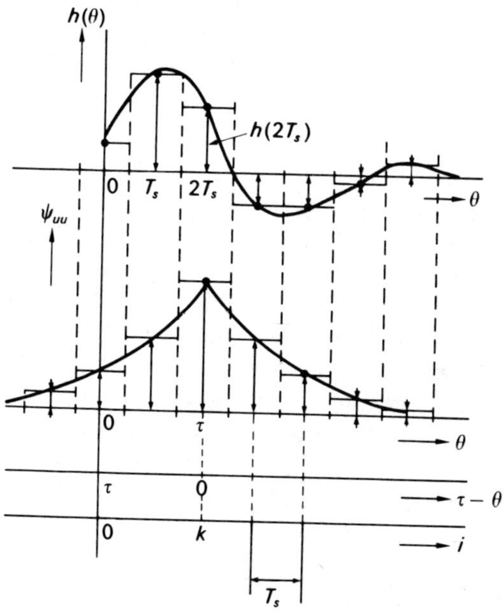


Figure 8.4

*Example.* Equation (8.11) is written in matrix form for three points of the impulse response:

$$\begin{bmatrix} \psi_{uy}(0) \\ \psi_{uy}(T_s) \\ \psi_{uy}(2T_s) \end{bmatrix} \approx T_s \begin{bmatrix} \psi_{uu}(0) & \psi_{uu}(-T_s) & \psi_{uu}(-2T_s) \\ \psi_{uu}(T_s) & \psi_{uu}(0) & \psi_{uu}(-T_s) \\ \psi_{uu}(2T_s) & \psi_{uu}(T_s) & \psi_{uu}(0) \end{bmatrix} \cdot \begin{bmatrix} \frac{1}{2}h(0) \\ h(T_s) \\ h(2T_s) \end{bmatrix}$$

Because $\psi_{uu}(\tau) = \psi_{uu}(-\tau)$, the matrix is symmetrical.

This formulation is closely related to those given in Chapter 6. Of course one can easily develop other and better approximations of the convolution integral, e.g. by using linear interpolations between sample points of the $h$ and $\psi$ functions.

A third type of approach towards solving for $h(\theta)$ is by means of the frequency domain. By taking the Fourier transform of equation (8.10), one obtains:

$$\phi_{uy}(j\omega) = H(j\omega)\phi_{uu}(j\omega) \tag{8.12}$$

from which one can determine

$$H(j\omega) = \frac{\phi_{uy}(j\omega)}{\phi_{uu}(j\omega)}$$

Due to the fact that only estimates of the power spectral density functions will be available, $H(j\omega)$ will be an estimate also. Numerical Fourier transformation can be performed by the celebrated Fast-Fourier Transform method (cf. Section 3.5).

### 8.1.2 *The Use of Orthogonal Filters*

So far, the discussion of correlation techniques has been restricted to instrumentations using a time-delay element. This, however, is not necessary. Another interesting choice of linear dynamic operators is that of orthogonal filters (Lampard, 1955; Kitamori, 1960). In this way, one circumvents the need for building an analogue delay line, still an expensive and critical component of a correlator. Yet, one may estimate the impulse response. The discussion of orthogonality properties can be done both in the time and in the frequency domain. The first approach can be compared with the discussion in Section 3.2. Here we consider the frequency domain.

The (conceptual) situation is that of Figure 8.5. For the mean-squared error one finds:

$$\begin{aligned}\overline{e^2(t)} &= \psi_{ee}(0) \\ &= \frac{1}{2\pi}\int_{-\infty}^{+\infty} \{H^*(j\omega) - G^*(j\omega)\}\{H(j\omega) - G(j\omega)\}\phi_{uu}(j\omega)\,d\omega + \psi_{nn}(0)\end{aligned} \tag{8.13}$$

where the integral represents an inverse Fourier transform with $\tau = 0$ and where $H$ and $G$ represent the dynamic properties of process $P$ and 'model' $M$. Apart from the noise contribution, this equation indicates the squared difference between $H$ and $G$, weighted with respect to $\phi_{uu}$. Now the model consists of some filters $F_i(j\omega)$, i.e.:

$$G(j\omega) = \sum_{i=0}^{m} \beta_i F_i(j\omega) \tag{8.14}$$

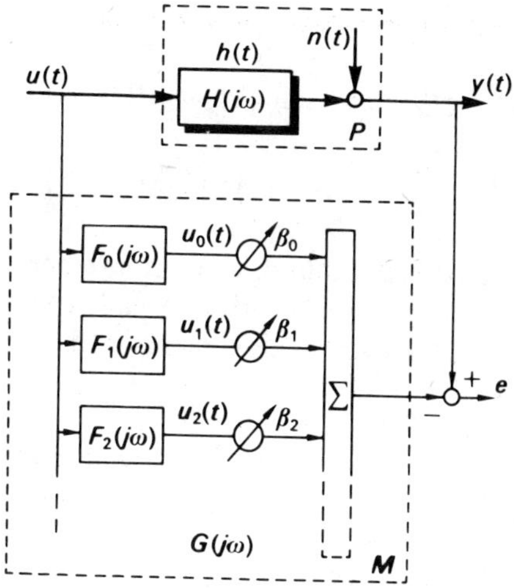


Figure 8.5

The coefficients $\beta_0, \ldots, \beta_m$ have to be 'adjusted' such that $\overline{e^2(t)}$ is a minimum. Substitution of equation (8.14) in (8.13), differentiation with respect to $\beta_k$ and equating this to zero for $\beta_k = \hat{\beta}_k$ yields:

$$\frac{1}{2\pi}\sum_{i=0}^{m}\hat{\beta}_i\int_{-\infty}^{+\infty}\{F_k^*(j\omega)F_i(j\omega) + F_k(j\omega)F_i^*(j\omega)\}\phi_{uu}(j\omega)\,d\omega$$
$$= \frac{1}{2\pi}\int_{-\infty}^{+\infty}\{F_k^*(j\omega)H(j\omega) + F_k(j\omega)H^*(j\omega)\}\phi_{uu}(j\omega)\,d\omega \qquad k = 0, 1, \ldots, m \tag{8.15}$$

or with symbols $I$ instead of the integrals:

$$\sum_{i=0}^{m}\hat{\beta}_i I_{ki} = I_{kH} \tag{8.16}$$

From this equation follows:

$$\hat{\beta}_k = \frac{1}{I_{kk}}\left\{I_{kH} - \sum_{i\neq k}\hat{\beta}_i I_{ki}\right\} \tag{8.17}$$

Consequently $\hat{\beta}_k$ is dependent on all other parameters; after adjustment of some other parameters the parameter $\beta_k$ has to be corrected again. This nuisance can be prevented by choosing

$$I_{ki} = \frac{1}{2\pi}\int_{-\infty}^{+\infty}\{F_k^*(j\omega)F_i(j\omega) + F_k(j\omega)F_i^*(j\omega)\}\phi_{uu}(j\omega)\,d\omega$$
$$= \delta_{ki} = \begin{cases}1 & k = i\\ 0 & k \neq i\end{cases} \tag{8.18}$$

i.e. the $F_k$ form an orthonormal system with respect to $\phi_{uu}(j\omega)$. In that case

$$\hat{\beta}_k = \frac{\displaystyle\int_{-\infty}^{+\infty} \{F_k^*(j\omega)H(j\omega) + F_k(j\omega)H^*(j\omega)\}\phi_{uu}(j\omega)\,d\omega}{\displaystyle 2\int_{-\infty}^{+\infty} F_k^*(j\omega)F_k(j\omega)\phi_{uu}(j\omega)\,d\omega} \tag{8.19}$$

which gives $\hat{\beta}_k$ as an explicit relation, without interdependence of the parameter values. The $F_k$'s have to be orthonormal with respect to $\phi_{uu}$. If $u$ represents white noise, then $\phi_{uu}(j\omega)$ is constant and as impulse responses of the filters $F_k$ the well-known Laguerre functions can be used (cf. Chapter 3 and Figure 4.17).

Note that numerator and denominator of equation (8.19) do not change if under the integral signs is included $\varepsilon^{j\omega\tau}$ with $\tau = 0$. Consequently, these integrals may be recognized as inverse $\mathscr{F}$-transforms, i.e. correlation functions (with argument $\tau = 0$):

$$\hat{\beta}_k = \frac{\psi_{u_k y}(0) + \psi_{y u_k}(0)}{2\psi_{u_k u_k}(0)} = \frac{\psi_{u_k y}(0)}{\psi_{u_k u_k}(0)} \qquad k = 0, 1, \ldots, m \tag{8.20}$$

This can be implemented according to Figure 8.6.

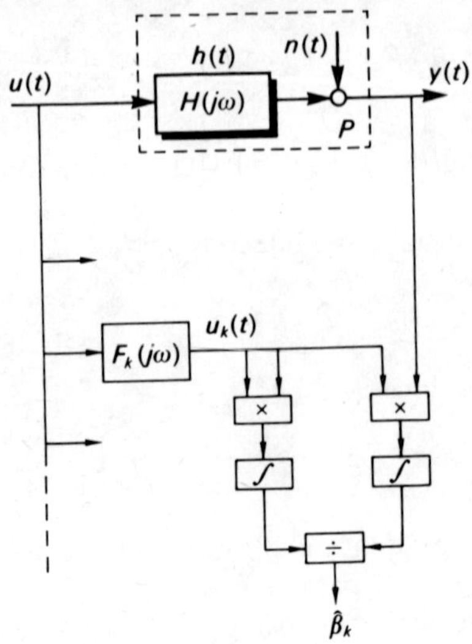


Figure 8.6

## 8.13 Statistical Errors of Correlation Measurements

The averaging operation over an infinite time interval is, of course, highly impractical. In actual applications one has to restrict the correlation time to a finite interval $[0, T]$. Consequently, one has to be satisfied with approximations of the correlation functions denoting them by $\tilde{\psi}_{uu}(\tau, T)$ and $\tilde{\psi}_{uy}(\tau, T)$. These measured 'correlation functions' show a variance (uncertainty) with respect to the theoretical correlation functions.

To discuss this aspect we consider Figure 8.7 where $x$ and $y$ are ergodic (stationary) stochastic signals and where the switch $S$ is closed at $t = 0$.

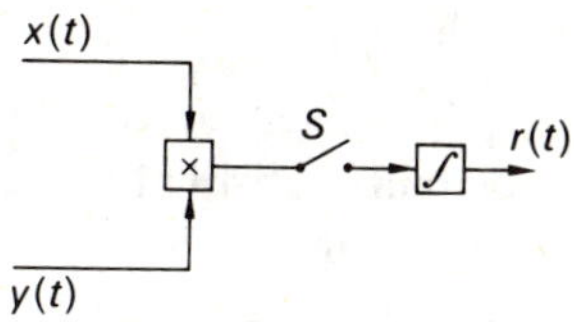


Figure 8.7

The initial condition $r(0) = 0$. In Appendix D it is shown that for this situation the following holds:

$$\mu_r(t) \stackrel{\text{def}}{=} \mathscr{E}[r(t)] = t\psi_{xy}(0) \tag{8.21}$$

and if $x$ and $y$ are jointly Gaussian:

$$\sigma_r^2(t) \stackrel{\text{def}}{=} \mathscr{E}[\{r(t) - \mu_r(t)\}^2] = 2\int_0^t (t - v)R(v)\,\mathrm{d}v \tag{8.22}$$

with

$$R(v) = \psi_{xx}(v)\psi_{yy}(v) + \psi_{xy}(v)\psi_{yx}(v) \tag{8.23}$$

By comparing Figures 8.3 and 8.7 one notices that these cases correspond if

$$x(t) \Rightarrow u(t - \tau)$$

$$y(t) \Rightarrow y(t) = x(t) + n(t)$$

Consequently,

$$\psi_{xy}(0) \Rightarrow \psi_{uy}(\tau) = \psi_{ux}(\tau) + \cancel{\psi_{un}(\tau)}$$

and

$$\psi_{xx}(v) \Rightarrow \psi_{uu}(v)$$

$$\psi_{yy}(v) \Rightarrow \psi_{yy}(v) = \psi_{xx}(v) + \psi_{nn}(v) + \cancel{\psi_{xn}(v)} + \cancel{\psi_{ux}(v)}$$

$$\psi_{xy}(v) \Rightarrow \psi_{uy}(v + \tau) = \psi_{ux}(v + \tau) + \cancel{\psi_{un}(v + \tau)}$$

$$\psi_{yx}(v) \Rightarrow \psi_{yu}(v - \tau) = \psi_{xu}(v - \tau) + \cancel{\psi_{nu}(v - \tau)}$$

where several terms drop out as the input and noise signals are assumed to be independent. This results in

$$\mu_r(t) = t\psi_{ux}(\tau) \tag{8.24}$$

i.e. the (independent) additive noise does not contribute to the expected value. For the variance, one can split $R(\nu)$ into two parts:

$$R(\nu) = R_1(\nu) + R_2(\nu)$$

in such a way that

$$R_1(\nu) = \psi_{uu}(\nu)\psi_{xx}(\nu) + \psi_{ux}(\nu + \tau)\psi_{xu}(\nu - \tau) \tag{8.25}$$

$$R_2(\nu) = \psi_{uu}(\nu)\psi_{nn}(\nu) \tag{8.26}$$

The contribution due to $R_1$ is present even if there is no additive noise; this might be called the **inherent statistical uncertainty.** The contribution due to $R_2$ is the **uncertainty due to the additive noise.**

This discussion of statistical errors can be applied to the experimentally determined points of the correlation function $\psi_{ux}(\tau)$, estimated over the finite time interval $[0, T]$:

$$\tilde{\psi}_{ux}(\tau, T) = \frac{1}{T}\int_0^T u(t - \tau)x(t)\,dt \tag{8.27}$$

The **inherent statistical uncertainty** can be expressed by the signal-to-noise ratio $S/N_1$ of the correlator output:

$$S = \mathscr{E}[\tilde{\psi}_{ux}(\tau, T)] = \frac{1}{T}\mu_r(T) = \psi_{ux}(\tau) \tag{8.28}$$

$$N_1^2 = \mathscr{E}[\tilde{\psi}_{ux}^2(\tau, T)] - \psi_{ux}^2(\tau) = \frac{1}{T^2}\sigma_r^2(T) \tag{8.29}$$

The expectation of the integrator output is the correct value of the desired correlation function independent of the length of the observation interval $T$. Of course the accuracy (variance or standard deviation) of the determined value does depend on the length of the observation interval. Because

$$\begin{aligned} N_1^2 &= \frac{1}{T^2}\sigma_r^2(T) = \frac{2}{T^2}\int_0^T (T - \nu)R_1(\nu)\,d\nu \\ &= \frac{2}{T^2}\left\{T\int_0^T R_1(\nu)\,d\nu - \int_0^T \nu R_1(\nu)\,d\nu\right\} \end{aligned} \tag{8.30}$$

this can be simplified for large $T$ if the second term can be neglected, i.e. if $R_1(\nu)$ approaches rapidly to zero. In that case

$$N_1^2 \approx \frac{2}{T}\int_0^T R_1(\nu)\,d\nu = \frac{2}{T}\int_0^T \{\psi_{uu}(\nu)\psi_{xx}(\nu) + \psi_{ux}(\nu + \tau)\psi_{xu}(\nu - \tau)\}\,d\nu \tag{8.31}$$

If $\psi_{ux}(\tau) \to 0$ for large $\tau$, then under that condition

$$N_1^2 \approx \frac{2}{T}\int_0^T \psi_{uu}(\nu)\psi_{xx}(\nu)\,d\nu \tag{8.32}$$

*Example.* The measurement of the autocorrelation function according to Figure 8.8 is considered, where $u(t) = x(t)$.

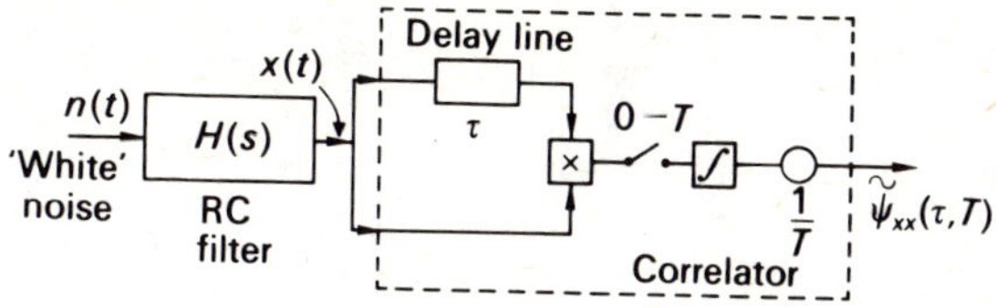


Figure 8.8

The white noise $n(t)$ has a power density: $k^2$;
the $RC$ filter has a transfer function: $H(s) = \alpha/(s + \alpha)$;
the power density of the signal $x(t)$ is: $\phi_{xx}(s) = k^2\alpha^2/(\alpha^2 - s^2)$;
the correlation function of $x(t)$ is:

$$\psi_{xx}(\tau) = \psi_{xx}(0)\varepsilon^{-\alpha|\tau|} \quad \text{with} \quad \psi_{xx}(0) = \alpha k^2/2$$

Consequently:

$$S = \psi_{xx}(\tau) = \psi_{xx}(0)\varepsilon^{-\alpha|\tau|}$$

$$N_1^2 = \frac{2}{T^2}\int_0^T (T - v)R_1(v)\,\mathrm{d}v$$

with

$$R_1(v) = \psi_{xx}^2(v) + \psi_{xx}(v + \tau)\psi_{xx}(v - \tau)$$
$$= \psi_{xx}^2(0)\{\varepsilon^{-2\alpha|v|} + \varepsilon^{-\alpha|v+\tau|}\varepsilon^{-\alpha|v-\tau|}\}$$

cf. Figure 8.9 which leads to

$$N_1^2 = \psi_{xx}^2(0)\frac{2\alpha T - 1 + 2\varepsilon^{-2\alpha T} + [(2\alpha\tau + 1)(2\alpha T - 1) - 2(\alpha\tau)^2]\varepsilon^{-2\alpha\tau}}{2(\alpha T)^2} \tag{8.33}$$

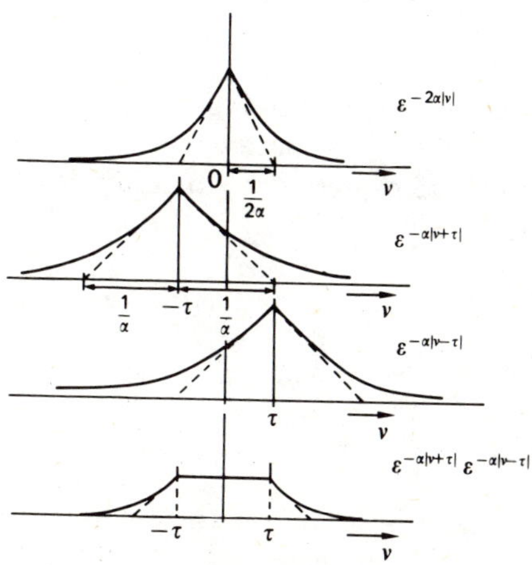


Figure 8.9

We introduce the dimensionless quantities $\alpha\tau = \beta$ and $\alpha T = \gamma$; then for $\beta \geqslant 0$

$$S = \psi_{xx}(0)\varepsilon^{-\beta} \tag{8.34}$$
$$N_1^2 = \psi_{xx}^2(0)\frac{2\gamma - 1 + 2\varepsilon^{-2\gamma} + [(2\beta + 1)(2\gamma - 1) - 2\beta^2]\varepsilon^{-2\beta}}{2\gamma^2}$$

Consider some special cases, where $\gamma \to \infty$ indicates long correlation intervals:

$$\beta = 0 \quad \begin{cases} N_1^2 = \psi_{xx}^2(0)\dfrac{2\gamma - 1 + \varepsilon^{-2\gamma}}{\gamma^2} & (8.35) \\ \gamma \to \infty \quad N_1^2 \to \psi_{xx}^2(0)\dfrac{2}{\gamma} & (8.36) \end{cases}$$

$$\beta \to \infty \qquad \gamma \to \infty \qquad N_1^2 \to \psi_{xx}^2(0)\frac{1}{\gamma} \tag{8.37}$$

Consequently, the variance of the measurements of points of the correlation function is inversely proportional to $\gamma$. Because $\gamma = \alpha T$, this means that the variance becomes smaller if $\alpha$ (the bandwidth of the filter) and/or $T$ (the observation interval) increase.

For the example given above, the signal-to-noise ratio is, in the case of a 'long' correlation interval:

$$\frac{S}{N_1} \approx \varepsilon^{-|\beta|}\sqrt{\frac{\gamma}{2}} \tag{8.38}$$

In terms of accuracy, one might specify e.g. $N_1 = 0{\cdot}05S$, i.e. a standard deviation equal to 5% of the measurement value. For $\beta = 0$ this results in $\gamma = \alpha T = 800$. Note that this means that the correlation interval is 800 times the time constant of the RC filter under consideration. Yet this rather long interval holds for a mediocre accuracy and for the most favourable point of the correlation function ($\beta = 0$). For large values of $\beta = \alpha\tau$, i.e. for points of the correlation function representing large arguments $\tau$, the $S/N$ ratio will be much worse. Correlation intervals of such lengths are dictated by the inherent statistical uncertainty only. Note that this $S/N$ ratio is independent of the level of the input signals to the correlator.

The uncertainty due to possible additive noise will put an even higher demand on correlation time. This **uncertainty due to additive noise** is of a simpler nature and can be expressed by an additional 'noise' term in the correlator output, viz.:

$$N_2^2 = \frac{2}{T^2}\int_0^T (T - v)R_2(v)\,\mathrm{d}v \tag{8.39}$$

with equation (8.26)

$$R_2(v) = \psi_{uu}(v)\psi_{nn}(v) \tag{8.40}$$

Of course, for this type of error, the signal-to-noise ratio of the correlator $S/N_2$ improves linearly with an increasing level of the input signal $u$. As an example, let us consider the same case as before, given by Figure 8.8, but now assuming that to **one** channel only additive noise is added. Let this noise $n(t)$ have the same band width as the signal $x(t)$, so:

$$\psi_{nn}(\tau) = \psi_{nn}(0)\varepsilon^{-\alpha|\tau|} \tag{8.41}$$

This leads to

$$N_2^2 = \psi_{xx}(0)\psi_{nn}(0)\frac{2\gamma - 1 + \varepsilon^{-2\gamma}}{2\gamma^2} \tag{8.42}$$

and for large $\gamma$:

$$N_2^2 \to \psi_{xx}(0)\psi_{nn}(0)\frac{1}{\gamma} \tag{8.43}$$

Now the total signal-to-noise ratio is found to be for long correlation intervals

$$\frac{S}{N} = \frac{S}{\sqrt{N_1^2 + N_2^2}} \approx \varepsilon^{-\beta}\sqrt{\frac{\gamma}{2 + \psi_{nn}(0)/\psi_{xx}(0)}} \tag{8.44}$$

Up to now, only the determination of one parameter (viz. one point of the impulse response) has been discussed. A new element enters the discussion if a number of parameters has to be determined. After one quantity has been estimated this estimate can be considered as being ***a priori* knowledge** for the next estimation cycle, etc. Intuitively, it is clear that the use of *a priori* knowledge may result in a more efficient estimation procedure. A crucial question, however, is in what way such knowledge can be incorporated into the instrumentation. This can be done by using a model of the process that is well adapted to the parameter description of the process. For example, if the attention is focussed on the impulse response, then a time-delay circuit with taps and potentiometers is adequate. Figure 8.10 gives an illustration of this situation (Eykhoff, 1967).

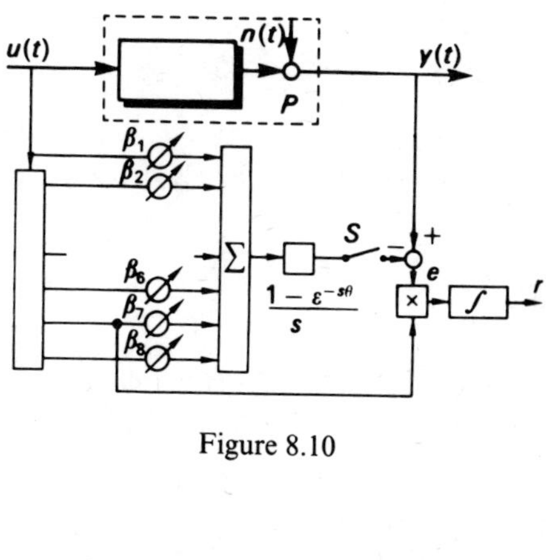


Figure 8.10

*Example.* In order to obtain some insight into the effect of the model, the relations (8.21) through (8.25) will be applied to an arbitrary simple example, viz. a process with impulse response:

$$h(\tau) = \begin{cases} 6\varepsilon^{-6\tau} & \text{for } \tau \geqslant 0 \\ 0 & \text{for } \tau < 0 \end{cases}$$

We will consider the determination of one point of this function, viz. for $\tau = 0{\cdot}225$ (Figure 8.11a) using a 'white' noise input signal.

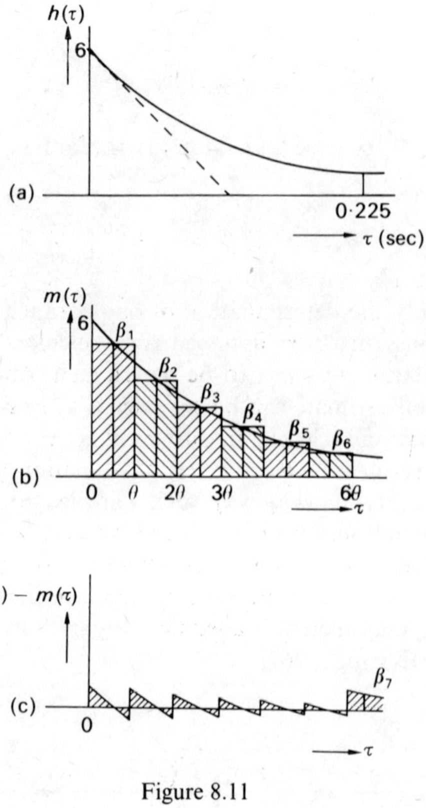


Figure 8.11

If the switch S in Figure 8.10 is open this method is identical to the correlation scheme of Figure 8.3. In that case we find for the standard deviation with respect to the expected value

$$\frac{1}{t}\mu_r = \psi_{uy}(\tau) \quad \text{with} \quad \tau = 0{\cdot}225$$

the relation given in Figure 8.12, curve *a*. From this diagram, the correlation interval needed for a certain accuracy (standard deviation) of the estimated parameter can be found.

Now we will assume that in Figure 8.10 the parameters $\beta_1$ through $\beta_6$ have already been estimated. The determined values are used for adjusting those potentiometers, the other potentiometers are set equal to zero and switch S is closed. The adjusted part of the model, with the hold circuit included, has the impulse response $m(\tau)$ shown in Figure 8.11. Consequently, the error signal $e$ is determined by the difference between $h(\tau)$ and $m(\tau)$ as shown in Figure 8.11c; yet, in this impulse response, $h(\tau) - m(\tau)$, the parameter to be estimated still has its original value. Now turning our attention to equation (8.25) we notice that in this case the index $x$ has to be replaced by the index $e$. This implies that the corresponding terms are smaller than the original ones, resulting in a smaller variance (or a reduction of the correlation interval that is needed for obtaining the same variance). For the example given we find the reduction of the correlation interval given by Figure 8.12, curve $b$; now the same standard deviation is obtained in about **one tenth** of the original time interval.

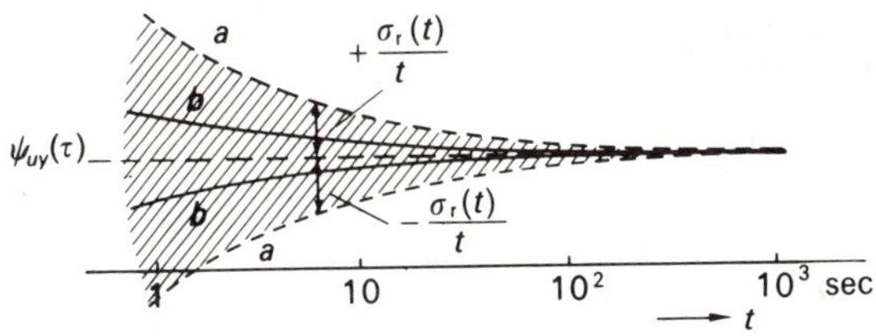


Figure 8.12

Figures 8.13a and b give a number of actual recordings of the integrator output $r(t)$ for the cases indicated above; S open viz. closed. The fluctuations represent the inherent statistical uncertainty of that correlator without and with a model. The influence of additive noise has been neglected; it can be taken into account using equation (8.26).

For further discussions on the accuracy of correlation techniques the reader is referred to Davenport and others (1952), Laning and Battin (1956), Green (1957), Solodovnikov (1960), Lee (1960), Bendat and Piersol (1966). There is an extensive literature on the use of correlation techniques; only a very small part of it can be mentioned as references (e.g. Balchen and Blandhol (1960), Rake (1968), Welfonder (1969), Buchta (1969), Hayashi (1969), Lange (1967), Rajbman et al (1970)) or as additional literature. Quite a number of correlator types is available commercially nowadays, some of them equipped with additional features such as probability-density-function meters.

Another development is that of correlators with a continuously changing time-delay $\tau$ and associated correcting filter (cf. Ball 1963).

Quite a popular type of additional instrumentation is that of Fourier transformation of the correlation functions to power spectral density curves. Often in such an instrument an appreciable number of points of the correlation function (e.g. 50 or 100) are determined simultaneously by parallel implementation or by Fast Fourier Transform.

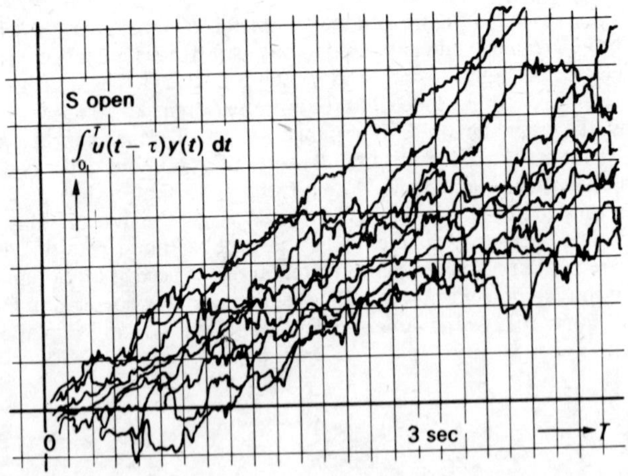


(a)

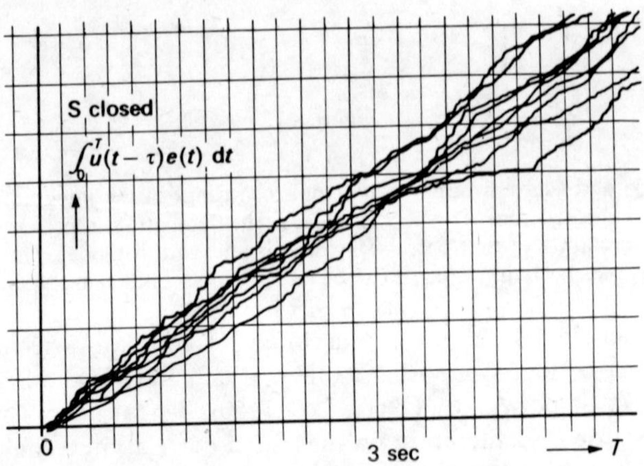


(b)

Figure 8.13

## 8.2 Operations on Quantized Signals

### *8.2.1 Amplitude Quantizing*

As indicated earlier, the quantizing operation $\mathcal{Q}[\ ]$ on 'analog' signals $x$ and $y$ is a non-linear operation, i.e. if

$$x_q = \mathcal{Q}[x] \quad \text{and} \quad y_q = \mathcal{Q}[y] \tag{8.45}$$

then

$$\alpha x_q + \beta y_q \quad \text{and} \quad \mathcal{Q}[\alpha x + \beta y] \tag{8.46}$$

are not necessarily equal for all values of $x$ and $y$. Consequently we will meet with the problems inherent to non-linear systems; that is they fail to obey the superposition principle.

For a discussion of the quantizing operation and its use in correlation determination the reader is referred to the papers by Widrow (1956) and Watts (1962). The following results are taken from Watts' paper. An 'analogue' input signal $x$ is transformed by the general equi-interval quantizer to a quantized signal $x_q$. The input–output characteristics may be that given in Figure 8.14a. This general quantizer $Q$ can be derived by means of a unit

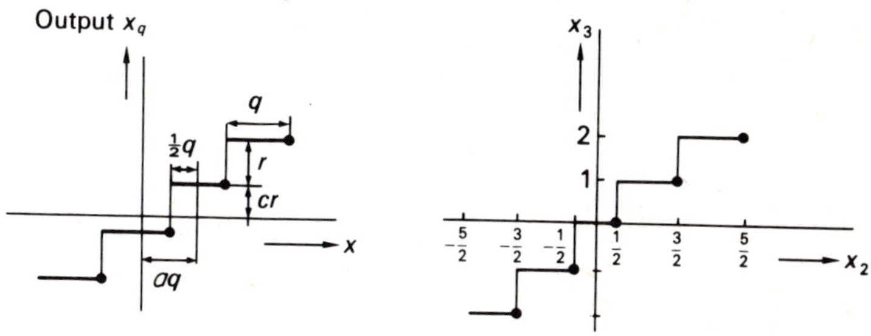


(a) General quantizer $Q$ (b) Unit quantizer $Q_1$

Figure 8.14

quantizer $Q_1$, the characteristics of which are shown in Figure 8.14b, as indicated in Figure 8.15. The corresponding relationships are the following:

gain $$x_1 = \frac{1}{q}x \tag{8.47}$$

shift $$x_2 = x_1 - a \tag{8.48}$$

unit quantizer $$x_3 = i \quad \text{for } (i - \tfrac{1}{2}) < x_2 \leqslant (i + \tfrac{1}{2}) \tag{8.49}$$

shift $$x_4 = x_3 + c \tag{8.50}$$

gain $$x_q = rx_4 \tag{8.51}$$

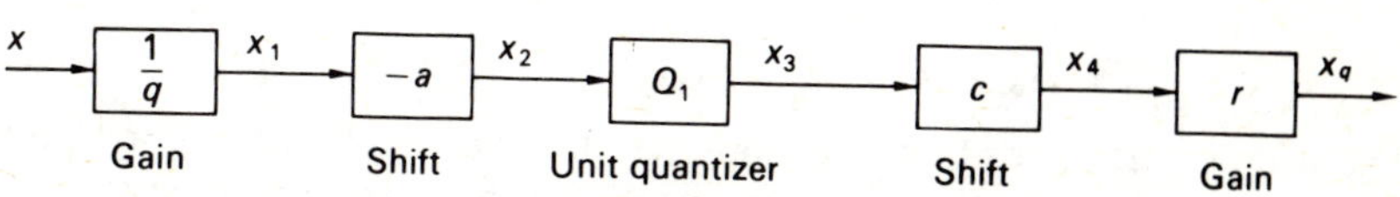


Figure 8.15

It is shown by Watts that the relationship between the probability-density functions at input $p(x)$ and output $p_q(x_q)$ is given by

$$p_q(x_q) = \frac{1}{r} \sum_{i=-\infty}^{+\infty} \left[ \int_{-\infty}^{+\infty} p(x) f_1 \left\{ \frac{x_q}{r} - c - \left( \frac{x}{q} - a \right) \right\} \mathrm{d}x \right] \delta\left( \frac{x_q}{r} - c - i \right) \tag{8.52}$$

with

$\delta(x)$ Dirac function

$$f_1\{x\} = \begin{cases} 1 & \text{for } -\frac{1}{2} < x \leqslant \frac{1}{2} \\ 0 & \text{elsewhere} \end{cases}$$

For $a = c = 0$, $q = r = 1$ this reduces to

$$p_q(x_q) = \sum_{i=-\infty}^{+\infty} \left[ \int_{-\infty}^{+\infty} p(x) f_1(x_q - x) \, \mathrm{d}x \right] \delta(x_q - i) \tag{8.53}$$

i.e. a convolution of the probability-density function with the window function $f_1(x)$, the result 'sampled' in the $x_q$ domain.

For further insight and ease of manipulations it pays to take the Fourier transform, by which the convolution changes into a multiplication; i.e. we determine the characteristic function

$$\varphi_{x_q}(u) = \mathscr{E}[\exp jux_q] = \int p_q(x_q) \exp jux_q \, \mathrm{d}x_q \tag{8.54}$$

and

$$\varphi_{x_q y_q}(u_1, u_2) = \mathscr{E}[\exp j(u_1 x_q + u_2 y_q)] = \int p_q(x_q, y_q) \exp j(u_1 x_q + u_2 y_q) \, \mathrm{d}x_q \, \mathrm{d}y_q \tag{8.55}$$

By differentiation of the characteristic function with respect to $u_1$ and $u_2$ one obtains the moment

$$\mathscr{E}[x_q y_q] = (j)^2 \left. \frac{\partial^2 \varphi_{x_q y_q}(u_1, u_2)}{\partial u_1 \, \partial u_2} \right|_{u_1 = u_2 = 0} \tag{8.56}$$

which is the correlation coefficient between these signals. In this way the correlation coefficient for the general form of a multiplier correlator can be found. One way to interpret the results is by considering the quantized signals as the original signal plus an additive quantization noise:

$$x_q = x + n_x \qquad y_q = y + n_y \tag{8.57}$$

$$\mathscr{E}[x_q y_q] = \mathscr{E}[xy] + \mathscr{E}[n_x y] + \mathscr{E}[x n_y] + \mathscr{E}[n_x n_y] \tag{8.58}$$

From this general form, expressions for the different types of correlators can be derived, including the ones operating with extreme quantization, resulting in binary signals.

### 8.2.2 *Types of Correlators*

From the previous discussion it may be clear, that the quantizing operation can be characterized by the following parameters:

$a$ shift along $x$ axis
$c$ shift along $x_q$ axis
$q$ step size along $x$ axis
$r$ step size along $x_q$ axis

Special cases are:

| | |
|---|---|
| $a = c = 0; q = r = 1$ | unit quantizer |
| $a = c = 0; q = r \to 0$ | analogue channel, no quantizer |
| $a = c = 0; q = r$ | general (symmetric) quantizer |
| $a = \frac{1}{2}; c = \frac{1}{2}; q \to \infty; r = 2$ | binary quantizer |

Depending on the types of quantization used in either or both channels of a correlator one may distinguish (cf. Figure 8.16):

(1) 'analogue' correlator;
(2) 'digital' correlator;
(3) polarity-coincidence correlator;
(4) relay or ring-modulator correlator;
(5) 'Stieltjes' correlator.

| $x$-Channel \ $y$-Channel | Binary | Quantized | Analogue |
|---|---|---|---|
| Analogue | (4) | (5) | (1) |
| Quantized | | (2) | |
| Binary | (3) | | |

A comparison of advantages and disadvantages of these correlators is given by Watts (1962). The reader is referred to that paper; here only some remarks are in order.

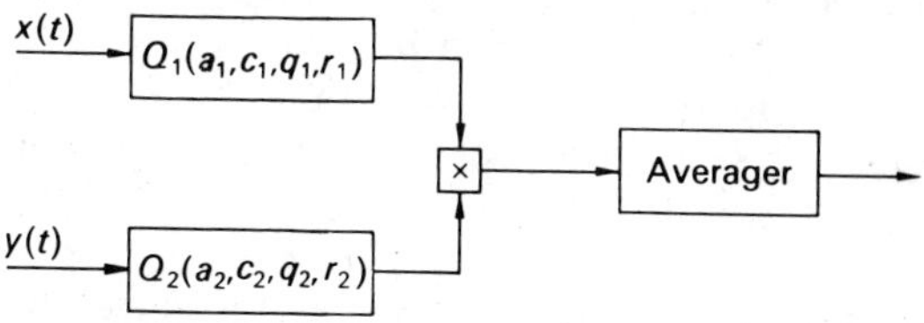


Figure 8.16

(*1*) '*Analogue*' *Correlator*. For this type of correlator, $x_q = x$ and $y_q = y$, i.e. there is no quantization noise and:

$$\mathscr{E}[x_q y_q] = \mathscr{E}[xy] = \mathscr{E}[\tilde{\psi}_{xy}(0, T)] = \psi_{xy}(0) \tag{8.59}$$

where as before

$$\psi_{xy}(0, T) = \frac{1}{T}\int_0^T x(t)y(t)\,\mathrm{d}t \tag{8.60}$$

This case was discussed extensively in Section 8.1.

(*2*) '*Digital*' *Correlator*. The signals in both channels are quantized. It turns out that even for rather coarse quantization there is an acceptable correspondence between $\mathscr{E}[x_q y_q]$ and $\mathscr{E}[xy]$. An estimate of the error can be obtained by considering equation (8.58). The worst case occurs in autocorrelation measurements. Then

$$\mathscr{E}[n_x n_y] = \mathscr{E}[n_x^2]$$

*Example*. Assuming a flat probability distribution $p(n_x) = q^{-1}$ over the quantization interval $(-q/2, q/2)$ around each quantization level, then this second moment of $n_x$ is:

$$\mathscr{E}[n_x^2] = \int_{-q/2}^{+q/2} n_x^2 p(n_x)\,\mathrm{d}n_x = \frac{1}{3q}n_x^3\Big|_{-q/2}^{+q/2} = \frac{1}{12}q^2$$

This reduces rapidly with diminishing $q$. If a signal amplitude interval $(-3\sigma, 3\sigma)$ is quantized equidistantly in 16 intervals, then $q = 6\sigma/16 = 3\sigma/8$ and

$$\mathscr{E}[n_x^2] = \frac{1}{12}\frac{9\sigma^2}{64} \approx \cdot 012\sigma^2$$

This implies that for digital correlators sometimes even three- or four-bit word lengths may suffice. Chang and Moore (1970) discuss the use of auxiliary random noise signals to unbias the output of this type of correlator, i.e. to obtain relatively high accuracy with a small number of quantization levels (cf. also Section 8.3).

(*3*) *Polarity-coincidence Correlator*. Cf. Section 8.3.

(*4*) *Relay or Ring-modulator Correlator*. In this case the multiplication is very simple to implement, e.g. $x \times \operatorname{sgn} y$ with $\operatorname{sgn} y = 1$ for $y > 0$, $\operatorname{sgn} 0 = 0$ and $\operatorname{sgn} y = -1$ for $y < 0$, which can be done by means of a mechanical or electronic relay. Of course the information about the power level of $y$ is lost. For interpretation of the measurement results, the probability-density function of the input signals has to be known.

(*5*) '*Stieltjes*' *Correlator*. This method was introduced by Watts (1962). Its name refers to the fact that the correlation integral for the analogue signal $x$ and the quantized signal $y$ can be recognized as a Stieltjes integral. It has interesting properties with respect to both theory and implementation, which can be quite simple. In spite of a rather coarse quantization, say four or five levels, it turns out that the error in the (expectation of the) measured correlation function is remarkably small. Figure 8.17 shows a way to implement this principle. For digital computer calculation of correlation functions, this principle is worth considering.

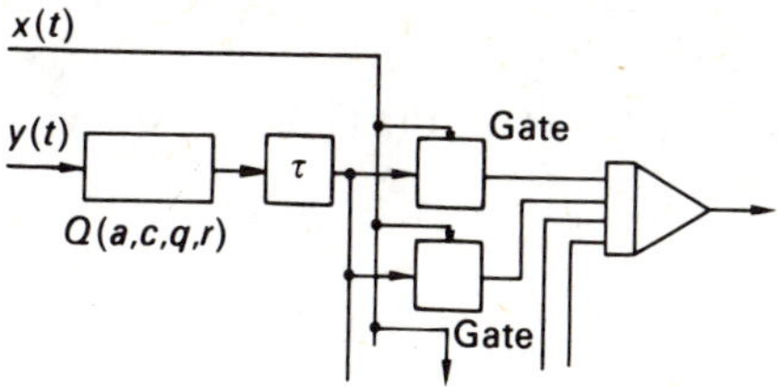


Figure 8.17

## 8.3 Operations on Binary Signals

It was recognized early (Faran and Hills, 1952) that for signals with a joint Gaussian probability-density function, the correlation measurement could well be performed using bivalent (binary) versions of these signals. This makes it possible to use purely digital means for building a correlator. As only the polarity of the signals comes into play, this principle is also called **polarity-coincidence correlation.**

### *8.3.1 The Correlation Function*

Assume the signals $x(t-\tau)$ and $y(t)$ have a Gaussian probability-density function with $\mathscr{E}[x] = \mathscr{E}[y] = 0$, i.e.:

$$p(x, y) = \frac{1}{2\pi\sigma_x\sigma_y\sqrt{1-\rho^2}} \exp\left[\frac{-1}{2(1-\rho^2)}\left\{\left(\frac{x}{\sigma_x}\right)^2 - 2\rho\left(\frac{x}{\sigma_x}\right)\left(\frac{y}{\sigma_y}\right) + \left(\frac{y}{\sigma_y}\right)^2\right\}\right] \tag{8.61}$$

with $\rho = \rho(\tau)$. The correlation function is defined as

$$\psi_{xy}(\tau) = \mathscr{E}[xy] = \int_{-\infty}^{+\infty}\int_{-\infty}^{+\infty} x \times y \times p(x, y)\,\mathrm{d}x\,\mathrm{d}y \tag{8.62}$$

For the density function given by equation (8.61) this results in

$$\psi_{xy}(\tau) = \sigma_x\sigma_y\rho(\tau) \tag{8.63}$$

When using binary signals sgn $x$ and sgn $y$ instead of the original signals, one finds:

$$\psi'_{xy}(\tau) = \mathscr{E}[\operatorname{sgn} x \times \operatorname{sgn} y]$$

$$= \int_{-\infty}^{+\infty} \int_{-\infty}^{+\infty} \operatorname{sgn} x \times \operatorname{sgn} y \times p(x, y)\,\mathrm{d}x\,\mathrm{d}y$$

$$= \underbrace{\int_{0}^{+\infty} \int_{0}^{+\infty} \cdots + \int_{-\infty}^{0} \int_{-\infty}^{0} \cdots}_{\geqslant 0} + \underbrace{\int_{0}^{\infty} \int_{-\infty}^{0} \cdots + \int_{-\infty}^{0} \int_{0}^{\infty} \cdots}_{\leqslant 0} \qquad (8.64)$$

Substitution of equation (8.61) for the Gaussian density function leads to:

$$\psi'_{xy}(\tau) = \frac{2}{\pi} \arcsin \rho(\tau) \qquad (8.65)$$

Comparing $\psi$ and $\psi'$ according to equations (8.63) and (8.65) one notices that:

as expected from the clipping operation, the 'powers' of the signals have disappeared;

the $\psi'$ is a monotonic function of $\rho(\tau)$ because $-1 \leqslant \rho \leqslant 1$

It can be shown that the variance of this correlation function is only slightly worse than that of the normalized correlation function calculated by means of an 'analogue' correlator (Veltman and Kwakernaak, 1961). It is interesting to note that (8.65) can be proved for a wider class of probability-density functions, viz. those with elliptic contours of equal height (elliptically symmetric distributions) as defined and discussed by McGraw and Wagner (1968).

### 8.3.2 *The Use of Auxiliary Signals*

The reasoning given above holds for signals with a Gaussian distribution. In practical situations, the amplitude distribution may be unknown. In those cases the relationship between $\psi'$ and $\psi$ is not known. There has been a number of attempts to alleviate this drawback of polarity-coincidence

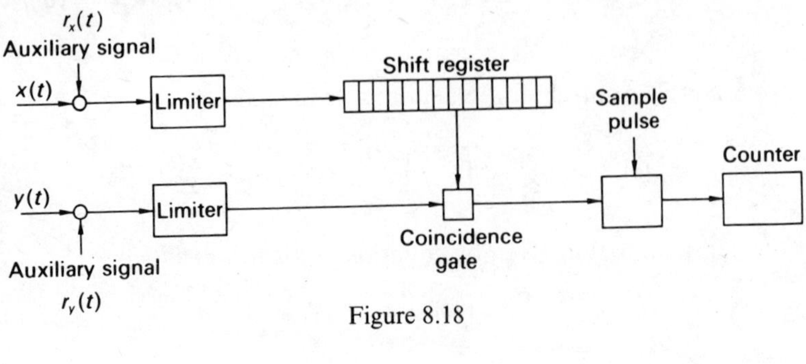


Figure 8.18

correlation ('linearizing'). All these attempts are based on the addition of auxiliary signals to the correlator input signals (cf. Figure 8.18). An analysis of this type of implementation indicates that the measured correlation functions correspond quite well with the proper correlation functions, in spite of the polarity multiplication. The auxiliary signals have been chosen as:

(*a*) Random numbers that are realizations of two statistically independent random processes, uniformly distributed (Veltman and Kwakernaak, 1961); also other distributions can be used (Chang and Moore, 1970).

(*b*) Random saw-tooth signals that are realizations of two independent processes, each of which has a uniform probability-density function (Jespers, Chu and Fettweis, 1962).

(*c*) Deterministic functions which, however, in a sense are mutually independent over a finite time interval, even under time-shift operations; 'shift-invariant independent functions' (Peek, 1967).

The difference between the latter method and the methods mentioned under (*a*) and (*b*), is the use of systematic instead of random auxiliary signals.

A kind of independence of functions is defined by: the set of functions $\{f_i(t)\}$, $i = 1, 2, \ldots, n$ is 'shift-invariant independent' in the interval $(0, T)$ if and only if

$$\frac{1}{T}\int_0^T f_1^{k_1}(t+\tau_1)\ldots f_n^{k_n}(t+\tau_n)\,\mathrm{d}t = \frac{1}{T}\prod_{s=1}^{n}\int_0^T f_s^{k_s}(t+\tau_s)\,\mathrm{d}t \qquad (8.66)$$

for every value of $\tau_1, \ldots, \tau_n$ and all positive integers $k_1, \ldots, k_n$.

A particularly interesting set of functions that exhibit this type of independence are the bivalent Rademacher functions defined by

$$r_i(t) = \operatorname{sgn}\left\{\sin\frac{2^i \pi t}{T_0}\right\} \qquad i = 1, 2, 3, \ldots \qquad (8.67)$$

These are orthonormal. The set is not complete, which is of no importance in this type of use. Figure 8.19 shows the first four elements of the set of

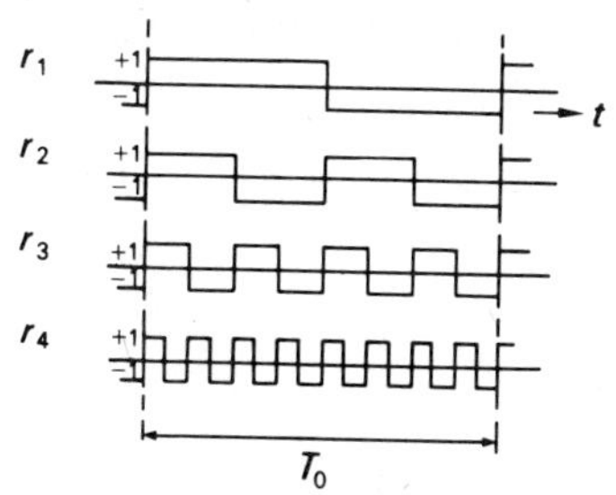


Figure 8.19

Rademacher functions. They can be generated by means of a binary counter, fed with periodic pulses. Linear combinations of these functions are also 'shift-invariant independent' if they do not contain a common Rademacher function. In this way signals with more than two levels may be constructed. Figure 8.20 shows two such three-level signals, having the desired independence property. Signals of such kind are added to the correlator input

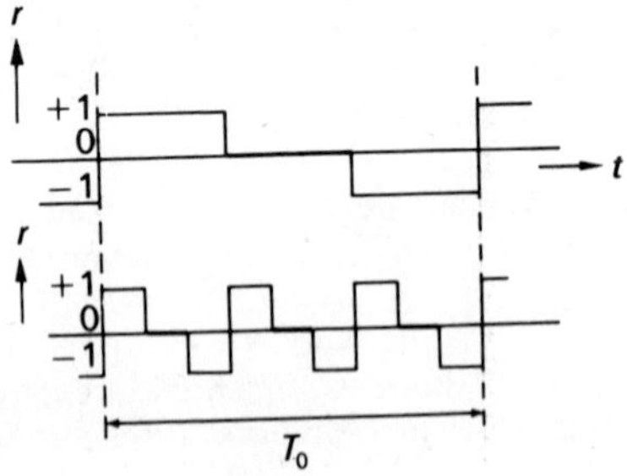


Figure 8.20

signals as indicated in Figure 8.18. The resulting sum signals are fed to hard limiters. It can be shown that the correlation function of these binary signals is the same as the correlation function of quantized input signals as discussed in Section 8.2. Consequently, the same analysis for distortion errors of the correlation function holds. Due to the periodic nature of the auxiliary signals their contribution to the correlator output noise, if measured over an integral multiple of periods, is favourable compared to random auxiliary signals. In Peek (1967) the output signal-to-noise ratio for several types of correlators is derived.

## 8.4 Differential Approximation and Related Methods

In essence the explicit methods depend on the use of operators, applied to the input and output signals, in such a way that there results a set of equations in the unknown parameters. Equations (8.11) and (8.16) may serve as illustrations. One may recognize a variety of such operators.

(*a*) In the previous sections we have studied operators on a time function $y(t)$ of the following kind:

$$\mathcal{O}_1[y(t)] = \frac{1}{T}\int_0^T y(t)u(t-\tau)\,dt$$

$$\mathcal{O}_2[y(t)] = \frac{1}{T}\int_0^T y(t)u_q(t-\tau)\,dt$$

$$\mathcal{O}_3[y(t)] = \frac{1}{T}\int_0^T y(t)\operatorname{sgn} u(t-\tau)\,dt$$

where also the quantizing of the time function $y(t)$ can be taken into consideration. From the resulting estimates of (points of the) correlation functions points of the process impulse response could be determined.

*(b) Derivative Method*

This method can be considered as a direct extension of the correlation techniques. It has been developed also under the name **differential approximation.** Consider the situation sketched in Figure 8.21. It is assumed that all

Figure 8.21

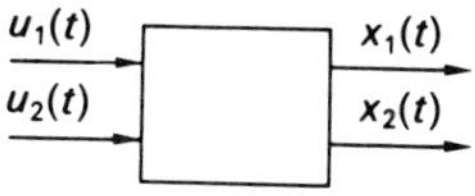


state variables $\mathbf{x}(t)$ and all input variables $\mathbf{u}(t)$ are measurable and that there is no noise present. The process dynamics are described by

$$\dot{\mathbf{x}} = \mathbf{f}(\mathbf{x}, \mathbf{u}, \mathbf{b}) \tag{8.68}$$

For a value of the unknown parameter vector $\mathbf{b}$ we may write

$$\dot{\mathbf{x}} - \mathbf{f}(\mathbf{x}, \mathbf{u}, \boldsymbol{\beta}) = \mathbf{e} \tag{8.69}$$

Again the criterion is to minimize $E(\boldsymbol{\beta})$ which is some functional quadratic in $\mathbf{e}$, e.g.

$$E(\boldsymbol{\beta}) = \int_0^T (e_1^2 + e_2^2)\, dt$$

or generally

$$E(\boldsymbol{\beta}) = \int_0^T \|\dot{\mathbf{x}} - \mathbf{f}(\mathbf{x}, \mathbf{u}, \boldsymbol{\beta})\|^2\, dt \tag{8.70}$$

The necessary condition for the minimum:

$$\nabla_{\boldsymbol{\beta}} E(\boldsymbol{\beta})|_{\boldsymbol{\beta}=\hat{\boldsymbol{\beta}}} = \mathbf{0} \tag{8.71}$$

leads to:

$$\int_0^T \frac{\partial \mathbf{f}'}{\partial \boldsymbol{\beta}} \dot{\mathbf{x}}\, dt = \int_0^T \frac{\partial \mathbf{f}'}{\partial \boldsymbol{\beta}} \mathbf{f}(\mathbf{x}, \mathbf{u}, \hat{\boldsymbol{\beta}})\, dt \tag{8.72}$$

where

$$\frac{\partial \mathbf{f}'}{\partial \boldsymbol{\beta}} = \left[\frac{\partial f_i}{\partial \beta_j}\right] = \begin{bmatrix} \dfrac{\partial \mathbf{f}'}{\partial \beta_0} \\ \vdots \\ \dfrac{\partial \mathbf{f}'}{\partial \beta_m} \end{bmatrix} = \begin{bmatrix} \dfrac{\partial f_1}{\partial \beta_0} & \cdots & \dfrac{\partial f_n}{\partial \beta_0} \\ \vdots & = & \vdots \\ \dfrac{\partial f_1}{\partial \beta_m} & \cdots & \dfrac{\partial f_n}{\partial \beta_m} \end{bmatrix} \tag{8.73}$$

From the set of equations (8.72) the unknown parameters $\hat{\boldsymbol{\beta}}$ have to be solved. If $\mathbf{f}$ is linear in $\hat{\boldsymbol{\beta}}$ then again this procedure is quite simple. For more details the reader may refer to the additional literature on this topic given at the end of this chapter.

Neither is it unavoidable to use the state–space description, nor does a non-linear process behaviour necessarily imply additional complications. (As pointed out in Chapter 4, linearity-in-the-parameters is of paramount importance.)

*Example.* The following simple case may be illustrative. The process is described by

$$\ddot{x} + a_1\dot{x} + a_0x^3 = bu \tag{8.74}$$

and the 'model' by:

$$\ddot{x} + \alpha_1\dot{x} + \alpha_0x^3 - \beta u = e \tag{8.75}$$

Consequently,

$$\frac{\partial e}{\partial \alpha_1} = \dot{x} \qquad \frac{\partial e}{\partial \alpha_0} = x^3 \qquad \frac{\partial e}{\partial \beta} = -u \tag{8.76}$$

Using min $\int_0^T e^2\,dt$ this leads to the set of equations, linear-in-the-unknown-parameters:

$$\int_0^T e\dot{x}\,dt = 0 \qquad \int_0^T ex^3\,dt = 0 \qquad \int_0^T eu\,dt = 0 \tag{8.77}$$

or:

$$\left.\begin{aligned}
\alpha_1 \int_0^T \dot{x}^2\,dt + \alpha_0 \int_0^T x^3\dot{x}\,dt - \beta \int_0^T u\dot{x}\,dt &= -\int_0^T \ddot{x}\dot{x}\,dt \\
\alpha_1 \int_0^T \dot{x}x^3\,dt + \alpha_0 \int_0^T x^6\,dt - \beta \int_0^T ux^3\,dt &= -\int_0^T \ddot{x}x^3\,dt \\
\alpha_1 \int_0^T \dot{x}u\,dt + \alpha_0 \int_0^T x^3u\,dt - \beta \int_0^T u^2\,dt &= -\int_0^T \ddot{x}u\,dt
\end{aligned}\right\} \tag{8.78}$$

from which the parameters may be solved.

So far it has been assumed that $x$ can be measured without additive noise. If only a contaminated output is available, e.g. $y(t) = x(t) + n(t)$, then one is tempted to apply the equations (8.78) using $y$ and $\dot{y}$ instead of $x$ and $\dot{x}$. The reader will note the intimate relationship with the generalized-model notion. From this relationship it will be clear that in this case as well, additive noise will lead to biased estimates of the parameters. Such a bias can be prevented by the so-called instrumental variable approach (cf. Section 9.4). This leads to a set of equations:

$$\int_0^T eg_1\,dt = 0 \qquad \int_0^T eg_2\,dt = 0 \qquad \int_0^T eg_3\,dt = 0 \tag{8.79}$$

where $g_1, g_2, g_3$ are the instrumental variables, i.e. time functions related to $u(t)$ but not contaminated with noise.

*(c) Laplace Transform Method*

In this case the operator is given by:

$$\mathcal{O}_L[y(t)] = \int_0^\infty y(t)\varepsilon^{-st}\,dt$$

This can be computed for a number of values $s_i$ of $s$. Again in this way a set of equations in the unknown parameters results.

*Example.* We take the same model as before, described by equation (8.75).

$$\mathcal{O}_L[\ddot{x} + \alpha_1\dot{x} + \alpha_0 x^3 - \beta u] = 0 \tag{8.80}$$

For the sake of convenience we assume the initial conditions to be zero. Taking into account the properties of the $\mathscr{L}$ operator we find:

$$s_i^2\int_0^\infty x(t)\varepsilon^{-s_it}\,dt + \alpha_1 s_i\int_0^\infty x(t)\varepsilon^{-s_it}\,dt + \alpha_0\int_0^\infty x^3(t)\varepsilon^{-s_it}\,dt = \beta\int_0^\infty u\varepsilon^{-s_it}\,dt \tag{8.81}$$

Using three different values $s_1, s_2, s_3$ one obtains three equations in the parameters $\alpha_1, \alpha_0, \beta$ which can be solved by applying matrix inversion. Of course one may use more than three different values of $s$ and find a least-squares estimate of the parameters from the resulting set of equations.

Note the advantage of this method; it is not necessary to differentiate the time functions. Also the initial conditions may be estimated if desired.

*(d) Fourier Transform Method*

This is analogous to the previous method; the operator is given by

$$\mathcal{O}_F[y(t)] = \int_0^\infty y(t)\varepsilon^{-j\omega t}\,dt \tag{8.82}$$

Note the convergence problems or the truncation errors that may occur due to the fact that $|\exp[-j\omega t]| = 1$ for all $\omega, t$.

From the previous three methods Shinbrot (1957) introduces the name **'method functions'** for the multiplying time functions, i.e. $[\dot{x}, x^3, u]$, $\exp[-s_i t]$, viz. $\exp[-j\omega_i t]$ and the name **equations-of-motion method** for the estimation procedure:

multiplication of the model equation by each of the selected method functions;

integration;

solving the resulting algebraic equations for the unknown parameters.

The method functions can be:

depending on the process input and output functions (e.g. differential approximation);

depending on the process input functions only (e.g. correlation techniques; instrumental variable method);

independent of the process input and output functions (e.g. Laplace and Fourier transform method).

*(e) General Equations-of-motion Method*

Now let us consider in a general way the method functions $g_i(t)$ that are independent of the process input and output functions. In this case of independence we will denote them by **modulating functions.** From an integration by parts with $\dot{x}(t)$ and $x(t)$ follows:

$$\left.\begin{aligned} \int_0^\infty \dot{x}g_i\,dt &= -x(0)g_i(0) - \int_0^\infty x\dot{g}_i\,dt \\ \int_0^\infty \ddot{x}g_i\,dt &= -\dot{x}(0)g_i(0) + x(0)\dot{g}_i(0) + \int_0^\infty x\ddot{g}_i\,dt \end{aligned}\right\} \tag{8.83}$$

This can be substituted in the model equation. It has the attractive feature that the derivatives of $x(t)$ need not be determined from the experimental data. It has, however, the disadvantage that the numerical values of $x(0)$ and $\dot{x}(0)$ play a very important role. This can be prevented by choosing the modulating functions such that $g_i(0) = \dot{g}_i(0) = 0$.

For practical situations one likes to restrict the observation interval to a finite one. The same type of reasoning can be used with respect to the end point of such a finite observation interval $0 \leqslant t \leqslant T$. Then it pays to chose the modulating functions in such a way that $g_i(T) = \dot{g}_i(T) = 0$. Among the functions that (approximately) fulfil these requirements are the Gaussian ones.

Shinbrot (1957) indicates that this type of approach can also be used for non-linear differential equations like

$$\ddot{x} + f(x)\dot{x} + g(x)x = 0 \tag{8.84}$$

where $f(x)$ and $g(x)$ can be approximated by polynomials, e.g.

$$f(x) = a_{10} + a_{11}x + a_{12}x^2 + \cdots$$

$$g(x) = a_{00} + a_{01}x + a_{02}x^2 + \cdots$$

The numerical values of these coefficients can be estimated along the schemes indicated.

*(f) Strejc's Method*

An approach, related to the methods described before, has been developed by Strejc (1961). He considers an engineering situation where the noise can be neglected and where records of process input and output signals are available. Assume the differential equation is known to be

$$a_2\ddot{y} + a_1\dot{y} + a_0 y = b_2\ddot{u} + b_1\dot{u} + u \tag{8.85}$$

where $u$ and $y$ are the input and the output signal respectively. The part of the recording used is that over the interval $(t_1, t_2)$. The method is based on repeated integration of equation (8.85) over the interval $(t, t_2)$. One integration yields:

$$\begin{aligned} &a_2[\dot{y}(t_2) - \dot{y}(t)] + a_1[y(t_2) - y(t)] + a_0 \int_t^{t_2} y(t)\,dt \\ &= b_2[\dot{u}(t_2) - \dot{u}(t)] + b_1[u(t_2) - u(t)] + \int_t^{t_2} u(t)\,dt \end{aligned} \tag{8.86}$$

Integrated again:

$$\begin{aligned} &a_2[\dot{y}(t_2)\{t_2 - t\} - y(t_2) + y(t)] + a_1[\cdots] + a_0[\cdots] \\ &= b_2[\cdots] + b_1[\cdots] + [\cdots] \end{aligned} \tag{8.87}$$

and integrated again:

$$\begin{aligned} &a_2\left[\dot{y}(t_2)\{t_2(t_2 - t_1) - \tfrac{1}{2}(t_2^2 - t^2)\} - y(t_2)\{t_2 - t\} + \int_t^{t_2} y(t)\,dt\right] \\ &+ a_1[\cdots] + a_0[\cdots] = \cdots \end{aligned} \tag{8.88}$$

By substituting $t = t_1$ in each of these equations the expressions in square brackets can be evaluated for the signal records over $(t_1, t_2)$. This can be done in a numerical or graphical way. Consequently, one obtains again a set of linear equations in the unknown coefficients:

$$\sum_{i=0}^{2} a_i Y_{ij} = \sum_{i=0}^{2} b_i U_{ij} \qquad b_0 = 1 \tag{8.89}$$

where the index $j$ refers to the $j$th equation. One needs the same number of independent equations as the number of unknown coefficients, in our case five. One may obtain this number by further integrations. In view of accuracy requirements, however, it would be better to take more signal recordings and integrate at most two times.

The best way to determine $\dot{u}(t_1)$, $\dot{u}(t_2)$, $\dot{y}(t_1)$, $\dot{y}(t_2)$ is at extremes or inflexion points of the recording, where some of these derivatives are zero. This may be used as a guideline in chosing the interval(s) $(t_1, t_2)$ on the signal records.

A comparatively simple case (Strejc, 1958; 1959) is obtained if the process is described by

$$a_n \overset{(n)}{y} + a_{n-1} \overset{(n-1)}{y} + \cdots + a_1\dot{y} + a_0 y = u \tag{8.90}$$

and if the input and output are constant for $t < 0$ and for $t \to \infty$ such that

$$\dot{y}(0) = \ddot{y}(0) = \cdots = \overset{(n-1)}{y}(0) = 0$$

$$\dot{y}(\infty) = \ddot{y}(\infty) = \cdots = \overset{(n-1)}{y}(\infty) = 0$$

*Example.* If in addition $y(0) = y(\infty)$ and $u(0) = u(\infty)$, then by repeated integration one can show that

$$\begin{aligned} a_0 &= c\int_0^\infty u(t)\,\mathrm{d}t \\ a_1 &= -c\int_0^\infty \int_t^\infty \{u(t) - a_0 y(t)\}\,\mathrm{d}t^2 \\ a_2 &= c\int_0^\infty \int_t^\infty \left[\int_t^\infty \{u(t) - a_0 y(t)\}\,\mathrm{d}t + a_1 y(t)\right]\mathrm{d}t^2 \end{aligned} \tag{8.91}$$

etc.

with $c^{-1} = \int_0^\infty y(t)\,\mathrm{d}t$.

Due to unavoidable measurement errors, this method is limited to differential equations of the order $\leqslant 3$. Calculation of the coefficients $a_0$, $a_1$, $a_2$ of an $n$th order differential equation, however, is sufficient to determine an approximation of that dynamics by supposedly $n$ equal time constants. For that situation in the Laplace notation the following holds:

$$a_0(sT + 1)^n X = a_0\{s^n T^n + \tbinom{n}{1}s^{n-1}T^{n-1} + \cdots + \tbinom{n}{2}s^2T^2 + \tbinom{n}{1}sT + 1\}X \tag{8.92}$$

Consequently:

$$nT = \frac{a_1}{a_0} \quad \text{and} \quad \binom{n}{2}T^2 = \frac{a_2}{a_0}$$

from which $n$ and $T$ follow; $n$ has to be an integer number.

## 8.5 Higher-order Correlation Functions

In the Sections 8.1–8.3 attention was restricted to correlation functions for two time functions; these might be called second-order correlation functions. In some cases, however, it is of interest to consider correlation functions of the order $n$, defined for ergodic signals by:

$$\psi_{x_1 \ldots x_n}(\tau_1, \ldots, \tau_n) = \lim_{T\to\infty} \frac{1}{T}\int_0^T x_1(t - \tau_1)\ldots x_n(t - \tau_n)\,\mathrm{d}t \tag{8.93}$$

These play a role in the characterization of non-linear signal and process relations, e.g. in the Wiener characterization of non-linear processes. As these methods still have found very little application in practice the reader is referred to the additional literature on this topic.

## 8.6 Dispersion Functions; Definitions and Properties*

From the previous sections it will be clear that correlation techniques are very useful for the characterization of linear processes. This can be attributed to the following reasons:

for a linear process a Gaussian input signal results in a Gaussian output signal;

for stationary Gaussian random signals their statistical properties are completely described by the (second-order) auto- and cross-correlation functions.

For non-linear processes the situation is much more complicated. The differential approximation and related techniques are acceptable for estimation purposes if the model can be made linear-in-the-parameters and if the noise does not deteriorate the estimation scheme. According to Section 8.5 for a Wiener type of process characterization one needs higher-order correlation functions.

*Example.* To illustrate the difficulties that may arise when trying to identify a non-linear process consider the situation where the relationship between input $u(t)$ and output $y(t)$ is given by

$$y(t) = u^2(t) \tag{8.94}$$

If for example $u(t)$ is a stationary Gaussian random signal with $\mathscr{E}[u(t)] = 0$ then the cross correlation is found to be:

$$\begin{aligned}\psi_{yu}(\tau) &= \int_{-\infty}^{+\infty}\int_{-\infty}^{+\infty} y\,u\,p(y, u; \tau)\,\mathrm{d}y\,\mathrm{d}u \\ &= \int_{-\infty}^{+\infty} u^3\,p(u; \tau)\,\mathrm{d}u = 0\end{aligned} \tag{8.95}$$

as under the integral there is the product of an odd and an even function. Although there is a definite relationship between $u$ and $y$ the cross-correlation function does not give any indication of this relation.

Some of these difficulties can be avoided by using the **dispersion method,** dealing with expectations, correlation and dispersion function. An exposition of the method and applications can be found in Rajbman (1970).

### *Definitions of Dispersion Functions*

The **cross-dispersion function** of the real random signal (process)† is a function of two arguments (Rajbman, 1966; 1967).

$$\Theta_{yu}(t_1, t_2) = \mathscr{E}_{u(t_2)}[\{\mathscr{E}_{y(t_1)}[y(t_1)|u(t_2)] - \mathscr{E}_{y(t_1)}[y(t_1)]\}^2] \tag{8.96}$$

* This section is based on notes by Mrs. J. Kregting-Jansen, University of Technology, Eindhoven, Netherlands.

† In view of the use of the term 'process' for the object under estimation we will use the word 'signal' for the random process from which the signal may be considered to be an ensemble element.

where $\mathscr{E}_{y(t_1)}[y(t_1)|u(t_2)]$ is the conditional mean of the random signal $y(t_1)$ given $u(t_2)$. If the necessary densities are known then it can be determined from:

$$\Theta_{yu}(t_1, t_2) = \int_{-\infty}^{+\infty} \left\{ \int_{-\infty}^{+\infty} y(t_1) p(y(t_1)|u(t_2)) \, \mathrm{d}y(t_1) - \int_{-\infty}^{+\infty} y(t_1) p(y(t_1)) \, \mathrm{d}y(t_1) \right\}^2 p(u(t_2)) \, \mathrm{d}u(t_2) \tag{8.97}$$

where $p(u(t_2))$ and $p(y(t_1))$ are the probability densities of $u(t_2)$ and $y(t_1)$ respectively; $p(y(t_1)|u(t_2))$ is the conditional probability density of $y(t_1)$ given $u(t_2)$.

The **autodispersion function** of the real random signals $u(t_1)$ and $u(t_2)$ is defined as:

$$\Theta_{uu}(t_1, t_2) = \mathscr{E}_{u(t_2)}[\{\mathscr{E}_{u(t_1)}[u(t_1)|u(t_2)] - \mathscr{E}_{u(t_1)}[u(t_1)]\}^2] \tag{8.98}$$

If the corresponding densities are known then the autodispersion function can be determined analogously to equation (8.97). The autodispersion functions $\Theta_{uu}(t_1, t_2)$ give quantitative characteristics of the random signals, similar to the autocorrelation functions $\psi_{uu}(t_1, t_2)$, and they characterize the internal structure of the random signal $u(t)$. The cross-dispersion functions $\Theta_{yu}(t_1, t_2)$ are, similar to the cross-correlation functions $\psi_{yu}(t_1, t_2)$, a measure of the relation of the two random signals $u(t)$ and $y(t)$. The correlation function $\psi_{yu}(t_1, t_2)$ serves as a measure of relation of **linear** dependence only, while the dispersion function $\Theta_{yu}(t_1, t_2)$ serves for an arbitrary dependence of $u(t_2)$ and $y(t_1)$.

In practical problems normalized values of dispersion functions are used to compare them with normalized correlation functions. The **normalised autodispersion function** is defined as:

$$\eta_{uu}(t_1, t_2) = \sqrt{\frac{\Theta_{uu}(t_1, t_2)}{\sigma_u^2(t_1)}} \tag{8.99}$$

where $\sigma_u^2$ is the variance; $\sigma_u^2(t_1) = \mathscr{E}_{u(t_1)}[\{u(t_1) - \mathscr{E}[u(t_1)]\}^2]$. The **normalized cross-dispersion function** is:

$$\eta_{yu}(t_1, t_2) = \sqrt{\frac{\Theta_{yu}(t_1, t_2)}{\sigma_y^2(t_1)}} \tag{8.100}$$

When $y(t_1)$ and $u(t_2)$ are independent then $\Theta_{yu}(t_1, t_2) = 0$ and if $\sigma_y^2(t_1) \neq 0$ we obtain $\eta_{yu}(t_1, t_2) = 0$. If $u(t_1)$ and $u(t_2)$ are independent then $\Theta_{uu}(t_1, t_2) = 0$ and if $\sigma_u^2(t_1) \neq 0$ we find $\eta_{uu}(t_1, t_2) = 0$. When there is a non-random functional dependence between $y(t_1)$ and $u(t_2)$ then $\Theta_{yu}(t_1, t_2) = \sigma_y^2(t_1)$ and

$\eta_{yu}(t_1, t_2) = 1$. If the cross-dispersion function or the normalized cross-dispersion function is zero for the arguments $t_1$ and $t_2$ of the random signals $y(t_1)$ and $u(t_2)$, then we call the random variables $y(t_1)$ and $u(t_2)$ **non-dispersed.** This notion corresponds to the concept of uncorrelated random variables in the linear case. If two random variables are non-dispersed, they are uncorrelated too. The converse, however, is not true.

*Some Properties of Dispersion Functions*

(1) Dispersion functions are non-negative functions;

$$\Theta_{uu}(t_1, t_2) \geqslant 0 \qquad \eta_{uu}(t_1, t_2) \geqslant 0$$

$$\Theta_{yu}(t_1, t_2) \geqslant 0 \qquad \eta_{yu}(t_1, t_2) \geqslant 0$$

(2) With Schwarz's inequality one can prove:

$$\sigma_y^2(t_1) \geqslant \Theta_{yu}(t_1, t_2)$$

(3) Combination of properties (1) and (2)

$$0 \leqslant \eta_{uu}(t_1, t_2) \leqslant 1$$

$$0 \leqslant \eta_{yu}(t_1, t_2) \leqslant 1$$

(4) In general the cross-dispersion function and the normalized cross-dispersion function are asymmetrical:

$$\Theta_{yu}(t_1, t_2) \neq \Theta_{uy}(t_2, t_1) \qquad \eta_{yu}(t_1, t_2) \neq \eta_{uy}(t_2, t_1)$$

(5) In general the normalized cross-dispersion function is larger than or equal to the absolute value of the normalized cross-correlation function $\Upsilon_{yu}$:

$$\eta_{yu}(t_1, t_2) \geqslant \left| \frac{\psi_{yu}(t_1, t_2)}{\sigma_y(t_1)\sigma_u(t_2)} \right| = |\Upsilon_{yu}(t_1, t_2)|$$

The equality holds only in linear cases.

In many problems the regression $u_1$ is required for an arbitrary value of $t_1$ of some values of $u_2, \ldots, u_n$ for arbitrary $t_2, \ldots, t_n$. Then the relation between $u(t_1)$ and $u_2(t_2), \ldots, u_n(t_n)$ is characterized by the **multiple dispersion function:**

$$\Theta_{\mathbf{u}}(t_1, \ldots, t_n) = \mathscr{E}_{u_2(t_2)\ldots u_n(t_n)}[\{\mathscr{E}[u_1(t_1)|u_2(t_2), \ldots, u_n(t_n)] - \mathscr{E}[u_1(t_1)]\}^2] \tag{8.101}$$

Similarly, the normalized multiple dispersion function is defined:

$$\eta_{\mathbf{u}}(t_1, \ldots, t_n) = \sqrt{\frac{\Theta_{\mathbf{u}}(t_1, \ldots, t_n)}{\sigma_{u_1}^2(t_1)}} \tag{8.102}$$

Consider as a measure of non-linearity of a regression curve $\mathscr{E}[y(t_1)|u(t_2)]$ the mean-square deviation of the curve from such a straight line that this deviation be the least as possible. We define the **degree of non-linearity** of the regression $y(t_1)$ on $u(t_2)$ as follows:

$$\nu_{yu}(t_1, t_2) = \sqrt{\frac{\min\limits_{a,b} \mathscr{E}_{u(t_2)}[\{\mathscr{E}_{y(t_1)}[y(t_1)|u(t_2)] - (a + bu(t_2))\}^2]}{\sigma_y^2(t_1)}} \tag{8.103}$$

The minimum for the above formula is attained for:

$$b = b(t_1, t_2) = \Upsilon_{yu}(t_1, t_2)\frac{\sigma_y(t_1)}{\sigma_u(t_2)} \tag{8.104}$$

and

$$a = a(t_1, t_2) = \mathscr{E}[y(t_1)] - \Upsilon_{yu}(t_1, t_2)\frac{\sigma_y(t_1)}{\sigma_u(t_2)}\mathscr{E}[u(t_2)] \tag{8.105}$$

Substituting $b(t_1, t_2)$ and $a(t_1, t_2)$ in the expression for the degree of non-linearity $y(t_1, t_2)$ we obtain for the square of the degree of non-linearity:

$$\nu_{yu}^2(t_1, t_2) = \eta_{yu}^2(t_1, t_2) - \Upsilon_{yu}^2(t_1, t_2) \tag{8.106}$$

*Examples.* In order to illustrate these ideas it may be helpful to consider a few illustrations. For the sake of notational convenience the arguments $t_1$ and $t_2$ are deleted.

(*a*) $y$ and $u$ are independent (cf. Figure 8.22a).

$$\mathscr{E}[y|u] = \mathscr{E}[y] = \text{constant}$$

$$\Theta_{yu} = \int \{\mathscr{E}[y] - \mathscr{E}[y]\}^2\, p(u)\,\mathrm{d}u = 0$$

$$\Psi_{yu} = \int\int u\, y\, p(u, y)\,\mathrm{d}u\,\mathrm{d}y = \mathscr{E}[u] \times \mathscr{E}[y]$$

(*b*) $y$ and $u$ are statistically related in linear sense (cf. Figure 8.22b).

$$\mathscr{E}[y|u] = a\,u \qquad a = \frac{\psi_{yu}}{\sigma_u^2} \qquad \mathscr{E}[u] = \mathscr{E}[y] = 0$$

$$\Theta_{yu} = \int \left\{\frac{\psi_{yu}}{\sigma_u^2}\right\}^2 u^2\, p(u)\,\mathrm{d}u = \frac{\psi_{yu}^2}{\sigma_u^2}$$

$$\psi_{yu} = \int\int u\, y\, p(u, y)\,\mathrm{d}u\,\mathrm{d}y$$

$$\nu_{yu}^2 = \eta_{yu}^2 - \gamma_{yu}^2 = \frac{\psi_{yu}^2}{\sigma_u^2\sigma_y^2} - \left(\frac{\psi_{yu}}{\sigma_u\sigma_y}\right)^2 = 0$$

(*c*) $y$ and $u$ are deterministically related in a linear sense (cf. Figure 8.22c).

$$y = a\,u \qquad \text{i.e. } \mathscr{E}[y|a] = a\,u \qquad \mathscr{E}[u] = \mathscr{E}[y] = 0$$

$$\Theta_{yu} = \int \{a\,u - \mathscr{E}[u]\}^2\, p(u)\,\mathrm{d}u = a^2\sigma_u^2$$

$$\psi_{yu} = \int\int u\, y\, p(u, y)\,\mathrm{d}u\,\mathrm{d}y = a\int u^2\, p(u)\,\mathrm{d}u = a\sigma_u^2$$

$$\nu_{yu}^2 = \eta_{yu}^2 - \gamma_{yu}^2 = a^2\frac{\sigma_u^2}{\sigma_y^2} - a^2\frac{\sigma_u^4}{\sigma_u^2\sigma_y^2} = 0$$

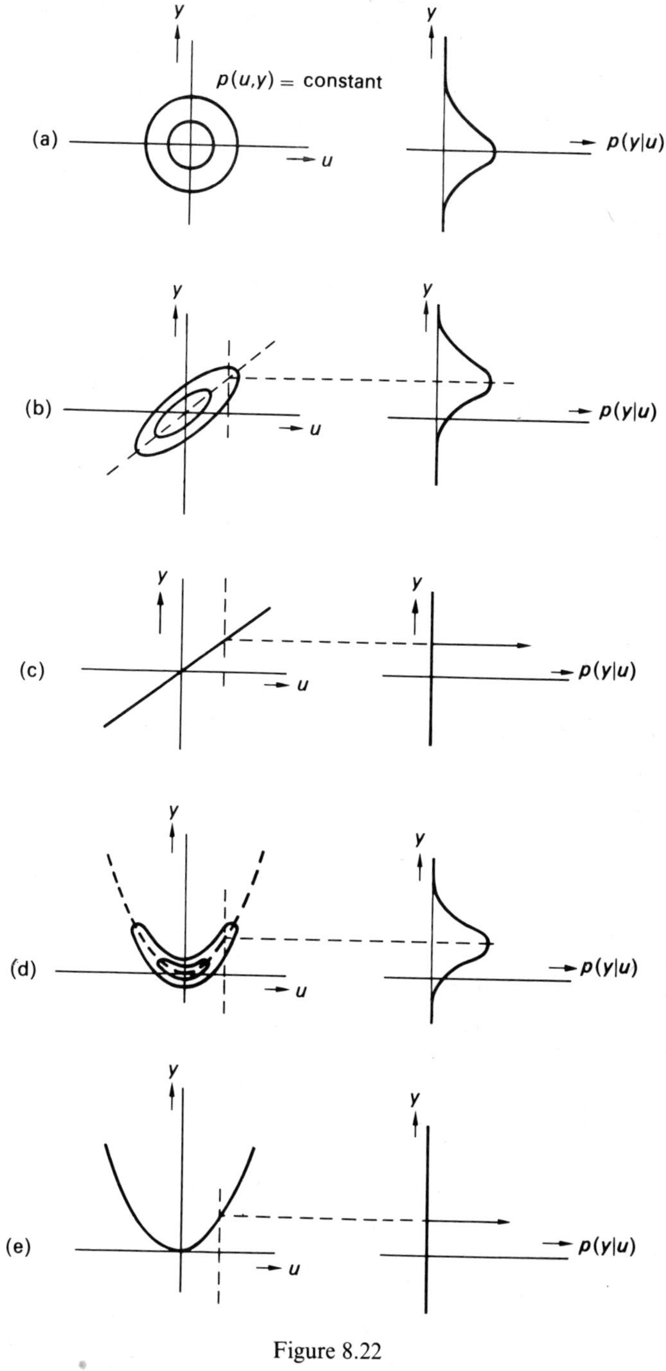


Figure 8.22

(*d*) $y$ and $u$ are statistically related in a quadratic sense (cf. Figure 8.22d).

$$\mathscr{E}[y|u] = u^2 \qquad \mathscr{E}[u] = 0 \qquad p(u) = p(-u)$$

$$\Theta_{yu} = \int \{u^2 - \mathscr{E}[y]\}^2\, p(u)\, du \geqslant 0$$

$$\psi_{yu} = \int\int u\, y\, p(u, y)\, du\, dy$$

$$= \int u \left\{ \int y\, p(y|u)\, dy \right\} p(u)\, du$$

$$= \int u^3\, p(u)\, du = 0$$

$$\nu_{yu}^2 = \eta_{yu}^2 - \gamma_{yu}^2 = \frac{\Theta_{yu}}{\sigma_y^2} - 0 \geqslant 0$$

(*e*) $y(t_1)$ and $u(t_2)$ are deterministically related in a quadratic sense (cf. Figure 8.22e).

$$y = u^2 \qquad \mathscr{E}[u] = 0 \qquad p(u) = p(-u)$$

$$\Theta_{yu} = \int \{u^2 - \mathscr{E}[u^2]\}^2\, p(u)\, du = \sigma_{u^2}^2 = \sigma_y^2 \neq 0$$

$$\psi_{yu} = \int\int u\, y\, p(u, y)\, du\, dy = \int u^3\, p(u)\, du = 0$$

$$\nu_{yu}^2 = \eta_{yu}^2 - \gamma_{yu}^2 = \frac{\sigma_y^2}{\sigma_y^2} - 0 = 1$$

Let the input $u(t)$ and output $y(t)$ of a process be random functions. The general characteristics of the process can be described by an operator $\mathcal{O}_P$

$$y(t_1) = \mathcal{O}_P[u(t_2)] \tag{8.107}$$

If $u(t)$ and $y(t)$ are measurable the identification problem is reduced to the problem of finding an operator $\mathcal{O}_M$ which serves as an approximation of the true operator $\mathcal{O}_P$. To approximate the true operator by $\mathcal{O}_M$ some criterion is required. A possible criterion is the mathematical expectation of the loss function

$$\mathscr{E}[C\{y(t_1), w(t_1)\}] \tag{8.108}$$

to be a minimum. When the loss function is: $C\{y(t), w(t)\} = \{y(t) - w(t)\}^2$ the minimum of the mathematical expectation of this loss function is reached if:

$$w(t_1) = \mathscr{E}_{y(t_1)}[y(t_1)|u(t_2)] \tag{8.109}$$

This situation is indicated in Figure 8.23. The dispersion of the output can be written as follows, where again for the sake of notational convenience

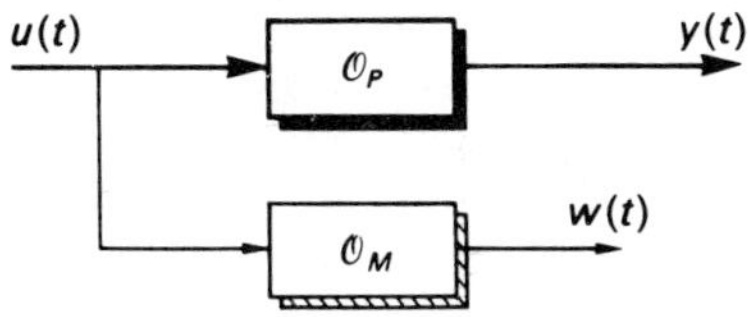


Figure 8.23

$u(t_2) = u$ and $y(t_1) = y$,

$$\sigma_y^2 = \mathscr{D}[y] = \mathscr{D}[\mathscr{E}_y[y|u]] + \mathscr{E}_{yu}[\mathscr{D}[y|u]] \tag{8.110}$$

where:

$$\mathscr{D}[y] \overset{\text{def}}{=} \int \{y - \mathscr{E}[y]\}^2 p(y)\, dy \tag{8.111}$$

$$\mathscr{D}[\mathscr{E}_y[y|u]] = \int \{\mathscr{E}_y[y|u] - \mathscr{E}_y[y]\}^2 p(u)\, du \tag{8.112}$$

$$\mathscr{E}_{yu}[\mathscr{D}[y|u]] = \int\int \{y - \mathscr{E}_y[y|u]\}^2 p(u, y)\, du\, dy \tag{8.113}$$

Having a quadratic loss function the expression for the dispersion of the output $y(t)$ will be:

$$\mathscr{D}[y] = \mathscr{E}_u[\{w - \mathscr{E}[y]\}^2] + \mathscr{E}[\{y - w\}^2] \tag{8.114}$$

where the first part of the right-hand side is the cross-dispersion function. For more information the reader is referred to the references given and to Pugachev (1965), Rajbman and Terekhin (1965), Rajbman (1967), Bunich and Rajbman (1972).

## References

Balchen, J. G. and E. Blandhol (1960). 'On the experimental determination of statistical properties of signals and disturbances in automatic control systems', *1st IFAC Congr., Moscow*, 788–796.

Ball, G. A. (1963). 'On a way of speeding up instrument correlation analysis', *Automn. Remote Control*, **24,** 1117–1122.

Bendat, J. S. and A. G. Piersol (1966). *Measurements and Analysis of Random Data*, Wiley, New York, 390 pp.

Buchta, H. (1969). 'Method of estimation of random errors in the determination of correlation functions of random infralow frequency signals in control systems', *4th IFAC Congr., Warsaw*, paper 33.4.

Bunich, A. L. and N. S. Rajbman (1972). 'A dispersion equation of nonlinear plant identification', *Problems of Control and Inform. Theory*, **1,** 29–36.

Chang, C. M., T. P. Goodman and J. B. Reswick (1956). 'Use of correlation functions to determine system characteristics without applying artificial disturbances', in *Regelungstechnik* (Ed. G. Müller), München, Oldenburg, 251–256.

Chang, K-Y. and A. D. Moore (1970). 'Modified digital correlator and its estimation errors', *IEEE Trans. Information Theory*, **IT-16,** 699–706.

Davenport, W. B. and others (1952). 'Statistical errors in measurement on random time functions', *J. Appl. Phys.*, **23,** 337–388.

Eykhoff, P. (1967). cf. references Chapter 2.

Faran, J. J. and R. Hills (1952). 'Correlators for signal reception', *Techn. Memo No. 27, Acoustics Research Laboratory, Harvard University*, Mass.

Goodman, T. P. and J. B. Reswick (1956). 'Determination of systems characteristics from normal operating records', *Trans. ASME*, **78,** 259–271.

Goodman, T. P. (1957). 'Determination of the characteristics of multi-input and nonlinear systems from normal operating records', *Trans. ASME*, **79,** 567–575.

Green Jr., P. E. (1957). 'The output signal-to-noise ratio of correlation detectors', *IRE Trans. Information Theory*, **IT-3,** 10–18.

Hayashi, N. (1969). 'On a method of correlation analysis for multivariate systems', *4th IFAC Congr., Warsaw*, paper 33.1.

Jespers, P., P. T. Chu and A. Fettweis (1962). 'A new method for computing correlation functions', *IRE Trans. Information Theory*, **IT-8,** 106–107.

Kitamori, T. (1960). Cf. references Chapter 3.

Lampard, D. G. (1955). Cf. references Chapter 3.

Lange, F. H. (1967). *Correlation Techniques*, Van Nostrand, Princeton, N.J., 464 pp.

Laning, J. H. and R. H. Battin (1956). *Random Processes in Automatic Control*, McGraw-Hill, New York.

Lee, Y. W. (1960). *Statistical Theory of Communication*, Wiley, New York, 509 pp.

McGraw, D. K. and J. F. Wagner (1968). 'Elliptically symmetric distributions', *IEEE Trans. Information Theory*, **IT-14,** 110–120.

Peek, J. B. H. (1967). 'The measurement of correlation functions in correlators using "shift-invariant independent" functions', *Doctoral dissert., Technological University, Eindhoven*, Netherlands.

Pugachev, V. S. (1965). *Theory of Random Functions and its Application to Control Problems*, Pergamon, Oxford, 833 pp.

Rajbman, N. S. and A. T. Terekhin (1965). 'Variance-ratio methods for random processes and their application to the analysis of nonlinear control plants', *Automn. Remote Control*, **26,** 496–506.

Rajbman, N. S. (1966). 'Nonlinear plants identification with the aid of dispersion functions', *3rd IFAC Congr., London*, paper 3C.

Rajbman, N. S. (1967). 'Identification of nonlinear controlled plants with dispersion functions', *IFAC Symp.—Identification in Autom. Control Systems, Prague*, paper 3.14.

Rajbman, N. S. et al. (1970). Cf. references Chapter 2.

Rake, H. (1968). 'The determination of frequency responses from stochastic signals by means of an electronic analog computer' (in German), *Regelungstechnik*, **16,** 258–260.

Shinbrot, M. (1957). 'On the analysis of linear and nonlinear systems', *Trans. ASME*, **79,** 547–552.

Solodovnikov, V. V. (1960). Cf. references Chapter 12.

Strejc, V. (1958). 'New methods for approximating differential equations of control loops using a general input signal' (in German), *Acta Techn.*, **3,** 241–261.

Strejc, V. (1959). 'Determination of dynamical properties of control loops using measured input and output signals of a general nature' (in German), *Tagungsbericht IV. Int. Kolloq. d. Hochschule für Elektrotechnik. Ilmenau:* 397–402; also in: *Z. Messen. Steuern. Regeln*, 7–11 (1960).

Strejc, V. (1961). 'Evaluation of general signals with non-zero initial conditions', *Acta Techn.* **6,** 378–391.

Veltman, B. P. Th. and H. Kwakernaak (1961). 'Theory and application of polarity correlation for analyzing low frequency signals and systems' (in German), *Regelungstechnik*, **9,** 337–364.

Wallman, H. (1950). 'Electronic integral-transform computer and practical solution of integral equations', *J. Franklin Inst.*, **250,** 45–61.

Watts, D. G. (1962). 'A general theory of amplitude quantization with application to correlation determination', *Proc. IEE*, **109C,** 209–218.

Welfonder, E. (1969). 'Stochastic signals; statistical averaging under real conditions' (in German), *Fortschritt-Bericht VDI-Z*, **8,** 165.

Widrow, B. (1956). 'A study of rough amplitude quantization by means of Nyquist sampling theory', *IRE Trans. Circuit Theory*, **CT-3,** 266–276.

## Additional Literature

### *8.1 Operations on 'Analogue Signals'*

Albert, G. E. (1960). 'Statistical methods in prediction, filtering and detection problems', *SIAM J.*, **8,** 640–653.

Alper, P. (1965). 'An open-loop procedure for process parameter estimation using a hybrid computer', *IFAC Symp.—The Theory of Self-adaptive Control Systems, Teddington*, paper 3.1.

Anderson, G. W., R. N. Buland and G. R. Cooper (1959). 'Use of crosscorrelation in an adaptive control system', *Proc. Nat. Electron. Conf.*, **15,** 34–35.

Bär, D. (1962). 'Practical method of determining the dynamic characteristic values of linear controlled plants with inherent control by means of the correlation analysis' (in German), *Z. Messen, Steuern, Regeln*, **5,** 251–256.

Bartels, E. (1966). 'Practical system analysis using correlation techniques' (in German), *Regelungstechnik*, **14,** 49–55.

Bastl, W. (1966). Cf. references Section 14.3.

Cooper, D. C. (1965). 'The probability density function for the output of a correlator with band-pass input waveforms', *IEEE Trans. Information Theory*, **IT-11,** 190–195.

Cowley, P. E. A., R. M. Fano and B. L. Basore (1951). 'A short-time correlator for speech waves', *Tech. Report 174. Res. Lab. Electron. MIT, Cambridge*, Mass.

Csáki, R. and J. Gyürki (1967). 'Some questions of the system identification by stochastic signals', *IFAC Symp.—Identification in Autom. Control Systems, Prague*, paper 3.21.

Elistratov, M. R. (1963). 'Determination of dynamic characteristics of non stationary systems during their normal operation', *Automn. Remote Control*, **24,** 852–862.

Elkind, J. I., D. M. Green and E. A. Starr (1963). 'Application of multiple regression analysis to identification of time-varying linear dynamic systems', *IEEE Trans. Autom. Control*, **AC-8,** 163–166.

Fife, D. W. (1963). 'Simrar: simulated receiver and recorder for statistical measurements', *Ann. Assoc. Int. Calcul. Analogue.*, **5,** 156–159.

Goodman, T. P. and R. H. Hillsley (1958). 'Continuous measurement of characteristics of systems with random inputs: a step toward selfoptimizing control', *Trans. ASME*, **80,** 1839–1848.

Goodman, N. R. and others (1961). 'Frequency response from stationary noise: two case histories', *Technometrics*, **3**, 245–268.

Henderson, J. G. and Pengilley (1958). 'Experimental determination of system transfer functions from normal operating data', *J. Br. IRE*, **18,** 149.

Hofstetter, M. (1965). 'Some results on stochastic signal parameter estimation problem', *IEEE Trans. Information Theory*, **IT-11,** 422–429.

Hughes, M. T. G. and A. R. M. Noton (1962). 'The measurement of control system characteristics by means of a cross correlator', *Proc. IEE*, **109B,** 77–83.

Janac, K. (1964). 'The solution of probability problems by analog computers', *Automn. Remote Control*, **25,** 79–84.

Jenkins, G. M. (1963). 'Cross spectral analysis and the estimation of linear open loop transfer functions', *Time Series Analysis* (Ed. Rosenblatt), Wiley, New York.

Kryze, J. (1963). 'MUSA-6, a universal statistical analyser' (in German), *Z. Messen, Steuern, Regeln*, **6,** 286, 386.

Kwakernaak, H. (1963). Cf. references Chapter 6.

Lange, F. H. and M. Zecha (1969). 'Stochastic error theory', *4th IFAC Congr., Warsaw*, paper 33.3.

Leonov, Y. P. and L. N. Lipatov (1961). 'Apparatus for statistical investigation of the dynamic characteristics of industrial processes in the presence of noise', *Instrum. Constr.*, **11,** 3.

Levey, E. C. (1959), 'Complex curve fitting', *IRE Trans. Autom. Control*, **AC-4,** 37–43.

Lindenlaub, J. C. and G. R. Cooper (1963). 'Noise limitations of system identification techniques', *IEEE Trans. Autom. Control*, **AC-8,** 43–48.

Mann, K. H., G. Otte and G. W. Werner (1963). 'Correlation measuring installation for scientific research and automatic control teaching' (in German), *Wiss. Z. Elektrotechn.*, **2,** 53–68.

Mesch, F. (1966). 'Non-ideal time delay circuits for correlation measurements', *Regelungstechnik*, **14,** 70–74.

Otterman, J. (1960). 'The properties and methods for computation of exponentially-mapped-past statistical variables', *IRE Trans. Autom. Control*, **AC-5,** 11–18.

Parks, P. C. and B. L. Clarkson (1965). 'Some theory and practice of random process analysis', *Proc. Inst. Mech. Engrs.*, **179,** part 34.

Perlis, H. J. (1963). 'Justification and limitation of a 2-parameter 2-adjustment self-adaptive system', *IEEE Trans. Applic. and Ind.*, **82,** 220–229.

Perlis, H. J. (1964). 'The minimization of measurement error in a general perturbation-correlation process identification system', *IEEE Trans. Autom. Control*, **AC-9,** 339–345.

Peschel, M. (1962). 'Investigation of linear systems by means of appropriate noise processes' (in German), *Z. Messen, Steuern, Regeln*, **5,** 257–260.

Qvarnström, B. (1960). 'Transfer function-determination in the presence of noise for a set of significant input functions', *Proc. 5th Inst. Instruments and Measurements Conf.*, Stockholm.

Rajbman, N. S. (1962). 'Correlation methods for determining multiple automatic line process approximate characteristics' (in Russian), *Izvest. Akad. Nauk SSSR*, Energetika i Avtomatika, Moscow, 72–91.

Resa, F. M. (1961). *An Introduction to Information Theory*, McGraw-Hill, New York, 317 pp.

Reswick, J. B. (1955). 'Determine system dynamics without upset', *Control Engng.*, **2,** 50.

Schetzen, M. (1965). 'Determination of optimum nonlinear systems for generalized error criteria based on the use of gate functions', *IEEE Trans. Information Theory*, **IT-11,** 117–125.

Simmons, G. J. (1966). 'A factorization technique for binary autocorrelation functions', *Proc. IEEE*, **54,** 794–795.

Solodownikov, V. V. and A. S. Uskov (1963). *Statistical Analysis of Control Loops* (in German), V.E.B. Verlag Technik, Berlin.
Sydow, A. (1963). 'Statistical analysis by means of the electronic analog computer' (in German), *Z. Messen, Steuern, Regeln*, **6,** 501–504.
Van der Grinten, P. M. E. M. (1962). 'Stochastic processes in measurement and control' (in Dutch, summary in English), *Doctoral Thesis, University of Technology, Eindhoven*, Netherlands, 108 pp.
Van der Grinten, P. M. E. M. and W. Krijger (1963). 'Processing of the auto and cross-correlation functions to step response', *Ann. Assoc. Int. Calcul. Analogue*, **5,** 160–161.
Werner, G. W. (1962). 'Contribution to correlation techniques in control engineering' (in German), *Z. Messen, Steuern, Regeln*, **5,** 347–354.
Werner, G. W. and W. Seide (1962). 'The determination of pole-zero distribution of an industrial controlled plant by means of correlation measurements' (in German), *Wiss. Z. Hochsch. Elektrotechn*, Ilmenau, D.D.R., **8,** 353–359.
Westcott, J. H. (1956). 'The determination of process-dynamics from normal disturbance records of a controlled process', in *Regelungstechnik* (Ed. G. Müller), München, Oldenbourg, 256–261.
Westcott, J. H. (1960). 'The parameter-estimation problem', *1st IFAC Congr., Moscow*, 779–787.
Wolf, A. A. (1962). 'On the significance of instantaneous and short term correlation functions for a class of stochastic processes', *Proc. IRE*, **50,** 1983.
Wolf, A. A. and J. H. Dietz (1962). 'A statistical theory for parameter identification in physical systems', *J. Franklin Inst.*, **274,** 369–400.
Woodrow, R. A. (1958). 'Closed loop dynamics from normal operating records', *Trans. Soc. Instrum. Technol.*, **10,** 101–105.

## *8.2 Operations on Quantized Signals*

Bennett, W. R. (1948). 'Spectra of quantized signals', *Bell Syst. Tech. J.*, **27,** 446–472.
Price, R. (1958). 'A useful theorem for nonlinear devices, having Gaussian inputs', *IRE Trans. Information Theory*, **IT-4,** 69–72.
Scanlon, W. W. and C. Lieberman (1959). 'Naval ordnance and electronic research', *Proc. IRE*, **47,** 910.
The, G. (1966). 'Theorem on Stieltjes correlation', *Control and Science Record*, **113,** 1091.

## *8.3 Operations on Binary Signals*

Anderson, G. W. and J. E. Murrin (1956). 'Autocorrelation analysis of binary noise signals', *IRE WESCON Conv. Rec.*,
Anderson, G. W., R. N. Buland and G. R. Cooper (1959). Cf. references Chapter 10.
Anderson, G. W. (1959). 'Use of crosscorrelation in an adaptive control system', *Proc. Nat. Electron. Conf.*, **15,** 34–35.
Bogner, R. E. (1965). 'New relay correlation method', *IEEE Trans. Electron. Letters*, **1,** 53.
Brown, R. F. (1969). 'Review and comparison of drift corrections schemes for periodic correlation', *Electron Letters*, **5,** 179–181.
Cheng, M. C. (1968). 'The clipping loss in correlation detectors for arbitrary input signal-to-noise ratios', *IEEE Trans. Information Theory*, **IT-14,** 382–389.
Diamessis, J. E. (1965). 'A new method for determining the parameters of physical systems' (correspondence), *Proc. IEEE*, **53,** 205–206.

Diamessis, J. E. (1965). 'A method for determining the parameters of certain time-varying systems' (correspondence), *Proc. IEEE*, **53,** 396–397.

Ekre, H. (1963). 'Polarity coincidence detection of a weak noise source', *IEEE Trans. Information Theory*, **IT-9,** 18–23.

Heins, W., A. Van den Bos and B. P. Th. Veltman (1965). 'On the moments of shifted Gaussian probability density functions', *Rep. Coop. Centre; Univ. of Technology, Delft*, Netherlands.

Kuz'michev, V. N. and V. F. Nesteruk (1964). 'Techniques and apparatus for determination of correlation functions of normally distributed and pulsed random processes', *Automn. Remote Control*, **25,** 660–664.

Liewers, P., and E. Buttler (1967). Cf. references Section 14.3.

Poortvliet, D. C. J. (1962). Cf. references Chapter 10.

Ruchkin, D. S. (1965). 'Error of correlation coefficient estimates from polarity coincidences' (correspondence), *IEEE Trans. Information Theory*, **IT-11,** 296–297.

Solodovnikov, V. V. and A. S. Uskov (1959). 'A frequency method for determining the dynamic characteristics of objects of automatic control from data on their normal usage', *Automn. Remote Control*, **20,** 1533–1542.

Stott, S. I. and T. I. Canning (1965). 'Method of correlating binary signals', *Proc. IEE*, **112,** 1835.

Van den Bos, A. (1965). 'The variance of auto-correlation estimates of clipped Gaussian noise', *Rep. Coop. Centre, Univ. of Technology, Delft*, Netherlands.

Van Vleck, J. and D. Middleton (1966). 'The spectrum of clipped noise', *Proc. IEEE*, **54,** 2–19.

Veltman, B. P. Th. and A. Van den Bos (1963). 'The applicability of the relay-correlator and the polarity-coincidence correlator in automatic control', *2nd IFAC Congr., Basel.*

Veltman, B. P. Th. (1966). 'Quantizing, sampling frequency and statistical dispersion of correlation measurements' (in German), *Regelungstechnik*, **14,** 151–158.

Wolff, S. S., J. B. Thomas and T. R. Williams (1962). 'The polarity-coincidence correlator: a nonparametric detection device', *IRE Trans. Information Theory*, **IT-8,** 5–9.

Womack, B. F. and J. T. Watt, Jr. (1965). 'A new method for determining system parameters' (correspondence), *Proc. IEEE*, **53,** 729–730.

## *8.4 Differential Approximation Method*

Bellman, R. E., R. E. Kalaba and B. Kotkin (1963). 'Differential approximation applied to the solution of convolution equations', *Rand Corp.*, RM-3601.

Bellman, R. E., B. Gluss and R. Roth (1964). 'Identification of differential systems with time-varying coefficients', *Rand Corp.*, RM-4288-PR.

Bellman, R. E., R. E. Kalaba and R. Sridhar (1966). 'Adaptive control via quasilinearization and differential approximation', *Computing*, **1,** 8–17.

Sage, A. P. and B. R. Eisenberg (1966). 'Experiments in nonlinear and nonstationary system identification via quasilinearization and differential approximation', *Joint Autom. Control Conf.*, 522–530.

## *8.5 Higher-order Correlation Functions*

Cf. also the additional literature Section 4.4.

Booton, R. C. (1954). 'The measurement and representation of nonlinear systems', *IRE Trans. Circuit Theory*, **CT-1**, 32–34.

Elkind, J. I. and D. M. Green (1961). 'A technique for measuring the characteristics of nonlinear systems', *Rep. 849, Bolt Beranek and Newman, Cambridge*, Mass.

Lee, Y. W. and M. Schetzen (1961). 'Measurement of the kernels of a nonlinear system by cross correlation', *Q. Prog. Rep. 60, Res. Lab. Electron., M.I.T.*, 118–130; also in *Int. J. Control*, (1965).

Lee, Y. W. and M. Schetzen (1965). 'Some aspects of the Wiener theory of nonlinear systems', *Proc. Nat. Electron. Conf.*, **21,** 759–764.

Wiener, N. (1958, 1962). Cf. references Chapter 4.

CHAPTER 9

# Continuous Signals; Implicit or Model-adjustment Techniques

As in Chapter 7, the model-adjustment techniques may be interpreted either as a control scheme for the parameters of a physical (analogue) model or as a computation scheme in which the 'mathematical model' can be clearly indicated. In Chapter 2 a 'generalized model' and 'error' have been introduced. In this case there is the advantage of a **linear** relationship between the error and the parameters to be adjusted. Some considerations of a general nature are given in Section 9.1. Some of the properties of these model-adjustment schemes are discussed. A distinction is made with respect to the use of an instantaneous error criterion, e.g.

$$E_1 = e^2(\boldsymbol{\beta}, t) \qquad \text{(cf. Section 9.2)}$$

and a time-average criterion, e.g.

$$E_2 = \frac{1}{T}\int_{t-T}^{t} e^2(\boldsymbol{\beta}, t)\,\mathrm{d}t \qquad \text{(cf. Section 9.3)}$$

In the last equation a certain averaging time $T$ is introduced. In practice this time must be chosen long enough to satisfy requirements as to accuracy of the estimate. The instantaneous criterion can be applied only when the corresponding adjustment loop is sufficiently slow. In this case, the adjustment loop, which always will be of a low-pass character, will provide the necessary averaging effect to obtain a certain accuracy.

Section 9.4 is devoted to the parameter-sensitivity functions and their use for continuous and for intermittent adjustment schemes. The approximation of the sensitivity functions through the use of two models with parameters $\beta$ and $(\beta + \Delta\beta)$ respectively is the subject of Section 9.5. Section 9.6 is devoted to the use of the relation

$$\frac{\mathrm{d}E}{\mathrm{d}t} = \frac{\partial E}{\partial \beta}\frac{\mathrm{d}\beta}{\mathrm{d}t} + \frac{\partial E}{\partial t}$$

where $\beta$ is varied, e.g. sinusoidally, and $\mathrm{d}E/\mathrm{d}t$ can be measured. If $\partial E/\partial t$ can be neglected then $\partial E/\partial \beta$ can be found. This provides us the direction to which the parameters have to be adjusted.

When studying this chapter, and especially the discussion of model adjustments, it should be kept in mind that the deterministic and the stochastic convergence schemes as discussed in Sections 5.2 and 5.3 have a direct applicability to this problem.

## 9.1 Models, Linear-in-the-parameters

### *9.1.1 Generalized error*

The generalized error was defined in Chapter 2 as

$$e = \sum_{i=0}^{n} \alpha_i y_i + \sum_{j=0}^{m} \beta_j u_j \tag{9.1}$$

where $u_j$ and $y_i$ are the outputs of the dynamic operators. The purpose of the arrangement of Figure 9.1 is to adjust the potentiometers $\alpha_0, \ldots, \alpha_n$ and

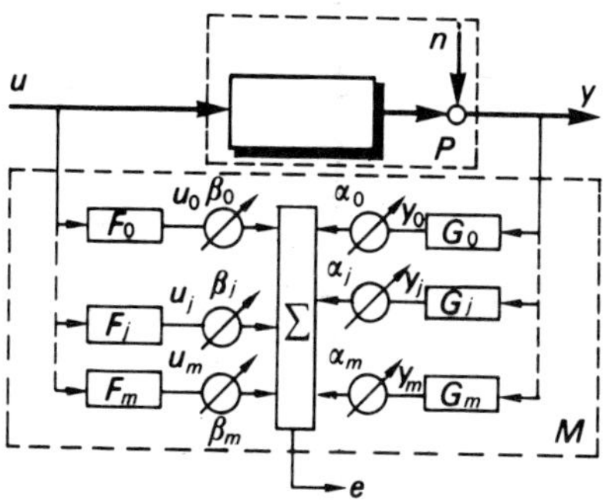


Figure 9.1

$\beta_0, \ldots, \beta_m$ in such a way that some functional of the generalized error $e$ is minimized. One of the 'branches' of the model can be chosen freely as reference, e.g. $\beta_0 F_0 = -1$. With this choice:

$$e = \sum_{i=0}^{n} \alpha_i y_i + \sum_{j=1}^{m} \beta_j u_j - u \tag{9.2}$$

As has already been pointed out, the essential feature of such a generalized model is the **linear** relation between $e$ and the parameters $\alpha_i$ and $\beta_j$. This implies that the partial derivatives (parameter-sensitivity functions; cf. Section 9.4) are given in a very simple way:

$$\frac{\partial e}{\partial \alpha_i} = y_i \qquad \frac{\partial e}{\partial \beta_j} = u_j \tag{9.3}$$

Consequently, no extra models are necessary to instrument the sensitivity equations and all parameter-sensitivity functions can be obtained from the one (generalized) model.

For the moment, let us choose as the error criterion a minimization of the squared error over an interval of length $T$. In order to eliminate the stochastic aspect we will take the expectation $\mathscr{E}$, thus:

$$E = \mathscr{E}\left[\int_{t-T}^{t} e^2\,\mathrm{d}t\right] \tag{9.4}$$

For this minimization we differentiate with respect to the parameters that have to be determined

$$\frac{1}{2}\zeta_i = \frac{1}{2}\frac{\partial E}{\partial \alpha_i} = \mathscr{E}\left[\int_{t-T}^{t} ey_i\,\mathrm{d}t\right] \stackrel{\text{def}}{=} \langle e, y_i\rangle \qquad i = 0, 1, \ldots, n \tag{9.5a}$$

$$\frac{1}{2}\eta_j = \frac{1}{2}\frac{\partial E}{\partial \beta_j} = \mathscr{E}\left[\int_{t-T}^{t} eu_j\,\mathrm{d}t\right] \stackrel{\text{def}}{=} \langle e, u_j\rangle \qquad j = 1, \ldots, m \tag{9.5b}$$

This leads to $m + n + 1$ equations. The symbol $\langle\ \rangle$ defined by equations (9.5a) and (9.5b) has been introduced for the sake of convenience. In the next section a wider interpretation will be given to it. For optimality $\zeta_0, \ldots, \zeta_n$ and $\eta_1, \ldots, \eta_m$ all have to be equal to zero. At the moment we will assume $\alpha_0, \ldots, \alpha_n, \beta_1, \ldots, \beta_m$ to be constant. If one substitutes the expression (9.2) for the generalized error $e$ in the equations (9.5a and b) then one will find in matrix notation:

$$\frac{1}{2}\begin{bmatrix}\zeta_0\\ \vdots\\ \zeta_n\\ \eta_1\\ \vdots\\ \eta_m\end{bmatrix} = \begin{bmatrix}\langle y_0, y_0\rangle & \cdots & \langle y_n, y_0\rangle\langle u_1, y_0\rangle & \cdots & \langle u_m, y_0\rangle\\ \vdots & & \vdots\quad\vdots & & \vdots\\ \langle y_0, y_n\rangle & \cdots & \langle y_n, y_n\rangle\langle u_1, y_n\rangle & \cdots & \langle u_m, y_n\rangle\\ \langle y_0, u_1\rangle & \cdots & \langle y_n, u_1\rangle\langle u_1, u_1\rangle & \cdots & \langle u_m, u_1\rangle\\ \vdots & & \vdots\quad\vdots & & \vdots\\ \langle y_0, u_m\rangle & \cdots & \langle y_n, u_m\rangle\langle u_1, u_m\rangle & \cdots & \langle u_m, u_m\rangle\end{bmatrix} \cdot \begin{bmatrix}\alpha_0\\ \vdots\\ \alpha_n\\ \beta_1\\ \vdots\\ \beta_m\end{bmatrix} - \begin{bmatrix}\langle u, y_0\rangle\\ \vdots\\ \langle u, y_n\rangle\\ \langle u, u_1\rangle\\ \vdots\\ \langle u, u_m\rangle\end{bmatrix}$$

or renaming the matrices in this order:

$$\tfrac{1}{2}\nabla E = \mathbf{M}\boldsymbol{\theta} - \boldsymbol{\psi} \tag{9.6}$$

The vector $\nabla E$ with elements $\zeta_i$ and $\eta_j$ will be called the **gradient vector,** the vector $\boldsymbol{\theta}$ with elements $\alpha_i$ and $\beta_j$ is the **parameter vector,** the vector $\boldsymbol{\psi}$ with elements $\langle u, y_i \rangle$ and $\langle u, u_j \rangle$ will be denoted by **reference vector.**

On the matrix **M** one can make observations analogously to those made in Chapter 6:

From the definition of **M** it follows that $\langle u_j, y_i \rangle = \langle y_i, u_j \rangle$, consequently **M** is symmetric about its main diagonal. In addition, it is clear that the elements on the main diagonal are non-negative.

The elements of the matrix depend only on the input signal $u$, the process and the linear operators of the model; they do not depend on the coefficient vector if, as it was assumed, this vector is constant.

In principle, each component of the gradient vector is dependent on all parameters; this leads to parameter interaction.

By use of orthogonality relations a number of matrix elements can be made zero for certain classes of input signals (cf. Chapters 3 and 4).

If **M** is not singular then, by multiplication of the matrix equation (9.6) with $\mathbf{M}^{-1}$, the right-hand side is simplified: $\mathbf{M}^{-1}\mathbf{M} - \mathbf{I}$ (with $\mathbf{I}$ – identity matrix). This means that if $\mathbf{M}^{-1}$ can be instrumented in the estimating scheme then there is no interdependence in the determination of the different coefficients.

The general expression of equation (9.6) looks rather formidable. One has to remember, however, that it governs the determination of not less than $m + n + 1$ unknown coefficients. Moreover, it will become clear that in the instrumentation of the problem it will not be necessary to determine all elements of matrix **M** explicitly. Starting from equation (9.6) there are two routes:

(*a*) Put $\nabla E = \mathbf{0}$ where $\mathbf{0}$ is the zero vector. Now equation (9.6) reduces to:

$$\mathbf{M}\boldsymbol{\theta} = \boldsymbol{\psi} \quad \text{and} \quad \boldsymbol{\theta} = \mathbf{M}^{-1}\boldsymbol{\psi} \tag{9.7}$$

Provided **M** is not singular, this matrix equation can be solved for the unknown elements of the vector $\boldsymbol{\theta}$ as all elements of **M** and $\boldsymbol{\psi}$ can be determined by a proper instrumentation. This leads us to the one class of Chapter 2 called: **'using an explicit mathematical relation',** which is discussed extensively in Chapter 8.

(*b*) Select an operational relation between the gradient vector $\nabla E$ and the parameter vector $\boldsymbol{\theta}$

$$\boldsymbol{\theta} = \mathscr{P}[\tfrac{1}{2}\nabla E] \tag{9.8}$$

where $\mathscr{P}$ is called the **policy operator.** Substituting equation (9.8) in equation (9.6) leads to:

$$\boldsymbol{\theta} = \mathscr{P}[\mathbf{M}\boldsymbol{\theta} - \boldsymbol{\psi}] \tag{9.9}$$

A proper choice of $\mathscr{P}$ will give a set of equations so that in time:

$$[\zeta_0, \ldots, \zeta_n, \eta_1, \ldots, \eta_m] \to [0, \ldots, 0, 0, \ldots, 0]$$

Except for pathological cases this implies:

$$[\alpha_0, \ldots, \alpha_n, \beta_1, \ldots, \beta_m] \to [a_0, \ldots, a_n, b_1, \ldots, b_m]$$

where $[a_i, b_j]$ are the parameters of the process.

As a result of the condition which requires the coefficient vector to be constant, the policy operator has to be one of the intermittent type, i.e. measuring over an interval with constant coefficients, correcting the model coefficients after the interval according to the results of the measurements, measuring again, correcting again, etc. Consequently, if the argument $i$ denotes the $i$th measuring and adjustment cycle then at the end of the measuring interval one finds for the error vector:

$$\tfrac{1}{2}\nabla E(i) = \mathbf{M}\boldsymbol{\theta}(i) - \boldsymbol{\psi}(i) \tag{9.10}$$

In the next adjustment interval the coefficient vector may be changed according to:

$$\boldsymbol{\theta}(i + 1) = \boldsymbol{\theta}(i) - \boldsymbol{\Gamma}(i)\nabla E(i) \tag{9.11}$$

where $\boldsymbol{\Gamma}(i)$ may be a matrix of real numbers, representing 'gain factors' of the adjustment loop. With this new value of $\boldsymbol{\theta}$ over the $(i + 1)$th measuring interval the error vector is determined again:

$$\tfrac{1}{2}\nabla E(i + 1) = \mathbf{M}\boldsymbol{\theta}(i + 1) - \boldsymbol{\psi}(i + 1)$$

This procedure leads to a set of difference equations.

When we choose as a policy operator

$$\dot{\boldsymbol{\theta}} = -\tfrac{1}{2}\boldsymbol{\Gamma}\nabla E \tag{9.12}$$

then we are led to a differential equation:

$$\dot{\boldsymbol{\theta}} + \boldsymbol{\Gamma}\mathbf{M}\boldsymbol{\theta} = \boldsymbol{\Gamma}\boldsymbol{\psi} \tag{9.13}$$

Strictly speaking, this does not hold as now $\boldsymbol{\theta}$ is changing whereas equation (9.6) was derived assuming $\boldsymbol{\theta}$ to be constant. In cases where the convergence is relatively slow, however, equation (9.13) leads to an approximation of the dynamic behaviour. The instrumentation of these equations leads to the class, mentioned in Chapter 2, **'using a model-adjustment technique'**; the model adjustment gives an approximation of the process characteristics.

If the model can describe the dynamics of the process completely and if there is no additive noise then there is a set of parameters $[a_0, \ldots, a_n, b_1, \ldots, b_m]$ for which the following holds:

$$0 = \sum_{i=0}^{n} a_i y_i + \sum_{j=1}^{m} b_j u_j - u \tag{9.14}$$

Subtraction of equation (9.14) from (9.2) leads us to

$$e = \sum_{i=0}^{n} (\alpha_i - a_i)y_i + \sum_{j=1}^{m} (\beta_j - b_j)u_j \tag{9.15}$$

Consequently, the error is linearly related to the deviation of the model parameters from their optimal values. Using again as the error criterion the minimization of (9.4) and differentiation according to (9.5a) and (9.5b), then this results in

$$\frac{1}{2}\begin{bmatrix} \zeta_1 \\ \vdots \\ \zeta_n \\ \eta_1 \\ \vdots \\ \eta_m \end{bmatrix} = \begin{bmatrix} \langle y_0, y_0 \rangle & \cdots & \langle y_n, y_0 \rangle & \langle u_1, y_0 \rangle & \cdots & \langle u_m, y_0 \rangle \\ \vdots & & \vdots & \vdots & & \vdots \\ \langle y_0, u_m \rangle & \cdots & \langle y_n, u_m \rangle & \langle u_1, u_m \rangle & \cdots & \langle u_m, u_m \rangle \end{bmatrix} \cdot \begin{bmatrix} \alpha_0 - a_0 \\ \vdots \\ \alpha_n - a_n \\ \beta_1 - b_1 \\ \vdots \\ \beta_m - b_m \end{bmatrix}$$

or renaming the vectors in this order:

$$\tfrac{1}{2}\nabla E = \mathbf{M}\boldsymbol{\lambda} \tag{9.16}$$

where $\nabla E$ and $\mathbf{M}$ are the same as defined for equation (9.8) and where the vector $\boldsymbol{\lambda}$ with elements $\alpha_i - a_i$ and $\beta_j - b_j$ is the **parameter-error vector.** As before different kinds of policy operators $\mathscr{P}$ can be used, e.g.

$$\Delta\boldsymbol{\lambda} = -\tfrac{1}{2}\boldsymbol{\Gamma}\nabla E \tag{9.17}$$

or

$$\Delta\boldsymbol{\lambda} = -\boldsymbol{\Gamma}\mathbf{M}\boldsymbol{\lambda}$$

This leads to a vector difference equation

$$\Delta\boldsymbol{\lambda} = \boldsymbol{\lambda}(i + 1) - \boldsymbol{\lambda}(i) = -\boldsymbol{\Gamma}\mathbf{M}\boldsymbol{\lambda}(i)$$

or

$$\boldsymbol{\lambda}(i + 1) = [\mathbf{I} - \boldsymbol{\Gamma}\mathbf{M}]\boldsymbol{\lambda}(i) \tag{9.18}$$

or

$$\boldsymbol{\lambda}(i) = [\mathbf{I} - \boldsymbol{\Gamma}\mathbf{M}]^i \boldsymbol{\lambda}(0) \tag{9.18a}$$

where $\boldsymbol{\lambda}(0)$ is the parameter-error vector at $t = 0$. Again the 'gain factors' $\boldsymbol{\Gamma}$ may be made dependent on the iteration number $i$. As before, we also may choose

$$\dot{\boldsymbol{\lambda}} = -\tfrac{1}{2}\boldsymbol{\Gamma}\nabla E$$

which leads to

$$\dot{\boldsymbol{\lambda}} = -\boldsymbol{\Gamma}\mathbf{M}\boldsymbol{\lambda} \tag{9.19}$$

or

$$\boldsymbol{\lambda}(t) = \boldsymbol{\lambda}(0) \exp[-\boldsymbol{\Gamma}\mathbf{M}t] \tag{9.19a}$$

As $\mathbf{M}$ is a real symmetric matrix its characteristic roots are real. According to the theory of matrix exponentials $\boldsymbol{\lambda}(t) \to \mathbf{0}$ for $t \to \infty$ if all characteristic roots of $\boldsymbol{\Gamma}\mathbf{M}$ have real positive parts.

### 9.1.2 *The Choice of Error Criterion and Type of Information Processing*

Let us consider the case where the parameter $b(t)$ in the process $P$ is the quantity that changes according to an arbitrary, unknown time function and $\beta(t)$ is the parameter in the model $M$ that has to follow that time function (parameter tracking). The quantities $b$ and $\beta$ may be coefficients of differential equations, of difference equations, of transfer functions or of sampled data functions. They may also represent the locations of poles and zeros of the transfer functions, coordinates of the impulse-response functions, moments of the impulse-response functions, coordinates of amplitude or phase characteristics.

The error in the correspondence of the transfer characteristics of $P$ and $M$ can be defined in many ways. One might propose using as an error criterion the minimization of

$$\int_{t-T}^{t} \{b(t) - \beta(t)\}^2 \, dt \tag{9.20}$$

or some corresponding relationship. A complete solution for $\beta(t)$ is explicitly obtainable only for the most simple cases. Further, $b(t)$ is not known in practical applications. Consequently, the error criterion of equation (9.20) can be used in simulation studies only. Therefore another type of **error,** $e = y - w$, was defined. As an **error criterion,** the squared error over an interval of time was used. There is, however, a considerable freedom in choosing error criteria, e.g.:

$$E_1 = \mathscr{E}\left[\int_{t-T}^{t} e^2 \, dt\right] \quad \overset{\text{def}}{=} \langle e, e\rangle \tag{9.21a}$$

$$E_2 = \mathscr{E}\left[\int_{t-T}^{t} |e| \, dt\right] \quad \overset{\text{def}}{=} \langle |e|, 1\rangle \tag{9.21b}$$

$$E_3 = \mathscr{E}\left[\int_{0}^{\infty} e^2(t-\theta)h(\theta) \, d\theta\right] \overset{\text{def}}{=} \langle e, e\rangle_h \tag{9.21c}$$

$$E_4 = \mathscr{E}\left[\int_{0}^{\infty} |e(t-\theta)|h(\theta) \, d\theta\right] \overset{\text{def}}{=} \langle |e|, 1\rangle_h \tag{9.21d}$$

In this order, the criteria may be called:

$E_1$ integral squared error
$E_2$ integral absolute error
$E_3$ time-weighted squared error
$E_4$ time-weighted absolute error

In $E_1$ and $E_2$ the integration interval is 'sliding' along the $t$-axis. Another possibility is the use of intervals that are prefixed on the $t$-axis (intermittent schemes). In $E_3$ and $E_4$ there is a weighting of the error with a function that puts less emphasis on older information, i.e. 'the past is gradually forgotten'. This can be achieved by choosing a low-pass filter with impulse response $h(t) \to 0$ for $t \to \infty$ instead of an integrator. As before, the bracket symbols defined by equation (9.21) are introduced for the sake of convenience. Differentiation of equations (9.21) with respect to $\alpha_i$ and $\beta_j$ analogously to equation (9.5) leads to:

$$\frac{1}{2}\zeta_i = \frac{1}{2}\frac{\partial E_1}{\partial a_i} = \langle e, y_i\rangle \qquad \frac{1}{2}\eta_j = \frac{1}{2}\frac{\partial E_1}{\partial \beta_j} = \langle e, u_j\rangle \tag{9.22a}$$

$$\zeta_i = \frac{\partial E_2}{\partial \alpha_i} = \left\langle \frac{e}{|e|}, y_i\right\rangle \qquad \eta_j = \frac{\partial E_2}{\partial \beta_j} = \left\langle \frac{e}{|e|}, u_j\right\rangle \tag{9.22b}$$

$$\frac{1}{2}\zeta_i = \frac{1}{2}\frac{\partial E_3}{\partial \alpha_i} = \langle e, y_i\rangle_h \qquad \frac{1}{2}\eta_j = \frac{1}{2}\frac{\partial E_3}{\partial \beta_j} = \langle e, u_j\rangle_k \tag{9.22c}$$

$$\zeta_i = \frac{\partial E_4}{\partial \alpha_i} = \left\langle \frac{e}{|e|}, y_i\right\rangle \qquad \eta_j = \frac{\partial E_4}{\partial \beta_j} = \left\langle \frac{e}{|e|}, u_j\right\rangle \tag{9.22d}$$

Of course, the error criterion used determines the implementation. The term

$$\frac{e}{|e|} = \operatorname{sgn} e = \pm 1 \qquad \text{signum function}$$

in equation (9.22) leads to a binary operation; for example a multiplication $x$ sgn $e$ can be performed by a type of relay driven by $e$ and switching between $+x$ and $-x$.

In Chapter 3, a distinction was made with regard to the type of signal used. The names **'continuous'** (a better term would be: non-sampled) and **sampled** refer to the information representation as a function of time. As a function of amplitude a distinction can be made as to: **analogue, quantized** and **binary.** In studying the following sections, it has to be kept in mind that many schemes can be implemented for the use of each of these signal types mentioned.

### *9.1.3 The Choice of the Dynamic Operators*

In the previous sections, nothing has been said about the dynamic operators $F_0, \ldots, F_m$ and $G_0, \ldots, G_n$ that are a part of the generalized model in Figure 9.1. It is clear that the choice of these operators influences the properties of the parameter-estimating scheme, e.g. the kind of information produced, the speed of convergence to the solution, the possible bias of the solution, the influence of additive noise, the possibilities of simplification of the implementation using orthogonality relations.

The following are a few possible choices for the **linear** operators:

| | | |
|---|---|---|
| $F_j$ | $G_i$ | differentiators |
| $F_j$ | $G_i$ | high-pass filters |
| $F_j$ | $G_i$ | integrators |
| $F_j$ | $G_i$ | low-pass filters |
| $F_j$ | $G_i$ | band-pass filters |
| $F_j$ | | pure time-delay units; $\alpha_0 = 1$, $G_0 = 1$ and $G_i = 0$ for $i > 0$ |
| $F_j$ | | orthogonal filters; $\alpha_0 = 1$, $G_0 = 1$ and $G_i = 0$ for $i > 0$ |

The last two choices lead to a scheme (cf. Figure 2.3) where the model is no longer a 'generalized' one. In such a case, the signal $e = y - w$ can be called the **error-in-the-dependent-variable,** as both $y$ and $w$ are dependent on the input variable $u$.

It has to be noted of course that a 'mixed' set of operators can also be used. A linear operator may consist also of a multiplier with the signal $u$ (or $y$) as one input and an **independent** time function as the other input. This arrangement may be used in development of the signal $u$ (or $y$) in orthogonal components. The choice of the linear operators hinges on the completeness and the simplicity of the process representation. It also depends on the type of information that is needed from the process (e.g. time-domain or frequency-domain information). It has to be remembered that the best fit in some sense to the observed data depends on the assumptions that are used for a particular type of description. The propagation of errors or uncertainty from one description to another one is also often difficult to analyze. Table 9.1

Table 9.1

| Model | Usable *a priori* knowledge? | Output linear-in-parameters? | Interaction of of the parameter adjustment? | Bias due to noise? |
|---|---|---|---|---|
| Time delay | Little (1) | Yes | Little | No |
| Orthogonal filter | Little (1) | Yes | Little | No |
| Generalized model | Much (2) | Yes | Some | Yes |
| Topologically identical model | Much (2) | No | Much | No |

(1) the process is (approximately) linear; the approximate length of the impulse response.
(2) order and form of the differential equation; order of magnitude of the parameter values that have to be estimated; known values of some of the parameters.

summarizes some of the properties of the different types of models (cf. Figure 2.13). Analogous considerations hold for non-linear models. The restriction that the error has to be linear-in-the-parameters limits the choice of such non-linear dynamic operators to a special class (cf. Chapter 4).

Apart from the previous considerations it may be important to consider the interaction of the model adjustments. For a minimal parameter interaction, one may choose orthogonal filters according to Sections 3.2 and 3.3. In that case

$$\int_0^T u_i u_j \, dt = \langle u_i, u_j \rangle = c\delta_{ij} \tag{9.23}$$

With the error criterion and the error

$$E = \int_0^T e^2 \, dt = \langle e, e \rangle \quad \text{and} \quad e = y - \sum_{j=0}^{m} \beta_j u_j$$

this leads to the simplified relationship:

$$-\tfrac{1}{2}\nabla E = \begin{bmatrix} \langle u_0, u_0 \rangle & \cdots & \langle u_m, u_0 \rangle \\ \vdots & & \vdots \\ \langle u_0, u_m \rangle & \cdots & \langle u_m, u_m \rangle \end{bmatrix} \begin{bmatrix} \beta_0 \\ \vdots \\ \beta_m \end{bmatrix} - \begin{bmatrix} \langle y, u_0 \rangle \\ \vdots \\ \langle y, u_m \rangle \end{bmatrix}$$

$$= c \begin{bmatrix} \beta_0 \\ \vdots \\ \beta_m \end{bmatrix} - \begin{bmatrix} \langle y, u_0 \rangle \\ \vdots \\ \langle y, u_m \rangle \end{bmatrix}$$

or

$$-\frac{1}{2}\frac{\partial E}{\partial \beta_j} = c\beta_j - \langle y, u_j \rangle \tag{9.24}$$

i.e. the gradient with respect to one parameter is now independent of all other parameters. The optimal value of the parameter is

$$\frac{\partial E}{\partial \beta_j} = 0 \quad \text{or} \quad \hat{\beta}_j = \frac{\langle y, u_j \rangle}{c}$$

From

$$E = \int_0^T e^2 \, dt = \int_0^T \{y - \sum \beta_j u_j\}^2 \, dt$$

it follows that the residual part of the error criterion is:

$$E_{res} = \int_0^T y^2 \, dt - c \sum_{j=0}^{m} \hat{\beta}_j^2 \tag{9.25}$$

This shows the contribution of each model parameter $\beta_j$ to the minimization of the error criterion. It gives us a means for judging the number of model parameters needed. Extending the number of model parameters with $m + 1, \ldots$ does not influence the optimal settings of the parameters $\beta_0, \ldots, \beta_m$.

The problem remains as to how one chooses the impulse responses of the dynamic operators in such a way that their output signals $u_j(t)$ are orthogonal. Assume we want

$$\frac{1}{T}\int_0^T u_i(t)u_j(t) \, dt = c\delta_{ij}$$

with

$$u_i(t) = \int_0^\infty u(t - \xi)g_i(\xi) \, d\xi$$

$$u_j(t) = \int_0^\infty u(t - \xi)g_j(\xi) \, d\xi$$

This leads to the condition

$$\begin{aligned} c\delta_{ij} &= \frac{1}{T}\int_0^T \int_0^\infty \int_0^\infty u(t - \xi)u(t - \nu)g_i(\xi)g_j(\nu) \, dt \, d\xi \, d\nu \\ &= \int_0^\infty \int_0^\infty \tilde{\psi}_{uu}(\nu - \xi, T)g_i(\xi)g_j(\nu) \, d\xi \, d\nu \end{aligned} \tag{9.26}$$

Consequently the impulse responses of the dynamic operators have to satisfy this condition in order to derive the orthogonal output signals $u_j(t)$

from $u(t)$. Therefore, these impulse responses are dependent on the input signal characteristics. For 'white' noise

$$\mathscr{E}[\tilde{\psi}_{uu}(v - \xi, T)] = \sigma_u^2 \delta(v - \xi)$$

Taking the expectation leads to the simple condition

$$\int_0^\infty g_i(\xi) g_j(\xi) \, d\xi = \frac{c}{\sigma_u^2} \delta_{ij}$$

For arbitrary input signals it is not a simple task to find the desired impulse responses (cf. Section 3.2).

It is evident that for a particular application one orthogonal series is 'better' than some other series. This connotation 'better' may be based on the number of terms needed for the approximation (with predetermined accuracy) of the process dynamics. This leads to the proposal of adjusting separately some parameter of the orthogonal filters in such a way that, given a limited number of filters, the residual error is minimized (cf. Chapter 4).

## 9.2 Minimization of the Instantaneous Error

In the previous section, a number of error criteria has been mentioned. The mathematical expectation was taken in order to remove the stochastic aspect. In a hardware realization of a model-adjustment system, one has only one ensemble element to operate on. In this section, the minimization of the **instantaneous** error is discussed; the minimization of an integral error is the topic of Section 9.3.

### *9.2.1 Basic Equations*

The generalized model is shown in Figure 9.1. For the moment the additive noise is neglected. The generalized error is defined before as:

$$e = \sum_{i=0}^{n} \alpha_i y_i + \sum_{j=1}^{m} \beta_j u_j - u \tag{9.27}$$

The error criterion is chosen as minimizing:

$$E = e^2 \tag{9.28}$$

From this it follows that

$$\left.\begin{aligned} \frac{1}{2}\zeta_i &= \frac{1}{2}\frac{\partial E}{\partial \alpha_i} = e\frac{\partial e}{\partial \alpha_i} = e y_i \\ \frac{1}{2}\eta_j &= \frac{1}{2}\frac{\partial E}{\partial \beta_j} = e\frac{\partial e}{\partial \beta_j} = e u_j \end{aligned}\right\} \tag{9.29}$$

and the vector equation is

$$\tfrac{1}{2}\nabla E = \mathbf{M}\boldsymbol{\theta} - \boldsymbol{\psi} \tag{9.30}$$

with

$$\begin{aligned}
\nabla E &= [\zeta_0, \ldots, \zeta_n, \eta_1, \ldots, \eta_m]' \\
\boldsymbol{\theta} &= [\alpha_0, \ldots, \alpha_n, \beta_1, \ldots, \beta_m]' \\
\boldsymbol{\psi} &= [uy_0, \ldots, uy_n, uu_1, \ldots, uu_m]'
\end{aligned}$$

and

$$\mathbf{M} = \begin{bmatrix}
y_0y_0 & \cdots & y_ny_0 & u_1y_0 & \cdots & u_my_0 \\
\vdots & & \vdots & \vdots & & \vdots \\
y_0y_n & \cdots & y_ny_n & u_1y_n & \cdots & u_my_n \\
y_0u_1 & \cdots & y_nu_1 & u_1u_1 & \cdots & u_mu_1 \\
\vdots & & \vdots & \vdots & & \vdots \\
y_0u_m & \cdots & y_nu_m & u_1u_m & \cdots & u_mu_m
\end{bmatrix}$$

The same reasoning holds for the error criterion

$$E = |e| \tag{9.28a}$$

Now

$$\left.\begin{aligned}
\zeta_i &= \frac{\partial E}{\partial \alpha_i} = \frac{e}{|e|}\frac{\partial e}{\partial \alpha_i} = \frac{e}{|e|}y_i \\
\eta_j &= \frac{\partial E}{\partial \beta_j} = \frac{e}{|e|}\frac{\partial e}{\partial \beta_j} = \frac{e}{|e|}u_j
\end{aligned}\right\} \tag{9.29a}$$

and a vector equation results of the form

$$\nabla E = \frac{\mathbf{M}\boldsymbol{\theta} - \boldsymbol{\psi}}{|e|} \tag{9.31}$$

where $\nabla E$, $\boldsymbol{\theta}$, $\boldsymbol{\psi}$, $\mathbf{M}$ are the same as given before.

It is interesting to note, that with these error criteria there is no need to stipulate that $\boldsymbol{\theta}$ has to be constant as was done in Section 9.1; equations (9.30) and (9.31) also hold exactly for varying $\boldsymbol{\theta}$. So one may choose the policy operator in the equation

$$\boldsymbol{\theta} = \mathscr{P}[\tfrac{1}{2}\nabla E] \tag{9.32}$$

for example as in equation (9.12)

$$\dot{\boldsymbol{\theta}} = -\tfrac{1}{2}\boldsymbol{\Gamma}\nabla E \tag{9.33}$$

With equation (9.30) this leads to the matrix-differential equation

$$\dot{\boldsymbol{\theta}} + \boldsymbol{\Gamma M \theta} = \boldsymbol{\Gamma \psi} \tag{9.34}$$

If the input $u$ is stochastic then the elements of $\mathbf{M}$ are also stochastic. This differential equation can be simplified again by assuming that the model is capable of an ideal representation (see Miller, 1962). This implies that

$$0 = \sum_{i=0}^{n} a_i y_i + \sum_{j=1}^{m} b_j u_j - u \tag{9.35}$$

and

$$e = \sum_{i=0}^{n} (\alpha_i - a_i) y_i + \sum_{j=1}^{m} (\beta_j - b_j) u_j \tag{9.36}$$

leads to

$$\dot{\boldsymbol{\lambda}} = -\boldsymbol{\Gamma M \lambda} \tag{9.37}$$

cf. equations (9.14), (9.15) and (9.19). For a stationary stochastic input signal the elements of the matrix $\mathbf{M}$ are stationary stochastic as well.

The vector equations (9.30) and (9.32) govern the **behaviour** of the adjusting system. For the **instrumentation scheme** the equations (9.27) and (9.33) are sufficient:

$$\dot{\boldsymbol{\theta}} = -\tfrac{1}{2}\boldsymbol{\Gamma}\nabla E \quad \text{or} \quad \begin{bmatrix} \dot{\alpha}_i \\ \dot{\beta}_j \end{bmatrix} = -\boldsymbol{\Gamma}\begin{bmatrix} e y_i \\ e u_j \end{bmatrix} \tag{9.38}$$

Using a simple 'gain matrix' $\boldsymbol{\Gamma} = \gamma \mathbf{I}$ then:

$$\begin{cases} \dot{\alpha}_i = -\gamma e y_i \\ \dot{\beta}_j = -\gamma e u_j \end{cases} \tag{9.38a}$$

*Example.* We will assume a process with a simple linear transfer function

$$H(s) = \frac{b}{1 + as}$$

where $a$ and $b$ have to be estimated. Written in Laplace notation and neglecting the initial conditions, the process obeys the equation

$$bU(s) - asY(s) - Y(s) = 0$$

or

$$W(s)\{bU(s) - asY(s) - Y(s)\} = 0$$

or

$$bU_0(s) - aY_1(s) - Y_0(s) = 0$$

where $W(s)$ is an arbitrary (filter) transfer function

$$U_0(s) = W(s)U(s)$$
$$Y_0(s) = W(s)Y(s)$$
$$Y_1(s) = W(s)sY(s)$$

By means of a suitable choice of $W(s)$ the transfer function $W(s)s$ may be more realizable practically, than without this filter, e.g.

$$W(s) = \frac{1}{1 + \tau s} \qquad W(s)s = \frac{s}{1 + \tau s}$$

Now the model equation is chosen as:

$$\beta U_0(s) - \alpha Y_1(s) - Y_0(s) = \mathscr{L}[e(t)] \tag{9.39}$$

This is represented in Figure 9.2 for that particular choice of $W(s)$.

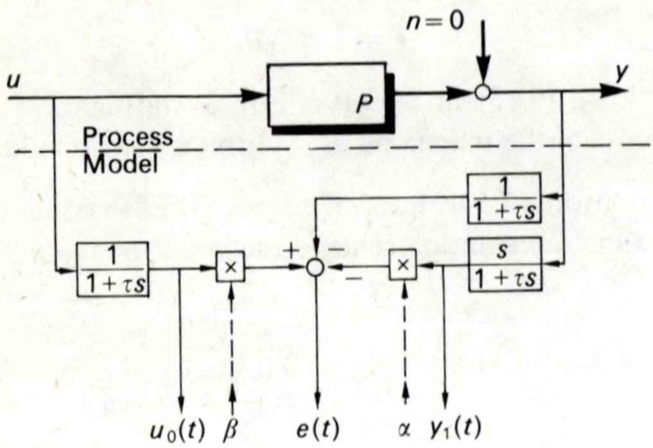


Figure 9.2

This example illustrates a general principle. By introducing an additional filter one may create the possibility of forming signals that are related to each other as, for example, time derivatives. It can be considered as a direct consequence of the notion of generalized model. Examples of specific choices can be found in Kohr (1963) and Young (1965).

Now one may choose from a variety of error criteria. For $E = e^2(t)$ one finds:

$$\left.\begin{aligned} \frac{\partial E}{\partial \alpha} &= \frac{\partial e^2}{\partial \alpha} = 2e\frac{\partial e}{\partial \alpha} = 2e(-y_1) \\ \frac{\partial E}{\partial \beta} &= \cdots\cdots\cdots\cdots = 2eu_0 \end{aligned}\right\} \tag{9.40}$$

and for the gradient equations

$$\left.\begin{aligned}\frac{d\alpha}{dt} &= -\gamma\frac{\partial E}{\partial \alpha} = -2\gamma e(-y_1)\\ \frac{d\beta}{dt} &= \cdots\cdots = -2\gamma e u_0\end{aligned}\right\} \tag{9.41}$$

From this it follows that Figure 9.2 has to be completed using the instrumentation of Figure 9.3a. Choosing the error criterion $E = |e|$, one finds

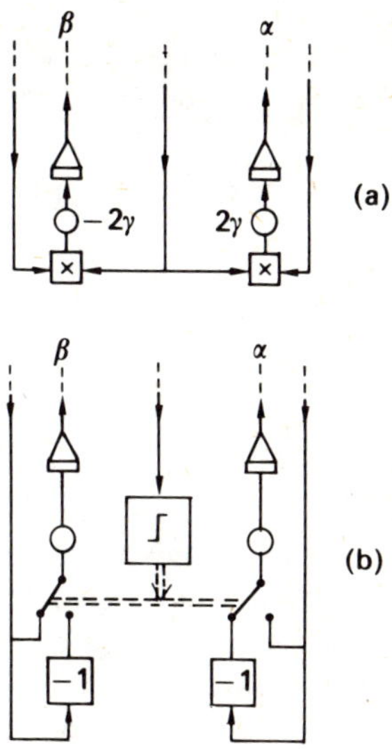


Figure 9.3

the instrumentation of Figure 9.3b, where a couple of analogue multipliers have been replaced by simple relays.

*9.2.2 Convergence Properties*

The convergence and stability properties of the parameter-estimation schemes, given in previous sections, are often difficult to establish. Therefore, another approach, where the basic requirement is not some type of error criterion but a monotonic model convergence, is given in this section (cf. Graupe, 1961).

The process is given by

$$g(t, b_1, \ldots, b_m) = 0 \tag{9.42}$$

where $b_1, \ldots, b_m$ are unknown parameters. This equation holds for a period of time, where the time dependence is given by several functions $u_j(t)$ and $y_i(t)$, which may include derivatives. The model is given by

$$g(t, \beta_1, \ldots, \beta_m) = e \tag{9.43}$$

where $e$ is the error that has to be minimized. Now the sense in which it will be minimized is not, however, determined in advance but follows from the following type of reasoning. It generally holds that:

$$\frac{\mathrm{d}e}{\mathrm{d}t} = \sum_{j=1}^{m} \frac{\partial e}{\partial \beta_j}\frac{\mathrm{d}\beta_j}{\mathrm{d}t} + \frac{\partial e}{\partial t} \tag{9.44}$$

Each term of this equation is multiplied by an odd function $-f$ of $e$; $f(-e) = -f(e)$. Now assume that the following condition is fulfilled:

$$-f(e)\sum_{j=1}^{m} \frac{\partial e}{\partial \beta_j}\frac{\mathrm{d}\beta_j}{\mathrm{d}t} > \left| -f(e)\frac{\partial e}{\partial t} \right| \qquad \text{for } 0 \leqslant e_{\min} < |e| \tag{9.45}$$

In that case

$$-f(e)\frac{\mathrm{d}e}{\mathrm{d}t} > 0 \qquad \text{for } 0 \leqslant e_{\min} < |e| \tag{9.46}$$

The last inequality assures monotonic convergence of the error towards zero as long as the condition posed by equation (9.45) is fulfilled. After the error is diminished in such a way that this condition no longer holds, then the error will fluctuate in the interval $-e_{\min} < e < e_{\min}$. For a certain choice of $f(e)$ there are several possible choices for $\mathrm{d}\beta/\mathrm{d}t$ in order to fulfil the requirements of equation (9.45). A few examples for the case of one parameter adjustment are given below. Take $f(e) = e^{-1}$, then

$$\left.\begin{array}{lll}
\dfrac{\mathrm{d}\beta}{\mathrm{d}t} = -\gamma e\dfrac{\partial e}{\partial \beta} & -\dfrac{1}{e}\dfrac{\partial e}{\partial \beta}\dfrac{\mathrm{d}\beta}{\mathrm{d}t} = \gamma\left(\dfrac{\partial e}{\partial \beta}\right)^2 & \geqslant 0 \\
\qquad -\gamma e \,\mathrm{sgn}\left[\dfrac{\partial e}{\partial \beta}\right] & \gamma\left|\dfrac{\partial e}{\partial \beta}\right| & \geqslant 0 \\
\qquad -\gamma\,\mathrm{sgn}\,[e]\dfrac{\partial e}{\partial \beta} & \gamma\left(\dfrac{\partial e}{\partial \beta}\right)^2\dfrac{1}{|e|} & \geqslant 0 \\
\qquad -\gamma\,\mathrm{sgn}\,[e]\,\mathrm{sgn}\left[\dfrac{\partial e}{\partial \beta}\right] & \gamma\left|\dfrac{\partial e}{\partial \beta}\right|\dfrac{1}{|e|} & \geqslant 0
\end{array}\right\} \text{for } \gamma > 0 \tag{9.47}$$

For one of these cases and for the simultaneous adjustment of $m$ parameters, Figure 9.4 shows the instrumentation schematically. The gains $\gamma_k$ of the integrators have to be chosen with respect to noise filtering ($\gamma_k$ low) and convergence speed ($\gamma_k$ high).

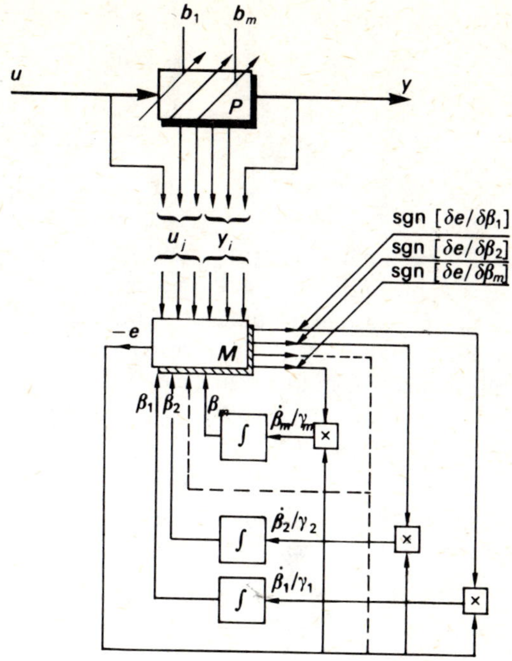


Figure 9.4

*Example.* Taking as a simple case a steepest-descent instrumentation for each parameter:

$$\frac{\mathrm{d}\beta_i}{\mathrm{d}t} = -\gamma \frac{\partial E}{\partial \beta_i} \tag{9.48}$$

then for

$$E = e^2 \Rightarrow \frac{\partial E}{\partial \beta_i} = 2e \frac{\partial e}{\partial \beta_i} \quad \text{and} \quad \frac{\mathrm{d}\beta_i}{\mathrm{d}t} = -2\gamma e \frac{\partial e}{\partial \beta_i} \tag{9.49a}$$

$$E = |e| \Rightarrow \frac{\partial E}{\partial \beta_i} = \operatorname{sgn}[e] \frac{\partial e}{\partial \beta_i} \quad \text{and} \quad \frac{\mathrm{d}\beta_i}{\mathrm{d}t} = -\gamma \operatorname{sgn}[e] \frac{\partial e}{\partial \beta_i} \tag{9.46b}$$

Thus, equation (9.47) guarantees the monotonic convergence of the steepest-descent adjustment using the error criterion given above for one parameter.

Using the error of equation (9.27) and the choices indicated by equation (9.49a) one finds for the condition of equation (9.45):

$$2\gamma\left\{\sum_{j=1}^{m} u_j^2 + \sum_{i=0}^{n} y_i^2\right\} > \left|-\frac{1}{e}\frac{\partial e}{\partial t}\right| \quad \text{for } 0 \leqslant e_{\min} < |e| \tag{9.50}$$

For the $E = |e|$ criterion one finds for the condition of equation (9.45):

$$\frac{\gamma}{|e|}\{\sum u_j^2 + \sum y_1^2\} > \left|-\frac{1}{e}\frac{\partial e}{\partial t}\right|$$

or

$$\gamma\{\sum u_j^2 + \sum y_i^2\} > \left|\frac{\partial e}{\partial t}\right| \qquad \text{for } 0 \leqslant e_{min} < |e| \tag{9.51}$$

If the input signal is stationary then the $u_j$ and $y_i$'s are stationary too. The coefficient $\gamma$ gives a contribution to the 'loop gain'.

An intuitive type of reasoning about the convergence of this type of system is given in the paper by Potts, Ornstein and Clymer (1961); the following example is taken from that reference.

*Example.* The process dynamics equation is given by:

$$a_1\dot{y} + a_0 y = u$$

where $a_1$ and $a_0$ have to be determined. For comparison with the previous discussion put $\dot{y} = y_1$; $y = y_0$ and $u = u_0$. The generalized error is:

$$e = \alpha_1\dot{y} + \alpha_0 y - u = \alpha_1 y_1 + \alpha_0 y_0 - u_0$$

where $\alpha_1$ and $\alpha_0$ are the computed values of the coefficients. Taking the absolute magnitude

$$|e| = |\alpha_1 y_1 + \alpha_0 y_0 - u_0|$$

gives the plane in the $|e(\alpha_1, \alpha_0)|$ space shown in Figure 9.5. This surface is tangent to the $(\alpha_1, \alpha_0)$ plane in line L given by:

$$\alpha_1 y_1 + \alpha_0 y_0 - u_0 = 0$$

The slopes of the surface along the $\alpha_1$ and $\alpha_0$ axis can be found as

$$\frac{\partial|e|}{\partial\alpha_1} = \frac{e}{|e|}y_1$$

$$\frac{\partial|e|}{\partial\alpha_0} = \frac{e}{|e|}y_0$$

The right-hand sides form the components of the gradient vector. If the coordinates of the running point are driven at rates proportional to the negatives of these slopes then the running point moves towards the line L along a path normal to that line. As the values of $y_1$, $y_0$ and $u_0$ change with time, the line L will rotate about the fixed point $(a_1, a_0)$. The running point will converge uniformly to the point $(a_1, a_0)$. The instrumentation in this case is like the one given already in Figure 9.3.

Some discussion of convergence properties can be found in the paper by Miller (1962), where the second method of Lyapunov is used. Define $R^2(t)$ as:

$$R^2(t) = \sum_{i=0}^{n} (\alpha_i - a_i)^2 + \sum_{j=1}^{m} (\beta_j - b_j)^2 \geqslant 0 \tag{9.52}$$

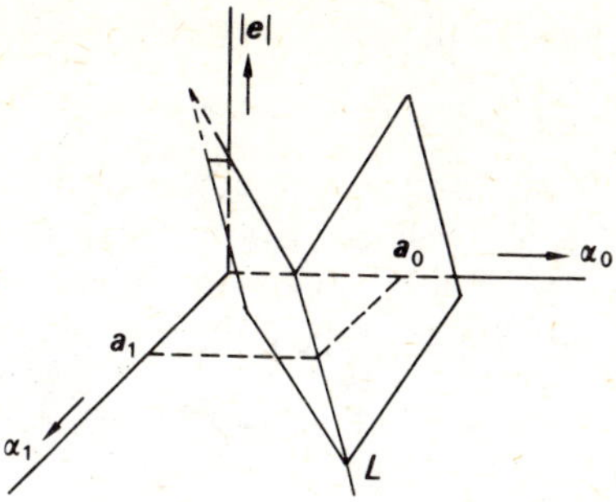


Figure 9.5

Differentiation of equation (9.52) with respect to time yields:

$$\frac{dR^2}{dt} = -2\gamma e\{\sum(\alpha_i - a_i)y_i + \sum(\beta_j - b_j)u_j\} = -2\gamma e^2$$

as a result of equations (9.38a) and (9.36). Thus, $R^2(t)$ will be a Lyapunov function.

### 9.2.3 *Influence of Additive Noise*

There has not yet been a discussion concerning the influence of additive noise. Consider, however, the simple example of implementation given in Figure 9.6. Assume the process is described by

$$\dot{x} + a_0 x = u \qquad (9.53)$$

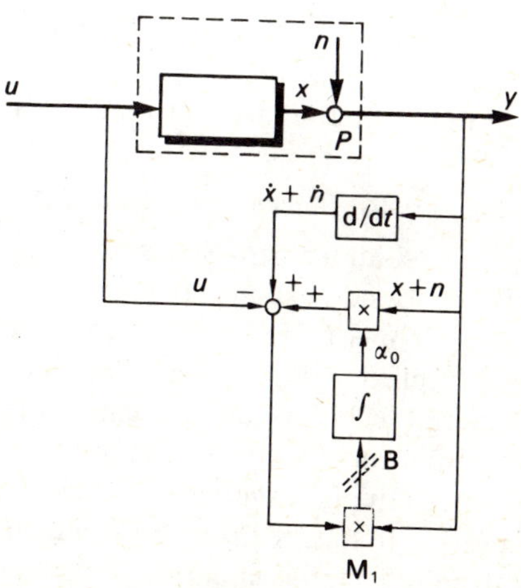


Figure 9.6

Without the additive noise $n$, the generalized error is:

$$e = \dot{x} + \alpha_0 x - u \tag{9.54}$$

Now assume that $\alpha_0 = a_0$ and that the adjustment loop is cut open at $B$. The signal and noise are not related; $\mathscr{E}[u, n] = 0$ and $\mathscr{E}[u, \dot{n}] = 0$. If $n \neq 0$, then the error signal $e$ has two components due to the noise signal, viz.: $\alpha_0 n$ and $\dot{n}$. The (negative) output of the multiplier $M_1$ consists of:

$$(x + n)\underbrace{(u - \dot{x} - a_0 x}_{0} - \dot{n} - a_0 n) = -(x + n)(\dot{n} + a_0 n)$$

Now

$$\mathscr{E}[x.n] = 0 \quad \text{and} \quad \mathscr{E}[x.\dot{n}] = 0$$

due to the assumption that the signal and noise are not related. Further it can be proved that

$$\mathscr{E}[n.\dot{n}] = 0$$

The fourth term, however, gives

$$\alpha_0 \mathscr{E}[n.n] > 0 \quad \text{for } n \not\equiv 0 \tag{9.55}$$

This amounts to a d.c. component at the output of the multiplier. So when the loop is closed at $B$ the system is not in equilibrium. Loosely speaking one would, therefore, expect an offset of the coefficient $\alpha_0$ such that:

$$\alpha_0 \mathscr{E}[n.n] = -(\alpha_0 - a_0)\mathscr{E}[x.x] \tag{9.56}$$

Because $x.x$ is the only combination of the inputs of $M_1$ that can provide an expectation $\neq 0$, this means that a bias of the coefficient will be found which is related to the signal/noise ratio. The input combination of $M_1$ which does not contribute to the expectation of the output does, however, add to the variance of that output and thus to the uncertainty of $\alpha_0$.

### 9.2.4 *Examples of Simulations, Implementations and Applications*

Examples of these model-adjustment schemes where the instantaneous error is minimized can be found in several references.

In pioneering papers, Clymer (1958; 1959) introduces the idea of 'system synthesis by dynamic implicit computation'. This line of thought has its origin in the implicit circuit technique of analogue computation, an example of which is the implementation of a divider circuit by the use of a multiplier and a high-gain amplifier. 'Implicit synthesis is thus a distinct mode of the application of computers. It differs from the usual mode of application, in which one knows the differential equation including all characteristics, and the excitation, while one wishes to determine the solution to the equation, which is the dependent variable as a function of the independent variable. By contrast, in implicit synthesis one knows the solution and the excitation, but one wishes to determine one or more unknown characteristics in the differential equation'.

Some computer results for simple systems are given in those papers, as well as an illustration based on chemical reaction rate equations. Further information on the development of this line of thought can be found in Potts, Ornstein, Clymer (1961), where application of the methods for the determination of aerodynamic coefficients and for the determination of human coefficients are indicated. Also cf. Ornstein (1963), Schoenemann (1965).

## 9.3 Minimization of a Time-averaged Error

### *9.3.1 Basic Equations*

Now the following notations of Section 9.1 are used: the generalized error according to equation (9.2)

$$e = \sum_{i=0}^{n} \alpha_i y_i + \sum_{j=1}^{m} \beta_j u_j - u \tag{9.57}$$

and the error criteria of the type according to equations (9.21)

$$\left.\begin{array}{lll}
\min \displaystyle\int_0^T e^2\,\mathrm{d}t & \mathscr{E}\left[\displaystyle\int_{t-T}^{t} e^2\,\mathrm{d}t\right] & = \langle e, e\rangle \\
\min \displaystyle\int_0^T |e|\,\mathrm{d}t & \mathscr{E}\left[\displaystyle\int_0^T |e|\,\mathrm{d}t\right] & = \langle |e|, 1\rangle \\
\min \displaystyle\int_0^\infty e^2(t-\theta)h(\theta)\,\mathrm{d}\theta & \mathscr{E}\left[\displaystyle\int_0^\infty e^2(t-\theta)h(\theta)\,\mathrm{d}\theta\right] & = \langle e, e\rangle_h \\
\min \displaystyle\int_0^\infty |e(t-\theta)|h(\theta)\,\mathrm{d}\theta & \mathscr{E}\left[\displaystyle\int_0^\infty |e(t-\theta)|h(\theta)\,\mathrm{d}\theta\right] & = \langle |e|, 1\rangle_h
\end{array}\right\} \tag{9.58}$$

with corresponding expressions for $\zeta_i$ and $\eta_j$ as components of the gradient vector $\nabla E$. In the following we neglect the factor 2 that appears when differentiating a quadratic error criterion. The vector equation

$$\nabla E = \mathbf{M}\boldsymbol{\theta} - \boldsymbol{\psi} \tag{9.59}$$

was obtained under the condition that the coefficient vector $\boldsymbol{\theta}' = [\alpha_0, \ldots, \alpha_n, \beta_1, \ldots, \beta_m]$ is constant. As has already been mentioned in Section 9.1, the latter condition implies that, strictly speaking, the policy equation

$$\dot{\boldsymbol{\theta}} = -\boldsymbol{\Gamma}\nabla E \tag{9.60}$$

leading to the differential equation

$$\dot{\boldsymbol{\theta}} + \boldsymbol{\Gamma}\mathbf{M}\boldsymbol{\theta} = \boldsymbol{\Gamma}\boldsymbol{\psi} \tag{9.61}$$

does not hold, because $\boldsymbol{\theta}$ is changing now. For a low loop gain $\boldsymbol{\Gamma}$ and a corresponding low convergence rate, however, equation (9.61) can be used for an approximation of the dynamic behaviour.

*Example.* Take the simple case where the process can be described by:

$$H(s) = \frac{1}{a_1 s + a_0}$$

and where $\boldsymbol{\Gamma} = \gamma\mathbf{I}$. Then the differential equation (9.61) is

$$-\begin{bmatrix} \dot{\alpha}_0 \\ \dot{\alpha}_1 \end{bmatrix} = \gamma \begin{bmatrix} \langle y_0, y_0 \rangle & \langle y_1, y_0 \rangle \\ \langle y_0, y_1 \rangle & \langle y_1, y_1 \rangle \end{bmatrix} \begin{bmatrix} \alpha_0 \\ \alpha_1 \end{bmatrix} - \gamma \begin{bmatrix} \langle u, y_0 \rangle \\ \langle u, y_1 \rangle \end{bmatrix} \tag{9.62}$$

which has the solution:

$$\begin{bmatrix} \alpha_0 \\ \alpha_1 \end{bmatrix} = \begin{bmatrix} \alpha_{00} \\ \alpha_{10} \end{bmatrix} \varepsilon^{-\gamma \mathbf{M} t} + \mathbf{M}^{-1} \begin{bmatrix} \langle u, y_0 \rangle \\ \langle u, y_1 \rangle \end{bmatrix} \tag{9.63}$$

where $\alpha_{00}$ and $\alpha_{10}$ are the offset components at $t = 0$.

Figure 9.7 shows a practical example and some of its results. The object of study is the automatic measurement of impedances by means of model adjustment. This has been indicated already by Figure 1.8. Now the three elements $C$, $R$ and $L$ of a circuit are being measured.* The respective elements are switched on and off in order to show the speed of response and the interaction between the different channels. The input is a triangle wave signal with a fundamental frequency of 500 Hz.

Indicated as well in Section 9.1 is the use of a policy operator of the intermittent type: during the $i$th measuring interval the vector $\nabla E(i)$ is determined:

$$\nabla E(i) = \mathbf{M}\boldsymbol{\theta}(i) - \boldsymbol{\psi}(i) \tag{9.64}$$

In the following adjustment interval the coefficient vector is changed to:

$$\boldsymbol{\theta}(i+1) = \boldsymbol{\theta}(i) - \boldsymbol{\Gamma}(i)\nabla E(i) \tag{9.65}$$

This coefficient vector is used during the $(i + 1)$th measuring interval, etc. Now the restriction of $\boldsymbol{\theta}$ to be constant is fulfilled during the measuring interval. In this type of operation the error criteria of the type

$$\int_{iT}^{(i+1)T} e^2 \, dt \quad \text{and} \quad \int_{iT}^{(i+1)T} |e| \, dt \tag{9.66}$$

where $T$ is the duration of the measuring interval are well suited.

Again reference can be made to Clymer (1959) for an early example of the 'dynamic least-squares' method. An application to a variable-parameter

* These results are taken from a project at the University of Technology, Eindhoven, Netherlands, cf. Van de Ven, et al. (1973).

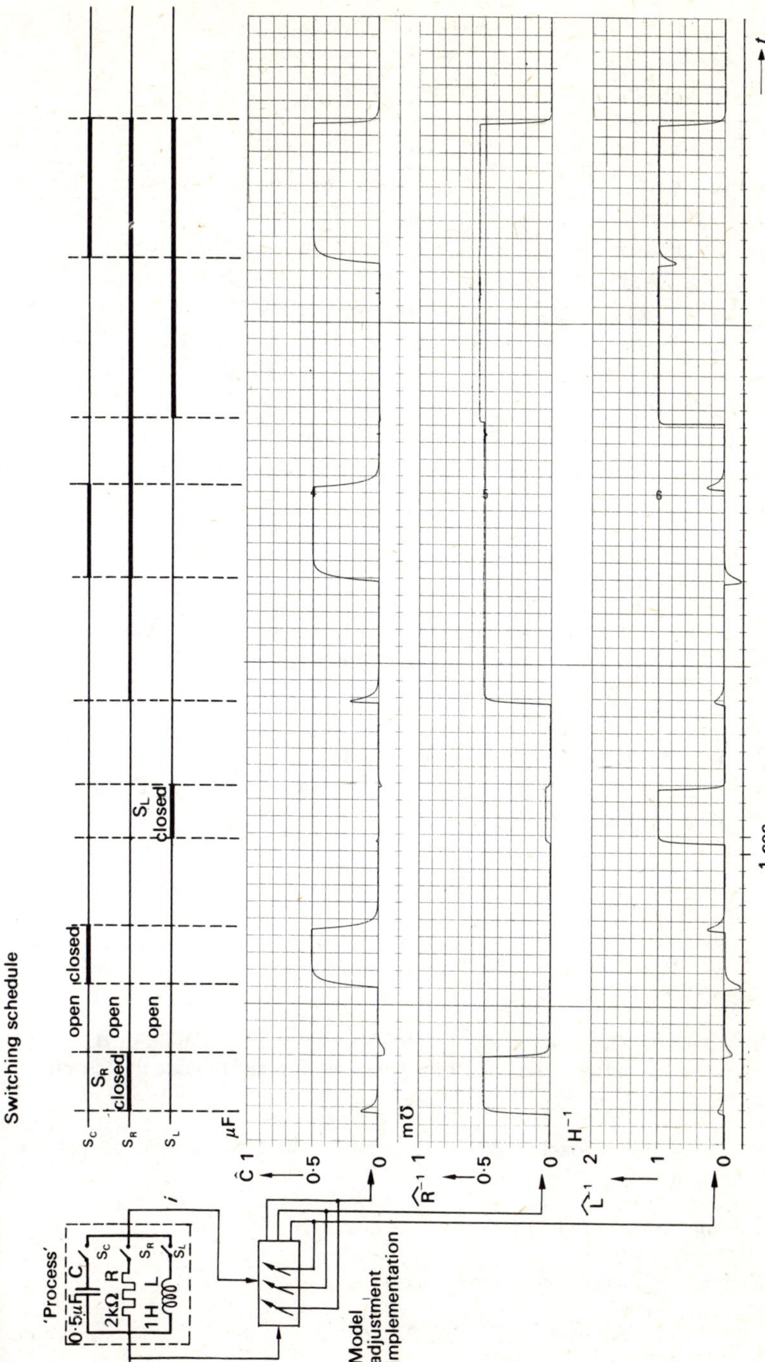

Switching schedule
'Process'
0·5μF
C
2kΩ
R
1H
L
Model adjustment implementation
open
closed
1 sec
t


Figure 9.7

PID controller is reported by Kaya and Yamamura (1962). They indicated that the model-adjustment loop is stable, provided the input signal is not constant over a long interval of time.

### *9.3.2 Convergence Properties*

The requirement that the closed loop is stable is a necessary design criterion. Since the system consisting of the adjustable model, the process and the adjustment mechanism is highly non-linear, the stability problem is not trivial. Using Lyapunov methods, Lion (1966), Shackcloth and Butchart (1965), Parks (1966), Pazdera and Pottinger (1969) have designed stable systems. Lion's results have been generalized to stochastic systems by Kushner (1969). The powerful stability tests developed by Popov (1962) and Zames (1966) have given new tools to design adjustment mechanisms which will result in stable systems. Initial efforts in this direction have been done by Landau (1970) who has proposed stable model reference systems using the Popov criterion.

## 9.4 Parameter-sensitivity Functions

In Section 2.4 the basic principles have been indicated of this computer technique which yields the partial derivatives (parameter-influence coefficients) with respect to the pertinent system parameters. These derivatives can be found simultaneously with the solution of the original differential equation.

The scope of the parameter-sensitivity or -influence technique is wider than that of parameter estimation alone. Meissinger has given the following list of possible applications:

Prediction of the solutions in the neighbourhood of a known solution by a linear extrapolation method.

Determination of parameter tolerances on the basis of linear prediction; detection of critical parameters.

Application to statistical studies: evaluation of the influence of randomly distributed system constants or initial conditions; extrapolation of results obtained with random-noise signal inputs.

Optimization of system parameters in accordance with the specified performance criteria using gradient techniques.

Analysis of solution sensitivity to computer errors.

Determination of system stability boundaries.

Variation of the time occurrence of specified conditions; variation of rise times, settling times.

Solution of boundary value problems in ordinary differential equations.

The discussion will be limited to the use of this technique for process-parameter estimation.

### 9.4.1 *Parameter-influence (Sensitivity) Techniques*

Now the basic principles of the parameter-influence technique will be amplified. The following inhomogeneous linear differential equation in $w$ will be used:

$$\frac{d^2w}{dt^2} + \alpha_1 \frac{dw}{dt} + \alpha_0 w = \beta_1 \frac{du}{dt} + \beta_0 u \tag{9.67}$$

with the initial conditions:

$$w(0) = \gamma_0 \quad \text{and} \quad \dot{w}(0) = \gamma_1$$

The solution $w$ is desired for specific values of the parameters $\alpha_1 = \alpha_{1,0}$, $\alpha_0 = \alpha_{0,0}$, $\beta_1 = \beta_{1,0}$, $\beta_0 = \beta_{0,0}$. If we consider only one parameter at a time we may look at $w$ as a function of two variables, e.g. $w = w(t; \alpha_1)$. From a known solution curve obtained for one parameter value $\alpha_{1,0}$ a neighbouring curve for $\alpha_1 = \alpha_{1,0} + \Delta\alpha_1$ can be determined by extrapolation in terms of $\Delta\alpha_1$

$$w(t; \alpha_1) = w(t; \alpha_{1,0}) + \left.\frac{\partial w}{\partial \alpha_1}\right|_{\alpha_{1,0}} \Delta\alpha_1 + \frac{1}{2!}\left.\frac{\partial^2 w}{\partial \alpha_1^2}\right|_{\alpha_{1,0}} (\Delta\alpha_1)^2 + \cdots \tag{9.68}$$

The number of terms needed in this expansion for a satisfactory approximation of $w(t; \alpha_{1,0} + \Delta\alpha_1)$ depends on the magnitude of $\Delta\alpha_1$ and on the behaviour of the solution $w$ and its partial derivatives with respect to $\alpha_1$ in the region of interest. Only first-order extrapolation will be considered here.

The partial derivative $\partial w/\partial \alpha_1$, itself a function of $t$ and $\alpha_1$, is called a **parameter-influence coefficient** or a **parameter-sensitivity function** of the first order. Other parameter influences pertaining to equation (9.67) are the derivatives

$$\frac{\partial w}{\partial \alpha_0}, \quad \frac{\partial w}{\partial \beta_1}, \quad \frac{\partial w}{\partial \beta_0}, \quad \frac{\partial w}{\partial \gamma_1}, \quad \frac{\partial w}{\partial \gamma_0}$$

The two latter terms represent the sensitivity changes in the initial conditions. Differentiation of equation (9.67) with respect to $\alpha_1$ yields

$$\frac{\partial^3 w}{\partial \alpha_1\, \partial t^2} + \alpha_1 \frac{\partial^2 w}{\partial \alpha_1\, \partial t} + \alpha_0 \frac{\partial w}{\partial \alpha_1} = -\frac{dw}{dt} \tag{9.69}$$

because $u$ and $\dot{u}$ do not depend on $\alpha_1$. Interchanging the order of differentiation and using the notation $v_1 = \partial w/\partial \alpha_1$ one obtains a differential equation in $v_1$:

$$\frac{\partial^2 v_1}{\partial t^2} + \alpha_1 \frac{\partial v_1}{\partial t} + \alpha_0 v_1 = -\frac{\mathrm{d}w}{\mathrm{d}t} \tag{9.70}$$

with the initial conditions

$$v_1(0) = \dot{v}_1(0) = 0$$

since the initial values of $w$ and $\dot{w}$ are constant and do not depend on $\alpha_1$. Equation (9.70) is known as the **sensitivity equation** of the system with respect to parameter $\alpha_1$. For small variations in $\alpha_1$ this equation gives information on the approximate gradient $v_1$. This equation can easily be simulated if the partial derivative is replaced by the total derivative:

$$\frac{\mathrm{d}^2 v_1}{\mathrm{d}t^2} + \alpha_1 \frac{\mathrm{d}v_1}{\mathrm{d}t} + \alpha_0 v_1 = -\frac{\mathrm{d}w}{\mathrm{d}t} \tag{9.71}$$

(**approximate sensitivity equation**). The reason that this is only an approximation is due to the fact that for the relation of the total and the partial derivative the following holds:

$$\frac{\mathrm{d}v_1}{\mathrm{d}t} = \frac{\partial v_1}{\partial t} + \sum_{i=0}^{1} \frac{\partial v_1}{\partial \alpha_i}\frac{\mathrm{d}\alpha_i}{\mathrm{d}t} + \sum_{j=0}^{1} \frac{\partial v_1}{\partial \beta_j}\frac{\mathrm{d}\beta_j}{\mathrm{d}t} \tag{9.72}$$

Consequently, equation (9.71) is a good approximation if the changes of the parameters in time are sufficiently small.

Similarly, one may derive the approximate sensitivity equations with respect to $\alpha_0$, $\beta_1$ and $\beta_0$. For the four parameters under consideration one finds:

$$\ddot{v}_1 + \alpha_1 \dot{v}_1 + \alpha_0 v_1 = -\dot{w} \quad \text{with} \quad v_1 = \frac{\partial w}{\partial \alpha_1} \qquad \dot{v}_1(0) = v_1(0) = 0 \tag{9.73a}$$

$$\ddot{v}_0 + \alpha_1 \dot{v}_0 + \alpha_0 v_0 = -w \quad \text{with} \quad v_0 = \frac{\partial w}{\partial \alpha_0} \qquad \dot{v}_0(0) = v_0(0) = 0 \tag{9.73b}$$

$$\ddot{u}_1 + \alpha_1 \dot{u}_1 + \alpha_0 u_1 = \dot{u} \quad \text{with} \quad u_1 = \frac{\partial w}{\partial \beta_1} \qquad \dot{u}_1(0) = u_1(0) = 0 \tag{9.73c}$$

$$\ddot{u}_0 + \alpha_1 \dot{u}_0 + \alpha_0 u_0 = u \quad \text{with} \quad u_0 = \frac{\partial w}{\partial \beta_0} \qquad \dot{u}_0(0) = u_0(0) = 0 \tag{9.73d}$$

The equations (9.73a–d) can be simulated using a separate sensitivity model for each sensitivity equation as shown in the block diagram Figure 9.8. It happens that in this linear case all approximate sensitivity equations are identical, except for the right-hand side. This implies that the parameter sensitivity functions can be calculated in successive operations using the

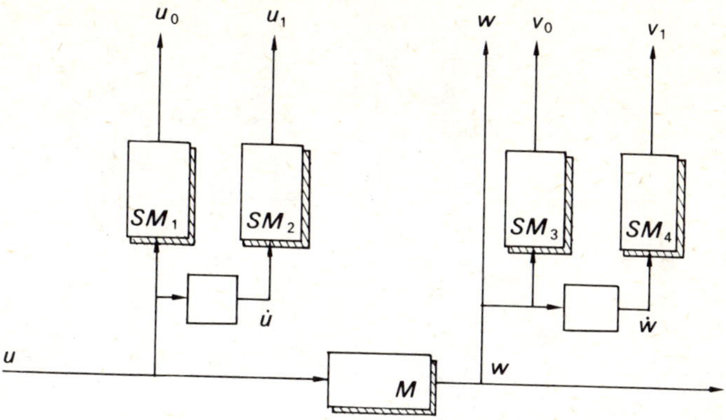


Figure 9.8

same model and the appropriate 'coupling term' $-\dot{w}$, $-w$, $\dot{u}$ or $u$. A further simplification can be made by recognizing:
from equations (9.73a and b):

$$v_1 = \dot{v}_0 \tag{9.74}$$

from equation (9.73c and d):

$$u_1 = \dot{u}_0 \tag{9.75}$$

and from the comparison of equation (9.67) with equations (9.73c and d):

$$w = \beta_1 u_1 + \beta_0 u_0 \tag{9.76}$$

Consequently, it suffices to instrument the equation (9.73b) and to use the relationships (9.74), (9.75) and (9.76) to obtain the four parameter-sensitivity functions simultaneously (cf. Figures 9.9a and b). This instrumentation is much less expensive than that according to Figure 9.8.

If the initial conditions $\gamma_1$ and $\gamma_0$ are the parameters of interest it can be seen that no coupling terms appear in the sensitivity equations. In the case of $\partial w/\partial \gamma_0 = v$ one obtains the homogeneous differential equation:

$$\frac{\partial^2 v}{\partial t^2} + \alpha_1 \frac{\partial v}{\partial t} + \alpha_0 = 0 \tag{9.77}$$

with initial conditions

$$v(0) = 1 \quad \text{and} \quad \dot{v}(0) = 0$$

This can be solved simply by operating the original computer set-up once more, but with the driving function $u$ identically zero, and with appropriately modified initial values.

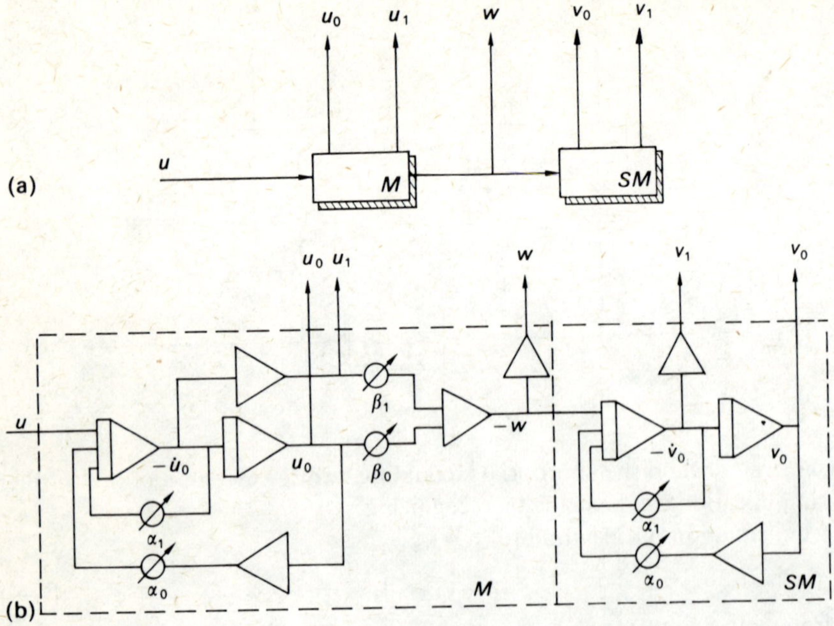


Figure 9.9

The application of the parameter-influence method is not restricted to linear systems. As an example of a non-linear system consider equation (9.68):

$$\ddot{w}_1 + \alpha_1\dot{w} + \alpha_0 w + \mu w^3 = u \tag{9.78}$$

The sensitivity equations are the following:

$$\ddot{v}_1 + \alpha_1\dot{v}_1 + \alpha_0 v_1 + 3\mu w^2 v_1 = -\dot{w} \quad \text{with} \quad v_1 = \frac{\partial w}{\partial \alpha_1} \tag{9.79a}$$

$$\ddot{v}_0 + \alpha_1\dot{v}_0 + \alpha_0 v_0 + 3\mu w^2 v_0 = -w \quad \text{with} \quad v_0 = \frac{\partial w}{\partial \alpha_0} \tag{9.79b}$$

$$\ddot{v}_\mu + \alpha_1\dot{v}_\mu + \alpha_0 v_\mu + 3\mu w^2 v_\mu = -w^3 \quad \text{with} \quad v_\mu = \frac{\partial w}{\partial \mu} \tag{9.79c}$$

Again the equations are the same except for the 'coupling term'. Consequently, the same model can be used with the driving functions $-\dot{w}$, $-w$ and $-w^3$ in succession.

The problem can be generalized to an $n$th-order system of differential equations with $m$ parameters $\alpha_1, \alpha_2, \ldots, \alpha_m$.

$$\frac{\mathrm{d}w_i}{\mathrm{d}t} = f_i(w_1, w_2, \ldots, w_n, t; \alpha_1, \alpha_2, \ldots, \alpha_m) \qquad i = 1, \ldots, n \tag{9.80}$$

The sensitivity equation with respect to $\alpha_k$ yielding the derivative $u_{ik} = \partial w_i/\partial\alpha_k$ may be formulated as follows:

$$\frac{\partial}{\partial\alpha_k}\left(\frac{dw_i}{dt}\right) = \frac{\partial u_{ik}}{\partial t} = \frac{\partial f_i}{\partial w_1}u_{1k} + \cdots + \frac{\partial f_i}{\partial w_n}u_{nk} + \frac{\partial f_i}{\partial\alpha_k} \tag{9.81}$$

The initial conditions are $u_{ik}(0) = 0$, provided none of the initial values of the original differential equation are considered as parameters. This formulation applies to linear as well as to non-linear systems. In order to study the influences of one parameter $\alpha_k$ it is necessary to programme a complete set of sensitivity equations according to equation (9.81) even though this parameter may appear explicitly in only one equation of the original system given by equation (9.80). If, for example, $\alpha_k$ occurs in the term $f_j$ only, then the $j$th sensitivity equation exhibits the coupling term $\partial f_j/\partial\alpha_k$ whereas $\partial f_i/\partial\alpha_k = 0$ for $i \neq j$. The remaining sensitivity equations nevertheless contain the influence of $\alpha_k$ implicitly in terms of the variables $u_{1k}, u_{2k}, \ldots,$ and they establish intercoupling with the $j$th equation.

Still another type of application may be found in studies of the effect of missing higher-order terms of a differential equation. Assume that the equation to be studied is

$$\ddot{w} + \alpha_1\dot{w} + \alpha_0 w = u \tag{9.82}$$

We wish to investigate the effect of a third-order term

$$\alpha_3\dddot{w} + \ddot{w} + \alpha_1\dot{w} + \alpha_0 w = u \tag{9.83}$$

The sensitivity equations with respect to $\alpha_3$ and $\alpha_1$ are:

$$\alpha_3\dddot{v}_3 + \ddot{v}_3 + \alpha_1\dot{v}_3 + \alpha_0 v_3 = -\dddot{w} \quad \text{with} \quad v_3 = \frac{\partial w}{\partial\alpha_3} \tag{9.84a}$$

$$\alpha_3\dddot{v}_1 + \ddot{v}_1 + \alpha_1\dot{v}_1 + \alpha_0 v_1 = -\dot{w} \quad \text{with} \quad v_1 = \frac{\partial w}{\partial\alpha_1} \tag{9.84b}$$

Consequently $v_3 = \ddot{v}_1$ and in the vicinity of $\alpha_3 = 0$ this parameter-influence coefficient can be obtained from a sensitivity model.

So far in this section we considered the (absolute) parameter-sensitivity function, e.g.: $v_1 = \partial w/\partial\alpha_1$. There may be an advantage in using the **relative** parameter sensitivity functions, e.g.

$$\frac{\partial w}{\partial \ln \alpha_1} = \alpha_1 \frac{\partial w}{\partial\alpha_1}$$

### *9.4.2 Sensitivity Points Method*

In the previous section, it turned out that for the determination of a number of sensitivity functions simultaneously, in addition to the (process) model, more than one sensitivity model is needed. This implies a complication of the analogue computer programming or the amount of digital computing time needed for handling such types of problems.

On the other hand, we learned in Section 9.1 that when using the generalized model no additional sensitivity model is needed as the sensitivity functions are measurable directly. This was due to the linearity-in-the-parameters of the generalized model.

In view of the desire to limit the (analogue) computer complications or the amount of (digital) computer time as much as possible it is worth while studying the types of models that can provide us with as many of the desired sensitivity functions as possible. This is done in literature under the name **sensitivity points method** (Kokotović, 1964).

The principle may be illustrated as follows. We consider a (linear) process with a transfer function $H(s, a_1, \ldots, a_m)$, depending on the parameters $a_i$, The input signal has the transform $U(s)$ and the process output is given by

$$Y(s, a_1, \ldots, a_m) = H(s, a_1, \ldots, a_m)U(s) \tag{9.85}$$

The output of the corresponding model is

$$W(a, \alpha_1, \ldots, \alpha_m) = H(s, \alpha_1, \ldots, \alpha_m)U(s) \tag{9.86}$$

By invoking the differentiability of the $\mathscr{L}$-transforms with respect to the parameters one obtains

(absolute) parameter-sensitivity functions $$\frac{\partial W(s, \alpha_1, \ldots, \alpha_m)}{\partial \alpha_i} = V_i(s) \tag{9.87}$$

relative parameter-sensitivity functions $$\frac{\partial W(s, \alpha_1, \ldots, \alpha_m)}{\partial \ln \alpha_i} = \alpha_i V_i(s) \tag{9.88}$$

The following example may illustrate the principles (cf. Figures 9.10a and b). For the model, the following holds:

$$W(s) = H(s, \alpha_1, \alpha_2)U(s) = \frac{\alpha_1 G_1(s)}{1 + \alpha_1 G_1(s)\alpha_2 G_2(s)} U(s) \tag{9.89}$$

For the relative sensitivity functions follow:

$$\begin{aligned} \frac{\partial W(s)}{\partial \ln \alpha_1} &= \alpha_1 \frac{\partial W(s)}{\partial \alpha_1} = \alpha_1 \frac{G_1(s)}{\{1 + \alpha_1 G_1(s)\alpha_2 G_2(s)\}^2} U(s) \\ &= \frac{1}{1 + \alpha_1 G_1(s)\alpha_2 G_2(s)} W(s) \end{aligned} \tag{9.90a}$$

$$\frac{\partial W(s)}{\partial \ln \alpha_2} = \alpha_2 \frac{\partial W(s)}{\partial \alpha_2} = \cdots = \frac{\alpha_1 G_1(s)\alpha_2 G_2(s)}{1 + \alpha_1 G_1(s)\alpha_2 G_2(s)} W(s) \tag{9.90b}$$

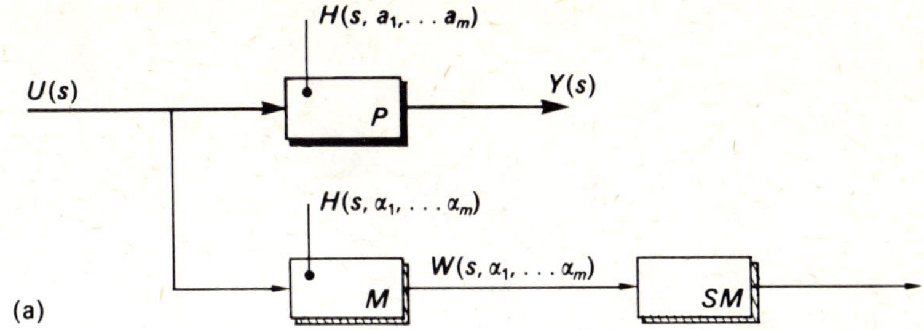


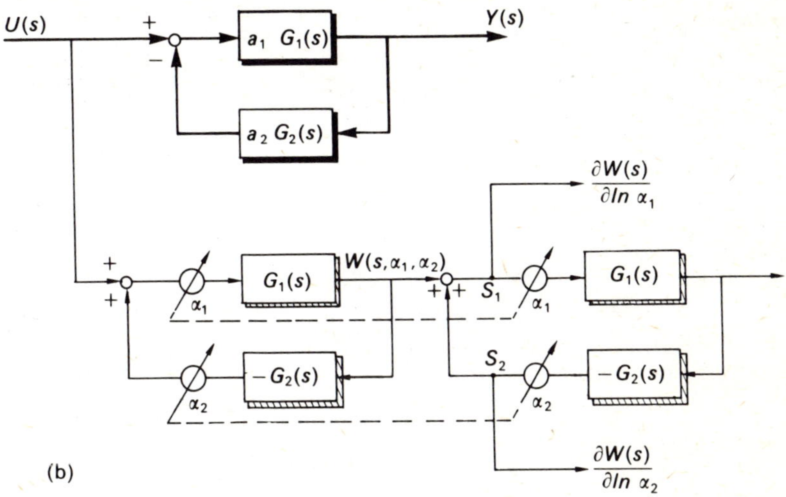


Figure 9.10

This leads to the structure given in Figure 9.10b. $S_1$ and $S_2$ are called the **sensitivity points.** In an analogue simulation both sensitivity functions can be measured simultaneously, in a digital programme both time functions can be derived from the same computation loop.

This idea can be generalized to multiloop feedback systems according to Figure 9.11. Here it is assumed that there are $2m$ elementary blocks and that each block has only one parameter for which the sensitivity function is of interest. Along the same lines as before it is not difficult to show that $S_j$ is a sensitivity point for the parameter in the block $G_j$. The only aspect that remains to be considered is the way in which parameter $\alpha_j$ is imbedded in the transfer function $G_j(s, \alpha_j)$. This is taken care of by an additional transfer function $T_j$, defined by

$$T_j(s) = \frac{\partial \ln G_j(s, \alpha_j)}{\partial \ln \alpha_j} \tag{9.91}$$

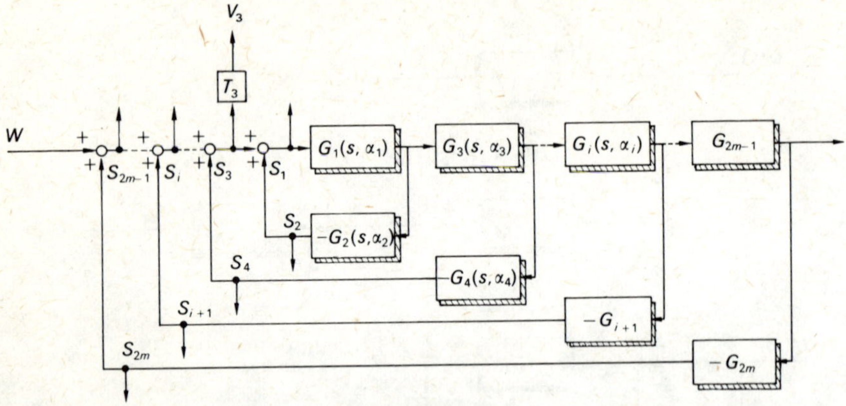


Figure 9.11

This is the logarithmic transfer-sensitivity function that was already defined by Bode (1945). The input of $T_j(s)$ is the signal from the sensitivity point $S_j$, its output is

$$V_j(s) = \frac{\partial W(s, \ldots, \alpha_j, \ldots)}{\partial \ln \alpha_j} \tag{9.92}$$

Some special cases of $T_j$:

If $G_j(s, \alpha_j) = \alpha_j F(s)$, then

$$T_j(s) = \frac{\alpha_j}{\alpha_j F(s)} \frac{\partial \alpha_j F(s)}{\partial \alpha_j} = 1 \tag{9.93}$$

In this case the signal from $S_j$ is the sensitivity function and there is no need to add more elements to the sensitivity model (cf. Figures 9.9b and 9.10b).

If $G_j(s, \alpha_j) = D(s, \alpha_j)F(s)$, i.e. the transfer function is the product of two transfer functions, but only one of which contains the parameter of interest:

$$T_j(s) = \frac{\alpha_j}{D(s, \alpha_j)} \frac{\partial D(s, \alpha_j)}{\partial \alpha_j} \tag{9.94}$$

i.e. just the transfer-sensitivity function of the part that contains $\alpha_j$.

These ideas can be generalized as well to sensitivity functions of higher order, e.g.

$$V_{12}(s) = \frac{\partial^2 W(s, \alpha_1, \alpha_2)}{\partial \ln \alpha_1 \, \partial \ln \alpha_2} \quad \text{etc.} \tag{9.95}$$

which can be derived in a similar way as the first-order sensitivity functions. It turns out that for this case still one more sensitivity model is needed.

Of course, the sensitivity analysis has been extended as well to processes in state–space description. A literature survey can be found in Kokotović and Rutman (1965). Many interesting papers are assembled in two symposium proceedings (IFAC Sensitivity, 1964 and 1968).

### 9.4.3 *Continuous Model-adjustment*

The scheme that will be studied is given in Figure 9.12. The error is defined as

$$e = y - w$$

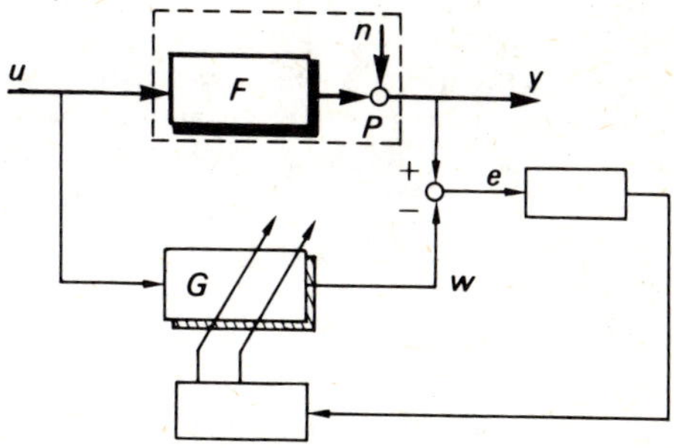


Figure 9.12

where one, using a functional notation $G$, can write

$$w = G[u; \boldsymbol{\alpha}] \tag{9.96}$$

The error criterion is to minimize $E$, which can be written as a functional $L$ of an even function $q(e)$:

$$E = L[q(e)] \tag{9.97}$$

The model $G$ is adjusted by changing the parameters $\boldsymbol{\alpha}' = (\alpha_1, \ldots, \alpha_m)$ according to the gradient of $E$

$$\dot{\boldsymbol{\alpha}} = -\gamma \boldsymbol{\nabla} E \tag{9.98}$$

The components of the gradient vector can be determined by a differentiation:

$$\frac{\partial E}{\partial \alpha_j} = L\left[\frac{\partial q(e)}{\partial e} \frac{\partial e}{\partial \alpha_j}\right] \tag{9.99}$$

where

$$\frac{\partial e}{\partial \alpha_j} = -\frac{\partial w}{\partial \alpha_j} = -\frac{\partial G[u; \boldsymbol{\alpha}]}{\partial \alpha_j} \tag{9.100}$$

$\partial w/\partial \alpha_j$ is the parameter-influence coefficient. Now an operator can be defined so that:

$$H_j[u;\boldsymbol{\alpha}] = \frac{\partial G[u;\boldsymbol{\alpha}]}{\partial \alpha_j} \tag{9.101}$$

which leads to:

$$\nabla E = -L\left[\frac{\partial q(e)}{\partial e}\mathbf{H}[u;\boldsymbol{\alpha}]\right] \tag{9.102}$$

The set of operators **H** with parameters $\boldsymbol{\alpha}$ and operating on $u$ generates the parameter-sensitivity functions as indicated in the previous section.

*Example.* Instead of developing this formalism the reader may wish to focus his attention on a simple case. This can be done by referring to Margolis and Leondes (1959). There the process and the model are described respectively by the equations

$$\frac{dy}{dt} + ay = u$$

$$\frac{dw}{dt} + \alpha w = u$$

The sensitivity equation follows by partial differentiation from the model equation:

$$\frac{\partial v}{\partial t} + \alpha v = -w$$

with $v = \partial w/\partial \alpha$ and $\alpha$ considered as a constant.
Using an error criterion the minimization of

$$E = e^2 = (y - w)^2$$

and the steepest descent as the adjustment criterion then

$$\frac{d\alpha}{dt} = -\gamma\frac{\partial E}{\partial \alpha} = -2\gamma v(y - w)$$

because only $w$ depends on $\alpha$. The behaviour of the model-adjustment scheme is given by the equations given. Because of the restriction that requires $\alpha$ to be constant these equations only approximately describe the behaviour if $\alpha$ is varying slowly.

In the paper cited, the convergence properties are studied for the cases where $u$ is a step input and where $u$ is a sinusoidal input. For the first case, stability of the equilibrium point $a = \alpha$ can be proved. The second case results in a Mathieu equation which may lead to stable, to periodic and to unstable solutions.

The second method of Lyapunov has been applied for stability studies; cf. Johnson and Rang (1961) and the references cited in the previous section. Note that the parameter-sensitivity function plays the role of the **instrumental variable,** as discussed in Chapters 6 and 7 for the sampled data case.

### 9.4.4 *Examples of Simulations, Implementations and Applications*

Although the work of Donalson and Leondes (1963) is not strictly related to parameter estimation it can be cited as another example of the use of parameter-influence coefficients. The system studied is shown in Figure 9.13.

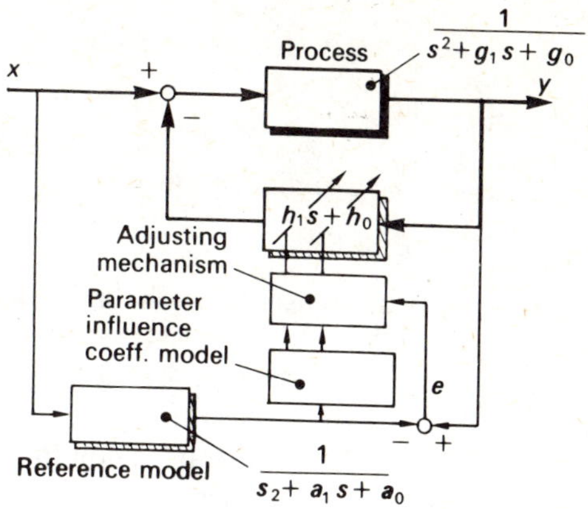


Figure 9.13

The parameters $g_1$ and $g_0$ of the process (e.g. the pitch axis angular rate response of an air-frame to control-surface deflexion) are changing. These changes have to be compensated for by an adjustment of the parameters $h_1$ and $h_0$ in the feedback loop. The desired values of the process feedback loop are set by the reference model which is a fixed analogue structure. The goal of the adjustment is to minimize an even functional of the error $e$. This implies:

$$g_1 + h_1 = \alpha_1 \rightarrow a_1$$

$$g_0 + h_0 = \alpha_0 \rightarrow a_0$$

This is done by generating the parameter-influence coefficients $\partial e/\partial a_1$ and $\partial e/\partial a_0$ of the reference model instead of those of the process + feedback loop. As $a_1$ and $a_0$ are fixed values, this has the advantage that the parameter-influence coefficients generated are exactly the partial derivatives that are desired. (This is not the case in the model-adjustment scheme mentioned previously.)

### 9.4.5 *Intermittent Model-adjustment*

In Section 9.2 it has been indicated that it is difficult to establish the convergence properties for continuous adjustment schemes. This is caused chiefly because it is difficult to determine the gradient while the model parameters are being changed (adjusted). Now we shall consider the schemes in which the model parameters are kept constant while the gradient is determined. After this measurement interval the model parameters are adjusted; a new measuring period starts, etc.

If two or more parameters have to be adjusted, for such an intermittent scheme one can distinguish between two cases:

all (or a number of) parameters are adjusted at the same time;

only one parameter at a time is adjusted.

Assume the error criterion to be a minimization of

$$E = E(\alpha_1, \ldots, \alpha_n)$$

where $\alpha_j$ are the parameters to be adjusted. For the simultaneous adjustment of a number of parameters, the partial derivatives with respect to these parameters have to be determined. The choice of the path towards the minimum has been studied in Chapter 5. In the following, the adjustment of one parameter at a time is considered. It draws upon the work of Brunner (1961), who describes a steepest-descent iterative approach for finding the model adjustment. The error criterion is to minimize

$$E = \sum_{k=1}^{K} g_k e_k^2$$

where $g_k$ represents weighting factors for the error samples. Differentiation with respect to $\alpha_j$ gives:

$$\frac{\partial E}{\partial \alpha_j} = 2 \sum_{k=1}^{K} g_k e_k \frac{\partial e_k}{\partial \alpha_j} \qquad j = 1, 2, \ldots, n \tag{9.103}$$

From here we want to derive an iteration procedure for the adjustment of parameter $\alpha_j$. As a single-step procedure is being studied this is the only parameter under consideration. The error can be written in vector form. Let

$$\mathbf{e}(\alpha_j) = [e_1(\alpha_j), \ldots, e_k(\alpha_j)]'$$

denote the error vector before $\alpha_j$ is updated and

$$\mathbf{e}(\alpha_j + \Delta\alpha_j)$$

is the error vector after $\alpha_j$ has been updated. To obtain a 'steepest-descent' path one has to try to make the updated vector $\mathbf{e}(\alpha_j + \Delta\alpha_j)$ orthogonal under the metric $g_k$ to the derivative of the $\mathbf{e}(\alpha_j)$ vector with respect to the parameter $\alpha_j$. This condition of orthogonality can be expressed as the following scalar product:

$$\sum_{k=1}^{K} g_k \frac{\partial e_k}{\partial \alpha_j} e_k(\alpha_j + \Delta\alpha_j) = 0 \tag{9.104}$$

which is the same as requiring the partial derivative $\partial E/\partial \alpha_j$ to be zero. If the components of the updated vector are now expanded in a Taylor series and if the first-order terms are preserved then equation (9.104) becomes:

$$\sum_{k=1}^{K} g_k \frac{\partial e_k}{\partial \alpha_j} \left\{ e_k(\alpha_j) + \frac{\partial e_k}{\partial \alpha_j} \Delta \alpha_j \right\} = 0 \tag{9.105}$$

Solving for $\Delta \alpha_j$ leads to:

$$\Delta \alpha_j = - \frac{\sum_{k=1}^{K} g_k e_k \frac{\partial e_k}{\partial \alpha_j}}{\sum_{k=1}^{K} g_k \left( \frac{\partial e_k}{\partial \alpha_j} \right)^2} \tag{9.106}$$

In the case where the $e_k$ are linearly dependent on the parameter $\alpha_j$ equations (9.104) and (9.105) are exactly equivalent. This is the case where a generalized model is introduced (cf. Chapters 2 and 7).

Next we will consider the adjustment of all parameters simultaneously after each measuring period. Let the process and model be given by

$$\ddot{y} + a_1 y + a_0 y = u \tag{9.107}$$

$$\ddot{w} + \alpha_1 \dot{w} + \alpha_0 w = u \tag{9.108}$$

The error criterion is given by

$$E(\alpha_1, \alpha_0) = \int_0^T e^2 \, dt = \int_0^T \{w(t; \alpha_1, \alpha_0) - y(t)\}^2 \, dt \tag{9.109}$$

To minimize this function, a gradient technique is used where the values of $\alpha_0$ and $\alpha_1$ are varied proportionally to the respective components of the gradient vector $\nabla E$. This technique assures a steepest descent to the minimum value of $E$ along a path perpendicular to the contours $E(\alpha_0, \alpha_1) =$ constant. In order to compute the gradient components

$$\frac{\partial E}{\partial \alpha_0} = 2 \int_0^T e \frac{\partial w}{\partial \alpha_0} \, dt \tag{9.110}$$

$$\frac{\partial E}{\partial \alpha_1} = 2 \int_0^T e \frac{\partial w}{\partial \alpha_1} \, dt \tag{9.111}$$

the parameter influences $v_0 = \partial w/\partial \alpha_0$ and $v_1 = \partial w/\partial \alpha_1$ are required. They are obtained as continuous functions of time together with the solution $w(t)$ (cf. Figure 9.14). The computer evaluates the integrals (9.110) and (9.111) simultaneously. The sequence of operations consists in a systematic step-by-step variation of parameter values $\alpha_0$ and $\alpha_1$ in accordance with the gradient direction established at the end of each computing cycle. The process

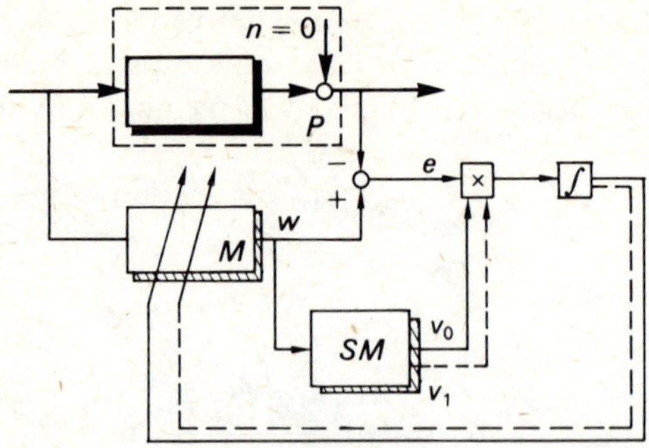


Figure 9.14

converges to the desired minimum of $E$. Figure 9.15 shows the typical contours and gradient lines obtained in the solution of the estimation problem.

A somewhat more refined technique uses quadratic prediction of the optimum. Assume for the moment that there are no noises, computation errors, etc. In that case the error $e$ is only due to

$$\begin{aligned} \Delta\alpha_1 &= a_1 - \alpha_1 = \lambda_1 \\ \Delta\alpha_0 &= a_0 - \alpha_0 = \lambda_0 \end{aligned} \tag{9.112}$$

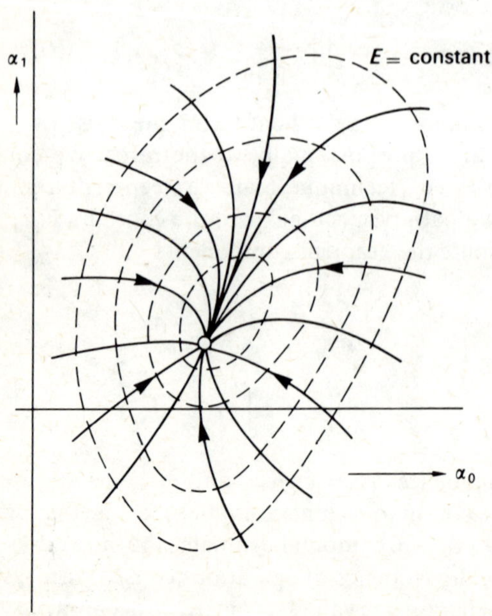


Figure 9.15

A development of $e$ in the vicinity of $(a_0, a_1)$ gives:

$$e(a_0, a_1) \approx e(\alpha_0, \alpha_1) + \frac{\partial e}{\partial \alpha_0}\lambda_0 + \frac{\partial e}{\partial \alpha_1}\lambda_1 + \cdots \tag{9.113}$$

Again the parameter-influence coefficients can be determined from the sensitivity equations in $v_0 = \partial e/\partial \alpha_0$ and $v_1 = \partial e/\partial \alpha_1$. As $e(a_0, a_1) = 0$ one finds:

$$e(\alpha_0, \alpha_1) + v_0\lambda_0 + v_1\lambda_1 \approx 0 \tag{9.114}$$

Because the assumption about the noises and computation errors is not completely correct in practical cases, the residual error $e_{\text{res}}$ will be introduced:

$$e(\alpha_0, \alpha_1) + v_0\lambda_0 + v_1\lambda_1 = e_{\text{res}} \tag{9.115}$$

Now the error criterion is

$$\min_{\alpha_0, \alpha_1} \int e^2 \, dt \tag{9.116}$$

Using the steepest-descent equation:

$$\begin{aligned} \Delta\alpha_0 &= -2\gamma \int_0^T v_0 e_{\text{res}} \, dt \\ \Delta\alpha_1 &= -2\gamma \int_0^T v_1 e_{\text{res}} \, dt \end{aligned} \tag{9.117}$$

## 9.5 Simultaneous Use of Two Models or Repetitive Use of One Model

By using two models with different parameter values it is possible to derive information on the direction of optimum adjustment. In this way, one can obtain a (rough) approximation of the derivative $\partial E/\partial \alpha$ by $\Delta E/\Delta \alpha$.

Two slightly different methods may be used. The relationship between the two methods is given by the following equation, which is the partial derivative of the error criterion with respect to the parameter to be adjusted:

$$\frac{\partial E}{\partial \alpha_i} = \frac{2}{T} \int_0^T e(\alpha_1, \alpha_2, \ldots, t) \frac{\partial e}{\partial \alpha_i} \, dt \tag{9.118}$$

In the first method an implementation is used to obtain $\Delta E/\Delta \alpha_i$ as an approximation of $\partial E/\partial \alpha_i$. In the second method $\Delta w/\Delta \alpha_i$ is used as an approximation of $\partial w/\partial \alpha_i = -\partial e/\partial \alpha_i$ if $e = y - w(\boldsymbol{\alpha})$.

The difference between these two methods is indicated in Figures 9.16a and b. If the input signal $u$ is stationary then the role of the two models in Figure 9.16a can fulfilled by one model by determining $E$, storing of this value, changing the parameter $\alpha_i$ and determining $E_2$, comparing this with $E_1$, etc. This type of approach is described by Balchen and Høsøien (1966).

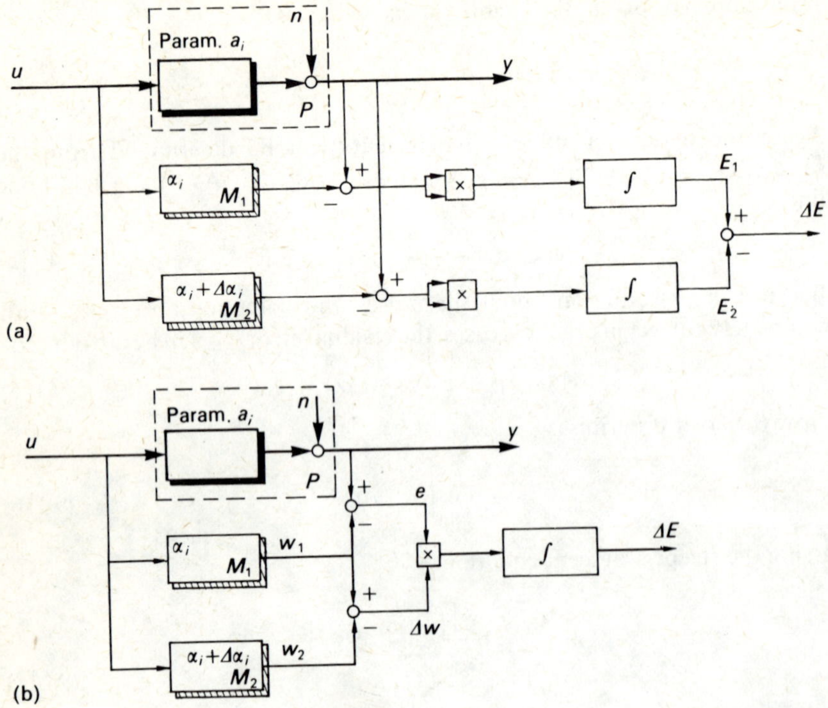


Figure 9.16

## 9.6 Model with Parameter Perturbation

The parameter perturbation method for obtaining partial derivatives is based upon modulation of the parameters of the model and upon detection of the respective responses in the error criterion.

Figure 9.17 indicates three different model conditions. In case (a) the parameter value $\alpha_1$ is not optimal. When this parameter is varied according to a sinusoidal modulation signal about a fixed value, a variation of the squared error is obtained if an instantaneous criterion is used. In the noiseless case this response is also sinusoidal and of a certain phase. By multiplying the response with a signal of the same frequency and phase and averaging, an approximation for the derivative $\partial E/\partial\alpha_1$ is obtained. The same holds for case (c), however with the phase of the response inverted. In case (b), where $\alpha_0$ is supposed to be optimal, the response in the squared error contains only higher-order harmonics of the modulation signal. By multiplying with the modulation frequency and averaging the result will be zero.

A possible adjustment system for one parameter is given in Figure 9.18. The process and the model have the same input signal. From the output signal of the process, including the additive noise, and the output signal of

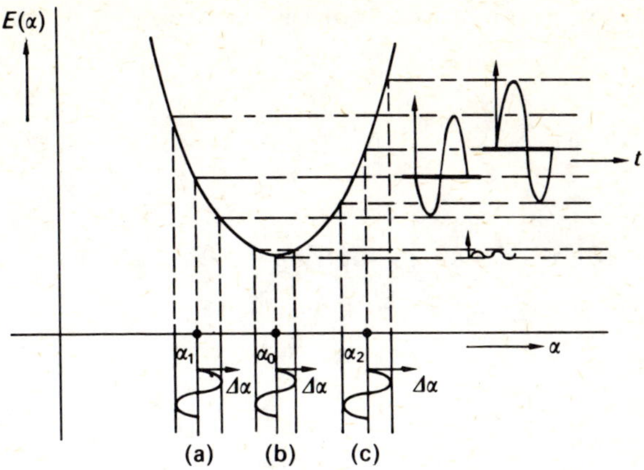


Figure 9.17

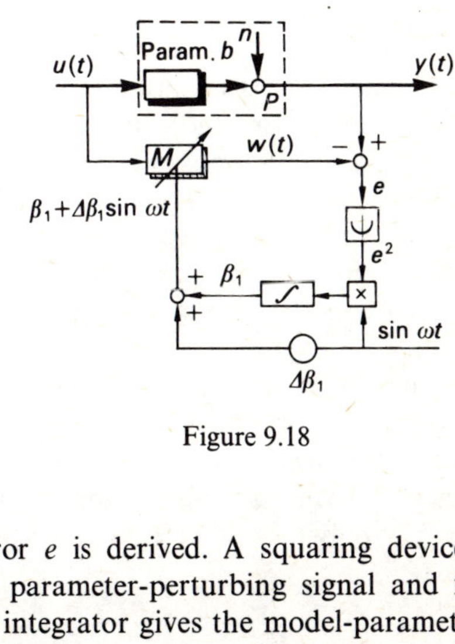


Figure 9.18

the model the error $e$ is derived. A squaring device forms $e^2$, which is multiplied by the parameter-perturbing signal and fed to an integrator. The output of the integrator gives the model-parameter value. This can be indicated by the following equations. Neglecting the noise we suppose our process equation to be:

$$a_2\ddot{y} + a_1\dot{y} + y - b_1\dot{u} - b_0u = 0 \tag{9.119}$$

Our model equation must then be of the form:

$$\alpha_2\ddot{w} + \alpha_1\dot{w} + w - \beta'_1\dot{u} - \beta_0u = 0 \tag{9.120}$$

Here $\beta'_1$ is the modulated parameter,

$$\beta'_1 = \beta_1 + \Delta\beta_1 \sin \omega t \tag{9.121}$$

By subtracting process and model equations we can form an error equation:

$$e = y - w$$

$$e = -(a_2\ddot{y} - \alpha_2\ddot{w}) - (a_1\dot{y} - \alpha_1\dot{w}) + (b_1 - \beta_1 - \Delta\beta_1 \sin\omega t)\dot{u} + (b_0 - \beta_0)u \tag{9.122}$$

After squaring this equation, multiplying it with $\sin\omega t$ and then averaging we arrive at:

$$\overline{e^2 \sin\omega t} = -\Delta\beta_1(b_1 - \beta_1)\overline{\dot{u}^2} + \Delta\beta_1\overline{(a_2\ddot{y} - \alpha_2\ddot{w})\dot{u}} + \Delta\beta_1\overline{(a_1\dot{y} - \alpha_1\dot{w})\dot{u}} \tag{9.123}$$

because $\overline{u\dot{u}} = 0$. Thus, we get an approximate gradient

$$\frac{\partial E}{\partial \beta_1} \approx \frac{\overline{e^2 \sin\omega t}}{\Delta\beta_1} = -(b_1 - \beta_1)\overline{\dot{u}^2} + \overline{(a_2\ddot{y} - \alpha_2\ddot{w})\dot{u}} + \overline{(a_1\dot{y} - \alpha_1\dot{w})\dot{u}} \tag{9.124}$$

The first right-hand member of this equation determines the sign of the gradient. The other right-hand terms indicate interaction with $\alpha_1$ and $\alpha_2$. For $\alpha_2 = a_2$ and $\alpha_1 = a_1$ these two terms will be very small and approach zero for $(\beta_1, \beta_0)$ approaching $(b_1, b_0)$. Consequently, we will get an optimum match for $(\alpha_2, \alpha_1, \beta_1, \beta_0) = (a_2, a_1, b_1, b_0)$. In an engineering sense, a crucial aspect of the instrumentation is the length of the averaging interval. Too long an interval may lead to a sluggish system, too short an interval may result in a 'noisy' system.

This discussion does not show that this technique can be used for multi-parameter adjustments. Realizing that the multiplicative method is a frequency selective one, it should easily be understood that by using different modulation frequencies a corresponding number of parameters can be adjusted simultaneously. Applying the method of steepest descent, we get the following equations:

$$\begin{aligned} \dot{\beta}_1 &= -\gamma\frac{\partial E}{\partial \beta_1} & \dot{\alpha}_2 &= -\gamma\frac{\partial E}{\partial \alpha_2} \\ \dot{\beta}_0 &= -\gamma\frac{\partial E}{\partial \beta_0} & \dot{\alpha}_1 &= -\gamma\frac{\partial E}{\partial \alpha_1} \end{aligned} \tag{9.125}$$

This results in four adjustment loops which will exhibit interaction. Depending upon the loop gain $\gamma$, we must consequently expect poor stability or instability. In developing the stability conditions for this type of problem we will arrive at the well-known Mathieu equation.

## 9.7 Concluding Remarks

Looking back, a central theme of this chapter can be summarized as follows. In model-adjustment techniques there is a definite need for determining partial derivatives with respect to the parameters to be adjusted. Some methods have been indicated; the need for additional implementation (sensitivity model) or extra observation time (model with parameter value $\beta_j$ viz. $\beta_j + \Delta\beta_j$) is apparent.

In this respect the generalized model is very attractive because, due to the linearity-in-the-parameters, the sensitivity functions are available without any need for additional implementation. The price that has to be paid for this elegant and inexpensive approach is the bias in the parameter estimates due to any additive noise.

When using a sensitivity model this problem does not arise. This is illustrated by Figures 9.19 and 9.20, where the influence of the noise is

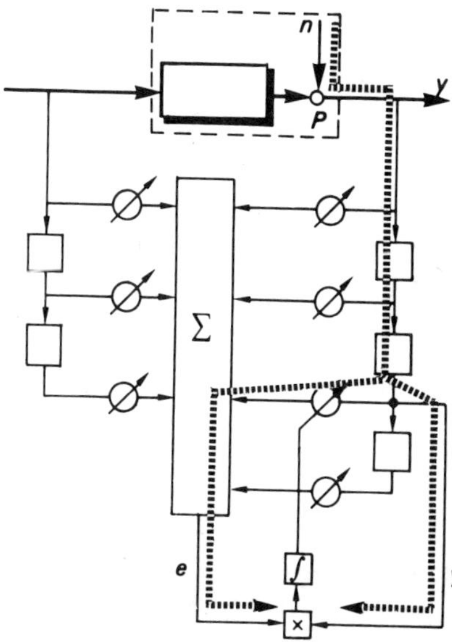


Figure 9.19

indicated for the generalized-model and for the sensitivity-model approach respectively. The bias problems originate in the multiplication of noise contribution in both the error signal $e$ and the sensitivity function $y_i$ of the generalized model.

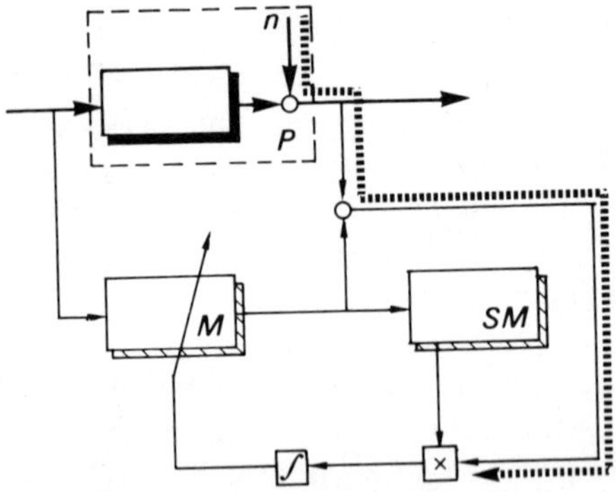


Figure 9.20

## References

Balchen, J. G. and C. Høsøien (1966). 'Adjustable models for determination of the statistical and dynamical behaviour of processes' (in German), *Regelungstechnik*, **14,** 145–150.

Bode, H. W. (1945). *Network Analysis and Feedback Amplifier Design*. Van Nostrand, Princeton, N.J., 551 pp.

Brunner, W. (1961). 'An iteration procedure for parametric model building and boundary value problems', *Proc. WJCC*, 517–533.

Clymer, A. B. (1958). 'System synthesis by dynamic implicit computation', *Paper: Meeting of the Midwestern and Central Simulation Councils*, St. Louis (Mo.).

Clymer, A. B. (1959). 'Direct system synthesis by means of computers', *Trans. AIEE*, **77,** part I (Communication and Electronics), 798–806.

Donalson, D. D. and C. T. Leondes (1963). 'A model referenced parameter tracking technique for adaptive control systems. Part I—The principle of adaption. Part II—Stability analysis by the second method of Lyapunov', *IEEE Trans. Applic. and Ind.*, **82,** No. 68, 241–262.

Graupe, K. K. (1961). 'The analog solution of some functional analysis problems', *Trans. AIEE*, **79,** part I (Communication and Electronics), 793–799.

IFAC Sensitivity (1964). *IFAC Symp.—Sensitivity Methods in Control Theory*, Dubrovnik, Yugoslavia.

IFAC Sensitivity (1968). *IFAC Symp.—System Sensitivity and Adaptivity*, Dubrovnik, Yugoslavia.

Johnson, C. W. and E. R. Rang (1961). 'The learning model approach to self-evaluation and the method of Lyapunov', *AIEE Paper DP 61–603*, Great Lakes District Meeting, Minneapolis.

Kaya, Y. and S. Yamamura (1962). 'A self-adaptive system with a variable-parameter PID controller', *AIEE Trans.* **80,** part II (Applic. and Ind., 58), 378–386.

Kohr, R. H. (1963). 'A method for the determination of a differential equation model for simple nonlinear systems', *IEEE Trans. Electron. Computers*, **EC-12,** 394–400.

Kokotović, P. (1964). 'Sensitivity points method in the analysis and optimization of linear control systems', *Automn. Remote Control*, **25,** 1512–1518.

Kokotović, P. and R. S. Rutman (1965). 'Sensitivity of automatic control systems' (survey), *Automn. Remote Control*, **26,** 727–749, 157 references.

Kushner, H. J. (1969). 'On the convergence of Lion's identification method with random inputs' (short paper), *IEEE Trans. Autom. Control*, **AC-15,** 652–654.

Landau, I. D. (1969). 'A hyperstability criterion for model reference adaptive control systems' (short paper), *IEEE Trans. Autom. Control*, **AC-14,** 552–555.

Lion, P. M. (1966). 'Rapid identification of linear and non-linear systems', *Joint Autom. Control Conf.*, 605.

Margolis, M. and C. T. Leondes (1959). 'A parameter tracking servo for adaptive control systems', *IRE Trans. Autom. Control*, **AC-4,** 100–111, and *IRE Wescon Conv. Rec.*, part 4, 104–115.

Miller, B. J. (1962). 'A general method of computing system parameters with an application to adaptive control', *Conf. Paper AIEE*, CP 62–96.

Ornstein, G. N. (1963). 'The automatic analog determination of human transfer function coefficients', *Medical Electronics, Biological Engineering*, Vol. 1, Pergamon Press, 377–387.

Parks, P. C. (1966). 'Liapunov redesign of model reference adaptive control systems', *IEEE Trans. Autom. Control*, **AC-11,** 362–367.

Pazdera, J. S. and H. J. Pottinger (1969). 'Linear system identification via Liapunov design techniques', *Joint Autom. Control Conf.*, 795–801.

Popov, V. M. (1962). 'Absolute stability of nonlinear systems of automatic control', *Automn. Remote Control*, **22,** 857–875.

Potts, T. F., G. N. Ornstein and A. B. Clymer (1961). 'The automatic determination of human and other system parameters', *Proc. West. Joint Computer Conf.*, Los Angeles, **19,** 645–660.

Schoenemann, P. T. (1965). 'An identification machine for dynamic biological data', *Ph.D. Thesis, E.E. Dept., Univ. of California*, Berkeley, Calif.

Shackcloth, B. and R. L. Butchart (1965). 'Synthesis of model reference adaptive systems by Liapunov's second method', *IFAC Symp.—Theory of Selfadaptive Control Systems*, Teddington, 145–152.

Van de Ven, H. H., A. P. Gelsing and P. Eykhoff (1973). 'The measurement of the impedance by automatic adjustment of a physical model', *Ann. de l'Assoc. Intern. pour le Calcul Analogique*, **15,** 8–14.

Young, P. C. (1965). Cf. references Section 14.1.

Zames, G. (1966). 'On the input–output stability of time-varying non-linear feedback systems', *IEEE Trans. Autom. Control*, **AC-11,** 228–238.

## Additional Literature

### *9.2 Minimization of the Instantaneous Error*

Butler, R. E. and E. V. Bohn (1966). 'An automatic identification technique for a class of nonlinear systems', *IEEE Trans. Autom. Control*, **AC-11,** 292–296.

Goodman, T. P. and J. B. Reswick (1956). 'Determination of system characteristics from normal operating records', *Trans. ASME*, **78,** 259–271.

Kovanits, P. (1966). 'Generalized discrete analogue of the Zadeh–Ragazzini problem', *Automn. Remote Control*, **27,** 215–225.

Kovanits, P. (1967). 'Identification of operators via static programming', *IFAC Symp.—Identification in Autom. Control Systems, Prague*, paper 1.3.

Narendra, K. S. and L. E. McBride (1964). 'Multiparameter self-optimization systems using correlation techniques', *IEEE Trans. Autom. Control*, **AC-9,** 31–38.

Roberts, J. D. (1962). 'A method of optimizing adjustable parameters in a control system', *Proc. IEE*, **109B,** 519–528.

Young, P. C. (1965). 'The determination of the parameters of a dynamic process', *Radio and Electron. Engr.*, **29,** 345–361.

### 9.3 *Minimization of the Time-average Error*

Balakrishnan, A. V. (1961). 'A general theory of nonlinear estimation problems in control systems', *Dept. Eng., Univ. of Calif.*, Los Angeles, Calif.

Hoberock, L. L. and R. H. Kohr (1967). 'An experimental determination of differential equations to describe simple nonlinear systems', *J. Basic Engng.* (*Trans. ASME*), **89,** 393–398.

Lekoff, K. (1967). 'On the problem of equivalence of model and object with regard the control of the object', *IFAC Symp.—Identification in Autom. Control Systems, Prague*, paper 5.2.

Maer, S. (1967). 'On the uniqueness of the adjustment in defining the dynamic characteristics with the help of selfadjusting model', *IFAC Symp.—Identification in Autom. Control Systems, Prague*, paper 5.6.

Maslov, E. P. (1967). 'An optimal statistical adaptive model', *IFAC Symp.—Identification in Autom. Control Systems, Prague*, paper 3.4.

Richalet, J., F. Lecamus and P. Hummel (1970). 'New trends in identification: structural distance and weak topology', *IFAC Symp.—Identification and Process Parameter Estimation, Prague*, paper 7.2.

Robinson, W. R. and A. C. Soudack (1970). 'A method for the identification of time delays in linear systems' (short paper), *IEEE Trans. Autom. Control*, **AC-15,** 97–101.

Rosen, J. B. (1960). 'The gradient projection method for nonlinear programming, part I, linear constraints', *SIAM J.*, **8,** 181–217.

Rosen, J. B. (1961). 'The gradient projection method for nonlinear programming, part II, nonlinear constraints', *SIAM J.*, **9,** 514–532.

Roy, R. and J. Sherman (1967). Cf. references chapter 4.

Skala, K. (1967). 'Experimental identification of systems by adapting of the parallel model', *IFAC Symp.—Identification in Autom. Control Systems, Prague*, paper 5.3.

### 9.4 *Parameter Sensitivity Functions*

Ariaratnam, S. T. and P. W. U. Graefe (1965). 'Linear systems with stochastic coefficients II', *Int. J. Control*, **2,** 161–169.

Åström, K. J. (1965). 'On a first order stochastic differential equation', *Int. J. Control*, **1,** 301–326.

Bekey, G. A. and R. B. McGhee (1964). Cf. references Chapter 5.

Brainin, S. M. (1967). 'The identification of random parameters', *IFAC Symp.—Identification in Autom. Control Systems, Prague*, paper 3.13.

Bogdanoff, J. L. and F. Kozin (1962). 'Moments of the output of linear random systems', *J. Acoust. Soc. Amer.*, **34,** 1063–1066.

Butchart, R. L. and B. Shackcloth (1965). 'Synthesis of model reference adaptive systems by Liapunov's second method', *IFAC Symp.—Theory of self-adaptive control systems, Teddington*, paper 3.4.

Chadeev, V. M. (1967). 'Adaptive models of plants', *IFAC Symp.—Identification in Autom. Control Systems, Prague*, paper 5.8.

Dymock, A. J., K. A. Helps and J. F. Meredith (1967). 'The identification of three parameters using a linear plant model', *IFAC Symp.—Identification in Autom. Control Systems, Prague*, paper 5.7.

Fishwick, W. and W. D. T. Davies (1965). 'The synthesis of dynamical models of plants and processes', *Proc. Inst. Mech. Engrs.*, **179,** Part 3H, 29–36.

Gabor, D., W. P. L. Wilby and R. Woodcock (1961). 'A universal nonlinear filter, predictor and simulator which optimizes itself by a learning process', *Proc. IEE*, **108B,** 422–438.

Graupe, D. and G. R. Cassir (1967). 'A predictive identification and mode-recognition technique for adaptive control and tracking systems', *IFAC Symp.—Identification in Autom. Control Systems, Prague*, paper 4.1.

Han, K. W. and G. J. Thaler (1963). 'Multiparameter self-adaptation using auxiliary models', *IEEE Trans. Applic. and Ind.*, **82,** 197–202.

Hughes, M. T. G. and R. May (1967). 'Estimation of frequency response from random data', *IFAC Symp.—Identification in Autom. Control Systems, Prague*, paper 6.4.

Kazakov, I. E. and L. G. Evlanov (1963). 'On the theory of self-tuning systems with a search of gradient by the method of auxiliary operator', *2nd IFAC Congr., Basle*, 510.

Kokotović, P. and others (1966). 'Sensitivity method in the experimental design of adaptive control systems', *3rd IFAC Congr., London*, paper 45B.

Kulick, V. T. and A. I. Sbitnev (1967). 'Interpolating and extrapolating methods of the statistical analysis of dynamic systems', *IFAC Symp.—Identification in Autom. Control Systems, Prague*, paper 3.18.

Kushner, H. J. (1963). 'A simple iterative procedure for the identification of the unknown parameters of a linear time varying discrete system', *J. Basic Eng.*, **85, D,** 227–235.

Lubbock, J. K. and H. A. Barker (1964). 'A solution of the identification problem', *IEEE Trans. Applic. and Ind.*, **83,** 166–173.

Margolis, M. and C. T. Leondes (1959). 'On the philosophy of adaptive control for process adaptive systems', *Proc. Nat. Electron. Conf.*, **15,** 27–34.

Margolis, M. and C. T. Leondes (1960). 'On the theory of adaptive control systems; the learning model approach', *1st IFAC Congr., Moscow*, **2,** 556–563.

Marsík, J. (1966). 'Experiments with a self-adjusting model for automatic parameter estimation of control loops' (in German), *Z. Messen, Steuern, Regeln*, **9,** 210–213.

Marsík, J. (1967). 'Quick-response adaptive identification', *IFAC Symp.—Identification in Autom. Control Systems, Prague*, paper 5.5.

Meissinger, H. F. (1960). Cf. references Chapter 2.

Meissinger, H. F. (1961). 'Parameter influence coefficients and weighting functions applied to perturbation analysis of dynamic systems', *Proc. 3rd Internat. Congr. on Analog Computation*, Opatija, Yugoslavia.

Meissinger, H. F. (1964). 'Parameter optimization by an automatic open-loop computing method', *Proc. 4th Internat. Congr. on Analog Computation*, Brighton, England, 268–274.

Meissinger, H. F. and G. A. Bekey (1964). 'Computation of sensitivity functions and their use in parameter optimization problems', *IFAC Conf. on Sensitivity Analysis*, Dubrovnik.

Meissinger, H. F. and G. A. Bekey (1966). 'An analysis of continuous parameter estimation methods', *Simulation*, **6,** 94–102.

Meredith, J. F. and A. J. Dymock (1965). 'A self-adaptive system employing high speed parameter identification', *IFAC Symp.—Theory of Self-adaptive Control Systems, Teddington*, paper 4.2.

Mesch, F. (1962). 'A self-adjusting controller with self-adjusting reference model' (in German), *Z. Messen, Steuern, Regeln*, **5,** 320–322.

Miller, K. S. and F. J. Murray (1958). 'A mathematical basis for an error analysis of differential analysers', *J. Math. Phys.*, **32,** No. 2, 3.

Norkin, K. B. (1967). 'Process Identification with the use of search controlled adaptive model', *IFAC Symp.—Identification in Autom. Control Systems, Prague*, paper 5.1.

Parks, P. C. (1967). 'Stability problems of model-reference and identification systems', *IFAC Symp.—Identification in Autom. Control Systems, Prague*, paper 5.4.

Plander, I. (1967). 'Automatic iteration method of process identification with cyclic adjustment of parameters using analog and hybrid computers', *IFAC Symp.—Identification in Autom. Control Systems, Prague*, paper 5.9.

Puri, N. N. and C. N. Weygandt (1963). 'Transfer function tracking of a linear time-varying system by means of orthogonal networks', *IEEE Trans. Applic. and Ind.*, **82,** No. 67, 234–238.

Puri, N. N. and C. N. Weygandt (1963). 'Transfer function tracking of a linear time-varying system by means of auxiliary simple lag network', *Joint Autom. Control Conf.*, 200–203.

Rubin, A. I., S. Driban and W. M. Meissner (1967). 'Regression analysis and parameter identification', *Simulation*, **7,** 39–47.

Sakrison, D. J. (1962). 'Application of stochastic approximation methods to system optimization', *Tech. Report 391, Res. Lab. Electron. M.I.T.*, Cambridge (Mass.).

Shapiro, I. I. (1958). *The Prediction of Ballistic Missile Trajectories from Radar Observations*, McGraw-Hill, New York.

Thal-Larsen, H. and Y. Takahashi (1960). 'Identification of process parameters by means of models', *ASME Paper 60-WA-21* (Winter Annual Meeting, New York).

Tomović, R. (1964). *Sensitivity Analysis of Dynamic Systems*, McGraw Hill, New York.

White, A. J. (1966). 'The analysis and design of model reference adaptive control systems', *Proc. IEE*, **113,** 175–184.

Young, P. C. (1968). 'Regression analysis and process parameter estimation: a cautionary message', *Simulation*, **8,** 125–128.

Zhivoglyadov, V. P. and V. K. Kaipov (1967). 'Identification on distributed parameter plants in the presence of noises', *IFAC Symp.—Identification in Autom. Control Systems, Prague*, paper 3.5.

## *9.5 Two Models*

Blandhol, E. (1962). 'On the use of adjustable models for determination of system dynamics', *Tech. Report Divis. Autom. Control, Techn. University, Trondheim*, Norway.

Blandhol, E. and J. G. Balchen (1963). 'Determination of system dynamics by the use of adjustable models', *2nd IFAC Congr., Basle*, 620.

## *9.6 Parameter Perturbation*

Alimov, Yu. I. (1963). 'The approximate calculation of a class of automatic systems with forced parameter optimization', *2nd IFAC Congr., Basle*, 315–323.

Anderson, G. W. et al. (1958). 'A self-adjusting system for optimum dynamic performance', *IRE Conv. Rec.*, part 4, 182–190.

Bogiorno Jr., J. J. (1962). 'Stability and convergence properties of model-reference adaptive control systems', *IRE Trans. Autom. Control*, **AC-7,** 30–41.

Dorf, R. C. and W. A. Byars (1963). 'A modulation method for model reference adaptive control systems', *IEEE internat. Conv.*

Douce, J. L. and K. C. Ng (1964). 'A six-channel adaptive computer', *Proc. IEE*, **111,** 1757.

Douce, J. L. and M. M. Gutpa (1966). 'Dynamics of the parameter-perturbation process', *Control and Science Record*, **113,** 1077 and *Proc. IEE*, **113,** 1077.

Eveleigh, V. W. (1963). 'General stability analysis of sinusoidal perturbation extrema searching adaptive systems', *2nd IFAC Congr., Basle*, 472.

Eykhoff, P. (1961). 'Process-parameter estimation using an analog model', *Proc. 3rd Internat. Analogue Computation Meetings*, Opatija, Yugoslavia, 276–290.

Eykhoff, P. and O. J. M. Smith (1962). 'Optimalizing control with process dynamics identification', *IRE Trans. Autom. Control*, **AC-7,** 140–155.
Gupta, M. M. (1968). 'Simultaneous generation of first and higher-order gradient functions of a cost function' (correspondence), *IEEE Trans. Autom. Control*, *AC*-**13,** 115–116.
Isobe, T. and T. Totani (1963). 'Analysis and design of a parameter-perturbation adaptive system for application to process control', *2nd IFAC Congr.*, *Basle*, 317.
Krasovski, A. A. (1960). Cf. references Section 14.1.
Lackey, R. B. (1962). 'An adaptive control system with sinusoidal parameter perturbation,' *AIEE Trans. Applic. and Ind.*, *61*, **81,** part I, 157–159.
McGrath, R. J. and V. C. Rideout (1961). 'A simulator study of a two-parameter adaptive system', *IRE Trans. Autom. Control*, **AC-6,** 35–42.
McGrath, R. J., V. Rajaraman and V. C. Rideout (1961). 'A parameter-perturbation adaptive control system', *IRE Trans. Autom. Control*, **AC-6,** 154–162.
Narenda, K. S. and T. S. Baker (1968). 'Gradient generation using a single perturbation signal' (correspondence), *IEEE Trans. Autom. Control*, **AC-13,** 298–299.
Rajamaran, V. (1962). 'Theory of a two-parameter adaptive control system', *IRE Trans. Autom. Control*, **AC-7,** 20–26.
Smith, K. C. (1962). 'Adaptive control through sinusoidal response', *IRE Trans. Autom. Control*, **AC-7,** 129–139.
Weygandt, C. N. and N. N. Puri (1961). 'Transfer function tracking and adaptive control systems', *IRE Trans. Autom. Control*, **AC-6,** 162–166.

*CHAPTER 10*

# *Test Signals; Periodic or Large Ones*

Fundamental for the estimation problem is the fact that the process has to be excited. It is often advantageous to use test signals, introduced from the outside if this is allowed; observations from spontaneous fluctuations alone give less accurate results. The use of disturbances, present already in the process, may suffer from the following disadvantages:

The power spectrum of the disturbances may be too narrow or it may contain some dominant frequencies, which require either prewhitening or periodic rejection in order to make the input spectrum approach that of white noise.

Because only finite records of input and output signals are available, the accuracy of correlation functions is limited.

The knowledge about the system can be used to find an optimal test signal, which can be defined as the signal with a prescribed maximum amplitude (allowable disturbance) that yields the required knowledge with specified accuracy in a minimum length of time. There is much choice in test signals, but if little *a priori* information is available about the system, the choice is seldom critical. For an extensive discussion of test signals the reader is referred to the excellent monograph by Strobel (1968).

In this chapter, several types of test signals are discussed. Generally speaking, the type of input signal determines the kind of knowledge that becomes available. Assume the linear process behaviour to be described by the equation:

$$a_2\ddot{y}(t) + a_1\dot{y}(t) + a_0y(t) = b_1\dot{u}(t) + b_0u(t) \quad \text{or} \quad a_2\ddot{y}(t) + a_1\dot{y}(t) + a_0y(t) = b_1\dot{u}(t-\tau) + b_0u(t-\tau) \tag{10.1}$$

where $u(t)$ is the input and $y(t)$ is the (uncorrupted) output signal. Description in the time domain leads to the convolution integral:

$$y(t) = \int_{-\infty}^{+\infty} h(\theta)u(t-\theta)\,d\theta \qquad y(t) = \int_{-\infty}^{+\infty} h(\theta)u(t-\tau-\theta)\,d\theta = \int_{-\infty}^{+\infty} h'(\theta)u(t-\theta)\,d\theta \tag{10.2}$$

$$h(t) = 0 \quad \text{for } t < 0 \qquad h'(t) = 0 \quad \text{for } t < \tau$$

and description in the frequency domain leads to:

$$Y(s) = H(s)U(s) \qquad Y(s) = H'(s)U(s)$$

with $\qquad$ with $\qquad$ (10.3)

$$H(s) = \frac{b_1 s + b_0}{a_2 s^2 + a_1 s + a_0} \qquad H'(s) = \frac{b_1 s + b_0}{a_2 s^2 + a_1 s + a_0} \varepsilon^{-s\tau}$$

Generally speaking, the measurement of transients is most useful for the first type of description, section 10.1, whereas for the second type of description periodic signals are mostly used; Section 10.2.

The relations between time- and frequency-domain description have been indicated in Chapters 3 and 4. For practical purpose, it is necessary to know how to go from the time-domain description to the frequency domain by means of Fourier transform methods (cf. Chapter 3).

Binary sequences used as test signals have several interesting properties. They are discussed in Section 10.3.

Some aspects of the influence of test signal properties on the accuracy of the estimates are discussed in Chapter 11.

## 10.1 Impulse and Step Signals

For information in the time domain there is the obvious solution of testing with impulse or step signals. Of theoretical interest is an input signal that approaches the properties of the Dirac impulse function, defined by:

$$u(t) = c\delta(t) \begin{cases} \delta(t) = 0 & \text{for } t \neq 0 \\ \displaystyle\int_{-\infty}^{+\infty} \delta(t)\, \mathrm{d}t = 1 \end{cases} \tag{10.4}$$

If this is substituted in the convolution integral equation (10.2) then $y(t) = ch(t)$. A somewhat more practical input test signal approximates a step signal:

$$u(t) = cr(t) \begin{cases} r(t) = 0 & \text{for } t < 0 \\ r(t) = 1 & \text{for } t \geqslant 0 \end{cases} \tag{10.5}$$

From the convolution integral follows

$$y(t) = c\int_0^t h(\theta)\, \mathrm{d}\theta \tag{10.6}$$

and by subsequent differentiation:

$$\frac{\mathrm{d}y}{\mathrm{d}t} = ch(t) \tag{10.7}$$

Step signals have decreasing power at high frequencies where processes usually show attenuation.

The impulse and step signals should, of course, not endanger the process, and not drive it into non-linear behaviour if a linear description of the behaviour around a set point condition is needed. In addition, ramp signals have some relevance in process testing.

From the response of the process to a step input, a number of practical parameters can be determined immediately, e.g. (cf. Figure 10.1)

$\tau$ dead time
$T_d$ delay time
$T_b$ build-up time.

For a more complete survey and nomenclature of such definitions, also in

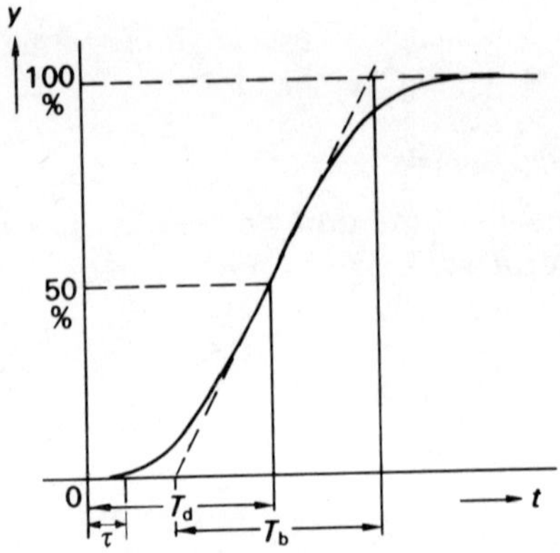


Figure 10.1

(ASA C85.1–1963). A rather common situation is that where the transfer function can be expected to have poles on the real axis only. Consequently,

$$H(s) = \frac{k\varepsilon^{-s\tau}}{\prod_{i=1}^{n} (s\tau_i + 1)}$$

If one may assume that the time constants are equal then from the step response curve, the number $n$ of those time constants and the numerical value of each of those time constants can be determined directly. For more general situations such rules for determining time constants have been reported (e.g., Schwarze, 1964; Strobel, 1968).

### 10.1.1 *The Influence of Additive Noise*

Now we have to consider the effect of additive noise $n(t)$. Instead of an uncorrupted output $x(t)$ there is available only $y(t) = x(t) + n(t)$. Assume $n(t)$ to be characterized by:

$$\mathscr{E}[n(t)] = 0 \qquad \mathscr{E}[n^2(t)] = \sigma^2$$

The uncertainty in one measurement introduced by the noise is given by the standard deviation $\sigma$. This uncertainty can be diminished by performing the same test a number (say $k$) of times, assuming the deterministic part of the test to be reproducible. The $k$ time functions are added and this sum is divided by $k$ (cf. Figure 10.2). The resulting time function has a smaller standard deviation, a fact which can be shown as follows:

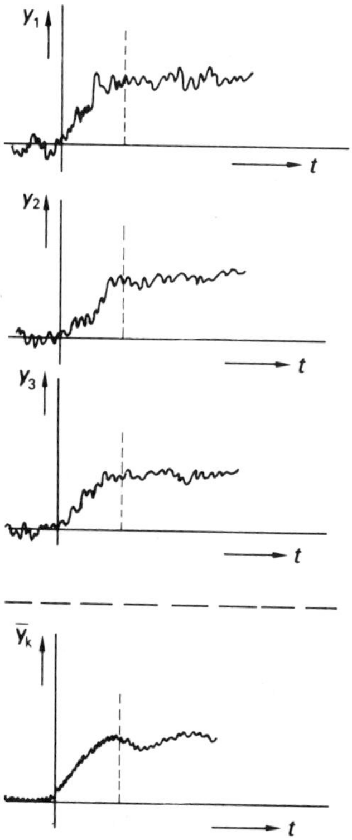


Figure 10.2

Assume the successive tests start at $t = t_1, t_2, \ldots, t_k$. Consider the samples of the output signal $\theta$ seconds after the start of each test and call

$$y(t_i + \theta) = x(t_i + \theta) + n(t_i + \theta)$$

or

$$y_i \quad = \quad x_i \quad + \quad n_i$$

Now the average over the $k$ tests is:

$$\bar{y}_k = \frac{1}{k}\sum_{i=1}^{k} y_i = x + \frac{1}{k}\sum_{i=1}^{k} n_i \tag{10.8}$$

because $x_i = x$ for all $i$ due to the reproducible deterministic part of the test. Consequently,

$$\mathscr{E}[\bar{y}_k] = x$$

$$\mathscr{E}[(\bar{y}_k - x)^2] = \frac{\sigma^2}{k}$$

The latter relation holds only if $n_i$ and $n_j$ are uncorrelated for $i \neq j$. This condition is sufficiently approximated if the autocorrelation function of the noise satisfies:

$$\psi_{nn}(t_i - t_j) \approx 0$$

that is, if the successive tests are sufficiently separated in time. Under that condition, the standard deviation diminishes inversely proportional to the square root of the number of tests (Clark, 1958; M.I.T., 1959; Clynes and Kohn, 1961; Clark, 1961). This principle, called **average response technique** is instrumented in commercially available equipment. Applications are in the field of the study of evoked brain potentials, nerve potentials, retinograms, cardiograms, etc. Furthermore, small digital computers with analogue-to-digital converters are being used for this purpose.

Another quantity that may be of interest to estimate is the variance $\sigma^2$. One might propose to do this as follows from $k$ observations:

$$\overline{(y - \bar{y}_k)^2_k} = \frac{1}{k}\sum_{i=1}^{k} (y_i - \bar{y}_k)^2 \tag{10.9}$$

For the expectation of this quantity we find:

$$\mathscr{E}[\overline{(y - \bar{y}_k)^2_k}] = \frac{1}{k}\mathscr{E}\left[\sum_{i=1}^{k} (y_i - x + x - \bar{y}_k)^2\right]$$

$$= \frac{1}{k}\mathscr{E}[\sum (y_i - x)^2 + 2\sum (y_i - x)(x - \bar{y}_k) + \sum (x - \bar{y}_k)^2]$$

$$= \sigma^2 - \frac{2}{k}\sigma^2 + \frac{1}{k}\sigma^2 = \frac{k-1}{k}\sigma^2$$

Hence as an unbiased estimate for $\sigma^2$ can be chosen:

$$s^2 = \frac{k}{k-1}\overline{(y - \bar{y}_k)^2_k} = \frac{1}{k-1}\sum_{i=1}^{k} (y_i - \bar{y}_k)^2 \tag{10.10}$$

(Note the case $k = 1$)

### *10.1.2 The Use of Correlating Filters*

For testing with impulse or step functions the input signal has to be of a considerable amplitude. This implies that one may drive the process into a region of non-linear operation for which the simple (impulse) response does not provide useful information. The process characteristics found in this way may differ from those under normal conditions. The problem may be circumvented by spreading the test signal energy over a longer time interval in such a way that the amplitude remains small. At any particular instant this can be arranged by using a so-called correlating filter (CF) and a corresponding test signal (cf. Figure 10.3a); Lichtenberger (1961). The

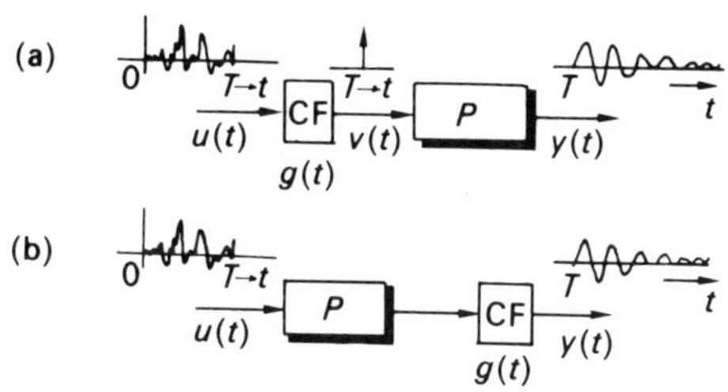


Figure 10.3

test signal $u(t)$ and the impulse-response function $g(t)$ are related to each other as $g(t) = u(T - t)$. Consequently, the output of the filter with this signal at its input is:

$$v(t) = \int_0^\infty u(t - \theta)g(\theta)\,d\theta = \int_0^\infty u(t - \theta)u(T - \theta)\,d\theta \tag{10.11}$$

This gives the autocorrelation function of the signal $u$. By a proper choice of $u(t)$ and the related $g(t)$ the signal $v(t)$ may approximate the impulse 'function' $c\delta(t)$ arbitrarily well. In that case the output $y(t)$ of the process $P$ is an approximation of its impulse response. If $P$ is linear, then CF and $P$ may be rearranged according to Figure 10.3b without changing $y(t)$ whereas now the energy of the signal $v(t) \approx c\delta(t)$ is spread over a time interval. Again the effect of additive noise can be diminished by repeating the experiment and by averaging over the responses.

The freedom of choice with respect to $g(t) = u(T - t)$ can be used to satisfy the criterion that the maximum amplitude of $u$ has to remain within acceptable bounds. This problem is identical with that of shaping radar signals for high resolution (cf. Woodward, 1964).

## 10.2 Sinusoidal Test Signals

Information in the frequency domain is most easily obtained by using sinusoidal or other periodic test signals. This method of testing processes has had a long history (Angström, 1861). There is a host of publications on this subject (Fuchs, 1954; Cowley, 1957). Sinusoidal signals have many advantages. One can limit the measurements to the frequencies of interest; orthogonality properties of periodic signals give labelling facilities; linearity assumptions are easily verified; the duration of the test can be chosen quite arbitrarily; the generation and processing of such signals has been studied extensively; in 'classical' feedback theory, stability studies are formulated in terms of sinusoidal signals. A disadvantage may be the possibility of dangerously exciting a resonance frequency of the process. The principles of this method can be summarized using the following example. Assume the process to be described by the differential equation

$$a_n\overset{(n)}{y} + \cdots + a_1\dot{y} + a_0 y = b_m\overset{(m)}{u} + \cdots + b_1\dot{u} + u \tag{10.12}$$

The input signal is chosen as:

$$u(t) = u_0 \sin \omega t = \frac{u_0}{2j}\{\varepsilon^{j\omega t} - \varepsilon^{-j\omega t}\} \tag{10.13}$$

Due to the linearity of the process its output consists of the sum of the responses of both components after the transients have disappeared:

$$y(t) = \frac{u_0}{2j}\{H(j\omega)\varepsilon^{j\omega t} - H(-j\omega)\varepsilon^{-j\omega t}\} \tag{10.14}$$

with

$$H(j\omega) = \frac{b_m(j\omega)^m + \cdots + b_1(j\omega) + 1}{a_n(j\omega)^n + \cdots + a_1(j\omega) + a_0} \tag{10.15}$$

and

$$\begin{aligned} H(j\omega) &= |H(j\omega)|\varepsilon^{j\varphi(\omega)} \\ H(-j\omega) &= |H(j\omega)|\varepsilon^{-j\varphi(\omega)} \end{aligned} \tag{10.16}$$

Consequently,

$$\begin{aligned} y(t) &= u_0|H(j\omega)| \sin\{\omega t + \varphi(\omega)\} \\ &\quad u_0\{c(\omega) \sin \omega t + \mathrm{d}(\omega) \cos \omega t\} \end{aligned} \tag{10.17}$$

with

$$c(\omega) = |H(j\omega)| \cos \varphi(\omega)$$
$$d(\omega) = |H(j\omega)| \sin \varphi(\omega) \tag{10.18}$$

and

$$|H(j\omega)| = \{c^2(\omega) + d^2(\omega)\}^{\frac{1}{2}}$$
$$\varphi(\omega) = \arctan \{d(\omega)/c(\omega)\} \tag{10.19}$$

The process parameters to be determined are $c(\omega_i)$ and $d(\omega_i)$ for a number of frequencies $\omega_i$. These can be determined by exciting the process with $u(t) = u_0 \sin \omega_i t$ and by applying **'Fourier filtering'** to the output signal $y(t)$. This filtering procedure consists of the multiplication of $y(t)$ respectively with $\sin \omega_i t$ and $\cos \omega_i t$ and averaging over an integer number of periods:

$$c(\omega_i) = \frac{1}{u_0 T} \int_0^T y(t) \sin \omega_i t \, dt$$
$$d(\omega_i) = \frac{2}{u_0 T} \int_0^T y(t) \cos \omega_i t \, dt \tag{10.20}$$

with

$$T = kT_i \quad \text{and} \quad T_i = \frac{2\pi}{\omega_i}$$

Comparison with Chapter 8 will illustrate the close relation to the correlation technique. In the same way, the harmonic content of $y(t)$ can be measured if $u(t) = u_0 \sin \omega_i t$ and the output is multiplied with frequencies $l\omega_i$; $l = 2, 3, 4, \ldots$.

The next problem is to derive, from the measured Nyquist plot, the parameters $a_i$ and $b_j$ in the transfer function given by equation (10.15) (cf. Strobel, 1968).

### 10.2.1 *The Influence of Additive Noise* (Sins, 1967)

Thus far a situation is considered in which there are no disturbances. If $y(t)$ is contaminated with a 'noise signal' $n(t)$ then the result of the Fourier filtering is

$$\tilde{c}(\omega_i, T) = c(\omega_i) + \frac{2}{u_0 T} \int_0^T n(t) \sin \omega_i t \, dt$$

and

$$\tilde{d}(\omega_i, T) = d(\omega_i) + \frac{2}{u_0 T} \int_0^T n(t) \cos \omega_i t \, dt \tag{10.21}$$

Now $\tilde{c}$ and $\tilde{d}$ are approximations which depend on $T = kT_i$, the time interval over which integration is performed. Because $n(t)$ is a stochastic signal, $c$ and $d$ will be of a stochastic nature. In the following we will assume $\mathscr{E}[n(t)] = 0$. By considering the expectation we find:

$$\mu_c(T) = \mathscr{E}[\tilde{c}(\omega_i, T)] = \mathscr{E}[c(\omega_i)] + \frac{2}{u_0 T}\int_0^T \mathscr{E}[n(t)] \sin \omega_i t \, \mathrm{d}t = c(\omega_i)$$
$$\mu_d(T) = \mathscr{E}[\tilde{d}(\omega_i, T)] = \qquad = d(\omega_i) \tag{10.22}$$

Consequently, the numerical results of the measurements are not biased. The variance of the measurements is also of interest

$$\sigma_c^2(T) = \mathscr{E}[\{\tilde{c}(\omega_i, T) - c(\omega_i)\}^2] = \mathscr{E}\left[\left\{\frac{2}{u_0 T}\int_0^T n(t) \sin \omega_i t \, \mathrm{d}t\right\}^2\right]$$
$$= \frac{4}{u_0^2 T^2}\int_0^T\int_0^T \mathscr{E}[n(u)n(v)] \sin \omega_i u \sin \omega_i v \, \mathrm{d}u \, \mathrm{d}v \tag{10.23}$$

where

$$\mathscr{E}[n(u)n(v)] = \psi_{nn}(v - u) = \psi_{nn}(\tau)$$

In Appendix D is derived that

$$\left.\begin{matrix}\sigma_c^2(T)\\ \sigma_d^2(T)\end{matrix}\right\} = \frac{4}{u_0^2 T}\int_0^T \psi_{nn}(\tau)\left\{\left(1 - \frac{\tau}{T}\right)\cos \omega_i \tau \pm \frac{1}{\omega_i T}\sin \omega_i \tau\right\} \mathrm{d}\tau \tag{10.24}$$

If the additive noise is 'white' then

$$\psi_{nn}(\tau) = C\delta(\tau)$$

and one finds*

$$\sigma_c^2(T) = \sigma_d^2(T) = \frac{2C}{u_0^2 T} = \frac{C}{\psi_{uu}(0)T} \tag{10.25}$$

because $\psi_{uu}(0) = u_0^2/2$ for $u = u_0 \sin \omega_i t$. As in cases discussed in previous chapters the standard deviation $\sigma$ is inversely proportional to $\sqrt{T}$ or $\sqrt{k}$. If the additive noise $n(t)$ can be considered as white noise $w(t)$, processed by low-pass filter with transfer function

$$H(s) = \frac{\alpha}{s + \alpha}$$

then

$$\varphi_{nn}(s) = C\frac{\alpha^2}{\alpha^2 - s^2} \Rightarrow \psi_{nn}(\tau) = \frac{C}{2}\alpha\varepsilon^{-\alpha|\tau|} \tag{10.26}$$

* Due to the relationship

$$\int_0^a \delta(\tau) \, \mathrm{d}\tau = \frac{1}{2}\int_{-a}^{+a} \delta(\tau) \, \mathrm{d}\tau = \frac{1}{2}$$

Substitution of this autocorrelation function in equation (10.24) leads to:

$$\sigma_c^2(T) = \frac{C}{\psi_{uu}(0)}\frac{\alpha}{T}\left\{\frac{\alpha}{\omega_i^2+\alpha^2} + \frac{2}{T}\frac{\omega_i^2}{(\omega_i^2+\alpha^2)^2}(1-\varepsilon^{-\alpha T})\right\} \tag{10.27a}$$

$$\sigma_d^2(T) = \frac{C}{\psi_{uu}(0)}\frac{\alpha}{T}\left\{\frac{\alpha}{\omega_i^2+\alpha^2} - \frac{2}{T}\frac{\alpha^2}{(\omega_i^2+\alpha^2)^2}(1-\varepsilon^{-\alpha T})\right\} \tag{10.27b}$$

The behaviour for $T \to \infty$ is found to be

$$\sigma_c^2(T) \approx \sigma_d^2(T) \approx \frac{C}{\psi_{nn}(0)T}\frac{\alpha^2}{\omega_i^2+\alpha^2} \tag{10.28}$$

which for $\alpha \to \infty$ ('white noise') reduces to equation (10.25). For white noise, using a certain interval, the variance is independent of $\omega_i$; for 'red-coloured' noise that variance improves with increasing $\omega_i$ (cf. Figure 10.4). It has been

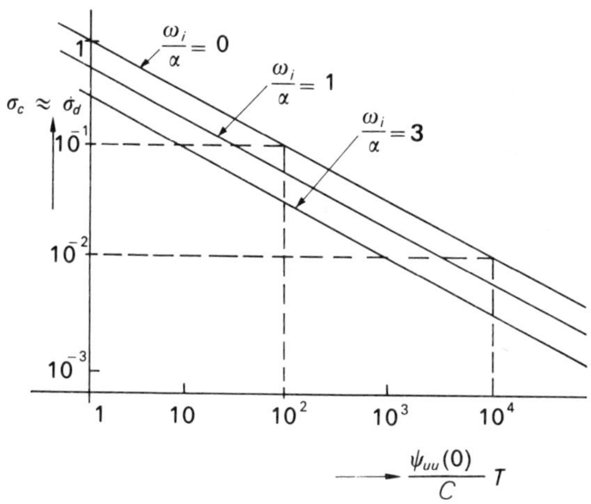


Figure 10.4

shown by Sins (1967) that the difference between equations (10.27a) and (10.27b) for $\sigma_c^2$ and $\sigma_d^2$ can be attributed to the particular choice of integration interval, viz. $(0, T) = (0, kT_i)$. One may as well use $\{(\varphi/2\pi)T_i, (k + \varphi/2\pi)T_i\}$. If $\varphi$ is stochastic and uniformly distributed over the interval $0 \leqslant \varphi < 2\pi$ then $\mathscr{E}_\varphi[\sigma_c^2(T)] = \mathscr{E}_\varphi[\sigma_d^2(T)]$.

Not only $\sigma_c$ and $\sigma_d$ are of interest, but also the correlation between the errors in the $c$ and $d$ direction. This can be expressed by

$$\sigma_{cd}^2(T) = \mathscr{E}[\{\tilde{c}(\omega_i, T) - c(\omega_i)\}\{\tilde{d}(\omega_i, T) - d(\omega_i)\}]. \tag{10.29}$$

It is not difficult to prove that $\sigma_{cd}^2(T) \equiv 0$ irrespective of $T$, $\omega_i$ and the noise correlation function $\psi_{nn}(\tau)$.

As an example, let us consider the measurement of the Nyquist or polar plot of a simple first-order process

$$H(j\omega) = \frac{\beta}{j\omega + \beta}$$

under the influence of additive noise given by equation (10.26); cf. Figure 10.5. Assume $\sigma_u^2 = C$ and $\beta = \alpha$. Then $\sigma_c$ and $\sigma_d$ can be calculated for different

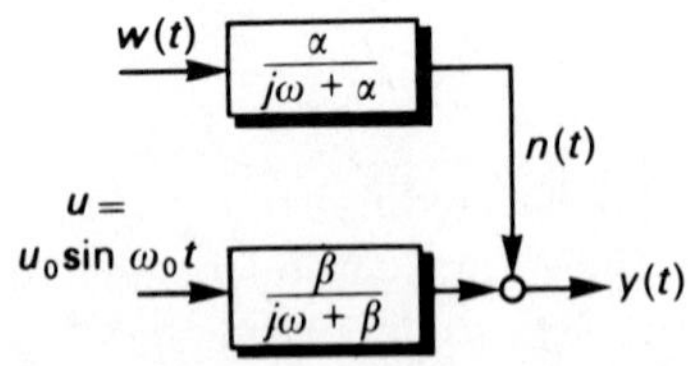


Figure 10.5

values of $\omega_i$ which, together with our knowledge $\sigma_{cd} \equiv 0$, leads to Figure 10.6. Here the circles around a point indicate the one-sigma regions for different lengths of integration intervals.

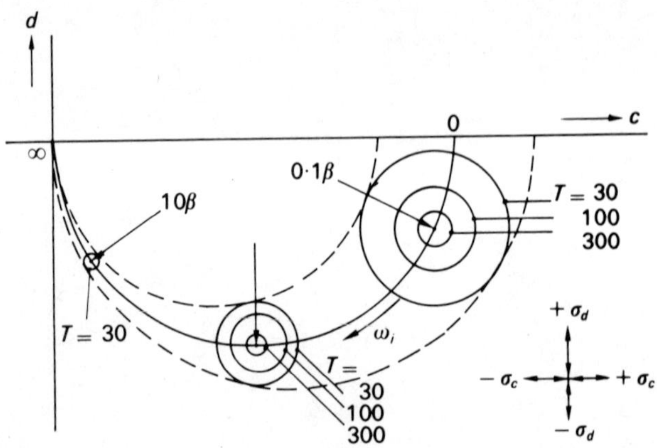


Figure 10.6

For further aspects of Fourier filtering see: Balchen (1962); Balchen and Blandhol (1960); Mesch (1964); Leonov and Lipatov (1959); Dwyer (1965); Werner (1962); Hennig (1963); Seifert (1962); Lange (1962).

### *10.2.2 Related Procedures* (Sins, 1967)

Considering the detrimental influence of the additive noise one may ask whether there are better techniques available. In this respect 'better' may be defined as: a method giving a smaller variance, using the same observation or integration interval, than the Fourier filter. One may consider:

The use of an **extra bandpass filter** with the Fourier filter in order to reduce the noise signal that interferes with our measurements (cf. Figure 10.7). Indeed, along these lines the correlation function $\psi_{nn}(\tau)$ in equation (10.24) can be influenced in a favourable way. There are, however, several disadvantages, viz.: the centre frequency of the band-pass

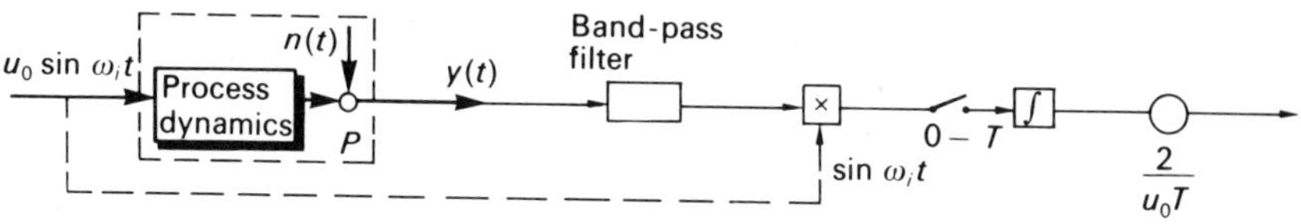


Figure 10.7

filter has to be slaved with the measuring frequency $\omega_i$; the adjustment of the band-pass filter with respect to $\omega_i$ is quite critical as a slight detuning manifests itself in a phase contribution $\neq 0$ which introduces errors in $c(\omega_i)$ and $d(\omega_i)$; when switching the sinusoidal signal to the band-pass filter there is a transient, during which the Fourier filter cannot yet be used. This transient time interval has to be added to the integration interval of the Fourier filter in order to obtain a fair comparison with the unmodified Fourier filter. It turns out that these disadvantages outweigh the possible advantages.

The use of a **compensation (model-adjustment) method** (cf. Figure 10.8). This can be compared with the Chapters 7 and 9. In this case, the model is very simple because the

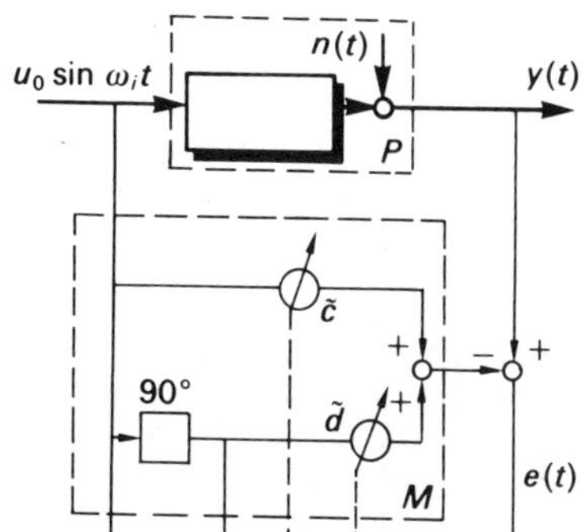


Figure 10.8

only parameters are $c$ and $d$. Again some type of error criterion containing $e(t)$ has to be minimized. It can be shown that the integral-squared-error-criterion is very closely related to the Fourier filter discussed already; cf. also Qvarnström (1961, 1963).

The use of **samplers.** This can be illustrated in the following way. Let

$$\begin{aligned} y(t) &= u_0|H(j\omega)| \sin \{\omega t + \varphi(\omega)\} \\ &= u_0|H(j\omega)| \sin \omega t \cos \varphi(\omega) + u_0|H(j\omega)| \cos \omega t \sin \varphi(\omega) \end{aligned} \tag{10.30}$$

If $\omega t_i = \pi/2 + i2\pi$:

$$y(t_i) = u_0|H(j\omega)| \cos \varphi(\omega) = u_0 c(\omega) \tag{10.31}$$

If $\omega t_i = i2\pi$:

$$y(t_i) = u_0|H(j\omega)| \sin \varphi(\omega) = u_0 d(\omega) \tag{10.32}$$

Consequently, the parameters to be determined can be found by sampling at the instants $t_i$. Therefore the method is extremely simple (Figure 10.9). It is clear, however, that this method is very sensitive to the presence of higher harmonics. The influence of additive

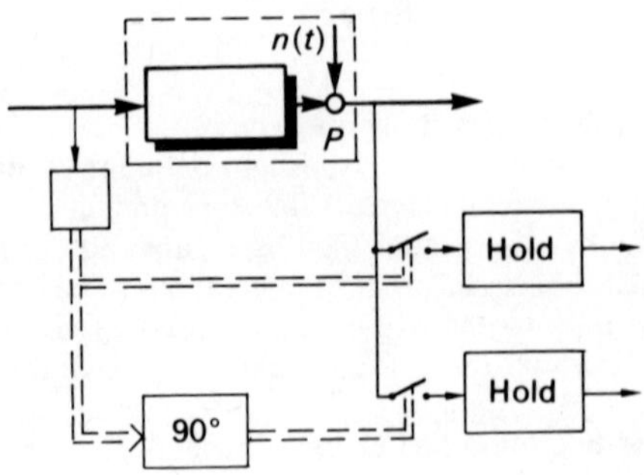


Figure 10.9

noise can be diminished again by averaging over a number of samples $t_i; i = 0, 1, 2, \ldots, k$. A comparison of the variance obtained by this method with that of Fourier filtering leads to the result that the latter is (much) better. Some improvement can be obtained, e.g. by sampling two times per period ($t_i = i\pi$) and taking care of inverting the proper observations. In spite of such added complications, the conclusion given remains valid.

The use of a **band filter**. In order to remove harmonics and noise, the output signal $y(t)$ is filtered by a band-pass filter and then used for making a Lissajous plot on a scope or $x$–$y$ plotter (cf. Figure 10.10). From this plot, both parameters can be determined as indicated in Figure 10.11. In principle the method is simple but, as with the previous application of a band-pass filter, there are some problems. From the theoretical consideration it follows that the Fourier filter is preferable (cf. Mesch 1964).

The use of an **optimal filter.** The problem can be described as estimating the unknown amplitude and phase (or real and imaginary component) of a signal of known frequency amidst contaminating noise. There is a multitude of publications on such types of optimal filter problems, that can be solved by using linear filters. From that point of view, the problem is sketched in Figure 10.12, where $g_r(t)$ and $g_i(t)$ indicate the impulse responses of the optimal filters for the real and imaginary component respectively. It turns out that for white additive noise the optimal filters are identical with the Fourier filter.

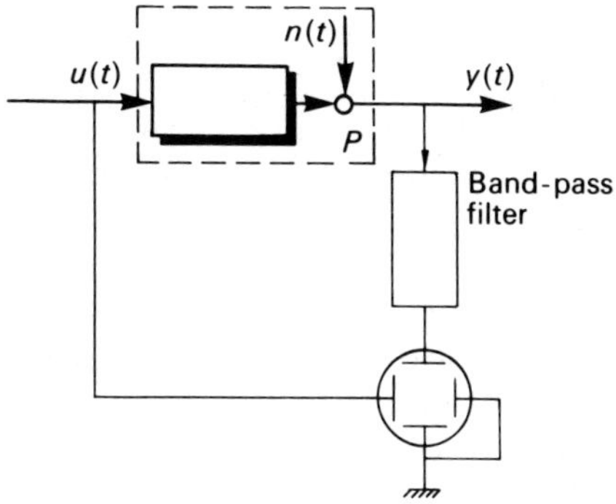


Figure 10.10

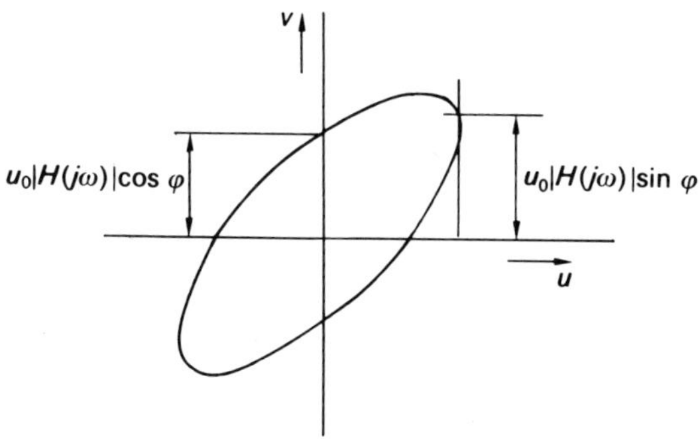


Figure 10.11

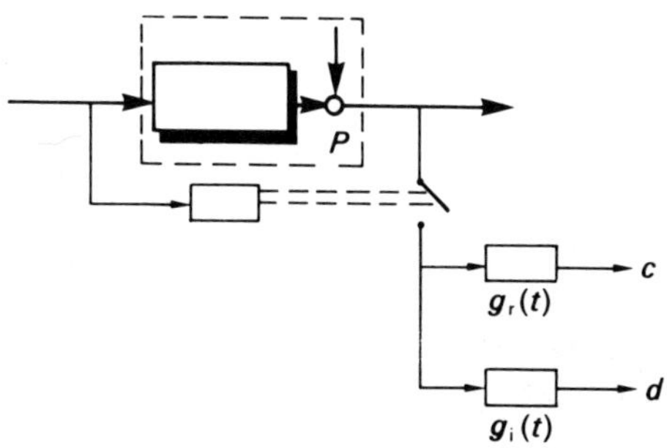


Figure 10.12

A disadvantage of the use of sine waves is the small amount of information that becomes available per frequency of the test signal. A means of diminishing the drawback of the sine-wave testing is the addition of several harmonics into one periodic test signal. In this way by an intelligent choice of phase relations of the sine-waves, the periodic signal can be made binary (i.e.

having only amplitude levels $+a$ and $-a$) without much disturbance of the original energy spectrum (cf. Jensen, 1959). If permitted, such a binary signal can easily be applied to many types of linear processes. The process output is filtered with respect to the harmonics which constitute the input signal. In this way, information regarding the amplitude transference of the process for several frequencies is obtained simultaneously. Another approach to determining time-dependent transfer characteristics (human operator) using the sum of a number of sine waves is described by Sheridan (1960).

There still is a need for a theory that enables us to construct on–off signals with a prescribed power spectrum. Trial and error methods, but also optimization procedures with on–off signals, can give good approximations to certain desired power spectra. If no *a priori* information is present, a pseudo-random sequence (Section 10.3) is very good for obtaining an overall picture of the response so that sharper measuring strategies can be devised. The foundation of optimum test signals is given in decision theory. As more knowledge is obtained from the measurements, sharper measuring strategies can be applied. If this is done with an automatic control system under normal operating conditions, the term 'dual control' is used (cf. Chapter 1).

## 10.3 Binary Signals

In Section 8.1 it was shown that 'white' noise has desirable properties with respect to the determination of the impulse response of a linear process. Its autocorrelation function, approximating a Dirac 'function', eliminates the need for deconvolution; the cross-correlation function of the process input and the output signal gives us directly the process impulse-response function. For sampled signals, the same idea holds: for 'white' noise, the matrix $[\mathbf{U}'\mathbf{U}]$ in the normal equations (Section 6.1) reduces to a diagonal matrix if the observation interval is of sufficient length. This alleviates the need for matrix inversion, corresponding to the deconvolution operation in the case of non-sampled signals. Therefore it is not surprising that a number of research workers have devoted time and energy to finding signals with autocorrelation properties corresponding to those of white noise. In spite of the fact that the same type of requirements occurs in the field of radar and sonar signal modulation (Woodward, 1964) there seems to have been only little interaction between this field and that of process-parameter estimation.

The signal or time sequence that we want to construct may be a periodic or an aperoidic one. Examples of the latter type are given by Huffman (1956). In the class of periodic signals, the subclass of binary test signals shows some interesting engineering properties:

– they can be generated using relay or digital type of instrumentation;

they reduce multiplication with an analogue signal to a relay-type of operation;

of all signals with amplitude limitation $-a \leqslant u \leqslant a$ the binary signal $u = \pm a$ transfers the maximum amount of 'power' to the process under test.

The following diagram shows some relationships

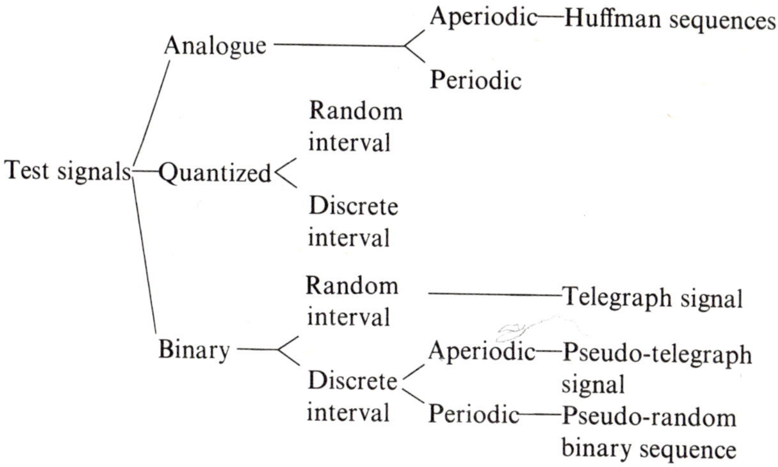


The random-interval (telegraph) signal may have a Poisson distribution for the number of switches from one amplitude to the other in a given time interval. It may be generated by using a radioactive source and a radiation detector, initiating a change of polarity at each detected particle. Its auto-correlation function is given by

$$\psi_{uu}(\tau) = \psi_{uu}(0) \exp[-2\lambda|\tau|] \tag{10.33}$$

where $\lambda$ is a real constant.

The switching intervals may be discretized by the use of a clock pulse with period $\theta$. In that case the change of signal polarity may occur only at the end of an integer number of clock-pulse periods. Such a signal has an auto-correlation function

$$\psi_{uu}(\tau) = \begin{cases} \psi_{uu}(0)\left\{1 - \left|\dfrac{\tau}{\theta}\right|\right\} & |\tau| \leqslant \theta \\ 0 & |\tau| > \theta \end{cases} \tag{10.34}$$

By a proper choice of implementation this correlation function can approximate the Dirac 'function' as closely as desired. This type of signal has the added advantage that it can easily be delayed by using a shift register driven by the same clock signal.

*10.3.1 m-Sequences*

Due to the random nature of the signals mentioned, one has to face the inherent statistical uncertainty in the correlation operation as explained in Section 6.1. Therefore, there is an advantage in using a **deterministic** signal that still has the desired autocorrelation function; pseudo-random binary sequence (*PRBS*). One of the early papers reporting on process dynamics estimation of this type is the one by Anderson, Buland and Cooper (1959). Here the generation of the binary test signal and the multiplication with the same time function were effected by using a code disc. Soon it was recognized, among others by Van Rede and Poortvliet (Poortvliet, 1962), that the so called **maximum-length sequences** (*m*-sequences), generated by feedback shift registers, have desirable autocorrelation characteristics. They have the following properties:

their autocorrelation function, if determined over an integer number of sequence periods, has no stochastic element (uncertainty) in it;

they can easily be generated by using digital components; cf. Figure 10.13 as an example;

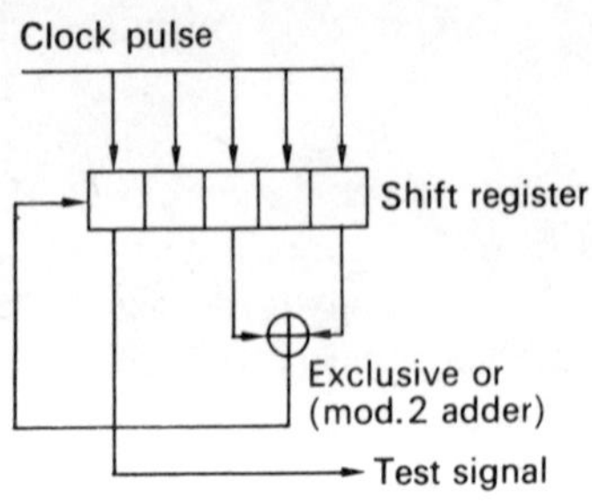


Figure 10.13

moreover the signal, delayed over any integer number of clock-signal periods $\theta$, can easily be generated;

the sequence period $T$ can easily be changed according to $T = (2^n - 1)\theta$, where $n$ = number of shift-register elements in use. Peterson (1961) gives a tabulation of feedback connexions around such shift registers up to $n = 33$;

the d.c. level of the sequence is only $a/N$; consequently, for large $N$ the normal operating point of the process need hardly be disturbed;

the autocorrelation function is shown in Figure 10.14; by choosing $N$ and $\theta$ a Dirac function can be well approximated;

the signal, consisting of amplitudes $+a$ and $-a$, can easily be applied to a variety of processes. Such signals represent the maximum 'power' under a constraint of amplitude; $\psi_{uu}(0) = a^2$.

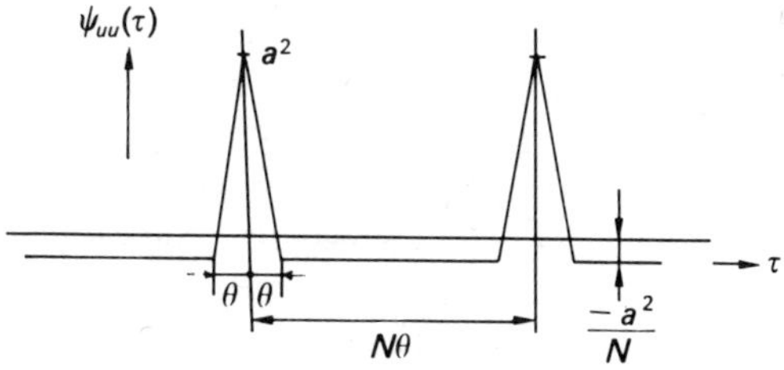


Figure 10.14

These signals and their properties are extensively discussed in Davies (1970). A disadvantage, due to the periodic character of these signals, is that the correlation functions necessarily also will be periodic.

The best way to illustrate these points is by using the simple example of Figure 10.15. The table shows the 'state' of the shift register during the intervals of time, marked by the clock pulses at $t = k\theta$. If the state of register element $\alpha$ is transformed to a test signal through $1 \rightarrow +a$ and $0 \rightarrow -a$, then

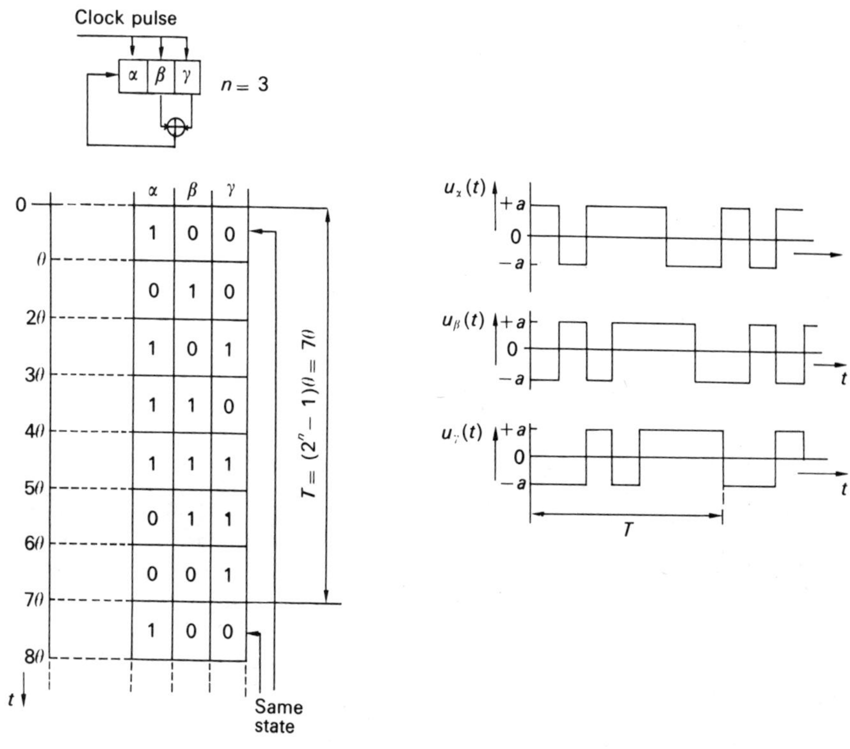


Figure 10.15

results the time function represented as $u_\alpha(t)$. Obviously, $u_\beta$ and $u_\gamma$ are delayed one, viz. two clock-pulse intervals. By direct multiplication the points of the autocorrelation function for $\tau = k\theta$ can easily be checked. For the intervals between those points, it is advantageous to prove first, that in each interval the correlation function consists of a linear interpolation between the end points of that interval. The small negative value of the correlation function is due to the fact that of the $2^n$ possible states there is only one not being used, viz. (0, 0, 0). Consequently, in averaging over the $N = 2^n - 1$ terms of the sequence there will remain one positive element uncompensated. A remedy may be the choice of slightly different values for positive and negative signal amplitudes, leading to correlation functions as sketched in Figure 10.16.

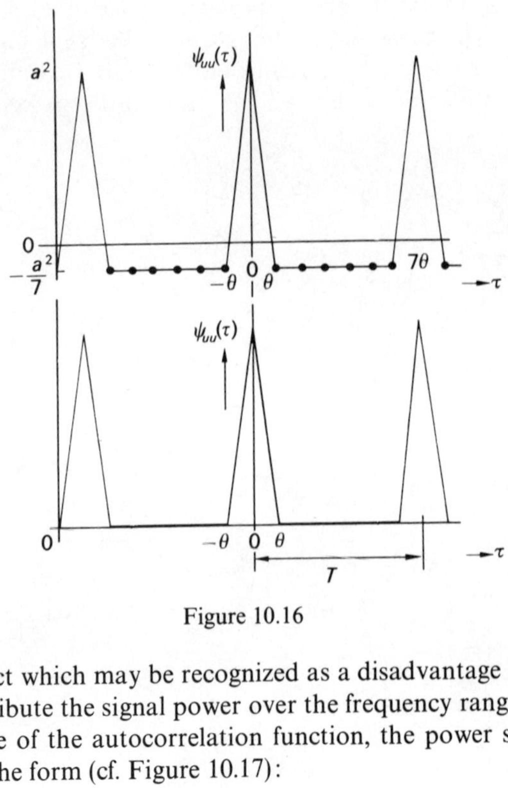


Figure 10.16

Another aspect which may be recognized as a disadvantage is the lack of freedom to distribute the signal power over the frequency range. Due to the triangular shape of the autocorrelation function, the power spectrum has an envelope of the form (cf. Figure 10.17):

$$\theta\left(\frac{\sin \omega\theta/2}{\omega\theta/2}\right)^2 \tag{10.35}$$

Due to the periodicity of the correlation function with a period $T = N\theta$, the frequency spectrum consists of lines at distances $2\pi/T$. (It is left to the

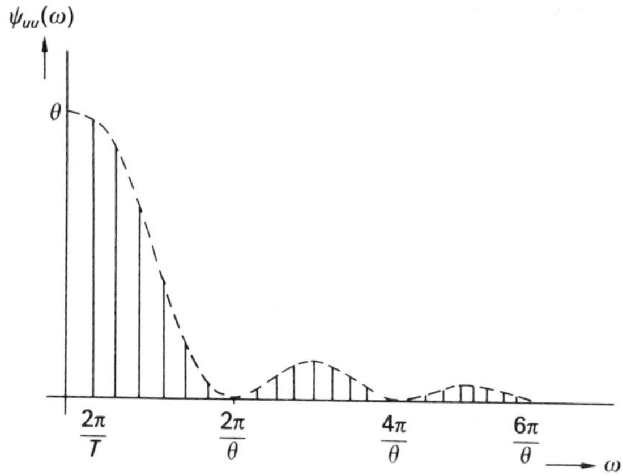


Figure 10.17

reader as an exercise to derive $\phi_{uu}(\omega)$ from $\psi_{uu}(\tau)$ either:

by finding the coefficients of the Fourier series for the periodic function or

by recognizing that

$$\psi_{uu}(\tau) = \frac{1}{\theta} f(\tau) * f(\tau) * \sum_{k=-\infty}^{\infty} \delta(\tau - kT)$$

$$f(\tau) = \begin{cases} 1 & |\tau| \leqslant \frac{\theta}{2} \\ 0 & |\tau| > \frac{\theta}{2} \end{cases}$$

* meaning convolution, which becomes multiplication in the time domain.

Consequently,

$$\phi_{uu}(\omega) = \frac{1}{\theta} F(\omega) \times F(\omega) \times \sum_{k=-\infty}^{\infty} \delta\left(\omega - k\frac{2\pi}{T}\right) \tag{10.36}$$

cf. Figure 10.18. Hence one has considerable freedom in determining the band width of the test signal (by choosing the clock-pulse frequency of the shift-register *m*-sequence generator) and the frequency resolution of its spectrum (by choosing $T = N\theta$, i.e. the length of the shift register and the feedback around).

The use of an m-squence as a test signal can be compared with the use of a large number of sinusoidal components and measuring at all frequencies simultaneously.

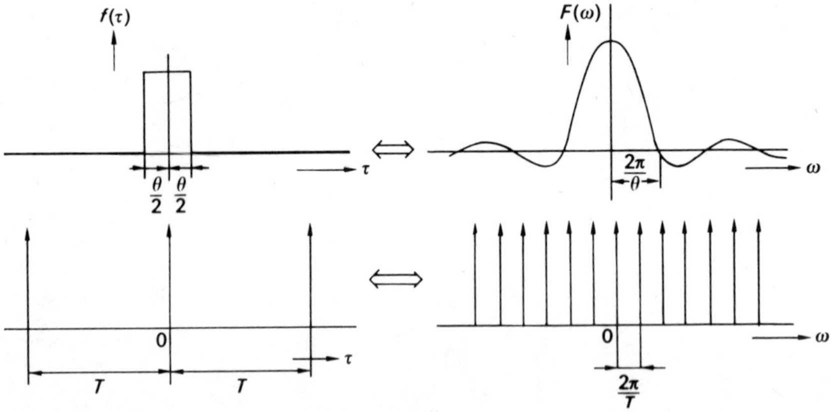


Figure 10.18

### *10.3.2 Errors when Using an m-Sequence Test Signal*

An important reason for choosing the m-sequence was the shape of its autocorrelation function. This may approximate the Dirac 'function' of white-noise test signals, eliminating the need for deconvolution when determining points of a process impulse response. Consequently, we start from the true relationship between the correlation functions

$$\psi_{uy}(\tau) = \int_0^\infty h(\theta)\psi_{uu}(\tau - \theta)\, d\theta \tag{10.37}$$

and compare the results of our approximation. The following types of error can be recognized (While, 1968):

(*a*) Periodicity of the signal.

(*b*) D.c. component of the signal.

(*c*) Use of a 'non-white' (band-limited) signal.

(*d*) Additive noise.

(*e*) Transducer and measurement error.

(*a*) *Periodicity of the Signal.* Because the m-sequence is periodic, its autocorrelation function will have the same property. This implies that when determining the impulse response $h(\tau)$ of a process, one has to be sure that $h(\tau) \approx 0$ for $\tau \geqslant T$, the sequence period. If that were not the case, then the effect of overlapping contributions due to different peaks of the $\psi_{uu}(\tau)$ (aliasing) will occur. Consequently, this technique cannot be used on processes having a pure integration in their dynamics.

When using m-sequences as process input signals for least-squares estimation one has to keep in mind that in spite of the periodicity of that signal the matrix $[\mathbf{U}'\mathbf{U}]$ has to be non-singular, i.e. its columns must be independent.

(*b*) *D.C. Component of the Signal.* The d.c. component in the test signal leads to a constant term in the autocorrelation function $\psi_{uu}(\tau)$, viz. $-a^2/N$. This will influence the cross correlation. To study this influence $\psi_{uu}(\tau)$ is split up into two parts, viz.

triangular functions with area $(N + 1)a^2\theta/N$;

a constant term $-a^2/N$.

The latter contributes to the cross correlation function

$$-\frac{a^2}{N}\int_0^\infty h(\tau)\,\mathrm{d}\tau \tag{10.38}$$

Provided $h(\tau) \to 0$ for $\tau \to T$ this contribution can be recognized from $\psi_{uy}(\tau)$ at large positive ($\tau \approx T$) or small negative values of $\tau$. This provides means to remove this error contribution.

(*c*) *Use of a 'Non white'* (*Band limited*) *Signal.* The use of this test signal was stimulated by its autocorrelation function, which may approximate a Dirac 'function'. Whether an approximation is fair enough depends on the situation at hand. As the fine details in a transfer function are caused by the smallest time constant $\tau_{\text{min}}$ of the process under study, it appears that $\theta \approx \tau_{\text{min}}/5$ is an acceptable choice. The amount of error depends of course on the situation, viz. on $h(\tau)$ of the process to be studied.

(*d*) *Additive Noise.* Due to the deterministic nature of these test signals, there is no inherent statistical uncertainty. The study of the uncertainty due to additive noise can be done as in previous cases. One finds for the variance in the estimate due to 'white' additive noise:

$$\sigma^2 = \frac{a^2\sigma_n^2}{iN\theta} \quad \text{or} \quad \sigma = \frac{a\sigma_n}{\sqrt{iN\theta}} \tag{10.39}$$

where $\sigma_n^2$ is the variance of the additive noise and $i$ is the (integer) number of sequence periods over which correlation has been performed. Note that the estimate $\psi_{uy}(\tau)$ is proportional to $a^2$; consequently the 'signal to noise' ratio of the estimate improves with increasing $a$; cf. Briggs and others (1964).

(*e*) *Transducer and Measurement Error*. Godfrey and Murgatroyd (1965) have studied the problem of the errors, introduced by the response of transducers used for applying m-sequence signals into processes.

An important extension is towards the use of such test signals for multi-input/multi-output processes. In the case where there are $p$ inputs and $q$ outputs, then the process can be characterized by $pq$ impulse responses. It is simple to show that the 'correlation' technique indicated in the previous lines applies also to this more general case and that the impulse response $h_{ij}$ from input $i$ to output $j$ can be found from

$$\psi_{u_i y_j}(\tau) = \int_0^{\infty} h_{ij}(\theta)\psi_{u_i u_i}(\tau - \theta)\, d\theta \tag{10.40}$$

if $u_i$ and the input signals at the other $p - 1$ inputs are statistically independent. If this condition is not fulfilled, then the result is a set of simultaneous integral equations in the unknown $h_{ij}$ which is difficult to solve.

The independence criterion for the $u$'s may be relaxed to: orthogonality over the $\tau$ interval of interest (which may be infinite for processes containing a pure integration).

The cross correlation between two m-sequences with the same $\theta$ and with periods of $M$ and $N$ (coprime) respectively, can be negligibly small for all shifts only if correlation is performed over the interval $MN\theta$.

For further information on the use of these test signals the reader is referred to Nikiforuk and Gupta (1969); Godfrey (1970); Davies (1970).

## References


Anderson, G. W., R. N. Buland and G. R. Cooper (1959). 'The Aeronutronic self-optimizing control system', in *Proc. of the Self-adaptive Flight Control Systems Symp.* (Ed. P. C. Gregory), WADC Techn. Report 59–49. ASTIA Document AD 209389: Wright-Patterson Air Force Base, Ohio, 349–406.

Angström, A. J. (1861). 'New methods to determine the heat conductivity of bodies' (in German), *Ann. Phys.*, **114,** 513–530.

ASA C85.1–1963 (1963). *American Standard Terminology for Automatic Control*, The American Society of Mechanical Engineers, N.Y.

Balchen, J. G. and E. Blandhol (1960). Cf. references Chapter 8.

Balchen, J. G. (1962). 'A simple instrument for experimental frequency response tests of control systems', *Regelungstechnik*, **33,** 200–205.

Briggs, P. A. N. and others (1964). 'Correlation analysis of process dynamics using pseudo-random binary test perturbations', *Proc. Inst. Mech. Engrs.*, **179,** pt. 3H, (Advances in Automatic Control), 37–51.

Clark, W. A. (1958). 'Average response computer (ARC-1)', *Quart. Progr. Rept., Res. Lab. Electron. M.I.T.*, 114–117.

Clark, W. A. (1961). 'The average response computer (ARC); a digital device for computing averages and amplitude and time histograms of electrophysiological response', *IRE Trans. Biomed. Electron.*, **BME-8,** 46–51.

Clynes, M. and M. Kohn (1961). 'Portable 4 channel online digital average response computer, CAT', *Proc. 4th Internat. Conf. on Med. Electronics*, New York.

Cowley, P. E. A. (1957). 'The application of an analog computer to the measurement of process dynamics', *ASME Trans.*, **79,** 823–832.

Davies, W. D. T. (1970). Cf. references Chapter 2.

Dwyer, B. (1965). 'A feedback transfer function analyser', *Control*, **9,** 676–678.

Fuchs, A. M. (1954). 'A bibliography of the frequency response method as applied to automatic feedback control systems', *ASME Trans.*, **76,** 1185–1194.

Godfrey, K. R. and M. Murgatroyd (1965). 'Input-transducer errors in binary cross-correlation experiments—1', *Proc. IEE*, **112,** 565–573.

Godfrey, K. R. (1970). 'The application of pseudo random sequences to industrial processes and nuclear power plant', *2nd IFAC Symp.—Identification and Process Parameter Estimation*, *Prague*, paper 7.1.

Hennig, T. (1963). 'Testing for plant transfer functions in presence of noise and non-linearity', *Control Engng.*, **10,** 67–70 and 119–124.

Huffman, D. A. (1962). 'The generation of impulse-equivalent pulse trains', *IRE Trans. Inform. Theory*, **IT-8,** 10–16.

Jensen, J. R. (1959). 'Notes on the measurement of dynamic characteristics of linear systems', *Part III*, *Report Servoteknisk Forskningslaboratorium*, Denmarks Tekniske Højskole, Copenhagen.

Lange, F. H. (1962). *Correlation Electronics* (in German), V.E.B. Verlag Technik, Berlin, 381 pp.

Leonov, Y. P. and L. N. Lipatov (1959). 'The use of statistical methods for determining the characteristics of objects', *Automn. Remote Control*, **20,** 1254–1268.

Lichtenberger, W. W. (1961). 'A technique of linear system identification using correlating filters', *IRE Trans. Autom. Control*, **AC-6,** 183–199.

Mesch, F. (1964). 'A comparison of frequency response measurements under stochastic disturbance' (in German), *Z. Messen, Steuern, Regeln*, **7,** 162–166.

M.I.T. (1959). Communications Biophysics Group and W. M. Siebert. 'Processing neuroelectric data', *Tech. Report 351. Res. Lab. of Electron.*, *MIT*, *Cambridge*, Mass.

Nikiforuk, P. N. and M. M. Gupta (1969). 'A bibliography of the properties, generation and control systems applications of shift-register sequences', *Int. J. Control*, **9,** 217–234.

Peterson, W. W. (1961). *Error Correcting Codes*, Wiley, New York, 285 pp.

Poortvliet, D. C. J. (1962). 'The measurement of system-impulse response by means of cross-correlation with binary signals', *Tech. Report Electronics Lab.*, *Univ. of Technology*, *Delft*, Netherlands; also in *Tijdschr. Nederl. Radio Genootschap*, **28,** 253–270 (1963).

Qvarnström, B. (1961). 'On the relative merits of methods to measure process dynamics', *Acta Imeko*, Budapest, 14-SW-126.

Qvarnström, B. (1963). 'An uncertainty relation for linear mathematical models', *2nd IFAC Congr.*, *Basle*, 634–639.

Schwarze, G. (1964). 'Algorithmic determination of the order and time constants of P-, I- and D-action elements with two different time constants and delays up to the 6th order' (in German), *Z. Messen, Steuern, Regeln*, **7,** 10–18.

Seifert, W. (1962). 'Commercial frequency-response measuring devices' (in German), *Regelungstechnik*, **10,** 350–353.

Sheridan, T. B. (1960). 'Experimental analysis of time-variation of the human operator's transfer function', *1st IFAC Congr.*, *Moscow*, 629–635.

Sins, A. W. (1967). 'The determination of a system transfer function in presence of output noise' (in Dutch), *Thesis*, *E.E. Dept.*, *University of Technology*, *Eindhoven*, Netherlands.

Strobel, H. (1968). *System Analysis Using Deterministic Test Signals* (in German), V.E.B. Verlag Technik, Berlin, 219 pp.

Werner, G. W. (1962). 'Frequency response measurement with low experimental expenditure in disturbed industrial control circuits' (in German), *Z. Messen, Steuern, Regeln*, **5,** 545–546.

While, C. (1968). 'System identification', *M.Sc. Thesis, Dept. of Electronic and Electrical Engng., Univ. of Birmingham*, Birmingham, England.

Woodward, P. M. (1964). *Probability and Information Theory with Applications to Radar*, Pergamon, London, 136 pp.

## Additional Literature

### *10.1 Impulse and Step Signals*

Adams, R. K. (1962). 'Digital computer analysis of closed-loop systems using number series approach', *Trans. Amer. Inst. Electr. Engrs*, **80** (Applic. and Industr., 58), 370–378.

Billingsley, D. S. and M. G. Rekoff (1961). 'Evaluation of transient coefficients', *IRE Trans. Autom. Control*, **AC-6,** 80–82.

Bux, D. and R. Isermann (1967). 'A comparison of aperiodical test signals for the measurement of the dynamic behaviour of control loops' (in German), *Fortschritt-Bericht VDI-Z.*, **8,** nr. 9.

Cuenod, M. and A. E. Durling (1969). *A Discrete-Time Approach for System-Analysis*, Academic Press, New York, 221 pp.

Dittman, H. (1963). 'Some extensions in determining characteristic values from the transient response' (in German), *Z. Messen, Steuern, Regeln*, **6,** 139–141.

Dreike, G. L. and J. O. Hougen (1963). 'Experimental determination of system dynamics by pulse methods', *Joint Autom. Control Conf.*, 608–633.

Ellington, J. P. and H. McCallion (1958). 'The determination of control system characteristics from a transient response', *Proc. IEE*, **105C,** 370–373.

Fuchs, H. (1963). 'On the evaluation of characteristic values of PID-units' (in German), *Z. Messen, Steuern, Regeln*, **6,** 385–386.

Gehrmann, E. and J. Schusta (1962). 'Method for determining the time constants of thermocouples' (in German), *Wiss. Z. Hochsch. Elektrotech., Ilmenau*, D.D.R., **8,** 419–421.

Gerhardt, G. (1966). 'Experimental parameter estimation in the time domain using aperiodic test signals' (in German), *Z. Messen, Steuern, Regeln*, **9,** 204–208.

Girshow, L. A., I. V. Gelman and L. A. Dobrin (1962). 'Certain engineering methods for analyzing controlled objects with monotonic transfer function', *Automn. Remote Control*, **23,** 1135–1139.

Gupta, S. C. (1966). 'Performance of process models', *3rd IFAC Congr., London*, paper 36D.

Hilscher, K. (1964). 'Parameter estimation of instruments and systems from the transient response that can be approximated by *n* equal time-constants and a dead time' (in German), *Z. Messen, Steuern, Regeln*, **7,** 272–273.

Hougen, J. O. and R. A. Walsh (1961). 'Pulse testing method', *Chem. Eng. Progr.*, **57,** 69–79.

John, R. M. and others (1962). 'System for processing and recording excitation function data', *Rev. Scient. Instrum.*, **33,** 1089–1094.

Lendaris, G. G. (1962). 'The identification of linear systems', *AIEE Trans.*, part II (Applics. Ind., 62), **81,** 231–242.

Mishkin, E. and R. A. Haddad (1959). 'Identification and command problems in adaptive systems', *IRE Wescon Conv. Rec.* part 4, 125–135; *IRE Trans. Autom. Control*, **AC-4,** 121–131.

Nyquist, J. K., R. N. Schindler and R. E. Gilbert (1963). 'Determination of frequency response from step response', *Chem. Eng. Progr.*, Symposium Series, **59,** 46.

Ormanns, G. (1956). 'On the processing of experimentally found transformations of control blocks' (in German), *Paper "Moderne Theorien und ihre Verwendbarkeit"*, Heidelberg, 130–133.

Radtke, M. and H. P. Seidel (1967). 'On estimation of static and dynamic parameters of nonlinear transfer elements from step responses', *IFAC Symp.—Identification in Autom. Control Systems, Prague*, paper 4.3.

Reinisch, K. (1962). 'Use of a control loop simulator in order to obtain simple dimensioning rules for controlled plants to find the characteristic values of controlled plants' (in German), *Z. Messen, Steuern, Regeln*, **5,** 245–251.

Reinisch, K. and J. Wernstedt (1967). 'Model method of analyzing disturbed linear and special nonlinear systems by means of aperiodic test signals', *Paper IFAC Symp.—Identification in Autom. Control Systems, Prague*, paper 4.2.

Sage, A. P. and W. C. Choate (1965). 'Minimum time identification of nonstationary dynamic processes', *Proc. Nat. Electron. Conf.*, **21,** 587–592.

Schechter, R. S. and E. H. Wissler (1959). 'Frequency response from step input response', *Ind. Eng. Chem.*, **51,** 945–948.

Stepan, J. (1965). 'Sensitivity studies of some methods for parameter estimation of control loops' (in German), *Z. Messen, Steuern, Regeln*, **8,** 35–38.

Strejc, V. (1959). 'Approximation process for aperiodic transfer characteristics' (in German), *Regelungstechnik*, **7,** 124–128.

Strejc, V. (1960). 'The approximation of aperiodic transient responses' (in German), *Z. Messen, Steuern, Regeln*, **3,** 115–124.

Strobel, H. (1967). 'On the limits imposed by random noise and measurement errors upon system identification in the time domain', *IFAC Symp.—Identification in Autom. Control Systems, Prague*, paper 4.4.

Turin, G. L. (1957). 'On the estimation in the presence of noise of the impulse response of a random linear filter', *IRE Trans. Inf. Theory*, **IT-3,** 5–10.

Unbehauen, H. (1963). 'A graphical-analytical computation scheme for the determination of the frequency response from the impulse response' (in German), *Regelungstechnik*, **11,** 551–555.

Unbehauen, H. (1966). 'Parameter estimation of control systems using the measured impulse response' (in German), *Z. Messen, Steuern, Regeln*, **9,** 188–191.

Verheyden, J. (1962). 'Measurement of controller characteristics', *Revue A*, **4,** 94–99.

Weller, W. (1962). 'The methods of determining control characteristics from the measured transient response' (in German), *Z. Messen, Steuern, Regeln*, **5,** 355–363.

Werner, G. W. (1966). 'Application of the impulse test procedure for parameter estimation' (in German), *Z. Messen, Steuern, Regeln*, **9,** 197–201.

Zadeh, L. A. (1950). 'The determination of impulse response of variable networks', *J. Appl. Phys.*, **21,** 642.

## *10.2 Sinusoidal Test Signals*

Atamanenko, V. G. (1962). Transfer function analyzer for experimental investigation of industrial automatic control system dynamic characteristics', *Instrum. Constr.*, 13–15.

Atherton, D. P. and N. K. Jain (1967). 'Identification of double-valued nonlinearities from their harmonic response', *Electron. Letters*, **3,** 485–486.

Bard, M. (1965). 'Implementation for automatic registration of frequency plots with the analog computer', *Z. Steuern, Regeln, Automatisieren*, **9,** 431–437.

Chen, C. F. and B. L. Philip (1965). 'Accurate determination of complex-root transfer functions from frequency response data' (short paper), *IEEE Trans. Autom. Control*, **AC-10,** 356–358.

Darovskikh, L. N. (1959). 'Experimental determination of automatic control system links, transfer functions by means of standard electronic models', *Automn. Remote Control*, **20,** 1180–1187.

Darovskikh, L. N. (1962). 'Finding a frequency characteristic by means of computing elements of an electronic analog computer', *Automn. Remote Control*, **23,** 219–222.

Dudnikov, E. E. (1959). 'Determination of transfer function coefficients of a linear system from the initial portion of an experimentally obtained amplitude phase characteristic', *Automn. Remote Control*, **20,** 552–558.

Dudnikov, E. E. (1963). 'On a method for the evaluation of transfer function coefficients of linear systems from the measured frequency response' (in German), *Z. Messen, Steuern, Regeln*, **6,** 453–456.

Ehret, R. J. (1953). 'An automatic transfer function measuring and recording system', *AIEE Trans.*, **72,** part I, 664–669.

Fraser, H. J. and W. V. P. Reece (1961). 'An automatic electronic Nyquist plotter', *Proc. IEE*, **108B,** 535–538.

Gochet, M. (1961). 'Device for measuring isochrone transmittance of servosystems' (in French), *Revue A*, **3,** 32–35.

Huber, E. A. (1962). 'A technique for the adaptive control of high order systems', *IRE Trans. Autom. Control*, **AC-7,** 22–29.

Hughes, M. T. G. and R. May (1967). 'Estimation of frequency response from random data', *IFAC Symp.—Identification in Autom. Control Systems, Prague*, paper 6.4.

Kardashov, A. A. and L. V. Karniushin (1958). 'Determination of system parameters from experimental frequency characteristics', *Automn. Remote Control*, **19,** 327–338.

Key, K. A. (1962). 'The analogue computer as a (one cycle) transfer function analyser', *Proc. 3rd Internat. Analogue Computation Meetings*, Presses Académiques Européennes, Brussels, 569–572.

Klein, V. (1966). 'The identification of linear system with a given structure by means of frequency responses', *3rd IFAC Congr., London*, paper 46A.

Lees, S. and others (1961). 'Experimental verification of a design basis for positional servomechanisms', *1st IFAC Congr., Moscow*, **2,** 998–1008.

Manton, F. E. S. (1962). 'Analog computer setup displays frequency response', *Control Engng.*, **9,** 83–85.

Martin, E. (1962). 'Instrument for automatic recording of the logarithmic amplitude and phase frequency response (Bode diagram)' (in German), *Z. Messen, Steuern, Regeln*, **5,** 1–6.

Minkwitz, G. (1966). 'On the application of digital techniques for phase measurements' (in German). *Nachrichtentechnik*, **16**, 97.

Nomoto, A. and others (1960). 'Automatisation of dynamic testing', *1st IFAC Congr., Moscow*, **3,** 304–309.

Ogar, G. W. (1961). 'Obtaining the frequency response of physical systems by analog computer techniques', *IRE Internat. Conv. Rec.*, part 2: Audio, Electronic Computers, 196–210.

Oppelt, W. (1972). *Kleines Handbuch technischer Regelvorgänge*, Verlag Chemie, Weinheim, 770 pp.

Parezanovic, N. (1962). 'Frequency characteristic determination from the system pulse response by the use of a repetitive differential analyser', *Proc. 3rd Internat. Analogue Computation Meeting*, Presses Académiques Européennes, Brussels, 573–575.

Reinisch, K. (1962). 'The use of a control-loop model for obtaining simple measurement rates for linear control loops and for parameter estimation on control systems' (in German), *Z. Messen, Steuern, Regeln*, **5,** 245–251.

Reinisch, K. (1965). 'A method for the measurement of continuous and sampled control loops including dead time and some nonlinearities' (in German), *Habilitationsschrift, Techn. Univ., Dresden.*

Sanathanan, C. K. and J. Koerner (1963). 'Transfer function synthesis as a ratio of two complex polynomials', *IEEE Trans. Autom. Control*, **8,** 56–58.

Schäfer, O. and W. Feissel (1955). 'An improved method for frequency response analysis of industrial control loops' (in German), *Regelungstechnik*, **3,** 225–229.

Senf, B. and H. Strobel (1961). 'Method for determining transfer functions of linear systems from measured values of frequency response' (in German), *Z. Messen, Steuern, Regeln*, **4,** 411–420.

Sourisse, C. and R. Wagner (1963). 'Pneumatic transferometer' (in French), *Automatisme*, **8,** 487–494.

Staffin, H. K. and R. Staffin (1965). 'Approximating transfer functions from frequency response data', *Instrum. and Control Syst.*, **38,** 137–144.

Strobel, H. (1964). 'On the problem of the essential frequency range when approximating a measured frequency response' (in German), *Z. Messen, Steuern, Regeln.*, **7,** 19–23.

Strobel, H. (1966). 'On a new method of determining the transfer function by simultaneous evaluation of the real and imaginary parts of the measured frequency response', *3rd IFAC Congr., London*, paper 1F.

Unbehauen, H. (1966). 'The determination of rational frequency responses from measurement data' (in German), *Regelungstechnik*, **14,** 268–273.

Vavilov, A. A. and A. I. Solodovnikov (1963). *Experimental Determination of Automatic Systems Frequency Response* (in Russian), Gosenergoizdat, Moscow and Leningrad, 252 pp.

Vichnevetsky, R. (1964). 'Analogue computer simulation of a time-dependent delay using the concept of generalized transfer function', *Ann. Assoc. Int. Calcul Analogue*, **6,** 105–109.

Webb, E. K. and N. E. Bacon (1962). 'A mechanical harmonic analyzer', *J. Sci. Instrum.*, **39,** 500–503.

Welfonder, E. (1966). 'Parameter estimation on disturbed control systems by means of correlation and periodic test signals' (in German), *Fortschritt-Bericht VDI-Z.*, **8,** Düsseldorf, VDI-Verlag.

Welfonder, E. (1967). 'Correlation method for identification on disturbed control systems by periodical test signal', *IFAC Symp.—Identification in Autom. Control System, Prague*, paper 3.8.

Weygandt., C. N. and N. N. Puri (1961). 'Transfer function tracking and adaptive control systems', *IRE Trans. Autom. Control*, **AC-6,** 162–166.

Weygandt, C. N. and N. N. Puri (1963). 'Multivariable adaptive control systems', *IEEE Trans. Applic. and Ind.*, **82,** 89–94.

Wierwille, W. W. (1961). 'Determination by frequency response of transfer function', *AIEE Trans.*, part 2 (Applics. Ind., 56), 183–189.

Wilfert, H. H. (1969). *Signal- and Frequency Response Analysis on Strongly-Disturbed Systems* (in German), V.E.B. Verlag Technik, Berlin, 195 pp.

Zoss, L. M. and H. S. Wilson (1962). 'Frequency response analysis', *Instrum. Soc. Amer. J.*, **2.**

*10.3 Binary Test Signals*


Barber, D. L. A. and P. H. Hammond (1965). 'The use of an on-line digital computer to evaluate the dynamic response of a pilot scale distillation column', *Trans. Soc. Instrum. Technol.*, **17** (3), 59.

Bell, D. A. (1966). 'Walsh functions and Hadamard matrices', *Electron. Letters*, **2,** 340.

Boardman, K. D. (1966). 'The measurement of the dynamic behaviour of a superheater by random signal testing', *IEE Conf. on Automatic Control in Electricity Supply*, Manchester, 198.

Briggs, P. A. N. and K. R. Godfrey (1966). 'Pseudo-random signals for the dynamic analysis of multivariable systems', *Proc. IEE*, **113,** 1259–1267.

Briggs, P. A. N., K. R. Godfrey and P. H. Hammond (1967). 'Estimation of process dynamic characteristics by correlation methods using pseudo-random signals', *IFAC Symp.—Identification in Autom. Control Systems, Prague*, paper 3.10.

Carter, G. A., J. A. Shepley and C. B. Roust (1967). 'A cross-correlation experiment on an experimental reactor using the method of binary input and output sequences', *IEE Colloquium—Pseudo-random Sequences Applied to Control Systems, London.*

Chow, P. E. K. and A. C. Davies (1969). 'The synthesis of cyclic code generators', *Electron. Technol.*, **41.**

Clarke, D. W. and K. R. Godfrey (1966). 'Simultaneous estimation of the first and second derivatives of a cost function', *Electron. Letters*, **2,** 338.

Cumming, I. G. (1970). 'Frequency of input signal in identification', *2nd IFAC Symp.—Identification and Process Parameter Estimation, Prague*, paper 7.8.

Darnell, M. (1966). 'Synthesis of pseudo-random signals derived from p-level m-sequences', *Electron. Letters*, **2,** 428.

Davies, A. C. (1965). 'Delayed versions of maximal-length linear binary sequences', *Electron. Letters*, **1,** 61. 'Further notes on delayed versions of linear binary sequences', *Electron. Letters*, **1,** 190.

Davies, W. A. (1966). 'Generation of delayed replicas of maximal-length linear binary sequences', *Proc. IEEE*, **113,** 295–296.

Davies, W. D. T. (1966). 'Generation and properties of maximum-length sequences 1, 2, 3', *Control and Autom. Progr.*, **10,** 302.

Davies, W. D. T. (1966). 'Identification of a system in the presence of low-frequency drift', *Electron. Letters*, **2,** 327.

Davies, W. D. T. and J. L. Douce (1967). 'On-line system identification in the presence of drift', *IFAC Symp.—Identification in Autom. Control systems, Prague*, paper 3.12.

Douce, J. L. and K. C. Ng (1965). 'The use of pseudo-random binary test signals in adaptive control', *IFAC Symp.—Theory of Self-adaptive Control Systems, Teddington*, Plenum Press, 1966.

Douce, J. L., K. C. Ng and A. E. G. Walker (1966). 'System identification in the presence of a ramp disturbance', *Electron. Letters*, **2,** 243.

Elispas, B. (1959). 'The theory of autonomous linear sequential networks', *IRE Trans. Circuit Theory*, **CT-6,** 45–60.

Everett, D. (1966). 'Periodic digital sequences with pseudo-noise properties', *GEC J. Sci. Technol.*, England, **33,** 115.

Gardiner, A. B. (1965). 'Logic pseudo-random binary sequence delay calculator and delayed version generator with automatic delay changing facility', *Electron. Letters*, **1,** 123.

Gardiner, A. B. (1966). 'Elimination of the effect of non-linearities on process cross-correlations', *Electron. Letters*, **2,** 164.

Godfrey, K. R. (1966). 'Three-level m-sequences', *Electron. Letters*, **2,** 241.

Godfrey, K. R. (1966). 'Input-transducer errors in binary crosscorrelation experiments—3', *Proc. IEE*, **113,** 1095–1102.

Godfrey, K. R. (1966). 'Theory and application of pseudo-random sequences', *Control and Autom. Progr.*, **10,** 305.

Godfrey, K. R., B. Everett and P. R. Bryant (1966). 'Input-transducer errors in binary crosscorrelation experiments—2', *Proc. IEE*, **113,** 185–189.

Hammond, P. H. and D. L. A. Barber (1965). 'The use of an on-line digital computer to evaluate the dynamic response of a pilot scale distillation column', *Trans. Soc. Inst. Techn.*, **59.**

Hampton, R. L. (1965). 'A hybrid analog-digital pseudo-random noise generator', *Simulation*, **4,** 179–190.

Havel, J. and K. Janáč (1967). 'Equipment for the generation of random and pseudo-random processes', *IFAC Symp.—Identification in Autom. Control Systems, Prague*, paper 6.3.

Hazlerigg, A. D. G. and A. R. M. Noton (1965). 'The application of crosscorrelating equipment to linear-system identification', *Proc. IEE*, **112,** 2385–2400.

Henderson, K. W. (1964). 'Some notes on the Walsh functions', *Trans. IEEE electron. computers*, **EC-13,** 50.

Hill, J. D. and G. J. McMurtry (1964). 'An application of digital computers to linear system identification', *IEEE Trans. Autom. Control*, **AC-9,** 536–538.

Isobe, T., T. Idogawa and H. Kashiwagi (1966). 'Impulse response determination with an on-line cross-correlator', *3rd IFAC Congr., London*, paper 15D.

Izawa, K. and K. Furuta (1966). 'Apparatus to measure dynamic characteristics', *Proc. 3rd IFAC Congr., London*, paper 15C.

Izawa, K. and K. Furuta (1966). 'Measurement of dynamic characteristics of heat exchanger using binary random signal', *Proc. Soc. Instrum. and Control Engng.*, **2,** 101–112.

Izawa, K. and K. Furuta (1967). Cf. references Chapter 4.

Kramer, C. (1965). 'A low frequency pseudo-random noise generator', *Electron. Engng.*, **37,** 465–467.

Krtolica, R. (1967). 'Some applications of pseudo-random maximum length binary sequences in the statistical identification', *IFAC Symp.—Identification in Autom. Control Systems, Prague*, paper 3.11.

Pitterman, F. and G. Schweizer (1966). 'Generation and use of binary noise in flight tests' (in German), *Regelungstechnik*, **14,** 63–70.

Rhihaczek, A. W. (1965). 'Radar signal design for target resolution', *Proc. IEEE*, **53,** 116–128.

Roberts, P. D. and R. H. Davis (1966). 'Statistical properties of smoothed maximal-length linear binary sequences', *Proc. IEE*, **113,** 190–196.

Sawaragi, Y. and K. Ogino (1972). 'Sensitivity approach to optimal input synthesis for parameter identification of bilinear systems', *5th IFAC Congr., Paris*, paper 31.2.

Scholefield, P. H. R. (1960). 'Shift registers generating maximum-length sequences', *Electron. Technol.*, **37,** 389–394.

Schwarze, G. (1966). 'Error discussions for the time-percentage parameter method when using non-ideal unit test signals' (in German), *Z. Messen, Steuern, Regeln*, **9,** 171–173.

Simpson, H. R. (1966). 'Statistical properties of a class of pseudorandom sequences', *Proc. IEE*, **113,** 2075–2080.

Smith, B. M. (1966). 'An asymmetrical property of binary pseudo-random noise generators' (correspondence), *Proc. IEEE*, **54,** 793–794.

Smith, K. D. and J. C. Hamilton (1966). 'The logical design of a digital pseudo-random noise generator', *IEE Trans. Nucl. Sci.*, **13,** 371–378.

COLLEGE LIBRARY CARDIFF

Stalder, J. E. and C. R. Cahn (1964). 'Bounds for correlation peaks of periodic digital sequences' (correspondence), *Proc. IEEE*, **52,** 1262–1263.

Tausworth, R. C. (1965). 'Random numbers generated by linear recurrence modulo two', *Math. of Comput.*, **19,** 201–209.

Turin, G. L. (1960). 'An introduction to matched filters', *IRE Trans. Inf. Theory*, **IT-6,** 311.

Unbehauen, H. and G. Schlegel (1967). 'Estimation of the accuracy in the identification of control systems using deterministic test signals', *IFAC Symp.—Identification in Autom. Control Systems*, *Prague*, paper 1.12.

Van Blerkom, R., R. E. Sears and D. G. Freeman (1965). 'Analysis and simulation of a digital matched filter receiver of pseudo-noise signals', *IBM J. Res. and Develop.*, **9,** 264–273.

Van den Bos, A. (1967). 'Construction of binary multifrequency test signals', *IFAC Symp.—Identification in Autom. Control Systems*, *Prague*, paper 4.6.

Van den Bos, A. (1970). 'Estimation of linear system coefficients from noisy responses to binary multi-frequency test signals', *2nd IFAC Symp.—Identification and Process Parameter Estimation*, paper 7.2.

Van der Grinten, P. M. (1963). 'The application of random test signals in process optimization', *2nd IFAC Congr.*, *Basle*, 551.

Zierler, N. (1960). 'Linear recurring sequences', *SIAM J.*, **7,** 31.

CHAPTER 11

# Bayes' and Maximum Likelihood Estimation

In Chapter 5 the relation between different types of estimates was indicated. It was shown that Markov and least-squares estimation have the advantage of requiring little or no *a priori* knowledge. It was also shown that for additive Gaussian noise these estimates are the same as the maximum likelihood ones.

Depending on the point of view, the strength of the least-squares estimator may turn into a weakness if *a priori* knowledge is available. Intuitively, it is clear that the use of the proper type of *a priori* knowledge may result in a 'better' estimate. In the least-squares model-adjustment approach it is possible to set the initial values of the model parameters according to our prior knowledge or guess. It is, however, not possible in a simple way to bring into the estimation scheme the amount of confidence that we have in our initial value. That is where the Bayes' estimation may be useful. This is discussed in Section 11.1.

Section 11.2 is devoted to the maximum likelihood estimation, which has the advantage of a well-developed theoretical background in statistics. Of great importance is the problem of maximum achievable accuracy. This is also dealt with in Section 11.2. Some instrumentation schemes for these types of estimates are discussed in 11.3. Requirements with respect to the process-input signal are discussed in Section 11.4.

## 11.1 Bayes' Estimation

The central aspects of the problem have been discussed in Section 5.1. It was shown that the crucial point in the estimation procedure is the determination of the (*a posteriori*) conditional probability-density function $p(\mathbf{b}|\mathbf{y})$, given the observations $\mathbf{y}$. For a certain sequence of observed process output variables $\mathbf{y}_k' = \{y(1), \ldots, y(k)\}$ this probability-density function provides all the knowledge that is of interest to the experimenter. From the development of $p(\mathbf{b}|\mathbf{y}_k)$, with $k$ increasing, one may learn about the characteristics of the estimate, such as bias, consistency, asymptotic normality (cf. Figure 5.2). It would be possible to determine $p(\mathbf{b}|\mathbf{y})$ and then let the experimenter decide on the basis of that knowledge what he considers to be the 'best' estimate $\hat{\boldsymbol{\beta}}$ of $\mathbf{b}$.

Several attempts to determine the development of conditional probability functions in time can be found in literature. There are approaches along the lines of solving Focker–Planck types of partial differential equations (Stratonovitch, 1960; Kushner, 1964; Bucy, 1965) or expansion of the process output in eigenfunctions (Kroy and Stubberud, 1968).

Objections to this procedure of first determining $p(\mathbf{b}|\mathbf{y})$ and then having the experimenter decide on the 'best' estimate may be of the following nature:

for a situation where $\mathbf{b}$ is an $m$-vector with $m > 2$ it is not a simple task to present $p(\mathbf{b}|\mathbf{y})$ to the experimenter in a form suitable for decision;

'best' has to be defined in some way or another, requiring the choice of a type of cost or loss function.

The choice of such a cost function remains more or less arbitrary; only very few problems clearly suggest a particular choice. Cost functions in connexion with probability-density functions that are discussed in the literature include:

$C(\beta, b) = (\beta - b)^2 c$ a quadratic function, leading to the conditional expectation or mean of $p(b|\mathbf{y})$ as in the one-dimensional case

$$\min \int (\beta - b)^2 c p(b|y)\, db$$

through differentiation with respect to $\beta$ and equating to zero results in

$$\hat{\beta} = \int b p(b|\mathbf{y})\, db$$

i.e. the **mean** of the probability-density functions.

$C(\beta, b) = |\beta - b| c$ an absolute value function which results in the **median** of $p(b|\mathbf{y})$.

$\beta = \max_b p(b|\mathbf{y})$, i.e. the most probable value, leading to the **mode** of $p(b|\mathbf{y})$.

In Figure 11.1 this is illustrated for several types of probability-density functions. Figure 11.2 indicates these and other types of cost functions. For symmetric unimodal probability functions the mean, median and mode coincide. For further discussion on the choice of cost functions cf. Deutsch (1965).

In Section 5.1 it was shown how the Bayes' estimate $\boldsymbol{\beta}$ depends on the *a priori* knowledge:

the conditional probability-density function of the output variable: $p(\mathbf{y}|\mathbf{b})$;

the (*a priori*) probability-density function of the parameter values $\mathbf{b}$: $\mathbf{q}(\mathbf{b})$;

the cost or loss function: $C(\boldsymbol{\beta}, \mathbf{b})$;

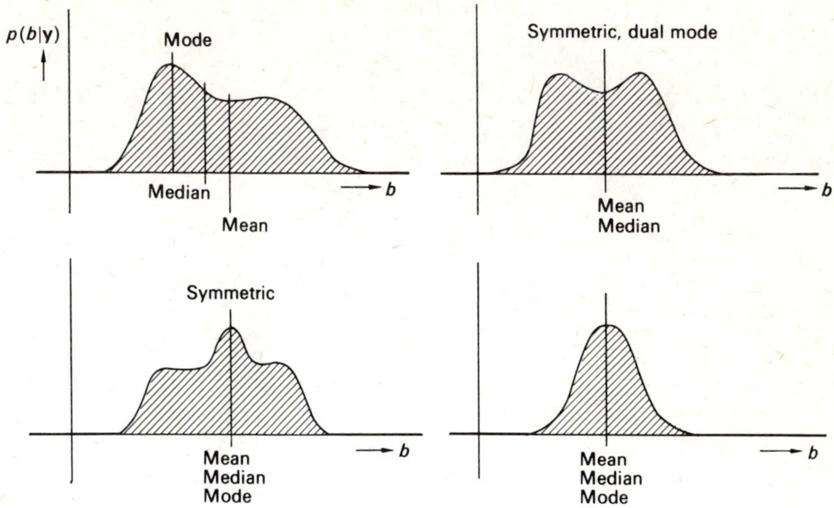


Figure 11.1

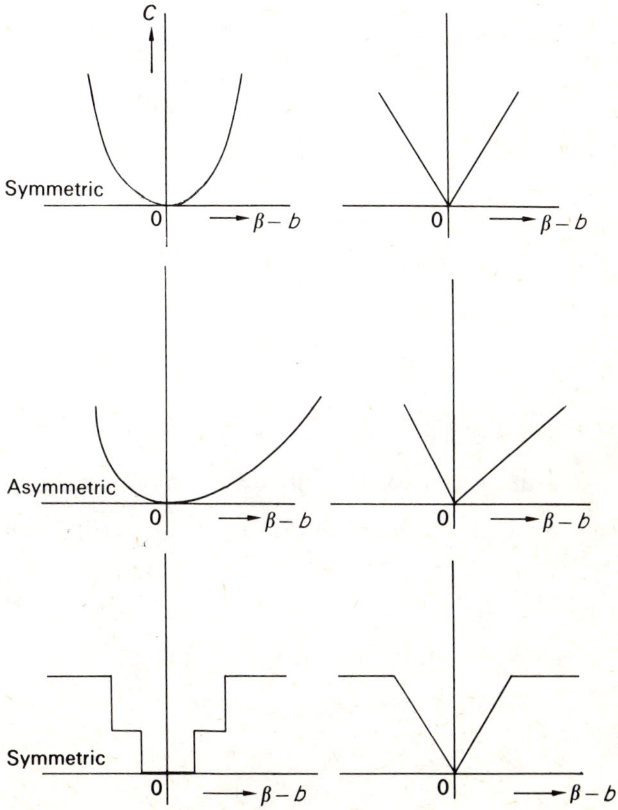


Figure 11.2

through the average risk

$$R(\boldsymbol{\beta}) = \mathscr{E}_{\mathbf{b}}[\mathscr{E}_{\mathbf{y}}[C(\boldsymbol{\beta}, \mathbf{b})|\mathbf{b}]] = \int_m \int_k C(\boldsymbol{\beta}, \mathbf{b}) p(\mathbf{y}, \mathbf{b}) \, \mathrm{d}^k\mathbf{y} \, \mathrm{d}^m\mathbf{b} \tag{11.1}$$

which has to be minimized. For a given observation **y** this reduces to

$$\min_{\boldsymbol{\beta}} \int_m C(\boldsymbol{\beta}, \mathbf{b}) p(\mathbf{b}|\mathbf{y}) \, \mathrm{d}^m\mathbf{b} \tag{11.2}$$

or the necessary condition

$$\frac{\partial}{\partial \boldsymbol{\beta}} \int_m C(\boldsymbol{\beta}, \mathbf{b}) p(\mathbf{b}|\mathbf{y}) \, \mathrm{d}^m\mathbf{b} \Big|_{\boldsymbol{\beta} = \hat{\boldsymbol{\beta}}} = \mathbf{0} \tag{11.3}$$

For one variable $\boldsymbol{\beta} = \beta$ and $C(\beta, b) = (\beta - b)^2 c$ we found

$$\hat{\beta} = \int_m b p(b|\mathbf{y}) \, \mathrm{d}b \tag{11.4}$$

In order to clarify the ideas presented we will consider the estimation of one parameter in the simple situation indicated by Figure 11.3:

$$\mathbf{y} = \mathbf{u}b + \mathbf{n}$$

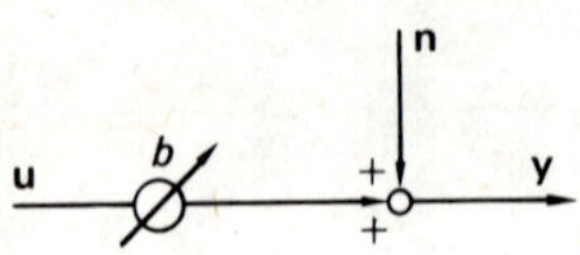


Figure 11.3

where

$$p_b(b) \quad \text{is Gaussian,} \quad \mathscr{E}[b] = \mu_b \quad \operatorname{var}[b] = \sigma_b^2$$

$$p_n(\mathbf{n}) \quad \text{is Gaussian,} \quad \mathscr{E}[\mathbf{n}] = \mathbf{0} \quad \operatorname{cov}[\mathbf{n}] = \sigma_n^2\mathbf{I}$$

$b$ and **n** are statistically independent. From this knowledge we also easily derive the probability-density function of **y**:

$$p_y(\mathbf{y}) \quad \text{is Gaussian,} \quad \mathscr{E}[\mathbf{y}] = \mathbf{u}\mu_b \quad \operatorname{cov}[\mathbf{y}] = \sigma_b^2\mathbf{u}\mathbf{u}' + \sigma_n^2\mathbf{I}$$

where the expression for the covariance follows from:

$$\begin{aligned} \operatorname{cov}[\mathbf{y}] &= \mathscr{E}[\{\mathbf{y} - \mathbf{u}\mu_b\}\{\mathbf{y} - \mathbf{u}\mu_b\}'] \\ &= \mathscr{E}[\{\mathbf{u}(b - \mu_b) + \mathbf{n}\}\{\mathbf{u}(b - \mu_b) + \mathbf{n}\}'] \\ &= \mathscr{E}[(b - \mu_b)^2]\mathbf{u}\mathbf{u}' + \mathscr{E}[\mathbf{n}\mathbf{n}'] \end{aligned} \tag{11.5}$$

Our interest is focussed on

$$p(b|\mathbf{y}) = \frac{p(\mathbf{y}, b)}{p_y(\mathbf{y})} = \frac{p_b(b)p_n(\mathbf{y} - \mathbf{u}b)}{p_y(\mathbf{y})} \tag{11.6}$$

as with $\mathbf{n} = \mathbf{y} - \mathbf{u}b$

$$p(\mathbf{y}, b) = p(\mathbf{y} - \mathbf{u}b, b)\underbrace{\left|\frac{\partial(\mathbf{y} - \mathbf{u}b)}{\partial \mathbf{y}}\right|}_{1} = p_b(b)p_n(\mathbf{y} - \mathbf{u}b)$$

Consequently,

$$\begin{array}{c} \quad p_b(b) \;\vdots\; p_n(\mathbf{y} - \mathbf{u}b) \;\vdots\; p_y^{-1}(\mathbf{y}) \\ p(b|\mathbf{y}) = C_1 \exp\left[-\frac{1}{2}\left\{\frac{(b-\mu_b)^2}{\sigma_b^2} + \frac{(\mathbf{y}-\mathbf{u}b)'(\mathbf{y}-\mathbf{u}b)}{\sigma_n^2} + \right.\right. \\ \left.\left. -(\mathbf{y} - \mathbf{u}\mu_b)'(\sigma_b^2\mathbf{u}\mathbf{u}' + \sigma_n^2\mathbf{I})^{-1}(\mathbf{y} - \mathbf{u}\mu_b)\right\}\right] \end{array} \tag{11.7}$$

with $C_1$ = normalizing constant.

From this point on, a student may easily spend an excessive amount of time trying to evaluate the integral of equation (11.4), keeping track of all parts of the exponent and the normalizing constant. The trouble may be drastically reduced by recalling some aspects of the Gaussian probability-density function:

$$p(x) = \frac{1}{\sqrt{2\pi}\sigma_x} \exp\left[-\frac{1}{2}\left\{\frac{x-\mu_x}{\sigma_x}\right\}^2\right]$$

viz.:

$$\int p(x)\,\mathrm{d}x = 1 \quad \text{and} \quad \int (x - \mu_x)p(x)\,\mathrm{d}x = 0$$

As we integrate with respect to $b$ the only thing of importance is to shape the exponent into a square in $b$:

$$\begin{aligned} &-\frac{1}{2}\left\{b^2\left(\frac{1}{\sigma_b^2} + \frac{\mathbf{u}'\mathbf{u}}{\sigma_n^2}\right) - 2b\left(\frac{\mu_b}{\sigma_b^2} + \frac{\mathbf{u}'\mathbf{y}}{\sigma_n^2}\right) + \cdots\right\} \\ &= -\frac{1}{2}\left(\frac{1}{\sigma_b^2} + \frac{\mathbf{u}'\mathbf{u}}{\sigma_n^2}\right)\left\{(b-\alpha)^2 + \cdots + \cdots\right\} \end{aligned} \tag{11.8}$$

with

$$\alpha = \frac{\dfrac{\mu_b}{\sigma_b^2} + \dfrac{\mathbf{u}'\mathbf{y}}{\sigma_n^2}}{\dfrac{1}{\sigma_b^2} + \dfrac{\mathbf{u}'\mathbf{u}}{\sigma_n^2}} \tag{11.9}$$

Consequently

$$\hat{\beta} = \int bp(b|\mathbf{y})\, db = \int (b - \alpha + \alpha)p(b|\mathbf{y})\, db$$

$$= \int (b - \alpha)p(b|\mathbf{y})\, db + \alpha \int p(b|\mathbf{y})\, db = 0 + \alpha$$

and

$$\hat{\beta} = \alpha = \frac{\sigma_n^2 \mu_b + \sigma_b^2 \mathbf{u}'\mathbf{y}}{\sigma_n^2 + \sigma_b^2 \mathbf{u}'\mathbf{u}} \tag{11.10}$$

Let us look at the two limiting cases:

$$\sigma_n^2 \neq 0 \qquad \sigma_b^2 = 0 \quad \text{then} \quad \hat{\beta} = \mu_b$$

$$\sigma_n^2 = 0 \qquad \sigma_b^2 \neq 0 \quad \text{then} \quad \hat{\beta} = \frac{\mathbf{u}'\mathbf{y}}{\mathbf{u}'\mathbf{u}}$$

In the first case there is no *a priori* uncertainty about the value of $b$, and our noisy measurements can not improve on that knowledge. In the second case of noiseless observations we can obtain a perfect estimate from our measurements.

In the general case $\sigma_n^2 \neq 0$, $\sigma_b^2 \neq 0$, we use both the *a priori* knowledge and the knowledge obtained by the measurement for estimation. The variances $\sigma_n^2$ and $\sigma_b^2$ determine the relative weight given to each of these pieces of knowledge. Note that

$$\mathscr{E}[\mathbf{u}'\mathbf{y}] = k\psi_{uy}$$
$$\mathscr{E}[\mathbf{u}'\mathbf{u}] = k\psi_{uu} \tag{11.11}$$

which implies that the weight given to the estimation measurements increases with the number of signal samples processed. If

$$\mu_b \neq \mathscr{E}[b]$$

then for $\sigma_n^2 \neq 0$, $\sigma_b^2 \neq 0$ the estimate is initially biased but becomes asymptotically unbiased.

As an exercise it is recommended to consider the same problem for the case that the noise $\mathbf{n}$ is not white, but is coloured according to a covariance matrix $\mathbf{N}$. Also the reader is advised to derive the equivalent of equation (11.10) for the case of a parameter vector $\hat{\boldsymbol{\beta}}$.

A still more general case is given by the interesting attempt to use decision theory to process-parameter estimation by Maslov (1963). The problem discussed is characterized by Figure 11.4 and

sampled input and output signals;

both input and output are contaminated with noises **v** and **w**;

**u**, **v**, **w**, **b** have known probability-density functions; they are stationary and independent;

the input–output operator is given by

$$x(k) = F(\mathbf{u}, \mathbf{b}, k)$$

the way the noises are added to the signal is known. These channels have no memory; $p(z(k)|u(k))$ and $p(y(k)|x(k))$ can be determined.

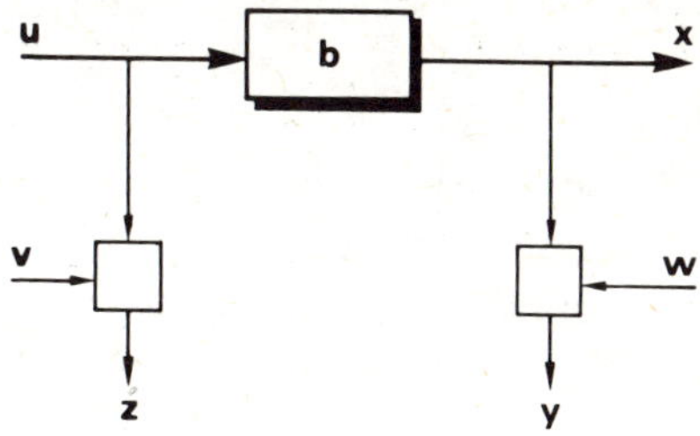


Figure 11.4

It is required to determine a 'best' estimate of **b**:

$$\boldsymbol{\beta} = \boldsymbol{\beta}(z, y) \tag{11.12}$$

Again statistically the best thing one can do is to determine the *a posteriori* density

$$p(\boldsymbol{\beta}|\mathbf{z}, \mathbf{y}) \tag{11.13}$$

for all $\boldsymbol{\beta}$. Taking into account a risk function, a decision rule

$$\boldsymbol{\beta} = \Gamma(\mathbf{z}, \mathbf{y}) \tag{11.14}$$

has to be constructed such that it fulfills the minimum risk criterion. This rule may be either deterministic or stochastic. A stochastic rule is necessary to minimize the risk in the case of a 'game' with an intelligent opponent. For the estimation situation with 'nature' as 'opponent' a deterministic rule has to be used.

Maslov discusses the cases: (1) where **z** and **y** are used for estimation, and: (2) where only **y** is being used. We restrict our summary to the first case. Now the conditional risk for given observations **z** and **y** is defined as:

$$\mathscr{E}[C(\boldsymbol{\beta}, \mathbf{b})|\mathbf{z}, \mathbf{y}] = \int_{\Omega(\boldsymbol{\beta},\mathbf{b})} C(\boldsymbol{\beta}, \mathbf{b})p(\boldsymbol{\beta}, \mathbf{b}|\mathbf{z}, \mathbf{y})\, \mathrm{d}\Omega(\boldsymbol{\beta}, \mathbf{b}) \tag{11.15}$$

The average risk for different experiments is given as

$$R(\boldsymbol{\beta}) = \int_{\Omega(\boldsymbol{\beta},\mathbf{b},\mathbf{z},\mathbf{y})} C(\boldsymbol{\beta}, \mathbf{b})p(\boldsymbol{\beta}, \mathbf{b}|\mathbf{z}, \mathbf{y})p(\mathbf{z}, \mathbf{y})\, \mathrm{d}\Omega(\boldsymbol{\beta}, \mathbf{b}, \mathbf{z}, \mathbf{y}) \tag{11.16}$$

Using Bayes' rule, the independence of the noises and the physical realizability of the process, Maslov derives a somewhat more general expression for the average risk. He shows that the optimal estimate $\hat{\boldsymbol{\beta}}$ is found by means of a deterministic decision rule. For further details and some examples the reader is referred to the paper cited.

Rather few publications have appeared on the use of Bayes' method in parameter-estimation techniques (Galtieri, 1963; Kerr, 1970). This is probably due to the computational problems when evaluating the conditional expectations. McGhee and Walford (1968) propose a solution by using a Monte Carlo approach.

## 11.2 Maximum Likelihood Estimation (M.L.E.)

Essential aspects of the M.L.E. have been summarized in Sections 2.2 and 5.2. A condition for its application is the availability of *a priori* knowledge of the probability-density function for the observable sample variables $\mathbf{y}$. This functional relationship will be denoted by

$p\{\mathbf{y};\mathbf{b}\}$ if $\mathbf{b}$ is known and our attention is focussed on $p(\mathbf{y})$

$L\{\mathbf{c};\boldsymbol{\beta}\}$ if $\mathbf{y}$ is known (observed), $\mathbf{y} = \mathbf{c}$, and an estimate $\boldsymbol{\beta}$ has to be found

The M.L. estimates $\hat{\boldsymbol{\beta}}$ of $\mathbf{b}$ are those for which the likelihood function $L$ is maximized, i.e. for which the necessary condition is fulfilled:

$$\frac{\partial}{\partial \mathbf{b}} \ln L\{\mathbf{y};\mathbf{b}\}|_{\mathbf{b}=\hat{\boldsymbol{\beta}}} = \mathbf{0} \tag{11.17}$$

One has to find the absolute maximum of $L$.

*Example.* The following case may illustrate this technique. Let $b$ be a parameter to be estimated and $n(i)$ additive noise samples which are independent and given by

$$p(n) = \frac{1}{(2\pi)^{\frac{1}{2}}\sigma} \exp\left[-\frac{1}{2}\frac{n^2}{\sigma^2}\right]$$
$$\mathscr{E}[n] = 0 \qquad \mathscr{E}[n^2] = \sigma^2 \tag{11.18}$$

The observed sample values are:

$$y(i) = b + n(i)$$

This leads to a likelihood function:

$$L\{\mathbf{y};b\} = \prod_{i=1}^{k} p\{y(i);b\} = \frac{1}{(2\pi)^{k/2}\sigma^k} \prod_{i=1}^{k} \exp\left[-\frac{1}{2}\left\{\frac{y(i)-b}{\sigma}\right\}^2\right] \tag{11.19}$$

and by taking the logarithm:

$$\ln L\{\mathbf{y};b\} = c - \frac{1}{2\sigma^2}\sum_{i=1}^{k} \{y(i) - b\}^2 \tag{11.20}$$

Differentiation and equating to zero yields:

$$\frac{\partial}{\partial b}[\ln L\{\mathbf{y}; b\}]_{b=\hat{\beta}} = -\frac{1}{\sigma^2}\sum_{i=1}^{k}\{y(i) - \hat{\beta}\} = 0 \tag{11.21}$$

or

$$\hat{\beta} = \frac{1}{k}\sum_{i=1}^{k} y(i) \tag{11.22}$$

as anticipated for this simple case. Just as an exercise and as a comparison with Bayes' method it pays to rework the example of the previous section as given by Figure 11.3 for the M.L.E.

In Section 5.2 the relation between the Markov and the least-squares estimators and the M.L. estimate has been indicated.

### *11.2.1 The Achievable Accuracy*

The accuracy obtainable from M.L.E. in terms of variances and covariances can be derived along the following lines (Kendall and Stuart, 1961). We consider the likelihood function of $k$ observations in the case of a one-dimensional parameter $b$:

$$L\{y(1), \ldots, y(k); b\} \tag{11.23}$$

and

$$\int_k L\,\mathrm{d}y(1), \ldots, \mathrm{d}y(k) = \int_k L\,\mathrm{d}^k y = 1 \tag{11.24}$$

Differentiation with respect to $b$ leads to:*

$$\int_k \frac{\partial L}{\partial b}\,\mathrm{d}^k\mathbf{y} = 0 \tag{11.25}$$

which can be rewritten as

$$\int_k \left(\frac{1}{L}\frac{\partial L}{\partial b}\right) L\,\mathrm{d}^k\mathbf{y} = \int_k \left(\frac{\partial \ln L}{\partial b}\right) L\,\mathrm{d}^k\mathbf{y} = \mathscr{E}\left[\frac{\partial \ln L}{\partial b}\right] = 0 \tag{11.25a}$$

Differentiating again yields:

$$\int_k \left\{\left(\frac{\partial \ln L}{\partial b}\right)^2 + \frac{\partial^2 \ln L}{\partial b^2}\right\} L\,\mathrm{d}^k y = 0 \tag{11.26}$$

or

$$\mathscr{E}\left[\left(\frac{\partial \ln L}{\partial b}\right)^2\right] = -\mathscr{E}\left[\frac{\partial^2 \ln L}{\partial b^2}\right] \tag{11.27}$$

* Here and in the following we assume the derivatives indicated do exist.

Now we consider $\hat{\beta} = \hat{\beta}(\mathbf{y})$ as an estimator for $b$:

$$\mathscr{E}[\hat{\beta}] = \int_k \hat{\beta} L \, \mathrm{d}^k\mathbf{y} = b + \lambda(b) = f(b) \tag{11.28}$$

where $\lambda(b)$ represents the bias that might be present in the estimate. Upon differentiation

$$\int_k \hat{\beta} \frac{\partial \ln L}{\partial b} L \, \mathrm{d}^k\mathbf{y} = 1 + \frac{\mathrm{d}\lambda(b)}{\mathrm{d}b} \tag{11.29}$$

or, using equations (11.25a) and (11.28):

$$\int_k (\hat{\beta} - f(b)) \frac{\partial \ln L}{\partial b} L \, \mathrm{d}^k\mathbf{y} = 1 + \frac{\mathrm{d}\lambda(b)}{\mathrm{d}b} \tag{11.30}$$

From the Cauchy–Schwarz inequality this can be written as

$$\left\{ \int_k \{\hat{\beta} - f(b)\}^2 L \, \mathrm{d}^k\mathbf{y} \right\} \left\{ \int_k \left( \frac{\partial \ln L}{\partial b} \right)^2 L \, \mathrm{d}^k\mathbf{y} \right\} \geqslant \left( 1 + \frac{\mathrm{d}\lambda(b)}{\mathrm{d}b} \right)^2 \tag{11.31}$$

or using equations (11.28) and (11.27):

$$\operatorname{var}[\hat{\beta}] = \mathscr{E}[\{\hat{\beta} - f(b)\}^2] \geqslant \frac{\left( 1 + \dfrac{\mathrm{d}\lambda(b)}{\mathrm{d}b} \right)^2}{\mathscr{E}\left[ \left( \dfrac{\partial \ln L}{\partial b} \right)^2 \right]} = - \frac{\left( 1 + \dfrac{\mathrm{d}\lambda(b)}{\mathrm{d}b} \right)^2}{\mathscr{E}\left[ \dfrac{\partial^2 \ln L}{\partial b^2} \right]} \tag{11.32}$$

This is called the **Cramér–Rao (in)equality** (Rao, 1945; Cramér, 1946). For unbiased estimators $\lambda(b) = 0$ and:

$$\mathscr{E}[\hat{\beta}] = f(b) = b \quad \operatorname{var}[\hat{\beta}] \geqslant - \frac{1}{\mathscr{E}\left[ \dfrac{\partial^2 \ln L}{\partial b^2} \right]} \tag{11.32a}$$

for biased estimators $\lambda(b) \neq 0$ and:

$$\mathscr{E}[\hat{\beta}] = b + \lambda(b) \quad \operatorname{var}[\hat{\beta}] \geqslant - \frac{\left( 1 + \dfrac{\mathrm{d}\lambda}{\mathrm{d}b} \right)^2}{\mathscr{E}\left[ \dfrac{\partial^2 \ln L}{\partial b^2} \right]} \tag{11.32b}$$

In the more-dimensional case of unbiased estimates one finds (cf. Appendix C):

$$\operatorname{cov}[\hat{\boldsymbol{\beta}}] = \mathscr{E}[(\hat{\boldsymbol{\beta}} - \mathbf{b})(\hat{\boldsymbol{\beta}} - \mathbf{b})'] \geqslant \mathbf{J}^{-1} \tag{11.32c}$$

with

$$\mathbf{J} = \mathscr{E}\left[\left(\frac{\partial \ln L}{\partial \mathbf{b}}\right)\left(\frac{\partial \ln L}{\partial \mathbf{b}}\right)'\right] = -\mathscr{E}\left[\frac{\partial^2 \ln L}{\partial \mathbf{b} \partial \mathbf{b}'}\right] \tag{11.33}$$

**J** is called the **information matrix** (Fisher).

As an application of the Cramér–Rao inequality consider the case given by

$$\mathbf{y} = \mathbf{U}\mathbf{b} + \mathbf{n}$$

$$p(\mathbf{n}) = \frac{1}{(2\pi)^{k/2}|\mathbf{N}|^{\frac{1}{2}}} \exp\left[-\tfrac{1}{2}\mathbf{n}'\mathbf{N}^{-1}\mathbf{n}\right]$$

which leads to the likelihood function

$$L\{\mathbf{y};\mathbf{b}\} = \frac{1}{(2\pi)^{k/2}|\mathbf{N}|^{\frac{1}{2}}} \exp\left[-\tfrac{1}{2}(\mathbf{y} - \mathbf{U}\mathbf{b})'\mathbf{N}^{-1}(\mathbf{y} - \mathbf{U}\mathbf{b})\right] \tag{11.34}$$

$$\ln L\{\mathbf{y};\mathbf{b}\} = c - \tfrac{1}{2}(\mathbf{y} - \mathbf{U}\mathbf{b})'\mathbf{N}^{-1}(\mathbf{y} - \mathbf{U}\mathbf{b})$$

$$\frac{\partial \ln L}{\partial \mathbf{b}} = +\mathbf{U}'\mathbf{N}^{-1}(\mathbf{y} - \mathbf{U}\mathbf{b})$$

$$\frac{\partial}{\partial \mathbf{b}'}\left\{\frac{\partial \ln L}{\partial \mathbf{b}}\right\} = -\mathbf{U}'\mathbf{N}^{-1}\mathbf{U}$$

$$\mathbf{J} = \mathscr{E}[\mathbf{U}'\mathbf{N}^{-1}\mathbf{U}] = \mathbf{U}'\mathbf{N}^{-1}\mathbf{U} \tag{11.35}$$

Consequently, the covariance is restricted to

$$\operatorname{cov}[\hat{\boldsymbol{\beta}}] \geqslant [\mathbf{U}'\mathbf{N}^{-1}\mathbf{U}]^{-1} \tag{11.36}$$

A comparison of this expression with equation (6.46), the covariance of the Markov estimator, is interesting. Considering the minimum variance properties of the Markov estimator and the correspondence between Markov and M.L.E. for Gaussian noise, this does not surprise us.

The $\mathscr{E}[\ ]$ in equation (11.35) has been dropped on the assumption that **u** can be measured without error. Consequently **J** is known *a posteriori*. If one wants to have an ***a priori*** guess of **J** then the expectation operator cannot be dropped if **u** is (*a priori*) a stochastic signal. If **n** is white noise then

$$\mathbf{N}^{-1} = \frac{1}{\sigma_n^2}\mathbf{I}$$

and

$$\mathbf{J} = \frac{1}{\sigma_n^2}\mathscr{E}[\mathbf{U}'\mathbf{U}] \tag{11.37}$$

We may continue this discussion of obtainable accuracy with a simple example viz. the estimation of two parameters $\boldsymbol{\beta}' = \{\beta_0, \beta_1\}$ for the case where $\mathbf{u}_0$ and $\mathbf{u}_1$ represent the same sequence of samples, shifted in time over one sampling interval. In this case, for ergodic $\mathbf{u}_0$:

$$\mathscr{E}[\mathbf{U}'\mathbf{U}] = k\begin{bmatrix} \psi_{uu}(0) & \psi_{uu}(1) \\ \psi_{uu}(1) & \psi_{uu}(0) \end{bmatrix} = k\sigma_u^2\begin{bmatrix} 1 & \rho \\ \rho & 1 \end{bmatrix} \tag{11.38}$$

and the minimum covariance is given by

$$\operatorname{cov}[\hat{\boldsymbol{\beta}}] = \frac{\sigma_n^2}{k\sigma_u^2}\frac{1}{1-\rho^2}\begin{bmatrix} 1 & -\rho \\ -\rho & 1 \end{bmatrix} = c\begin{bmatrix} 1 & -\rho \\ -\rho & 1 \end{bmatrix} \tag{11.39}$$

For $\rho \neq 0$ the off-diagonal terms represent a correlation between the errors in $\beta_0$ and $\beta_1$. By means of a simple transformation one may study the variance of the vector components $\mathbf{T}\boldsymbol{\beta}$, e.g.

$$\mathbf{T}\boldsymbol{\beta} = \frac{1}{\sqrt{2}}\begin{bmatrix} 1 & 1 \\ -1 & 1 \end{bmatrix}\begin{bmatrix} \beta_0 \\ \beta_1 \end{bmatrix} = \frac{1}{\sqrt{2}}\begin{bmatrix} \beta_0 + \beta_1 \\ -\beta_0 + \beta_1 \end{bmatrix} = \boldsymbol{\gamma}$$

Consequently

$$\operatorname{cov}[\hat{\boldsymbol{\gamma}}] = \frac{c}{2}\begin{bmatrix} 1 & 1 \\ -1 & 1 \end{bmatrix}\begin{bmatrix} 1 & -\rho \\ -\rho & 1 \end{bmatrix}\begin{bmatrix} 1 & -1 \\ 1 & 1 \end{bmatrix} = c\begin{bmatrix} 1-\rho & 0 \\ 0 & 1+\rho \end{bmatrix} \tag{11.40}$$

Now the off-diagonal terms are zero and the components of $\hat{\boldsymbol{\gamma}}$ are not correlated. This can be illustrated by Figure 11.5, where the so-called **concentration**

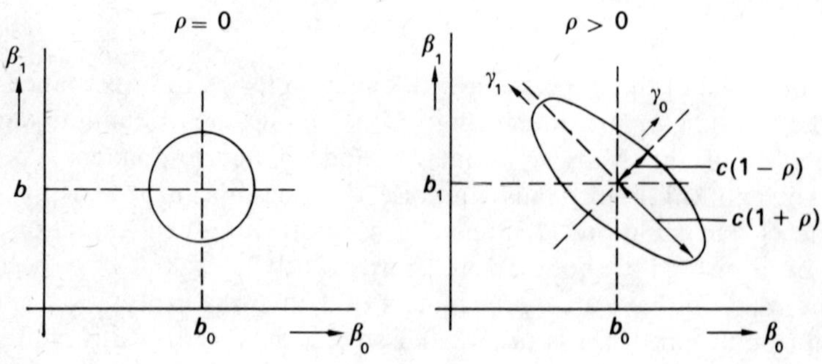


Figure 11.5

**ellipsoids** are given. These provide us with a pictorial presentation of the accuracy with which the parameters can be determined, taking into account the correlation of the errors. With increasing number of samples $k$, the concentration ellipsoids shrink to the point $(b_0, b_1)$ as $k$ is contained in the denominator of equation (11.39).

### *11.2.2 Some of the Most Important Properties*

For the case that $k$, the number of samples available, is constant an unbiased estimate is called **efficient** if in the Cramér–Rao relation (equation 11.32) the equal sign holds. The **efficiency** is defined as:

$$\eta(\beta) = \frac{\min \operatorname{var}[\hat{\beta}]}{\operatorname{var}[\beta]} \tag{11.41}$$

An estimate $\hat{\beta}$ is called **sufficient** if the following holds:

$$p\{y(1), \ldots, y(k); b\} = g\{\hat{\beta}; b\} H\{y(1), \ldots, y(k)\} \tag{11.42}$$

From Cramér (1946, p. 499):

if an efficient estimate $\hat{\beta}$ of $b$ exists, then the likelihood equation will have a unique solution equal to $\hat{\beta}$;

if a sufficient estimate $\hat{\beta}$ of $b$ exists, then any solution of the likelihood equation will be a function of $\hat{\beta}$.

Not all maximum likelihood estimates are either unbiased or efficient.

For the case that $k$, the number of samples available, increases one can study the asymptotic behaviour of $\hat{\beta}\{y(1), \ldots, y(k)\}$. If $\hat{\beta}$ converges in probability to $b$ as $k \to \infty$ then $\hat{\beta}$ is called a **consistent estimate;**

$$\lim_{k \to \infty} P[|\hat{\beta}\{y(1), \ldots, y(k)\} - b| \leqslant \epsilon] = 1 \tag{11.43}$$

with $\epsilon$ arbitrarily small. Each M.L.E. is consistent. In Åström and Bohlin (1966), a process is called identifiable if the estimate is consistent. A necessary condition is that the information matrix, associated with the estimation problem, is positive definite. If the efficiency $\eta(\hat{\beta}) \to 1$ as $k \to \infty$ then the estimate is called **asymptotically efficient.** It can be proved that the maximum likelihood estimates are asymptotically efficient. In addition, they have the property of being **asymptotically normal,** i.e. they approximate the normal distribution with mean $\mathbf{b}$ and variance $\mathbf{J}^{-1}$.

## 11.3 Some Implementation Schemes

An interesting application of maximum likelihood estimation to process-parameter estimation is given by Åström and Bohlin (1966). Their process is characterized by a linear, time-invariant difference equation. The process is of the single input, single output type:

$$A^*(z^{-1})y(k) = B^*(z^{-1})u(k) + \lambda C^*(z^{-1})n(k) \tag{11.44}$$

with

$$z = \text{shift operator}: z^{-1}x(k) = x(k-1)$$

$$u(k) = \text{process input}$$

$$y(k) = \text{process output}$$

$$n(k) = \text{additive noise, independent, with a Gaussian distribution, } \sigma_n^2 = 1$$

$$\lambda = \text{level of the noise signal.}$$

$$A^*(z^{-1}) = 1 + a_1 z^{-1} + \cdots + a_n z^{-n}$$

$$B^*(z^{-1}) = b + b_1 z^{-1} + \cdots + b_n z^{-n}$$

$$C^*(z^{-1}) = 1 + c_1 z^{-1} + \cdots + c_n z^{-n}$$

In Figure 11.6a this relation is represented as a block diagram. The polynomials $A^*$, $B^*$ and $C^*$ are assumed to have no common factors; some

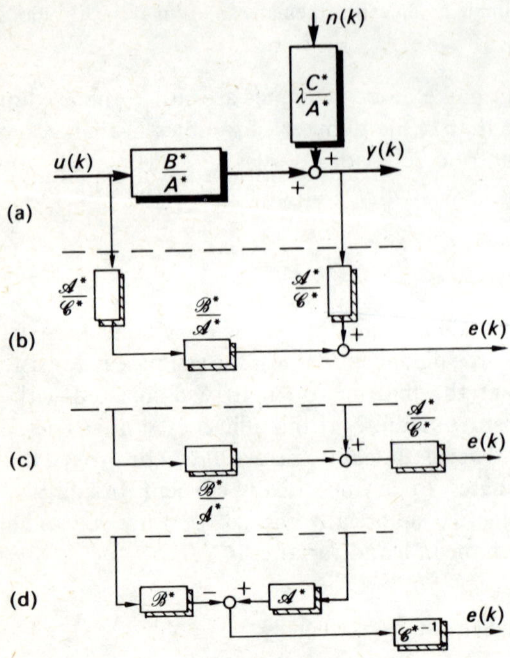


Figure 11.6

coefficients may be zero; the zeros of $A^*$ and $C^*$ all lie inside the unit circle. As a 'model', one may choose the relation

$$\mathscr{C}^*(z^{-1})e(k) = \mathscr{A}^*(z^{-1})y(k) - \mathscr{B}^*(z^{-1})u(k) \tag{11.45}$$

which may be represented by Figures 11.6b, c or d. The blocks $\mathscr{A}^*/\mathscr{C}^*$ may be recognized as noise-whitening filters. It is worth noting that the equations (11.44) and (11.45) also can be written in vector form:

$$\mathbf{ay} = \mathbf{bu} + \lambda\mathbf{cn} \tag{11.46}$$

$$\boldsymbol{\alpha}\mathbf{y} = \boldsymbol{\beta}\mathbf{u} + \boldsymbol{\gamma}\mathbf{e} \quad \text{or} \quad \boldsymbol{\gamma}\mathbf{e} = \boldsymbol{\alpha}\mathbf{y} - \boldsymbol{\beta}\mathbf{u} \tag{11.47}$$

In passing it may be mentioned that this scheme includes quite a number of distinguishable problems (Åström), for example:

| | | |
|---|---|---|
| $\mathbf{y} = \mathbf{bu}$ | $+ \lambda\mathbf{n}$ | regression |
| $\mathbf{ay} =$ | $\lambda\mathbf{n}$ | autoregression |
| $\mathbf{y} =$ | $\lambda\mathbf{cn}$ | estimation of a parameter in a moving average |
| $\mathbf{ay} =$ | $\lambda\mathbf{cn}$ | estimation of a rational power spectrum |
| $\mathbf{ay} = \mathbf{bu}$ | $+ \mathbf{n}$ | least-squares estimation |
| $\mathbf{ay} = \mathbf{bu}$ | $+ \lambda\mathbf{cn}$ | Markov estimation |
| $\mathbf{a}(\mathbf{y} - \lambda\mathbf{n}) = \mathbf{bu}$ | | estimation of a noise-free process with measurement errors |

Based on the knowledge that the noise $\mathbf{n}$ has a Gaussian distribution, the likelihood function for $\mathbf{e}' = \{e(1), \ldots, e(k)\}$ can be found as:

$$L(\mathbf{e}; \mathbf{u}, \boldsymbol{\alpha}, \boldsymbol{\beta}, \boldsymbol{\gamma}, \lambda) = \frac{1}{(2\pi)^{k/2}\lambda^k} \exp\left[-\frac{\mathbf{e}'\mathbf{e}}{2\lambda^2}\right] \tag{11.48}$$

or

$$\ln L = -\frac{k}{2}\ln 2\pi - k\ln\lambda - \frac{\mathbf{e}'\mathbf{e}}{2\lambda^2} \tag{11.49}$$

The M.L. estimate $\hat{\lambda}$ follows from

$$\frac{\partial \ln L}{\partial \lambda} = \left[-\frac{k}{\lambda} + \frac{\mathbf{e}'\mathbf{e}}{\lambda^3}\right]\Bigg|_{\lambda=\hat{\lambda}} = 0 \quad \text{or} \quad \hat{\lambda}^2 = \frac{\mathbf{e}'\mathbf{e}}{k} \tag{11.50}$$

For the following discussion there is an advantage in putting

$$\boldsymbol{\theta} = \begin{bmatrix} \boldsymbol{\alpha} \\ \hdashline \boldsymbol{\beta} \\ \hdashline \boldsymbol{\gamma} \end{bmatrix}, \qquad \mathbf{e} = \mathbf{e}(\boldsymbol{\theta}) \quad \text{and} \quad \mathbf{e}'\mathbf{e} = 2E(\boldsymbol{\theta})$$

where $\boldsymbol{\theta}$ represents the vector of parameters to be determined. From equation (11.47) and from Figure 11.6 it is clear that $\mathbf{e}$ is linear in $\boldsymbol{\alpha}$ and $\boldsymbol{\beta}$ and non-linear in $\gamma$. Consequently, finding the maximum of $L$ or $\ln L$ by differentiation and equating to zero is not necessarily a simple procedure. Åström has chosen the Newton–Raphson procedure:

$$\boldsymbol{\theta}(i+1) = \boldsymbol{\theta}(i) - \left[\left[\frac{\partial^2 E(\boldsymbol{\theta})}{\partial \boldsymbol{\theta}' \, \partial \boldsymbol{\theta}}\right]^{-1} \frac{\partial E(\boldsymbol{\theta})}{\partial \boldsymbol{\theta}}\right]_{\boldsymbol{\theta} = \boldsymbol{\theta}(i)} \tag{11.51}$$

For further details the reader is referred to the paper cited.

Other references on the use of the maximum likelihood method for parameter estimation are: Åström (1964), Åström, Bohlin and Wensmark (1965), Åström and Bohlin (1966), Gustavsson (1969), Woo (1970), Tretter and Steiglitz (1967).

The maximum likelihood procedure can also be interpreted as finding the coefficients of the prediction model

$$\begin{aligned} y_M(k) &= \hat{y}(k|k-1) \\ &= \frac{\mathscr{B}^*(z^{-1})}{\mathscr{C}^*(z^{-1})} u(k) - \frac{\mathscr{A}^*(z^{-1}) - \mathscr{C}^*(z^{-1})}{\mathscr{C}^*(z^{-1})} y(k) \end{aligned} \tag{11.52}$$

in such a way that the criterion

$$E = \sum_{i=1}^{k} \{y(i) - y_M(i)\}^2 = \sum_{i=1}^{k} e^2(i) \tag{11.53}$$

is as small as possible.

Notice, that (11.44) can also be written as

$$A^*(z^{-1}) \left[\frac{1}{C^*(z^{-1})} y(k)\right] = B^*(z^{-1}) \left[\frac{1}{C^*(z^{-1})} u(k)\right] + \lambda n(k) \tag{11.54}$$

This means that the maximum likelihood method can also be interpreted as a generalized least-squares method where the filter function $G = 1/C$ is determined from the model (11.44) (cf. Åström, 1967; 1970). The maximum likelihood method has been applied extensively to industrial measurements. See, for example, Åström (1964; 1967), and Gustavsson (1970). In the last-mentioned paper, comparisons with other techniques such as correlation methods and generalized least squares are also given. The maximum likelihood method has also been applied to time-series analysis (put $B = 0$). The maximum likelihood estimate is a strongly non-linear function of the parameters. Since time-series analysis is concerned mostly with quadratic functions, such as covariances and spectral densities, one might expect that the estimates can be expressed as non-linear functions of the sample covariances. Estimates of this nature which are asymptotically equivalent to the maximum likelihood estimates for parametric time-series analysis have been given by Zetterberg (1969). The generalization of the maximum likelihood method to the multivariable case has been done by Woo (1970) and Caines (1970).

## 11.4 Requirements with Respect to the Process-input Signal

In Section 11.2, the information matrix proved to be an important measure for the accuracy obtainable by the estimation method (cf. equation 11.33). From equations (11.37) and (11.38) it is clear how for the given process description the process-input signal **u** affects that accuracy. The question can be asked as to what conditions **u** has to satisfy in order to get consistent estimates. This problem has been studied by Åström and Bohlin (1966), Åström (1968), Aoki and Staley (1969a). It follows that for those conditions, the limits

$$\bar{u} = \lim_{k \to \infty} \frac{1}{k} \sum_{i=1}^{k} u(i)$$

and

$$R_{uu}(j) = \lim_{k \to \infty} \frac{1}{k} \sum_{i=1}^{k} \{u(i) - \bar{u}\}\{u(i+j) - \bar{u}\} \tag{11.55}$$

have to exist and the matrix $A_n$ defined by

$$A_n = [a_{jl} = R_{uu}(j - l)] \qquad j, l = 1, \ldots, n \tag{11.56}$$

has to be positive definite. These conditions are called **persistent excitation** (of the order $n$). Notice, that if the 'mean-square Fourier transform' of the input, i.e.

$$f_u(\omega) = \lim_{k \to \infty} \frac{1}{\sqrt{k}} \sum_{i=1}^{k} \{u(i) - \bar{u}\}\varepsilon^{j\omega i} \tag{11.57}$$

exists, then

$$R_{uu}(\tau) = \int |f_u(\omega)|^2 \varepsilon^{j\omega\tau}\, d\omega \tag{11.58}$$

and the matrix $A_n$ given by (11.56) is automatically non-negative definite for arbitrary $n$. Moreover, if

$$f_u(\omega) > 0 \tag{11.59}$$

the matrix $A_n$ is also positive definite for arbitrary $n$. The condition (11.59) is thus sufficient to guarantee that an input signal is persistently exciting of arbitrary order. Notice as well that if the input signal is an ergodic stationary stochastic process then its spectral density is given by

$$\phi_{uu}(\omega) = |f_u(\omega)|^2 \tag{11.60}$$

The condition (11.59) then implies that the spectral density of the input signal does not vanish for any $\omega$, a condition which is well known for correlation analysis.

Apart from persistent excitation many applications will require that the output is kept within specified limits during the experiment. The problem of designing input signals, energy and time constrained, which are optimal e.g. in the sense that they minimize the variances of the estimates, has been discussed by Levadi (1966), Aoki and Staley (1969a), (1969b), Nahi and Wallis (1969), Rault et al. (1969), Keviczky and Bányász (1973). It is closely related to the problem of optimal signal design in communication theory; see e.g. Middleton (1960).

## References

Aoki, M. and R. M. Staley (1969a). 'On input signal synthesis in parameter identification', *4th IFAC Congr., Warsaw*, paper 26.2. Also: *Automatica*, **6,** 431–440.

Aoki, M. and R. M. Staley (1969b). 'Some computational considerations in input signal synthesis problems', in *Computing Methods in Optimization Problems*, Vol. 2 (Eds. Zadeh, Neustadt and Balakrishnan), Academic Press, New York.

Åström, K. J. (1964). Cf. references Chapter 4.

Åström, K. J., T. Bohlin and S. Wensmark (1965). Cf. references Chapter 4.

Åström, K. J. and T. Bohlin (1966). 'Numerical identification of linear dynamic systems from normal operating records', *IFAC Symp.—Theory of Self-adaptive Control Systems, Teddington*, England, in *Theory of Self-Adaptive Control Systems* (Ed. P. H. Hammond), Plenum Press, New York.

Åström, K. J. (1967). 'On the achievable accuracy in identification problems', *IFAC Symp.—Identification in Automatic Control Systems, Prague*, paper 1.8.

Åström, K. J. (1967). 'Computer control of a paper machine—an application of linear stochastic control theory', *IBM J. Res. and Develop.*, **11,** 389–405.

Åström, K. J. (1968). Cf. references Chapter 6.

Åström, K. J. (1970). Cf. References Chapter 4.

Bucy, R. S. (1965). 'Nonlinear filtering theory', *IEEE Trans. Autom. Control,* **AC-10**, 198.

Caines, P. (1970). 'Maximum likelihood identification of multivariable systems', *Ph.D. Thesis, Imperial College, London.*

Cramér, H. (1946). Cf. references Chapter 6.

Deutsch, R. (1965). Cf. references Chapter 5.

Galtieri, C. A. (1963). 'The problem of identification of discrete time processes', *Report, Electronics Res. Lab., Univ. of Calif., Berkeley.*

Gustavsson, I. (1969). 'Maximum likelihood identification of dynamics of the Ågesta reactor and comparison with results of spectral analysis', *Report 6903, Div. Autom. Control, Inst. Technol., Lund, Sweden.*

Gustavsson, I. (1970). 'Comparison of different methods for identification of linear models for industrial processes', *2nd IFAC Symp.—Identification and Process Parameter Estimation, Prague*, paper 11.4. Also in: *Automatica*, **8,** 127–142.

Kendall, M. G. and A. Stuart (1961, 1962). Cf. references Chapter 6.

Kerr, R. B. (1970). 'Bayesian identification of system parameters', *J. Eng. Math.*, **4,** 273–281.

Keviczky, L. and Cs. Bányász (1973). 'On input signal synthesis for linear discrete-time systems', *Proc. 3rd IFAC Symp.—Identification and System Parometer Estimation*, 1011–1014, North-Holland Publ. Co., Amsterdam.

Kroy, W. H. and A. R. Stubberud (1968). 'Identification via nonlinear filtering', *Joint Autom. Control Conf.*, 394–412. Also in *Int. J. Control*, **6,** 499–522.

Kushner, H. J. (1964). 'On the differential equations satisfied by conditional probability densities of Markov processes with applications', *SIAM J. Control*, ser. A, **2,** 106–119.

Levadi, V. S. (1966). 'Design of input signals for parameter estimation', *IEEE Trans. Autom. Control*, **AC-11,** 205–211.

Maslov, E. P. (1963). Cf. references Chapter 5.

McGhee, R. B. and R. B. Walford (1968). 'A Monte Carlo approach to the evaluation of conditional expectation parameter estimates for non-linear dynamic systems', *IEEE Trans. Autom. Control*, **AC-13,** 29–37.

Middleton, D. (1960). *An Introduction to Statistical Communication Theory*, McGraw-Hill, New York, 1140 pp.

Nahi, N. E. and D. E. Wallis Jr. (1969). 'Optimal inputs for parameter estimation in dynamic systems with white observation noise', *Joint Autom. Control Conf.*, 506–513.

Rao, C. R. (1945). 'Information and accuracy attainable in the estimation of statistical parameters', *Bull. Calcutta Math. Soc.*, **37,** 81.

Rault, A., R. Pouliquen and J. Richalet (1969). 'Sensibilizing input and identification', *Proc. 4th IFAC Congr., Warsaw*, paper 5.2.

Stratonovich, R. L. (1960). 'Conditional Markov processes', *Theory Probab. Applic.*, **5,** 156–178.

Tretter, S. A. and K. Steiglitz (1967). 'Power spectrum identification in terms of rational models', *IEEE Trans. Autom. Control*, **AC-12,** 185–188.

Woo, K. T. (1970). 'Maximum likelihood identification of noisy systems', *2nd IFAC Symp.—Identification and Process Parameter Estimation, Prague*, paper 3.1.

Zetterberg, L. H. (1969). 'Estimation of parameters for a linear difference equation with application to EEG analysis', *Math. Biosci.*, **5,** 227–275.

## Additional Literature

Alspach, D. L. and H. W. Sorenson (1972). 'Nonlinear Bayesian estimation using Gaussian sum approximations', *IEEE Trans. Autom. Control*, **AC-17,** 439–448.

Aoki, M. (1965). 'On some convergence questions in Bayesian optimization problems', *IEEE Trans. Autom. Control*, **AC-10,**

Bányász, Cs., R. Haber and L. Keviczky (1973). 'Some estimation methods for nonlinear discrete time identification', *Proc. 3rd IFAC Symp.—Identification and System Parameter Estimation*, 798–802, North-Holland Publ. Co., Amsterdam.

Bohlin, T. (1970). 'On the maximum likelihood method of identification', *IBM J. Res. and Develop.*, **14,** 41–51.

Cramér, H. and M. R. Leadbetter (1967). *Stationary and Related Stochastic Processes*, Wiley, New York, 348 pp.

Davies, W. D. T. and J. L. Douce (1967). 'On-line system identification in the presence of drift', *IFAC Symp.—Identification in Autom. Control Systems, Prague*, paper 3.12.

Florentin, J. J. (1962). 'Optimal, probing, adaptive control of a simple Bayesian system', *J. Electron. Control*, ser. **I,** 13.

Helstrom, C. W. (1960). *Statistical Theory of Signal Detection*, Pergamon Press, New York, 364 pp.

Ho, Y. C. and R. C. K. Lee (1964). 'A Bayesian approach to problems in stochastic estimation and control', *IEEE Trans. Autom. Control*, **AC-9,** 333–339. Also in *Joint Autom. Control Conf.*, 382–387.

Kale, B. K. (1962). 'On the solution of likelihood equations by iteration processes—the multiparametric case', *Biometrica*, **49,** 479.

Lainiotis, D. G. (1968). 'A nonlinear adaptive estimation recursive algorithm' (correspondence), *IEEE Trans. Autom. Control*, **AC-13,** 197–198.

Liporace, L. A. (1971). 'Variance of Bayes estimates', *IEEE Trans. Int. Theory*, **IT-17,** 665–669.

Miller, R. W. and R. Roy (1964). 'Nonlinear process identification using decision theory' (short paper), *IEEE Trans. Autom. Control*, **AC-9,** 538–540.

Qvarnström, B. (1963). Cf. references Chapter 10.

Rajbman, N. S., S. A. Anisimov and F. A. Hovsepian (1969). 'Some problems of control plants identification', *4th IFAC Congr., Warsaw*, paper 5.1.

Rogers, A. E. and K. Steiglitz (1967). 'Maximum likelihood estimation of rational transfer function parameters' (short paper), *IEEE Trans. Autom. Control*, **AC-12,** 594–597; cf. also same journal, **AC-13,** 304.

Segerståhl, B. (1969). 'A theory of parameter tracking applied to slowly varying nonlinear systems', *4th IFAC Congr., Warsaw*, paper 5.4.

Seidman, L. P. (1970). 'Performance limitations and error calculations for parameter estimation', *Proc. IEEE*, **58,** 644–652.

Štěphán, J. (1970). 'The uncertainty problem in system identification', *2nd IFAC Symp.—Identification and Process Parameter Estimation, Prague*, paper 1.8.

Swerling, P. (1964). 'Parameter estimation accuracy formulas', *IEEE Trans. Inf. Theory*, **IT-10,** 302–314.

Wald, A. (1950). *Statistical Decision Functions*, Wiley, New York, 179 pp.

Ziv, J. and M. Zakai (1969). 'Some lower bounds on signal parameter estimation', *IEEE Trans. Inf. Theory*, **IT-15,** 386–391.

*CHAPTER 12*

# *Process State Estimation; A Review*

In Chapter 1 it was mentioned that the problem of parameter estimation is closely related with that of optimal filtering and state estimation. In the next chapter methods will be discussed which provide parameter and state estimates together. In this chapter we summarize some state estimation (Kalman filter) methods in order to put the next chapter in the proper perspective. Data processing schemes which separate desired signals from unwanted disturbances (noise) are referred to as filters.

The Kalman filter can be described as a method (implementation, computer programme) for combining measurement data, which are contaminated with noise; they have to be combined in such a way that an 'optimal' estimate will be obtained of some desired variable. It can be applied to different types of problems:

Filtering; interpolation, smoothing, extrapolation (prediction) of time series or signals (cf. Section 12.1).

State estimation as needed for implementation of optimal-control algorithms in cases where not all, or even no elements of the process state vector can be measured directly.

Combination of measurement data on some desired variable(s) provided by different measuring devices, each of which has its own type of errors (e.g. in navigation problems: gyroscopes, accelerometers, doppler radar, etc.).

Parameter estimation, by restating the problem with a parameter instead of a state vector.

For a highly interesting discussion of the historic perspective of these techniques the reader is referred to Sorenson (1970), who traces back the use of the least-squares method to applications made by K. F. Gauss in 1795 and shows its intimate relationship with Kalman filtering.

## 12.1 The Wiener Filtering Problem

During the second world war Wiener did his fundamental and, by now, classical studies on interpolation, extrapolation and smoothing of time series (Wiener, 1949). For our purpose we can restrict the attention to the situation sketched in Figure 12.1.

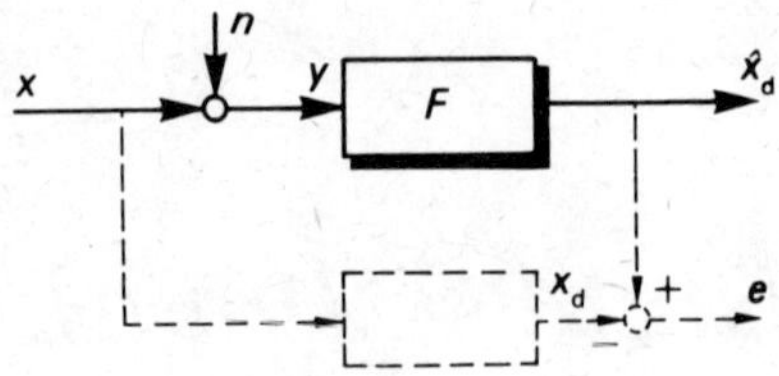


Figure 12.1

| | Continuous case | Sampled case |
|---|---|---|
| $x$ = information-carrying signal | $x(t)$ | $x(k)$ |
| $n$ = disturbing signal (observation noise) | $n(t)$ | $n(k)$ |
| $x_d$ = desired signal: | | |
| interpolated | $x(t - \tau)$ | $x(k - m)$ |
| filtered | $x(t)$ | $x(k)$ |
| extrapolated, predicted | $x(t + \tau)$ | $x(k + m)$ |
| | $\tau > 0$ | $m > 0$ |
| $\hat{x}_d$ = estimation of $x_d$, obtained by means of the filter $F$ from the signal $x + n$ | $\hat{x}_d(t)$ | $\hat{x}_d(k)$ |
| $e = \hat{x}_d - x_d$ = error signal that, by the choice of $F$, has to be minimized in the least-squares sense. | $e(t)$ | $e(k)$ |

Of course the choice of $F$ has to be based on the *a priori* knowledge about both $x$ and $n$. The problem can be solved in the time domain as well as in the frequency domain (cf. e.g. Solodovnikov, 1960; Sage, 1968).

Wiener assumed the signals $x(t)$ and $n(t)$ to be stationary, i.e. their statistical properties do not change with time, and to have known power spectra. The filter $F$ is restricted to be linear, time invariant and physically realizable. Consequently, it can be characterized by an impulse response $h(t)$. Assuming that $x(t)$ and $n(t)$ are sample functions of an ergodic stochastic process, a suitable criterion of 'optimality' is $\min \mathscr{E}[e^2(t)]$. For $x_d(t) = x(t)$ and $\hat{x}_d(t) = \hat{x}(t)$ this results in a minimization of

$$\mathscr{E}[e^2(t)] = \mathscr{E}[\{\hat{x}(t) - x(t)\}^2]$$

with

$$\hat{x}(t) = \int_{-\infty}^{\infty} h(\theta) y(t - \theta)\, d\theta \tag{12.1}$$

The minimization proceeds along the same lines as given before in Section 8.1 and it results in the necessary relationship expressed by the Fredholm integral equation:

$$\psi_{xy}(\tau) = \int_{-\infty}^{\infty} h(\theta)\psi_{yy}(\tau - \theta)\, d\theta \tag{12.2}$$

For physical realizability of the filter, one has to restrict the impulse response to $h(t) = 0$ for $t < 0$. With this condition we have again the Wiener–Hopf (W.H.) equation. The solution of this equation for $h(\theta)$ is in general not simple. Usually it is done by transformation to the frequency domain; here the condition of physical realizability adds a serious complication. The resulting transfer function can be written as:

$$H(s) = \frac{1}{\phi_{yy}^{+}(s)}\left[\frac{\phi_{xx}(s)}{\phi_{yy}^{-}(s)}\right]_{+} \tag{12.3}$$

where

$\phi_{yy}(s) = \phi_{yy}^{+}(s)\phi_{yy}^{-}(s)$

$\phi_{yy}^{+}(s)$ represents the poles and zeros in the left half of the $s$-plane

$\phi_{yy}^{-}(s)$ represents the poles and zeros in the right half of the $s$-plane

and

$[G(s)] = [G(s)]_{+} + [G(s)]_{-}$

$[\quad]_{+}$ indicates that only the poles in the left half of the $s$-plane are considered

The Wiener filter thus found is optimal for signals that have a Gaussian distribution. Such filters can also be derived for sampled data (cf. e.g. Ragazzini and Franklin, 1958; Jury, 1964). Some characteristics of the Wiener filtering problem are the following:

the *a priori* knowledge of the signal and noise is assumed to be given in terms of correlation functions or power spectra. This brings us to the problem of spectral factorization, i.e. the problem under what condition a spectral density function $\phi(j\omega)$ can be written as

$$\phi(j\omega) = F(j\omega)F(-j\omega)$$

cf. e.g. Åström (1970);

the determination of the optimal filter does not lend itself very well to digital computation;

the determination of the optimal filter ends with an impulse response $h(t)$ or a transfer function $H(s)$; the synthesis of the actual filter may still be quite cumbersome;

it cannot be adapted easily to the case of non-stationary signals;

it can be adapted to the case of multivariable processes, but it requires much computational effort.

## 12.2 The Kalman–Bucy Filter

The Kalman (or Kalman–Bucy) filter provides the solution for the same problem, but described in terms of state variables. Unlike the Wiener type of description it can easily be adapted to the case of non-stationary signals (Kalman, 1960; Kalman and Bucy, 1961). Usually the resulting filter is a time-varying one. In addition, the Kalman filter has a recursive form that is well suited for programming on a digital computer. The new estimate is found by updating the old one based on new observations; there is no need to store the whole string of observation data, consequently the storage and computing requirements are modest.

Even if the interest is focussed purely on signals it pays to assume these (stochastic) signals as being derived from a 'white' noise by means of a so called 'shaping filter'. In what follows such a filter will be also referred to as a 'process'.

In the general case the dynamic behaviour of a process can be given by:

$$\left.\begin{aligned} \hat{\mathbf{x}} &= \mathbf{f}(\mathbf{x}, \mathbf{u}, \mathbf{v}, t) \\ \mathbf{y} &= \mathbf{g}(\mathbf{x}, \mathbf{u}, \mathbf{n}, t) \end{aligned}\right\} \qquad (12.4)$$

cf. Figure 12.2. In the case of a linear(ized) process this reduces to

$$\left.\begin{aligned} \dot{\mathbf{x}} &= \mathbf{A}\mathbf{x} + \mathbf{B}\mathbf{u} + \mathbf{v} \\ \mathbf{y} &= \mathbf{C}\mathbf{x} + \mathbf{D}\mathbf{u} + \mathbf{n} \end{aligned}\right\}$$

cf. Figure 12.3. The functional relationships between the state variables and the other 'signals' are assumed to be known, including the numerical values of the parameters. Based on the observation of $\mathbf{y}$ over a time interval $(0, T)$ we want to find some estimate for $\mathbf{x}(t)$, denoted by $\hat{\mathbf{x}}(t|T)$. As before there are three cases:

$t > T$ prediction problem

$t = T$ filtering problem

$t < T$ smoothing problem

In line with Chapter 5 we can approach the problem by starting from the minimum risk or Bayes estimation, the maximum likelihood estimation or the least-squares estimation. For this survey we will take the first one as being the most general.

Again from the statistical point of view the crucial information is provided by the conditional probability-density function

$$p(\mathbf{x}|\mathbf{y})$$

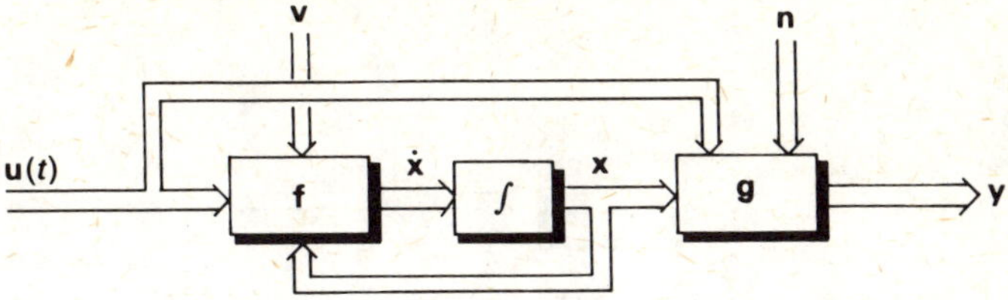


Figure 12.2

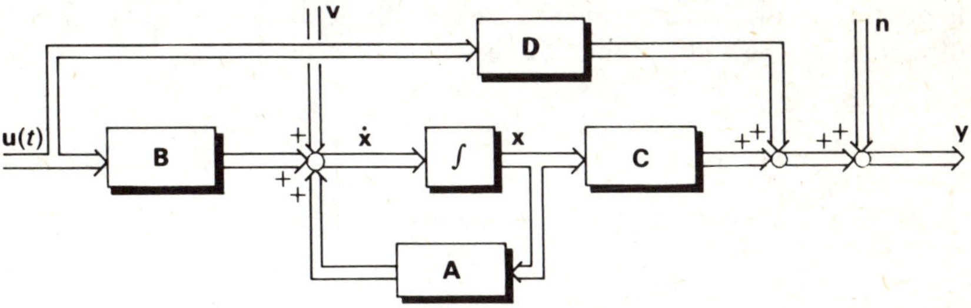


Figure 12.3

where **y** are the observed sample values of the process output. As in Chapter 5 from this probability function on can derive different estimates. Using Bayes' rule

$$p(\mathbf{x}|\mathbf{y}) = \frac{p(\mathbf{x}, \mathbf{y})}{p(\mathbf{y})} = \frac{p(\mathbf{y}|\mathbf{x})p(\mathbf{x})}{p(\mathbf{y})} \tag{12.5}$$

one can use the *a priori* knowledge about **x** and, after the observation of **y**, come up with a new *a posteriori* probability-density function for **x**. This leads quite naturally to an iterative estimation procedure, in which the outcome of the previous calculation is the *a priori* knowledge for the next calculation, using the new observations.

First we will consider the simple one-stage filter and after that the multi-stage version.

### *12.2.1 One-stage Filter*

Given is the following *a priori* knowledge

| | | |
|---|---|---|
| $\mathbf{y} = \mathbf{C}\mathbf{x} + \mathbf{n}$ | process, | cf. Figure 12.4 |
| $p(\mathbf{x})$ Gaussian; | $\mathscr{E}[\mathbf{x}] = \mathbf{x}_0$; | $\operatorname{cov}[\mathbf{x}] = \mathbf{P}_0$ |
| $p(\mathbf{n})$ Gaussian; | $\mathscr{E}[\mathbf{n}] = \mathbf{0}$; | $\operatorname{cov}[\mathbf{n}] = \mathbf{N}$ |

**x** and **n** statistically independent.

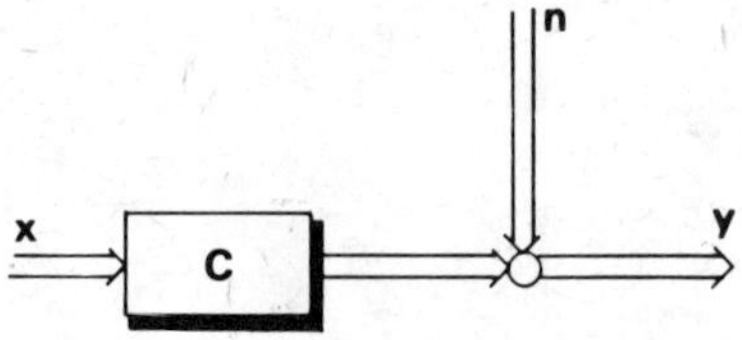


Figure 12.4

From this knowledge we can easily derive the probability-density function of $\mathbf{y}$, although we will not actually need it:

$$p(\mathbf{y}) \quad \text{Gaussian}; \qquad \mathscr{E}[\mathbf{y}] = \mathbf{C}\mathbf{x}_0; \qquad \text{cov}\,[\mathbf{y}] = \mathbf{C}\mathbf{P}_0\mathbf{C}' + \mathbf{N}$$

as

$$\begin{aligned}\text{cov}\,[\mathbf{y}] &= \mathscr{E}[\{\mathbf{y} - \mathbf{C}\mathbf{x}_0\}\{\mathbf{y} - \mathbf{C}\mathbf{x}_0\}'] \\ &= \mathscr{E}[\{\mathbf{C}(\mathbf{x} - \mathbf{x}_0) + \mathbf{n}\}\{\mathbf{C}(\mathbf{x} - \mathbf{x}_0) + \mathbf{n}\}']\end{aligned}$$

Now we want to derive an expression for $p(\mathbf{x}|\mathbf{y})$:

$$\begin{aligned}p(\mathbf{x}|\mathbf{y}) &= \frac{p(\mathbf{x}, \mathbf{y})}{p(\mathbf{y})} = \frac{p(\mathbf{x}, \mathbf{n})\,|\partial\mathbf{n}/\partial\mathbf{y}|}{p(\mathbf{y})} \\ &= \frac{p(\mathbf{x})p(\mathbf{n})}{p(\mathbf{y})} = \frac{p(\mathbf{x})p_n(\mathbf{y} - \mathbf{C}\mathbf{x})}{p(\mathbf{y})}\end{aligned} \tag{12.6}$$

The expressions for each of the probability-density functions can be substituted. Then one has to decide on what kind of estimate one wants by choosing a cost function. It has already been pointed out in Chapter 5 that in the case of a Gaussian distribution the use of any even cost function and the maximization of the density function leads to the same estimate. We will chose the maximization of the density function.

In order to save as much manipulation as possible it is worth while to note that $p(\mathbf{y})$ is known completely from our *a priori* knowledge. Consequently for our purpose it is considered as being part of the normalizing constant only and

$$p(\mathbf{x}|\mathbf{y}) = \text{const.}\exp\,[-\tfrac{1}{2}\{(\mathbf{x} - \mathbf{x}_0)'\mathbf{P}_0^{-1}(\mathbf{x} - \mathbf{x}_0) + (\mathbf{y} - C\mathbf{x})'\mathbf{N}^{-1}(\mathbf{y} - C\mathbf{x})\}] \tag{12.7}$$

This expression has to be maximized with respect to $\mathbf{x}$, which can be done by completing the square of the exponent (cf. Aoki 1967). Another way is by differentiating $\ln p(\mathbf{x}|\mathbf{y})$ and equating to $\mathbf{0}$ for the optimal estimate:

$$\{\mathbf{P}_0^{-1}(\mathbf{x} - \mathbf{x}_0) - \mathbf{C}'\mathbf{N}^{-1}(\mathbf{y} - \mathbf{C}\mathbf{x})\}|_{\mathbf{x}=\hat{\mathbf{x}}} = \mathbf{0} \tag{12.8}$$

or

$$\underbrace{[\mathbf{P}_0^{-1} + \mathbf{C}'\mathbf{N}^{-1}\mathbf{C}]}_{\mathbf{P}_1^{-1}}\hat{\mathbf{x}} = \mathbf{P}_0^{-1}\mathbf{x}_0 + \mathbf{C}'\mathbf{N}^{-1}\mathbf{y} \tag{12.9}$$

or the following expressions:

$$\hat{\mathbf{x}} = \mathbf{P}_1\mathbf{P}_0^{-1}\mathbf{x}_0 + \mathbf{P}_1\mathbf{C}'\mathbf{N}^{-1}\mathbf{y} \tag{12.10a}$$

$$\hat{\mathbf{x}} = \mathbf{x}_0 + \mathbf{P}_1\mathbf{C}'\mathbf{N}^{-1}\{\mathbf{y} - \mathbf{C}\mathbf{x}_0\} \tag{12.10b}$$

$$\hat{\mathbf{x}} = \mathbf{x}_0 + \mathbf{P}_0\mathbf{C}'[\mathbf{C}\mathbf{P}_0\mathbf{C}' + \mathbf{N}]^{-1}\{\mathbf{y} - \mathbf{C}\mathbf{x}_0\} \tag{12.10c}$$

where

$$\mathbf{P}_1 = \mathbf{P}_0 - \mathbf{P}_0\mathbf{C}'[\mathbf{C}\mathbf{P}_0\mathbf{C}' + \mathbf{N}]^{-1}\mathbf{C}\mathbf{P}_0 \tag{12.11}$$

These expressions follow from application of the matrix identities summarized in Appendix C. Equation (12.10) can be represented by Figure 12.5.

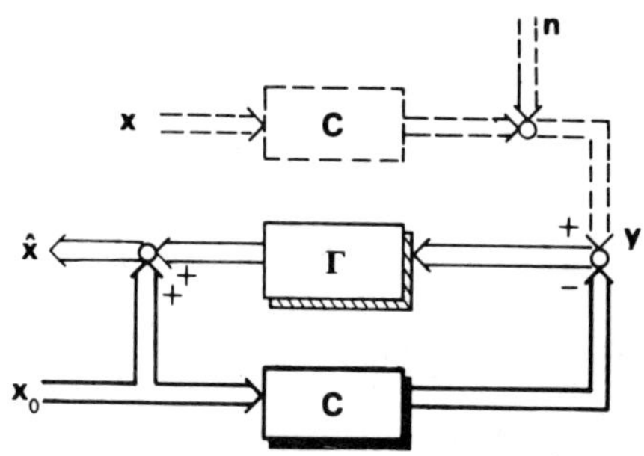


Figure 12.5

Some remarks are in order here:

Note that $\mathbf{y}$ are the only observations; all other quantities are given *a priori*. The 'gain factor' $\mathbf{\Gamma} = \mathbf{P}_0\mathbf{C}'[\mathbf{C}\mathbf{P}_0\mathbf{C}' + \mathbf{N}]^{-1}$ does not depend on the observation and can be determined beforehand.

It is easily shown that $\mathscr{E}[\hat{\mathbf{x}}] = \mathbf{x}_0$ and

$$\mathscr{E}[(\hat{\mathbf{x}} - \mathbf{x}_0)(\hat{\mathbf{x}} - \mathbf{x}_0)'] = \mathbf{P}_0\mathbf{C}'[\mathbf{C}\mathbf{P}_0\mathbf{C}' + \mathbf{N}]^{-1}\mathbf{C}\mathbf{P}_0 \tag{12.12}$$

which suggests an interpretation of equation (12.11).

Some simplifications that may aid to the insight:

$$\mathbf{C} = \mathbf{I} \qquad \hat{\mathbf{x}} = \mathbf{x}_0 + \mathbf{P}_0[\mathbf{P}_0 + \mathbf{N}]^{-1}\{\mathbf{y} - \mathbf{x}_0\}$$

$$\mathbf{P}_1 = \mathbf{P}_0 - \mathbf{P}_0[\mathbf{P}_0 + \mathbf{N}]^{-1}\mathbf{P}_0 \tag{12.13}$$

or, if $\mathbf{C}^{-1}$ exists:

$$\mathbf{N} = \mathbf{0} \qquad \begin{aligned} \hat{\mathbf{x}} &= \mathbf{x}_0 + \mathbf{C}^{-1}\mathbf{C}\mathbf{P}_0\mathbf{C}'[\mathbf{C}\mathbf{P}_0\mathbf{C}']^{-1}\{\mathbf{y} - \mathbf{C}\mathbf{x}_0\} \\ &= \mathbf{x}_0 + \mathbf{C}^{-1}\{\mathbf{y} - \mathbf{C}\mathbf{x}_0\} = \mathbf{C}^{-1}\mathbf{y} \end{aligned} \tag{12.14}$$

*Example.* Some more insight is provided if we simplify these equations to a one-dimensional case: $\mathbf{x} = x$, $\mathbf{y} = y$ such that $\mathbf{P}$, $\mathbf{N}$ and $\mathbf{C}$ become scalars: If $C = 1$

$$\hat{x} = x_0 + \frac{P_0}{P_0 + N}\{y - x_0\} \tag{12.15}$$

$$P_1 = P_0 - \frac{P_0^2}{P_0 + N} = \frac{P_0 N}{P_0 + N} \quad \text{or} \quad \frac{1}{P_1} = \frac{1}{P_0} + \frac{1}{N} \tag{12.16}$$

The terms $P_0$ and $N$ remind us of signal and noise powers in communication problems. If $N$ is comparatively large then $y - x_0$ has little influence.

### *12.2.2 Multi-stage Filter*

Given the following *a priori* knowledge:

$$\left.\begin{aligned} \mathbf{x}(i+1) &= \mathbf{A}\mathbf{x}(i) + \mathbf{v}(i) \\ \mathbf{y}(i) &= \mathbf{C}\mathbf{x}(i) + \mathbf{n}(i) \end{aligned}\right\} \text{ process, cf. Figure 12.6a} \tag{12.17}$$

$$p\{\mathbf{x}(k)|\mathbf{y}(1), \ldots, \mathbf{y}(k)\} = p\{\mathbf{x}(k)|\mathbf{Y}_k\}$$

$$p\{\mathbf{x}(k)|\mathbf{Y}_k\} \Rightarrow \mathscr{E}[\mathbf{x}(k)|\mathbf{Y}_k] = \hat{\mathbf{x}}(k|k) \qquad \operatorname{cov}[\mathbf{x}(k)|\mathbf{Y}_k] = \mathbf{P}(k)$$

$$p\{\mathbf{v}(i)\} \quad \Rightarrow \mathscr{E}[\mathbf{v}(i)] = \mathbf{0} \qquad \operatorname{cov}[\mathbf{v}(i)] = \mathbf{V}$$

$$p\{\mathbf{n}(i)\} \quad \Rightarrow \mathscr{E}[\mathbf{n}(i)] = \mathbf{0} \qquad \operatorname{cov}[\mathbf{n}(i)] = \mathbf{N}$$

$\mathbf{v}$ and $\mathbf{n}$ statistically independent.

The notation $\hat{\mathbf{x}}(k|k)$ refers to the estimate of $\mathbf{x}(k)$, based on the $k$ observations $\mathbf{y}(1), \ldots, \mathbf{y}(k)$. This type of notation can also be used for the prediction and the smoothing problem. As our attention is limited to the filtering problem it suffices to simplify the notation to $\hat{\mathbf{x}}(k)$. From this follows:

$$p\{\mathbf{x}(k+1)|\mathbf{Y}_k\} \Rightarrow \mathscr{E}[\mathbf{x}(k+1)|\mathbf{Y}_k] = \mathbf{A}\hat{\mathbf{x}}(k)$$

$$\operatorname{cov}[\mathbf{x}(k+1)|\mathbf{Y}_k] = \mathbf{A}\mathbf{P}(k)\mathbf{A}' + \mathbf{V} = \mathbf{Q}(k+1)$$

We can also derive the probability-density function for $\mathbf{y}(k+1)$ but, as before, we will not need it:

$$p\{\mathbf{y}(k+1)|\mathbf{Y}_k\} \Rightarrow \mathscr{E}[\mathbf{y}(k+1)|\mathbf{Y}_k] = \mathbf{C}\mathbf{A}\hat{\mathbf{x}}(k)$$

$$\operatorname{cov}[\mathbf{y}(k+1)|\mathbf{Y}_k] = \mathbf{C}\mathbf{Q}(k+1)\mathbf{C}' + \mathbf{N}$$

Due to the dependence on $\mathbf{Y}_k$ the formulation of Bayes' rule needed is a little more general:

$$p(a, b|c)p(c) = p(a, b, c) = p(a|b, c)p(b|c)p(c) \tag{12.18}$$

from which follows:

$$p(a|b, c) = \frac{p(a, b|c)}{p(b|c)} \tag{12.19}$$

Hence we notice a direct parallel with the previous case:

one-stage filter

$$p(\mathbf{x}|\mathbf{y}) = \frac{p(\mathbf{x})p_n(\mathbf{y} - \mathbf{C}\mathbf{x})}{p(\mathbf{y})}$$

multistage filter

$$p\{\mathbf{x}(k+1)|\mathbf{y}(k+1), \mathbf{Y}_k\} = \frac{p\{\mathbf{x}(k+1)|\mathbf{Y}_k\}p_n\{\mathbf{Y}(k+1) - \mathbf{C}\mathbf{x}(k+1)\}}{p\{\mathbf{y}(k+1)|\mathbf{Y}_k\}}$$

and for Gaussian distributions:

$$p\{\mathbf{x}(k+1)|\mathbf{y}(k+1), \mathbf{Y}_k\} = \text{const.}\exp\left[-\tfrac{1}{2}\{\|\mathbf{x}(k+1) - \mathbf{A}\hat{\mathbf{x}}(k)\|^2_{\mathbf{Q}^{-1}(k+1)} + \|\mathbf{y}(k+1) - \mathbf{C}\mathbf{x}(k+1)\|^2_{\mathbf{N}^{-1}}\}\right] \tag{12.20}$$

Differentiation of the logarithm of this function and equating it to **0** for the optimal estimate $\hat{\mathbf{x}}(k+1)$ yields:

$$\mathbf{Q}^{-1}(k+1)[\hat{\mathbf{x}}(k+1) - \mathbf{A}\hat{\mathbf{x}}(k)] - \mathbf{C}'\mathbf{N}^{-1}[\mathbf{y}(k+1) - \mathbf{C}\hat{\mathbf{x}}(k+1)] = \mathbf{0} \tag{12.21}$$

or

$$\underbrace{[\mathbf{Q}^{-1}(k+1) + \mathbf{C}'\mathbf{N}^{-1}\mathbf{C}]}_{\mathbf{P}^{-1}(k+1)}\hat{\mathbf{x}}(k+1) = \mathbf{Q}^{-1}(k+1)\mathbf{A}\hat{\mathbf{x}}(k) + \mathbf{C}'\mathbf{N}^{-1}\mathbf{C}\mathbf{y}(k+1) \tag{12.22}$$

or

$$\hat{\mathbf{x}}(k+1) = \mathbf{A}\hat{\mathbf{x}}(k) + \mathbf{Q}(k+1)\mathbf{C}'[\mathbf{C}\mathbf{Q}(k+1)\mathbf{C}' + \mathbf{N}]^{-1}[\mathbf{y}(k+1) - \mathbf{C}\mathbf{A}\hat{\mathbf{x}}(k)] \tag{12.23}$$

which is closely parallel to equation (12.10c). Both process and filter are shown in Figure 12.6. Note the similarity of the process and part of the filter.

Some characteristics of the K.-filter, to be compared with those of the Wiener filter, are the following:

the *a priori* knowledge of the signal (and noise) has to be given in terms of state equations;

the filtering algorithms are in a recursive form and lend themselves very well to digital computation;

the filter algorithm is a straightforward description of the filter realization;

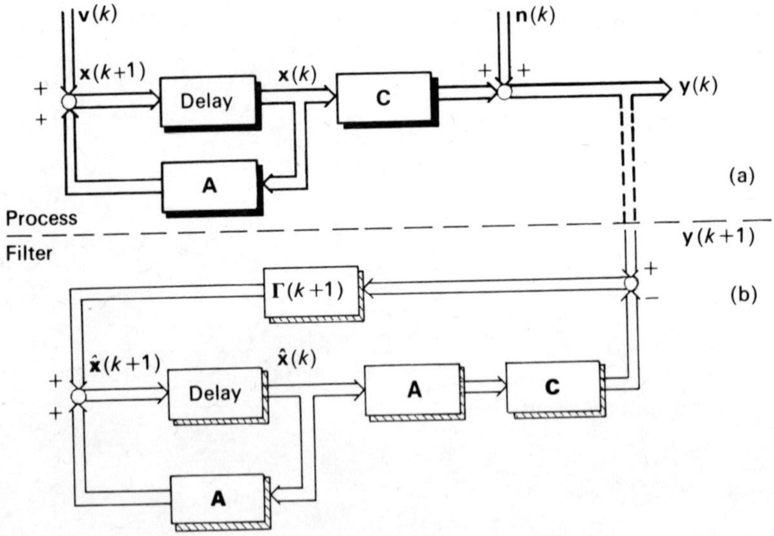


Figure 12.6

non-stationary signals can be included quite easily; this also implies the starting of observations at an arbitrary moment;

also multivariable cases can be included easily.

Among the many references cf. Meditch (1969); Sage and Melsa (1971).

If a measurable input vector $\mathbf{u}(k)$ is added:

$$\begin{aligned}\mathbf{x}(k+1) &= \mathbf{A}\mathbf{x}(k) + \mathbf{B}\mathbf{u}(k) + \mathbf{v}(k)\\ \mathbf{y}(k) &= \mathbf{C}\mathbf{x}(k) + \mathbf{n}(k)\end{aligned} \tag{12.24}$$

one finds in exactly the same way:

$$\hat{\mathbf{x}}(k+1) = \mathbf{A}\hat{\mathbf{x}}(k) + \mathbf{B}\mathbf{u}(k) + \mathbf{\Gamma}(k+1)[\mathbf{y}(k+1) - \mathbf{C}\mathbf{A}\hat{\mathbf{x}}(k) - \mathbf{C}\mathbf{B}\mathbf{u}(k)] \tag{12.25}$$

$$\mathbf{\Gamma}(k+1) = \mathbf{Q}(k+1)\mathbf{C}'[\mathbf{C}\mathbf{Q}(k+1)\mathbf{C}' + \mathbf{N}]^{-1} = \mathbf{P}(k+1)\mathbf{C}'\mathbf{N}^{-1} \tag{12.26}$$

$$\mathbf{Q}(k+1) = \mathbf{A}\mathbf{P}(k)\mathbf{A}' + \mathbf{V} \tag{12.27}$$

$$\begin{aligned}\mathbf{P}(k+1) &= [\mathbf{Q}^{-1}(k+1) + \mathbf{C}'\mathbf{N}^{-1}\mathbf{C}]^{-1} =\\ &\quad \mathbf{Q}(k+1) - \mathbf{Q}(k+1)\mathbf{C}'[\mathbf{C}\mathbf{Q}(k+1)\mathbf{C}' + \mathbf{N}]^{-1}\mathbf{C}\mathbf{Q}(k+1)\end{aligned} \tag{12.28}$$

where $\hat{\mathbf{x}}(0)$ and $\mathbf{P}_0$ are given.

In this scheme one may recognize:

| | |
|---|---|
| $\mathbf{A}\hat{\mathbf{x}}(k) + \mathbf{B}\mathbf{u}(k)$ | extrapolated state |
| $\mathbf{C}[\mathbf{A}\hat{\mathbf{x}}(k) + \mathbf{B}\mathbf{u}(k)]$ | observation of extrapolated state |
| $\mathbf{y}(k+1) - \mathbf{C}\mathbf{A}\hat{\mathbf{x}}(k) - \mathbf{C}\mathbf{B}\mathbf{u}(k)$ | 'error' between observed and predicted value |
| $\mathbf{Q}(k+1)$ | *a priori* covariance matrix of $\hat{\mathbf{x}}(k+1)$ based on $k$ observations |
| $\mathbf{P}(k+1)$ | *a posteriori* covariance matrix of $\hat{\mathbf{x}}(k+1)$ based on $k+1$ observations. |

Also the process noise may be coloured and expressed by $\mathbf{G}\mathbf{v}(k)$, where $\mathbf{v}(k)$ is again 'white' noise.

Some remarks, closely parallel to those made earlier, include:

Note that the 'gain factor' $\mathbf{\Gamma}(k+1)$ does not depend on the observation and can be determined beforehand for the whole estimation procedure.

For a one-dimensional case and $C = 1$

$$\mathbf{P}(k+1) = P(k+1) = \frac{Q(k+1)N}{Q(k+1)+N} \tag{12.29}$$

$$\mathbf{\Gamma}(k+1) = \Gamma(k+1) = \frac{P(k+1)}{N} = \frac{Q(k+1)}{Q(k+1)+N} \tag{12.30}$$

Note also that a time dependence of, for example, matrix $\mathbf{A}$ does not change the essence of the problem, at least if this time dependence is known completely.

A comparison of these recursive equations for state estimation, representing the K.-filter, with those derived in Section 7.1 for the recursive technique for parameter estimation indicates a close correspondence. A study of the relation between those techniques can be found in Ho (1962).

The estimate is linear in the observations. This is due to the assumption of Gaussian-distributed variables. For those cases this filter is optimal in the least-squares sense. For the non-Gaussian situation it provides the **optimal-linear** filter.

Instead of the derivation along the lines of Bayes' estimation as given before, it is also possible to obtain those filter equations along different lines. Consider the simplest case. Assume the observations are given by

$$y(i) = x + n(i)$$

An estimator, linear in the observations $y(1), \ldots, y(k)$ has to be found:

$$\hat{x} = \mathbf{q}'\mathbf{y} = q_1 y(1) + \cdots + q_k y(k)$$

such that $E = \mathscr{E}[e^2] = \mathscr{E}[(\hat{x} - x)^2]$ is minimized. By differentiation of $E$ with respect to $q_i$ one obtains the necessary conditions:

$$\frac{\partial E}{\partial q_i} = 2\mathscr{E}[(\hat{x} - x)y(i)] = 0 \qquad i = 1, \ldots, k \tag{12.31}$$

These equations formulate the so-called **orthogonality conditions** indicating that $(\hat{x} - x)$, the error in the estimate, has to be orthogonal to the observations. From the set of equations (12.31) one can determine the coefficients $q_1, \ldots, q_k$, which constitute the filter. In a straightforward way these orthogonality conditions can be derived for more general and complex situations. Based on these expressions equations for the different filtering jobs can be derived.

Still another way to look at the filter equations is by recognizing the close parallel between the process and the filter (model) structure. This parallel had already been recognized with respect to Wiener filtering (e.g. cf. Smith, 1958). It has been developed into a theory on **'observers'** (Luenberger, 1966; 1971).

In the previous sections state estimation procedures have been considered for simple cases: linear process, Gaussian noise. For a non-linear process the techniques remain valid through linearizing around a known approximate solution of the process equations. This may be indicated by equations (12.3) and (12.4), where the latter expression can be used instead of the non-linear one if the matrices **A**, **B**, **C** and **D** consist of the proper partial derivatives, taken with the approximate solution in mind.

Another approach is called **extended Kalman filter.** In this method the linearizing is being done around the best estimate $\hat{\mathbf{x}}$ of the state vector (and the measurable input vector $\mathbf{u}$ if present). If the process is described by for example:

$$\begin{aligned} \mathbf{x}(i + 1) &= \mathbf{f}(\mathbf{x}(i), i) + \mathbf{v}(i) \\ \mathbf{y}(i) &= \mathbf{g}(\mathbf{x}(i), i) + \mathbf{n}(i) \end{aligned} \tag{12.32}$$

then the filter equations can be:

$$\hat{\mathbf{x}}(k + 1|k + 1) = \hat{\mathbf{x}}(k + 1|k) + \mathbf{\Gamma}(k + 1)[\mathbf{y}(k + 1) - \mathbf{g}(\hat{\mathbf{x}}(k + 1|k), k + 1)] \tag{12.33}$$

where, as before,

$$\hat{\mathbf{x}}(k + 1|k) = \mathbf{f}(\hat{\mathbf{x}}(k|k), k) \tag{12.34}$$

is the best estimate of $\mathbf{x}(k + 1)$ based on $k$ observations.

The close relationship between recursive parameter and state estimation has been noted before. For a process that is linear-in-the-parameters the correspondence is easily illustrated. Assume the process is described by:

$$\begin{aligned} y(k) + a_1 y(k-1) + \cdots + a_n y(k-n) \\ = b_1 u(k-1) + \cdots + b_n u(k-n) + n(k) \end{aligned} \tag{12.35}$$

Consequently the process and observation equations are:

$$\begin{aligned} \mathbf{x}(k+1) &= \mathbf{x}(k) \\ y(k) &= \mathbf{c}'(k)\mathbf{x}(k) + n(k) \end{aligned} \tag{12.36}$$

with:

$$\mathbf{x}(k) = \begin{bmatrix} a_1 \\ \vdots \\ a_n \\ b_1 \\ \vdots \\ b_n \end{bmatrix} \qquad \mathbf{c}(k) = \begin{bmatrix} -y(k-1) \\ \vdots \\ -y(k-n) \\ u(k-1) \\ \vdots \\ u(k-n) \end{bmatrix}$$

In this way the parameter-estimation problem can be solved along the lines of the filter algorithms given in this chapter.

## 12.3 Further Aspects, Applications and Examples

There remain many aspects that should be discussed, e.g.

the possibilities of divergence of the estimation results, i.e. occurrence of a covariance matrix which is much smaller than the actual errors in the estimate, and its possible causes;

the investigation of the residuals

$$\mathbf{y}(k+1) - \mathbf{CA}\hat{\mathbf{x}}(k)$$

which should have the statistical characteristics of a white-noise sequence and which may be used to verify the *a priori* knowledge, e.g. about the level of the observation noise;

the case of absence of observation noise $\mathbf{n}(k)$, which implies $\mathbf{N} = \mathbf{0}$ and consequently $\mathbf{N}$ cannot be inverted;

the case of non-white observation noise which can be taken care of by including the noise-colouring filter into the process model;

the case of correlated signal and noise which also can be treated by extending the process model;

the case of more than one sampling rate for the different input and output signals and/or for the estimation sequence;

the effect of inaccurate or incomplete *a priori* knowledge; process parameters that are only partially known may be incorporated in the estimation scheme as state variables to be estimated (Jazwinsky, 1970); the possibilities of constructing adaptive filters for those cases;

the case of processes with continuous-time signals, which can be described by the following equations (cf. e.g. Kailath and Frost, 1968):
process:

$$\begin{aligned} \dot{\mathbf{x}} &= \mathbf{A}\mathbf{x} + \mathbf{G}\mathbf{v} \\ \mathbf{y} &= \mathbf{C}\mathbf{x} + \mathbf{n} \end{aligned} \tag{12.37}$$

$$\mathscr{E}[\mathbf{v}] = \mathbf{0} \qquad \mathscr{E}[\mathbf{v}(t)\mathbf{v}'(\theta)] = \mathbf{V}\delta(t-\theta)$$

$$\mathscr{E}[\mathbf{n}] = \mathbf{0} \qquad \mathscr{E}[\mathbf{n}(t)\mathbf{n}'(\theta)] = \mathbf{N}\delta(t-\theta)$$

$$\mathscr{E}[\mathbf{v}(t)\mathbf{n}'(\theta)] = \mathbf{0}$$

Of course the use of continuous 'white' noises **v** and **n** requires caution for mathematical and physical reasons.
filter:

$$\begin{aligned} \dot{\hat{\mathbf{x}}} &= \mathbf{A}\hat{\mathbf{x}} + \mathbf{\Gamma}[\mathbf{y} - \mathbf{C}\hat{\mathbf{x}}] \\ \mathbf{\Gamma} &= \mathbf{P}\mathbf{C}'\mathbf{N}^{-1} \\ \dot{\mathbf{P}} &= \mathbf{A}\mathbf{P} + \mathbf{P}\mathbf{A}' - \mathbf{P}\mathbf{C}'\mathbf{N}^{-1}\mathbf{C}\mathbf{P} + \mathbf{G}\mathbf{V}\mathbf{G}' \\ \hat{\mathbf{x}}(0) &= \hat{\mathbf{x}}_0 \qquad \mathbf{P}(0) = \mathbf{P}_0 \end{aligned} \tag{12.38}$$

*Example.* In order to illustrate the ideas exposed a very simple example will be discussed. A stationary stochastic signal $x(k)$ with known statistical properties is disturbed by a stationary 'white' noise. The assignment is to estimate $x(k)$ as well as possible. The process and observations are given by:

$$x(k+1) = ax(k) + v(k)$$

$$y(k) = x(k) + n(k)$$

$$\mathscr{E}[v(i)] = 0 \qquad \mathscr{E}[v(i)v(j)] = V\delta_{ij}$$

$$\mathscr{E}[n(i)] = 0 \qquad \mathscr{E}[n(i)n(j)] = N\delta_{ij}$$

$$\mathscr{E}[n(i)v(j)] = 0$$

The filter is described by:

$$\hat{x}(k+1) = a\hat{x}(k) + \gamma(k+1)\{y(k+1) - ax(k)\}$$

$$\gamma(k+1) = \frac{Q(k+1)}{Q(k+1)+N} \qquad x(0) = x_0$$

$$Q(k+1) = a^2P(k) + V \qquad P(0) = P_0$$

$$P(k+1) = Q(k+1) - \frac{Q^2(k+1)}{Q(k+1)+N} = \frac{Q(k+1)N}{Q(k+1)+N}$$

The filter may also be represented by

$$\hat{x}(k+1) = \underbrace{\frac{N}{Q(k+1)+N}}_{\alpha(k+1)} a\hat{x}(k) + \underbrace{\frac{Q(k+1)}{Q(k+1)+N}}_{\beta(k+1)} y(k+1)$$

with

$$\alpha(k+1) + \beta(k+1) = 1$$

From the *a priori* knowledge $P(k+1)$, $Q(k+1)$ and $\gamma(k+1)$ are completely determined beforehand. We consider two cases:

No *a priori* knowledge on $x_0$;
e.g. $x_0 = 0$ and $P_0 = \infty$
$N = V$; $a = \sqrt{0{\cdot}5} \approx 0{\cdot}7$

| $k$ | $Q(k)$ | $\gamma(k)$ | $P(k)$ |
|---|---|---|---|
| 0 | | | $\infty$ |
| 1 | $\infty$ | 1 | $N$ |
| 2 | $1{\cdot}5N$ | 0·6 | $0{\cdot}6N$ |
| 3 | $1{\cdot}3N$ | 0·565 | $0{\cdot}565N$ |
| 4 | $1{\cdot}283N$ | 0·56 | |
| 5 | $1{\cdot}28N$ | 0·56 | |

Full *a priori* knowledge on $x_0$;
exact $x_0$ and $P_0 = 0$
$N = V$; $a = \sqrt{0{\cdot}5} \approx 0{\cdot}7$

| $k$ | $Q(k)$ | $\gamma(k)$ | $P(k)$ |
|---|---|---|---|
| 0 | | | 0 |
| 1 | $N$ | 0·5 | $0{\cdot}5N$ |
| 2 | $1{\cdot}25N$ | 0·555 | $0{\cdot}555N$ |
| 3 | $1{\cdot}278N$ | 0·56 | $0{\cdot}56N$ |
| 4 | $1{\cdot}28N$ | 0·56 | |

After a few iterations $\gamma(k)$ is a constant and consequently the filter has become time invariant. Figure 12.7 shows the *a posteriori* covariance (the 'uncertainty') of the

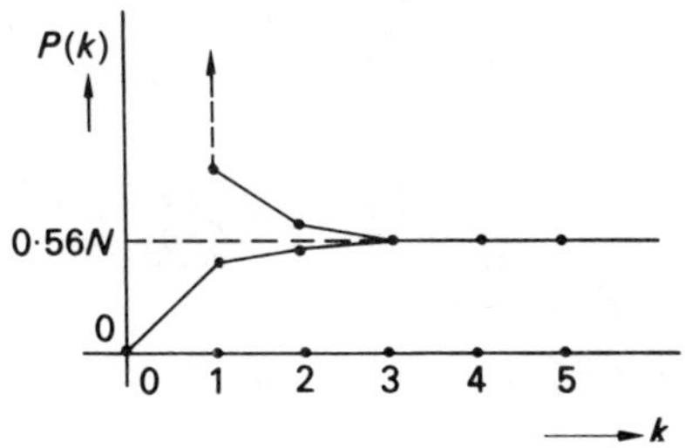


Figure 12.7

estimate. From the values $\infty$ viz. 0 it rapidly approaches an asymptotic value. The speed of reaching stationarity depends on the parameters. The condition for stationarity is

$$P(k) = P(k-1) = P$$

Consequently:

$$\frac{1}{P(k)} = \frac{1}{Q(k)} + \frac{1}{N} = \frac{1}{a^2P(k-1)+V} + \frac{1}{N}$$

$$(a^2P + V)N = PN + P(a^2P + V)$$

$$P^2 + \frac{N + V - a^2N}{a^2}P - \frac{VN}{a^2} = 0$$

from which $P$ can be determined. In the case discussed:

$$P^2 + 3NP - 2N^2 = 0$$

The positive root is: $P = 0{\cdot}56N$.

*Examples.* As illustrations of these filter algorithms a few elementary examples are given in Figures 12.8–12.10.*

Figure 12.8 represents a static case, i.e. the estimation $\hat{x}(k)$ of a constant signal from observed samples that are contaminated with 'white' noise. The pictures illustrate the

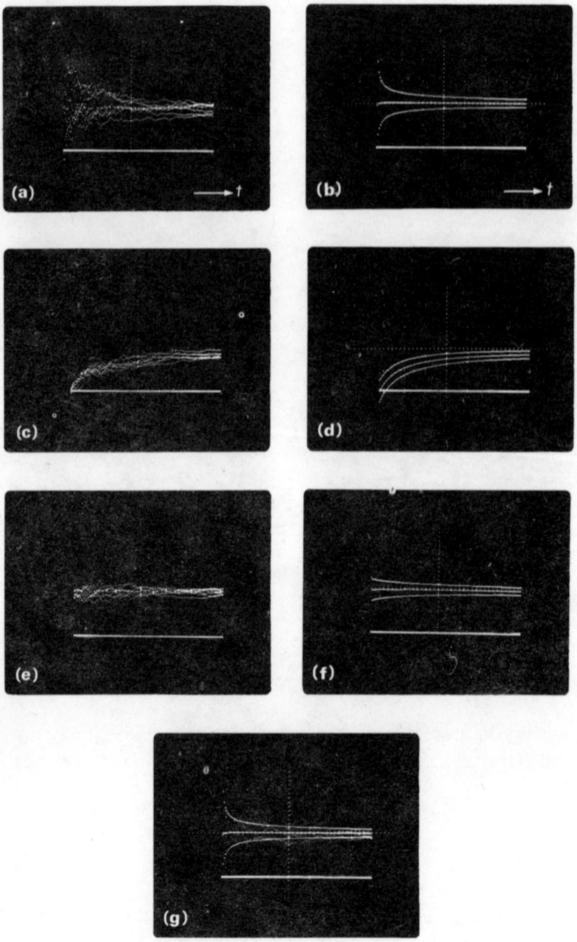


Figure 12.8

* These figures are taken from a simulation on a PDP-8 computer by Mr. G. J. Schoenmakers at the University of Technology, Eindhoven, Netherlands.

influence of (assumed) *a priori* knowledge about the quantity to be estimated. In (a) and (b) it is assumed that $\hat{x}(0) = 0$, which is not correct, but with a large value of $P(0)$, expressing a large degree of uncertainty with respect to that value of $\hat{x}(0)$. In (a) seven runs are given for different white-noise sequences, in (b) the development of the standard deviation around the expected value of $\hat{x}(k)$ is given. In (c) and (d) again $\hat{x}(0) = 0$, but now with a small valuc of $P(0)$, i.e. an unjustified certainty about the $x(0)$ chosen. Again (c) shows seven runs and in (d) the development of the standard deviation around the expected value of $\hat{x}(k)$ is given. Note that the estimate approaches its true value slowly due to the wrong values that were assumed for $\hat{x}(0)$ and $P(0)$. In (e) and (f) the value of $\hat{x}(0)$ was chosen as the right value with a small $P(0)$; (e) gives seven runs, (f) shows the standard deviation around the expected value. In (g) the pictures (a) and (b) are combined.

Figures 12.9 and 12.10 show a one-dimensional dynamic case; the estimation of a low-frequency signal from observations that are also contaminated with white noise.

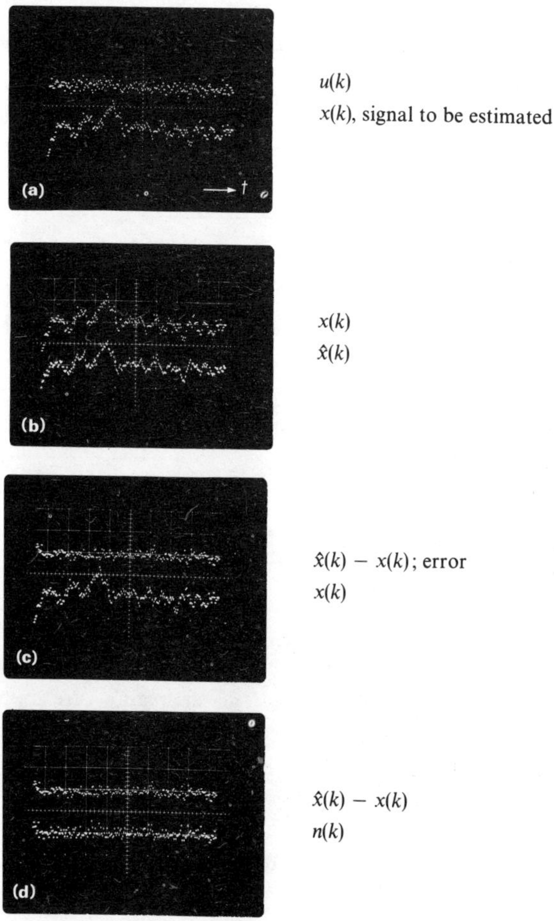


Figure 12.9

Figure 12.9a shows the process noise $u(k)$ and the resulting signal $x(k)$, that has to be estimated; (b) shows both $x(k)$ and its estimate $\hat{x}(k)$; (c) shows the error $\hat{x}(k) - x(k)$ and $x(k)$, the signal to be estimated; (d) shows the error $\hat{x}(k) - x(k)$ and the additive noise that was used in this example.

Figure 12.10 shows the influence of the (assumed) *a priori* (un)certainty about $\hat{x}(0)$.

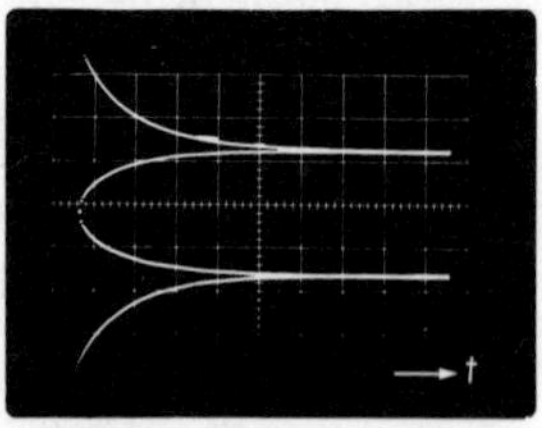


Figure 12.10

Figure 12.11 represents again a static one, but now for continuous signals. The upper diagram represents the filtering by an ordinary simple (first-order, constant coefficient) low-pass filter, the lower diagram gives the output of a (first-order) K.-filter.

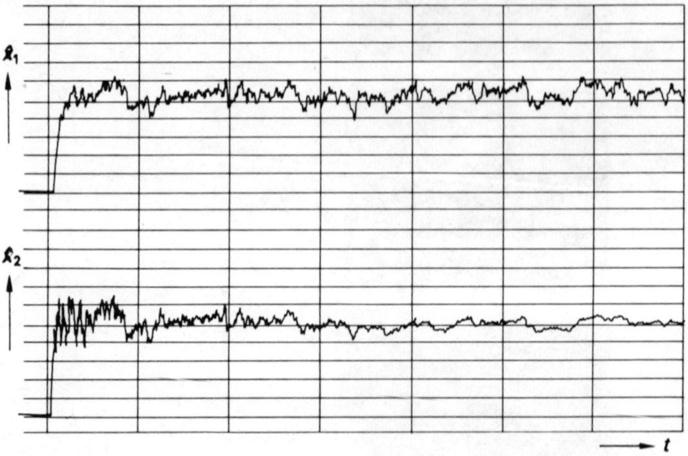


Figure 12.11

Both filters have the same input. The process and observations are given by:

$$\dot{x}(t) = 0 \qquad y(t) = x(t) + n(t)$$

$n(t)$ is stationary 'white' noise with zero mean and variance $\sigma_n^2$. The filter is described by:

$$\dot{\hat{x}}(t) = \frac{P(t)}{\sigma_n^2}\{y(t) - \hat{x}(t)\} \qquad \hat{x}(0) = \hat{x}_0$$

$$\dot{P}(t) = -\frac{P^2(t)}{\sigma_n^2} \qquad P(0) = P_0$$

This simple non-linear differential equation is easily solved:

$$\frac{dP}{P^2} = -\frac{dt}{\sigma_n^2} \quad \text{or} \quad -\frac{1}{P} = -\frac{t}{\sigma_n^2} + c = -\frac{t}{\sigma_n^2} - \frac{1}{P_0}$$

$$P(t) = \frac{\sigma_n^2}{(\sigma_n^2/P_0) + t}$$

Consequently the filter is described by:

$$\dot{\hat{x}}(t) = \frac{1}{(\sigma_n^2/P_0) + t}\{y(t) - \hat{x}(t)\}$$

or

$$\left(\frac{\sigma_n^2}{P_0} + t\right)\dot{\hat{x}}(t) + \hat{x}(t) = y(t)$$

The K.-filter has a time-varying gain; cf. Figure 12.12. It can also be described as a low-pass filter with a time constant that increases with time. Note from Figure 12.11 that in the K.-case the variance of the output diminishes whereas for the constant-coefficient filter the variance tends to a constant value.

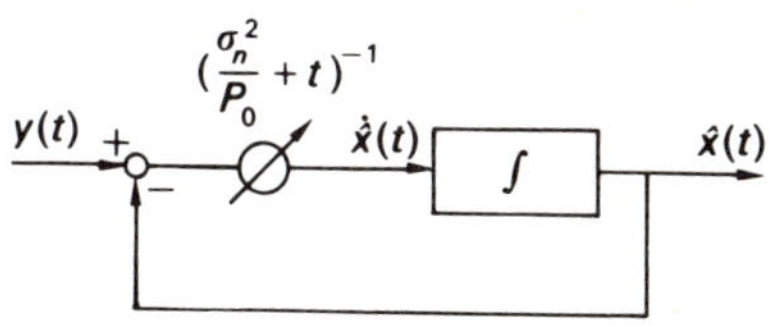


Figure 12.12

*Relation Between the Kalman and the Wiener Filter*

As both types of filters are optimal in the sense indicated before, they should lead to the same result under the appropriate conditions. This will be illustrated by means of a very simple continuous case.

*Example.* The process and observation are given by

$$\dot{x}(t) = -ax(t) + v(t) \qquad a > 0$$

$$y(t) = nx(t) + n(t)$$

$v(t)$ and $n(t)$ are uncorrelated stationary 'white' noises, with zero means and variances $V = \sigma_v^2$ and $N = \sigma_n^2$.

The **K.-filter** is described by:

$$\dot{\hat{x}}(t) = -a\hat{x}(t) + \frac{P(t)}{\sigma_n^2}\{y(t) - \hat{x}(t)\}$$

$$\dot{P}(t) = -2aP(t) - \frac{P^2(t)}{\sigma_n^2} + \sigma_v^2$$

For $t \to \infty$ we expect the time-varying filter to become constant. Consequently $P(t) = P$, $\dot{P}(t) = 0$ and

$$P^2 + 2a\sigma_n^2 P + \sigma_n^2\sigma_v^2 = 0$$

or

$$P_{1,2} = -a\sigma_n^2 \pm \sqrt{a^2\sigma_n^4 + \sigma_n^2\sigma_v^2}$$

$P$ has to be non-negative. Denoting $\sigma_v^2/\sigma_n^2 = \lambda^2$ then the constant filter is given by

$$\dot{\hat{x}}(t) + \{a - a + \sqrt{a^2 + \lambda^2}\}\hat{x}(t) = \{-a + \sqrt{a^2 + \lambda^2}\}y(t)$$

or

$$H_K(s) = \frac{\hat{X}(s)}{Y(s)} = \frac{-a + \sqrt{a^2 + \lambda^2}}{s + \sqrt{a^2 + \lambda^2}}$$

The **W.-filter** is given by equation (12.3), where for this simple case:

$$\phi_{xx}(s) = \frac{\sigma_v^2}{(a + s)(a - s)}$$

$$\phi_{nn}(s) = \sigma_n^2$$

$$\phi_{yy}(s) = \sigma_n^2 + \frac{\sigma_v^2}{(a + s)(a - s)} = \sigma_n^2 \frac{(\sqrt{a^2 + \lambda^2} + s)(\sqrt{a^2 + \lambda^2} - s)}{(a + s)(a - s)}$$

Consequently:

$$\frac{1}{\phi_{yy}^{+}(s)} = \frac{a + s}{\sigma_n(\sqrt{a^2 + \lambda^2} + s)}$$

$$G(s) = \frac{\phi_{xx}(s)}{\phi_{yy}^{-}(s)} = \frac{\sigma_v^2}{(a + s)(a - s)} \times \frac{a - s}{\sigma_n(\sqrt{a^2 + \lambda^2} - s)}$$

$$[G(s)]_+ = \frac{r}{a + s} \qquad \text{with } r = \text{residue at } s = -a$$

$$r = \lim_{s \to -a} (a + s)G(s) = \frac{\sigma_v^2}{\sigma_n} \frac{1}{a + \sqrt{a^2 + \lambda^2}}$$

This results in

$$H_W(s) = \lambda^2 \frac{1}{a + \sqrt{a^2 + \lambda^2}} \frac{1}{s + \sqrt{a^2 + \lambda^2}} = \frac{-a + \sqrt{a^2 + \lambda^2}}{s + \sqrt{a^2 + \lambda^2}}$$

The equality of both types of filters follows from $H_K(s) = H_W(s)$.

In view of the limited goal of this chapter, to be an introductory review for the next one, the reader is referred to the references given, for more details about these filter procedures, e.g. Rhodes (1971). Some papers on applications of the K.-filter in navigation can be found in Richman and Thau (1966), Schmidt et al. (1968), Toda et al. (1969), Leondes (1970).

Applications in communication are given in Irwin and O'Neal (1968), Snijder (1969). References to applications in industrial control and with respect to biological objects can be found in Chapter 14.

## References

Aoki, M. (1967). Cf. references Chapter 4.

Åström, K. J. (1970). Cf. references Chapter 4.

Ho, V. C. (1962). 'The methods of least squares and optimal filtering', *Memo 3329-PR, RAND Corp.*

Irwin, J. D. and J. B. O'Neal (1968). 'The design of optimum dpcm encoding systems via the Kalman predictor', *Joint Autom. Control Conf.*, 130–136.

Jazwinsky, A. H. (1970). *Stochastic Processes and Filtering Theory*, Academic Press, New York, 376 pp.

Jury, E. I. (1964). *Theory and Application of the Z-Transform Method*, Wiley, New York.

Kailath, T. and P. Frost (1968). Cf. references Chapter 4.

Kalman, R. E. (1960). 'A new approach to linear filtering and prediction problems', *J. Basic Engng.*, **82D,** 35–45.

Kalman, R. E. and R. S. Bucy (1961). 'New results in linear filtering and prediction theory', *J. Basic Engng.*, **83D,** 95–108.

Leondes, C. T. (Ed..) (1970). *Theory and Applications of Kalman Filtering*, AGARDOgraph 139.

Luenberger, D. G. (1966). 'Observers for multivariable systems', *IEEE Trans. Autom. Control*, **AC-11,** 190–197.

Luenberger, D. G. (1971). 'An introduction to observers', *IEEE Trans. Autom. Control*, **AC-16,** 596–602.

Meditch, J. S. (1969). *Stochastic Optimal Linear Estimation and Control*, McGraw-Hill, New York, 394 pp.

Ragazzini, J. and G. Franklin (1958). *Sampled-Data Control Systems*, McGraw-Hill, New York.

Rhodes, I. B. (1971). 'A tutorial introduction to estimation and filtering', *IEEE trans. Autom. Control*, **AC-16,** 608–706.

Richman, J. and F. Thau (1966). 'Elements of the Kalman filter theory'; part I—theory, part II—Applications, *Techn. News Bull.*, General Precision, Aerospace Group, **9,** No. 3,4.

Sage, A. P. (1968). *Optimum Systems Control*, Prentice-Hall, Englewood Cliffs, N.J.

Sage, A. P. and J. L. Melsa (1971). *Estimation Theory with Applications to Communications and Control*, McGraw-Hill, New York.

Schmidt, S. F. et al. (1968). 'Case study of Kalman filtering in the C-5 aircraft navigation system', *Case Studies in System Control*, workshop JACC, IEEE group on autom. control.

Smith, O. J. M. (1958). *Feedback Control Systems*, McGraw-Hill, New York, 694 pp.

Snijder, D. L. (1969). *The State-variable Approach to Continuous Estimation with Applications to Analog Communication Theory*, M.I.T. Press, Cambridge, Mass.; research monogr. 51, 114 pp.

Solodovnikov, V. V. (1960). *Introduction to the Statistical Dynamics of Automatic Control Systems*, Dover Publ., New York.

Sorenson, H. W. (1970). 'Least-squares estimation: from Gauss to Kalman', *IEEE Spectrum*, **7,** 63–68.

Toda, N. F. et al. (1969). 'Region of Kalman filter convergence for several autonomous navigation modes', *AIEE J.* **7,** No. 4.

Wiener, N. (1949). *The Extrapolation, Interpolation and Smoothing of Stationary Time Series*, Wiley, New York.

## Additional Literature

Athans, M., R. P. Wishner and A. Bertolini (1968). 'Suboptimal state estimation for continuous-time nonlinear systems from discrete noisy measurements', *IEEE Trans. Autom. Control*, **AC-13,** 504–514.

Breipohl, A. M. (1969). 'Kalman filtering and its application to reliability', *IEEE Trans. Reliability*, **R-18,** 127–130.

Bryson Jr., A. E. and D. E. Johansen (1965). 'Linear filtering for time-varying systems using measurements containing colored noise', *IEEE Trans. Autom. Control*, **AC-10,** 4–10.

Bryson Jr., A. E. and Y. C. Ho (1969). *Applied Optimal Control; Optimization, Estimation and Control*, Blaisdell, Waltham, Mass.

Bucy, R. S. and P. D. Joseph (1968). *Filtering for Stochastic Processes with Applications to Guidance*, Interscience, New York.

Bucy, R. S. (1970). 'Linear and nonlinear filtering' (invited paper), *Proc. IEEE*, **58,** 854–864.

Deutsch, R. (1965). Cf. references Chapter 5.

Deyst Jr., J. J. and C. F. Price (1968). 'Conditions for asymptotic stability of the discrete minimum-variance linear estimator' (short paper), *IEEE Trans. Autom. Control*, **AC-13,** 702–705.

Fischer, J. R. (1967). 'Optimal nonlinear filtering', in *Advances in Control Systems* (Ed. C. T. Leondes), **5,** 197–300, 57 ref.

Griffin, R. E. and A. P. Sage (1968). 'Large and small scale sensitivity analysis of optimum estimation algorithms', *IEEE Trans. Autom. Control*, **AC-13,** 320–328.

Griffin, R. E. and A. P. Sage (1969). 'Sensitivity analysis of discrete filtering and smoothing algorithms', *AIAA J.*, **7,** 1890–1897.

Gura, I. A. (1969). 'An algebraic solution of the state estimation problem', *AIAA J.*, **7,** 1242–1247.

Heffer, H. (1966). 'The effect of erroneous models on the Kalman filter response' (short paper), *IEEE Trans. Autom. Control*, **AC-11,** 541–543.

Huddle, J. R. and D. A. Wismer (1968). Degradation of linear filter performance due to modeling error', *IEEE Trans. Autom. Control*, **AC-13,** 421–423.

Klinger, A. (1968). Cf. references Chapter 7.

Kwakernaak, H. (1967). 'Optimal filtering in linear system with time delay', *IEEE Trans. Autom. Control*, **AC-12,** 169–173.

Lee, R. C. K. (1964). Cf. references Chapter 4.

Leondes, C. T. and L. M. Novak (1972). 'Optimal minimal-order observers for discrete-time systems—a unified theory. *5th IFAC Congr., Paris*, paper 38.6.

Liebelt, P. B. (1967). *An Introduction to Optimal Estimation*, Addison-Wesley, Reading, Mass., 273 pp.

Magne, D. Q. (1966). 'A solution of the smoothing problem for linear dynamic systems', *Automatica*, **44,** 73–92.

Nahi, N. (1969). Cf. references Chapter 5.

Nash, R. A. and F. B. Tuteur (1968). 'The effect of uncertainties in the noise covariance matrices on the maximum likelihood estimate of a vector' (short paper), *IEEE Trans. Autom. Control*, **AC-13,** 86–88.

Nishimura, T. (1966). 'On the *a priori* information in sequential estimation problems', *IEEE Trans. Autom. Control*, **AC-11,** 197–204. Cf. also same journal, **AC-12,** 123.

Ohap, R. F. and A. R. Stubberud (1965). Cf. references Chapter 13.

Phillipson, G. A. and S. K. Mitter (1969). 'State identification of a class of linear distributed systems', *4th IFAC Congr., Warsaw*, paper 12.3.

Schwartz, L. and E. B. Stear (1968). 'A computational comparison of several nonlinear filters' (short paper), *IEEE Trans. Autom. Control*, **AC-13**, 83–86.

Shaw, L. G. (1965). 'Optimum stochastic control', *Disciplines and Techniques of Systems Control* (Ed. J. Peschon), Blaisdell, New York, 125–185.

Sorensen, H. W. (1966). 'Kalman filtering techniques', in *Advances in Control Systems*, Vol. 3 (Ed. C. T. Leondes), Academic Press, New York, pp. 219–292.

Tait, G. R. and P. R. Bélanger (1970). 'A comparison of some parameter identification schemes using first and second order extended Kalman Bucy filters and sensitivity functions', *2nd IFAC Symp.—Identification and Process Parameter Estimation, Prague*, paper 10.3.

Wells, C. H. (1968). 'An approximation of the Kalman filter equations' (correspondence), *IEEE Trans. Autom. Control*, **AC-13**, 445.

*CHAPTER 13*

# *Parameter and State Estimation*

Many aspects of modern control theory are centred around the notion of state. For applications of these theories one has to have available, explicitly or implicitly, a measurement or estimate of the state variable. This can be found along the lines given in the previous chapter if both the structure and the parameters of the process are known. If the parameters are not known then in many schemes a combination of parameter and state estimation is needed.

In Section 13.1 it is pointed out that this combined parameter- and state-estimation problem is unavoidably of a non-linear nature.

The assumption that the noise is additive makes it feasible to derive an 'error function', which is illustrated in Section 13.2.

Section 13.3 indicates that the minimization of such an error function may lead to the celebrated two-point boundary value problem, the solution of which is illustrated in the next section by means of quasilinearization and invariant imbedding. Section 13.3 is concluded by an illustration. This chapter may introduce the reader to the wide variety of publications on (non-linear) estimation methods.

## 13.1 Non-linear Nature of the Problem

A straightforward approach advocated in a number of papers is the extension of the state vector with the unknown parameters. As before the process is described by

$$\left.\begin{aligned} \dot{\mathbf{x}} &= \mathbf{f}(\mathbf{x}, \mathbf{u}, \mathbf{a}, \mathbf{v}, t) \\ \mathbf{y} &= \mathbf{g}(\mathbf{x}, \mathbf{u}, \mathbf{c}, \mathbf{n}, t) \end{aligned}\right\} \tag{13.1}$$

cf. equation (4.64). The process-parameter vector $\mathbf{a}$, if it is known or expected to be constant, is treated as a function of time by the equation

$$\dot{\mathbf{a}} = \mathbf{0} \tag{13.2}$$

This leads to a set of equations of the following form:

$$\begin{bmatrix} \dot{\mathbf{x}} \\ \dot{\mathbf{a}} \end{bmatrix} = \begin{bmatrix} \mathbf{f}(\mathbf{x}, \mathbf{u}, \mathbf{a}, \mathbf{v}, t) \\ \mathbf{0} \end{bmatrix} \tag{13.3}$$

It will be clear that, even if the dynamics of the process is linear, this equation can not properly be separated into a form

$$\begin{bmatrix} \dot{\mathbf{x}} \\ \dot{\mathbf{a}} \end{bmatrix} = \mathbf{M} \begin{bmatrix} \mathbf{x} \\ \mathbf{a} \end{bmatrix} \tag{13.4}$$

where **M** does not depend on either **x** or **a**. Consequently even for a process with linear dynamics and which is linear in the parameters the problem of combined parameter and state estimation is a **non-linear** one in the 'state/parameter vector' $[\mathbf{x}', \mathbf{a}']'$.

This implies that all approaches to this problem will have the characteristics of the model-adjustment technique, i.e. they will be of an iterative type. Equation (13.1) provides a general representation of the behaviour of (non-linear) continuous processes. For sampled processes a corresponding representation is:

$$\left.\begin{aligned} \mathbf{x}(k+1) &= \mathbf{f}(\mathbf{x}(k), \mathbf{u}(k), \mathbf{a}(k), \mathbf{v}(k), k) \\ \mathbf{y}(k) &= \mathbf{g}(\mathbf{x}(k), \mathbf{u}(k), \mathbf{c}(k), \mathbf{n}(k), k) \end{aligned}\right\} \tag{13.5}$$

Based on *a priori* knowledge about the process and about the way it is operated or controlled the representation may be simplified. In the pertinent literature such simplifications are introduced frequently in order to make the estimation problem more tractable. This may result in many process descriptions*, some of which are indicated in the diagram below and are illustrated in Figure 13.1.

| | | |
|---|---|---|
| $\mathbf{A}\mathbf{x}(k) + \mathbf{B}\mathbf{u}(k)$ | $\Rightarrow$ | process dynamics are constant in time |
| $\Uparrow$ | | |
| $\mathbf{A}(k)\mathbf{x}(k) + \mathbf{B}(k)\mathbf{u}(k)$ | $\Rightarrow$ | process dynamics are linear |
| $\Uparrow$ | | |
| $\mathbf{f}(\mathbf{x}(k), \mathbf{u}(k), \mathbf{a}(k), k) + \mathbf{G}(k)\mathbf{v}(k)$ | $\Rightarrow$ | process noise is additive |
| $\Uparrow$ | | |
| $\mathbf{x}(k+1) = \mathbf{f}(\mathbf{x}(k), \mathbf{u}(k), \mathbf{a}(k), \mathbf{v}(k), k)$ <br> $\mathbf{y}(k) = \mathbf{g}(\mathbf{x}(k), \mathbf{u}(k), \mathbf{c}(k), \mathbf{n}(k), k)$ | $\Rightarrow$ | non-linear, time varying |
| $\Downarrow$ | | |
| $\mathbf{g}(\mathbf{x}(k), \mathbf{u}(k), \mathbf{c}(k), k) + \mathbf{n}(k)$ | $\Rightarrow$ | observation noise is additive |
| $\Downarrow$ | | |
| $\mathbf{C}(k)\mathbf{x}(k) + \mathbf{D}(k)\mathbf{u}(k)$ | $\Rightarrow$ | observations are linear in the state variable |
| $\Downarrow$ | | |
| $\mathbf{C}\mathbf{x}(k) + \mathbf{D}\mathbf{u}(k)$ | $\Rightarrow$ | observation matrices are constant in time |

* In many cases the equivalent continuous type of representation can be derived from the sampled type by careful limiting procedures, taking into account the mathematical complications in describing the effect of 'white'-noise sources (Ito; Stratonovich).

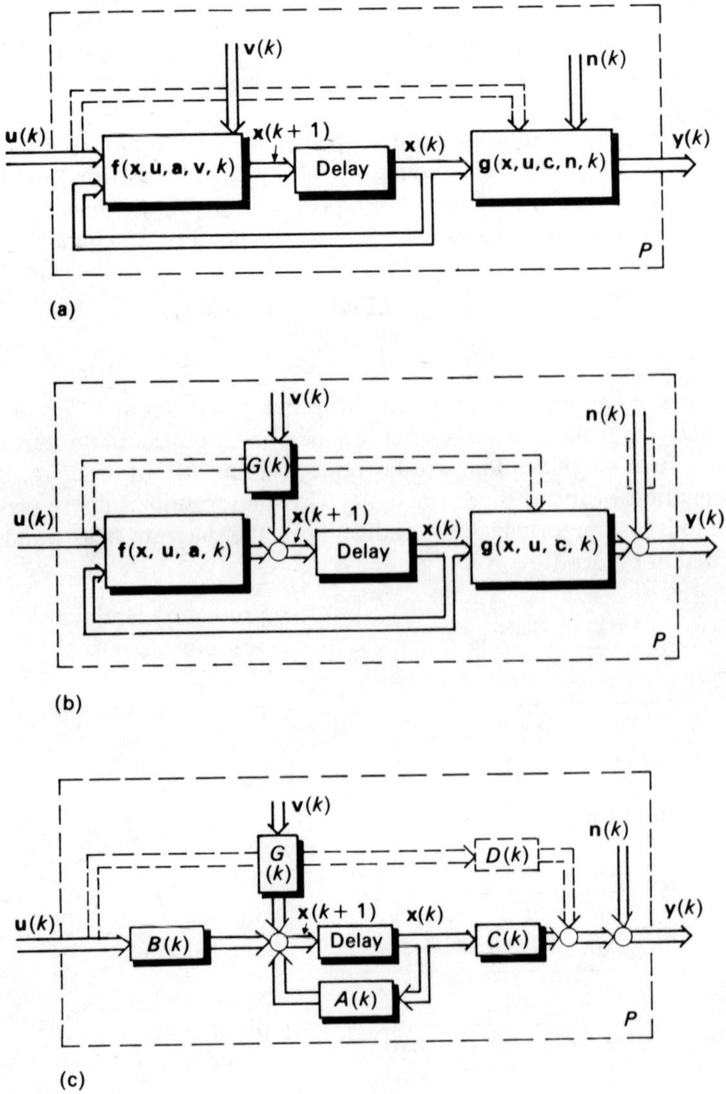


Figure 13.1

Among the other alterations that may hold are:

$$\mathbf{u}(k) = \mathbf{0} \quad \Rightarrow \begin{cases} \text{no measurable input} \\ \text{signal available} \end{cases}$$

$$\left.\begin{aligned} \mathbf{a}(k) &= \mathbf{a} \\ \mathbf{c}(k) &= \mathbf{c} \end{aligned}\right\} \text{constant} \quad \Rightarrow \begin{cases} \text{dynamics are not time} \\ \text{varying} \end{cases}$$

The condition that the noise is **additive** is a very convenient one. From a simple example:

$$\mathbf{y}(k) = \mathbf{g}(\mathbf{x}(k)) + \mathbf{n}(k)$$

it follows that

$$p(\mathbf{y}(k)|\mathbf{x}(k)) = p(\mathbf{y}(k) - \mathbf{g}(\mathbf{x}(k)) = p(\mathbf{n}(k)) \tag{13.6}$$

Consequently that conditional probability function is equal to the (assumed or known) probability-density function of the noise. Particularly if this function is Gaussian then this way of approach leads to acceptable expressions as has already been demonstrated in the previous chapter and as will be noticed again in what follows.

## 13.2 Derivation of an 'Error Function'

For the moment we consider a 'state/parameter vector' $\mathbf{x}$ which may consist of:

state variables and parameters together or

state variables (for a non-linear process) or

parameters for a process that is non-linear-in-the-parameters.

In this way these different types of estimation problems may be discussed together. Thus the numerous publications on non-linear K.-filtering, for example by means of the extended K.-filter, are relevant to our problem.

Again the problem can be approached from different points of view:

Bayes' estimation;

maximum likelihood estimation;

least-squares estimation.

Again one may discuss the relations between the different estimates for special cases (even cost function; Gaussian probability-density functions; 'white' noise) as has been done in Chapter 5. In all cases the methods lead to a minimization of an 'error function' $E$ with respect to the state/parameter vector $\mathbf{x}(i)$ over an interval $i = 0, \ldots, k$. First we summarize two derivations of such 'error functions'.

(a) The process description chosen is:

$$\mathbf{x}(i + 1) = \mathbf{f}(\mathbf{x}(i), i) + \mathbf{G}(i)\mathbf{v}(i) \tag{13.7}$$

$$\mathbf{y}(i) = \mathbf{g}(\mathbf{x}(i), i) + \mathbf{n}(i) \tag{13.8}$$

If no *a priori* knowledge of the process and the observation noise is available a straightforward approach is the following (cf. Detchmendy and Sridhar, 1966). Define:

$$\mathbf{e}_1(i) = \mathbf{x}^*(i+1) - \mathbf{f}(\mathbf{x}^*(i), i) \tag{13.9}$$

$$\mathbf{e}_2(i) = \mathbf{y}(i) - \mathbf{g}(\mathbf{x}^*(i), i) \tag{13.10}$$

where $\mathbf{x}^*(i)$, $i = 0, \ldots, k$ stands for a nominal trajectory. If $\mathbf{x}^*(i)$ were the true trajectory and if $\mathbf{v}(i) \equiv \mathbf{0}$ and $\mathbf{n}(i) \equiv \mathbf{0}$ for $i = 0, \ldots, k$ then $\mathbf{e}_1(i) \equiv \mathbf{e}_2(i) \equiv \mathbf{0}$. Consequently the following 'error function' could be chosen to be minimized.

$$E = \sum_{i=0}^{k} [\mathbf{e}_1'(i)\mathbf{Q}\mathbf{e}_1(i) + \mathbf{e}_2'(i)\mathbf{R}\mathbf{e}_2(i)] \tag{13.11}$$

$\mathbf{Q}$ and $\mathbf{R}$ are non-negative definite weighting matrices; apparently $E$ is a weighted least-squares criterion. If $\hat{\mathbf{x}}(i)$ results in a minimal value of $E$ then this estimate is optimal in the sense defined.

(b) The process chosen is again given by equations (13.7) and (13.8). Now it is assumed (cf. Cox, 1964) that $\mathbf{v}(i)$ and $\mathbf{n}(i)$ are Gaussian random vectors, with zero means and covariance matrices:

$$\left.\begin{aligned} \mathscr{E}[\mathbf{v}(i)\mathbf{v}'(j)] &= \mathbf{V}(i)\delta_{ij} \\ \mathscr{E}[\mathbf{n}(i)\mathbf{n}'(j)] &= \mathbf{N}(i)\delta_{ij} \\ \mathscr{E}[\mathbf{v}(i)\mathbf{n}'(j)] &= \mathbf{0} \end{aligned}\right\} \quad \text{all } i, j \tag{13.12}$$

i.e. the noise sequences are assumed to be 'white'. Again we select the Bayes' estimation approach. The following notation is used:

$$\mathbf{X}_k = [\mathbf{x}(0), \ldots, \mathbf{x}(k)]$$

$$\mathbf{Y}_k = [\mathbf{y}(0), \ldots, \mathbf{y}(k)]$$

Again the conditional probability-density function $P(\mathbf{X}_k|\mathbf{Y}_k)$ of $\mathbf{X}_k$, given the observations $\mathbf{Y}_k$, has to be found. From Bayes' rule:

$$p(\mathbf{X}_k|\mathbf{Y}_k) = \frac{p(\mathbf{Y}_k|\mathbf{X}_k)p(\mathbf{X}_k)}{p(\mathbf{Y}_k)} \tag{13.13}$$

Just as in Chapter 5, a cost or loss function can be used to reduce that *a posteriori* probability-density function to simple numerical values, i.e. a minimum cost or minimum loss estimation. On the other hand one might be satisfied in determining those numerical values of the estimate for which the conditional probability function equation (13.13) obtains its maximum value. For Gaussian probability-density functions and even cost or loss functions both methods will result in the same estimate.

Under the conditions given by (13.7), (13.8) and (13.12) let us therefore consider the maximization of equation (13.13) and discuss each of the terms of the right-hand side of that equation.

$p(\mathbf{Y}_k)$ is the most simple part. As it contains only the numerical values of the observations $\mathbf{Y}_k$ it just acts as a normalizing constant in order to make the integral

$$\int p(\mathbf{X}_k|\mathbf{Y}_k)\,d\mathbf{X}_k \qquad \text{equal to one}$$

For maximization purposes it does not play any role and thus can be neglected.

$p(\mathbf{Y}_k|\mathbf{X}_k)$ also results in a simple expression due to the analogy with equation (13.6) and the fact that the sequence $\{n(k)\}$ is non-correlated in time ('white'):

$$p(\mathbf{Y}_k|\mathbf{X}_k) = \prod_{i=0}^{k} p\{\mathbf{y}(i) - \mathbf{g}(\mathbf{x}(i), i)\}$$

$$= c_1 \prod_{i=0}^{k} \exp\{-\tfrac{1}{2}[\mathbf{y}(i) - \mathbf{g}(\mathbf{x}(i), i)]'\mathbf{N}^{-1}(i)[\mathbf{y}(i) - \mathbf{g}(\mathbf{x}(i), i)]\}$$

where again normalizing factors are taken care of by the constant $c_1$. Using the shorthand notation $\mathbf{z}'\mathbf{A}\mathbf{z} = \|\mathbf{z}\|^2_{\mathbf{A}}$ it follows:

$$p(\mathbf{Y}_k|\mathbf{X}_k) = c_1 \exp\left\{-\tfrac{1}{2}\sum_{i=0}^{k} \|\mathbf{y}(i) - \mathbf{g}(\mathbf{x}(i), i)\|^2_{\mathbf{N}^{-1}(i)}\right\}$$

$p(\mathbf{X}_k)$ can be rewritten by recognizing that

$$p(\mathbf{X}_i) = p(\mathbf{x}(i), \mathbf{X}_{i-1}) = p(\mathbf{x}(i)|\mathbf{X}_{i-1})p(\mathbf{X}_{i-1})$$

Due to equation (13.7) and the assumption that the sequence $\{\mathbf{v}(k)\}$ is non-correlated in time ('white') it follows that

$$p(\mathbf{x}(i)|\mathbf{X}_{i-1}) = p(\mathbf{x}(i)|\mathbf{x}(i-1), \ldots, \mathbf{x}(0)) = p(\mathbf{x}(i)|\mathbf{x}(i-1))$$

Consequently

$$p(\mathbf{X}_k) = p(\mathbf{x}(0)) \prod_{i=1}^{k} p(\mathbf{x}(i)|\mathbf{x}(i-1))$$

Also due to equation (13.7) $p(\mathbf{x}(i+1)|\mathbf{x}(i))$ is Gaussian with:

| | |
|---|---|
| mean | $\mathbf{f}(\mathbf{x}(i), i)$ |
| covariance matrix | $\mathbf{G}(i)\mathbf{V}(i)\mathbf{G}'(i)$ |

The *a priori* knowledge $p(\mathbf{x}(0))$ is also assumed to be Gaussian with:

| | |
|---|---|
| mean | $\mathbf{x}_0$ |
| covariance matrix | $\mathbf{P}_0$ |

Consequently the term under discussion can be written as

$$p(\mathbf{X}_k) = c_2 \exp\{-\tfrac{1}{2}[\mathbf{x}(0) - \mathbf{x}_0]'\mathbf{P}_0^{-1}[\mathbf{x}(0) - \mathbf{x}_0]\}$$

$$\times \prod_{i=0}^{k-1} \exp\{-\tfrac{1}{2}[\mathbf{x}(i+1) - \mathbf{f}(\mathbf{x}(i), i)]'[\mathbf{G}(i)\mathbf{V}(i)\mathbf{G}'(i)]^{-1}[\mathbf{x}(i+1) - \mathbf{f}(\mathbf{x}(i), i)]\}$$

provided the matrix $\mathbf{GVG}'$ can be inverted. Again $c_2$ is a normalizing constant. In the shorthand notation

$$p(\mathbf{X}_k) = c_2 \exp\left\{-\tfrac{1}{2}\|\mathbf{x}(0) - \mathbf{x}_0\|^2_{\mathbf{P}_0^{-1}} - \tfrac{1}{2}\sum_{i=0}^{k-1} \|\mathbf{x}(i+1) - \mathbf{f}(\mathbf{x}(i), i)\|^2_{[\mathbf{GVG}']^{-1}}\right\}$$

If we combine these terms there results:

$$p(\mathbf{X}_k|\mathbf{Y}_k) = c_3 \exp\{-E_k\} \tag{13.14}$$

where

$$\left.\begin{aligned} E_k = {} & \tfrac{1}{2}\|\mathbf{x}(0) - \mathbf{x}_0\|^2_{\mathbf{P}_0^{-1}} \\ & + \tfrac{1}{2}\sum_{i=0}^{k} \|\mathbf{y}(i) - \mathbf{g}(\mathbf{x}(i), i)\|^2_{\mathbf{N}^{-1}(i)} \\ & + \tfrac{1}{2}\sum_{i=0}^{k-1} \|\mathbf{x}(i+1) - \mathbf{f}(\mathbf{x}(i), i)\|^2_{[\mathbf{GVG}']^{-1}} \end{aligned}\right\} \tag{13.15}$$

Maximizing equation (13.14) with respect to $\mathbf{X}_k = [\mathbf{x}(0), \ldots, \mathbf{x}(k)]$ implies minimization of $E_k$. The last term of (13.15) can be replaced by

$$\tfrac{1}{2}\sum_{i=0}^{k-1} \|\mathbf{v}(i)\|^2_{\mathbf{V}^{-1}(i)}$$

and the constraint expressed by equation (13.7). Consequently one may also minimize:

$$\begin{aligned} E'_k = {} & \tfrac{1}{2}\|\mathbf{x}(0) - \mathbf{x}_0\|^2_{\mathbf{P}_0^{-1}} \\ & + \tfrac{1}{2}\sum_{i=0}^{k} \|\mathbf{y}(i) - \mathbf{g}(\mathbf{x}(i), i)\|^2_{\mathbf{N}^{-1}(i)} + \tfrac{1}{2}\sum_{i=0}^{k-1} \|\mathbf{v}(i)\|^2_{\mathbf{V}^{-1}(i)} \\ & + \boldsymbol{\lambda}'(i)[\mathbf{x}(i+1) - \mathbf{f}(\mathbf{x}(i), i) - \mathbf{G}(i)\mathbf{v}(i)] \\ & \qquad i = 0, \ldots, k-1 \end{aligned} \tag{13.16}$$

where $\boldsymbol{\lambda}(i)$ is a vector of Lagrange multipliers. In this formulation $\mathbf{GVG}'$ does not have to be invertible.

Note that $E'_k$ consists of three components, which can be recognized as being related to:

the assumption on the state/parameter vector at $t = 0$;

the additive 'observation noise' $\mathbf{n}$;

the 'process noise' $\mathbf{v}$ (together with the constraints given by the process description).

The weighting matrices $\mathbf{P}_0^{-1}$, $\mathbf{N}^{-1}(i)$ and $\mathbf{V}^{-1}(i)$ provide a measure for the relative weight of these contributions to the 'error' $E'_k$. Note also that $\mathbf{y}(i)$ for $i = 0, \ldots, k$ are the numerical values of the observations on the process. If no *a priori* knowledge on $\mathbf{x}(0)$ is available, then $\mathbf{P}_0^{-1} \to \mathbf{0}$ and the first term of $E'_k$ drops out.

Note that in the process description (13.7) it has been assumed that there is no measurable input sequence $\{\mathbf{u}(i)\}$ $i = 0, 1, \ldots, k$. If such an input is present then that description and the subsequent derivation can easily be adapted to incorporate that extra information.

## 13.3 Minimization of the 'Error Function'

The error function (13.16) was derived under the following conditions:

'observation noise' $\mathbf{n}$: additive, Gaussian, 'white';

'process noise' $\mathbf{v}$: additive, Gaussian.

For noises with unknown probability distribution the expression (13.16) may be assumed to be an acceptable error function on intuitive grounds. In that case its minimization leads to a least-squares estimate.

By minimizing (13.16) one finds the estimate $\hat{\mathbf{x}}(j|k)$, i.e. the estimation of the state/parameter vector at time $j$ based on the observations $0, 1, \ldots, k$. Again one may recognize the cases:

$0 \leqslant j < k$ interpolation (smoothing)
$j = k$ estimation of the present state (filtering)
$j > k$ extrapolation (if a simple extension of the expression for $E_k$ is used).

Several approaches to this minimization problem are available e.g.:

the dynamic programming formulation leading to a sequential procedure with the possibility of using new observations which become available, but also having its inherent problems with respect to the memory capacity needed in the computer (Bellman: curse of dimensionality);

the discrete maximum principle;

the discrete Euler–Lagrange equations;

These two methods result in a non linear **two-point boundary value problem** (TPBVP)

the same, but linearizing the process equations about the point $\mathbf{x}^*(k|k)$. This leads to approximate estimates, but may result in recursive schemes in which new observations can be processed as soon as they become available.

A straightforward derivation of the necessary conditions for a minimum of $E'_k$ can be found by differentiation:

$$\mathbf{0} = \frac{\partial E'_k}{\partial \mathbf{x}(i)} = -[\mathbf{y}(i) - \mathbf{g}(\mathbf{x}(i), i)]'\mathbf{N}^{-1}(i)\frac{\partial \mathbf{g}(\mathbf{x}(i), i)}{\partial \mathbf{x}(i)} + \boldsymbol{\lambda}'(i-1) - \boldsymbol{\lambda}'(i)\frac{\partial \mathbf{f}(\mathbf{x}(i), i)}{\partial \mathbf{x}(i)} \tag{13.17}$$

$$\mathbf{0} = \frac{\partial E'_k}{\partial \boldsymbol{\lambda}(i)} = \mathbf{x}(i+1) - \mathbf{f}(\mathbf{x}(i), i) - \mathbf{G}(i)\mathbf{v}(i) \tag{13.18}$$

$$\mathbf{0} = \frac{\partial E'_k}{\partial \mathbf{v}(i)} = \mathbf{v}'(i)\mathbf{V}^{-1}(i) - \boldsymbol{\lambda}'(i)\mathbf{G}(i) \tag{13.19}$$

By elimination of $\mathbf{v}(i)$ and noting that these are the conditions for the optimal estimates $\hat{\mathbf{x}}(i|k)$ and $\hat{\boldsymbol{\lambda}}(i)$ there results:

$$\hat{\mathbf{x}}(i+1|k) = \mathbf{f}(\hat{\mathbf{x}}(i|k), i) - \mathbf{G}(i)\mathbf{V}(i)\mathbf{G}'(i)\hat{\boldsymbol{\lambda}}(i) \qquad i = 0, \ldots, k-1 \tag{13.20}$$

$$\hat{\boldsymbol{\lambda}}(i-1) = \left.\frac{\partial \mathbf{f}'(\mathbf{x}(i), i)}{\partial \mathbf{x}(i)}\right|_{\mathbf{x}(i)=\hat{\mathbf{x}}(i|k)} \hat{\boldsymbol{\lambda}}(i) + \left.\frac{\partial \mathbf{g}'(\mathbf{x}(i), i)}{\partial \mathbf{x}(i)}\right|_{\mathbf{x}(i)=\hat{\mathbf{x}}(i|k)} \mathbf{N}^{-1}(i)[\mathbf{y}(i) - \mathbf{g}(\hat{\mathbf{x}}(i|k), i)] \qquad i = 1, \ldots, k \tag{13.21}$$

The expressions (13.20) and (13.21) are called the **canonic equations.** In the same way follow the boundary conditions for $i = 0$ and $i = k$:

$$\hat{\mathbf{x}}(0|k) = \mathbf{x}_0 + \mathbf{P}_0 \left.\frac{\partial \mathbf{g}'(\mathbf{x}(i), i)}{\partial \mathbf{x}(i)}\right|_{\mathbf{x}(i)=\hat{\mathbf{x}}(0|k)} \mathbf{N}^{-1}(0)[\mathbf{y}(0) - \mathbf{g}(\hat{\mathbf{x}}(0|k), 0)] + \mathbf{P}_0 \left.\frac{\partial \mathbf{f}'(\mathbf{x}(i), i)}{\partial \mathbf{x}(i)}\right|_{\mathbf{x}(i)=\hat{\mathbf{x}}(0|k)} \hat{\boldsymbol{\lambda}}(0) \tag{13.22}$$

$$\hat{\boldsymbol{\lambda}}(k) = \mathbf{0} \tag{13.23}$$

The expressions (13.20) to (13.23) define a type of the celebrated TPBVP. The boundary conditions are split; for $\mathbf{x}$ it is given at the beginning of the interval under consideration, for $\boldsymbol{\lambda}$ it is given at the end of that interval. If both were given at the same end of the interval then a straightforward computation in forward or backward direction would be possible. Due to the split conditions the solution of the TPBVP is not that straightforward (cf. Section 13.4).

Expressions analogous to (13.20) to (13.23) can be derived for the continuous type of process representations.

*Simplifications*

In the preceding discussions the process was assumed to have non-linear dynamics. In Section 13.1 it was indicated already that, in spite of linear dynamics, the combined parameter- and state-estimation problem is of a non-linear nature. Yet, one may expect that a simpler type of process will result in a simpler estimation problem. Only partial insight into this relationship between complexity of the process dynamics and the complexity of the estimation problem seems to be available.

As an interesting case the following example is cited (Anderson et al, 1969). Assume a process with linear dynamics that is constant in time, to be described by equation (13.24)

$$\left.\begin{aligned} \mathbf{x}(i+1) &= \mathbf{A}\mathbf{x}(i) + \mathbf{v}(i) \\ \mathbf{y}(i) &= \phantom{\mathbf{A}}\mathbf{x}(i) + \mathbf{n}(i) \end{aligned}\right\} \tag{13.24}$$

$\mathbf{v}(i)$ and $\mathbf{n}(i)$ are assumed to be 'white'. Note that the entire state is observed. The authors show that the following estimates for $\mathbf{A}$

$$\hat{\mathscr{A}}_n = \left[\sum_{i=3}^{n} \mathbf{y}(i)\mathbf{y}'(i-2)\right]\left[\sum_{i=3}^{n} \mathbf{y}(i-1)\mathbf{y}'(i-2)\right]^{(-1)}$$

converge with probability one to $\mathbf{A}$ and that a Kalman filter updated by these $\hat{\mathscr{A}}_n$ has an estimate which converges with probability one to that of the correct Kalman filter. Consequently, both parameters and states are properly estimated.

For a model-adjustment type of approach to this type of problem cf. Baggerud and Balchen (1970).

## 13.4 Some Algorithms

Of the host of literature on optimal-control problems many publications are relevant to our problem. It is far beyond the scope of this chapter to try to give an extensive discussion of the many methods discussed or proposed We just give some notes on:

quasilinearization;

invariant imbedding.

### 13.4.1 *Quasilinearization*

This is an interesting and powerful technique. Its development can be traced from the Newton–Raphson method for determining the root of an algebraic equation, through the generalization of this method by Kantorovich for functional equations. Under the name quasilinearization it was developed by Bellman and Kalaba (1965).

For this exposition we use the continuous type of process description given before, but the output $\mathbf{y}$ is assumed to be linearly dependent on the state $\mathbf{x}$:

$$\left.\begin{aligned} \begin{bmatrix} \dot{\mathbf{x}} \\ \dot{\mathbf{a}} \end{bmatrix} &= \begin{bmatrix} \mathbf{f}(\mathbf{x}, \mathbf{u}, \mathbf{a}, \mathbf{v}, t) \\ \mathbf{0} \end{bmatrix} \\ \mathbf{y}(t) &= \mathbf{C}\mathbf{x}(t) \end{aligned}\right\} \tag{13.25}$$

$\mathbf{x}$ and $\mathbf{a}$ have the dimension $n$ and $m$. Over the interval of process observation there are available

$$\mathbf{y}(j) = \mathbf{y}_j \qquad j = 1, 2, \ldots, m + n \tag{13.26}$$

i.e. numerical values for the measurements at $(m + n)$ instances of time. The expressions (13.25) provide $(m + n)$ equations that have to satisfy the boundary conditions given by (13.26). For this reason this problem is called a multipoint boundary value problem (MPBVP). This boundary value problem can in the following way be tackled by means of the method of quasilinearization. The $(i + 1)$th approximation of the time functions $\mathbf{x}(t)$ and $\mathbf{a}(t)$ over the time interval of observation are

$$\left.\begin{aligned} \dot{\mathbf{x}}^{(i+1)} &= \mathbf{f}(\mathbf{x}^{(i)}, \mathbf{u}, \mathbf{a}^{(i)}) + \mathbf{F}_1^{(i)}[\mathbf{x}^{(i+1)} - \mathbf{x}^{(i)}] + \mathbf{F}_2^{(i)}[\mathbf{a}^{(i+1)} - \mathbf{a}^{(i)}] \\ \dot{\mathbf{a}}^{(i+1)} &= \mathbf{0} \end{aligned}\right\} \tag{13.27}$$

with the conditions:

$$\mathbf{C}\mathbf{x}^{(i+1)} = \mathbf{y}^{(i+1)} = \mathbf{y}_j \tag{13.28}$$

The matrices $\mathbf{F}_1^{(i)}$ and $\mathbf{F}_2^{(i)}$ are the Jacobians:

$$\mathbf{F}_1^{(i)} = \left[\frac{\partial \mathbf{f}'}{\partial \mathbf{x}}\right]_{\mathbf{x}^{(i)}, \mathbf{a}^{(i)}} \quad \text{and} \quad \mathbf{F}_2^{(i)} = \left[\frac{\partial \mathbf{f}'}{\partial \mathbf{a}}\right]_{\mathbf{x}^{(i)}, \mathbf{a}^{(i)}} \tag{13.29}$$

For the $(i + 1)$th iteration the problem statement is:

$$\left.\begin{aligned} \dot{\mathbf{x}}^{(i+1)} &= \mathbf{F}_1^{(i)}\mathbf{x}^{(i+1)} + \mathbf{F}_2^{(i)}\mathbf{a}^{(i+1)} + \mathbf{w}^{(i)} \\ \mathbf{y}(j) &= \mathbf{C}\mathbf{x}^{(i+1)} \end{aligned}\right\} \tag{13.30}$$

where $\mathbf{w}^{(i)}$ is available from the previous iteration:

$$\mathbf{w}^{(i)} = \mathbf{f}(\mathbf{x}^{(i)}, \mathbf{u}, \mathbf{a}^{(i)}) - \mathbf{F}_1^{(i)}\mathbf{x}^{(i)} - \mathbf{F}_2^{(i)}\mathbf{a}^{(i)} \tag{13.31}$$

For the first iteration one starts with an initial guess $\mathbf{x}^{(0)}$ and $\mathbf{a}^{(0)}$. In order to obtain convergence of the iteration process the initial guesses sometimes have to be quite good. Differential (or difference) approximation (Section 9.4) can be used to obtain such guesses. The iteration process can be terminated when the difference between the solutions of two successive iterations, i.e. $\mathbf{x}^{(i+1)}(t) - \mathbf{x}^{(i)}(t)$, $\mathbf{a}^{(i+1)}(t) - \mathbf{a}^{(i)}(t)$, fulfil preassigned conditions. Bellman and

Kalaba (1965) show that the sequence of functions $\mathbf{x}^{(0)}(t), \ldots, \mathbf{x}^{(i+1)}(t)$ converges to the true solution in a quadratic way, if $\mathbf{f}(\mathbf{x}, \mathbf{u}, \mathbf{a})$ is a convex function and if the elements of the Jacobian above and below the main diagonal are positive. Ohap and Stubberud (1965) have shown that in cases where these conditions are not fulfilled convergence may also be obtained.

So far it has been assumed that the number of measurements equals the number of equations $(n + m)$. In view of disturbances on the observations in practice, the number of measurements $k$ will be chosen $k > (n + m)$. In that case one may minimize the squared error between the measurements $\mathbf{y}_j$ and the 'model' output $\mathbf{y}(j)$. Detchmendy and Sridhar (1965) have discussed the application of this technique to parameter and state estimation. The same type of approach can be used for the sampled type of process description.

### *13.4.2 Invariant Imbedding*

Again for a simple type of exposition of the basic ideas we use the following type of process description:

$$\left.\begin{aligned} \dot{\mathbf{x}}(t) &= \mathbf{f}(\mathbf{x}(t), t) + \mathbf{G}(t)\mathbf{v}(t) \\ \mathbf{y}(t) &= \mathbf{g}(\mathbf{x}(t), t) + \qquad \mathbf{n}(t) \end{aligned}\right\} \tag{13.32}$$

As before the parameters to be estimated are incorporated in the augmented 'state/parameter vector' $\mathbf{x}(t)$. In a way analogous to the one used in Section 13.3 we find the corresponding TPBVP over the interval $0 \leqslant t \leqslant T$ to be:

$$\dot{\mathbf{x}}(t) = \mathbf{f}(\mathbf{x}(t), t) - \mathbf{G}(t)\mathbf{V}(t)\mathbf{G}'(t)\boldsymbol{\lambda}(t) \tag{13.33}$$

$$\dot{\boldsymbol{\lambda}}(t) = \frac{\partial \mathbf{g}'(\mathbf{x}(t), t)}{\partial \mathbf{x}(t)}\mathbf{N}^{-1}(t)[\mathbf{y}(t) - \mathbf{g}(\mathbf{x}(t), t)] + \frac{\partial \mathbf{f}'(\mathbf{x}(t), t)}{\partial \mathbf{x}(t)}\boldsymbol{\lambda}(t) \tag{13.34}$$

with boundary conditions:

$$\mathbf{x}(0) = \mathbf{x}_0 - \mathbf{P}_0\boldsymbol{\lambda}(0) \tag{13.35}$$

$$\boldsymbol{\lambda}(T) = \mathbf{0} \tag{13.36}$$

For the sake of notational convenience equations (13.33) and (13.34) are written as:

$$\dot{\mathbf{x}}(t) = \boldsymbol{\gamma}(\mathbf{x}(t), \boldsymbol{\lambda}(t), t) \tag{13.37}$$

$$\dot{\boldsymbol{\lambda}}(t) = \boldsymbol{\eta}(\mathbf{x}(t), \boldsymbol{\lambda}(t), t) \tag{13.38}$$

We devote our attention to the boundary condition $\boldsymbol{\lambda}(T) = \mathbf{0}$. The problem is generalized to (imbedded into) the more general case (cf. e.g. Detchmendy and Sridhar, 1966):

$$\boldsymbol{\lambda}(T) = \mathbf{c} \tag{13.39}$$

Now $\mathbf{c}$ and $T$ are assumed to be variable. A change of $\mathbf{c}$ will influence the terminal value of $\boldsymbol{\lambda}$. Also a change of $T$ to $T + \Delta T$ will influence the value of $\boldsymbol{\lambda}$. Consequently one finds the relations:

$$\boldsymbol{\lambda}(T) = \mathbf{c} \quad \Rightarrow \quad \mathbf{x}(T) = \mathbf{r}(\mathbf{c}, T)$$

$$\lambda(T + \Delta T) = \mathbf{c} + \Delta\mathbf{c} \Rightarrow \mathbf{x}(T + \Delta T) = \mathbf{r}(\mathbf{c}, T) + \Delta\mathbf{x}$$

or

$$\mathbf{x}(T + \Delta T) = \mathbf{r}(\mathbf{c} + \Delta\mathbf{c}, T + \Delta T)$$

Through a Taylor series expansion of the last expression one finds:

$$\Delta\mathbf{x} = \mathbf{r}(\mathbf{c} + \Delta\mathbf{c}, T + \Delta T) - \mathbf{r}(\mathbf{c}, T)$$

$$\doteq \frac{\partial\mathbf{r}(\mathbf{c}, T)}{\partial\mathbf{c}}\Delta\mathbf{c} + \frac{\partial\mathbf{r}(\mathbf{c}, T)}{\partial T}\Delta T + \mathcal{O}[\Delta T^2] \tag{13.40}$$

where $\mathcal{O}[\Delta T^2]$ indicates that the remaining terms are at least of the second degree in $\Delta T$. From equations (13.37) and (13.38) one notices that

$$\Delta\mathbf{x} = \boldsymbol{\gamma}(\mathbf{r}(\mathbf{c}, T), \mathbf{c}, T)\Delta T + \mathcal{O}[\Delta T^2] \tag{13.41}$$

$$\Delta\mathbf{c} = \boldsymbol{\eta}(\mathbf{r}(\mathbf{c}, T), \mathbf{c}, T)\Delta T + \mathcal{O}[\Delta T^2] \tag{13.42}$$

Substituting in equation (13.40), dividing by $\Delta T$ and letting $\Delta T \to 0$ one obtains:

$$\boldsymbol{\gamma}(\mathbf{r}, \mathbf{c}, T) = \frac{\partial\mathbf{r}(\mathbf{c}, T)}{\partial\mathbf{c}}\boldsymbol{\eta}(\mathbf{r}, \mathbf{c}, T) + \frac{\partial\mathbf{r}(\mathbf{c}, T)}{\partial T} \tag{13.43}$$

which is called the **invariant imbedding equation.** As an approximate solution for this partial differential equation in the neighbourhood of $\mathbf{c} = \mathbf{0}$ is chosen:

$$\mathbf{r}(\mathbf{c}, T) = \hat{\mathbf{x}}(T) - \mathbf{P}(T)\mathbf{c} \tag{13.44}$$

According to the TPBVP if $\mathbf{c} = \mathbf{0}$ one should find indeed $\hat{\mathbf{x}}(T)$, the optimal value of the state/parameter vector at $t = T$. Substitution of this approximate solution equation (13.44) into equation (13.43) gives:

$$\boldsymbol{\gamma}(\hat{\mathbf{x}}(T) - \mathbf{P}(T)\mathbf{c}, \mathbf{c}, T) = -\mathbf{P}(T)\boldsymbol{\eta}(\hat{\mathbf{x}}(T) - \mathbf{P}(T)\mathbf{c}, \mathbf{c}, T) + \frac{\partial\hat{\mathbf{x}}(T)}{\partial T} - \frac{\partial\mathbf{P}(T)}{\partial T}\mathbf{c} \tag{13.45}$$

Now we return to our original equations for $\gamma$ and $\boldsymbol{\eta}$, viz equations (13.33) and (13.34) and substitute them into equation (13.45). There results:

$$\begin{aligned}
&\mathbf{f}(\hat{\mathbf{x}}(T) - \mathbf{P}(T)\mathbf{c}, T) - \mathbf{G}(T)\mathbf{V}(T)\mathbf{G}'(T)\mathbf{c} \\
&= -\mathbf{P}(T)\frac{\partial \mathbf{g}'(\hat{\mathbf{x}}(T) - \mathbf{P}(T)\mathbf{c}, T)}{\partial \hat{\mathbf{x}}}\mathbf{N}^{-1}(T)[\mathbf{y}(T) - \mathbf{g}(\hat{\mathbf{x}}(T) - \mathbf{P}(T)\mathbf{c}, T)] \qquad (13.46)\\
&\quad + \mathbf{P}(T)\frac{\partial \mathbf{f}'(\hat{\mathbf{x}}(T) - \mathbf{P}(T)\mathbf{c}, T)}{\partial \hat{x}}\mathbf{c} + \frac{\partial \hat{\mathbf{x}}(T)}{\partial T} - \frac{\partial \mathbf{P}(T)}{\partial T}\mathbf{c}
\end{aligned}$$

As $\hat{\mathbf{x}}(T) = \mathbf{r}(0, T)$ we write down an expansion around this optimal estimate, retaining only zero and first-order terms in $\mathbf{c}$:

$$\begin{aligned}
&\mathbf{f}(\hat{\mathbf{x}}(T), T) - \frac{\partial \mathbf{f}'(\hat{\mathbf{x}}(T), T)}{\partial \hat{\mathbf{x}}}\mathbf{P}(T)\mathbf{c} - \mathbf{G}(T)\mathbf{V}(T)\mathbf{G}'(T)\mathbf{c} \\
&= -\mathbf{P}(T)\frac{\partial \mathbf{g}'(\hat{\mathbf{x}}(T), T)}{\partial \hat{\mathbf{x}}}\mathbf{N}^{-1}(T)[\mathbf{y}(T) - \mathbf{g}(\hat{\mathbf{x}}, T)] \\
&\quad + \mathbf{P}(T)\frac{\partial}{\partial \hat{\mathbf{x}}}\left[\frac{\partial \mathbf{g}'(\hat{\mathbf{x}}(T), T)}{\partial \hat{\mathbf{x}}}\mathbf{N}^{-1}(T)[\mathbf{y}(T) - \mathbf{g}(\hat{\mathbf{x}}, T)]\right]\mathbf{P}(T)\mathbf{c} \qquad (13.47)\\
&\quad + \mathbf{P}(T)\frac{\partial \mathbf{f}'(\hat{\mathbf{x}}(T), T)}{\partial \hat{\mathbf{x}}}\mathbf{c} + \frac{\partial \hat{\mathbf{x}}(T)}{\partial T} - \frac{\partial \mathbf{P}(T)}{\partial T}\mathbf{c}
\end{aligned}$$

As this must hold for (small) arbitrary values of $\mathbf{c}$ expression (13.47) may be separated in zero and first-order terms in $\mathbf{c}$, which leads to the following equations:

$$\frac{\partial \hat{\mathbf{x}}(T)}{\partial T} = \mathbf{f}(\hat{\mathbf{x}}(T), T) + \mathbf{P}(T)\frac{\partial \mathbf{g}'(\hat{\mathbf{x}}(T), T)}{\partial \hat{\mathbf{x}}}\mathbf{N}^{-1}(T)[\mathbf{y}(T) - \mathbf{g}(\hat{\mathbf{x}}(T), T)] \qquad (13.48)$$

$$\begin{aligned}
\frac{\partial \mathbf{P}(T)}{\partial T} &= \mathbf{G}(T)\mathbf{V}(T)\mathbf{G}'(T) \\
&\quad + \mathbf{P}(T)\frac{\partial \mathbf{f}'(\hat{\mathbf{x}}(T), T)}{\partial \hat{\mathbf{x}}} + \frac{\partial \mathbf{f}(\hat{\mathbf{x}}(T), T)}{\partial \hat{\mathbf{x}}}\mathbf{P}(T) \qquad (13.49)\\
&\quad + \mathbf{P}(T)\frac{\partial}{\partial \hat{\mathbf{x}}}\left[\frac{\partial \mathbf{g}'(\hat{\mathbf{x}}(T), T)}{\partial \hat{\mathbf{x}}}\mathbf{N}^{-1}(T)[\mathbf{y}(T) - \mathbf{g}(\hat{\mathbf{x}}(T), T)]\right]\mathbf{P}(T)
\end{aligned}$$

These equations describe a sequential least-squares estimator, as $T$ can now be considered as the running variable $t$.

The initial conditions follow from equation (13.35)

$$\mathbf{x}(0) = \mathbf{x}_0 - \mathbf{P}_0\boldsymbol{\lambda}(0)$$

and equation (13.44). Consequently

$$\hat{\mathbf{x}}(0) = \mathbf{x}_0 \qquad \mathbf{P}(0) = \mathbf{P}_0$$

*Example.* The following illustration is an example given by Detchmendy and Sridhar (1966). They assume a second-order process:

$$\dot{x}_1 = x_2$$
$$\dot{x}_2 = -2x_1 - a(t)x_1^3 - 3x_2 + 5\sin t$$

with a time-varying parameter $a(t) = 2\exp[-0{\cdot}1t]$ and output observations given by

$$y = x_1 + n$$

If the kind of variation of $a(t)$ is known, but not its initial value nor its time constant, then it can be represented by

$$\dot{a} = -ba \quad \text{or} \quad \dot{x}_3 = -x_4x_3$$
$$\dot{b} = 0 \quad \text{or} \quad \dot{x}_4 = 0$$

Consequently

$$\dot{\mathbf{x}} = \begin{bmatrix} x_1 \\ x_2 \\ x_3 \\ x_4 \end{bmatrix} = \begin{bmatrix} x_2 \\ -2x_1 - x_3x_1^3 - 3x_2 + 5\sin t \\ -x_4x_3 \\ 0 \end{bmatrix} = \mathbf{f}(\mathbf{x}, t)$$

$$\mathbf{y} = \begin{bmatrix} y \\ 0 \\ 0 \\ 0 \end{bmatrix} = \begin{bmatrix} x_1 \\ 0 \\ 0 \\ 0 \end{bmatrix} + \begin{bmatrix} n \\ 0 \\ 0 \\ 0 \end{bmatrix} = \mathbf{g}(\mathbf{x}, t) + \mathbf{n}$$

Also given is:

$$\mathbf{N}(t) = 1 \quad \text{and} \quad \mathbf{GVG}' = \mathbf{I}$$

Note that

$$\mathbf{P}(t)\frac{\partial \mathbf{g}'}{\partial \hat{\mathbf{x}}} = [P_{11}, P_{21}, P_{31}, P_{41}]'$$

Consequently the estimator is found to be:

$$\begin{bmatrix} \dot{\hat{x}}_1 \\ \dot{\hat{x}}_2 \\ \dot{\hat{x}}_3 \\ \dot{\hat{x}}_4 \end{bmatrix} = \begin{bmatrix} \hat{x}_2 \\ -2\hat{x}_1 - \hat{x}_3\hat{x}_1^3 - 3\hat{x}_2 + 5\sin t \\ -\hat{x}_4\hat{x}_3 \\ 0 \end{bmatrix} + \begin{bmatrix} P_{11} \\ P_{21} \\ P_{31} \\ P_{41} \end{bmatrix}(y - \hat{x}_1)$$

and the equation for $\dot{\mathbf{P}}$ which also can be found in a straightforward way. Given the initial conditions $\mathbf{x}_0 = \mathbf{0}$ and $\mathbf{P}_0$ the estimation procedure can be started. Figure 13.2 shows the development in time of such an estimate.

The invariant imbedding method can be derived also for the sampled type of representation along analogous ways.

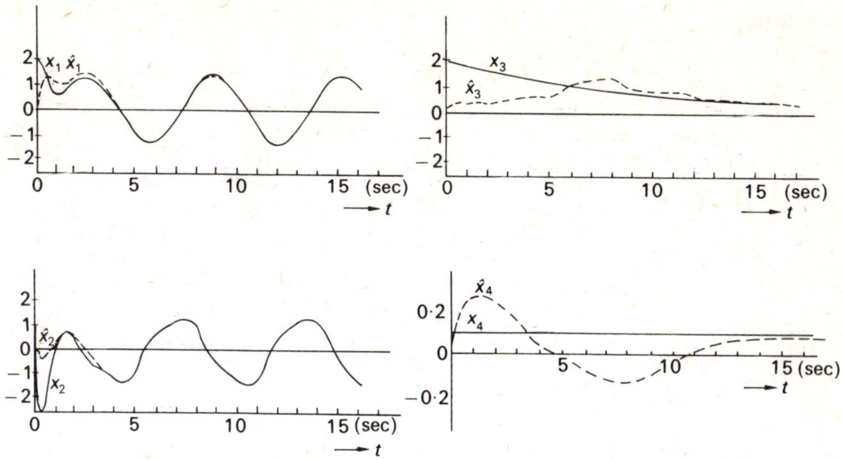


Figure 13.2

## 13.5 Concluding Remarks

As it is virtually impossible to do justice to the many interesting publications on relevant methods for minimizing the 'error functions' and solving TPBVP the reader is referred to:

the monograph by Sage and Melsa (1971) on this topic;

the papers mentioned under 'additional literature'.

## References


Anderson, W. N. et al. (1969). 'Consistent estimates of the parameters of a linear system', *Ann. Math. Stat.*, **40,** 2064–2075.

Baggerud, A. and J. G. Balchen (1970). 'An adaptive state estimator', *2nd IFAC Symp.—Identification and Process Parameter Estimation, Prague*, paper 10.1.

Bellman, R. E. and R. E. Kalaba (1965). *Quasilinearization and Nonlinear Boundary-Value Problems*, American Elsevier Publ., New York, 206 pp.

Cox, H. (1964). 'On the estimation of state variables and parameters for noisy dynamic systems', *IEEE Trans. Autom. Control*, **AC-9,** 5–12.

Detchmendy, D. M. and R. Sridhar (1965). 'On the experimental determination of the dynamical characteristics of physical systems', *Proc. Nat. Electron. Conf.*, Chicago, Ill., **21,** 575–580.

Detchmendy, D. M. and R. Sridhar (1966). 'Sequential estimation of states and parameters in noisy nonlinear dynamical systems', *J. Basic Engng.* (*Trans. ASME*), **88D,** 362–368.

Ohap, R. F. and A. R. Stubberud (1965). 'A technique for estimating the state of a nonlinear system', *IEEE Autom. Control*, **AC-10,** 150–155.

Sage, A. P. and J. L. Melsa (1971). Cf. references Chapter 2.

## Additional Literature

Cf. also the additional literature of Section 8.4.


Aoki, M. (1967). Cf. references Chapter 4.

Balakrishnan, A. V. and L. W. Neustadt (Eds.) (1964). *Computing Methods in Optimization Problems*, Academic Press, New York.

Balakrishnan, A. V. (1967). 'Identification of control systems from input–output data', *IFAC Symp.—Identification in Autom. Control Systems, Prague*, paper 1.1.

Bar-Shalom, Y. (1972). 'Optimal simultaneous state estimation and parameter identification in linear discrete-time systems', *IEEE—Trans. Autom. Control*, **AC-17,** 308–319.

Bellman, R. E., R. E. Kalaba and G. M. Wing (1960). 'Invariant imbedding and mathematical physics I—particle processes', *J. Math. Phys.*, **1,** 280–308.

Bellman, R. E., R. E. Kalaba and G. M. Wing (1960). 'Invariant imbedding and the reduction of two-point boundary value problems to initial-value problems', *Proc. U.S. Nat. Acad. Sci.*, **46,** 1646–1649.

Bellman, R. E. (1961). *Adaptive Control Processes, A Guided Tour*, University Press, New Jersey, Princeton.

Bellman, R. E. and R. E. Kalaba (1961). 'On the fundamental equations of invariant imbedding—I', *Proc. U.S. Nat. Acad. Sci.*, **47,** 336–338.

Bellman, R. E., H. Kagiwada and R. E. Kalaba (1963). 'Quasilinearization, system identification, and prediction', *Report Rand Corp.*, RM-3812-PR.

Bellman, R. E., H. Kagiwada and R. E. Kalaba (1964). 'Numerical studies of a two-point nonlinear boundary value problem using dynamic programming, invariant imbedding and quasilinearization', *Report Rand Corp.*, RM-4069-PR.

Bellman, R. E. and R. E. Kalaba (1964). 'Dynamic programming, invariant imbedding, and quasilinearization: comparisons and interconnections', *Report Rand Corp.*, RM-4038-PR.

Bellman, R. E., H. H. Kagiwada and R. E. Kalaba (1965). 'Quasilinearization, boundary-value problems, and linear programming', *IEEE Trans. Autom. Control*, **AC-10,** 2.

Bellman, R. E. (1965). 'Dynamic programming, system identification and suboptimization', *Report Rand Corp.*, RM-4593-PR.

Berkovec, J. W. (1969). 'A method for the validation of Kalman filter models', *Proc. Joint Autom. Control Conf.*, 488–493.

Coasert, R. (1964. 'On the estimation of the plant state and plant parameters and external disturbances from observations of the output of a control system', *Joint Autom. Control Conf.*, 340–348.

Coggan, G. C. and A. R. M. Noton (1970). 'Discrete-time sequential state and parameter estimation in chemical engineering', *Trans. Inst. Chem. Eng.*, **40,** 255–64.

Cox, H. (1966). 'Sequential minimix estimation', *IEEE Trans. Autom. Control*, **AC-11,** 323–324.

Ellis, T. W. and A. P. Sage (1968). 'Application of a method for on-line combined estimation and control', *Proc. Southwest IEEE Conf.*

Fan, L. T. and C. S. Wing (1964). *The Discrete Maximum Principle*, Wiley, New York.

Giese, C. and R. B. McGhee (1965). 'Estimation of nonlinear system states and parameters by regression methods', *Joint Autom. Control Conf.*, 46–53.

Habegger, L. J. and R. E. Bailey (1969). Cf. references Section 14.4.

Henrici, P. (1962). *Discrete Variable Methods in Ordinary Differential Equations*, Wiley, New York.

Hilborn, C. G. and D. G. Lainiotis (1969). 'Optimal estimation in the presence of unknown parameters', *IEEE Trans. Systems Sci. Cybernet.*, **SSC-5,** 38–43.

Jazwinski, A. H. (1966). 'Filtering for nonlinear dynamical systems', *IEEE Trans. Autom. Control*, **AC-11,** 765.

Jazwinski, A. H. (1969). 'Adaptive filtering', *Automatica*, **5,** 475–485.

Jazwinski, A. H. (1970). Cf. references Chapter 12.

Kailath, T. and P. Frost (1968). Cf. references Chapter 4.

Kalaba, R. E. (1959). 'On nonlinear differential equations, the maximum operation, and monotone convergence', *J. Math. Mech.*, **8,** 519–574.

Kalaba, R. (1963). 'Invariant imbedding and the analysis of processes', *Report Rand Corp.*, RM-3806-PR.

Kalaba, R. (1963). 'Some aspects of quasilinearization', in *Nonlinear Differential Equations and Nonlinear Mechanics* (Eds. J. P. LaSalle and S. Lefschetz), Academic Press, New York.

Kenneth, P. and R. McGill (1966). 'Two point boundary value problem techniques', *Advan. Contr. Syst.*, **3,** 69–110.

Kopp, R. E. and R. J. Orford (1963). 'Linear regression applied to system identification for adaptive control systems', *ATAA J.* **1,** 2300–2306.

Kumar, K. P. S. and R. Sridhar (1964). On the identification of control systems by the quasi-linearization method', *IEEE Trans. Autom. Control*, **AC-9,** 151–154.

Kushner, H. J. (1967). 'Approximations to optimal non-linear filters', *Joint Autom. Control Conf.*, 613–623.

Kushner, H. J. (1967). 'Non-linear filtering: the exact dynamical equations satisfied by the conditional mode', *IEEE Trans. Autom. Control*, **AC-12,** 262–267.

Larminat, Ph. de and M. H. Tallec (1970). 'On-line identifications of time varying systems', *2nd IFAC Symp.—Identification and Process Parameter Estimation, Prague*, paper 3.3.

Lavi, A. and J. C. Strauss (1965). 'Parameter identification in continuous dynamic systems', *IEEE Int. Conv. Rec.*, Symposium on Automatic Control; Systems Science; Cybernetics; Human Factors, Part VI, 49–61.

Lee R. C. K. (1964). Cf. references Chapter 4.

Mowery, V. O. (1965). 'Least squares recursive differential-correction estimation in nonlinear problems', *IEEE Trans. Autom. Control*, **AC-10,** 399–407.

Neal, S. R. (1968). 'Nonlinear estimation techniques', *IEEE Trans. Autom. Control*, **AC-13,** 705–708.

Noton, A. R. M. (1965). *Introduction to Variational Methods in Control Engineering*, Pergamon Press, Oxford, 122 pp.

Pearson, J. B. (1968). 'A note on nonlinear filtering' (correspondence), *IEEE Trans. Autom. Control*, **AC-13,** 103–105.

Sage, A. P. and G. W. Masters (1966). 'On-line estimation states and parameters for discrete nonlinear dynamic systems', *Proc. Nat. Electron. Conf.*, **22,** 677–682.

Sage, A. P. and G. W. Masters (1967). 'Identification and modeling of states and parameters of nuclear reactor systems', *IEEE Trans. Nucl. Sci.*, **NS-14,** 1, 279–285.

Schley, C. H. Jr. and I. Lee (1967). 'Optimal control computation by the Newton–Raphson method and the Riccati transformation', *IEEE Trans. Autom. Control*, **AC-12,** 139–144.

Schwartz, L. and E. B. Stear (1968). 'A computational comparison of several nonlinear filters', *IEEE Trans. Autom. Control*, **AC-13,** 83–86.

Schweppe, F. C. (1968). 'Recursive state estimation: unknown but bounded errors and system inputs', *IEEE Trans. Autom. Control*, **AC-13,** 22–28.

Sunahara, Y. (1969). 'An approximate method of state estimation for non-linear dynamical systems', *Joint Autom. Control Conf.*, 161–172.

Sylvester, R. J. and F. Meyer (1965). 'Two point boundary value problems by quasi-linearization', *SIAM J.*, **13,** 586–602.

Tait, G. R. and P. R. Bélanger (1970). 'A comparison of some parameter identification schemes using first and second order extended Kalman–Bucy filters and sensitivity functions', *2nd IFAC Symp.—Identification and Process Parameter Estimation, Prague*, paper 10.3.

Wishner, R. P., J. A. Tabaczynski and M. Athans (1969). 'A comparison of three non-linear filters', *Automatica*, **5,** 487–496.

Wonham, W. M. (1963). 'Stochastic problems in optimal control', *IEEE Conv. Rec.*, part 2, 114–124.

Wonham, W. M. (1964). 'Some applications of stochastic differential equations to non-linear filtering', *SIAM J. Control*, **2,** 347–369.

*CHAPTER 14*

# *Survey of Applications from Different Fields*

As mentioned in Section 1.6 the parameter- and state-estimation techniques can be used:

in different ways of application, e.g.: 'diagnostic' applications, control application, automatic (industrial) measurements, automatic (industrial) decisions, automatic (industrial) adjustments, pattern recognition.

in different fields of human endeavour, e.g.: communication engineering, power engineering, mechanical engineering, aeronautical engineering, chemical engineering, physics, geology, economy, biology/medicine.

In applications directed to automatic control there is the problem of how the estimation procedures fit into that control problem. This is briefly discussed in Section 14.1. The remaining part of the chapter is devoted to particular fields of applications, illustrated by specific examples. These are taken from the fields of:

| | |
|---|---|
| physical/mechanical/chemical processes | Section 14.2 |
| nuclear reactors | Section 14.3 |
| power plants and power distribution systems | Section 14.4 |
| telecommunication | Section 14.5 |
| air and space vehicles | Section 14.6 |
| biological objects | Section 14.7 |
| socioeconomic systems | Section 14.8 |
| learning systems and pattern recognition | Section 14.9 |

This list of 'fields' does not constitute a unique division of types of applications, nor are the examples given necessarily the most important or the most spectacular ones. By use of the literature references the picture of applications can be adapted to the interest of the reader. With this chapter the list of literature is divided according to the topics of the sections.

## 14.1 Automatic Control

### *14.1.1 Optimum Control; The Separation Hypothesis*

In Chapter 1 it was stressed that the application of feedback has been recognized as an effective expedient for combating uncertainties, but that there are limitations:

very strict requirements for optimization, leading to **optimal** or **self-optimizing** control;

very large parameter variations, leading to the use of **adaptive** control (cf. Jacobs, 1964).

In such cases the need for very good or for permanently updated knowledge of the process under control is of crucial importance. In Section 1.4 it was mentioned that Fel'dbaum's **dual control** principle implies two functions, performed simultaneously:

to study or learn the characteristics of the process;

to direct the process to the required state.

It is unfortunate that such a coherent treatment of the problems at hand leads to a theory that as yet results in exorbitant computational problems, even for simple cases (cf. Fel'dbaum, 1960, 1961; 1965; Mendes, 1970; cf. also Snegur, Tomashov and Dubina, 1967). Consequently attempts have been made to simplify this general problem. This is being done by an *a priori* assumption that the design of a control scheme can be divided into two parts: identification and control. We refer to this assumption as the **separation hypothesis** (cf. Åström, 1970). The approach is very natural, in particular if we consider the multitude of techniques which have been developed for the design of systems with known process dynamics and known environments. However, it is seldom true that optimum solutions are obtained if a process is identified and the results of the identification are used in a design procedure, developed under the assumption that the process and its environment are known precisely. It can be necessary to modify the control strategy to take into account the fact that the identification is not precise. Apart from the obvious fact that it is desirable to choose a class of models $\mathscr{S}$ for which there is a control theory available, there are also many other interesting questions in the area of identification and control, e.g.

Is it possible to obtain rational choices of model structures and criteria for the identification if we know that the results of identification will be used to design control strategies?

What 'accuracy' is required of the solution of an identification problem if the separation hypothesis should be valid at least with a specified error?

Partial answers to these questions are given by Åström and Wittenmark (1969) for a restricted class of problems.

An extensive survey of applications of identification methods in the U.S.S.R. has been written by Rajbman (1973).

### *14.1.2 Self-optimizing and Self-adaptive Systems*

Since the early contribution of Draper and Li (1951) there have been many attempts and suggestions for self-adaptive and -optimizing systems (cf. Kalman, 1958; Aseltine and others, 1958; Gibson, 1960; Krasovski, 1960; Mathias and Van Nice, 1960; Jacobs, 1961; Mishkin and Braun, 1961; Pun, 1961; Rutman, 1962).

Many of the methods proposed rely on estimation techniques. It is, however, also possible to mechanize the self-optimizing and (self-)adaptive principles in a direct way. As a typical example we mention the on-line hunting for the optimum or the on-line tuning of PID controllers. In such cases the introduction of test signals on the quantities of interest (set points, viz. coefficients) may be needed in order to determine the direction of the optimum. One of the classes of adaptive systems is indicated by the name **'model-reference'** approach. One of the early contributions along this line was by Whitaker et al. (1958). This method can be illustrated for an airplane control system indicated in Figure 14.1. At different altitudes, due to the

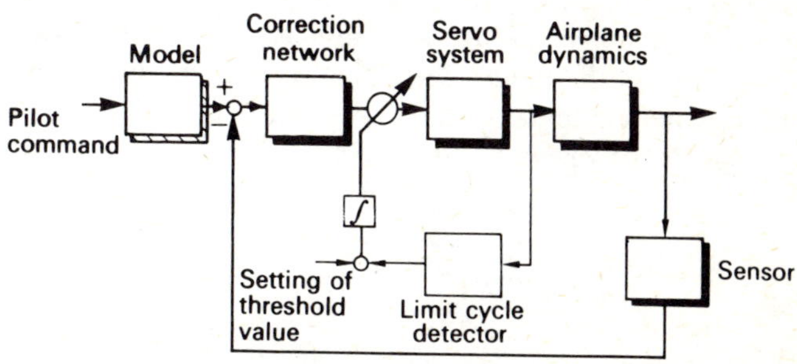


Figure 14.1

difference in air density, the effectiveness of the rudders change. This results in a change of the airplane dynamics and consequently in the command properties of the plane. On the pilot commands the (second-order) model gives a response as it is desired for the airplane as a whole. By automatic adjustment of a gain factor $K$ the loop gain of the control system is made as high as permissible, such that this closed loop forms a 'stiff' system. The adjustment of the maximum permissible gain is done by detection of a small-limit cycle. If no limit cycle is detected, then $K$ is increased, if the amplitude of the limit cycle gets too large then $K$ is diminished. Consequently no explicit use of estimation techniques is needed.

Now we consider two simple examples of control systems in which use is made of **estimation techniques** for adaptive purposes. The first one is a simple servo system as described by Young (1965, 1969). The principle is indicated in Figure 14.2. The purpose of the block 'estimation scheme' is to

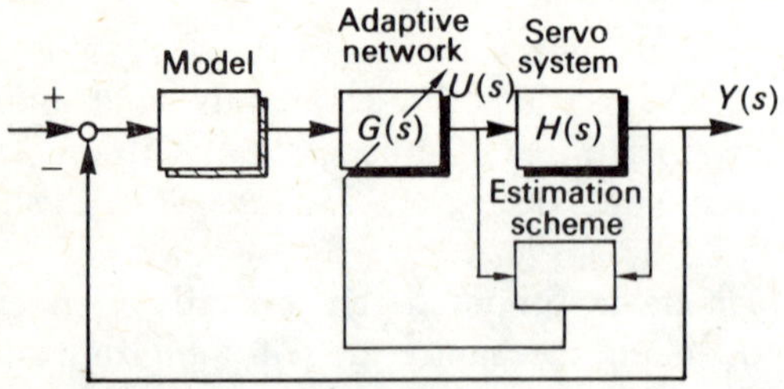


Figure 14.2

determine the coefficients of the transfer function

$$H(s) = \frac{1}{a_2 s^2 + a_1 s + a_0}$$

Based on these estimates the transfer function $G(s)$ of an adaptive network is adjusted in such a way that, as well as possible, $G(s) = H^{-1}(s)$. If this could be realized completely then the characteristics of the feedback system would be determined by the model, which can be chosen properly. For the estimation scheme analogue filters were used that enable one to derive a signal and its derivatives without using pure differentiators (cf. Figure 14.3

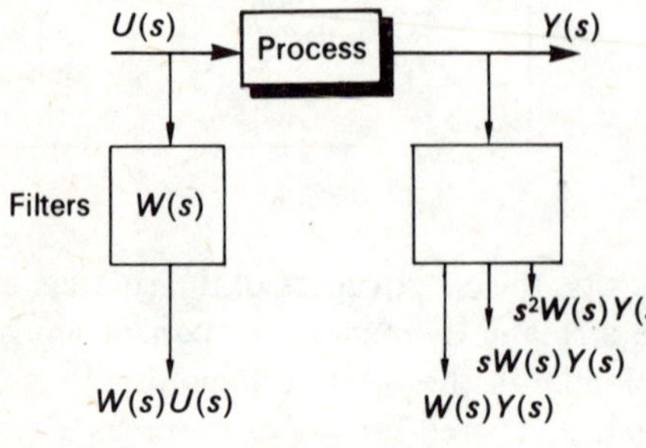


Figure 14.3

and Section 9.2). By sampling and analogue/digital conversion a set of simultaneous equations in the three unknown parameters is made, from which the estimate $\boldsymbol{\alpha}$ for $\mathbf{a}' = [a_2, a_1, a_0]$ can be determined by **explicit** methods.

The second example of a control system in which use is made of estimation techniques is taken from Rumold and Speth (1968) (cf. also Speth, 1969, 1971). In the case described the process consists of an electric speed-control

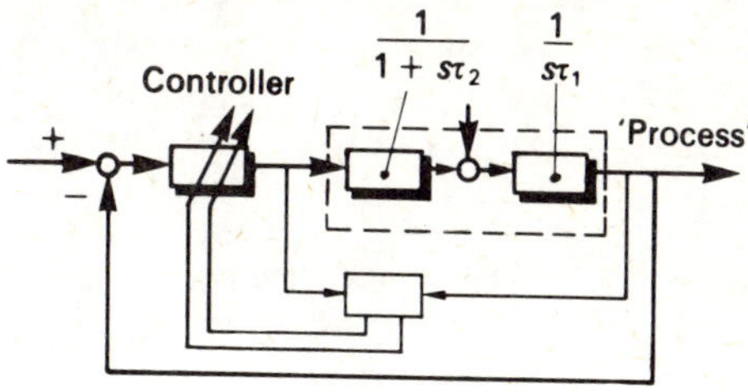


Figure 14.4

system with the transfer functions indicated in Figure 14.4:

$$\frac{1}{1 + s\tau_2} \quad \text{viz.} \quad \frac{1}{s\tau_1}$$

where $\tau_1$ is fixed by the moment of inertia and $\tau_2$ is the time constant of a torque control unit. Both the parameters $a_1 = \tau_1$ and $a_2 = \tau_1\tau_2$ are estimated by an **implicit** (model-adjustment) method. In order to prevent differentation again use is made of analogue filters that provide a signal and its derivative cf. Figure 14.5 where $\alpha_1$ and $\alpha_2$ are adjusted according to the gradient

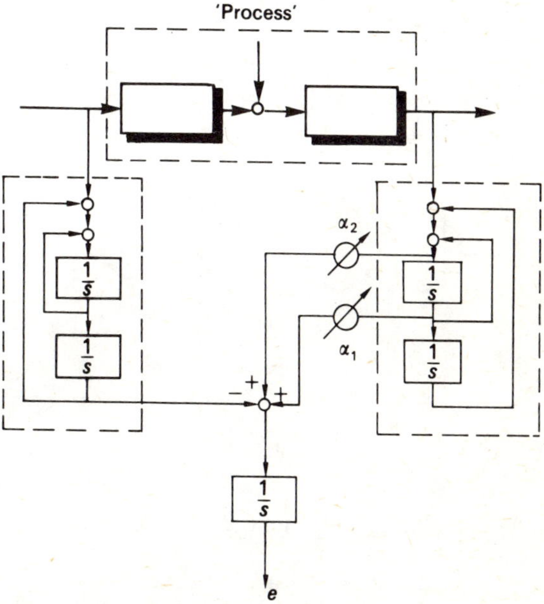


Figure 14.5

method based on an integral squared-error criterion. Based on these parameter values the controller can be adjusted by using explicit rules, based on a 'generalized symmetrical optimum', as referred to in the cited publications. It has been demonstrated that this type of scheme works satisfactorily; it has been implemented in practical applications. Cf. also Spiegel (1969).

In concluding this section it is worth mentioning the different structures of self-adaptive systems according to Figure 14.6 as given by Speth (1971).

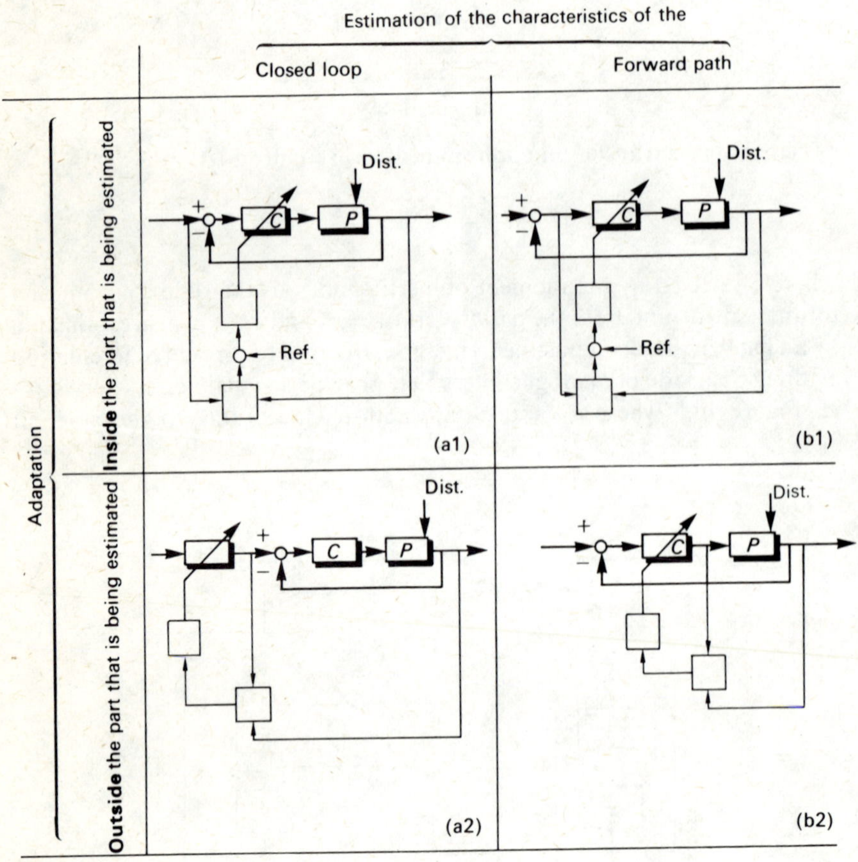


Figure 14.6

Among the differences and resulting properties of these schemes there are:

(a) First column—**estimation of the closed-loop characteristics.** The estimation results can be made independent of the disturbances, if those disturbances are not related to the system input function.

(b) Second column—**estimation of the forward path.** If the control loop to be adjusted is unstable then the estimation is fast due to the large signal amplitudes. Under ordinary conditions, however, one of the signals used for the estimation will be small and the estimation will be slow. Disturbance signals will bias the estimation results due to the 'circulating noise'.

(1) First row—**adaptation inside the part being estimated.** As the controller $C$ belongs to the part being estimated it is not necessary to determine the unknown parameters completely; it is sufficient to test whether they are bigger or smaller than the desired value.

(2) Second row—**adaptation outside the part being estimated.** A fast change of the adaptive element does not interfere with the estimation procedure.

## 14.2 Physical/Mechanical/Chemical Processes

Much has been written on the choice, derivation, use and evaluation of models for different purposes. Introductory remarks on the use of industrial models

for stabilizing control and

for optimizing control

together with some examples of dynamic and state models by Van der Grinten can be found in the survey paper Eykhoff, Van der Grinten, Kwakernaak and Veltman (1966). A survey of applications in chemical and physical processes in given in Gustavsson (1973).

There is an abundant literature on modelling of the great variety of industrial processes. Some general references: Stout (1961) and IEE (1969). Again it is far beyond the scope of this section to try to provide a coherent and complete picture of the many applications. In addition to the references mentioned explicitly the reader is referred to the additional literature. A series of notes on models and testing methods have been written by Murrill, Pike and Smith (1968, 1969).

An interesting table listing experiments that were made using pseudo-random binary time functions as perturbation signals and using correlation techniques was given by Godfrey (1970). A detailed discussion of such techniques applied to an oil refinery can be found in Godfrey (1969).

A few references to specific applications are:

distillation column: Rees (1966). A case study in this field is Foulard and Bornard (1973);

chemical reactions: Sawaragi et al. (1965), Paiziev (1968), Burovoi and Mamedov (1970), Vallerschamp and Perlmutter (1971), Bremermann (1970), Nesbit and Engel (1967), Seinfeld (1970);

water gas shift reactor: Wise, Price and Rippin (1967), Price and Rippin (1968);

paper machine: Gerdin (1970);

evaporators: Manczak (1967), Zavorka et al. (1967);

furnaces: Brown (1969), Bailey (1968), Carlson (1966), Loosemore and Bray (1967), El'kinson (1969). A case study on the identification of a glass furnace is Richalet (1973);

heat exchanger: Isermann (1965), Corlis and Luus (1969);

cement kiln: Rocco (1968), Bailey (1969);

gas chromatographs: Booth (1967), Godfrey and Devenish (1969), Ng and Moss (1970), Kelly and Harris (1971).

As an example we consider the topic of the last-mentioned paper. Schematically a gas chromatograph can be represented by Figure 14.7. It consists of the following parts: (1) a cylinder filled with gas, e.g. hydrogen; (2) a control valve which regulates a constant flow of gas; (3) an injection

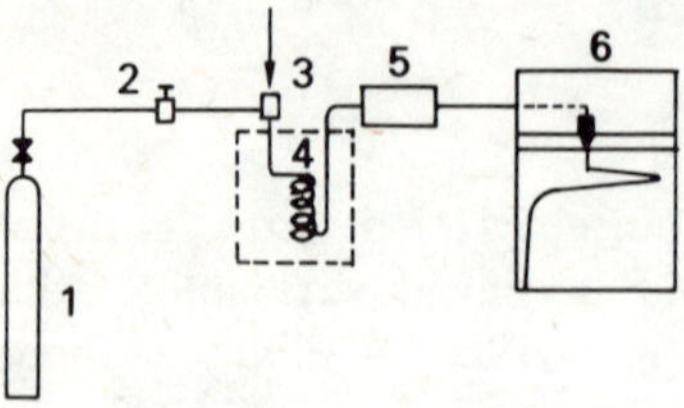


Figure 14.7

device by which the sample to be tested is brought into the gas stream; (4) the 'column', a tube in which the separation of the sample components takes place. This column is enclosed in a thermostat; (5) the detector, measuring some physical property of the gas, e.g. the heat conductivity; (6) the recorder.

The very small sample, mostly liquid, evaporates in the carrier gas flow. Due to the physical properties of the column the sample molecules are delayed with respect to the carrier gas; this delay is specific for each kind of molecule. Consequently, at the end of the column the components of the sample are detected as a sequence of peaks in the recorded chromatogram (cf. Figure 14.8). The time delay since the time of injection (retention time) $t_R$ is characteristic for a component, the area $A$ under the peak provides a measure for the quantity of that component in the sample. The width (standard deviation $\sigma$) of the peak is also of importance.

Although the attention is focussed on the properties of the input sample, these properties can also be attributed to the process (measuring device) which causes the separation. Consequently the problem can be formulated

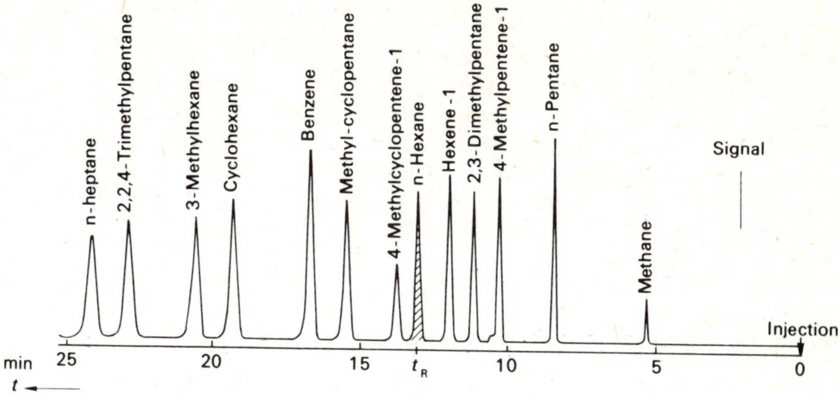


Figure 14.8

as that of estimating (for each peak) the parameter vector $\boldsymbol{\beta} = [t_R, A, \sigma]'$. In practice the chromatogram may be disturbed by noise, e.g. due to pollution in the carrier gas, changes of temperature, electrical noise of detector and amplifier. A technique described in the reference cited, assuming the noise to be stationary and additive, is: estimate the properties of the noise when there is no measurement signal; filter the signal by a 'noise-whitening filter'; estimate $\hat{\boldsymbol{\beta}}$.

## 14.3 Nuclear Reactors

Also in this field a variety of estimation schemes has been proposed and implemented. As a few specific examples we may cite:

The measurement of the under-critical (neutron) multiplication factor (e.g. Bastl, 1966). This may be approached by introducing a neutron source which is modulated by a pseudo-random binary sequence and by correlation of that (delayed) signal with measurements of the neutron density level.

The determination of kinetic parameters of the reactor from correlation measurements based on the statistical fluctuations of the neutron chain reaction (cf. Wilkie, 1967; Liewers and Buttler, 1967 who use the polarity correlation technique; Roman, Hsu and Habegger, 1971 who applied quasilinearization).

Parameter estimation of coupled reactors (cf. Hendrikson and Murphy, 1968; Schwalm, 1971).

State estimation, where the time-dependent reactivity is considered to be one of the state variables, using Kalman–Bucy filter theory (cf. Venerus and Bullock, 1970), or combined parameter and state estimation by means of a TPBVP (cf. Sage and Masters, 1966).

## 14.4 Power Plants and Power Distribution Systems

A survey of application methods in power generation and distribution is given in Baeyens and Jacquet (1973).

In order to deal with frequent load changes in an efficient way, much knowledge and even parametric control may be necessary in power plants. Some examples of literature on estimation problems: Stanton (1965), Suleimenova (1965), Debelle, Foureau and Vazquez (1966), Thompson (1967), Eklund (1972).

For the load dispatching problem, use can be made of state (and parameter) estimation, as it is discussed in e.g. Habegger and Bailey (1969), Debs and Larson (1970), Schweppe et al. (1970), Toyoda et al. (1970), Uemura (1972).

For the adjustment of load frequency controller parameters one needs to know the transfer function $G$ as indicated in Figure 14.9, where $u$ is the control signal of the

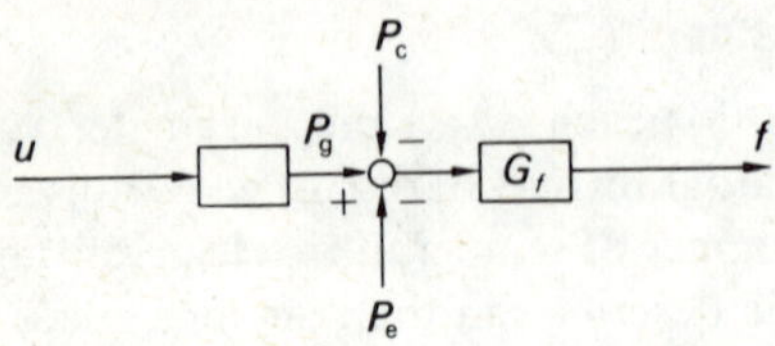


Figure 14.9

generating unit; $P_g$, $P_c$ and $P_e$ are the generated, the consumed and the 'exported' power; and where $f$ is the frequency. Cf. Cuenod and Quazza (1969), Koszelnik, Malkiewicz and Trybula (1969), Donati et al. (1970). A source of general interest is Handschin (1972). A case study on the identification of drum boiler dynamics is Eklund and Gustavsson (1973).

## 14.5 Telecommunication

It is quite evident that there is an intimate relationship between many problems in the telecommunication area and the field of parameter and state estimation, e.g. the dominating role of disturbances (noise), causing fundamental limitations to the possible achievements. Such relationships are clear in the problem of separating signal and noise, in estimating parameters of signals (e.g. distance and velocity of the object in radar observations), in the estimation of bit synchronization, etc. (cf. Middleton, 1960). Some special applications are indicated briefly.

The attenuation and phase characteristics of transmission lines change with length, type of cable, temperature, humidity, etc. Equalizers are used to compensate as well as possible for undesirable characteristics and for changes that may occur in those properties. Variable equalizers have been discussed already by Bode (1938). Automatic equalization has become a popular subject of research and application. Many papers have been written on adaptive filters that adjust automatically to the (changes in) characteristics of the transmission line in such a way that e.g. the intersymbol interference is minimized. The techniques used can be considered as model-adjustment procedures where the 'model' (equalizer) is placed in series with the changing object (cf. Figure 14.10). A popular choice of equalizer is a tapped delay line with an adjustable attenuator at every tap. A survey of automatic equalization is given in Kanal (1969); many papers on this topic are mentioned as additional literature.

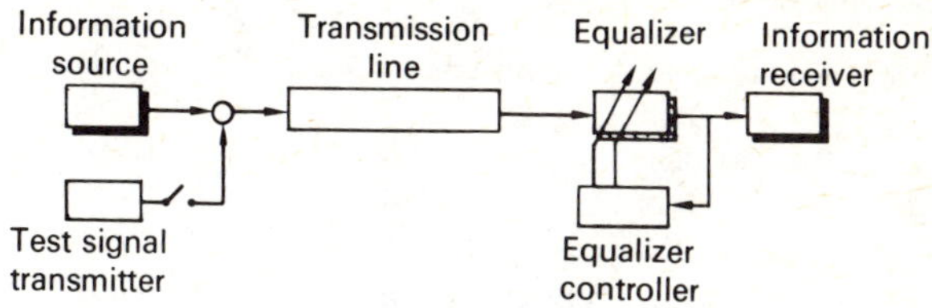


Figure 14.10

Adaptive filters used as frequency-selective limiters (cf. Orth, 1968).

Adaptive redundancy removal in data transmission (cf. Lucky, 1968).

Adaptive antenna systems (cf. Widrow et al., 1967).

Estimation of the characteristics of multipath communication channels (cf. Goldman, 1968).

Estimation or measurement of electr(on)ical components by self-adjusting bridges (cf. Lückers and Mulders, 1965) or by the model-adjustment technique (cf. Figure 1.7 and Figure 9.7; Van de Ven et al., 1973).

## 14.6 Air and Space Vehicles

In Section 14.1 control-type of applications have been indicated that have been used on high-performance airplanes. Such adaptive systems may be directed towards compensation of:

variations of vehicle dynamics, i.e. providing a closed-loop control system with good stability and manoeuvering properties in spite of drastic changes that the open-loop characteristics of the vehicle may undergo;

external disturbances due to atmospheric turbulence and outer-loop command signals.

For surveys of such applications to aerospace vehicles cf. Smyth and Ehlers (1970). In AGARD (1970) a number of papers on specific adaptive control schemes can be found. Case studies in this field are: Mehra and Tyler (1973) and Mulder (1973). Of the abundance of literature only a few publications are mentioned under additional references.

For determining unknown air and spacecraft parameters several techniques have been applied (cf. Rault, Pouliquen and Lefèvre, 1970; Rault, 1973). See also:

An iterative weighted least-squares estimation of satellite dynamics (Farrell et al., 1969).

An integral least-squares estimation of aerodynamic coefficients for missiles by use of a sensitivity analysis (Huff et al., 1970).

Estimation by means of frequency-response data (Dale and Cohen, 1970).

Estimation of aircraft parameters by quasilinearization techniques (Larson and Fleck, 1969) and by maximum likelihood methods (Mehra, 1970).

In line with the earliest application of least-squares estimation techniques by Gauss for determining celestial state variables, these types of procedures have been used extensively for orbit calculations of space vehicles. Cf. the literature cited in Chapters 12 and 13.

## 14.7 Biological Objects

Due to the complexity of living creatures, the difficulty of measuring crucial quantities in a direct way, the emphasis on more accurate diagnoses and a better treatment of diseases, the estimation techniques may be regarded as (potentially) valuable aids. This development is closely connected to the recognition of system/control aspects in biological objects, as is exemplified by a number of books (e.g. Bayliss, 1966; Milsum, 1966; Mesarovic, 1968; Stark, 1968). A survey paper on parameter estimation in biological systems is Bekey (1973).

### *Human Operator*

For many years a very popular object of study has been the human being as an element of a control system; cf. Figure 1.5. One of the early contributions is from Tustin (1947). The interest in identification and estimation studies on such human operators stems primarily from assignments of controlling various kinds of land-, water-, air-, space- and lunar vehicles. Crucial issues in such studies are: the difficulty of the assignment, the variations in time of the dynamic properties due to learning and fatigue, the adaptive capability shown in reactions to sudden changes in the vehicle dynamics, e.g. due to failure, etc. Very much attention has been paid to these types of problems; early surveys are: Clymer and Ax (1960), Clymer (1960) and Cosgriff and Briggs (1960). An extensive bibliography has been published by Costello and Higgins (1966); many interesting contributions are published in the proceedings of the annual NASA-university conferences on manual control (NASA).

The wide variety of papers published is caused by the many aspects that can be discerned, e.g.

the type of assignment: pursuit or compensatory (cf. Figure 14.11); single or more axes (degree of freedom);

the type of information desired: impulse response; transfer-function coefficients, either constant or stochastically time varying; non-linearities; sampling frequency;

the type of input characteristics: deterministic multifrequency signal; stochastic signal with prescribed probability-density and power spectrum curves;

the process dynamics; unstable; non-linear; sudden changes/failures;

the type of signal preprocessing, e.g. quickening, prediction;

influence of the environment on the human operator;

the type of estimation techniques used.

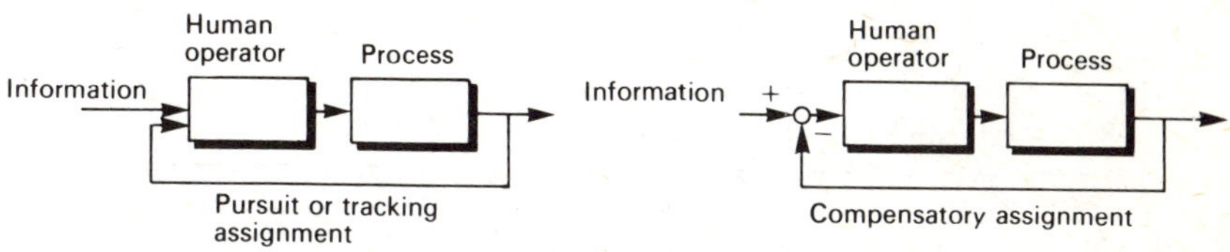


Figure 14.11

Only a very limited number of publications can be mentioned under additional literature.

The fact that the human operator is embedded in a closed loop (cf. Figure 14.12) results in the crucial problem of circulating noise, i.e. the signals $n$ and

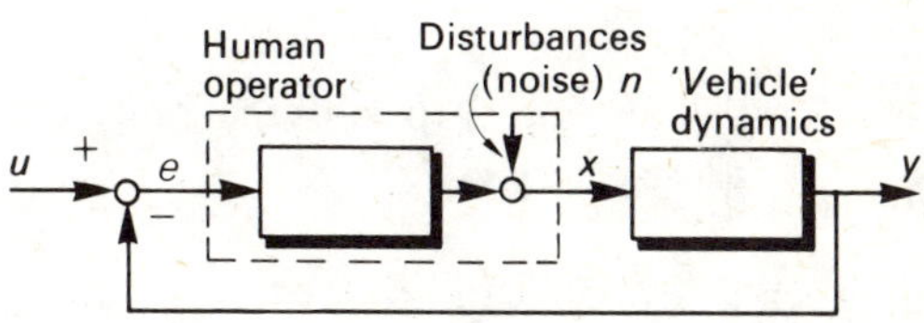


Figure 14.12

$e$ are correlated. Consequently the use of only the signals $e$ and $x$ for estimation purposes may lead to biased estimates. One of the methods to deal with this problem is to have recourse to, for example, the technique where instrumental variables are derived from the signal $u$.

Most NASA studies on the human operator have been directed to highly qualified and trained men in simulations of extraordinary missions. We will briefly summarize an interesting experiment directed to the man-in-the-street, viz. a bicycle rider. This description is based on a private communication* and Stassen (1969). The simulator consists of a bicycle frame at a fixed position on which forces can be applied by an hydraulic servo system. The task of the subject is to 'ride' this bicycle, i.e. to keep the inherently unstable system in balance. The hydraulic system takes care of the natural forces that occur due to the gyroscopic moment of the front wheel and the centrifugal moment of the bike (due to small deviations from a straight course). The sign conventions defined by Figure 14.13 and the following notations are used:

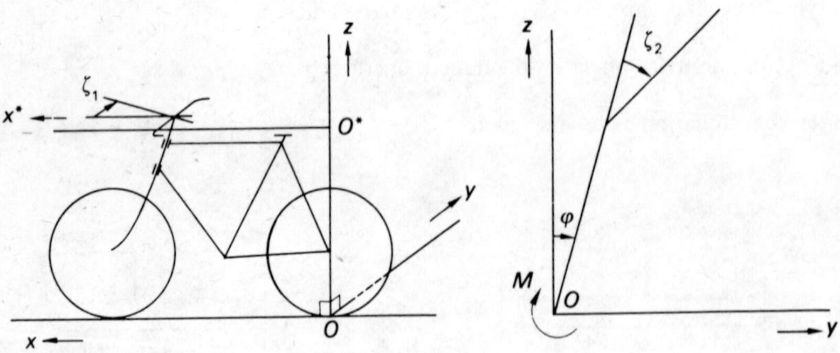


Figure 14.13

$\zeta_1(t)$ angle between handlebar and frame

$\zeta_2(t)$ angle between upper (part of the) body and frame

$\varphi(t)$ angle between $O$–$z$ axis and frame

$J$ moment of inertia }
$M$ static moment } of bike and body with respect to the $O$–$x$ axis.

$J^*$ moment of inertia }
$M^*$ static moment } of the upper (part of the) body with respect to the $O^*$–$x^*$ axis

$z_1$ distance between $O^*$–$x^*$ axis and $O$–$x$ axis

$v$ foreward velocity

$A, B$ constants

* Delft University of Technology, Mech. E. Dept., Lab. for Measurement and Control; H. G. Stassen, A. van Lunteren; Annual report 1969 of the Man–Machine Systems Group.

With a number of simplifying assumptions the moments around the $O-x$ axis can be equated as follows:

$$\overbrace{Av\dot{\zeta}_1(t)}^{\substack{\text{gyroscopic moment}\\ \text{due to the front wheel}}} + \overbrace{Bv^2\zeta_2(t)}^{\substack{\text{centrifugal moment}\\ \text{when riding a curved path}}} =$$

$$= \underbrace{\{J\ddot{\varphi}(t) - Mg\varphi(t)\}}_{\substack{\text{moment due to a}\\ \text{deviation of the frame}\\ \text{and the body from the}\\ \text{equilibrium position}}} + \underbrace{\{(J^* + z_1M^*)\ddot{\zeta}_2(t) - M^*g\zeta_2(t)\}}_{\substack{\text{moment due to a deviation}\\ \text{of the upper part of the body}\\ \text{with respect to the frame}}}$$

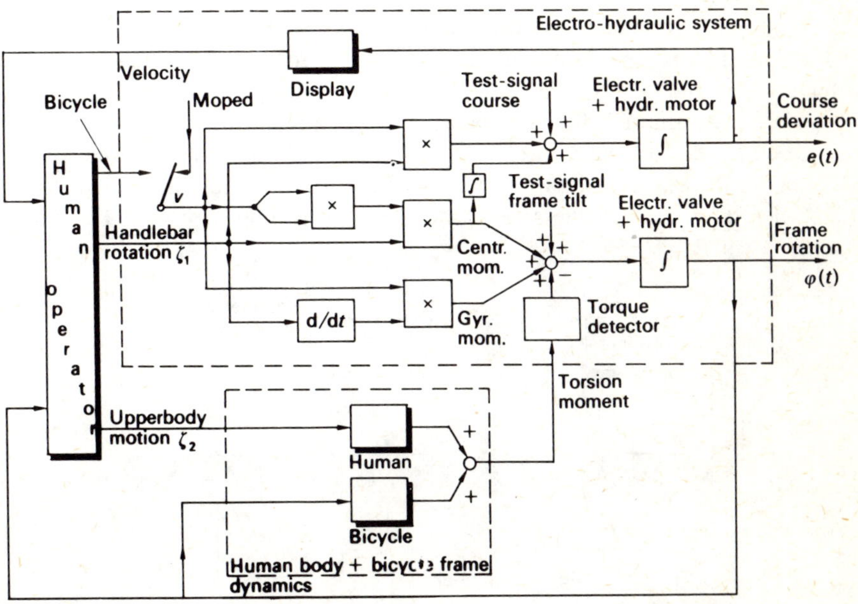


Figure 14.14

The diagram of Figure 14.14 represents the experimental set up as a block diagram. By means of a test signal course the human operator is given an assignment; the course deviation $e(t)$ acts as an error signal. In this experimental situation the interest is focussed on the dynamic properties of the human operator. As models for the human operator are chosen:

for the relation of frame rotation → handlebar rotation:

$$H_1(s) = \left(\alpha_{11} + \frac{\alpha_{12}}{s} + \alpha_{13}s\right)\varepsilon^{-\tau_1 s}$$

for the relation of frame rotation → upper body motion

$$H_2(s) = \left(\alpha_{21} + \frac{\alpha_{22}}{s} + \alpha_{23}s\right)\varepsilon^{-\tau_2 s}$$

Use is made of a model-adjustment technique by means of a small digital computer. Again one has to be aware of the bias problems due to circulating noise. One of the lines of research was the testing of the effects of drugs and alcohol on the different parameters.

*Blood Circulation*

Numerous attempts have been made to build (electrical) models of (parts of) the (human) blood circulation. The cardio-vascular system is a distributed system, the dynamic behaviour of which has to be described by partial differential equations. Different kinds of approximations by lumped circuits have been studied; Noordergraaf et al. (1963), Beneken (1965), De Pater (1966), e.g. by simulations on an analogue computer.

One of the objectives of such studies may be the wish to determine parameters of the blood circulation that are difficult to measure in a direct way, e.g. the compliance of the arteries or veins. According to Figure 14.15 the

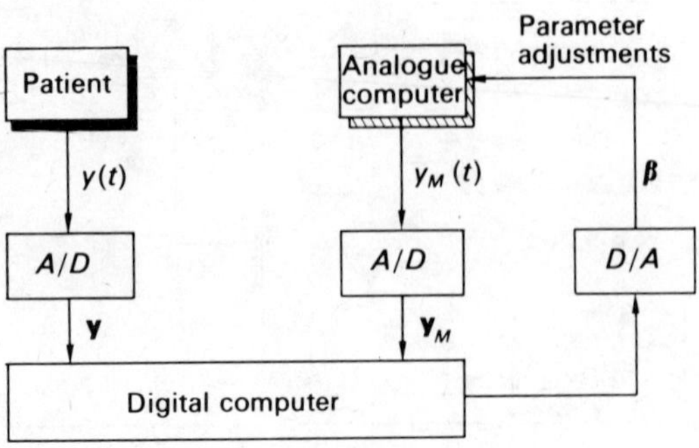


Figure 14.15

analogue computer model may be adjusted by a digital computer in such a way that the measurements from the model compare as well as possible with those from the patients. An example of a model of the human aorta is

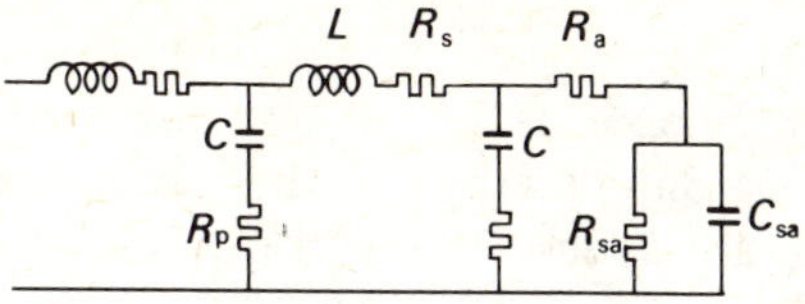


Figure 14.16

depicted in Figure 14.16.* Inertia of the blood is represented by inductances, vessel wall compliance by capacities. The resistors stand for hydraulic resistance and viscous effects of the wall. Needless to say, this is only one of the numerous possible ways of modelling the aorta.

Figure 14.17 shows the error function or performance index of this model dependent on the variation of inductance and capacitance parameters

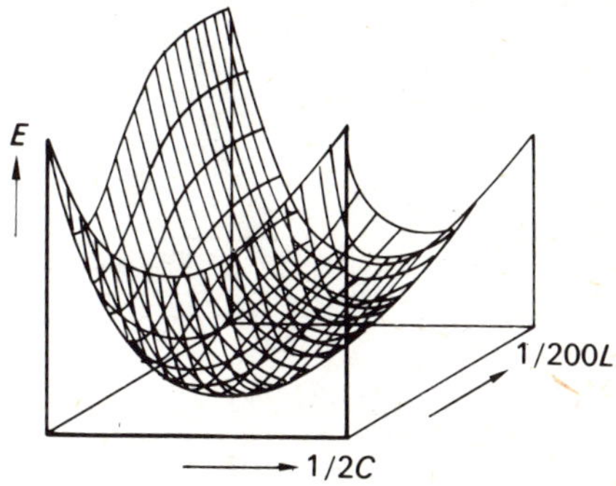


Figure 14.17

around a nominal value. An interesting phenomenon, which may serve as well as a warning, is shown in Figure 14.18. The essential change in the error function's landscape is merely caused by a larger variation of the same parameters around the same nominal values as in Figure 14.17.

Under the additional literature a few publications are mentioned.

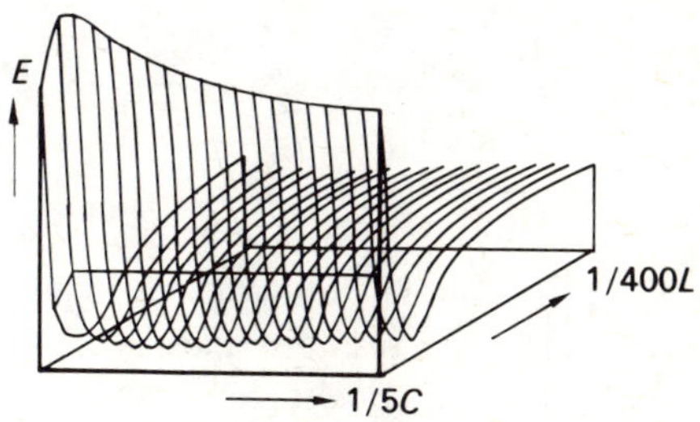


Figure 14.18

* Personal communication by Messrs. C. A. M. Van den Brekel and H. W. Timmer, Group Measurement and Control, EE Dept., University of Technology, Eindhoven, Netherlands.

*Electric Heart Action*

From the electrical phenomena of the heart muscle (depolarization wave) the resulting electrocardiogram (e.c.g.) can be measured at the skin of the body. Given an adequate model of the heart and properties of the surrounding parts of the body the e.c.g. can be simulated. Such a model may consist of say $m$ current dipoles, assigned to the different segments of the heart muscle (Figure 14.19) where each dipole is characterized by the time function of its intensity (Figure 14.20). If such a time function can be characterized by

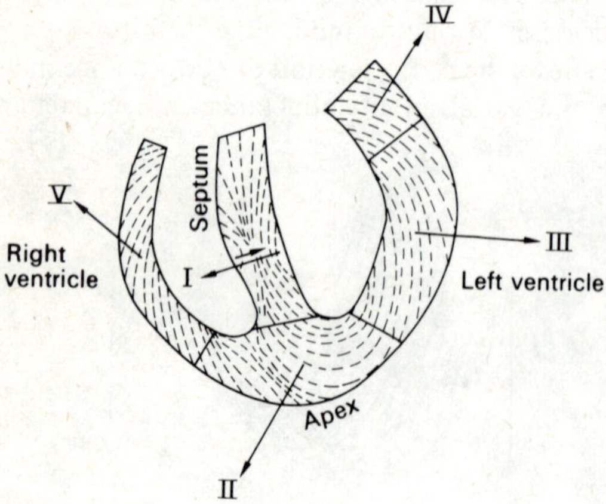


Figure 14.19

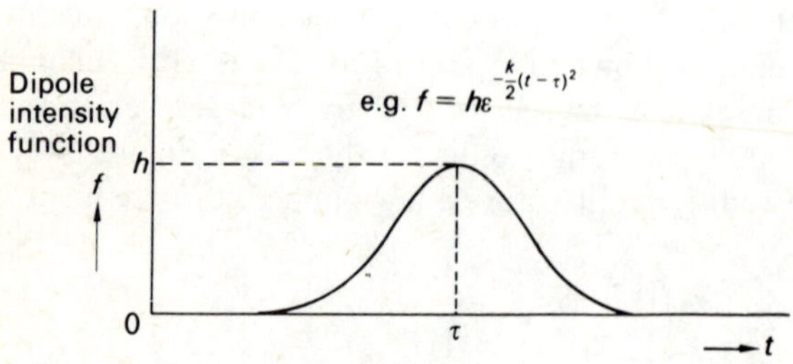


Figure 14.20

$n$ parameters, then the whole model contains $m \times n$ parameters. Combinations of electrode potential signals result in spatial orthogonal signals, which are approximately algebraic sums of the dipole intensity functions (Figure 14.21).* A crucial problem now is whether it is possible to determine those

* Personal Communication by Mr. A. A. H. Damen, group 'Measurement and Control', EE Dept., University of Technology, Eindhoven, Netherlands.

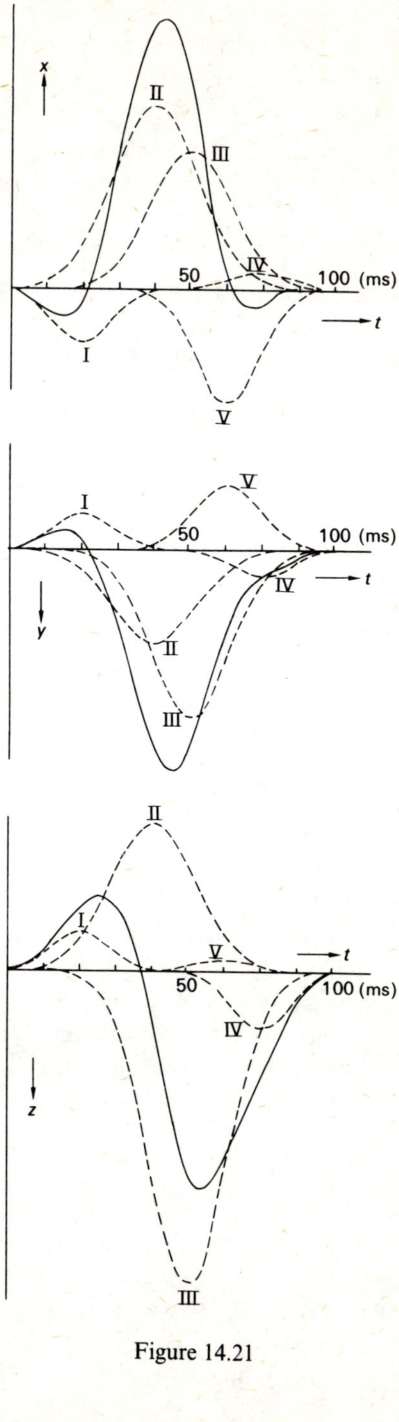
x
II
III
IV
50
100 (ms)
t
I
V
I
V
50
100 (ms)
t
IV
y
II
III
II
I
V
t
50
100 (ms)
IV
z
III

Figure 14.21

dipole parameters from the measured e.c.g. If that were the case then the location and severity of heart defects could be determined by estimation, viz. model-adjustment techniques. Relevant references are: Bellman et al. (1964), Damen and Piceni (1971).

Of the other applications to biological objects we mention only the model building:

with respect to muscles (Bahler, 1968; Gottlieb, Agarwal and Stark, 1969);

with respect to e.e.g. analysis (cf. Zetterberg, 1969; Gersch, 1970);

with respect to the human lens accommodation (cf. O'Neill, Sanathanan and Brodkey, 1969);

with respect to drug metabolism (Buel and Kalaba, 1969).

## 14.8 Socioeconomic Systems

For many years econometricians have been working on the development of models of essential parts of national economies and other sectors of economic activities. Again it is outside the scope of this text to discuss examples of these projects. Only a few general references to pertinent literature may suffice (Koopmans, 1937; Fisher, 1966; Manetsch, 1966; Bergstrom, 1967; Naylor, 1971; Young et al. 1972; Johnston, 1972). For a case study cf. Young et al. (1973).

Extensions of these techniques have been made to so-called industrial dynamics, urban dynamics and world dynamics (Forrester, 1961; 1969; 1971).

## 14.9 Learning Systems and Pattern Recognition

The reader is reminded of the close relationships that there are between the parameter-estimation procedures and the problems that are discussed under the terms learning systems and pattern recognition. Among those relationships there are the choice of model structures and the ways in which weighting coefficients (parameters) may be adjusted. Again only a few publications are mentioned as additional literature.

Due to the nature of this chapter it contains either too much or far too little, depending on the interests of the reader. It is only meant to demonstrate the wide variety of goals and means of identification, parameter- and state-estimation that have been proposed and used. The quite extensive list of literature can be instrumental in providing more information. The reader is reminded of the possibilities offered by e.g. *Computer and Control Abstracts*, *Electrical and Electronics Abstracts*, *Cybernetics Abstracts*, *The Engineering Index* to get access to the world literature. He is reminded of the advantages of using the *Citation Index* for finding more recent publications that refer to a particular paper.

Originated by definite needs in particular areas, stimulated by the potential possibilities and the demonstrated capabilities, and supported by the decreasing costs of information-processing equipment the field of parameter- and state estimation is developing rapidly. It can clearly be recognized as a means for improving human well-being in many respects; e.g. by economic gains, by increased safety of airplanes through applying these techniques for control and for a checking of maintenance, by providing better models for socioeconomic processes, by contributing to the feasibility of mass health screening, etc. etc.

At this moment neither the whole spectrum of possible applications nor the ultimate limitations can yet be foreseen adequately. The need for further theoretical foundations and more experience with practical applications is apparent from this book and from literature. May these efforts be with human well-being in mind ....

> Not enjoyment, and not sorrow,
> Is our destined end of way;
> But to act, that each to-morrow
> Brings us farther than to-day.
>
> H. W. Longfellow, A Psalm of Life.

## Literature

Please note: as this chapter is divided into well-defined quite homogeneous sections so is the literature. For each section first the references and then the additional literature is given.

### 14.1 Automatic Control

*References*

Aseltine, J. A. and others (1958). 'A survey of adaptive control systems', *IRE Trans. Autom. Control*, **PGAC-6,** 102–108.

Åström, K. J. and B. Wittenmark (1969). 'Problems of identification and control'. Report Electronic Sciences Laboratory, University of South. Calif., Los Angeles, Calif., *J. Math. Anal. Appl.*

Åström, K. J. (1970). Cf. references Chapter 4.

Draper, C. S. and V. T. Li (1951). 'Principles of optimalizing control systems and an application to the internal combustion engine', *ASME Publ.*, New York.

Fel'dbaum, A. A. (1960, 1961; 1965). Cf. references Chapter 1.

Gibson, J. E. (1960). 'Mechanising the adaptive principle', *Control Engng.*, **7,** 109.

Jacobs, O. L. R. (1961). 'A review of self-adjusting systems in automatic control', *J. Electron. Control*, **10,** 311–322.

Jacobs, O. L. R. (1964). 'Two uses of the term "adaptive" in automatic control', *IEEE Trans. Autom. Control*, **AC-9,** 574–575.

Kalman, R. E. (1958). Cf. references Chapter 7.

Krasovski, A. A. (1960). 'The dynamics of continuous automatic control systems with extremal self-adjustment of the correcting devices', *1st IFAC Congr., Moscow*, 679–685.

Mathias, R. A. and R. I. Van Nice (1960). 'Adaptive control systems; a survey', *Electro Technology*, **66,** 116–125.

Mendes, M. (1970). 'An on line adaptive control method', *IFAC Symp.—Identification and Process Parameter Estimation, Prague*, paper 6.5; see also *Automatica*, **7,** 323–332, (1971), and: 'Dual adaptive control of time-varying systems' (in German), *Regelungstechnik*, **18,** 127–164 (1970).

Mishkin, E. and L. Braun (1961). *Adaptive Control System*, McGraw-Hill, New York, 533 pp.

Pun, L. (1961). 'The evolution of optimal control' (in French), *Automatisme*, **6,** 47–53, 103–111, 148–152.

Rajbman, N. S. (1973). 'The application of identification methods (survey)', *Proc. 3rd IFAC Symp.—Identification and System Parameter Estimation*, North-Holland Publ. Comp., Amsterdam, 1–48.

Rumold, G. and W. Speth (1968). 'Self-adjusting PI controller' (in German), *Siemens Z.*, **42,** 765–768.

Rutman, R. S. (1962). 'Adaptive systems with dynamic characteristic adjustment' (survey of estim.), *Automn. Remote Control*, **23,** 602–625.

Snegur, A., V. Tomashov and L. Dubina (1967). 'Synthesis of chemical reactor adaptive control system and analysis of system stability with inaccurate *a priori* knowledge on reactor model', *IFAC Symp.—Identification in Autom. Control, Prague*, paper 7.14.

Speth, W. (1969). 'Simple methods for the rapid self-adaptation of automatic controllers in drive applications', *4th IFAC Congr., Warsaw*, paper 50.4.

Speth, W. (1971). 'Self-adjusting control systems in electrical-drive technology' (in German), *Dr.-Ing. Dissert., Techn. Univ., Braunschweig*, Germany.

Spiegel, P. (1969). 'Parameter control in electrical drives' (in German), *Z. Messen, Steuern, Regeln*, **12,** 197–202.

Whitaker, H. P. et al. (1958). 'Design of a model reference adaptive control system for aircraft', *Report R-164, Instrum. Lab., MIT*, Cambridge (Mass.).

Young, P. C. (1965). 'Process parameter estimation and self-adaptive control' in *Proc. 2nd IFAC Symp.—The Theory of Self-Adaptive Control Systems* (Ed. P. Hammond), Plenum Press, New York, 118–139.

Young, P. C. (1969). 'Applying parameter estimation to dynamic systems', *Control Engng.*, **16,** Oct. 119–125; Nov. 118–124.

*Additional Literature*

Bakke, R. M. (1965). 'Adaptive direct digital control with multi-parameter adjustment', *Paper IFAC Symp.—Theory of Self-adaptive control Systems*, Teddington.

Bohn, E. V., R. E. Butler and M. R. Mukherji (1966). 'Parameter-tracking models for adaptive control systems', *Proc. IEE*, **113,** 378.

Box, G. E. P. and G. M. Jenkins (1965). 'Mathematical models for adaptive control and optimization', *Univ. of Wisconsin, Dept. Stat.*, Techn. report 45.

Braun Jr., L. (1959). 'On adaptive control systems', *IRE Nat. Conv. Rec.*, part 4, 32–44; *IRE Trans. Autom. Control.*, **AC-4,** 30–42.

Braun, L. E., E. Mishkin and J. G. Truxal (1960). 'Approximate identification of process dynamics in computer controlled adaptive systems', *1st IFAC Congr., Moscow*.

Butkovskii, A. G. and A. Ya. Lerner (1960). 'The optimal control of systems with distributed parameters', *Automn. Remote Control*, **21,** 472–477.

Chang, S. S. L. (1965). 'On the relative time of adaptive processes', *IEEE Trans.*, **AC-10,** 11–17.

Choquette, P., A. R. M. Noton and C. A. G. Watson (1970). 'Remote computer control of an industrial process', *Proc. IEEE*, **58,** 10–16.

Eveleigh, V. W. (1967). *Adaptive Control and Optimization Techniques*, McGraw-Hill, New York.

Gibson, J. E. (1963). *Nonlinear Automatic Control*, McGraw-Hill, New York.

Graupe, D. and G. R. Cassir (1967). 'Adaptive control by predictive identification and optimization' (short paper), *IEEE Trans. Autom. Control*, **AC-12,** 191–199.

Haimes, Y. Y. and D. A. Wismer (1972). 'A computational approach to the combined problem of optimization and parameter estimation', *Automatica*, **8,** 337–237.

Hougen, J. O. (1966). 'Experimental techniques and the synthesis of control systems', *Progr. Engng.*, Vol. 3, Heyward Books, London.

Kershow, W. V. (1965). 'Adaptive control by plant identification', *Control Engng.*, **12,** 103.

Kondo, B. and T. Suzuki (1964). 'Statistical consideration of an optimal control system based on the alternate parameter-adjustment method', *Electrical Engng., Japan*, **84,** 88.

Kondo, B. and T. Suzuki (1965). 'Adaptive control of correlation-type optimalizing control system', *Electrical Engng., Japan*, **85,** 1.

Kumar, K. S. P. and R. Sridhar (1964). 'A note on combined identification and control' (correspondence), *IEEE Trans. Autom. Control*, **AC-9,** 118.

Kuntsevich, V. M. (1965). 'Investigation of the stability of the simplest adaptive system', *Automn. Remote Control*, **26,** 1367.

Macmillan, R. H. and N. W. Rees (1960). 'Automatic control systems', *Process Control and Automn.*, 576–580; 623–629.

Maslov, E. P. and L. M. Osovskii (1966). 'Adaptive control systems with models', *Automn. Remote Control*, **27,** 1116–1136.

Moe, M. L. and G. J. Murphy (1962). 'An approach to self-adaptive control based on the use of time moments and a model reference', *IRE Trans. Autom. Control*, **AC-7,** 82–92.

Narendra, K. S. and L. E. McBride (1964). 'Multiparameter self optimizing systems using correlation techniques., *IEEE Trans. Autom. Control*, **AC-9,** 31–38.

Nikiforuk, P. N. and E. K. Lashyn (1963). 'Self-zeroing self-adaptive control system', *Proc. IEE*, **110,** 2280–2284.

Petersen, I. and Y. Jaaksoo (1967). 'On the optimal adaptive control of linear multi-dimensional plant', *IFAC Symp.—Identification in Autom. Control Systems, Prague*, paper 5.11.

Reinisch, K. (1966). 'The effect of errors in process analysis on the synthesis of control loops and the consequences on the model method for the analysis of control loops' (in German), *Z. Messen, Steuern, Regeln*, **9,** 183–187.

Sage, A. P. (1968). Cf. references Chapter 12.

Sano, A. and M. Terao (1969). 'Measurement optimization in batch process optimal control', *4th IFAC Congr., Warsaw*, paper 12.4; Also *Automatica*, **6,** 705–713 (1970).

Sarachik, P. E. (1965). 'Identification of the steady state operator for discrete self-optimizing systems', *IEEE Trans.*, **AC-10,** 83–85.

Sawaragi, Y., Y. Sunahara and T. Nakamizo (1967). *Statistical Decision Theory in Adaptive Control Systems*, Academic Press, New York, 216 pp.

Schmidt, G. (1962). 'A self-adjusting control loop with a reference model' (in German), *Regelungstechnik*, **10,** 145–151.

Sklansky, J. (1966). 'Learning systems for automatic control', *IEEE Trans. Autom. Control*, **AC-11,** 6–19.

Staffin, W. (1960). 'Executive-controlled adaptive systems', *AIEE Trans. Part II* (*Applics. & Ind.*, *46*), **78,** 523–530.

Štěpán, J. (1967). 'The identification of controlled systems and the limit of stability', *IFAC Symp.—Identification in Autom. Control Systems*, *Prague*, paper 1.11.

Stout, T. M. (Ed.) (1970). 'Computers in industrial process control', *Proc. IEEE*, **58,** 1–132.

Sworder, D. D. (1966). 'A study of the relationship between identification and optimization in adaptive control problems', *J. Franklin Inst.*, **281,** 198.

Takahashi, Y., M. J. Rabens and D. M. Auslander (1970). *Control and Dynamic Systems*, Addison-Wesley, London.

Tou, J. T. (1959). *Digital and Sampled Data Control Systems*, McGraw-Hill, New York.

Vasu, G. (1960). 'Self-adaptive system for automatic control of dynamic performance by controlling gain, phase shift, gain margin, phase margin or slope', *Paper AIEE*, 60–1381 (Fall General Meeting, Chicago).

Wells, C. H. and R. E. Larson (1970). 'Application of combined optimum control and estimation theory to direct digital control', *Proc. IEEE*, **58,** 16–22.

## 14.2 Physical/Mechanical/chemical Processes

*References*

Bailey, A. J. (1968). 'Modelling of a steel reheating furnace', *IEE*, *London*, 5 pp.

Bailey, M. J. (1969). 'On line dynamic modelling for improved control', cf. reference IEE (1969), 117–184.

Booth, A. D. (1967). 'Resolution of (complex) gas chromatograms using a digital computer', *Trans. Soc. Instrum. Technol.*, **19,** 12–16.

Bremermann, H. (1970). 'Identification of rate constants in chemical cycles of known structure', *IEEE Symp.—Adaptive Processes, Decision and Control*, *Austin*, *Tex.*, *U.S.A.*, 3pp.

Brown, P. (1969). 'The adaptive control of electric arc furnaces', cf. reference IEE (1969).

Burovoi, I. A. and A. A. Mamedov (1970). 'Formulating the structure of a model for identifying a certain class of heterogeneous catalytic processes', *Autom. Remote Control*, **31,** 116–1123.

Carlson, N. R. (1966). 'Model for the basic oxygen process in steel refining', *IEEE Trans. System. Sci. Cybernet.*, **SCC-2,** 41–44.

Corlis, R. G. and R. Luus (1969). 'Use of residuals in the identification and control of two-input, single-output systems', *Ind. Eng. Chem. Fundamentals*, **8,** 246–253.

El'kinson, L. Z. (1969). 'Mathematical model of a glass furnace with respect to control parameters' (in Russian), *Steklo i Keramika.*, **25,** 148–154. Transl. in *Glass and Ceram.*

Eykhoff, P., P. M. E. M. Van der Grinten, H. Kwakernaak and B. P. Th. Veltman (1966). Cf. references Chapter 1.

Foulard, C. and G. Bornard (1973). 'Identification and system parameter estimation of distillation processes—a case study', *Proc. 3rd IFAC Symp.—Identification and System Parameter Estimation*, North-Holland Publ. Comp., Amsterdam, 167–186.

Gerdin, K. (1970). 'Markov modelling via correlation analysis applied to identification of a paper machine in the absence of test signal', *2nd IFAC Symp.—Identification and Process Parameter Estimation, Prague*, paper 2.8.

Godfrey, K. R. (1969). 'Dynamic analysis of an oil-refinery unit under normal operating conditions', *Proc. IEE*, **116,** 879–888.

Godfrey, K. R. and M. Devenish (1969). 'An experimental investigation of continuous gas chromatography using pseudo-random-binary sequences', *Measurement and Control*, **2,** 228–233.

Godfrey, K. R. (1970). 'The application of pseudo random sequences to industrial processes and nuclear power plant', *2nd IFAC Symp.—Identification and Process Parameter Estimation, Prague*, paper 7.1. Cf. also: 'The theory of the correlation method of dynamic analysis and its application to industrial processes and nuclear power plant', *Measurement and Control*, **2,** 165–172 (1969).

Gustavsson, I. (1973). 'Survey of applications of identification in chemical and physical processes', *Proc. 3rd IFAC Symp.—Identification and System Parameter Estimation*, North-Holland Publ. Comp., Amsterdam, 67–86.

IEE (1969). 'Industrial application of dynamic modelling', *Conf. held at Durham, England, 16–18 Sept.*

Isermann, R. (1965). 'Measurement and calculation of the control behaviour of a super heater', *Fortschritt-Bericht VDI-Z.* **6,** VDI-Verlag, Düsseldorf.

Kelly, P. C. and W. E. Harris (1971). 'Application of method of maximum posterior probability to estimation of gas-chromatographic peak parameters', *Anal. Chem.*, **43,** 1184–1195.

Loosemore, M. J. and J. W. Bray (1967). 'Optimisation model for a lead–zinc blast furnace', *I.Ch.E. Symp. Nottingham, London*, 19–31.

Manczak, K. (1967). 'Identification of an evaporator plant as the exemplification of the analytical identification of complex technological units', *Paper IFAC Symp.—Identification in Autom. Control Systems, Prague*, paper 2.2.

Murrill, P. W., R. W. Pike and C. L. Smith (1968, 1969). *Chem. Engng.*, **75**; No. 23, 117; No. 25, 165; No. 27, 103. **76**; No. 2, 167; No. 4, 105; No. 5, 111; No. 7, 151; No. 11, 195; No. 13, 97; No. 18, 125.

Nesbit, R. A. and R. D. Engel (1967). 'An example program for the determination of chemical rate coefficients from experimental data', *Simulation*, **8,** 133–137.

Ng, K. C. and G. C. Moss (1970). 'Continuous gas chromatography', *Proc. Conf. Lab. Autom., London*, 197–208.

Paiziev, E. (1968). 'Identification of complex controlled plant by means of partial correlation and dispersion functions of random functions', *Automn. Remote Control*, **29,** 1683–1689.

Price, R. J. and D. W. T. Rippin (1968). 'Single variable on-line adaptive optimisation of a water gas shift reactor', *Chem. Eng. Sci.*, **23,** 593–617.

Rees, N. W. (1966). 'Dynamic identification and optimum control studies on a pilot distillation column', *3rd IFAC Congr., London*, 19 pp.

Richalet, J. (1973). 'Identification of a glass furnace', *Proc. 3rd IFAC Symp.—Identification and System Parameter Estimation*, North-Holland Publ. Comp., Amsterdam, 109–116.

Rocco, V. (1968). 'Building a dynamic model to simulate in real time the evolution of a rotating cement kiln' (in Italian), *Automaz. e Strum.*, **16,** 146–152.

Sawaragi, Y., T. Takamatsu, H. Fukwa and H. Sayama (1965). 'Fundamental approach to the identification and optimization of chemical reaction processes', *Proc. IFAC Symp.—Systems Engineering for Control System Design, Tokyo*, 415–423.

Seinfeld, J. H. (1970). 'Nonlinear estimation theory', *Ind. Eng. Chem.*, **62,** 32–42.

Stout, T. M. (1961). 'Mathematical models for computer control systems', *1st IFAC Congr., Moscow*, 991–997.

Vallerschamp, R. E. and D. D. Perlmutter (1971). 'Constrained and weighted least squares procedures for parameter estimation', *Ind. Eng. Chem. Fundamentals*, **10,** 150–153.

Wise, H. E., R. J. Price and D. W. T. Rippin (1967). 'A comparison of the use of pseudo random binary sequences and frequency response methods for the determination of the dynamics of a pilot scale water gas shift reactor', *5th Int. Congr. AICA, Lausanne*, 171.

Závorka, J., L. Sutek, A. Aguado and E. Delgado (1967). 'The dynamics of multiple-effect evaporators in the sugar industry', *IFAC Symp.—Identification in Autom. Control Systems, Prague*, paper 2.2.

## *Additional Literature*

Åström, K. J. (1964). 'Control problems in papermaking', *Proc. IBM Sci. Computing Symp. on Control Theory and Applic.*, Yorktown Heights, N.Y.

Åström, K. J. (1967). 'Computer control of a paper machine—an application of linear stochastic control theory', *IBM J. Res. and Develop.*, **11,** 389–405.

Buyanovskii, L. A. and A. D. Grishunin (1970). 'Representing the dynamic characteristics of a controlled plant by the method of stochastic approximation' (in Russian), *Izv. VUZ Neft. Gaz*, **1,** 101–104.

Coggan, G. C. and A. R. M. Noton (1970). 'Discrete-time sequential state and parameter estimation in chemical engineering', *Trans. Inst. Chem. Eng.*, **40,** p. 255–264.

Damert, K. (1969). 'A standard program for determining kinetic constants from differential equations by non-linear regression' (in German), *Chem. Techn.*, **21,** 458–459.

Fitzgerald, T. J. (1967). 'Determination of the average impulse response for a fluidized bed', *Dissert. Abstr. B*, **27,** paper 67–7024.

Hougen, J. O. (1964). 'Experiences and experiments with process dynamics', *Chem. Eng. Progr., Monograph*, Series 4, 60.

Ibragimov, I. A. and M. S. Meth (1967). 'Investigation of a multi-phase oil-refining process by non-linear correlation method' (in Russian), *Izv. VUZ Neft. Gaz*, 83–87.

Kaplick, K. (1967; 1968). 'Investigation of process by regression analysis' (in German), *Chem. Techn.*, **19,** 466–407, 538–542; **20,** 30–34.

Klyuer, Yu. B. (1967). 'A method of preliminary analysis of continuous technological processes as objects of control (in Russian), *Mekh. Avtomat. Upravleniya*, 7–9.

Michalski, B. (1970). 'Mathematical model and optimization of industrial manufacturing process of phthalic anhydride by fluidization method', *Pomiary Automat. Kontrola*, **16,** 30–33.

Paul, R. J. A., T. T. Al-Naimi and D. K. Das-gupta (1970). 'Dynamics of a gas-solid fluidized bed', *Int. J. Contr.*, **12,** 817–834.

Pell Jr., T. M. and R. Aris (1970). 'Problems in chemical reactor analysis with stochastic features. Control of linearized distributed systems on discrete and corrupted observations', *Indust. Eng. Chem., Process Design Develop.*, **9,** 15–20.

Ray, W. H. and R. Aris (1967). 'An adaptive control of the batch reactor. IV. A more sophisticated controller', *Automatica*, **4,** 139–161.

Reid, K. N. (1969). 'Lateral dynamics of an idealized moving web', *Joint Autom. Control Conf.*, 955–963.

Reid, K. N. (1969). 'Lateral dynamics of a real moving web', *Joint Autom. Control Conf.*, 964–976.

Sastry, V. A., W. J. Vetter and R. Caston (1969). 'The application of identification methods to an industrial process', *Joint Autom. Control Conf.*, 787–794.

Schaefer, W. (1967). 'Mathématical problems of simulation and optimisation of processes in chemical engineering' (in German), *XII Int. Sci. Colloquium, Ilmenau*, 7–11.

Vaclavek, V. (1969). 'Studies on system engineering. III. Optimal choice of the balance measurements in complicated chemical engineering systems', *Chem. Eng. Sci.*, **24,** 947–956.

Welfonder, E. and O. Hasenkopf (1970). Comparison of deterministic and statistical methods for system identification by test-measurements at a disturbed industrial control system', *2nd IFAC Symp.—Identification and Process Parameter Estimation, Prague*, paper 7.5.

Wells, C. H. (1969). 'Application of the extended Kalman estimator to the non-linear well stirred reactor problem', *Joint Autom. Control Conf.*, 473–482.

## 14.3 Nuclear Reactors

*References*

Bastl, W. (1966). 'Correlation procedure applied to measuring in nuclear reactor systems' (in German), *Regelungstechnik*, **14,** 56–63.

Hendrikson, R. A. and G. Murphy (1968). 'Cross spectral density measurements in a coupled-core reactor', *Nucl. Sci. Eng.*, **31,** 215–221.

Liewers, P. and E. Buttler (1967). 'Measurement of correlation functions of reactor noise by means of the polarity correlation method', *IFAC Symp.—Identification in Automatic Control Systems, Prague*, paper 3.7.

Roman, W. S., C. Hsu and L. J. Habegger (1971). 'Parameter identification in a non-linear reactor system', *IEEE Trans. Nucl. Sci.*, **NS-18,** 426–429.

Sage, A. P. and G. W. Masters (1966). 'Identification and modeling of states and parameters of nuclear reactor systems', *IEEE Trans. Nucl. Sci.*, **NS-14,** 279–285.

Schwalm, D. (1971). 'Parameter estimation of coupled reactors by correlation methods', *Atomkernenergie*, **17,** 140–142.

Venerus, J. C. and T. E. Bullock (1970). 'Estimation of the dynamic reactivity using digital Kalman filtering', *Nucl. Sci. Eng.*, **40,** 199–105.

Wilkie, J. D. F. (1967). 'Off-line statistical identification of the kinetics of a nuclear reactor' (a comparison of digital models), *IFAC Symp.—Identification in Autom. Control Systems, Prague*, paper 3.20.

*Additional Literature*

Balcomb, J. D., H. B. Demuth and E. P. Gyftopoulos (1961). 'A cross-correlation method for measuring the impulse response of reactor systems', *Nucl. Sci. Eng.*, **11,** 159–166.

Cohn, C. E. (1959). 'Determination of reactor kinetic parameters by pile noise analysis', *Nucl. Sci. Eng.*, **5,** 331.

Dang Van Ba, M. and A. Roch (1967). 'Control of a nuclear reactor: measurement of transfer functions' (in French), *Bull. Tech. Suisse Romande*, **93,** 335–339.

Dragt, J. B. (1966). 'Accurate reactor noise measurements in a low power critical reactor', *Nukleonik*, **6,** 188.

Kemeny, L. G. (1968). 'Hybrid simulation of nuclear reactor system stochastic models', *Proc. Int. Symp. on Analog and Hybrid Computation Applied to Nuclear Energy, Versailles.*

Kovanic, P. (1967). 'Identification of operators in the reactor physics' (in German), *Atomkernenergie*, **12,** 404–408.
Larssen, B. (1967). 'The use of statistical methods in the determination of reactor dynamics', *Neue Tech.*, **9,** 141–145.
Robinson, E. (1968). 'Correlation experiments in nuclear reactors by digital measurement of counting rates' (in German), *Atomkernenergie*, **13,** 8–12.
Sumner, A. (1966). 'Optimal identification of some parameters of a nuclear reactor dynamical system', *3rd IFAC Congr., London*, paper 26A.
Walker, P. A. W., E. J. Ward and D. Maudsley (1970). 'Identification of a nuclear reactor control system by cross-correlation', *J. Brit. Nucl. Energy Soc.*, **9,** 85–93.

## 14.4 Power Plants and Power Distribution Systems

*References*

Baeyens, R. and B. Jacquet (1973). 'Applications of identification methods in power generation and distribution; *Proc. 3rd IFAC Symp.—Identification and System Parameter Estimation*, North-Holland Publ. Comp., Amsterdam, 1107–1121.
Cuenod, M. and G. Quazza (1969). 'Dynamic statistical analysis of electric power system controls', *Inst. Engrs. Austral. Elect. Eng. Trans.*, **EE-5,** 125–135.
Debelle, J., A. Foureau and R. Vazquez (1966). 'Parametric control of boilers and identification of furnace perturbation', *3rd IFAC Congr., London*, paper 10C.
Debs, A. S. and R. E. Larson (1970). 'A dynamic estimator for tracking the state of a power system', *IEEE Trans. Power Apparat. and Systems*, **PAS-89,** 1670–1678.
Donati, F., G. Fiorio, R. Genesio and V. Mauro (1970). 'Statistical identification of interconnected power systems', *Ricerche di Automatica*, **1,** 30–56.
Eklund, K. (1972). 'A comparison of a drum boiler-turbine model to measurements and models obtained by identification', *5th IFAC Congr., Paris*, paper 5.3.
Eklund, K. and I. Gustavsson (1973). 'Identification of drum boiler dynamics', *Proc. 3rd IFAC Symp.—Identification and System Parameter Estimation*, North-Holland Publ. Comp., Amsterdam, 87–108.
Habegger, L. J. and R. E. Bailey (1969). 'Minimum variance estimation of parameters and states in nuclear power systems', *4th IFAC Congr., Wcrsaw*, paper 12.2.
Handschin, E. (Ed.) (1972). *Real-time Control of Electric Power Systems*, Elsevier, Amsterdam, 320 pp.
Koszelnik, M., J. Malkiewicz and St. Trybula (1969). 'A method to determine the transfer functions of power systems', *4th IFAC Congr., Warsaw*, paper 33.2.
Schweppe, F. C. et al. (1970). 'Power system static state estimation. Part I—exact model; part II—approximate model; part III—implementation', *IEEE Trans. Power Apparat. and Systems*, **PAS-89,** 120–135.
Stanton, K. N. (1965). 'Estimation of turboalternator transfer functions using normal operating data', *Proc. IEE*, **112,** 1713–1720.
Suleimenova, L. K. (1965). 'Experimental determination by statistical methods of dynamic characteristics of a boiler-turbine-block', *Automn. Remote Control*, **26,** 262–267.
Thompson, F. T. (1967). 'A dynamic model of a drum-type boiler system', *IEEE Trans. Power Apparat. and Systems*, **PAS-86,** 625–635.
Toyoda, J. et al. (1970). 'An application of state estimation to short-term load forecasting', *IEEE Trans. Power Apparat. and Systems*, **PAS-89,** 1678.
Uemura, K. (1972). 'State estimation of large-scale electric power systems by decomposition methods', *5th IFAC Congr., Paris*, paper 7.5.

## 14.5 Telecommunication

*References*

Bode, H. W. (1938). 'Variable equalisers', *Bell Syst. Tech. J.*, **17,** 229–244.

Goldman, J. (1968). 'Estimation of a random time-varying transfer function', *IEEE Trans. Inform. Theory*, **IT-13,** 598–599.

Kanal, L. (1969). 'Survey of automatic equalization', *IEEE Computers and Communic. Conf., Rome*, 23–33.

Lückers, J. and C. E. Mulders (1965). 'An automatic A.C. bridge with phase-sensitive detectors', *Revue A*, **7,** 88–92.

Lucky, R. W. (1968). 'Adaptive redundancy removal in data transmission', *Bell Syst. Tech. J.*, **47,** 549–573.

Middleton, D. (1960). Cf. references Chapter 11.

Orth, R. W. (1968). 'Frequency-selective limiters and their application', *IEEE Trans. Electromagn. Compatibility*, **EMC-10,** 273–283.

Van de Ven, H. H. et al. (1973). Cf. references Chapter 9.

Widrow, B., P. E. Mantey, L. J. Griffiths and B. B. Goode (1967). 'Adaptive antenna systems', *Proc. IEEE*, **55,** 2143–2159.

*Additional Literature*

Arnon, E. (1969). 'An adaptive equalizer for television channels', *IEEE Trans. Commun. Technol.*, **17,** 726–734.

Becker, F. K., L. N. Holzman, R. W. Lucky and E. Port (1965). 'Automatic equalisation for digital communications', *Proc. IEEE*, **53,** 96–97.

Boyd, R. T. and F. C. Monds (1970). 'Adaptive equaliser for multipath channels', *Electron. Letters*, **6,** 556–558.

Gersho, A. (1969). 'Adaptive equalization of highly dispersive channels for data transmission', *Bell Syst. Tech. J.*, **48,** 55–70.

Gonsalves, A. and D. W. Tufts (1968). 'Data transmission through a random noisy channel by PAM', *IEEE Trans. Commun. Technol.*, **COM-16,** 439–444.

Hirsch, D. and W. J. Wolf (1970). 'A simple adaptive equalizer for efficient data transmission', *IEEE Trans. Commun. Technol.*, **COM-18,** 5–12.

Lucky, R. W. (1965). 'Automatic equalization for digital communication', *Bell Syst. Tech. J.*, **44,** 547–588.

Lucky, R. W. and H. R. Rudin (1966). 'Generalized automatic equalization for communication channels', *Proc. IEEE*, **54,** 439–440.

Mohrmann, K. (1970). 'Implementation of an adaptive equalizer for fast data transmission involving a minimum of complexity', *Nachrichtentech. Z.*, **23,** 36–42.

Niessen, C. W. (1967). 'Automatic channel equalization algorithms', *Proc. IEEE*, **55,** 700–701.

Potter, J. B. (1969). 'The adaptive equalisation of communication systems, particularly those using waveform fidelity criteria', *Proc. Inst. Radio Electron. Engrs. Australia*, **30,** 326–336.

Proakis, J. G. and J. H. Miller (1969). 'An adaptive receiver for digital signalling through channels with intersymbol interference', *IEEE Trans. Inform. Theory*, **IT-15,** 484–497.

Proakis, J. G. (1970). 'Adaptive digital filters for equalization of telephone channels', *IEEE Trans. Audio Electroacoust.*, **AU-18,** 195–200.

Rudin Jr., H. R. (1967). 'Automatic equalization using transversal filters', *IEEE Spectrum*, **4,** 53–60.

Rudin Jr., H. R. (1969). 'A continuously adaptive equalizer for general-purpose communication channels', *Bell Syst. Tech. J.*, **48,** 1865–1884.

Schonfeld, T. J. (1970). 'Rapidly converging second order tracking algorithms for adaptive equalization', *Proc. Internat. Symp. Inform. Theory, Noordwijk*, Netherlands.

Ungerboeck, G. (1972). 'Theory on the speed of convergence in adaptive equalizers for digital communication', *IBM J. Res. Develop.*, **16,** 546–555.

## 14.6 Air and Space Vehicles

*References*

AGARD (1970). *Advanced Control System Concepts.*

Dale, O. E. and R. Cohen (1970). 'Multi-parameter identification in linear continuous vibratory systems', *Joint Autom. Control Conf.*, 556–564.

Farrell, J. L., J. K. Newton, J. A. Miller and J. J. Connely (1969). 'Optimal estimation of rotation-coupled flexural oscillations', *Joint Autom. Control Conf.*, 433–441.

Huff, C. F., R. B. Asher and H. R. Sebesta (1970). 'Parameter sensitivity of aerodynamic coefficients determined from experimental data', *J. Spacecraft and Rockets*, **7,** 1265–1267.

Larson, D. B. and J. T. Fleck (1969). 'Quasilinearization techniques', *Proc. Nat. Electron. Conf., Chicago*, Ill., 95–100.

Mehra, R. K. (1970). 'Maximum likelihood identification of aircraft parameters', *Joint Autom. Control Conf.*, 442–444.

Mehra, R. K. and J. S. Tyler (1973). 'Case studies in aircraft parameter identification', Proc. *3rd IFAC Symp.—Identification and System Parameter Estimation*, North-Holland Publ. Comp., Amsterdam, 117–144.

Mulder, J. A. (1973). 'Aircraft performance measurements in nonsteady flights', *Proc. 3rd IFAC Symp.—Identification and system Parameter Estimation*, North-Holland Publ. Comp., Amsterdam, 1131–1145.

Rault, A., R. Pouliquen and C. Lefèvre (1970). 'Applications of identification methods to aeronautics', *2nd IFAC Symp.—Identification and Process Parameter Estimation, Prague*, paper 11.2.

Rault, A. (1973). 'Identification applications to aeronautics', *Proc. 3rd IFAC Symp.—Identification and System Parameter Estimation*, North-Holland Publ. Comp., Amsterdam, 49–66.

Smyth, R. K. and H. L. Ehlers (1970). 'Survey of adaptive control application to aerospace vehicles', *AGARD Advanced Control System Concepts*, 1–21.

*Additional Literature*

Andeen, R. E. and P. P. Shipley (1963). 'Digital adaptive flight control systems for aerospace vehicles', *AIAA J.*, 1.

Buel, J. D., H. H. Kagiwada and R. E. Kalaba (1967). Cf. references Chapter 7.

Dymock, A. J., K. A. Helps and J. F. Meredith (1970). 'The place of adaptive control techniques in the future development of automatic fiight control systems', *AGARD Advan. Control Syst. Concepts*, 87–100.

Grusas, P. A. (1969). 'Satellite attitude determination form celestial sightings', *J. Spacecraft and Rockets*, **6,** 1007–1012.

Hollstien, R. B. and W. H. Gruber (1970). 'A response regulation technique for self-adaptive control of flexible aircraft', *AGARD Advan. Control Syst. Concepts*, 121–138.

Howell, G. C. and G. T. Shanks (1970). 'Self-adaptive manoeuvre demand control systems: some practical design considerations', *AGARD Advan. Control Syst. Concepts*, 149–168.

Kishi, F. H. (1964). 'On-line computer control techniques and their application to re-entry aerospace vehicle control', in *Advances in Control Systems* (Ed. C. T. Leondes), Academic Press, New York.

Nash Jr., R. A. (1968). 'The estimation and control of terrestrial inertial navigation system errors due to vertical deflections', *IEEE Trans. Autom. Control*, **AC-13,** 329–338.

Nikiforuk, P. N. and R. J. Pronovost (1967). 'A three variable self-adaptive flight control system using discrete interval binary noise perturbation signals', *IFAC Symp.—Identification in Autom. Control Systems, Prague*, paper 3.2.

Taylor, L. W., K. W. Iliff and B. G. Powers (1969). 'A comparison of Newton–Raphson and other methods for determining stability derivatives from flight data' (paper), *AIAA Third Flight Test Simulation and Support Conf.*, Houston, Texas.

## 14.7 Biological Objects

*References*

Bahler, A. S. (1968). 'Modeling of mammalian skeletal muscle', *IEEE Trans. Biomed. Eng.*, **15,** 249–257.

Bayliss, L. E. (1966). *Living Control Systems*, Freeman, San Francisco, 189 pp.

Bekey, G. A. (1973). 'Parameter estimation in biological systems; a survey', *Proc. 3rd IFAC Symp.—Identification and System Parameter Estimation*, North-Holland Publ. Comp., Amsterdam, 1123–1130.

Bellman, R. et al. (1964). 'Estimation of heart parameters using skin potential measurements', *Communic. ACM*, **7,** 666–668.

Beneken, J. E. W. (1965). 'A mathematical approach to cardio-vascular function', *Doctoral Dissert., Univ. Utrecht, Netherlands.*

Buel, J. D. and Kalaba (1969). Cf. references Chapter 7.

Clymer, A. B. (1960). 'Accomplishment in human simulation', *ASME Paper*, 60-AV-26, Aviation Conference, Dallas.

Clymer, A. B. and A. F. Ax (1960). 'Possibilities for human simulation', *ASME Paper*, 60-AV-37, Aviation Conference, Dallas.

Cosgriff, R. L. and G. E. Briggs (1960). 'Accomplishments in human operator simulations', *ASME Paper*, 60-AV-40, Aviation Conference, Dallas.

Costello, R. G. and T. J. Higgins (1966). 'An inclusive classified bibliography pertaining to modeling the human operator as an element in an automatic control system', *IEEE Trans. Human Factors Electr.*, **HFE-7,** 174–181.

Damen, A. A. H. and H. A. L. Piceni (1971). 'The multiple dipole model of the ventricular depolarisation', Report, *Group Measurement and Control, EE Dept., University of Technology, Eindhoven, Netherlands.*

De Pater, L. (1966). 'An electrical analogue of the human circulatory system', *Doctoral Dissert., Univ. Groningen, Netherlands.*

Gersch, W. (1970). 'Spectral analysis of e.e.g. by autoregressive decomposition of time series', *Math. Biosci.*, **7,** 205–222.

Gottlieb, G. L., G. C. Agarwal and L. Stark (1969). 'Stretch receptor models. I. Single-efferent single-afferent innervation', *IEEE Trans. Man-Mach. Syst.*, **MMS-10,** 17–27.

Mesarovic, M. D. (Ed.) (1968). *Systems Theory and Biology*, Springer.

Milsum, J. H. (1966). *Biological Control System Analysis*, McGraw-Hill, New York, 466 pp.

NASA. *Annual NASA-Univ. Conf. Manual Control.* Scient. and Technical Info. Division, Office of Technology Utilization, NASA, Washington, DC 20546.

Noordergraaf, A. et al. (1963). 'Analog of the arterial bed', in *Pulsatile Blood Flow* (Ed. E. O. Attinger), McGraw-Hill, New York.

O'Neill, W. D., C. K. Sanathanan and J. S. Brodkey (1969). 'A minimum variance, time optimal control system model of human lens accomodation', *IEEE Trans. Systems Sci. Cybernet.*, **SSC-5,** 290–299.

Stark, L. (1968). *Neurological Control Systems Studies in Bioengineering*, Plenum Press, New York, 428 pp.

Stassen, H. G. (1969). 'The polarity coincidence correlation technique—a useful tool in the analysis of human operator dynamics', *IEEE Trans. Man-Machine Syst.*, **MMS-10,** 34–39.

Tustin, A. (1947). 'The nature of the operator's response in manual control and its implications for controller design', *J. IEE*, **94,** part IIA, 190–207.

Zetterberg, L. H. (1969). Cf. references Chapter 11.

*Additional Literature (Human Operator)*

Angel, E. S. and G. A. Bekey (1968). 'Adaptive finite-state models of manual control systems', *IEEE Trans. Man-Machine Syst.*, **MMS-9,** 15–20.

Baron, S. and D. W. Kleinman (1969). 'The human as an optimal control and information processor', *IEEE Trans. Man-Machine Syst.*, **MMS-10,** 9–17.

Bekey, G. E. (1963). 'Discrete models of the human operator in a control system', *2nd IFAC Congr., Basle*, 430.

Bekey, G. A., H. F. Meissinger and R. E. Rose (1965). 'Mathematical models of human operators in simple two-axis manual control systems', *IEEE Trans. Human Factors Electron.*, **HFE-6,** 42–52.

Bekey, G. A. and C. B. Neal (1968). 'Identification of sampling intervals in sampled-data models of human operators', *IEEE Trans. Man-Machine Syst.*, **MMS-9,** 138–142.

Blank, G. L. and P. J. Schumacher, Jr. (1968). 'The human operator as an optimal controller', *Proc. 6th Allerton Conf. Circuit and System Theory, Monticello, Ill.*, 680–689.

Costello, R. G. (1968). 'The surge model of the well-trained human operator in simple manual control', *IEEE Trans. Man-Machine Syst.*, **MMS-9,** 2–9.

Donaldson, P. E. K. (1964). 'Error decorrelation studies on a human operator performing a balancing task', *Med. Electron. Biol. Engng.*, **2,** 393.

Elkind, J. I. and C. D. Forgie (1959). 'Characteristics of the human operator in simple manual control systems', *IRE Trans. Autom. Control*, **AC-4,** 44–55.

Elkind, J. I. and D. C. Miller (1966). 'On the process of adaptation by the human controller', *3rd IFAC Congr., London.*

Kahn, A. and A. C. Beare (1966). 'A critical re-evaluation of the human transfer function problem., *IEEE Trans. Aerospace Electron. Syst.*, AES-2, 711–718.

Kreifeldt, J. G. (1965). 'A sampled-data pursuit tracking model', *IEEE Trans. Human Factors Electron.*, **HFE-6,** 65–73.

Kreil, W. and G. Schweizer (1968). 'Man acting as a controller. A contribution to the dynamical transfer model' (in German), *Regelungstechnik*, **16,** 49–56.

Lang, R. P. and A. G. Vacroux (1969). 'Operator-centred adaptive compensation of continuous manual control systems', *IEEE Trans. Man-Machine Syst.*, **MMS-10,** 53–59.

McRuer, D. T., D. Graham, E. S. Krendel and W. C. Reisener, Jr. (1966). 'System performance and operator stationarity in manual control systems', *3rd IFAC Congr., London.*

Naslin, P. and J. C. Raoult (1963). 'Continuous and sampled-data models of the human operator in a control loop' (in French), *2nd IFAC Congr., Basle*, 421.

Ornstein, G. N. (1963). Cf. references Chapter 9.

Paskin, H. M. (1970). 'A discrete stochastic optimal control model of the human operator', *Proc. Nat. Aerospace Electron. Conf., Dayton, Ohio.*

Phatak, A. V. and G. A. Bekey (1969). 'Model of the adaptive behavior of the human operator in response to a sudden change in the control situation', *IEEE Trans. Man-Machine Syst.*, **MMS-10,** 72–80.

Potts, T. E., G. N. Ornstein and A. B. Clymer (1961). Cf. references Chapter 9.

Preyss, A. E. and J. L. Meiry (1968). 'Stochastic modeling of human learning behavior', *IEEE Trans. Man-machine Syst.*, **MMS-9,** 36–46.

Sheridan, T. B. (1960). Cf. references Chapter 10.

Smith, H. J. (1967). 'Human describing functions measured in flight and on similators', *IEEE Trans. Human Factors Electron.*, **HFE-8,** 264–268.

Snyder, D. L. (1964). 'A nonlinear study of compensatory manual control systems', *IEEE Trans. Human Factors Electron.*, **HFE-5,** 25–28.

Taylor, L. W. and A. V. Balakrishnan (1967). 'Identification of human response models in manual control systems', *IFAC Symp.—Identification in Autom. Control Systems, Prague*, paper 1.7.

Tsibulevskii, I. E. (1967). 'Adaption of the (human) operator to changes in gain factor of a series-connected amplifier', *Automn. Remote Control*, **28,** 461–467.

Wieringa, R. D. (1969). 'An evaluation of a pilot model based on Kalman filtering and optimal control', *IEEE Trans. Man-Machine Syst.*, **MMS-10,** 108–117.

Wierwille, W. W. (1964). 'Improvement of the human operator's tracking performance by means of optimum filtering and prediction', *IEEE Trans. Human Factors Electron.*, **HFE-5,** 20–24.

Wierwille, W. W. (1965). 'A theory for optimal deterministic characterization of time-varying human operator dynamics', *IEEE Trans. Human Factors Electron.*, **HFE-6,** 53–61.

Young, L. R., D. M. Green, J. I. Elkind and J. A. Kelly (1964). 'Adaptive dynamic response characteristics of the human operator in simple manual control', *IEEE Trans. Human Factors Electron.*, **HFE-5,** 6–13.

### *Additional Literature (Blood Circulation)*

Abel, F. L. (1966). 'An analysis of the left ventricle as a pressure and flow generator in the intact systemic circulation', *IEEE Trans. Biomed. Engng.*, **BME-13,** 182–188.

Bekey, G. A., M. J. Merritt and N. S. Assali (1964). 'Synthesis and optimization of a model of cardiovascular dynamics', *Proc. 4th Int. Analogue Computation Meetings, Brighton*, 383–388.

Burrus, C. S., T. W. Parks and T. B. Watt Jr. (1971). 'A digital parameter-identification technique applied to biological signals', *IEEE Trans. Biomed. Engng.*, **BME-18,** 35–37.

Ellis, M. F., P. J. Seaba, F. D. Alton and D. M. Levy (1966). 'A hybrid computer simulation of the mammalian cardiovascular system', *19th Annual Conf. Engng. Med. and Biol., San Francisco*, **18,** 27.

Goldwyn, R. M. and T. B. Watt Jr. (1966). 'Modal analysis of the dicrotic portion of the human blood pressure curve', *IEEE Int. Conv. Rec.*, **14,** 17–26.

Goldwyn, R. M. and T. B. Watt, Jr. (1967). 'Arterial pressure pulse contour analysis via a mathematical model for the clinical quantification of human vascular properties', *IEEE Trans. Biomed. Engng.*, **BME-14,** 11–17.

Granick, J. L. (1966). 'An electrical analog of the cardiovascular system and its use in simulating pathological cardiology', *19th Ann. Conf. Eng. Med. and Biol., San Francisco*, **8,** 130.

King, R. E. and R. E. Blodgett (1966). 'A network realization of aortal haemodynamics', *Symp. Generalized Networks*, New York, 599–606.

Morehouse, F., J. T. Apter and L. Amiot (1970). 'Use of hybrid computers to analyze behavior of detailed models for biological systems. I. Model for aortic blood flow', *Comput. Biol. and Med.*, **1,** 29–40.

Pickering, W. D., P. N. Nikiforuk and J. F. Merriman (1969). 'Analogue computer model of the human cardiovascular control system', *Med. Biol. Engng.*, **7,** 401–410.

Pollack, G. H., R. V. Reddy and A. Noordergraaf (1968). 'Input impedance, wave travel and reflections in the human pulmonary arterial tree: studies using an electrical analog', *IEEE Trans. Biomed. Engng.*, **BME-15,** 151–164.

Rideout, V. C. (1969). 'Computer simulation study of the pulmonary circulation', *Simulation*, **12,** 239–245.

Rideout, V. C. and J. B. Sims (1969). 'Computer study of the effects of small nonlinearities in the arterial system', *Math. Biosci.*, **4,** 411–426.

Strano, J. J., W. Welkowitz and S. Fich (1972). 'Measurement and utilization of *in vivo* blood-pressure transfer functions of dog and chicken aortas', *IEEE Trans. Biomed. Engng.*, **BME-19,** 261–271.

Topham, W. S. (1967). 'An analog model of the control of cardiac output', *Simulation*, **8,** 49–56.

Verdouw, P. D. and A. Noordergraaf (1966). 'A critical consideration of the design of electrical analogs for the pressure-flow relationship in arteries', *19th Ann. Conf. Engng. Med. and Biol., San Francisco*, **8,** 181.

Watt Jr., T. B. and R. M. Goldwyn (1966). 'Clinical estimation of human vascular properties via a mathematical model describing the contour of the arterial pressure pulse', *19th Ann. Conf. Engng. Med. and Biol., San Francisco*, **8,** 182.

Westerhof, N. and A. Noordergraaf (1966). 'Wave travel and input impedance for various types of networks in an electrical model of the human systemic arterial tree, with and without elastic tapering', *19th Ann. Conf. Engng. Med. and Biol., San Francisco*, **8,** 90.

Willard, P. W. and H. R. Beck (1969). 'A system for measurement of cardiovascular parameters', *Med. Res. Engng.*, **8,** 33–36.

## 14.8 Socioeconomic Systems

Bergstrom, A. R. (1967). *The Construction and Use of Economic Models*, English Universities Press, London, 127 pp.

Fisher, F. M. (1966). *The Identification in Econometrics*, McGraw-Hill, New York.

Forrester, J. W. (1961). *Industrial Dynamics*, MIT Press, Cambridge, 464 pp.

Forrester, J. W. (1969). *Urban Dynamics*, MIT Press, Cambridge, 285 pp.

Forrester, J. W. (1971). *World Dynamics*, Wright-Allen Press, Cambridge, 142 pp.

Johnston, J. (1972). *Econometric Methods*, McGraw-Hill, New York, 437 pp.

Koopmans, T. C. (1937). *Linear Regression Analysis of Economic Time Series*, Haarlem, Netherlands, Bohn.

Manetsch, T. J. (1966). 'Transfer function representation of the aggregate behaviour of a class of economic processes', *IEEE Trans. Autom. Control*, **AC-11,** 693–698.

Naylor, T. H. (1971). *Computer Simulation Experiments with Models of Economic Systems*, Wiley, New York, 502 pp.

Young, J. W., W. F. Arnold and J. W. Brewer (1972). 'Parameter identification and dynamic models of socioeconomic phenomena', *IEEE Trans. Syst., Man. Cybernet.*, **SMC-2,** 460–467.

Young, P., J. Naughton, C. Neethling and S. Shellswell (1973). 'Macro-economic modelling—a case study', *Proc. 3rd IFAC Symp.—Identification and System Parameter Estimation*, North-Holland Publ. Comp., Amsterdam, 145–166.

## 14.9 Learning Systems and Pattern Recognition

*Additional Literature*

Barát, J. and Gy. Muszély (1967). 'Pattern recognition algorithm adapting with searching method', *IFAC Symp.—Identification in Autom. Control Systems, Prague*, paper 1.4.

Biglow, J. W., E. S. McVey and J. W. Moore (1967). 'Pattern recognition in a learning automatic control system', *IFAC Symp.—Identification in Autom. Control Systems, Prague*, paper 1.5.

Gaines, B. R. (1967). 'Techniques of identification with the stochastic computer', *IFAC Symp.—Identification in Autom. Control Systems, Prague*, paper 6.1.

Hawkins, J. (1961). 'Self-organizing systems: a review and commentary', *Proc. IRE*, **49,** 31–48.

Ho, Y. C. and A. K. Agrawala (1968). 'On pattern classification algorithms—introduction and survey' (expository paper), *IEEE Trans. Autom. Control*, **AC-13,** 676–690; also *Proc. IEEE*, **56,** 2101–2114.

Nilsson, N. J. (1965). *Learning Machines*, McGraw-Hill, New York, 137 pp.

Tsypkin, Ya. Z. (1969; 1967; 1968; 1970). Cf. references Chapter 1.

# *Appendixes*

## APPENDIX A

### List of Symbols Used

Only the most important symbols are given. General rules (with a few exceptions) are:

—bold face lower-case symbols indicate vectors

—bold face capitals indicate matrices

—where applicable capitals denote the $\mathscr{F}$-, $\mathscr{L}$- or $\mathscr{Z}$-transform of lower-case variables.

| | |
|---|---|
| $A$, $B$, $C$, $D$ | —transfer function; dynamic operator |
| $C$ | —cost or loss function |
| $D$ | —differentiation operator |
| $E$ | —even function(al) of the error signal; error or loss function |
| $F$ | —function(al), $F[\ ]$ |
| $F$, $G$, $H$ | —transfer function; dynamic operator, $F(\ )$ |
| $J$ | —criterion function |
| $L$ | —likelihood function |
| $M$ | —model |
| $N$ | —normal distribution function; $N(\mu, \sigma^2)$, mean $\mu$, variance $\sigma^2$ |
| $N$ | —noise power, e.g. at correlator output or process output |
| $P$ | —process |
| $P$ | —probability, $P[\ ]$ |
| $R$ | —Lyapunov function |
| $S$ | —system |
| $S$ | —signal power, e.g. at correlator output or process output |
| $T$ | —time, integration interval |
| $U$, $V$, $W$, $X$, $Y$ | —transform of the corresponding time function |

| | |
|---|---|
| $\mathbf{A}$ | —system matrix |
| $\mathbf{B}$ | —distribution matrix |
| $\mathbf{C}$ | —output or measurement matrix |
| $\mathbf{D}$ | —lower triangular matrix; input–output matrix |
| $\mathbf{F}$ | —matrix of partial derivatives of vector function $\mathbf{f}$ |
| $\mathbf{I}$ | —identity matrix |
| $\mathbf{J}$ | —information matrix |
| $\mathbf{M}$ | —matrix |
| $\mathbf{N}$ | —covariance of the additive noise |
| $\mathbf{P}$ | —'covariance matrix' of an estimate |
| $\mathbf{Q}$ | —estimation matrix |
| $\mathbf{R}$ | —positive definite (weighting) matrix |
| $\mathbf{R}$ | —matrix of samples of noise |
| $\mathbf{T}$ | —(similarity) transformation matrix |
| $\mathbf{U}$ | —matrix of sample values of signals $[\mathbf{u}_0, \ldots, \mathbf{u}_m]$ |
| $\mathbf{W}$ | —matrix of samples of 'white' noise |
| $\mathbf{Y}$ | —matrix of sample values of output signals |

| | |
|---|---|
| $a$ $\mathbf{a}$, $b$ $\mathbf{b}$ | —coefficients of differential equation, difference equation, transfer function of the *process*; (varying) parameters of the *process* |
| $c$, $d$ | —coefficients in (orthogonal) developments |
| $e$ $\mathbf{e}$ | —(generalized) error signal |
| $\mathbf{e}$ | —basis vectors |
| $f$ | —frequency |
| $f$ $\mathbf{f}$, $g$ $\mathbf{g}$, $h$ $\mathbf{h}$ | —function; weighting function; impulse response; sample points of the impulse response; unit-sample-response function |
| $i$ | —integer numbers; current |
| $j$ | —integer numbers; $j^2 = -1$ |
| $k$, $l$, $m$, $n$ | —integer numbers |
| $n$ $\mathbf{n}$ | —additive noise (vector) |
| $p$, $q$ | —probability density function |
| $r$ | —weighting factor |
| $\mathbf{r}$ | —noise vector |
| $s$ | —Laplace transform variable |
| $t$ | —time |
| $u$ $\mathbf{u}$ | —measurable input (vector) of the process (and model); voltage |

| | |
|---|---|
| $u$ $\mathbf{u}_j$; $v$ $\mathbf{v}_i$ | —output of the linear operators in the (generalized) model; parameter sensitivity functions |
| $v$ $\mathbf{v}$; $w$ $\mathbf{w}$ | —unmeasurable noise (vectors) of the process |
| $w$ $\mathbf{w}$ | —output of the model; weighting function |
| $x$ $\mathbf{x}$ | —process state (vector) |
| $y$ $\mathbf{y}$ | —output (vector) of the process, either or not corrupted with noise |
| $z$ | —shift operator; $\mathscr{Z}$-transform variable |
| $\mathscr{A}$ $\boldsymbol{\mathscr{A}}$; $\mathscr{B}$ $\boldsymbol{\mathscr{B}}$; $\mathscr{C}$; $\mathscr{D}$ | —estimates of $A$, $B$, $C$, $D$, $\mathbf{A}$, $\mathbf{B}$ |
| $\mathscr{D}$ | —dispersion operator; $\mathscr{D}[\ ]$ |
| $\mathscr{E}$ | —expectation operator; ensemble average; $\mathscr{E}[\ ]$ |
| $\mathscr{F}$; $\mathscr{F}^{-1}$ | —Fourier transform; inverse Fourier transform; $\mathscr{F}[\ ]$ |
| $\mathscr{F}^{+}$ | —Fourier transform acting on positive time components only |
| $\mathscr{F}^{-}$ | —Fourier transform acting on negative time components only |
| $\mathscr{L}$; $\mathscr{L}^{-1}$ | —one-sided Laplace transform; inverse Laplace transform; $\mathscr{L}[\ ]$ |
| $\mathscr{L}_2$; $\mathscr{L}_2^{-1}$ | —two-sided Laplace transform; inverse Laplace transform; $\mathscr{L}_2[\ ]$ |
| $\mathscr{O}$ | —dynamic operator; $\mathscr{O}[\ ]$ |
| $\mathscr{P}$ | —policy operator; $\mathscr{P}[\ ]$ |
| $\mathscr{Q}$ | —quantizing operator; $\mathscr{Q}[\ ]$ |
| $\mathscr{S}$ | —sampling operator; $\mathscr{S}[\ ]$ |
| $\mathscr{S}$ | —class of systems; |
| $\mathscr{T}$ | —mapping; |
| $\mathscr{U}$ | —class of input signals; |
| $\mathscr{Z}$; $\mathscr{Z}^{-1}$ | —$z$-transform; inverse $z$-transform; $\mathscr{Z}[\ ]$ |
| $\Gamma$ (gamma) | —gain factor |
| $\boldsymbol{\Gamma}$ | —matrix of (gain) factors |
| $\Delta$ (delta) | —variation |
| $\Theta$ (theta) | —dispersion function |
| $\boldsymbol{\Lambda}$ (lambda) | —diagonal matrix |
| $\Pi$ (pi) | —continued product |
| $\Pi_i$ | —orthogonal polynomials |
| $\Sigma$ (sigma) | —continued sum |
| $\Upsilon$ (upsilon) | —normalized correlation function |
| $\boldsymbol{\Phi}$ (phi) | —transition or principal matrix; $\boldsymbol{\Phi}(t)$, $\boldsymbol{\Phi}(k)$ |

$\phi$ —power spectral density function; $\phi(\omega)$
$\psi$ (psi) —correlation function; $\psi(\tau)$
$\mathbf{\Omega}$ (omega) —matrix of process input and output samples;
$\nabla$ —gradient; $\nabla_{\boldsymbol{\beta}}$ gradient with respect to $\boldsymbol{\beta}$

$\alpha$ $\boldsymbol{\alpha}$ (alpha), $\beta$ $\boldsymbol{\beta}$ (beta) —estimates of **a**, **b**; coefficients of differential equation, difference equation, transfer function of the *model*; adjustable parameters of the *model*; components of the coefficient vector
$\gamma$ —gain factor; coefficient of orthogonal development
$\delta$ (delta) —Dirac's delta function; $\delta(t) = 0$ for $t \neq 0$ and

$$\int_{-\infty}^{+\infty} f(t)\,\delta(t-\theta)\,\mathrm{d}t = f(\theta)$$

$\delta_{mn}$ —Kronecker delta
$\varepsilon$ (epsilon) —base of Naperian logarithms
$\epsilon$ —small real constant $>0$
$\zeta$ (zeta), $\eta$ (eta) —components of the gradient vector
$\eta$ —normalized dispersion function
$\theta$ (theta) —time delay interval, sample interval, integration variable
$\boldsymbol{\theta}$ —parameter vector with components $[\boldsymbol{\beta}', \boldsymbol{\alpha}']'$ or $[\boldsymbol{\beta}', \boldsymbol{\alpha}', \boldsymbol{\gamma}', \boldsymbol{\delta}']'$
$\kappa$ (kappa), $\lambda$ (lambda) —scalar quantities
$\boldsymbol{\lambda}$ —parameter error vector with components $[\boldsymbol{\beta}', \boldsymbol{\alpha}']' - [\mathbf{b}', \mathbf{a}']'$
$\lambda$ —eigen value
$\mu$ (mu) —first (general) moment of a random variable; e.g. $\mu_x = \mathscr{E}[x]$
$\nu$ (nu), $\xi$ (xi) —dummy integration variables
$\boldsymbol{\xi}$ —model state (vector)
$\pi$ (pi) —polynomials
$\pi$ —3·14159
$\rho$ (rho) —correlation coefficient
$\sigma$ (sigma) —standard deviation
$\sigma^2$ —second central moment of random variable; variance e.g. $\sigma_x^2 = \mathscr{E}[(x-\mu)^2]$
$\tau$ (tau) —time (age) variable; time constant
$\varphi$ (phi) —characteristic function
$\phi$ (phi) —phase angle
$\chi$ (chi) —e.g. $\chi^2$ distribution
$\boldsymbol{\psi}$ (psi) —reference vector
$\omega$ (omega) —radian frequency variable
$\boldsymbol{\omega}$ —vector with components $[\mathbf{u}', \mathbf{y}', \mathbf{w}', \mathbf{r}']'$

Subscripts, superscripts, etc.

| | |
|---|---|
| $\overset{(n)}{u}$ ; $\dot{u}$ | —$n$th; first derivative |
| $u(k)$ | —sample value at $t = k\theta$ or $t = kT_s$ |
| $u_i$ | —$i$th component of the vector $\mathbf{u}$ |
| $\mathbf{u}_i$ | —vector during the $i$th measurement interval |
| $\mathbf{U}_k$ | —matrix containing the observations $1, \ldots, k$ |
| $u^{(i)}$ | —$i$th approximation |
| $\mathbf{u}(k+1)$ | —new observation on $\mathbf{U}$ at time $k+1$ |
| $\boldsymbol{\beta}(i)$ | —vector after the $i$th adjustment |
| $E(i)$ | —value after the $i$th measurement |
| * | —e.g. $u^*$ = sampled version of time-function $u(t)$ |
| * | —e.g. $G^*$ = pulse transfer function |
| ‾‾‾ | —e.g. $\overline{G(j\omega)}$ = complex conjugate of $G(j\omega)$ |
| ‾ | —e.g. $\bar{x}$ = ensemble average over infinite ensemble<br>—e.g. $\bar{x}_N$ = ensemble average over finite ensemble $= \frac{1}{N}\sum_{i=1}^{N} x_i$ |
| ⟨ ⟩ | —e.g. $\langle x \rangle$ = time average over infinite time $= \lim_{T\to\infty} \frac{1}{T}\int_0^T x(t)\,\mathrm{d}t$<br>—e.g. $\langle x \rangle_T$ = time average over finite time $= \frac{1}{T}\int_0^T x(t)\,\mathrm{d}t$ |
| ⟨ ⟩ | —e.g. $\langle x, y \rangle$ = some type of (nonlinear) signal processing operation: $= \int_0^T x(t)y(t)\,\mathrm{d}t$ |
| * | —e.g. $x * h$ = convolution |
| ∠ | —e.g. $\angle H(j\omega)$ = phase angle contribution |
| \| \| | —absolute value; determinant, e.g. Jacobian |
| ≡ | —identity |
| ‖ ‖ | —norm |
| $\| \ \|_{\mathbf{R}}^2$ | —positive definite form with weighting matrix $\mathbf{R}$ |
| $\overset{\text{def}}{=}$ | —equal by definition |
| sgn | —signum, e.g. $\operatorname{sgn} e = \frac{e}{\|e\|}$ |
| var[ ] | —variance |

| | |
|---|---|
| cov[ ] | —covariance |
| $^{(-1)}$ | —e.g. $A^{(-1)}$: generalized inverse |
| $\mid$ | —e.g. $p(z\mid x)$: conditional probability |
| $\hat{}$ | —e.g. $\hat{\beta}$: optimum estimate |
| $\sim$ | —e.g. $\tilde{\psi}$: approximation of correlation function |
| Re[ ] | —real part of [ ] |
| Im[ ] | —imaginary part of [ ] |
| exp[ ] | —$\varepsilon$ to the power [ ] |
| $'$ | —transpose of matrix or vector, e.g. $\mathbf{U}'$ |
| $\int_k \cdots d^k y$ | —$k$ fold integral |
| $_M$ | —model, e.g. $y_M$ |
| $_P$ | —process, e.g. $y_P$ |

# APPENDIX B

## Summary of Notions from Probability Theory and Stochastic Signals

In writing the text it is supposed that the reader has some knowledge of probability theory and stochastic signals. This summary is designed only for easy reference of properties and notation. The contents of this appendix will be used freely in this text without further reference.

### *Sets*

The relations between 'events' are governed by the rules of set theory:
Notions: set, subset, element, empty set.
Operations; cf. Figure B.1:

| | | |
|---|---|---|
| $\bar{}$ | e.g. $\bar{A}$ | complement |
| $\cup$ | e.g. $A \cup B$ | union |
| $\cap$ | e.g. $A \cap B$ | intersection |

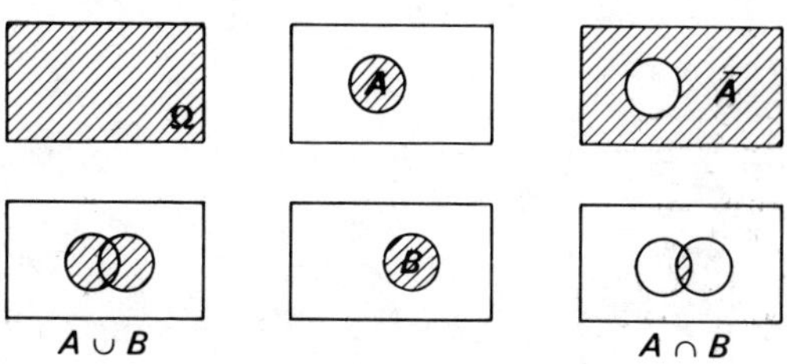


Figure B.1

Properties:

| Commutative | Associative |
|---|---|
| $A \cup B = B \cup A$ | $(A \cup B) \cup C = A \cup (B \cup C)$ |
| $A \cap B = B \cap A$ | $(A \cap B) \cap C = A \cap (B \cap C)$ |

Distributive

$$A \cap (B \cup C) = (A \cap B) \cup (A \cap C)$$
$$A \cup (B \cap C) = (A \cup B) \cap (A \cup C)$$

etc.

*Probability*

To each 'event' $A$ of a class of possible events in a simple experiment a number $P[A]$ is assigned. This number is called probability.

axiom I $\quad P[A] \geqslant 0$

axiom II $\quad P[\Omega] = 1 \quad$ if $\Omega$ is a certain event

$P[\phi] = 0 \quad$ if $\phi$ is an impossible event

axiom III $\quad P[A \cup B] = P[A] + P[B] \quad$ if $A \cap B = \phi$;

i.e. if the events are mutually exclusive.

For cases where there is an infinite number of events:

axiom IV $\quad P[A_1 \cup A_2 \cup A_3 \cup \cdots] = \sum_{i=1}^{\infty} P[A_i]$

if $A_i \cap A_j = \phi$ for each $i \neq j$.

*Combined Experiments*

The outcomes of two simple experiments, viz. $A_i$ and $B_j$, are considered as a (combined) event $[A_i, B_j]$. *

$$0 \leqslant P[A_i, B_j] \leqslant 1$$

$$\sum_{i,j} P[A_i, B_j] = 1$$

$$P[A_i] = \sum_j P[A_i, B_j] \qquad \text{if } B_k \cap B_j = \phi \qquad \text{for each } k \neq j$$

$$P[B_j] = \sum_i P[A_i, B_j] \qquad \text{if } A_k \cap A_i = \phi \qquad \text{for each } k \neq i$$

* The symbols $\sum_i$ and $\sum_{i,j}$ mean summation over all $i$, viz. all combinations $i, j$.

*Conditional Probability*

Definition:

$$P[A_i|B_j] = \frac{P[A_i, B_j]}{P[B_j]} \qquad \text{if } P[B_j] \neq 0$$

consequently

$$P[A_i|B_j]P[B_j] = P[A_i, B_j] = P[B_j|A_i]P[A_i]$$

For independent events:

$$P[A_i|B_j] = P[A_i]$$
$$P[A_i, B_j] = P[A_i]P[B_j]$$

In most cases the event is (or can be interpreted by means of) a numerical value or a set of numerical values.

A (real) **random variable** $X$* is a quantity that can have different values in such a way that for each given real number $x$ the probability $P[X \leqslant x]$ is defined.

The random variable can be **discrete,** if it can have a finite or denumerable set of distinct values, or it can be **continuous.** Based on these ideas the following notions can be recognized.

| | Random variable $X$ | |
|---|---|---|
| | Discrete | Continuous |
| **One-dimensional case** | | |
| Distribution function (cf. Figure B.2) | $F(x) = P[X \leqslant x]$ | $F(x) = P[X \leqslant x]$ |
| Probability (density) function | $P_i = P[X = x_i]$ | $p(x) = \frac{dF(x)}{dx}$ |
| Probability | | $p(x)\,\Delta x \approx P[x < X \leqslant x + \Delta x]$ |
| | $\sum_{\text{some } i} P_i$ | $\int_a^b p(\xi)\,d\xi = P[a < X \leqslant b]$ |

* In this appendix a random variable is denoted by an upper-case type. This convention is not adhered to in the main text.

| | Random variable $X$ | |
|---|---|---|
| | Discrete | Continuous |
| Properties | $0 \leqslant P_i \leqslant 1$ | $0 \leqslant p(x)$ |
| | $\sum_i P_i = 1$ | $\int_{-\infty}^{+\infty} p(\xi)\, d\xi = 1$ |
| | $F(x) = \sum_{\text{all } x_i \leqslant x} P_i$ | $F(x) = \int_{-\infty}^{x} p(\xi)\, d\xi$ |
| | | $F(-\infty) = 0$ |
| | | $F(\infty) = 1$ |
| Examples of density functions | Binomial: | Normal (cf. Figure B-3): |
| | $P[X = x] = \binom{n}{x} p^x (1 - p)^{n-x}$ | $p(x) = \frac{1}{\sigma\sqrt{2\pi}} \exp\left[-\frac{1}{2}\left(\frac{x - \mu}{\sigma}\right)^2\right]$ |
| | $x = 0, 1, \ldots, n$ | $-\infty \leqslant x \leqslant \infty$ |
| | Poisson: | Exponential: |
| | $P[X = x] = \frac{\lambda^x \varepsilon^{-\lambda}}{x!}$ | $p(x) = \begin{cases} \lambda\varepsilon^{-\lambda x} & 0 \leqslant x < \infty \\ 0 & -\infty < x < 0 \end{cases}$ |
| | $x = 0, 1, 2 \ldots$ | |
| **Two-dimensional case** | | |
| (Joint) distribution function | $F(x, y) = P[X \leqslant x, Y \leqslant y]$ | $F(x, y) = P[X \leqslant x, Y \leqslant y]$ |
| (Joint) probability (density) function | $P_{ij} = P[X = x_i, Y = y_j]$ | $p(x, y) = \frac{\partial^2 F(x, y)}{\partial x\, \partial y}$ |
| (Joint) probability | $\sum_{\substack{\text{some } i \\ \text{some } j}} P_{ij}$ | $p(x, y)\, \Delta x\, \Delta y \approx P[x < X \leqslant x + \Delta x, y < y \leqslant y + \Delta y]$ |
| Properties | $0 \leqslant P_{ij} \leqslant 1$ | $0 \leqslant p(x, y)$ |
| | $\sum_{i,j} P_{ij} = 1$ | $\int_{-\infty}^{+\infty} \int_{-\infty}^{+\infty} p(\xi, \eta)\, d\xi\, d\eta = 1$ |
| | $F(x, y) = \sum_{\substack{\text{all } x_i \leqslant x \\ \text{all } y_i \leqslant y}} P_{ij}$ | $F(x, y) = \int_{-\infty}^{x} \int_{-\infty}^{y} p(\xi, \eta)\, d\xi\, d\eta$ |
| | | $F(-\infty, y) = 0$ |
| | | $F(x, -\infty) = 0$ |
| | | $F(\infty, \infty) = 1$ |

| | Random variable $X$ | |
|---|---|---|
| | Discrete * | Continuous * |
| Marginal distribution function | $F(x, \infty) = P[X \leqslant x, Y \leqslant \infty]$ $= P[X \leqslant x]$ <br> $F(\infty, y) = P[X \leqslant \infty, Y \leqslant y]$ $= P[Y \leqslant y]$ | $F(x, \infty) = P[X \leqslant x, Y \leqslant \infty]$ $= P[X \leqslant x]$ <br> $F(\infty, y) = P[X \leqslant \infty, Y \leqslant y]$ $= P[Y \leqslant y]$ |
| Marginal probability (density function) | $P_i = \sum_j P_{ij}$ <br> $P_j = \sum_i P_{ij}$ | $p(x) = \dfrac{dF(x, \infty)}{dx} = \int_{-\infty}^{+\infty} p(x, \eta)\, d\eta$ <br> $p(y) = \dfrac{dF(\infty, y)}{dy} = \int_{-\infty}^{+\infty} (\xi, y)\, d\xi$ |
| Conditional distribution function | | $F(x \mid y) = P[X \leqslant x \mid Y = y]$ <br> $F(y \mid x) = P[Y \leqslant y \mid X = x]$ |
| Conditional probability (density) function | $P(X = x_i \mid Y = y_j) = \dfrac{P_{ij}}{P_j}$ <br> $P(Y = y_j \mid X = x_i) = \dfrac{P_{ij}}{P_i}$ | $p(x \mid y) = \dfrac{p(x, y)}{p(y)}$ <br> $p(y \mid x) = \dfrac{p(x, y)}{p(x)}$ <br> $p(x \mid y)p(y) = p(x, y)$ $= p(y \mid x)p(x)$ |
| Independence of $X$, $Y$ | $F(x, y) = F(x)F(y)$ <br> $P_{ij} = P_i P_j$ | $F(x, y) = F(x)F(y)$ <br> $p(x, y) = p(x)p(y)$ |
| Most important example of a bivariate density function | | Bivariate normal (cf. Figure B.4): <br> $p(x, y) = \dfrac{1}{2\pi\sigma_x\sigma_y(1 - \rho^2)^{\frac{1}{2}}} \exp[-\tfrac{1}{2}Q]$ <br> $Q = \dfrac{1}{1 - \rho^2}\left\{\left(\dfrac{x - \mu_x}{\sigma_x}\right)^2 - 2\rho\left(\dfrac{x - \mu_x}{\sigma_x}\right)\left(\dfrac{y - \mu_y}{\sigma_y}\right) + \left(\dfrac{y - \mu_y}{\sigma_y}\right)^2\right\}$ |

**More-dimensional case**

| | | |
|---|---|---|
| (Joint) distribution function | $F(x, y, z, \ldots)$ $= P[X \leqslant x, Y \leqslant y, Z \leqslant z, \ldots]$ | $F(x, y, z, \ldots)$ $= P[X \leqslant x, Y \leqslant y, Z \leqslant z, \ldots]$ |

etc., analogous to the previous case.

* The assumption will be used that a change of argument indicates a different function.

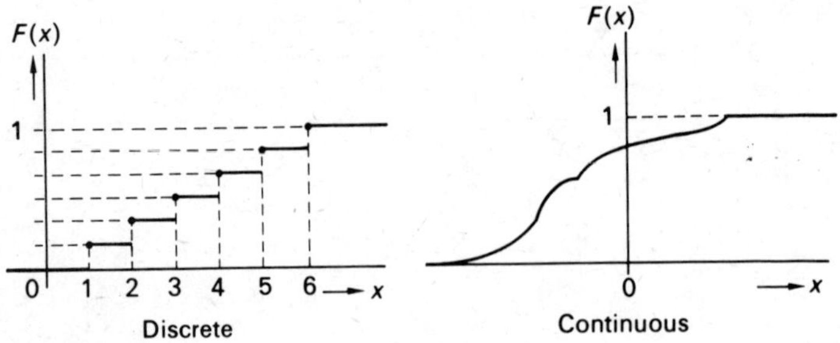


Figure B.2

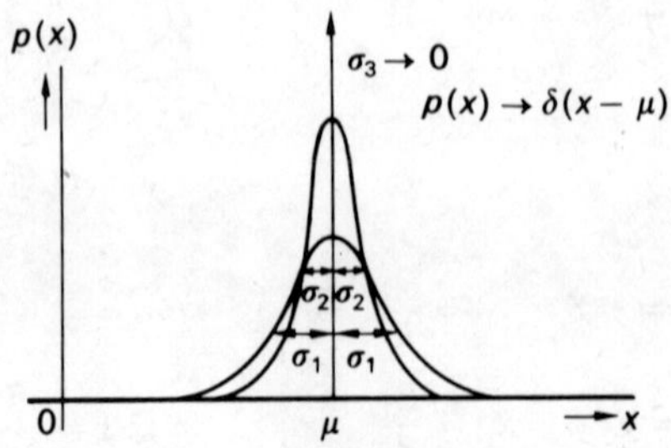


Figure B.3

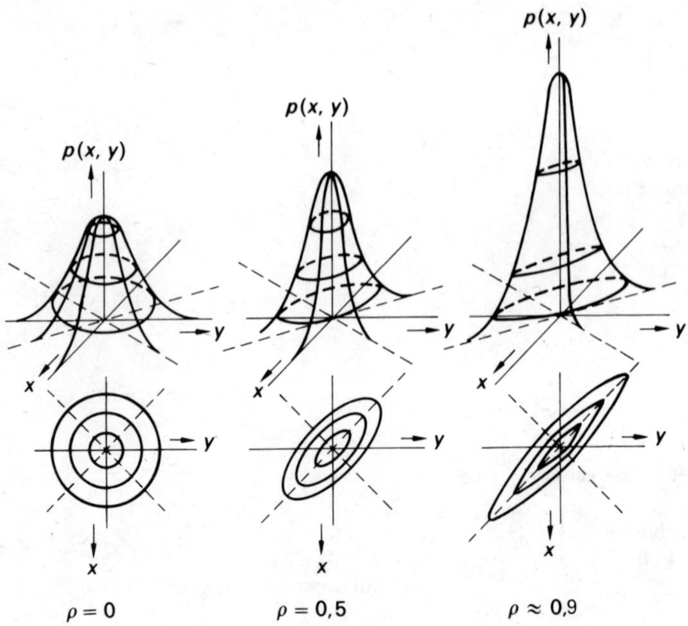


Figure B.4

In many cases it is a quite elaborate procedure to work with the expressions for these (probability) functions. Fortunately quite often it is permissible (and more convenient) to work with **parameters** instead of such **functions.** This leads to the following notions.

| | Random variable $X$ | |
|---|---|---|
| | Discrete | Continuous |
| **One-dimensional case** | | |
| Expectation | $\mathscr{E}[\cdots] = \sum \cdots P_i$ | $\mathscr{E}[\cdots] = \int_{-\infty}^{+\infty} \cdots p(x)\,dx$ |
| | $\mathscr{E}[f(X)] = \sum_i f(x_i)P_i$ | $\mathscr{E}[f(X)] = \int_{-\infty}^{+\infty} f(x)p(x)\,dx$ |
| Property: linearity | $\mathscr{E}[\alpha X + \beta\gamma] = \alpha\mathscr{E}[X] + \beta\mathscr{E}[Y]$ | |
| $n$th moment | $\mathscr{E}[X^n] = \sum_i x_i^n P_i$ | $\mathscr{E}[X^n] = \int_{-\infty}^{+\infty} x^n p(x)\,dx$ |
| First moment; | $\mu = \mathscr{E}[X]$ | |
| mean: expectation | $\mu = \sum_i x_i P_i$ | $\mu = \int_{-\infty}^{+\infty} xp(x)\,dx$ |
| $n$th central moment | $\mathscr{E}[(X - \mu)^n]$ | |
| First central moment | $\mathscr{E}[(X - \mu)] = 0$ | |
| Second central moment; | $\sigma^2 = \mathscr{E}[(X - \mu)^2] = \mathscr{E}[X^2] - \{\mathscr{E}[X]\}^2$ | |
| variance | $\sigma^2 = \sum_i x_i^2 P_i - \mu^2$ | $\sigma^2 = \int_{-\infty}^{+\infty} x^2 p(x)\,dx - \mu^2$ |
| Spread; standard deviation | $\sigma$ | |
| **Two-dimensional case** | | |
| Expectation | $\mathscr{E}[\cdots] = \sum_{i,j} \cdots P_{ij}$ | $\mathscr{E}[\cdots] = \int_{-\infty}^{+\infty}\int_{-\infty}^{+\infty} \cdots p(x, y)\,dx\,dy$ |
| | $\mathscr{E}[f(X, Y)] = \sum_{i,j} f(x_i, y_i)P_{ij}$ | $\mathscr{E}[f(X, Y)] = \int\int f(x, y)p(x, y) \times dx\,dy$ |
| Mean | $\mu_x = \mathscr{E}(X)$ | |
| | $\mu_x = \sum_{i,j} x_i P_{ij}$ | $\mu_x = \int\int xp(x, y)\,dx\,dy$ |

<table>
<tr><th></th><th colspan="2">Random variable $X$</th></tr>
<tr><th></th><th>Discrete</th><th>Continuous</th></tr>
<tr><td>Variance</td><td colspan="2">$\sigma_x^2 = \mathscr{E}[(X - \mu_x)^2]$</td></tr>
<tr><td></td><td>$\sigma_x^2 = \sum_{i,j} x_i^2 P_{ij} - \mu_x^2$</td><td>$\sigma_x^2 = \int\int x^2 p(x, y)\,dx\,dy - \mu_x^2$</td></tr>
<tr><td>Covariance</td><td colspan="2">$\sigma_{xy}^2 = \text{cov}\,[XY] = \mathscr{E}[(X - \mu_x)(Y - \mu_y)]$<br>$= \mathscr{E}[XY] - \mu_x\mu_y$</td></tr>
<tr><td>Correlation coefficient</td><td colspan="2">$\rho_{xy} = \dfrac{\sigma_{xy}^2}{\sigma_x\sigma_y}$</td></tr>
<tr><td>Conditional expectation</td><td>$\mathscr{E}[X \mid Y = y_j] = \dfrac{\sum_i x_i P_{ij}}{P_j}$</td><td>$\mathscr{E}[X \mid Y] = \dfrac{\int xp(x, y)\,dx}{p(y)}$</td></tr>
<tr><td>Property</td><td colspan="2">$\mathscr{E}[\mathscr{E}[X \mid Y]] = \mathscr{E}[X]$</td></tr>
<tr><td>Independence of of $X$, $Y$</td><td colspan="2">$\mathscr{E}[XY] = \mathscr{E}[X]\mathscr{E}[Y]$<br>$\rho_{xy} = 0$</td></tr>
<tr><td></td><td>$\sum_{i,j} x_i y_j P_{ij}$<br>$= \left\{\sum_i x_i P_i\right\}\left\{\sum_j y_j P_j\right\}$<br>$= \mu_x\mu_y$</td><td>$\int\int xyp(x, y)\,dx\,dy$<br>$= \left\{\int xp(x)\,dx\right\}\left\{\int yp(y)\,dy\right\}$<br>$= \mu_x\mu_y$</td></tr>
</table>

A **stochastic process** is a family of random variables $X(t)$, where $t$ is a parameter running over a set $T$.

A **realization** or **sample function** of a stochastic process is an assignment of a possible value to $X(t)$ for each $t$ of $T$.

In this text $t$ = time and $T$ = the time interval of observation. In that case we may choose:

as a sample function—a (stochastic) signal

as a stochastic process—an ensemble of stochastic signals.

Conceptually such an ensemble might consist of a (large) number of physical systems of (macroscopically) identical structure and under identical environmental conditions (cf. Figure B.5). If each system produces a stochastic signal then we may assume that the statistical properties of all these signals

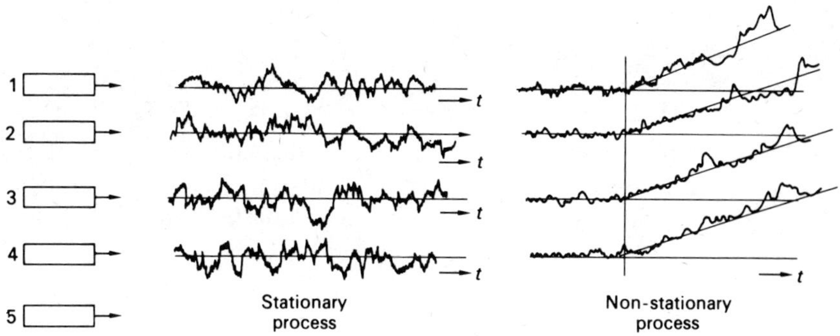


Figure B.5

are the same. In that situation one can define the probability of the amplitude $X(t_1)$ at time $t_1$:

$$F_1(x_1, t_1) = P[X(t_1) \leqslant x_1]$$

and

$$p_1(x_1, t_1) = \frac{\mathrm{d}F_1(x_1, t_1)}{\mathrm{d}x_1}$$

or

$$p_1(x_1, t_1)\Delta x_1 \approx P[x_1 < X(t_1) \leqslant x_1 + \Delta x_1]$$

(if $F_1$ can be differentiated). Figure B.6 shows some signals (ensemble elements) that pass through the interval $(x_1, x_1 + \Delta x_1]$ at time $t_1$.

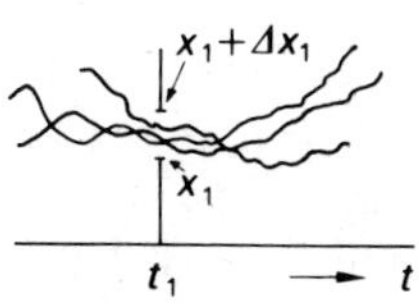


Figure B.6

A more detailed description of the random process can be given by studying the ensemble at **two** instants of time:

$$F_2(x_1, t_1; x_2, t_2) = P[X(t_1) \leqslant x_1; X(t_2) \leqslant x_2]$$

and

$$p_2(x_1, t_1; x_2, t_2) = \frac{\partial^2}{\partial x_1\,\partial x_2} F_2(\ldots)$$

or

$$p_2(x_1, t_1; x_2, t_2)\Delta x_1 \Delta x_2$$
$$\approx P[x_1 < X(t_1) \leqslant x_1 + \Delta x_1; x_2 < X(t_2) \leqslant x_2 + \Delta x_2]$$

cf. Figure B.7. In the same way one may find probability-density functions $p_3, p_4, \ldots$, relating the amplitude probabilities at 3, 4, ..., instances of time, that give a more detailed description of the stochastic process. As done

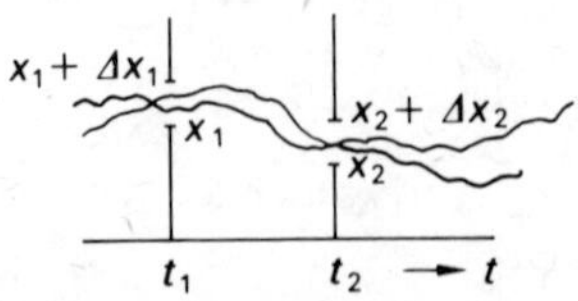


Figure B.7

previously for random variables, one may look for parameters that can be used, instead of probability-density functions, for characterizing a stochastic process. These are called **ensemble moments.** Some examples:

$$\mu_x(t_1) = \overline{X(t_1)}^{\text{ens}} = \mathscr{E}[X(t_1)] = \int_{-\infty}^{+\infty} x_1 p(x_1, t_1)\, dx_1$$

$$\overline{X^2(t_1)}^{\text{ens}} = \mathscr{E}[X^2(t_1)] = \int_{-\infty}^{+\infty} x_1^2 p(x_1, t_1)\, dx_1$$

$$\psi_{xx}(t_1, t_2) = \overline{X(t_1 X(t_2)}^{\text{ens}} = \mathscr{E}[X(t_1)X(t_2)]$$

$$= \int_{-\infty}^{+\infty} \int_{-\infty}^{+\infty} x_1 x_2 p(x_1, t_1; x_2, t_2)\, dx_1\, dx_2$$

The latter expression, **the autocorrelation function,** is of great importance for describing the properties of a stochastic process. Analogously one may describe the relation between two stochastic processes $X$ and $Y$ by the **cross-correlation function** $\psi_{xy}(t_1, t_2)$. These definitions hold for general stochastic processes, which may be non-stationary. A very important subclass is that of the **stationary** processes, i.e. for which it is true that the statistical properties are invariant with respect to shifting along the time axis (cf. Figure B.5):

$$F_n(x_1, t_1 + \theta; \ldots; x_n, t_n + \theta) = F_n(x_1, t_1; \ldots; x_n, t_n) \qquad \text{each } \theta$$

One consequence is, that for each stationary processes all parameters are invariant under a time shift:

$$\mu_x(t_1) = \mu_x(t) = \mu_x$$

$$\psi_{xx}(t_1, t_2) = \psi_{xx}(t, t + \tau) = \psi_{xx}(\tau) \qquad \text{with } \tau = t_2 - t_1$$

For such processes it makes sense to use the notion of time average or **time moments.** If the corresponding time and ensemble moments are equal:

$$\lim_{T\to\infty}\frac{1}{T}\int_0^T X(t)\,dt = \overline{X(t)}^{\text{time}} = \overline{X(t)}^{\text{ens}} = \int_{-\infty}^{+\infty} xp(x)\,dx$$

$$\lim_{T\to\infty}\frac{1}{T}\int_0^T X^2(t)\,dt = \overline{X^2(t)}^{\text{time}} = \overline{X^2(t)}^{\text{ens}} = \int_{-\infty}^{+\infty} x^2p(x)\,dx$$

$$\psi_{xy}(\tau) = \lim_{T\to\infty}\frac{1}{T}\int_0^T X(t)Y(t+\tau)\,dt = \overline{X(t)Y(t+\tau)}^{\text{time}} = \overline{X(t)Y(t+\tau)}^{\text{ens}}$$

$$= \int_{-\infty}^{+\infty}\int_{-\infty}^{+\infty} xyp(x, y, \tau)\,dx\,dy$$

etc.

then the process is called **ergodic.**

Some properties of autocorrelation functions for stationary processes:

$\psi_{xx}(t)$ exists if $\mathscr{E}[X^2(t)]$ exists

$\psi_{xx}(0) =$ average 'power' $> 0$ if $\mathscr{E}[X^2(t)] \neq 0$

$\psi_{xx}(\tau) = \psi_{xx}(-\tau)$

$|\psi_{xx}(\tau)| \leqslant \psi_{xx}(0)$

if $X(t)$ has a periodic component, so has $\psi_{xx}(\tau)$
if $X(t)$ has no periodic component and no d.c. component then $\psi_{xx}(\tau) \to 0$ for $\tau \to \infty$.

For a presentation of a stochastic process in the frequency domain one may use the **auto-** and **cross-spectral density** functions (power spectrum) $\phi_{xx}(j\omega)$ and $\phi_{xy}(j\omega)$ which can be found through Fourier or two-sided Laplace transform of the respective correlation functions (Wiener–Khintchin relations):

$$\begin{cases} \phi_{xy}(j\omega) = \mathscr{F}[\psi_{xy}(\tau)] = \displaystyle\int_{-\infty}^{+\infty} \psi_{xy}(\tau)\varepsilon^{-j\omega\tau}\,d\tau \\ \psi_{xy}(\tau) = \mathscr{F}^{-1}[\phi_{xy}(j\omega)] = \dfrac{1}{2\pi}\displaystyle\int_{-\infty}^{+\infty} \phi_{xy}(j\omega)\varepsilon^{j\omega\tau}\,d\omega \end{cases}$$

or

$$\begin{cases} \phi_{xy}(s) = \mathscr{L}_2[\psi_{xy}(\tau)] = \displaystyle\int_{-\infty}^{+\infty} \psi_{xy}(\tau)\varepsilon^{-s\tau}\,d\tau & \alpha < \mathrm{Re}\,[s] < \beta \\ \psi_{xy}(\tau) = \mathscr{L}_2^{-1}[\phi_{xy}(s)] = \dfrac{1}{2\pi j}\displaystyle\int_{c-j\infty}^{c+j\infty} \phi_{xy}(s)\varepsilon^{s\tau}\,ds & \alpha < c < \beta \end{cases}$$

Figure B.8 shows some transform pairs. Note the use of the $j$-operator in the argument of the $\mathscr{F}$-transformed variable; this supports the parallelism between $\mathscr{F}$- and $\mathscr{L}_2$-transformation through $j\omega \leftrightarrow s$.

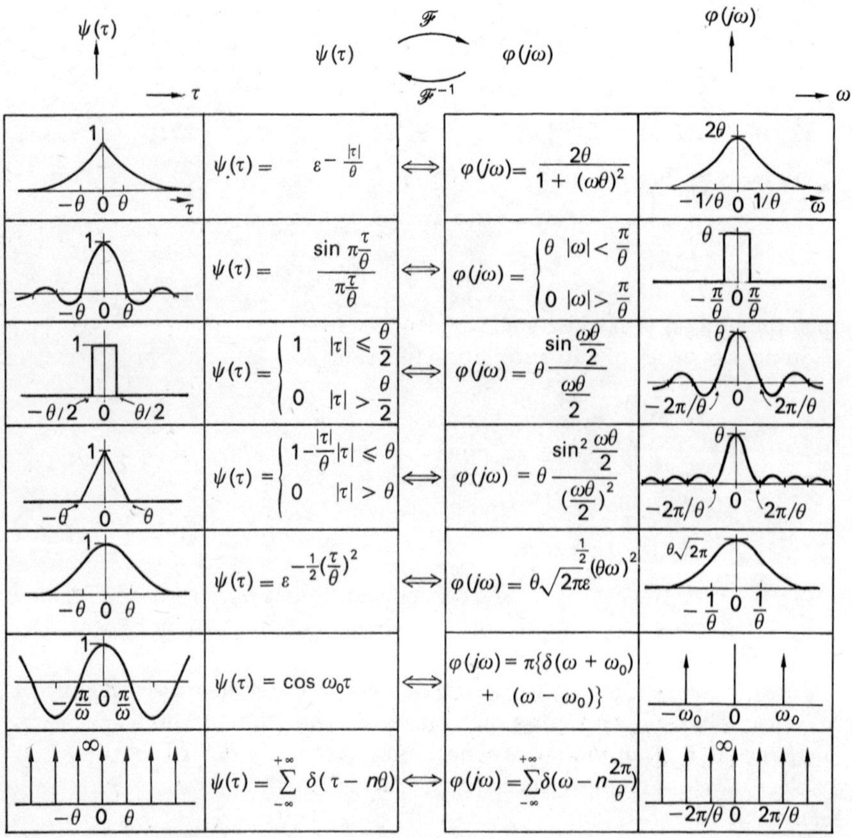


Figure B.8

One aspect of the $\mathscr{L}_2$-transformation needs special attention. For the transformation of correlation functions one may choose $c = 0$, i.e. the convergence strip contains the $j\omega$ axis. This axis divides the $s$-plane:

the left-hand part, $\mathrm{Re}\,[s] < 0$, corresponds to $\tau > 0$

the right-hand part, $\mathrm{Re}\,[s] > 0$, corresponds to $\tau < 0$

cf. Figure B.9.

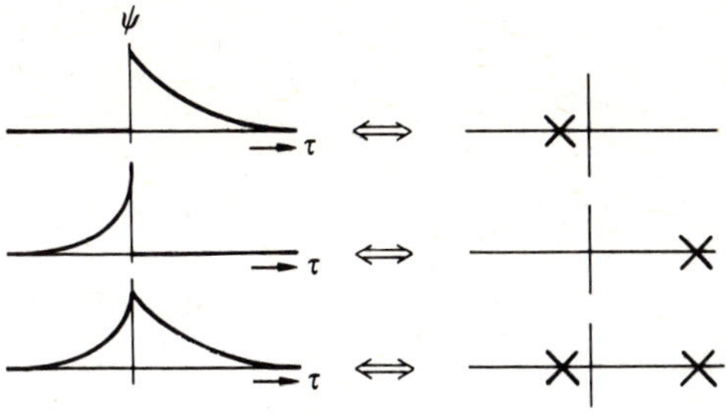


Figure B.9

As an application of signal characterization in time and frequency domain consider Figure B.10. We want to determine the cross-correlation function for $Y$ and $Z$, as well as the cross-power spectral density (ergodic case).

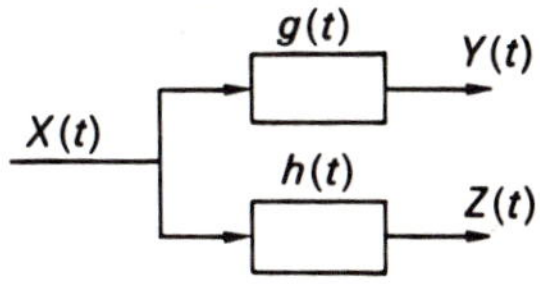


Figure B.10

$$\psi_{yz}(\tau) = \overline{Y(t)Z(t+\tau)}$$

$$= \overline{\left\{\int_{-\infty}^{+\infty} g(\theta)X(t-\theta)\,\mathrm{d}\theta\right\}\left\{\int_{-\infty}^{+\infty} h(\xi)X(t+\tau-\xi)\,\mathrm{d}\xi\right\}}$$

$$= \int_{-\infty}^{+\infty} \mathrm{d}\theta\, g(\theta) \int_{-\infty}^{+\infty} \mathrm{d}\xi\, h(\xi)\overline{X(t-\theta)X(t+\tau-\xi)}$$

$$= \int_{-\infty}^{+\infty} \mathrm{d}\theta\, g(\theta) \int_{-\infty}^{+\infty} \mathrm{d}\xi\, h(\xi)\psi_{xx}(\tau-\xi+\theta)$$

This represents two convolutions, one after the other. By two-sided Laplace transforms one obtains the cross-power spectral density function:

$$\phi_{yz}(s) = \mathscr{L}_2[\psi_{yz}(\tau)] = \int_{-\infty}^{+\infty} \psi_{yz}(\tau)\varepsilon^{-s\tau}\,\mathrm{d}\tau$$

$$= \int_{-\infty}^{+\infty} \mathrm{d}\theta\, g(\theta)\varepsilon^{s\theta} \int_{-\infty}^{+\infty} \mathrm{d}\xi\, h(\xi)\varepsilon^{-s\xi} \int_{-\infty}^{+\infty} \psi(\tau-\xi+\theta)\varepsilon^{-s(\tau-\xi+\theta)}\,\mathrm{d}\tau$$

$$= G(-s)H(s)\phi_{xx}(s)$$

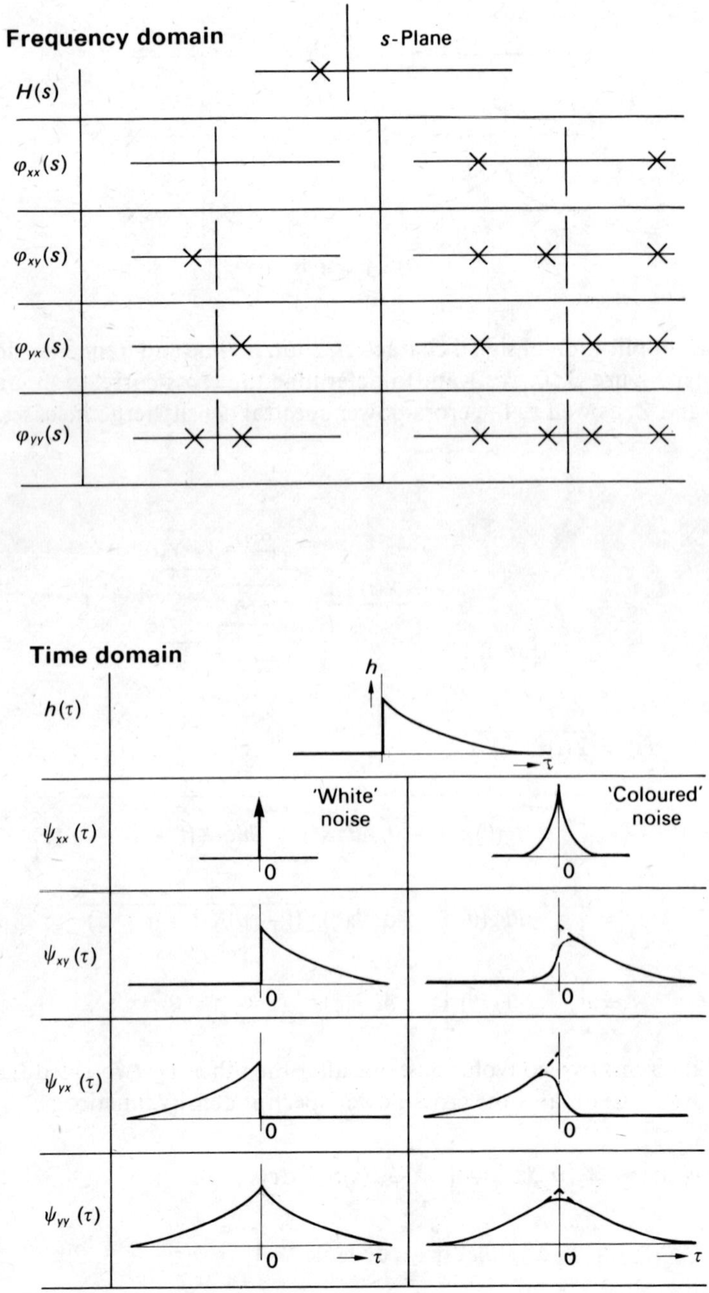

Frequency domain
s-Plane
H(s)
φxx(s)
φxy(s)
φyx(s)
φyy(s)
Time domain
h(τ)
h
τ
'White' noise
'Coloured' noise
ψxx (τ)
ψxy (τ)
ψyx (τ)
ψyy (τ)
0
τ


Figure B.11

with

$$\left.\begin{aligned} G(s) &= \mathscr{L}_2[g(t)] \\ H(s) &= \mathscr{L}_2[h(t)] \end{aligned}\right\} \quad \begin{gathered} g(t) = h(t) = 0 \text{ for } t < 0 \\ \text{(physical realizability)} \end{gathered}$$

$$\phi_{xx}(s) = \mathscr{L}_2[\psi_{xx}(\tau)]$$

From this derivation the formalism is clear:

$$\phi_{yz}(s) = G(-s)H(s)\phi_{xx}(s)$$

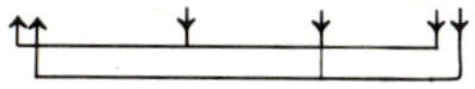

The transfer function with $-s$ changes the first subscript, the transfer function with $+s$ changes the second subscript of the power spectrum. Figure B.11 illustrates this procedure for two very simple examples. A corresponding notation $G(s) \to G(j\omega)$ and $G(-s) \to G^*(j\omega)$ can be used for the respective Fourier transforms.

# APPENDIX C

## Summary of Matrix Operations

*Notions*

Vector: Matrix:

$$\mathbf{x} = \begin{bmatrix} x_1 \\ \vdots \\ x_k \end{bmatrix} \qquad \mathbf{A} = [a_{ij}] = \begin{bmatrix} a_{11} & \cdots & a_{1n} \\ \vdots & & \vdots \\ a_{m1} & \cdots & a_{mn} \end{bmatrix}$$

$k \times 1$ matrix $\qquad$ $m \times n$ matrix

dimensions, elements, row, column, cofactor, trace, diagonal matrix, unit matrix.

*Matrix Algebra*

Transpose:

$$\mathbf{x}' = [x_1, \ldots, x_k] \qquad \mathbf{A}' = [a_{ji}]$$

$$[\mathbf{A}']' = \mathbf{A}$$

$$[\mathbf{A} + \mathbf{B}]' = \mathbf{A}' + \mathbf{B}'$$

$$[\mathbf{ABC}]' = \mathbf{C}'\mathbf{B}'\mathbf{A}'$$

square matrix $\mathbf{A}$:
if $\mathbf{A}' = \mathbf{A}$ symmetric;
if $\mathbf{A}' = -\mathbf{A}$ skewsymmetric;

if $\mathbf{A}'\mathbf{A} = \mathbf{A}\mathbf{A}' = \mathbf{I}$ orthogonal; $\mathbf{I}$ identity or unit matrix
if $\mathbf{A}'\mathbf{A} = \mathbf{A}$ idempotent

Determinant:

$$|\mathbf{A}| = \sum_{j=1}^{n} a_{ij} \text{ cofactor } (a_{ij}) \qquad \text{if } \mathbf{A} \text{ is an } n \times n \text{ matrix}$$

$$|\mathbf{A}| = 0 \Rightarrow \text{matrix singular}$$

$$|\mathbf{AB}| = |\mathbf{A}|\,|\mathbf{B}|$$

Rank:

order of its largest non-singular square submatrix;

number of linearly independent row (column) vectors.

$$\text{rank } C = \text{rank } ACB \qquad \text{if } A \text{ and } B \text{ non-singular; } C \text{ and } ACB \text{ equivalent matrices.}$$

Adjoint:

$$\tilde{\mathbf{A}} = \begin{bmatrix} \text{matrix of} \\ \text{cofactors} \end{bmatrix}'$$

$$\tilde{\mathbf{A}}\mathbf{A} = \mathbf{A}\tilde{\mathbf{A}} = |\mathbf{A}|\mathbf{I}$$

Addition:

associative
commutative

Multiplication:

| | |
|---|---|
| associative | $\mathbf{A}[\mathbf{BC}] = [\mathbf{AB}]\mathbf{C}$ |
| distributive | $\mathbf{A}[\mathbf{B} + \mathbf{C}] = \mathbf{AB} + \mathbf{AC}$ |
| not commutative, except | $\mathbf{CA} = \mathbf{AC}$ if $\mathbf{C} = c\mathbf{I}$ |
| scalar multiplication | $c\mathbf{AB} = \mathbf{A}c\mathbf{B} = \mathbf{AB}c$ |
| vector dot (inner) product | $\mathbf{x}'\mathbf{y} \rightarrow$ scalar |
| dyadic (outer) product | $\mathbf{xy}' \rightarrow$ matrix |

Matrix multiplication:

an expression ending with column vector is col. vector $\mathbf{ABx} = \mathbf{y}$

an expression beginning with row vector is row vector $\mathbf{y}'\mathbf{BC} = \mathbf{d}'$

an expression beginning with row vector and ending with column vector } is scalar $\mathbf{x}'\mathbf{Ay} = c$

Quadratic forms:

$$[x_1, x_2]\begin{bmatrix} a & b \\ b & c \end{bmatrix}\begin{bmatrix} x_1 \\ x_2 \end{bmatrix} = ax_1^2 + 2bx_1x_2 + cx_2^2$$

$$\begin{matrix} \mathbf{x}'\mathbf{A}_1\mathbf{x} & \text{quadratic form} \\ \mathbf{x}'\mathbf{A}_2\mathbf{y} & \text{bilinear form} \end{matrix} \quad \begin{cases} >0 \text{ positive definite} \\ <0 \text{ negative definite} \\ \geqslant 0 \text{ non-negative definite} \\ \leqslant 0 \text{ non-positive definite} \\ \quad \text{for each } \mathbf{x} \neq \mathbf{0},\ \mathbf{y} \neq \mathbf{0} \end{cases}$$

$\mathbf{A}_1$ can always be made symmetric, i.e. $\mathbf{A}_1 = \mathbf{A}_1'$; if $\mathbf{Q}$ is positive definite then

$$[\mathbf{x} - \mathbf{x}_0]'\mathbf{Q}^{-1}[\mathbf{x} - \mathbf{x}_0] = c \text{ ellipsoid, center } \mathbf{x}_0$$

$$|\text{semi-axes}| = \sqrt{\text{charact. roots of } \mathbf{Q}}$$

Singularity:

$$\mathbf{A} \text{ singular} \qquad \left.\begin{matrix} \mathbf{A}\mathbf{x} = \mathbf{0} \\ \text{or } \mathbf{A}'\mathbf{x} = \mathbf{0} \end{matrix}\right\} \quad \text{for any } \mathbf{x} \neq \mathbf{0}$$

Inverse:

$$\mathbf{A}^{-1}\mathbf{A} = \mathbf{I} = \mathbf{A}\mathbf{A}^{-1}$$

$$\mathbf{A}^{-1} = |\mathbf{A}|^{-1}\tilde{\mathbf{A}}$$

$$[\mathbf{A}^{-1}]' = [\mathbf{A}']^{-1}$$

$$[\mathbf{A}^{-1}]^{-1} = \mathbf{A}$$

$$[\mathbf{ABC}]^{-1} = \mathbf{C}^{-1}\mathbf{B}^{-1}\mathbf{A}^{-1}$$

if $\mathbf{A}' = \mathbf{A}^{-1}$ orthogonal;

for $\mathbf{A}' = \mathbf{A},\ \mathbf{C}' = \mathbf{C}$;

$$[\mathbf{A} + \mathbf{BCB}']^{-1} = \mathbf{A}^{-1} - \mathbf{A}^{-1}\mathbf{B}[\mathbf{C}^{-1} + \mathbf{B}'\mathbf{A}^{-1}\mathbf{B}]^{-1}\mathbf{B}'\mathbf{A}^{-1}$$

$$= [\mathbf{I} + \mathbf{A}^{-1}\mathbf{BCB}']^{-1}\mathbf{A}^{-1} \qquad *$$

$$\left.\begin{matrix} \mathbf{CB}'[\mathbf{A} + \mathbf{BCB}']^{-1} = [\mathbf{C}^{-1} + \mathbf{B}'\mathbf{A}^{-1}\mathbf{B}]^{-1}\mathbf{B}'\mathbf{A}^{-1} \\ [\mathbf{A} + \mathbf{BCB}']^{-1}\mathbf{BC} = \mathbf{A}^{-1}\mathbf{B}[\mathbf{C}^{-1} + \mathbf{B}'\mathbf{A}^{-1}\mathbf{B}]^{-1} \end{matrix}\right\} \text{transpose}$$

* This can be proved by multiplying both sides of the equation with $[\mathbf{A} + \mathbf{BCB}']$:

$$\begin{aligned} \mathbf{I} &= [\mathbf{A} + \mathbf{BCB}'][\mathbf{A}^{-1} - \mathbf{A}^{-1}\mathbf{B}[\mathbf{C}^{-1} + \mathbf{B}'\mathbf{A}^{-1}\mathbf{B}]^{-1}\mathbf{B}'\mathbf{A}^{-1}] \\ &= \mathbf{I} - \mathbf{B}[\mathbf{C}^{-1} + \mathbf{B}'\mathbf{A}^{-1}\mathbf{B}]^{-1}\mathbf{B}'\mathbf{A}^{-1} \\ &\quad + \mathbf{BC} \qquad\qquad\qquad \mathbf{B}'\mathbf{A}^{-1} \\ &\quad - \mathbf{BCB}'\mathbf{A}^{-1}\mathbf{B}[\mathbf{C}^{-1} + \mathbf{B}'\mathbf{A}^{-1}\mathbf{B}]^{-1}\mathbf{B}'\mathbf{A}^{-1} \\ &= \mathbf{I} + \mathbf{B}[\mathbf{C} - [\mathbf{C}^{-1} + \mathbf{B}'\mathbf{A}^{-1}\mathbf{B}]^{-1}[\mathbf{I} + \mathbf{CB}'\mathbf{A}^{-1}\mathbf{B}]]\mathbf{B}'\mathbf{A}^{-1} \\ &= \mathbf{I} + \mathbf{BC}[\mathbf{I} - \mathbf{I}]\mathbf{B}'\mathbf{A}^{-1} = \mathbf{I} \end{aligned}$$

possible modifications $\mathbf{C} \to -\mathbf{C}$

$$\mathbf{A} \to -\mathbf{A}$$

$$\mathbf{A} \to -\mathbf{A}^{-1}$$

$$\mathbf{C} \to \mathbf{C}^{-1}$$

$$\mathbf{B} \to \mathbf{B}'$$

Characteristic roots:

| | |
|---|---|
| $\mathbf{A} - \lambda\mathbf{I}$ | characteristic matrix; **A** square |
| $\lvert\mathbf{A} - \lambda\mathbf{I}\rvert = f(\lambda)$ | characteristic function |
| $\lvert\mathbf{A} - \lambda\mathbf{I}\rvert = f(\lambda) = 0$ | characteristic equation |
| $\lambda_1, \ldots, \lambda_n$ | characteristic roots; eigenvalues |
| $[\mathbf{A} - \lambda_i\mathbf{I}]\mathbf{x}_i = \mathbf{0} \Rightarrow \mathbf{x}_1, \ldots, \mathbf{x}_n$ | characteristic vectors; eigenvectors |

$$\mathbf{r}_i = \frac{\mathbf{x}_i}{|\mathbf{x}_i|}$$

$$\mathbf{R} = [\mathbf{r}_1, \mathbf{r}_2, \ldots, \mathbf{r}_n]$$

**A** real symmetric then there is at least one **R** such that:

$$\left.\begin{aligned} \mathbf{R}'\mathbf{A}\mathbf{R} &= \mathbf{D} \\ \mathbf{A} &= \mathbf{R}\mathbf{D}\mathbf{R}' \end{aligned}\right\} \mathbf{D}\text{—diagonal}$$

if **A** real symmetric $\Rightarrow \lambda_i$ real

if **A** real symmetric $\lambda_i > 0 \Rightarrow$ positive definite

if **A** real symmetric $\lambda_i \geqslant 0 \Rightarrow$ non-negative definite

Similarity transformation:

$$\left.\begin{aligned} \mathbf{B} &= \mathbf{T}^{-1}\mathbf{A}\mathbf{T} \\ \mathbf{T}\mathbf{T}^{-1} &= \mathbf{T}^{-1}\mathbf{T} = \mathbf{I} \end{aligned}\right\} \Rightarrow \mathbf{A} \text{ and } \mathbf{B} \text{ similar matrices}$$

$$\mathbf{A} = \mathbf{T}\mathbf{B}\mathbf{T}^{-1}$$

$$\mathbf{A}^i = \mathbf{T}\mathbf{B}^i\mathbf{T}^{-1}$$

Congruence transformation:

$$\left.\begin{aligned} &\mathbf{B} = \mathbf{T}'\mathbf{A}\mathbf{T} \\ &\mathbf{A} \text{ and } \mathbf{B} \text{ symmetric} \end{aligned}\right\} \Rightarrow \mathbf{A} \text{ and } \mathbf{B} \text{ congruent matrices}$$

**B** can be made diagonal with elements $-1, 0, 1$;
numberof 1's is rank (invariant)

Orthogonal transformation:

$$\mathbf{B} = \mathbf{T}^{-1}\mathbf{A}\mathbf{T}$$

$$\mathbf{T}^{-1} = \mathbf{T}' \quad \Rightarrow \mathbf{T} \text{ orthogonal matrix}$$

Symmetric idempotent matrix:

$$\mathbf{A}^2 = \mathbf{A} \Rightarrow \quad \text{characteristic vector } \mathbf{x}_i:$$

$$\left.\begin{aligned} \mathbf{A}^2\mathbf{x}_i &= \mathbf{A}\mathbf{x}_i = \lambda_i \mathbf{x}_i \\ \mathbf{A}^2\mathbf{x}_i &= \mathbf{A}\lambda_i\mathbf{x}_i = \lambda_i^2\mathbf{x}_i \end{aligned}\right\} \lambda_i = 0 \text{ or } 1$$

$$[\mathbf{I} - \mathbf{A}]^2 = \mathbf{I}^2 - 2\mathbf{A} + \mathbf{A}^2 = \mathbf{I} - \mathbf{A} \quad \text{also idempotent}$$

Generalized inverse:

$$\mathbf{A}^{(-1)}$$

$$\mathbf{A}\mathbf{A}^{(-1)}\mathbf{A} = \mathbf{A} \qquad \text{definition } \mathbf{A}^{(-1)}$$

$$\mathbf{A}^{(-1)} = \mathbf{A}^{-1} \qquad \text{if } \mathbf{A} \text{ square, non-singular}$$

$$\mathbf{A}^{(-1)}\mathbf{A} = \mathbf{I} \qquad \Rightarrow \text{inverse from the left}$$

$$\mathbf{A}\mathbf{A}^{(-1)} = \mathbf{I} \qquad \Rightarrow \text{inverse to the right}$$

*Matrix Calculus*

Differentiation with respect to scalar:

$$\frac{d\mathbf{A}}{dt} = \left[\frac{da_{ij}}{dt}\right]$$

$$\frac{d\mathbf{x}}{dt} = \begin{bmatrix} \dfrac{dx_1}{dt} \\ \vdots \\ \dfrac{dx_k}{dt} \end{bmatrix} \qquad \frac{d\mathbf{x}'}{dt} = \left[\frac{dx_1}{dt}, \ldots, \frac{dx_k}{dt}\right]$$

$$\frac{d}{dt}(\mathbf{A} + \mathbf{B}) = \frac{d\mathbf{A}}{dt} + \frac{d\mathbf{B}}{dt}$$

$$\frac{d}{dt}(\mathbf{ABC}) = \frac{d\mathbf{A}}{dt}\mathbf{BC} + \mathbf{A}\frac{d\mathbf{B}}{dt}\mathbf{C} + \mathbf{AB}\frac{d\mathbf{C}}{dt}$$

$$\frac{d}{dt}\mathbf{A}^n = \frac{d\mathbf{A}}{dt}\mathbf{A}^{n-1} + \mathbf{A}\frac{d\mathbf{A}}{dt}\mathbf{A}^{n-2} + \cdots + \mathbf{A}^{n-1}\frac{d\mathbf{A}}{dt}$$

$$\frac{d}{dt}\mathbf{A}^{-1} = -\mathbf{A}^{-1}\frac{d\mathbf{A}}{dt}\mathbf{A}^{-1}$$

$$\frac{d}{dt}|\mathbf{A}| = \text{trace}\left[\frac{d\mathbf{A}}{dt}\tilde{\mathbf{A}}\right]$$

Differentiation with respect to vector:

$$\frac{\partial}{\partial \mathbf{x}} = \nabla_{\mathbf{x}} \overset{\text{def}}{=} \begin{bmatrix} \dfrac{\partial}{\partial x_1} \\ \vdots \\ \dfrac{\partial}{\partial x_k} \end{bmatrix} \quad \text{gradient operator}$$

$$\frac{\partial}{\partial \mathbf{x}'} = \nabla_{\mathbf{x}'} \overset{\text{def}}{=} \left[ \frac{\partial}{\partial x_1}, \ldots, \frac{\partial}{\partial x_k} \right]$$

scalar $c$

$$\nabla_{\mathbf{x}} c = \frac{\partial c}{\partial \mathbf{x}} = \begin{bmatrix} \dfrac{\partial c}{\partial x_1} \\ \vdots \\ \dfrac{\partial c}{\partial x_k} \end{bmatrix} \rightarrow \text{vector}; \qquad \nabla_{\mathbf{x}\mathbf{x}'} c = \nabla_{\mathbf{x}} \nabla_{\mathbf{x}'} c = \begin{bmatrix} \dfrac{\partial^2 c}{\partial x_1 \partial x_1} & \cdots & \dfrac{\partial^2 c}{\partial x_1 \partial x_k} \\ \vdots & & \vdots \\ \dfrac{\partial^2 c}{\partial x_k \partial x_1} & \cdots & \dfrac{\partial^2 c}{\partial x_k \partial x_k} \end{bmatrix}$$

vector $\mathbf{y}$

$$\nabla_{\mathbf{x}} \mathbf{y}' = \begin{bmatrix} \dfrac{\partial y_1}{\partial x_1} & \cdots & \dfrac{\partial y_m}{\partial x_1} \\ \vdots & & \vdots \\ \dfrac{\partial y_1}{\partial x_k} & \cdots & \dfrac{\partial y_m}{\partial x_k} \end{bmatrix}; \qquad \nabla_{\mathbf{x}'} \mathbf{y} = \begin{bmatrix} \dfrac{\partial y_1}{\partial x_1} & \cdots & \dfrac{\partial y_1}{\partial x_k} \\ \vdots & & \vdots \\ \dfrac{\partial y_m}{\partial x_1} & \cdots & \dfrac{\partial y_m}{\partial x_k} \end{bmatrix}$$

$$\nabla_{\mathbf{x}'} \mathbf{x} = \mathbf{I}$$

$$\nabla_{\mathbf{x}} \mathbf{x}' = \mathbf{I}$$

$$\nabla_{\mathbf{x}}[\mathbf{y}'\mathbf{x}] = \mathbf{y} = \nabla_{\mathbf{x}}[\mathbf{x}'\mathbf{y}]$$

$$\nabla_{\mathbf{x}}[\mathbf{x}'\mathbf{x}] = 2\mathbf{x}$$

$$\nabla_{\mathbf{x}}[\mathbf{x}'\mathbf{A}\mathbf{x}] = \mathbf{A}\mathbf{x} + \mathbf{A}'\mathbf{x}$$

$$\nabla_{\mathbf{x}}[\mathbf{x}'\mathbf{A}] = \mathbf{A}$$

both $\mathbf{z}$ and $\mathbf{y}$ a function of $\mathbf{x}$

$$\nabla_{\mathbf{x}}[\mathbf{y}'\mathbf{z}] = [\nabla_{\mathbf{x}}\mathbf{y}]\mathbf{z} + [\nabla_{\mathbf{x}}\mathbf{z}']\mathbf{y}$$

**A** symmetric, non-negative definite, not a function of **x**:

$$\nabla_{\mathbf{x}}[\mathbf{y}'\mathbf{A}\mathbf{y}] = \nabla_{\mathbf{x}}[[\mathbf{y}'\mathbf{A}'^{\frac{1}{2}}][\mathbf{A}^{\frac{1}{2}}\mathbf{y}]] = 2[\nabla_{\mathbf{x}}[\mathbf{y}'\mathbf{A}'^{\frac{1}{2}}]]\mathbf{W}^{\frac{1}{2}}\mathbf{y} = 2[\nabla_{\mathbf{x}}\mathbf{y}']\mathbf{A}\mathbf{y}$$

where

$$\mathbf{A} = \mathbf{R}'\mathbf{D}\mathbf{R} = [\mathbf{R}'\mathbf{D}^{\frac{1}{2}}][\mathbf{D}^{\frac{1}{2}}\mathbf{R}] \qquad \mathbf{A}^{\frac{1}{2}} = \mathbf{D}^{\frac{1}{2}}\mathbf{R}$$

$|\nabla_{\mathbf{x}'}\mathbf{y}|$—Jacobian or functional determinant

Stationary value:

scalar $c$

$$\left.\frac{\partial c}{\partial \mathbf{x}}\right|_{\mathbf{x}=\hat{\mathbf{x}}} = \mathbf{0} \qquad \text{min if } \left.\frac{\partial}{\partial \mathbf{x}'}\frac{\partial c}{\partial \mathbf{x}}\right|_{\mathbf{x}=\hat{\mathbf{x}}} \quad \text{positive definite}$$

scalar $\mathbf{e}'\mathbf{e}$ with $\mathbf{e} = \mathbf{z} - \mathbf{U}\boldsymbol{\beta}$

$$\frac{\partial \mathbf{e}'\mathbf{e}}{\partial \boldsymbol{\beta}} = -2\mathbf{U}'\mathbf{e} = -2\mathbf{U}'[\mathbf{z} - \mathbf{U}\boldsymbol{\beta}]$$

$$\frac{\partial}{\partial \boldsymbol{\beta}'}\frac{\partial \mathbf{e}'\mathbf{e}}{\partial \boldsymbol{\beta}} = 2\mathbf{U}'\mathbf{U}$$

Matrix power series and exponential:

$$f(\mathbf{A}) \stackrel{\text{def}}{=} \sum_{i=0}^{\infty} \alpha_i \mathbf{A}^i$$

$$\varepsilon^{\mathbf{A}t} \stackrel{\text{def}}{=} \mathbf{I} + \mathbf{A}t + \frac{[\mathbf{A}t]^2}{2!} + \frac{[\mathbf{A}t]^3}{3!} + \cdots$$

Cayley–Hamilton ⇒ each matrix polynomial and matrix power series that converges can be expressed in a linear combination of $\mathbf{A}^{n-1}, \ldots, \mathbf{A}, \mathbf{I}$, if **A** is an $n \times n$ matrix. Consequently

$$\varepsilon^{\mathbf{A}t} = \alpha_0\mathbf{I} + \alpha_1\mathbf{A} + \alpha_2\mathbf{A}^2 \qquad \text{if } \mathbf{A} \text{ is a } 3 \times 3 \text{ matrix}$$

Schwarz's inequality (outline of proof):

$$[\mathbf{A}\mathbf{x} + \mathbf{B}\mathbf{y}]'[\mathbf{A}\mathbf{x} + \mathbf{B}\mathbf{y}] \geqslant 0$$

$$\mathbf{x}'\underbrace{\mathbf{A}'\mathbf{A}}_{\mathbf{C}}\mathbf{x} + \mathbf{x}'\underbrace{\mathbf{A}'\mathbf{B}}_{\mathbf{D}}\mathbf{y} + \mathbf{y}'\underbrace{\mathbf{B}'\mathbf{A}}_{\mathbf{D}'}\mathbf{x} + \mathbf{y}'\underbrace{\mathbf{B}'\mathbf{B}}_{\mathbf{E}}\mathbf{y} \geqslant 0$$

$$\underbrace{[\mathbf{C}^{\frac{1}{2}}\mathbf{x} + \mathbf{C}^{-\frac{1}{2}}\mathbf{D}\mathbf{y}]'[\mathbf{C}^{\frac{1}{2}}\mathbf{x} + \mathbf{C}^{-\frac{1}{2}}\mathbf{D}\mathbf{y}]}_{=\,\mathbf{0} \quad \text{if one chooses} \quad \mathbf{x} = -\mathbf{C}^{-1}\mathbf{D}\mathbf{y}} + \mathbf{y}'[\mathbf{E} - \mathbf{D}'\mathbf{C}^{-1}\mathbf{D}]\mathbf{y} \geqslant 0$$

Consequently

$$\left.\begin{aligned} \mathbf{y}'[\mathbf{E} - \mathbf{D}'\mathbf{C}^{-1}\mathbf{D}]\mathbf{y} \geqslant 0 \\ \mathbf{E} - \mathbf{D}'\mathbf{C}^{-1}\mathbf{D} \geqslant \mathbf{0} \end{aligned}\right\} \quad \text{all } \mathbf{y}$$

or

$$[\mathbf{B}'\mathbf{B}] \geqslant [\mathbf{B}'\mathbf{A}][\mathbf{A}'\mathbf{A}]^{-1}[\mathbf{A}'\mathbf{B}]$$

## References


Bellman, R. (1960). *Introduction to Matrix Analysis*, McGraw-Hill, New York, 328 pp.

Bodewig, E. (1956). *Matrix Calculus*, North-Holland Publ. Comp., Amsterdam, 334 pp.

Derusso, P. M., R. J. Roy and C. M. Close (1965). *State Variables for Engineers*, Wiley, New York, 608 pp.

Frame, J. S. (1964). 'Matrix functions and applications', *IEEE Spectrum*, **1**, part I: 'Matrix operations and generalized inverses', March, 208–220; part II: 'Functions of a matrix', April, 102–108; part III: Application of matrices to systems analysis', May, 100–109; part IV: 'Matrix functions and constituent matrices', June, 123–131; part V: 'Similarity reductions by rational or orthogonal matrices', July, 103–109.

Gantmacher, F. R. (1959). *Application of the Theory of Matrices*, Interscience, New York, 317 pp.

Householder, A. S. (1964). *The Theory of Matrices in Numerical Analysis*, Blaisdell Publ. Comp., New York, 257 pp.

Zurmühl, R. (1964). *Matrices and Their Engineering Applications* (in German), Springer, Berlin, 452 pp.

*On Generalized Inverses*

Albert, A. (1972). *Regression and the Moore–Penrose Pseudoinverse*, Academic Press, New York, 192 pp.

Manherz, R. K., B. W. Jordan and S. L. Hakimi (1968). 'Analog methods for computation of the generalized inverse', *IEEE Trans. Autom. Control*, **AC-13**, 582–585.

Penrose, R. (1955). 'A generalized inverse for matrices', *Proc. Phil. Soc.*, **51**, 406–413.

Penrose, R. (1956). 'On best approximate solutions of linear matrix equations', *Proc. Camb. Phil. Soc.*, **52**, 17–19.


# APPENDIX D

## Transients in Statistical Average and Variance of Correlation Measurements

### *Product of Two Random Signals*

The system of Figure D.1 is considered. This system is slightly more general than is necessary for our purpose. Both $x$ and $y$ are stationary random signals, elements of ergodic ensembles, which are multiplied together. The potentiometer 'modulates' the product with a time function $f(t)$ which may, for example, be a step function due to the closing of a

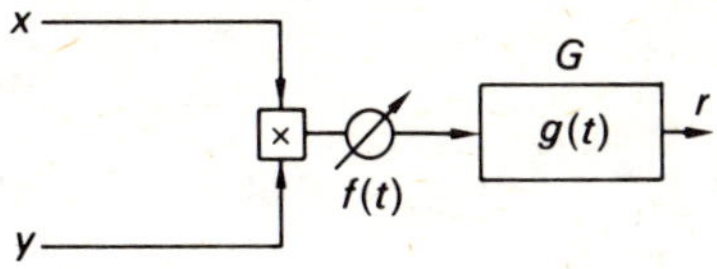


Figure D.1

switch. (If one or both of the time functions $x$ and $y$ are 'modulated' in power, then the time function $f(t)$ can represent that modulation.) This modulated product is fed to a block G with an impulse response $g(t)$.

Consequently assuming $f(t) = 0$ for $t < 0$ and G having initial conditions zero;

$$r(t) = \int_0^t f(t-\theta)x(t-\theta)y(t-\theta)g(\theta)\,\mathrm{d}\theta \tag{D.1}$$

If $f(t)$ is a unit step function at $t = 0$ then:

$$r(t) = \int_0^t x(t-\theta)y(t-\theta)g(\theta)\,\mathrm{d}\theta \tag{D.2}$$

If the block denoted by G is a pure integrator then $g(t)$ is a step function as well and by change of variable one finds:

$$r(t) = \int_0^t x(\tau)y(\tau)\,\mathrm{d}\tau \tag{D.3}$$

Both the low-pass filter and the pure-integrator case can be represented by the same formula, equation (D.2).

One wants to obtain an expression for the statistical average and for the variance of the output $r$ as a function of time.

*a. Statistical Average.* The statistical average of $r$ is:

$$\begin{aligned}\mu_r(t) = \mathscr{E}[r(t)] &= \mathscr{E}\left[\int_0^t f(t-\theta)x(t-\theta)y(t-\theta)g(\theta)\,\mathrm{d}\theta\right]\\ &= \int_0^t f(t-\theta)\mathscr{E}[x(t-\theta)y(t-\theta)]g(\theta)\,\mathrm{d}\theta \qquad \text{(D.4)}\\ &= \psi_{xy}(0)\int_0^t f(t-\theta)g(\theta)\,\mathrm{d}\theta\end{aligned}$$

as both signals have been assumed to be stationary.

If $f(t)$ is a step function:

$$f(t) = \begin{cases}1 & \text{for } t \geqslant 0\\ 0 & \text{for } t < 0\end{cases} \quad \text{then} \quad \mu_r(t) = \psi_{xy}(0)\int_0^t g(\theta)\,\mathrm{d}\theta \tag{D.5}$$

If an integrator is used:

$$g(t) = \begin{cases} 1 & \text{for } t \geqslant 0 \\ 0 & \text{for } t < 0 \end{cases} \quad \text{then} \quad \mu_r(t) = \psi_{xy}(0)t \tag{D.6}$$

*b. Variance.* The variance of $r$ is:

$$\begin{aligned} \sigma_r^2(t) &= \mathscr{E}[\{r(t) - \mu_r(t)\}^2] \\ &= \mathscr{E}[r^2(t)] - \mu_r^2(t)] \end{aligned} \tag{D.7}$$

This can be written as follows:

$$\begin{aligned} \sigma_r^2(t) = \mathscr{E}\bigg[ &\int_0^t f(t-\theta)x(t-\theta)y(t-\theta)g(\theta)\,\mathrm{d}\theta \\ &\times \int_0^t f(t-\xi)x(t-\xi)y(t-\xi)g(\xi)\,\mathrm{d}\xi \bigg] \\ &- \left\{ \psi_{xy}(0) \int_0^t f(t-\theta)g(\theta)\,\mathrm{d}\theta \right\}^2 \\ = &\int_0^t \int_0^t f(t-\theta)f(t-\xi)\mathscr{E}[x(t-\theta)y(t-\theta)x(t-\xi)y(t-\xi)] \\ &\times g(\theta)g(\xi)\,\mathrm{d}\theta\,\mathrm{d}\xi - \left\{ \psi_{xy}(0) \int_0^t f(t-\theta)g(\theta)\,\mathrm{d}\theta \right\}^2 \end{aligned} \tag{D.8}$$

Now we have to determine a fourth-order moment. Consider the moment generating function $P(\mathbf{s})$ defined by:

$$\begin{aligned} P(\mathbf{s}) &= \mathscr{E}[\exp \mathbf{s}'\mathbf{x}] \\ &= \iiiint_{-\infty}^{+\infty} \exp\{s_1x_1 + s_2x_2 + s_3x_3 + s_4x_4\} p(\mathbf{x})\,\mathrm{d}x_1\,\mathrm{d}x_2\,\mathrm{d}x_3\,\mathrm{d}x_4 \end{aligned} \tag{D.9}$$

where

$$\mathbf{s} = \begin{bmatrix} s_1 \\ s_2 \\ s_3 \\ s_4 \end{bmatrix} \quad \text{and} \quad \mathbf{x} = \begin{bmatrix} x_1 \\ x_2 \\ x_3 \\ x_4 \end{bmatrix}$$

and $p(\mathbf{x})$ is the four-dimensional probability-density function of the random process under consideration.

Differentiation of $P(\mathbf{s})$ with respect to $s_1, s_2, s_3$ and $s_4$ results in the type of fourth-order moment that we need:

$$\frac{\partial^4 P(\mathbf{s})}{\partial s_1\, \partial s_2\, \partial s_3\, \partial s_4} = \mathscr{E}[x_1 x_2 x_3 x_4] \tag{D.10}$$

Let us assume that the random variables have zero mean values and that the joint probability-density function is Gaussian:

$$p(\mathbf{x}) = \frac{1}{(2\pi)^2 |M|^{\frac{1}{2}}} \exp\{-\tfrac{1}{2}\mathbf{x}'\mathbf{M}^{-1}\mathbf{x}\}$$

with

$$\mathbf{M} = \begin{bmatrix} \psi_{11} \cdots \psi_{1n} \\ \vdots \qquad \vdots \\ \psi_{n1} \cdots \psi_{nn} \end{bmatrix} \quad \text{and} \quad \psi_{ij} = \mathscr{E}[x_i x_j] = \psi_{ji}$$

Consequently its moment generating function can be found to be:

$$P(\mathbf{s}) = \frac{1}{(2\pi)^2 |M|^{\frac{1}{2}}} \iiiint_{-\infty}^{+\infty} \exp\{-\tfrac{1}{2}\mathbf{x}'\mathbf{M}^{-1}\mathbf{x} + \tfrac{1}{2}\mathbf{x}'\mathbf{s} + \tfrac{1}{2}\mathbf{s}'\mathbf{x}\}\, dx_1\, dx_2\, dx_3\, dx_4$$

By completing the square:

$$-\tfrac{1}{2}\{\mathbf{x}'\mathbf{M}^{-1}\mathbf{x} - \mathbf{x}'\mathbf{s} - \mathbf{s}'\mathbf{x}\} = -\tfrac{1}{2}\{(\mathbf{x}'\mathbf{M}^{-\frac{1}{2}} - \mathbf{s}'\mathbf{M}^{\frac{1}{2}})(\mathbf{M}^{-\frac{1}{2}}\mathbf{x} - \mathbf{M}^{\frac{1}{2}}\mathbf{s}) - \mathbf{s}'\mathbf{M}\mathbf{s}\}$$

and taking into account the properties of integrals of probability-density functions, the resulting expression is:

$$P(\mathbf{s}) = \exp\{\tfrac{1}{2}\mathbf{s}'\mathbf{M}\mathbf{s}\} \tag{D.11}$$

Differentiation according to equation (D.10) leads to:

$$\mathscr{E}[x_1 x_2 x_3 x_4] = \psi_{12}\psi_{34} + \psi_{13}\psi_{24} + \psi_{14}\psi_{23} \tag{D.12}$$

For the case under consideration one finds:

$$\begin{aligned} &\mathscr{E}[x(t-\theta)y(t-\theta)x(t-\xi)y(t-\xi)] \\ &= \psi_{xy}^2(0) + \psi_{xx}(\theta-\xi)\psi_{yy}(\theta-\xi) + \psi_{xy}(\theta-\xi)\psi_{yx}(\theta-\xi) \\ &= \psi_{xy}^2(0) + R(\theta-\xi) \end{aligned} \tag{D.13}$$

where the term $R(\theta - \xi)$ is substituted for the sake of convenience. As

$$\psi_{xx}(-v) = \psi_{xx}(v) \qquad \psi_{yy}(-v) = \psi_{yy}(v)$$

$$\psi_{xy}(-v) = \psi_{yx}(v)$$

it follows that $R(-\nu) = R(\nu)$. As the first term $\psi_{xy}^2(0)$, gives rise to

$$\int_0^t \int_0^t f(t-\theta) f(t-\xi) \psi_{xy}^2(0) g(\theta) g(\xi) \, d\theta \, d\xi = \left\{ \psi_{xy}(0) \int_0^t f(t-\theta) g(\theta) \, d\theta \right\}^2 \tag{D.14}$$

this term cancels the corresponding one in equation (D.8). Thus a substitution of equation (D.13) in equation (D.8) gives:

$$\sigma_r^2(t) = \int_0^t \int_0^t f(t-\theta) f(t-\xi) R(\theta-\xi) g(\theta) g(\xi) \, d\theta \, d\xi \tag{D.15}$$

By putting $\theta - \xi = \nu$ and by dividing the region of integration of Figure D.2 into two parts according to Figure D.3 one obtains

$$|\text{Jacobian}| = 1$$

$$\text{For } I_1 : 0 \leqslant \xi \leqslant \theta \leqslant t \Rightarrow 0 \leqslant \xi \leqslant (\xi + \nu) \leqslant t$$

$$\text{For } I_2 : 0 \leqslant \theta \leqslant \xi \leqslant t \Rightarrow 0 \leqslant \theta \leqslant (\theta - \nu) \leqslant t$$

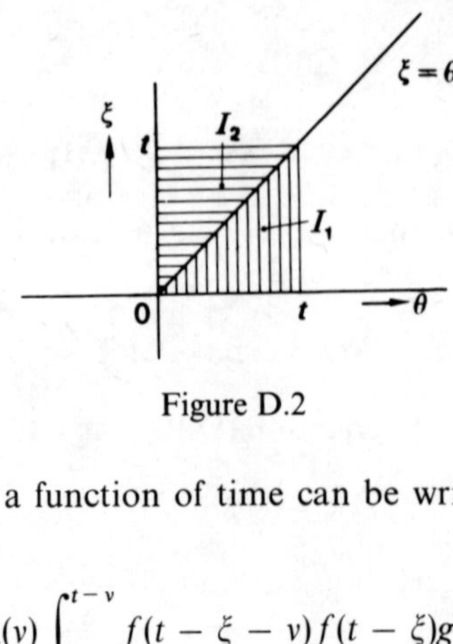


Figure D.2

Thus the variance as a function of time can be written as the sum of two integrals:

$$\sigma_r^2(t) = \int_0^t d\nu \, R(\nu) \int_0^{t-\nu} f(t-\xi-\nu) f(t-\xi) g(\xi+\nu) g(\xi) \, d\xi + \int_0^t d\nu \, R(\nu) \int_0^{t+\nu} f(t-\theta) f(t-\theta+\nu) g(\theta) g(\theta-\nu) \, d\theta$$

As $R(-\nu) = R(\nu)$, both integrals are equal and:

$$\underbrace{\sigma_r^2(t) = 2 \int_0^t d\nu \, R(\nu) \int_0^{t-\nu} f(t-\nu-\xi) f(t-\xi) g(\nu+\xi) g(\xi) \, d\xi}_{0 \leqslant \nu \leqslant t} \tag{D.16}$$

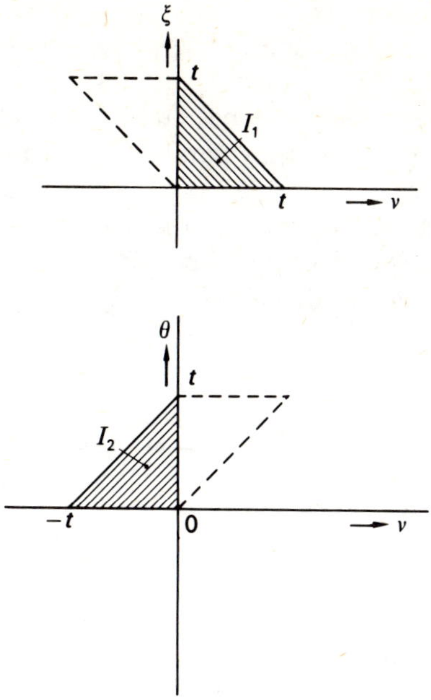


Figure D.3

with

$$R(v) = \psi_{xx}(v)\psi_{yy}(v) + \psi_{xy}(v)\psi_{yx}(v)$$

If $f(t)$ is a step function:

$$f(t) = \begin{cases} 1 & \text{for } t \geqslant 0 \\ 0 & \text{for } t < 0 \end{cases}$$

then

$$\underline{\sigma_r^2(t) = 2\int_0^t \mathrm{d}v\, R(v) \int_0^{t-v} g(v+\xi)g(\xi)\,\mathrm{d}\xi} \qquad \text{(D.17)}$$

If an integrator is used then

$$g(t) = \begin{cases} 1 & \text{for } t \geqslant 0 \\ 0 & \text{for } t < 0 \end{cases}$$

then

$$\underline{\sigma_r^2(t) = 2\int_0^t (t-v)R(v)\,\mathrm{d}v} \qquad \text{(D.18)}$$

*c. Discussion of These Results.* First it has to be noted that the results hold in general for stationary random processes for which one can determine $\psi_{xy}(0)$ and the fourth-order moment $\mathscr{E}[x(t-\theta)y(t-\theta)x(t-\xi)y(t-\xi)]$. As an example a Gaussian process has been taken. Secondly, it has to be mentioned that although the time-delay parameter $\tau = 0$ has been chosen this does not restrict the results.

If one wishes to determine the $\mu_r(t)$ and $\sigma_r^2(t)$ for another argument say $\tau = \tau_1$ then this can be obtained by introducing a virtual time-delay element in the process-generating block diagram (Figure D.4).

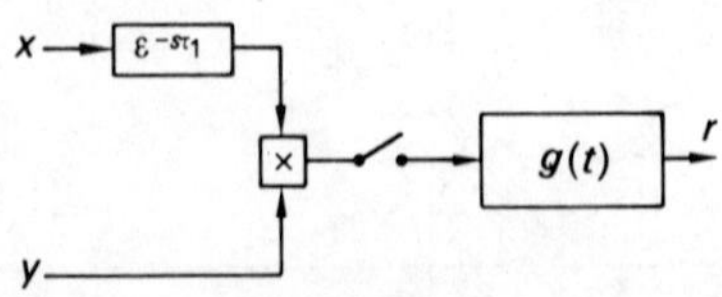


Figure D.4

Finally it can be seen that the formulas that have been derived can also be used for the study of the transients in statistical average and variance for the circuit shown in Figure D.5, where a stationary signal is modulated by a time function $f(t)$ and then fed to a linear operator with $g(t)$ as impulse response.

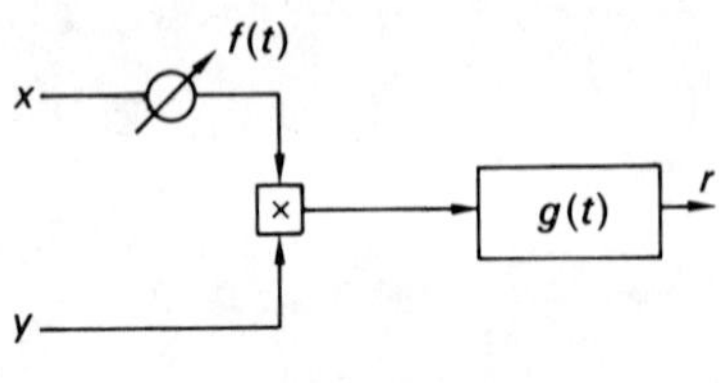


Figure D.5

*Product of a Sinusoidal and a Random Signal*

Now the system under consideration is given by Figure D.6. Consequently we restrict ourselves to a discussion of the influence of the additive noise.

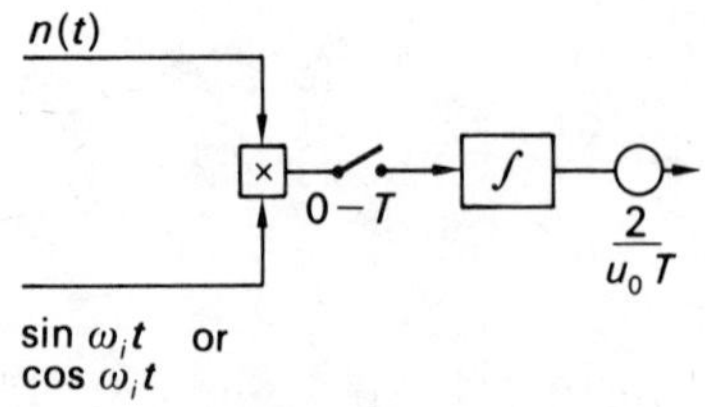


Figure D.6

As in Chapter 10 the integration interval is restricted to $(0, T)$ with $T = kT_i$, $k$ integer and $T_i = 2\pi/\omega_i$.

*a. Statistical Average.* In Chapter 10 it was mentioned that due to the independence of signal and noise the noise does not influence the value of the statistical average.

*b. Variance.* In Chapter 10 the expressions for the variance were found to be:

$$\left.\begin{matrix}\sigma_c^2(T)\\ \sigma_d^2(T)\\ \sigma_{cd}^2(T)\end{matrix}\right\} = \frac{4}{u_0^2T^2}\int_0^T\int_0^T \psi_{nn}(\theta-\xi)\begin{Bmatrix}\sin\omega_i\xi & \sin\omega_i\theta\\ \cos\omega_i\xi & \cos\omega_i\theta\\ \sin\omega_i\xi & \cos\omega_i\theta\end{Bmatrix} d\xi\, d\theta \qquad \text{(D.19)}$$

These expressions can be reworked just as in equation (D.15). Again by substitution of $\theta - \xi = v$ and by dividing the region of integration of Figure D.2 into two parts according to Figure D.3 (with $t = T$) one obtains:

$$\sigma_c^2(T) = \frac{4}{u_0^2T^2}\int_0^T dv\,\psi_{nn}(v)\int_0^{T-v}\sin\omega_i\xi\sin\omega_i(\xi+v)\,d\xi$$
$$+\frac{4}{u_0^2T^2}\int_0^T dv\,\psi_{nn}(v)\int_0^{T-v}\sin\omega_i(\theta-v)\sin\omega_i\theta\,d\theta$$

As $\psi_{nn}(-v) = \psi_{nn}(v)$ both integrals are equal and

$$\sigma_c^2(T) = \frac{8}{u^2T^2}\int_0^T dv\,\psi_{nn}(v)\int_0^{T-v}\sin\omega_i\xi\sin\omega_i(\xi+v)\,d\xi$$

where

$$\sin\omega_i\xi\sin\omega_i(\xi+v) = \tfrac{1}{2}\cos\omega_i v - \tfrac{1}{2}\cos\omega_i(2\xi+v)$$

which leads to

$$\sigma_c^2(T) = \frac{4}{u_0^2T^2}\int_0^T \psi_{nn}(v)$$
$$+\left\{(T-v)\cos\omega_i v - \frac{1}{2\omega_i}\sin\omega_i(2T-v) + \frac{1}{2\omega_i}\sin\omega_i v\right\}dv$$
$$= \frac{4}{u_0^2T}\int_0^T \psi_{nn}(v)\left\{\left(1-\frac{v}{T}\right)\cos\omega_i v + \frac{1}{\omega_i T}\sin\omega_i v\right\}dv \qquad \text{(D.20)}$$

as given in equation (10.24). In the same way the expressions for $\sigma_d^2(T)$ and $\sigma_{cd}^2(T)$ can be derived.

# *Author Index*

The page numbers set in *italics* indicate the pages where the full references are found. An arrow → points from an incomplete to the complete reference at the end of another chapter.

The other numbers indicate the pages where the references are used in the text.

# *Subject Index*

UNIVERSITY COLLEGE LIBRARY CARDIFF

WITHDRAWN FROM
STOCK